THE PAPERS OF THOMAS A. EDISON

FINANCIAL CONTRIBUTORS

Public Foundations

National Science Foundation
National Endowment for the Humanities
National Historical Publications and Records Commission

Private Foundations

The Alfred P. Sloan Foundation
Charles Edison Fund
The Hyde and Watson Foundation
Geraldine R. Dodge Foundation

Private Corporations and Individuals

Alabama Power Company
Amerada Hess Corporation
Anonymous
AT&T
Atlantic Electric
Association of Edison Illuminating Companies, Inc.
Battelle Memorial Institute
The Boston Edison Foundation
Cabot Corporation Foundation, Inc.
Carolina Power & Light Company
Consolidated Edison Company of New York, Inc.
Consumers Power Company
Corning Glass Works Foundation
Duke Power Company
Entergy Corporation (Middle South Electric System)
Exxon Corporation
Florida Power & Light Company
General Electric Foundation
Gould Inc. Foundation
Gulf States Utilities Company
Idaho Power Company
International Brotherhood of Electrical Workers
Iowa Power and Light Company
Mr. and Mrs. Stanley H. Katz
Matsushita Electric Industrial Co., Ltd.
McGraw-Edison Company
Minnesota Power
New Jersey Bell
New York State Electric & Gas Corporation
North American Philips Corporation
Philadelphia Electric Company
Philips International B.V.
Public Service Electric and Gas Company
RCA Corporation
Robert Bosch GmbH
Rochester Gas and Electric Corporation
San Diego Gas & Electric
Savannah Electric and Power Company
Schering-Plough Foundation
Texas Utilities Company
Thomas & Betts Corporation
Thomson Grand Public
Transamerica Delaval Inc.
Westinghouse Educational Foundation
Wisconsin Public Service Corporation

THE PAPERS OF THOMAS A. EDISON

Volume 2

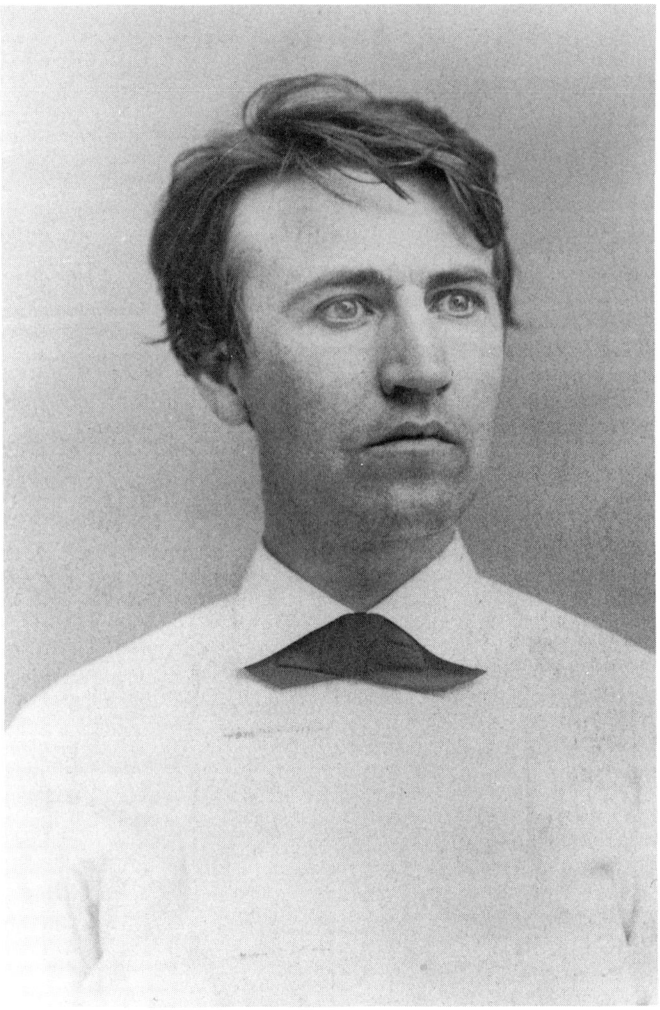

Thomas A. Edison in his mid to late twenties.

Volume 2

The Papers of Thomas A. Edison

FROM WORKSHOP TO LABORATORY

June 1873–March 1876

VOLUME EDITORS
Robert A. Rosenberg
Paul B. Israel
Keith A. Nier
Melodie Andrews

EDITORIAL STAFF
Grace Kurkowski
Helen Endick

DIRECTOR AND EDITOR
Reese V. Jenkins

SPONSORS
Rutgers, The State University of New Jersey
National Park Service, Edison National Historic Site
New Jersey Historical Commission
Smithsonian Institution

THE JOHNS HOPKINS UNIVERSITY PRESS
BALTIMORE AND LONDON

Volume 2 of *The Papers of Thomas A. Edison* appears through the generous financial assistance of the National Historical Publications and Records Commission, the National Endowment for the Humanities, and Rutgers, The State University.

The Johns Hopkins University Press
701 West 40th Street
Baltimore, Maryland 21211-2190
The Johns Hopkins Press Ltd., London

Printed in the United States of America

The paper used in this book meets the minimum requirements of the American National Standard for Information Sciences—Permanence of Paper for Printed Library Materials, ANSI Z39.48-1984.

Library of Congress Cataloging-in-Publication Data

Edison, Thomas A. (Thomas Alva), 1847–1931.
The papers of Thomas A. Edison.
Includes bibliographies and indexes.
Contents: v. 1. The making of an inventor, February 1847–June 1873 — v. 2. From workshop to laboratory, June 1873–March 1876.
1. Edison, Thomas A. (Thomas Alva), 1847–1931. 2. Edison, Thomas A. (Thomas Alva), 1847–1931—Archives. 3. Inventors—United States—Biography. I. Jenkins, Reese. II. Title.
TK140.E3A2 1989 600 88-9017
ISBN 0-8018-3100-8 (v. 1. : alk. paper)
ISBN 0-8018-3101-6 (v. 2. : alk. paper)

TO THE MEMORY OF DR. EDWARD J. BLOUSTEIN

Former President
Rutgers, The State University of New Jersey

Contents

Calendar of Documents

List of Editorial Headnotes

Preface

The mid-1870s were heady years for Thomas Alva Edison. He commanded the respect of the telegraph industry's corporate leaders; watched his own financial fortunes rise, sink, and rise again; fathered Thomas, Jr., his second child; and moved increasingly into the public eye—reviled as a perfidious mountebank, praised as an inventive genius, and dismissed as an incompetent scientific investigator. He left the running of his manufacturing shop largely to his partner Joseph Murray and broadened his own horizons in several directions, assuming at various times the roles of student, author, and designer/manufacturer of consumer goods. By 1875 he felt financially sophisticated enough to offer marketing advice to the notorious financier Jay Gould. As the nation wrestled with a prolonged economic depression, Edison took a bold initiative. Determined to live primarily by his wits, he left his shop-floor, experimental workshop to start an independent laboratory devoted to invention. One year later he moved the laboratory, its staff, and his family out of Newark.

The three years documented in this volume were important ones in Edison's development. During these years he was increasingly following his own lines of research. Moreover, with several years of intensive experimentation behind him, he drew on his previous work for ideas and solutions with growing frequency. By 1876 Edison was widely recognized as a leading electrical inventor. His principal assistant, Charles Batchelor, called him America's "foremost inventor and electrician . . . by far," and a credit report acknowledged him to be "a genius." Just back from England in 1873, he was a promising young man with more backers than notable inventive accomplishments; three years later, he moved to Menlo Park a seasoned, successful inventor.

Because of the wide range of documentary evidence available, the years covered by Volume Two of the Edison Papers have been almost as interesting and diverse for the volume editors to study as they were for Edison to live through. Edison tried several methods of recording laboratory notes, one being his "Experimental Researches," a book of entries self-consciously patterned on that of Michael Faraday. As Edison assembled an experimental team, beginning with Charles Batchelor and James Adams, laboratory notebook entries became collective documents, unassignable to a single author. Patent-related materials for this period include thirty-two successful applications, thirty-six caveats, and all the associated correspondence and legal documents. Edison provided public accounts of his work in interviews with newspaper reporters and in articles he wrote for the technical press. He started a book manuscript that was as much a description of his own explorations as a discussion of telegraph technology, and he published articles as science editor of the *Operator*. Edison's often-complex business arrangements generated a variety of agreements and correspondence. The technical success of his quadruplex telegraph, combined with his 1874 financial straits, led to public acclaim and abuse, labyrinthine intrigues in the U.S. Patent Office, and proceedings in several courts, each with its own documentary legacy.

Quantity is no measure of historical significance. Edison's accomplishments, not the simple vastness of the Edison archive, are the motivation for this edition. But quantity may be the most editorially challenging aspect of the documents available for Volume Two and the volumes to come. In assembling Volume One the editors selected roughly ninety percent of the relevant material in the archive at the Edison National Historic Site; for Volume Two the selection approximates only fifteen percent. Partly as a result of this change—and partly as a reflection of Edison's growing intellectual independence—the editors' task in annotating documents has become one of alerting the reader to relevant unselected archival material as much as referring to developments outside Edison's ambit. Almost all of the unselected documents cited are in the microfilm edition of Edison's papers. The film and its accompanying guides not only serve general researchers but are also fundamental research tools of the book edition's editorial staff.

The Thomas A. Edison Papers project continues to benefit from substantial assistance from many quarters, including the Sponsors, other financial contributors, scholars, archivists, curators, collectors, and students. All of the representatives from the four sponsoring institutions have given generously of their time and energies, but the project is especially indebted to the late Edward J. Bloustein (to whom this volume is dedicated), T. Alexander Pond, Fahy Whitaker, Bernard Bush, Howard Green, and Bernard Finn for their sustained personal interest and institutional support. Likewise, the members of the Corporate Associates and the Editorial Advisory Board have responded to particular needs of the project. Especially notable for their contributions are William C. Hittinger and Nathan Reingold. At the Rutgers University Foundation, Bruce Newman and Dick Lloyd have made the project a priority.

Public and private foundation grants continue to play a vital role. The preparation of this volume was made possible in part by grants from the Division of Research Programs (Editions-Texts) of the National Endowment for the Humanities, an independent federal agency, by grants from the Program in History and Philosophy of Science and the Engineering Division of the National Science Foundation (DIR-8521524), and with the assistance of the National Park Service, Edison National Historic Site. Katherine Fuller and Douglas M. Arnold of the National Endowment for the Humanities and Ronald Overmann and Nam Suh of the National Science Foundation have provided valuable counsel. Grants from the Alfred P. Sloan Foundation and the Charles Edison Fund also have contributed significantly to this volume. Albert Rees, Stewart F. Campbell, Paul Christiansen, John Venable, David Shantz, John P. Keegan, Robert W. Parsons, Jr., Scott McVay, and Thomas L. Morrissey have offered key counsel. Any opinions, findings, and conclusions or recommendations expressed in this publication are solely those of the editors and do not necessarily reflect the views of the National Science Foundation, the National Endowment for the Humanities, the United States Government, or any other financial contributor.

The editors are grateful to many scholars for their publications, professional presentations, teaching, and private discussions. For this volume the editors are especially indebted to Brian Bowers, James Brittain, John Bryant, Aaron Ihde, and Frank W. Dittman.

The project is indebted to many institutions and their staffs for providing documents, photographs, photocopies, and research assistance. Staff members of the Henry Ford Museum & Greenfield Village have extended themselves, especially William S. Pretzer, John Bowditch, Tim Binkley, and Rudy Ruzicska. Similar help came from Joyce Bedi of the History Center of the Institute of Electrical and Electronics Engineers and from Jeff Sturchio of the Corporate Research Archives at AT&T. The editors also received assistance from Florence Bartechevski of the Baker Library at Harvard University, Irene Coffee of the Franklin Institute, Mrs. C. I. Constantinides of the Post Office Archives in London, Jane Hardy of the Stevens Institute, Margaret Jasko of Western Union Telegraph Company, Janice Pickering of the U.S. Patent Office, Lenore Symons of the Institute of Electrical Engineers in London, and Steven Wheeler of the New York Stock Exchange Archives. The project is indebted to several persons in the Federal Archives and Record Services: John Butler, Marjorie Ciarlante, and Sue McDonough in Washington, D.C.; and Joel Buchwald and Donna Cabarle in Bayonne, N.J. Others who have provided resources or special assistance include Charles Hummel, Theodore Edison, and John Edison Sloane.

Many people associated with the National Park Service have contributed significantly to this project. Worthy of special mention are Herbert S. Cables, Jr., Gerald R. Patten, John Maounis, and Edward Dallop of the Boston Regional Office and Fahy Whitaker, Nancy Waters, Mary Bowling, George Tselos, and Edward Pershey of the Edison National Historic Site.

The project benefits from many friends within Rutgers University who have given assistance. The editors are appreciative of the efforts of colleagues in the History Department, especially James W. Reed, Susan R. Schrepfer, Philip J. Pauly, Gerald Grob, Michael Adas, Richard L. McCormick, Rudy Bell, and John Gillis. All of the staff of the Rutgers University Libraries have given fine service, but noteworthy is the help given by Ruth Simmons, Ronald L. Becker, and Clark L. Beck, Jr. Administrative people who are special contributors to this project include Paul L. Leath, Peter D. Klein, Robert F. Pack, John Salapatas, David A. Cayer, Andrew B. Rudczynski, C. Fred Main, Francis A. Baran, David A. Rumbo, Barbara A. Reis, Albert Hanna, Donna Estler, Muriel Wilson, Cynthia Jucks, and Joseph Harrigan.

Several staff members, interns, and students not named on the title page have contributed in many ways to this volume, including Thomas E. Jeffrey, Mary Ann Hellrigel, Douglas C. Tarr, Dennis Madden, David Hutchings, Leonard DeGraaf, Gregory Field, Jacqueline Miller, Gregory Jankunis, Paul Garelick, Linhhue Duong, Jatinder Sharma, Beth Smollen, and James Spiller. Patricia Matteson and Richard W. Mitchell of the Agricultural Museum at Rutgers University performed a significant portion of the extensive photographic work for this volume.

The Johns Hopkins University Press, under the direction of Jack Goellner, has heroically undertaken the unusually complex job of publishing the Edison Papers. The editorial and design staff at the press distinguished themselves with Volume One, demonstrating their skill, imagination, and devotion. The project editors are especially grateful for the privilege of working with Penny Moudrianakis, Barbara Lamb, Martha Farlow, Jim Johnston, Robert Brugger, and Henry Y. K. Tom.

Chronology of Thomas A. Edison

June 1873–March 1876

1873	
25 June	Returns to Newark from England.
2 August	Drafts caveat that contains the basis of the quadruplex.
25 August	Executes patent application for his roman-letter automatic-telegraph perforator.
2 September	Sells British rights for his automatic telegraph to a London company.
6 September	A carbon rheostat for use in artificial-cable telegraph experiments is finished in the shop.
Summer–Fall	Devises and experiments with circuit designs for duplex, automatic, and cable telegraphy using electromagnets and storage batteries to counter inductive effects.
	Studies chemistry handbooks, prepares automatic-telegraph experiments using different metal styli, and executes four patent applications for chemical solutions for telegraph recording paper.
	Resumes experiments with three-key automatic-telegraph perforators.
1 October	Demonstrates quadruplex telegraph for several important telegraph officials.
28 October	Executes first caveats for roman-letter automatic-telegraph circuits.
Fall	Makes Charles Batchelor his principal laboratory assistant and conducts experiments on electrolytic solutions in batteries.
	Works with telegraph inventor-manufacturer Jesse Bunnell on a railway signal device.
Late December	Conducts test of his automatic telegraph system for several important telegraph officials.

1874	
27 January	The Automatic Telegraph Co. publicly demonstrates Edison's automatic telegraph system.
1 February	Drafts a new roman-letter automatic-telegraph caveat and begins experiments based on its designs.
Winter	Continues experiments on battery and automatic-telegraph recording solutions and on cable telegraphy.
	British investors urge his return to England to supervise continuing tests of his automatic telegraph on British Post Office and private cable lines.
	Works in earnest on his proposed book on telegraphy.
Early March	Begins development of a district telegraph and fire alarm system.
26 March	Agrees with Joseph Murray and Jarvis Edson to incorporate the Domestic Telegraph Co. to exploit his district-telegraph and fire alarm inventions.
Late March	Seeks increased support for duplex telegraph research from Western Union president William Orton.
10 April	Discovers electromotograph phenomenon.
c. 16 April	Demonstrates roman-letter automatic to George Ward, American manager of the French and Anglo-American cable lines.
18 April	Begins experiments on induction coils for electromedical devices.
28 April	Shop builds models of the inductorium and of a student telegraph operator's instrument for proposed Scientific Toy Co.
19 May	Seeks cooperation of Western Union electrician George Prescott in duplex experiments in exchange for a half interest in resulting patents.
22 May	Begins selling inductorium.
May	Searches for a source of tellurium for use as recording stylus in automatic and domestic telegraphy.
1 June	Executes a patent application for a tellurium recording stylus for automatic telegraphy.
c. 1 June	George Prescott accepts Edison's proposal with approval of President Orton; Edison and Prescott begin quadruplex experiments on Western Union lines and have apparatus made at the Western Union shop.
21 June	Drafts partnership agreement with George Prescott regarding duplex and quadruplex.
June	Edison and Murray prepares instruments for a July demonstration of Edison's fire alarm system in Utica, N.Y.
Spring	George Harrington and Josiah Reiff negotiate to form a new

	new company to exploit Edison's automatic-telegraph inventions.
3 July	Mortgage and notes due William Unger; payments made during month by Automatic Telegraph Co. investors in exchange for Edison's promissory notes to Harrington.
8 July	Demonstrates quadruplex to Western Union officials on the company's lines.
9 July	Signs first partnership agreement with George Prescott.
10 July	Western Union announces the quadruplex through an article in the *New York Times*.
18 July	Former partner James Ashley begins vituperative attacks on Edison in the *Telegrapher*.
24 July	Finishes several automatic-telegraph patent models for George Harrington, including modifications of the British Wheatstone system.
3 August	Resumes electromotograph experiments.
7 August	Executes patent application for a telegraph device employing the electromotograph.
19 August	Signs a revised partnership agreement with George Prescott and executes several patent applications for duplex and quadruplex telegraphy.
August	Becomes science editor of the *Operator*. James Adams joins Edison's experimental staff.
1 September	Publishes first of several articles in the *Operator*.
5 September	Announces electromotograph discovery in the *Scientific American*.
21 September	Begins one month of chemistry lessons under Robert Spice, Brooklyn High School professor of chemistry and natural philosophy .
Late September	Successfully tests quadruplex on Western Union line between New York and Boston; commercial operation begins 3 October.
October–November	Installs and modifies quadruplex on Western Union lines.
c. 1 November	Moves his family to an apartment on Bank St.
2 November	Conducts experiments looking for new forces.
c. 6 November	Exhibits electromotograph at the National Academy of Sciences meeting in Philadelphia.
c. 1 December	Sells his house on Wright St.
3 December	Agrees to establish a Domestic Telegraph Co. in Canada and begins manufacturing new private-line printer for the domestic system.
4 December	Successfully establishes quadruplex operation on Western Union line between New York and Chicago with a repeater at Buffalo.

	Executes first four of twelve multiple-telegraph caveats and patent applications prepared during December and January.
9 December	Society of Telegraph Engineers in London elects Edison and Charles Batchelor members.
10 December	Receives $5,000 advance from Western Union for quadruplex.
Mid-December	Tries to negotiate sale of quadruplex to Western Union; President Orton departs on business trip.
30 December	Josiah Reiff, John McManus, and Jay Gould reach agreement on Atlantic and Pacific Telegraph Co. takeover of Automatic Telegraph Co., with Edison to be company electrician.
	Gould, Thomas Eckert, and Albert Chandler visit Edison at night to see automatic telegraph and quadruplex.
31 December	Quadruplex patent application suspended for possible interference.
1875	
4 January	Sells his quadruplex patent rights to Jay Gould for $30,000.
5–8 January	Discusses strategies with Gould for Atlantic and Pacific competition with Western Union.
	Purchases books and scientific equipment.
c. 11 January	Visits his family in Port Huron, Mich.
19 January	William Orton accepts terms for quadruplex purchase suggested by Edison in December.
20–21 January	Files arguments and amendments for applications examined by Patent Office on 31 December 1874 and requests immediate approval of all multiple-telegraphy cases that were not suspended.
23 January	Writes Commissioner of Patents to ask that quadruplex patents be assigned jointly with Harrington rather than Prescott.
	Writes William Orton, saying he cannot accept Western Union's terms.
28 January	Western Union begins legal proceedings against Edison in New Jersey over the quadruplex.
January	Begins serving as electrician of Atlantic and Pacific; inspects New York–Boston line.
	Edison and Murray begins manufacturing domestic-telegraph instruments for New York City lines and automatic instruments for Atlantic and Pacific.
11 February	With his wife, Mary, gives a masquerade party on his birthday.
24 February	Executes a patent application with Charles Batchelor for the domestic telegraph system.

27 February	Jay Gould agrees to pay expenses for roman-letter automatic experiments.
20 March	Commissioner of Patents orders that quadruplex patents be assigned jointly with Prescott rather than Harrington.
23 March	Files two patent applications (for a quadruplex repeater and for the domestic telegraph system designed with Batchelor); does not file another application for a year.
Late March	Demonstrates an automatic telegraph system employing a spectroscope to the secretary of the German telegraph system.
March–Early April	Moves family to a house on South Orange Ave. in Newark.
c. 1 April	Ezra Gilliland begins selling Edison and Murray products at new store in New York City.
16 April	Transfers to George Harrington and Josiah Reiff part of his interest in automatic-telegraph patents in expectation of a settlement from Atlantic and Pacific.
20 April	Agrees with Gold and Stock Telegraph Co. to settle their outstanding accounts.
	Assesses his financial situation by preparing a summary account of his assets and liabilities.
27 April	Swears affidavit presenting his argument to Secretary of the Interior Delano appealing the Patent Commissioner's decision on quadruplex applications.
30 April	Begins experiments to develop a new copying process.
7 May	Begins experiments on electromotograph repeater for automatic telegraphy.
16 May	Agrees with Joseph Murray to dissolve their manufacturing partnership; sets up separate laboratory at Ward St. shop.
31 May	With Charles Batchelor prepares list of proprosed experimental topics and conducts first experiments in new laboratory.
2 June	Experiments to find new force for use in telegraphy.
30 June	Conceives electric-pen copying system.
15 July	Signs formal agreement dissolving Edison and Murray.
26 July	Proposes lease of Atlantic and Pacific wires for night-letter business to be run from laboratory.
July	Meets with William Orton to discuss acoustic telegraphy and draws alternative designs based on article about Philip Reis's telephone.
	Furnishes Ezra Gilliland with space and machinery in the Ward St. shop for the manufacturing enterprise Gilliland & Co.
5 August	Prepares list of inventions to be displayed at 1876 Centennial Exhibition in Philadelphia.

August	Suggests various terms for settling disputes over automatic and quadruplex patents during negotiations over possible merger of Atlantic and Pacific and Western Union.
16 September	Proposes duplexing Atlantic cable of Direct United States Cable Co.
Summer	Develops basic electric-pen copying system; begins manufacture and sale of system by early September.
23 September	Executes first caveat for electric-pen copying system.
	Leaves for Port Huron to visit family.
1 October	Signs first of several agreements for electric-pen agencies; this one includes his nephew Charles Edison.
2 October	Agrees to give Charles Batchelor and James Adams a percentage of profits from electric-pen sales.
c. 5 October	Proposes a system for transmitting stockbrokers' private messages to the floor of the New York Stock Exchange.
14–21 October	Develops new battery for electric pen.
30 October	Drafts caveat for facsimile telegraph system employing ideas from his other inventions—e.g., the electromotograph and electric pen.
16 November	Begins acoustic-telegraph experiments for Western Union.
22 November	Executes his first acoustic-telegraph caveat and develops new quadruplex designs.
	Notices phenomena he ascribes to "etheric force" while working with acoustic-telegraph apparatus and begins two weeks of intensive experiments on new force.
29 November	Gives newspaper reporters first interviews on etheric force.
	Again employs Robert Spice, this time as consultant and assistant on acoustic-telegraph experiments.
1 December	Agrees to assign rights in automatic telegraphy to the new American Automatic Telegraph Co. as part of strategy to force Atlantic and Pacific and Jay Gould to settle with those interested in the old Automatic Telegraph Co.
14 December	Signs contracts with Western Union to settle their mutual claims regarding the quadruplex controversy and to formalize the company's support for his work in acoustic telegraphy in exchange for control of his inventions.
	Conducts etheric-force experiments with Professor Henry Morton at the Stevens Institute of Technology in Hoboken, N.J.
16 December	Demonstrates etheric force with Dr. George Beard at a meeting of the Polytechnic Association of the American Institute of New York.
23 December	Demonstrates etheric force at meeting of the Newark Scientific Association.

29 December	Purchases property in Menlo Park, N.J., for new laboratory and home.
December	Conducts numerous acoustic-telegraph experiments.
1876	
10 January	Edison's first son, Thomas Alva, Jr. ("Dash"), is born in Newark.
Mid-January	Executes five acoustic-telegraph caveats.
January–February	Continues acoustic-telegraph experiments in laboratory and on Western Union lines.
	Demonstrates a new quadruplex on Western Union lines.
	Continues construction of new laboratory at Menlo Park under the supervision of his father, Samuel.
7 February	Improves design of electric pen; laboratory machinists begin altering existing pens for manufacturer Gilliland & Co.
8 February	Assigns a three-tenths interest in electric pen to Ezra Gilliland's father, Robert.
7 March	Executes a patent application for his electric-pen copying system; files it on 13 March as his first application in one year.
	Arranges public sale of stock in new electric-pen company.
c. 13 March	Agrees to share space at Centennial Exhibition with Western Union.
16 March	Mary Edison is given a surprise party at the Edison home.
21 March	Attempts to withdraw his quadruplex- and automatic-telegraph patent applications and associated powers of attorney in order to negate reactivation of his quadruplex applications by Interior Secretary Chandler.
Winter	Continues public dispute over nature of etheric force.
	With Charles Batchelor arranges foreign agencies for electric pen.
c. 26–28 March	Moves laboratory and family to Menlo Park.

Editorial Policy

The editorial policy for the book edition of Thomas Edison's papers remains essentially as stated in Volume One. The additions that follow include new editorial situations that have arisen in assembling Volume Two and certain minor points that were overlooked in the initial statement.

Selection

Most of Edison's documents exist in one, original version only. However, some documents—especially those associated with court or Patent Office proceedings—exist in multiple transcriptions, often printed, the original having been lost. In such cases the editors have let reader convenience guide their selection, choosing the text with the fewest obvious errors. The general textnote for such a document indicates the existence of the other copies, and specific textnotes to the document flag any significant variations in the text.

Most legal or Patent Office testimony is too lengthy and diffuse to supply useful documents in an edition like this one. Occasionally, however, a party to a proceeding enters a summary statement that contains interpretative analyses of prior events, presenting a great deal of information in relatively brief compass. Although such statements may be generated within the period covered by a given volume, their retrospective nature makes them akin to the autobiographical reminiscences presented in volume appendixes. In order to provide the reader with the perspectives and the significant, otherwise unavailable material that these documents contain, the editors will include particularly valuable ones in appendixes; two, dealing with Edison's quadruplex patents, are printed in this volume.

Annotation

A document included in the book edition may bear a different date in the microfilm index. This only indicates that further information has allowed the editors to date it more precisely. Occasionally an endnote refers to an archival document that has not yet been microfilmed. In this volume, that includes any document at the Edison National Historic Site in files for the years after 1886. Such references provide an identification that will allow the reader to find the document when the appropriate part of the microfilm edition is published.

With this volume, chemical compounds appear with increasing frequency in experimental records. The editors have identified only those names that are not readily found in such references as *Hawley's Condensed Chemical Dictionary.*[1]

Form

Notebook entries and technical notes. Edison's laboratory notebook entries and technical notes do not readily submit to traditional editorial procedures in many respects. Because these documents served as a legal record of invention, Edison and his staff often signed and dated individual pages or drawings of a single experimental process. In order to avoid unduly fragmenting the record of Edison's work, the editors have presented such continuous, one-day records as single documents. In these cases, the date of the document appears only once, on the dateline, in the form in which it was first written. Other points at which the date appears are textnoted. Similarly, signatures appear only at the end of the document, and other occurrences are textnoted to indicate who signed at that place. If all participants did not sign in all places, names appear at the end of the document in the order in which they were written.

Authorship. Authorship of a document is usually a relatively straightforward determination. However, the increasingly collaborative nature of the research represented by Edison's laboratory records gives rise to an unusual situation. It is clear that the greatest part of the laboratory work was directed, inspired, designed, or performed by Edison. Nevertheless, it is often not evident who played what role in a particular experiment. The person writing an entry may have been the sole experimenter, one of several experimenters, or simply a scribe. The editors are treating such research entries and technical notes as having no specific author. The physical description in the textnote is simply "X"; these by definition are not autograph documents, nor can they be considered signed by an author.

Transcription principles. In many patent applications and caveats, Edison or a secretary underlined letters designating parts of a drawing, often preceding or following these letters with dots to set them off from the text. The editors have underlined the dots as well except for the dots that serve as grammatical periods or commas. In documents with several independent drawings and no text, the editors have added capital letters in brackets as labels. These allow placement of endnote references, facilitate cross-reference, and indicate original order when small drawings are rearranged on the published page.

1. *Hawley's Condensed Chemical Dictionary,* 11th ed. (New York: Van Nostrand Reinhold Co., 1987).

Editorial Symbols

~~Newark~~ Overstruck letters
Legible manuscript cancellations; crossed-out or overwritten letters are placed before corrections

[Newark] Text in brackets
Material supplied by editors

[Newark?] Text with a question mark in brackets
Conjecture

[Newark?][a] Text with a question mark in brackets followed by a textnote
Conjecture of illegible text

⟨Newark⟩ Text in angle brackets
Marginalia; in Edison's hand unless otherwise noted

[ ] Empty brackets
Text missing from damaged manuscript

[---] Hyphens in brackets
Conjecture of number of characters in illegible material

Superscript numbers in editors' headnotes and in the documents refer to endnotes, which are grouped at the end of each headnote and after the textnote of each document.

Superscript lowercase letters in the documents refer to textnotes, which appear at the end of each document.

List of Abbreviations

ABBREVIATIONS USED TO DESCRIBE DOCUMENTS

The following abbreviations describe the basic nature of the documents included in the second volume of *The Papers of Thomas A. Edison:*

AD	Autograph Document
ADf	Autograph Draft
ADfS	Autograph Draft Signed
ADS	Autograph Document Signed
AL	Autograph Letter
ALS	Autograph Letter Signed
AX	Autograph Experimental Note
AXS	Autograph Experimental Note Signed
D	Document
Df	Draft
DS	Document Signed
L	Letter
LS	Letter Signed
M	Model
PD	Printed Document
PDS	Printed Document Signed
PL	Printed Letter
TD	Typed Document
TL	Typed Letter
X	Experimental Note

In these descriptions the following meanings are assumed:

Document Accounts, agreements and contracts, bills and receipts, legal documents, memoranda, patent applications, and published material, but excluding letters, models, and experimental notes

Draft A preliminary or unfinished version of a document or letter

Experimental Note Technical notes or drawings not included in letters, legal documents, and the like

Letter Correspondence

Model An artifact, whether a patent model, production model, or other

The symbols may be followed in parentheses by one of these descriptive terms:

abstract A condensation of a document

copy A version of a document made at the time of the creation of the document by the author or other associated party

electric-pen copy A copy made with Edison's electric-pen copying system

fragment Incomplete document, the missing part not found by editors

historic drawing A drawing of an artifact no longer extant or no longer in its original form

historic photograph A photograph of an artifact no longer extant or no longer in its original form

letterpress copy A transfer copy made by pressing the original under a sheet of damp tissue paper

photographic transcript A transcript of a document made photographically

telegram A received telegraph message

transcript A version of a document made at a substantially later date than that of the original, by someone not directly associated with the creation of the document

STANDARD REFERENCES AND JOURNALS

Standard References

BDAS	*Biographical Dictionary of American Science*
DAB	*Dictionary of American Biography*
DNB	*Dictionary of National Biography*
DSB	*Dictionary of Scientific Biography*

MEB	*Modern English Biography*
NCAB	*National Cyclopedia of American Biography*
OED	*Oxford English Dictionary*
TAEB	*The Papers of Thomas A. Edison* (book edition)
TAEM	*Thomas A. Edison Papers: A Selective Microfilm Edition*
WWW-HV	*Who Was Who in America, Historical Volume*

Journals

Elec. W.	*Electrical World*
J. Frank. Inst.	*Journal of the Franklin Institute*
J. Soc. Teleg. Eng.	*Journal of the Society of Telegraph Engineers*
J. Teleg.	*Journal of the Telegraph*
Phil. Mag.	*Philosophical Magazine*
Sci. Am.	*Scientific American*
Teleg. and Tel.	*Telegraph and Telephone Age;* formerly *Telegraph Age*
Telegr.	*The Telegrapher*
Teleg. J. and Elec. Rev.	*Telegraphic Journal and Electrical Review;* formerly *Telegraphic Journal*

ARCHIVES AND REPOSITORIES

In general, repositories are identified according to the Library of Congress system of abbreviations. Parenthetical letters added to Library of Congress abbreviations have been supplied by the editors. Abbreviations contained entirely within parentheses have been created by the editors and appear without parentheses in citations.

DLC	Library of Congress, Washington, D.C.
DNA	National Archives, Washington, D.C.
DSI-NMAH	Archives, National Museum of American History, Smithsonian Institution, Washington, D.C.
MdSuFR	Washington National Records Center, Suitland, Md.
MH-BA	Baker Library, Graduate School of Business Administration, Harvard University, Boston, Mass.

MiD-B	Burton Historical Collection, Detroit Public Library, Detroit, Mich.
MiDbEI	Library and Archives, Henry Ford Museum & Greenfield Village, Dearborn, Mich.
MiDbEI(H)	Henry Ford Museum & Greenfield Village, Dearborn, Mich.
(MiPHStCo)	St. Clair County Building, Port Huron, Mich.
MoSHi	Missouri Historical Society, St. Louis, Mo.
NjBaFAR	Federal Archives and Records Center, Bayonne, N.J.
(NjHum)	Private Collection, Charles Hummel, Wayne, N.J.
(NjNECo)	Essex County Hall of Records, Newark, N.J.
NjWAT	Corporate Research Archives, American Telephone and Telegraph Company, Warren, N.J.
NjWOE	Edison National Historic Site, West Orange, N.J.
(NjWU)	Western Union Corporate Records, Western Union Corporation, Upper Saddle River, N.J.
NNC	Rare Books and Manuscripts, Butler Library, Columbia University, New York, N.Y.
NNHi	New-York Historical Society, New York, N.Y.
(NNYCo)	New York County Hall of Records, New York, N.Y.
PHi	Historical Society of Pennsylvania, Philadelphia, Pa.
PU	Archives, University of Pennsylvania, Philadelphia, Pa.
(UKLPO)	Archives, Post Office, London, United Kingdom

MANUSCRIPT COLLECTIONS AND COURT CASES

References to documents included in *Thomas A. Edison Papers: A Selective Microfilm Edition* (Frederick, Md.: University Publications of America, 1985–) are followed by parenthetical citations of reel and frame of that work; for example, Cat. 1185:34, Accts. (*TAEM* 22:562). Documents found at NjWOE after contemporaneous material had been microfilmed will be filmed as a supplement of the next part of the microfilm edition. For *TAEM* Supp. III, see *Guide to Thomas A. Edison Papers, Part III.*

Accts.	Accounts, NjWOE
Anders v. Warner	*Anders v. Warner,* Patent Interference File 5603, RG-241, MdSuFR
ATF	Automatic Telegraphy File, UKLPO
Batchelor	Charles Batchelor Collection, NjWOE
BC	Biographical Collection, NjWOE
Buell v. Martin	*Buell v. Martin,* Patent Interference File 10,110, RG-241, MdSuFR
DF	Document File, NjWOE
EBC	Edison Biographical Collection, NjWOE
Edison Caveat	Edison Caveats, RG-241, DNA
Edison Coll.	Edison Manuscript Collection, NNC
Edison v. Lane	*Edison v. Lane v. Gray v. Rose v. Gilliland and Edison v. Lane v. Gray v. Edison and Johnson,* Patent Interference Files 8027–28, RG-241, MdSuFR
EP&RI	Edison Papers and Related Items, MiDbEI.
Field v. Lugo	*Field v. Lugo,* Patent Interference File 8087, RG-241, MdSuFR
GF	General File, NjWOE
Harrington v. A&P	*Harrington, Edison, and Reiff v. Atlantic & Pacific, and George Gould et al.,* Equity Case File 3980, RG-276, and Equity Case File 4940, RG-21, NjBaFAR
Kellow	Richard W. Kellow File, Legal Series, NjWOE
Lab.	Laboratory notebooks and scrapbooks, NjWOE
Lbk.	Letterbooks, NjWOE
LBO	William Orton Letterbooks, NjWU
Lefferts	Marshall Lefferts Papers, NNHi
Libers Pat.	Libers of Patent Assignments, U.S. Patent Office Records, MdSuFR
Lit.	Litigation Series, NjWOE
Meadowcroft	William H. Meadowcroft Collection, NjWOE

Miller	Harry F. Miller File, Legal Series, NjWOE
Nicholson v. Edison	*Nicholson v. Edison,* Patent Interference Files 8689–90, RG-241, MdSuFR
Nipher	Francis E. Nipher Papers, MoSHi
Painter	Uriah H. Painter Papers, PHi
Pat. App.	Patent Application Files, RG-241, MdSuFR
Pioneers Bio.	Edison Pioneers Biographical File, NjWOE
PS	Patent Series, NjWOE
Quad.	Quadruplex Case, NjWOE (see list below)
RGD	R. G. Dun Collection, MH-BA
Scraps.	Scrapbooks, NjWOE
Supp. III	Supplement to *TAEM,* Part III
TI	Telephone Interferences, NjWOE
Wiley v. Field	*Wiley v. Field,* Patent Interference File 9219, RG-241, MdSuFR
WU Coll.	Western Union Collection, DSI-NMAH
WUTAE	Envelope of TAE letters, Box 8, Secretary's Office, NjWU

Quadruplex Case Citations

Volume 70 (*Atlantic & Pacific v. Prescott & others*)

1 Amended Bill of Complaint
2 Answer of Prescott
3 Answer of Edison
4 Answer of Western Union
5 Answer of Prescott to Amended Complaint
6 Plaintiff's Exhibits 1 through Z.3
7 Testimony for Plaintiff (with index)
8 Exhibits Z.4 through Z.8
9 Plaintiff's Rebuttal Witnesses (with index)

Volume 71 (*Atlantic & Pacific v. Prescott & others*)

1 Testimony for Defendants (with index)
2 Defendant's Exhibits

Volume 72

1 "Priority of Invention," *American Law Register* (Oct.–Nov. 1872)
2 *Harrington & Edison v. Prescott,* Prescott to Commissioner of Patents

3 *Prescott against Wilber,* Prescott's Preferral of Charges
4 *Prescott against Wilber,* Prescott's Brief
5 *Prescott against Wilber,* Statement and Brief of Wilber
6 *Harrington & Edison v. Prescott,* Argument for Prescott
7 *Harrington & Edison v. Prescott,* Argument for Edison
8 *Harrington & Edison v. Prescott,* Confidential Argument for Edison
9 *Harrington & Edison v. Prescott,* Argument for Edison
10 *Harrington & Edison v. Prescott,* Statement of Harrington
11 *Harrington & Edison v. Prescott,* Brief for Edison
12 *Harrington & Edison v. Prescott,* Brief for Harrington
13 *Harrington & Edison v. Prescott,* Reply for Harrington
14 *Western Union v. Edison & Prescott,* Answer of Edison
15 *Rules of Practice* (U.S. Patent Office, Sept. 1875)
16 *Harrington & Edison v. Prescott,* Edison's Affidavit
17 Telepost prospectus
18 *Harrington & Edison v. Prescott,* Argument for Prescott

Volume 73 (*Atlantic & Pacific v. Prescott & others*)
1 Findings Submitted by Defendant (annotated by judge)
2 Findings Submitted by Plaintiff (annotated by judge)
3 Plaintiff's General Argument
4 Points for Plaintiff
5 Argument for Plaintiff (by Wheeler)
6 Brief for Plaintiff (by Latrobe)
7 Brief for Plaintiff (by Butler)
8 Brief for Plaintiff (by Hodges)
9 Argument for Plaintiff (by Myers)
10 Brief for Plaintiff (by Russell)
11 Defendant's Proposed Findings
12 Suggested Points by Defendant
13 Defendant's Brief
14 Argument for Defendant (by Lowrey)
15 Argument for Defendant (by Dickerson)
16 Decision by Judge Sanford

Telegraph Law Cases (TLC)
1 *Western Union v. Harrington & others,* Affidavits and Injunction
2 *Western Union v. Harrington & others,* Summons and Complaint
3 *Edison & Harrington v. Western Union & others,* Bill of Complaint
4 *Edison & Harrington v. Western Union & Prescott,* Bill of Complaint

5 *Edison & Harrington v. Western Union & others*, Brief for Complainant
6 *Harrington & Edison v. Prescott*, Reply of Prescott
7 *Western Union v. Edison & Prescott*, Bill of Complaint and Injunction

FROM WORKSHOP TO LABORATORY, JUNE 1873–MARCH 1876

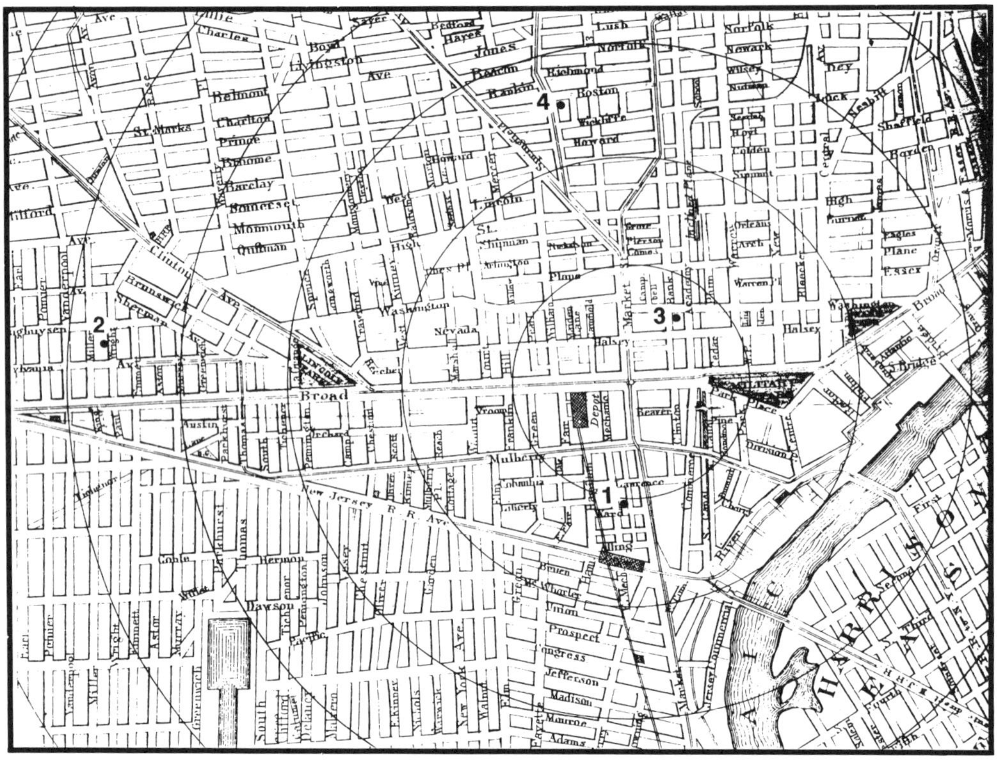

This 1875 map of downtown Newark, N.J., shows the locations of Edison's business and residences between 1873 and 1876. The concentric circles are spaced at quarter-mile intervals. (1) *The Edison and Murray shop was located at 10–12 Ward St.* (2) *Edison, his wife, Mary, and his first child, Marion, lived at 97 Wright St. until November 1874.* (3) *The Edisons lived over a pharmacy at 65 Bank St. for about five months in late 1874 and early 1875.* (4) *Early in the spring of 1875 the family moved to 89 South Orange Ave., where they lived until relocating to Menlo Park the next year.*

-1- June–September 1873

At the end of June 1873 Edison returned from England, where he had demonstrated his automatic telegraph system to the British Post Office. During these demonstrations Edison successfully met the conditions set by British telegraph officials (under which the Post Office would cover the cost of the trials), but the Post Office made no offer to purchase rights to use his system. In early September, however, a group of British capitalists bought a one-fifth share of the worldwide rights (except for the United States) for overland use of Edison's automatic.[1] Edison had also experimented with his automatic system on a coiled submarine telegraph cable stored at the Greenwich works of the Telegraph and Construction Maintenance Company.[2] The results of these experiments had puzzled Edison; nevertheless, his associates attempted to interest British cable telegraph companies in the system.[3]

The experimental problems Edison encountered in England piqued his interest, and he quickly filled several small notebooks with ideas for experiments on automatic and cable systems. On his return Edison began a period of intense experimentation on these telegraph technologies and a few related topics; for almost a year he executed few patent applications. He set up an artificial cable line at his shop in Newark, New Jersey. Edison continued to work on the electrical and electromechanical problems in automatic telegraphy that had engaged his interest before his trip to England, but he also experimented increasingly with chemical solutions both for batteries and for automatic telegraph recording.

Inductive effects plagued both automatic and cable telegraphy, and Edison used several circuit components to cope with them.[4] He created electromagnets with cores as long as ten feet and introduced condensers and secondary (storage) batteries in many combinations. In order to work on cable problems, he designed several circuits that replicated the resistance and capacitance of a deep-sea line. In the process, he built a rheostat that used compressed carbon in glass tubes, an arrangement whose sensitivity to vibrations made it useless for its intended purpose but which years later became the basis for his successful telephone transmitter. Induction also hampered multiple telegraphy, and Edison transferred techniques between his automatic and cable work and his duplex and diplex circuits.

As Edison plunged into experimental work on automatic and cable telegraphy, he expanded his knowledge of chemistry considerably. Beginning with less rigorous texts, he read widely, selecting experiments to repeat in the laboratory. Battery problems, such as longevity and diffusion of solutions, were an important prod for this research. He also focused his attention on improving solutions used for treating the receiving paper of his electrochemical telegraph, and complemented that work with research into the efficacy of different metals in the receiving styli.[5]

In this period, Edison began to develop a quadruplex system, a multiple telegraph system combining a new diplex circuit (which sent two messages in the same direction) with standard duplex designs (which sent two messages in opposite directions) to transmit four messages simultaneously over one wire, two in each direction. In early August he drew up a caveat that contained the key elements of his successful quadruplex, and he continued to test apparatus both in his shop and at night on Western Union lines. His new design, based on an insight he reached in England, circumvented the principal difficulty in his diplex. The problem was that a necessary concomitant of transmitting one message could mutilate the other. Unable to eradicate the disturbance at its source, he found a way to isolate it electromechanically, thus preventing the interference. This solution—Edison referred to it as a "bug trap"—illustrates one of his inventive strategies; when he could not eliminate the cause of a problem, he sought an arrangement that rendered the effects insignificant.[6]

Although Edison no longer concentrated on inventions in printing telegraphy,[7] he and his partner Joseph Murray continued to manufacture at their Newark shop Edison's universal stock and private-line printers for the Gold and Stock Telegraph Company. They also produced instruments for the Automatic Telegraph Company, which was using Edison's system on its line between New York and Washington, and manufactured most of the instruments Edison used in his experimental work.[8] Additional income came from renting a room in the shop to the inventor of a new rifle.[9] Edison's personal income came from both Gold and Stock and Automatic, each of which paid him a $2,000 annual base salary.[10] With the sale of the overseas rights to his automatic system, he expected to receive an additional $17,000. But all financial arrangements became somewhat uncertain later in September with the onset of the Panic of 1873.

1. They also agreed to work, on commission, as agents selling Edison's automatic to both submarine and overland telegraph companies or departments.

2. The Telegraph and Construction Maintenance Co. manufactured and laid most of the world's oceanic telegraph cables; John Pender had organized it in 1864 (Haigh 1968, 41; Lawford and Nicholson 1950, 53–74). John Pender, MP (later knighted), was a textile merchant who also organized, ran, and was the leading investor in most of the world's major oceanic cable telegraph companies. *DNB* (Supp.), s.v. "Pender, Sir John (1815–1896)"; Barty-King 1979, pt. 1 passim.

3. Apparently some arrangement was reached during the few months with John Pender's Eastern Telegraph Co. or some of its backers; see Doc. 381 n. 4.

4. Inductive effects are notable in high-speed signaling on long lines and also in underground cables, a feature of many British lines.

5. At the end of September, Edison filed two patent applications for electrochemical telegraph solutions (U.S. Pats. 160,402 and 160,404). On the styli see Doc. 370 n. 2.

6. Edison used this approach, for example, in developing the universal stock printer, in which key parts were sufficiently adjustable to compensate for inexact manufacturing tolerances, thus making all parts interchangeable (*TAEB* 1:347); in those printing telegraphs where the printing and typewheel magnets were in series, one being made insensitive to the weaker currents that activated the other (e.g., *TAEB* 1:161, 292); and in an automatic telegraphy caveat, where an unavoidable current between two receiving pens was made negligible by increasing the pens' separation (Edison Caveat 43).

7. An account entry of 8 August 1873 shows work on a new experimental transmitter for stock quotations (Cat. 1219:88, Accts. [*TAEM* 21:749]), and in one notebook entry Edison suggests using a condenser for a circuit of twenty universal stock printers (Doc. 343).

8. Manufacturing accounts are found in Cat. 1219:78–114, Accts. (*TAEM* 21:744–62).

9. Account records indicate that Edison and Murray rented space between September and December 1873 to a man identifed as "Mr. Keene alias Gun Man" (Cat. 1219:89–141, Accts. [*TAEM* 21:749–76]). In his reminiscences Edison stated, "I think it was the Berdan rifle. In

any event it was a rifle, which was subsequently adopted in the British Army" (*TAEB* 1:644). Edison and Murray also seem to have rented space to a man named Prosser between September 1873 and February 1874 (Cat. 1219:99, 149; Cat. 1184:13, 27; both Accts. [*TAEM* 21:754, 780, 789, 796]).

10. See Docs. 164 and 274.

FRAGMENTARY POCKET NOTEBOOK

Docs. 341–346

Edison used this notebook around the time of his return from England for noting ideas and planning experiments. The entries can be dated only approximately. They forecast some experimental work under way at the Edison and Murray shop during the summer and fall of 1873. Similarities of style and subject suggest that Edison made them all in a short time. They are closely related in subject to the notebooks he used in England but are prospective rather than observational in character.[1] The entries relate primarily to arrangements for automatic telegraphy on undersea cables or to apparatus that would be used in that work. Some of the entries are cut off at either the beginning or the end; only a small number of leaves of the original notebook survive.[2] Some of the material repeats notes found in the Greenwich Pocket Notebook (Docs. 321–36).

1. It is likely that Edison recorded the entries just after his English trip, but he could have written them as early as his last days in England or as late as mid-July.

2. The covers are missing, as are most of the pages; some of the remainder are damaged. The surviving leaves were long separated in the archive; they have been reassembled and ordered by alignment of signature holes and page-edge marbling. Some of the disassembled leaves were kept with the other materials from the English trip, and were filmed in that way (interspersed in PN-73-00-00.1, Lab. [*TAEM* 6:822–71]) before the other leaves were found in a mislabeled crate of documents. The reassembled notebook will be filmed as a whole in *TAEM* Supp. III.

–341–

Notebook Entry: Automatic and Cable Telegraphy[1]

[June 1873?]

American Cable Experime~~r~~nts

First= Transmitter & Recvr The transmitter to go by Electric Eng. Worm on nice Rosewood base on hard rubber pillars.

Receiver—ditto base. to take narrow paper less than quarter inch to have great surface on drum to reduce resistance to have convenient slide for different pens. to have English binders[2]

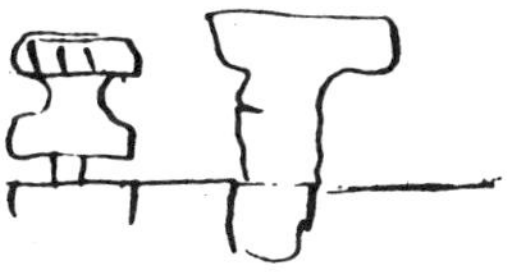

plugs etc to get any desired connection

The paper to be drawn forward by steel tooth wheels nickled so that small pressure is required

Two nice boxes with handles, drawers in base for spare pens expmtl pens etc =

Wheatstone puncher all new = [3]

AX, NjWOE, Lab., PN-73-00-00.3 (*TAEM* Supp. III).

1. See headnote above.

2. In English binding posts the conducting wire was wrapped around a threaded post and clamped with a nut; in American binders the wire was inserted into a hole and clamped with a nut. English posts made better permanent contact. Maver 1892, 559.

3. Edison may have intended to design a new perforator or arrangement of perforations for the Wheatstone automatic telegraph system. No record has been found of work on such a perforator, but for modifications of other parts of the Wheatstone system see Doc. 457; see also *TAEB* 1:592 n. 3.

-342-

Notebook Entry: Galvanometer[1]

[June 1873?]

Sensitive Recording Galvanometer[2]

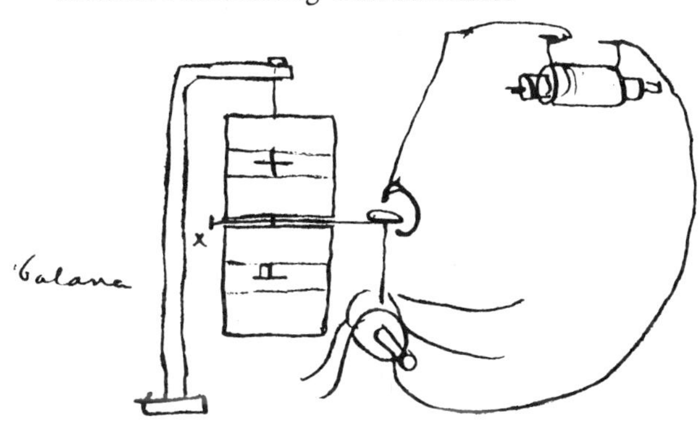

Aluminum pointer with disk over Iron paper jump spark[3] = ozone = in tube etc = [a]

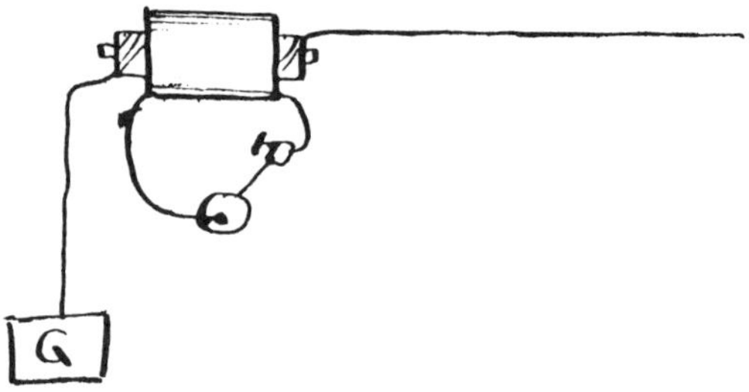

~~T~~Experiment on the thinnest & thickest paper for reduction of R. & sensativeness

Organic & Inorganic colors paper bleached by ozone from pen.

[A]

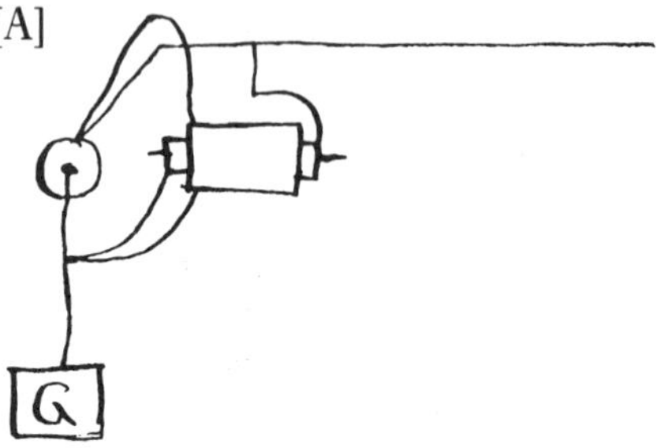

[B]

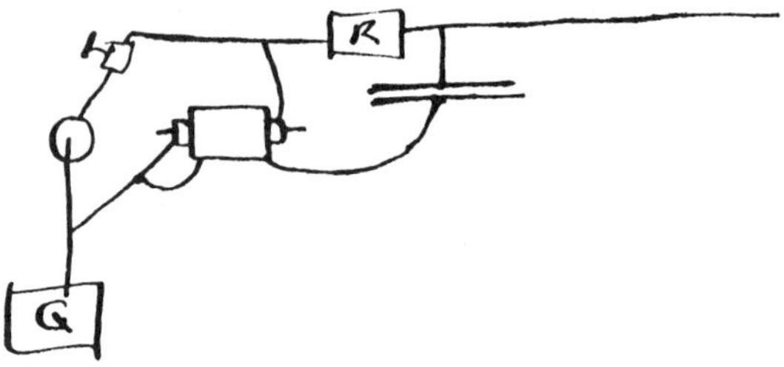

[C]

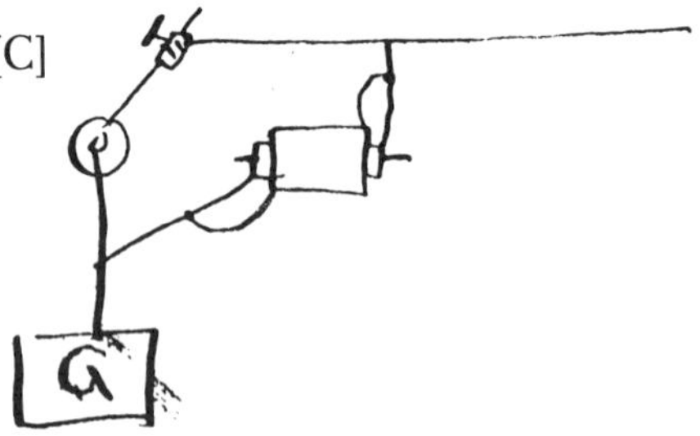

[D]

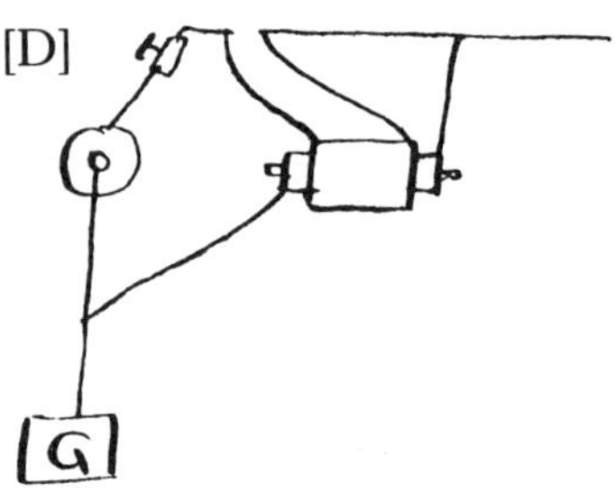

[E]

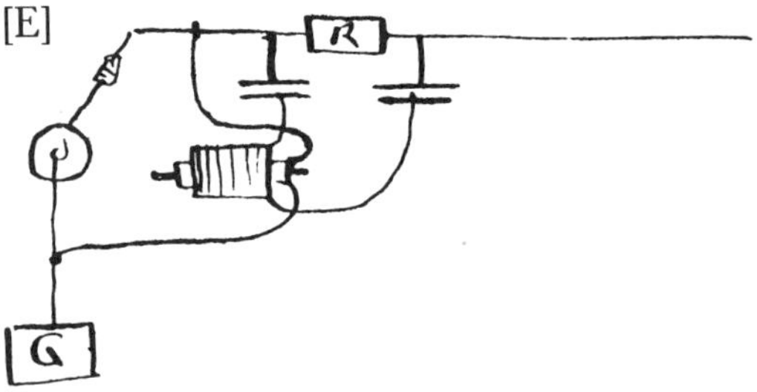

[F]

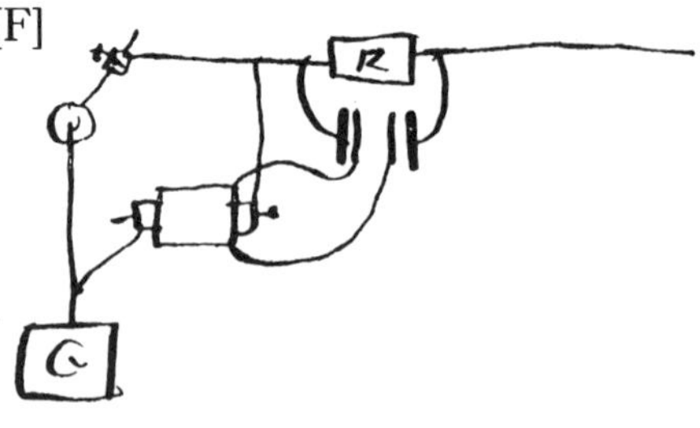

AX, NjWOE, Lab., PN-73-00-00.3 (*TAEM* Supp. III). [a]Page ends here.

1. See headnote, p. 6.
2. A galvanometer measures current strength.
3. The "jump spark" from the pointer to the paper would be gener-

ated by the secondary winding of the induction coil, at upper right. The circuit to be measured, through the primary winding, is not shown.

–343–

Notebook Entry: Automatic and Cable Telegraphy[1]

[June 1873?]

Apply a small condenser to each printer in a circuit of twenty instruments to stop spark, also to all relays in W.U. to Universals[2] to stop spark and make them act quick.

Make Model—no=

Attend to Quantizing[3] Automatic batteries, and Shunts attended to[a]

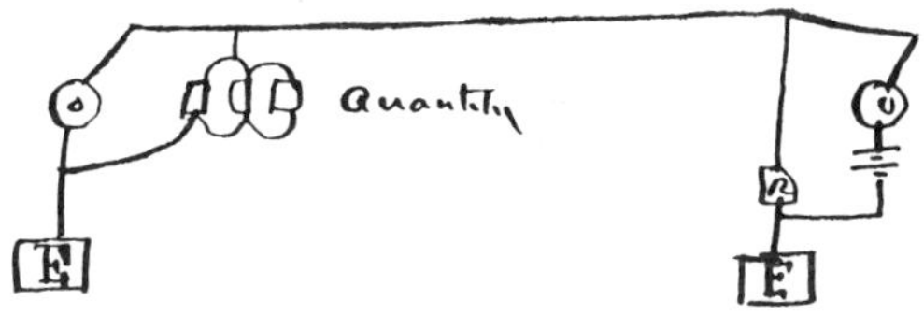

connect a Relay spool opposite so that the magnet will pull nothing see if there is any charge and if it is not in the opposite direction to that when connected properly.

Test the difference in the amount of the discharge from a magnet with the armature off and the armature on.

test the length of the discharge on a fine wire spool & Coarse ditto with same battery

Test quantilizing a number of very fine wire spools with seperate cores. Then all on one core.

Test iron wire spools & magnets

Magnetic[4] shunt of high Res & use low battery power.

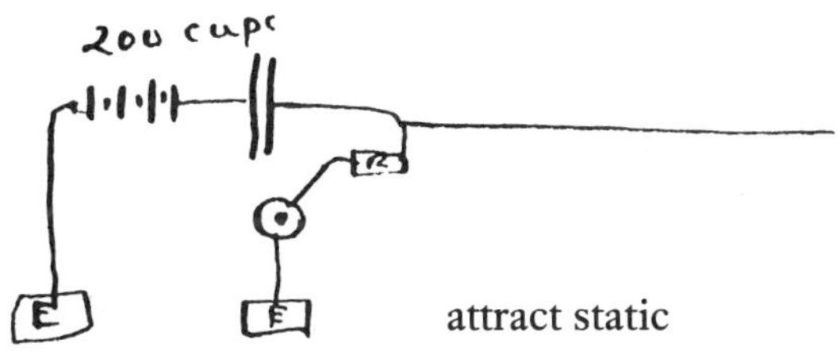

attract static

attraction of currents

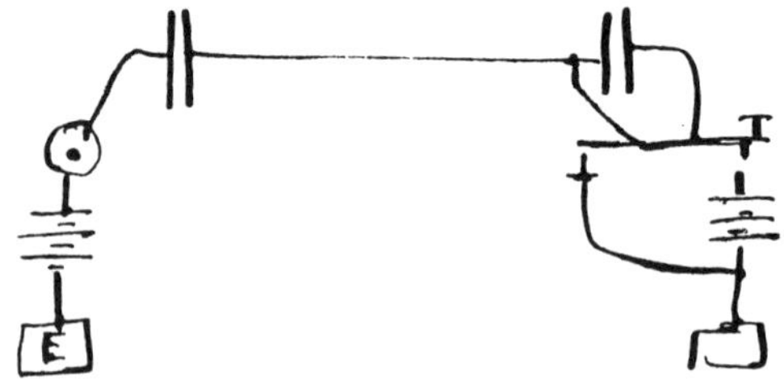

Magnetic Res with adjustbl Cores to vary discharge.

AX, NjWOE, Lab., PN-73-00-00.3 (*TAEM* Supp. III). [a]Page ends here.

1. See headnote, p. 6.
2. Universal stock printer (Doc. 195).
3. By "Quantizing" ("quantilizing" below), Edison means arranging in a parallel (quantity) circuit. See *TAEB* 1:96 n. 6.
4. Inductive.

–344– [June 1873?]

Notebook Entry: Automatic and Cable Telegraphy[1]

Cable Perforations[2]

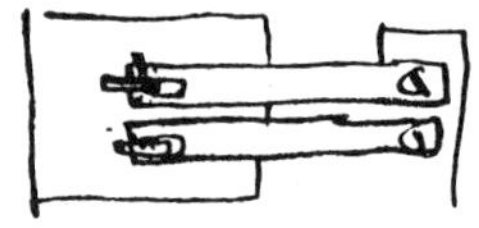

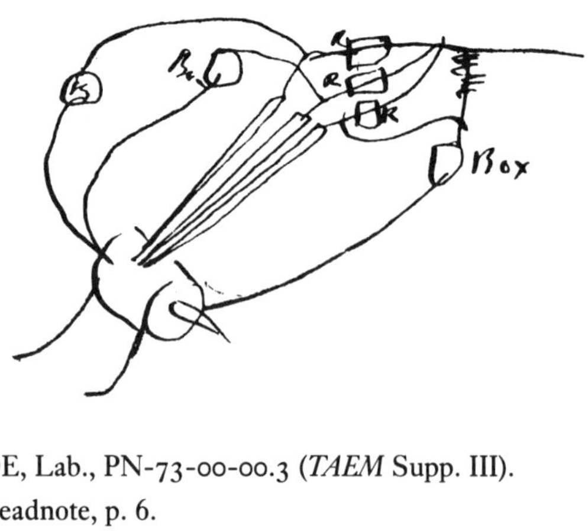

AX, NjWOE, Lab., PN-73-00-00.3 (*TAEM* Supp. III).

1. See headnote, p. 6.
2. In two undated fragments, probably written before his trip to England, Edison indicated the use of very small holes in the perforations

to keep the line statically charged in order to overcome tailing problems (Cat. 297:12[3–4], Lab. [*TAEM* 5:479–80]; on tailing see *TAEB* 1:149). See also Doc. 329 for other perforation designs and Doc. 328 for a similar drawing of recording pens.

"Box" in the drawing refers to Edison's magnet box. *TAEB* 1:608 n. 2.

–345–

Notebook Entry: Automatic and Cable Telegraphy[1]

[June 1873?]

instrument can be made to work. possibly several local cells added to recg paper <u>N</u> would allow N waves to preponderate. this would give the required instrument.

Test the effacy of glass tubes filled tightly with flour of gas retort carbon, for constant resistance & high[2] ⟨no g⟩ use heavy glass tubes with 1/16 or 1/32 bore, & platinum terminals—& well sealed with hard rubber & Gutta P. Have Dr B[3] Measure them & have a box of 1 million ohms made of ~~1~~50 000 ohms each & one of 100,000 of ~~3~~5000 ohms each, with plugs Pol Mahog. ⟨No good⟩[4a]

Try this

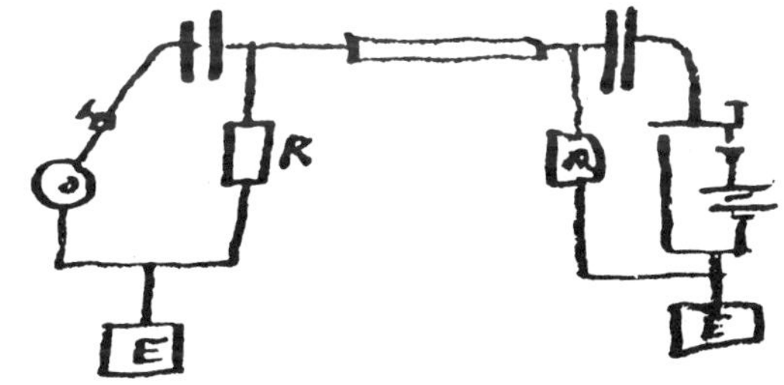

Varleys being this way

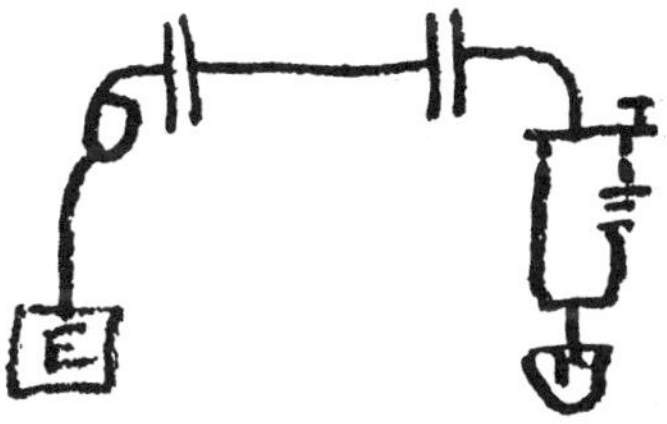

obtain some metal that will not decompose iodide, then[b] atta~~t~~ch it to platina pen to reduce resistance

see through 500,000 ohms with 3 cups of battery, if anything can be put into the Iodide that will reduce its resistance or improve its delicacy

Try and see which two cups of battery D or Grove[5] will give the mark the quickest

It may be that signals could be given by 20 cup Grove & spaces by twenty cup Daniels both P to line.

Secondary batteries have the advantage over a Condr that they send a longer current have six large ones made of Lead. test. line, or shunt

have a magnet made Thus of 50 foot in length wound with one layer of 36 silk cov wire total length Thus

back and forth an put in a nice box

Test fine iron wire bundle cores in compensating magnets. also in the secondary coil compensatn

Keeping the cable statically charged

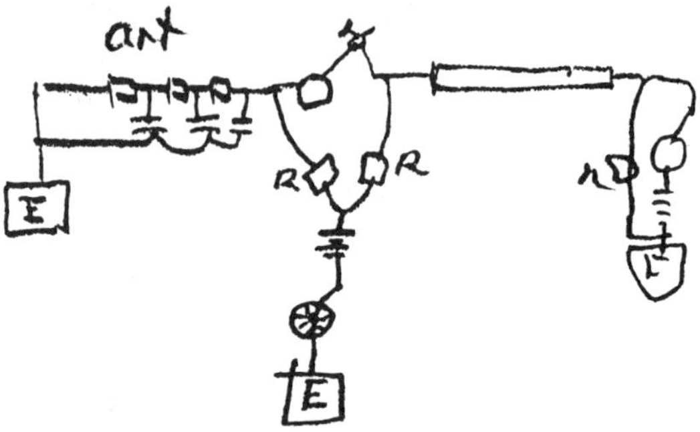

have a clockwork or Magnetic engine vibrator[6] made with break wheel thus.

Double crown platined so single or P & N can be sent.

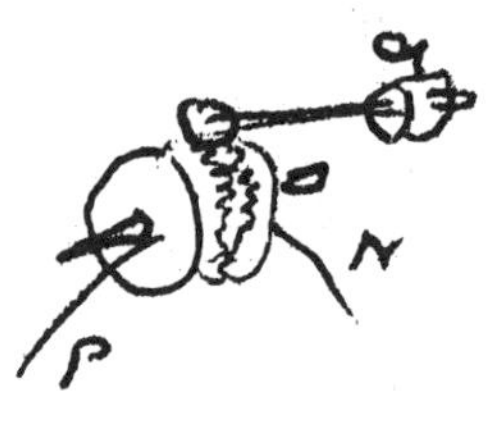

with brush to keep clean

Make a Condenser of ½ inch strip ½ mile long connected thus

AX (fragment), NjWOE, Lab., PN-73-00-00.3 (*TAEM* Supp. III). The middle pages of this signature are missing. [a]Cancels first half of paragraph. [b]"n" interlined above.

1. See headnote, p. 6. This document contains versions of Docs. 334 and 335.

2. Edison here sought a convenient way of matching the very large

resistances of oceanic cables in an instrument for experiments in his workshop.

3. Leverett Bradley was known for his patented magnetic helices and electrical measuring instruments. See *TAEB* 1:141.

4. Edison had such a resistance instrument built in his shop after his return to Newark and found that the resistance produced was not constant. See Doc. 351.

5. On the Daniell and Grove batteries, see *TAEB* 1:615 nn. 3–4.

6. Cf. Docs. 326 and 327.

–346–

Notebook Entry: Miscellaneous[1]

[June 1873?]

[ ]ms[a]
[ ]no =[a]

Arrange one or more Condensers ~~w~~connected to Earth with delicate revolving armature Engine having high resistance spools to that the condrs will be discharged ~~from~~ through the engine & give a constant rotation

~~M~~Small magneto—no perm Gramme Compound Int[erest] prin[ciple][2] applied WU wires 50 seperate in a rack
~~S~~Coal chpr than Zinc[3]

ascertain if some magnetic arrangement could be made to attach to every Knot or 10 K of a cable which should counter equal the electrostatic capacity of that number of Knots, if they are constant, & increase & decrease with each other in the same prop[or]t[io]n

Disks Vulcanite faced 1/8 circle zinc other 1/8—1/8 C[arbon] each equal to zinc, revolve rapidly very close together. connect C with one side Gal[vanometer] [ ][a] paper Z with other [ ]ing.[a] get indicatns

AX, NjWOE, Lab., PN-73-00-00.3 (*TAEM* Supp. III). [a]Corner of paper missing.

1. See headnote, p. 6. Cf. Doc. 331.

2. This is Edison's initial impression of a Gramme dynamo, which he saw for the first time while in England. A generator working on the dynamo principle—discovered in the late 1860s—does not have permanent field magnets and generates only a feeble current when it starts. However, that current strengthens the field magnets, thereby generating a stronger current, which further strengthens the field, and so on, eventually generating far more electricity than can a permanent-magnet (magnetoelectric) machine. Edison likened this positive-feedback loop to compound interest.

3. Burning coal to produce steam power (which can drive an electric generator) is cheaper than producing the equivalent electric power with primary batteries. This had been the subject of considerable study. See, for example, Hunt 1857.

–347–

To Norman Miller[1]

[Newark, c. August 1, 1873[2]]

After an almost infinite amount of experimenting I have a duplex working in shop two messages in the same direction, way stations, etc.[3] I want the loan of three duplex sounders and one Phelps ordinary relay for a week.[4]

I am ready to put the new duplex in operation first between New York and Phila.

I have been very sick in bed—have had the most interesting features of 4,000 nightmares in the day time. Cause—root beer and duplex.

PL (transcript), Quad. 71.2, p. 13 (*TAEM* 10:228).

1. At this time Miller was secretary and treasurer of the Gold and Stock Telegraph Co. For his relationship with Edison see *TAEB* 1:519.

2. Edison testified that he wrote this note to Miller, but he could not specify the date (Quad. 70.7, pp. 269, 338, 345–46; testimony of William Orton, Quad. 71.1, p. 235 [*TAEM* 9:501, 536, 540; and 10:124]). Since this note announces the completion of the diplex invention described in Doc. 348 on 2 August, it probably dates from approximately the same time. Miller brought the requested equipment to Newark when he witnessed Doc. 348 on 21 August 1873. Equipment list, Quad. 71.2, following p. 14 (*TAEM* 10:230); see also testimony of George Phelps, Jr., Quad. 71.1, pp. 306–7, 314 (*TAEM* 10:160, 164).

3. In telegraphy "duplex" generally meant one wire carrying two simultaneous messages in opposite directions. Eventually "diplex" became the preferred label for two simultaneous messages in the same direction.

4. On "Duplex sounders" see *TAEB* 1:101 n. 3; and Doc. 348 n. 21. A Phelps relay was a standard relay manufactured at the Western Union factory, which was managed by George Phelps (*TAEB* 1:30, 135 n. 2).

–348–

Draft Caveat: Multiple Telegraphy

Newark August 2nd 1873.

Caveat.[1] Duplex Telegraph.[2]

I Thos A Edison of Newark etc etc.

The object of this invention is to transmit two or more telegrams in the same direction over one telegraph wire at the same time.[3]

The invention consist of combining a receiving magnet which works by a reversal of the direction of the current in the wire ~~wi~~the directive force being any magnetic constant or electric constant, with an electromagnet which will respond to either current and worked by incease and decrease of the currents of opposite polarity sent over the wire.[4]

It further consists in ~~arran~~ combining two local circuits with sounders with the relay which works by increase and decrease of the current, also in working the~~s~~ first ~~e~~local circuit by the back point of the relay and the second circuit by the back point of the sounder in that circuit[5]

My invention further consists in the manner of reversing the direction of the current and increasing and decreasing the same[6] It further consists in places the same receiving apparatus as used at the terminal stations at intermediate stations so the both despatches may be received by ~~any~~ as many stations as is desired.[7]

It further consists in placing in the circuit at the transmitting end a relay worked either by increase or decrease or by reversals and a key at the receiving end which totally breaks the circuit for stopping the transmitting operators. the opening of the key at any station stops the relay responding to the signals of the transmitting operator[8]

I will mention here four messages may be sent by this apparatus two in one direction and two in the other, by winding the bobbins of all the magnets double.[9] The increase & decrease and reversed current at one station ~~t~~are sent through the wire without effecting either relay at that station upon thc same plan as ~~usually~~ generally used, balancing the effect by the double coils and secondary artificial circuit.[10]

figure 1 shows the relay[a] and local circuits, worked[a] by increase and decrease of the P & N[b] currents.[11]

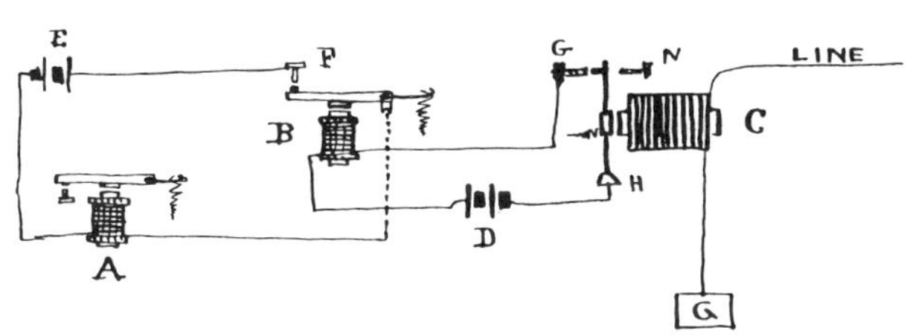

C is the relay which responds to either a positive or negative current[12] as either one current or the other is always on the line the relay lever remains drawn to the face of the magnet when the battery is full strength. When the lever is away from the point G the repeating sounder B is open ie[c] its lever remains in contact with the point F and the receiving sounder A is closed = just the same as if the point N were used and the sounder A worked direct in the usual way When the lever H is in contact with the point G the lever of B is drawn away from the point F. and the circuit of A opens.

The object of using the back point[13] and ~~see~~extra circuit[14] is that when the polarity of the current is changed at the moment of change the lever H will fly away from the face of the c[--]ores ~~no r~~ but be quickly reattracted after its cores have changed their polarity. This opening takes place always no matter how light the retractile force of the adjusting spring may be or how weak or strong the battery is.[15] This would interrupt the local circuit continually and prevent the reception of signals were the ~~e~~local circuit arranged in the usual way but by referring to the drawings it will be seen that the lever H must come in contact with the point G and the magnet B charge and the lever be drawn away from F before the receiving sounder will be interrupted As the time which this takes is greater (when the spring of B is strong) than the time which it takes to reverse the polarity of the cores of the relay .C. the alternating of the currents upon the line has no effect upon ~~this receiever~~ the recieption of the signals. If reversals are being sent over the line with a half battery, the spring of the lever H draws it away from the cores and the sounder A is open ready to receive signals When the battery is doubled by a depression of the ~~so~~key at the transmitting end the lever H is attracted and A closes. While thus closed the other operator should be sending reversals they will have no effect on A, but vibrate the lever H slightly.[16] The reversed currents are received upon a polarized relay of any form or construction placed in the same circuit as .C. Any increase or decrease in the strength of the current does not ~~e~~affect it, the tongue being thrown to the right or to the left by a ~~e~~positive or negative irregardless of its strength. figure two shows the connections at the receiving station.

fig 2.

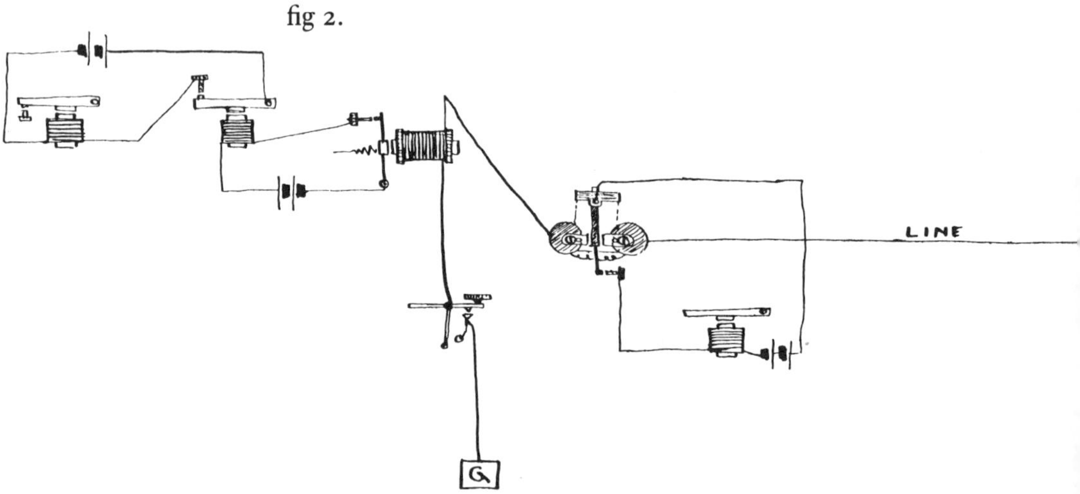

~~fig~~

figure 3 shows the connections at the transmitting station[17] fig 3.

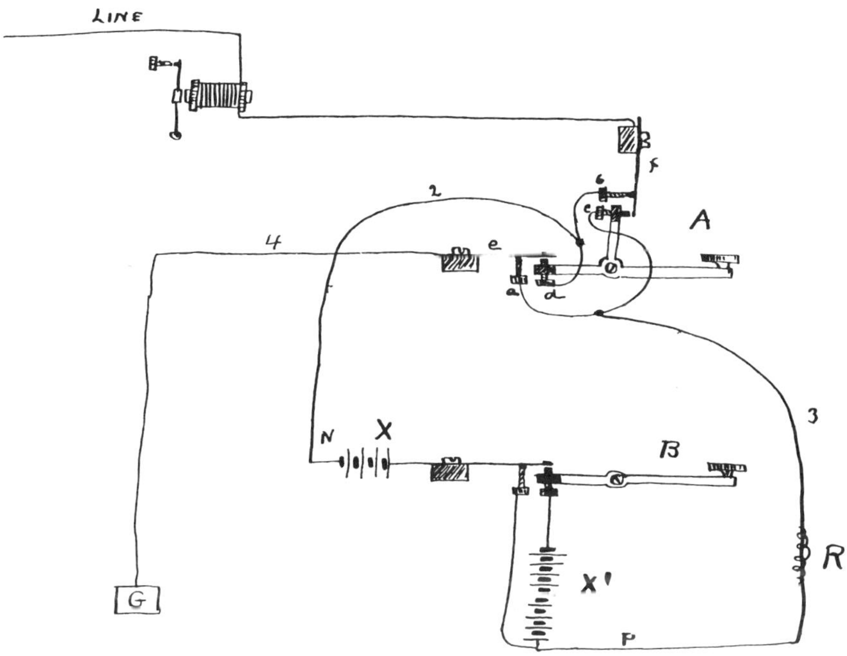

The key A reverses the direction of the current upon the line. the line is connected to the spring f and the ground wire to the spring .e. The P pole of the battery is connected to the points. a & .c. and the N pole to the points b and .d. When the key is raised the springs f and e rest upon the points a.b. then the circuit is completed by line to wire 2 through battery and secondary key to wire 3 thence to ground with N to line. When the key is depressed the line passes down through wire 3 through secondary key & battery thence by wire 2 to e and ground with P. to line.[18] A resestance R is inserted to regulate the spark at the contact points of the key A. in the act of reversing.

The number of elements of the battery is increased or decreased by the key B. When it is open the battery X alone remains in circuit. When depressed the battery X′ is added which has nearly twice the number of elements The relay is ~~a~~inserted in the line and so adjusted that it will respond to the working ~~of either key~~ of the key B and be interrupted when the circuit is broken ~~at~~by the receiving station, the object being to notify the transmitting stations that the signals are not being received or a repetition of a word is desired.[19]

I do not wish to confine myself to any particular method of

increasing of decreasing the current or reversing it, as it may be done as well ~~on~~by the plan shown in figure 4

fig 4

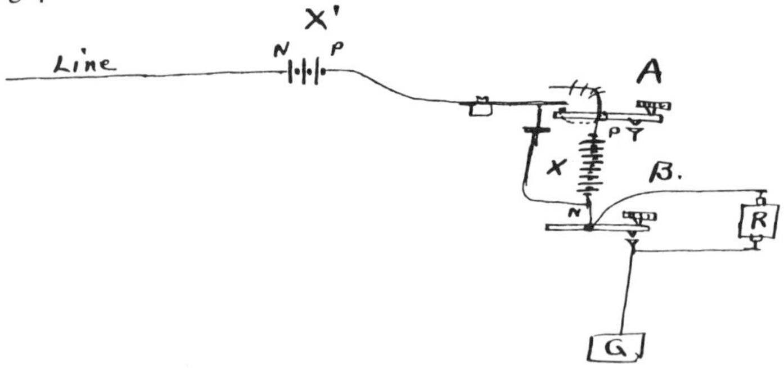

When the key A is open the battery X is out of circuit, but X′ is in circuit and with its N pole to line. This battery has but half the number of elements as X. When the key A is depressed the battery X is added to the line half of its cups goes to neutralize the battery X′ and the remainder give a current equal to that generated by X′ but in the opposite direction (ie) P to line. One advantage of this plan is that the current from the batteries cannot be "short-circuited" at the contact points at the moment of reversals and draw~~n~~ the current[d] from the line for an instant, ~~as when but one battery~~ like that shown in fig .3. as one battery remains constantly in the line and the change in polarity is done electrically not mechanacally.

The current in this second plan is increased or decreased by placing the resistance in & out of circuit by the key .B.[20] Another plan is shown in figure 5

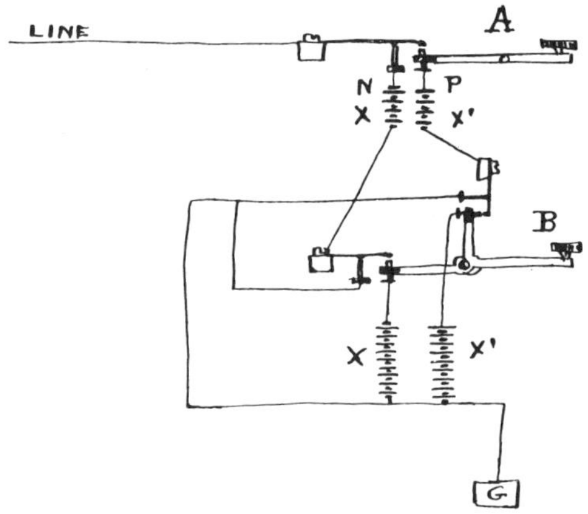

The current upon the line is reversed by .A. when A is open the battery X gives an N current when depressed the battery X′ gives a P current.

The key B increases and decreases the number of elements of the two batteres.

All these methods of reversing the direction of the current and increasing and decreasing its strength have been arranged with keys but in practice I prefer to attach them to a local magnet operated by a key.[21]

My claims will be learned from the first part of the specification.[22]

Thomas A Edison

⟨Witness, Augt 21/73 Norman C Miller⟩[23]

ADDENDA

[Newark, August 1873?[24]][e]

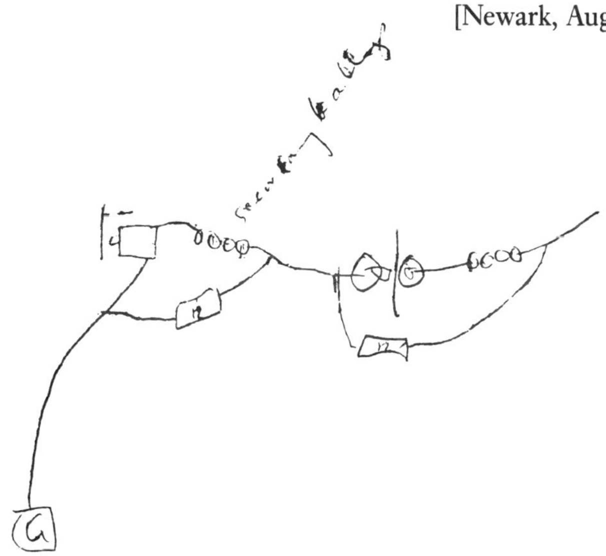

[Newark, Summer 1873?[25]][f]

Fig. 1^{A}

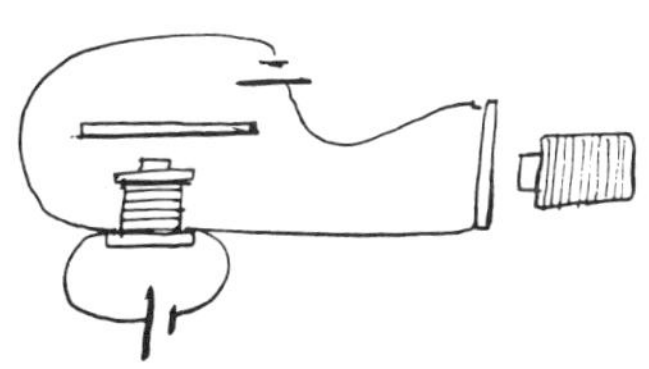

Fig. 1^{A} is a modification illustrating two batteries of equal power that are ordinarily opposed to each other so that the magnet is inoperative except when the relay breaks one circuit and allows the other to operate in the magnet.

ADS, MdSuFR, RG-241, *Nicholson v. Edison,* Exhibit draft Caveat 45. [a]Preceded by superscript "x" in unknown hand. [b]"P & N" interlined above. [c]"ie" circled. [d]"the current" interlined above. [e]Addendum is an

AX. [f]Addendum is a D (copy), in the same hand as the finished caveat filed at the Patent Office.

1. Edison drafted this in the beginning of August to record his priority while still experimentally testing his concept. He worked intensively on it as soon as he returned from his trip to England, producing the design presented here and in Doc. 356 (testimony of Edison and Charles Batchelor, Testimony and Exhibits on Behalf of T. A. Edison, pp. 24–25, 65–66, *Nicholson v. Edison*), but almost three weeks passed before he had it witnessed (see n. 23). Lemuel Serrell, Edison's patent attorney, later marked portions of the text for a clerk to follow in drafting a declaration of probable claims. At some time an alternative design for part of the apparatus was inserted (see nn. 11, 24) and the text was edited at a few points; a fair copy of the revised whole was signed on 28 October 1873 though not filed until the summer of 1874 (Edison Caveat 45; a printed copy is on pp. 155–63c, Testimony and Exhibits on Behalf of T. A. Edison, *Nicholson v. Edison*). Apparently due to an oversight, the original draft of 2 August 1873 was not put in evidence during the extensive litigation surrounding the resulting patent application (the Quadruplex Case). Edison's and Serrell's testimony, pp. 13–17, 60–63, 101–6, ibid.

2. See Doc. 347 n. 3.

3. This became the first claim of the revised caveat.

4. This became the second claim. The descriptions generally would have been understood as referring to a polarized relay and a standard relay, respectively (as used in the "Case H" duplex design, Doc. 315), but are cast in somewhat broader language, possibly to try to include application of the basic approach to modes of telegraphy other than the American "Morse" system (for ordinary relays see *TAEB* 1:10, 30; for polarized relays see Doc. 12; and for "receiving magnet" see Page 1866). Edison may have initially thought of using this plan of diplexing in connection with an automatic system as the British later did (Uriah Painter's testimony, pp. 89–92, Testimony and Exhibits on Behalf of T. A. Edison, *Nicholson v. Edison*; Preece and Sivewright 1891, 200). He may have tried it experimentally with Wheatstone automatic equipment in the fall of 1874, and with Hughes-Phelps combination printing telegraph equipment as well (Reid 1886, 735; Georges d'Infreville's testimony, Quad. 70.7, p. 101 [*TAEM* 9:415]; George Prescott's testimony, Quad. 71.1, p. 327 [*TAEM* 10:170]).

5. This became the third claim.

6. This became the fourth claim.

7. This became the fifth claim.

8. This paragraph became the sixth claim.

9. The value of the innovation depended on this possibility of elaborating the design into one for quadruplex telegraphy, though the list of probable claims did not include this point (see also Docs. 446 and 472). As Edison knew, telegraph engineers and scientists had recognized for decades that in principle an effective diplex arrangement could be combined with a practical duplex method to carry four messages on a wire at one time (Prescott 1877, 829, 833–35; Sabine 1867, 163; Heaviside 1873, 431–32; Doc. 300; testimony of Edison and Charles Batchelor, Testimony and Exhibits on Behalf of T. A. Edison, pp. 23–24, 64, *Nicholson v. Edison;* Edison's testimony, Quad. 70.7., p. 268 [*TAEM* 9:501];

Orton's testimony, Quad. 71.1, p. 127 *TAEM* 10:68]). The first really practical duplex methods for long lines had been developed by Joseph Stearns and widely introduced by Western Union in 1872, but diplex methods up to this point were judged "beautiful in their way, but quite useless in a practical point of view" (Docs. 50 n. 3 and 283 n. 3; Sabine 1867, 163).

Doubly wound electromagnets were the basis of all differential methods of duplex telegraphy (see Doc. 446 and *TAEB* 1:32); however, Edison's diplex design could also be combined with bridge duplex arrangements as shown in the accompanying illustration. See also Docs. 285 n. 17, 309 n. 1, 449, 485, 494 n. 1, 512, and 513; Edison's testimony, pp. 32–33, Testimony and Exhibits on Behalf of T. A. Edison, *Nicholson v. Edison*.

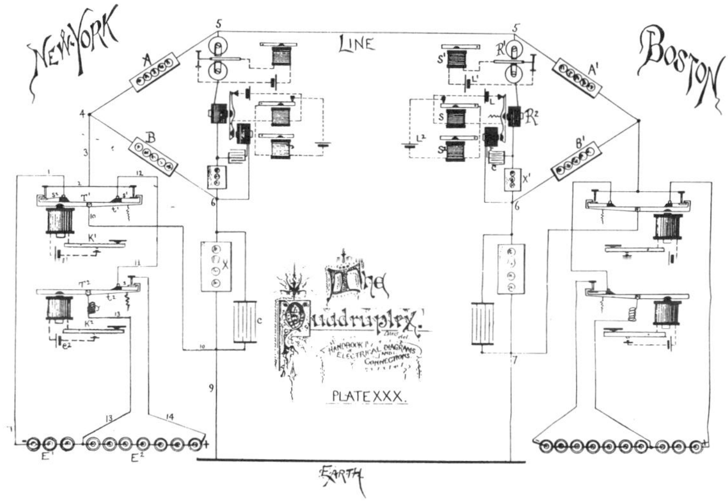

This 1875 quadruplex design combines Edison's diplex with the standard bridge duplex arrangement (Davis and Rae 1876).

10. This description applies the general idea of differential duplex telegraphy to Edison's diplex design. See Docs. 23, 28, 285, and 301.

11. A brief description of an alternative design, with an accompanying drawing (fig. 1^{A}), was later inserted here. See second addendum to this document.

12. That is, a standard or "neutral" relay, which responds in the same way to either kind of current, as distinct from a polarized relay. Edison's testimony and exhibit drawing, pp. 4–6, 199, Testimony and Exhibits on Behalf of T. A. Edison, *Nicholson v. Edison*.

13. The back point of the relay **C** is **G**; normally the front point of a relay (here **N**) was the contact point.

14. The "extra circuit" (**B G H D**) is the one with the "repeating" rather than the "receiving" sounder in it.

15. This break in the magnetism was the basic problem for any design along these lines; Edison had been grappling with it intermittently for some time. See *TAEB* 1:143–44, 508; and Docs. 88, 275–77, 285, 294, 297, 300, and 315.

16. This is the first "bug trap," the unwanted movement of the ordinary relay's armature lever **H** being the "bug" (Edison's testimony, Quad. 70.7, p. 269 [*TAEM* 9:501]). Edison needed to eliminate the false breaks caused by current reversals in order to make a practical diplex or quadruplex telegraph system of this type. The design in the second ad-

dendum to this document (fig. 1[A]) is another bug trap; eventually many more were invented. Prescott 1877, 8, 38–53; Maver 1892, chap. 12.

Edison conducted many experiments of limited success late in 1872 and early in 1873 trying to use both reversals of current direction and variations in current strength to send two messages the same way simultaneously on one wire (*TAEB* 1:507–8; Docs. 275–79, 285, 294, 300, and 315; Edison's testimony, Quad. 70.7, pp. 228–29, 302 [*TAEM* 9:481–82, 518]; Edison's testimony and preliminary statements, Testimony and Exhibits on Behalf of T. A. Edison, pp. 3–4, 10–13, 24, 29–30, 43, 147–54, *Nicholson v. Edison*). See also Docs. 36, 48, 50, 63, and 88 for his earlier efforts.

Edison said he thought of a new approach to the problem—that is, using a relay's back point combined with a repeating sounder to get over the moment of no magnetism—while he was in England (Edison's testimony, Testimony and Exhibits on Behalf of T. A. Edison, p. 24, *Nicholson v. Edison*). He had already used back point contacts and other aspects of this arrangement, notably in some printing telegraphy designs also done in the first half of 1873 (U.S. Pats. 139,128, 139,129, 140,487, and 140,488). The arrangement used here apparently evolved in the context of Edison's continuing work with another of the general types of circuitry that he considered extensively and experimented with in his work on multiple telegraphy (see the early material cited in Doc. 285 n. 12; Docs. 286, 297, 298, 312, 485, 512, and 533; U.S. Pat. 147,917; Cat. 297:37(2), 81(1), 81(3) back, 85(3), 90(4) back, 125(2), 127(2), 144(2), Lab. [*TAEM* 5:570, 757, 760, 784, 817, 897, 908, 960]; and Cat. 299:69, Lab. [*TAEM* 6:131]). The accompanying illustration, probably from the spring or summer of 1873, shows such a multiple telegraph design, in which Edison replaced two front point contacts with back point contacts.

A circuit design with back point contacts.

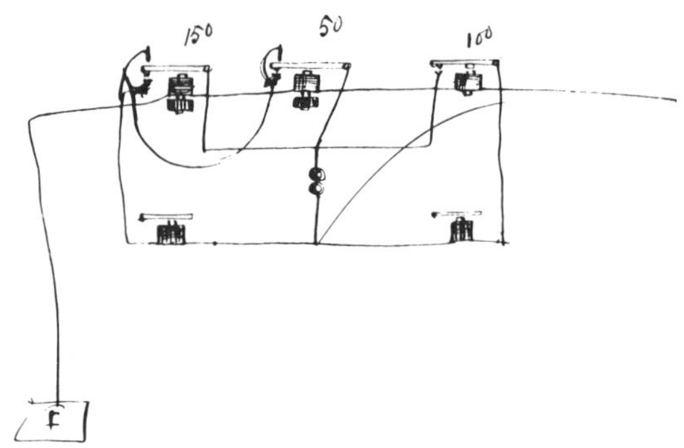

17. Figures 2 and 3 combined provide a basis for comparison with drawings in Docs. 285, 300, 315, and 472. Although Edison had not yet worked out most of the practical details of the design, the general principle is close to that of the design in Doc. 472, Edison's key patent application for quadruplex telegraphy.

18. The arrangements of the transmitting keys in figures 3, 4, and 5 had to maintain a connection to the ground to keep the circuit unbroken and thus not disturb the signals made by the other sender.

19. The accompanying illustration shows an earlier stage in Edison's work on this apparatus.

An earlier design of one of Edison's quadruplex circuits.

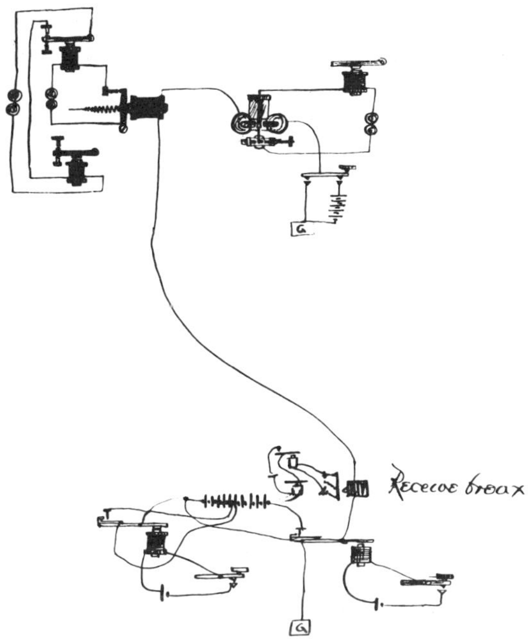

20. This and the next variation occur frequently in Edison's designs. See Docs. 24, 275–78, 285, and 297.

21. This refers to using a local circuit with a special transmitter in duplex operation, an arrangement pioneered by Joseph Stearns in 1868. The key would activate the transmitter's magnet to move an armature lever that served both as the switch in the main circuit, thus making and breaking contact more crisply than would a purely hand operation, and as the lever of a sounder, thus allowing the sender to monitor a message as it was keyed. Since the duplex apparatus was designed to be unresponsive to outgoing signals, this sounder was particularly important for American operators, who normally worked by sound. Each transmitter also needed connections to ensure that it never broke the main circuit and interfered with the other signal; the design depended on whether the transmitter switched current polarity or just varied its strength. Prescott 1877, 792–96, 853–58; Maver 1892, 180; see also Docs. 23 n. 2, 28, 50 n. 3 (and illustration), and 285 n. 7.

22. In the completed caveat the declaration of probable claims replaced this sentence.

23. Miller witnessed this caveat draft when he brought to Newark equipment that Edison had requested from Phelps's Western Union shop in New York (Doc. 347 n. 2; testimony of George Phelps, Jr., Quad. 71.1, pp. 306–7, 315 [*TAEM* 10:160, 164]). Miller later recalled that he had taken this draft with him, showed it to President Orton of Western Union, and then delivered it to Edison's patent attorney, Lemuel Serrell. Uriah Painter had already seen the system in operation, and Edison told Miller the design had been successfully tried, but Miller did not recall having seen the instruments set up. Miller's and Painter's testimony, Testimony and Exhibits on Behalf of T. A. Edison, pp. 90–91, 97, *Nicholson v. Edison.*

24. This drawing was done below the signatures, in pencil rather than the blue ink Edison used in the rest of the draft. Since Miller took the document with him on 21 October, the drawing was probably done by then, although it could date from 28 October, when Edison was at Serrell's office in New York and signed the finished version of the caveat along with Docs. 373–75. This design uses secondary batteries and a shunt around the neutral relay either to compensate for inductive effects or as an alternative bug trap. Cf. Cat. 1171:97–99 (*TAEM* 6:256); and Docs. 357, 359, 392, 513, and 531.

25. This design probably postdates Miller's witnessing of the draft caveat and predates the preparation of the finished version (executed on 28 October 1873), which included it. Cf. also Doc. 472.

ROMAN-LETTER PERFORATOR Doc. 349

Edison had been working since the summer of 1871 on an automatic telegraph system that would record a message in roman letters. Such a system would save the time and cost of translating messages from Morse code.[1] Edison originally undertook the design of a roman-letter system on his own initiative, but on 31 March 1873 he entered into an agreement with George Harrington and Josiah Reiff under which they agreed to pay development costs.[2]

The first perforator with twenty-five punches was built between 12 and 30 March 1873.[3] Nothing is known of this design, but it could be the same as that embodied in Edison's

The puncher and paper tape of Edison's standard perforator design, described in U.S. Patent 121,601.

One of the production models of Edison's roman-letter perforator.

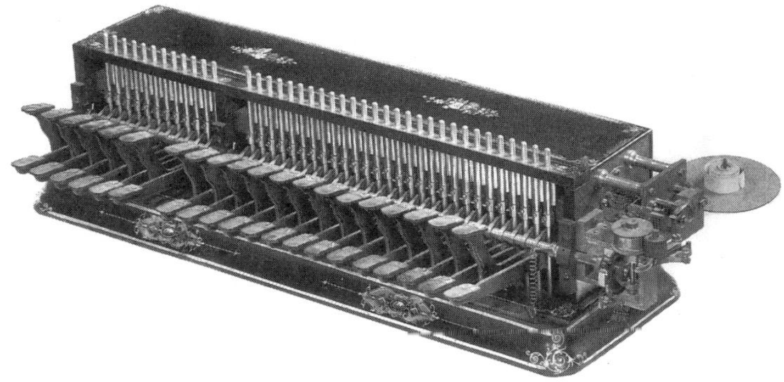

U.S. Patent 151,209. In the patented perforator a square matrix of twenty-five punches replaced the two rows of dot-and-dash punches used in his original large perforator.[4] At least two of these perforators were manufactured in October 1873, one of which may still be extant.[5] In his roman-letter system employing this perforator, transmission required five transmitting rollers, five line wires, and five pens at the receiving end, which recorded letters made up of dots that matched the perforations.[6] Edison subsequently experimented with a roman-letter system using a perforator with a square matrix of forty-nine punches, a transmitter with seven rollers, and a receiver with seven pens.[7]

1. See Docs. 183, 184, and 186.

2. Doc. 295.

3. Cat. 1214:31, Accts. (*TAEM* 21:579).

4. U.S. Pat. 121,601; see also *TAEB* 1:474. One of Edison's photograph albums (Cat. 558) shows the original punches and the tape they produced.

5. Cat. 1214:34, Accts. (*TAEM* 21:581); Acc. 29.1980.294, MiDbEI(H).

6. The perforations (fig. 1) and received letters (fig. 2) are shown in the patent drawings. U.S. Pat. 151,209.

The perforated tape and received message for Edison's roman-letter automatic system (U.S. Pat. 151,209).

Fig. 1.

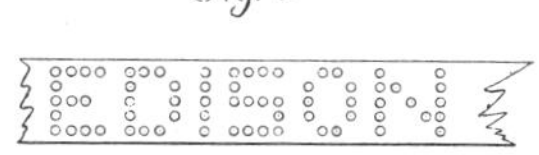

Fig. 2.

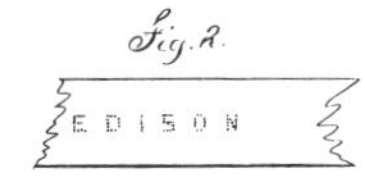

7. No record of this experimental work is found in the notebooks. However, a perforator (Cat. 270) and a modified transmitter-receiver (Cat. 1527) for such a system can be found at the Edison National Historic Site, and a transmitter (Acc. 31.1217.129) can be found at the Edison Institute, Dearborn, Mich.

Patent Model: Automatic Telegraphy[1]

M (30 cm × 14 cm × 14 cm), MiDbEI(H), Acc. 29.1980.1350.

1. See headnote above.

2. Edison executed the covering patent application on this date (U.S. Pat. 151,209).

–350–

Agreement with George Harrington, Josiah Reiff, John Fleming, Robert McIlwraith, William Nicol, and John Puleston

September 2, 1873[a]

This[b] Deed or Agreement is made the second day of September One thousand eight hundred and seventy three between Thomas Alva Edison of Newark in the State of New Jersey in the United States of America of the first part, George Harrington[1] of Washington in the District of Columbia and Josiah Custer Keiff[2] of the City and State of New York both in the said United States of the second part, John Fleming, Robert McIlwraith and William Nicol of 17 and 18 Leadenhall Street in the City of London sometimes hereafter referred to as "Smith, Fleming & Co.,"[3] of the third part and the said John Fleming and I[4c] H. Puleston[5] of in the City of London sometimes hereafter referred to as the Trustees of the fourth part.

Whereas it has been agreed between the said parties hereto of the first and second parts and the said Messrs Smith, Fleming & Co., that the latter shall purchase and acquire from the parties of the first and second parts in consideration of the sum of Ten thousand pounds Sterling a one fifth share or interest in all the inventions patent rights and privileges as well British as Foreign and as well existing as future as mentioned or referred to in the Schedule hereto so far as the same relate to or are or can be made applicable to Land Telegraphs as destinguished from Submarine or Subaqueous Telegraphs.

And whereas consequent upon such Agreement it has been arranged between the parties hereto that all such Inventions and patent rights and privileges as mentioned or referred to in the said Schedule hereto shall be from time to time assigned to the Trustees of thcsc presents and shall be held, worked, used and exercised by them for the benefit of the parties hereto of the first second and third parts respectively upon the terms and in the manner hereinafter expressed.

And whereas in part performance of such agreement and arrangements an indenture has already been prepared and will be executed contemporaneously with the execution of these presents whereby the said parties hereto of the first and second parts together with George Edward Gourand[6] will assign to the said Trustees the two British Letters Patent first and secondly mentioned in the said Schedule and will enter into covenants with the said Trustees for the completion of British Letters Patent in respect of the provisionally protected invention thirdly mentioned in the same Schedule and for the assignment of such Letters Patent when completed to the said Trustees.[7]

Now these presents witness that it is hereby mutually covenanted and agreed by and between the said parties hereto (each party covenanting for the acts and observances of each party respectively and where consisting of more than one person covenanting jointly and severally with the other parties respectively) in manner following that is to say:—

1. In consideration of the payment of the sum of Ten thousand Pounds Sterling to the parties hereto of the first and second parts by Messrs. Smith, Fleming & Co., (the receipt whereof is hereby acknowledged) Messrs. Smith, Fleming and Co., shall be treated as having acquired and there is hereby assigned to them by the parties hereto of the first and second parts two undivided tenth shares or interests in all monies, profits, emoluments advantages or benefits arising or accruing now or hereafter from the making, working, sale, disposition, use or exercise of all or any of the Inventions Patent rights and privileges (existing or future and as well British or Foreign) mentioned or referred to in the Schedule hereto so far as the same relate to or are or can or may be applicable to Law[8] Telegraphs, and also so far as may be consistent with the provisions of these presents there is also hereby assigned to Messrs. Smith, Fleming & Co., two undivided tenth shares or interests in all the said Inventions Letters Patent, Patent rights and privileges themselves and for the like consideration Messrs. Smith, Fleming & Co., shall also be entitled to the several commissions or payments hereinafter provided for and also to the option as hereinafter expressed of acquiring a share or interest in the said Inventions, Letters Patent Patent rights and privileges so far as the same relate to or are or can or may be applicable to Submarine or Subaqueous Telegraphs.

2. The parties of the first and second parts engage at their own cost and expense from time to time to apply for and obtain in respect of the said inventions as mentioned or referred to in the said Schedule all British and Foreign letters Patent, Brevets d'Invention or other authorities granting exclusive rights and privileges in all countries and places (except the United States of America) where such exclusive rights and privileges are or may be granted by the Governments of such Countries and places and they engage to assign or cause the same respectively to be from time to time assigned to or vested in the Trustees of these Presents, and they also engage to maintain, uphold and keep the same on foot under all circumstances and for the longest periods which grants of exclusive rights can be obtained from such Governments respectively.

3. The parties hereto of the first and second parts hereby warrant and guarantee the legal validity of all the said British and Foreign Letters Patent, Patent rights and privileges mentioned or refered to in the said Schedule the Agreement for such purchase as aforesaid by Messrs. Smith, Fleming and Co. having been made upon the express condition and understanding that the validity and sufficiency of all such British and Foreign Letters Patent should at all times be guaranteed maintained and upheld by the said parties hereto of the first and second parts under all circumstances.

4. The inventions, Letters Patent rights and privileges mentioned or refered to in the said Schedule and assigned or to be assigned to the Trustees in accordance with the provisions of these presents shall as and when so assigned be held by the Trustees for the benefit of all the parties hereto of the first second and third parts and be dealt with, sold, disposed of, used, exercised or put in practice by the Trustees in such manner and in such manner only as shall be approved by the parties hereto of the first second and third parts or in the event of a difference of opinion between them as shall be decided by a majority of votes each of the said parties having or being entitled to one vote in respect of each one tenth share or interest possessed by each and in case of an equality of votes then as shall be determined by arbitration in the manner hereinafter mentioned.

5. there shall be only one vote in respect of each one tenth share though the firm or persons interested therein may be composed of several individuals

6. Whenever a firm or two or more persons shall be interested in one or more share or shares then the vote or decision in respect thereof may be given by such individual of the number as may be from time to time appointed to act as the representative of such firm or of two or more persons interested in such share or shares or by the lawfully authorized Attorney or Agent of the parties respectively.

7. The decision or votes of the parties may be given on any occasion on which the parties shall meet or by writing under the hands of the parties or of their representative as aforesaid or of their lawful Attorney or Agent as aforesaid.

8. Messrs. Smith, Fleming and Co. shall use their best endeavors to find a purchaser on terms to be approved by the parties hereto of the first and second parts of such parts, shares or interests of or in the Inventions, Letters Patent rights and privileges mentioned or referred to in the said

Schedule as the parties hereto of the first and second parts may from time to time sanction or approve in writing under the hands of some of one of them or of their authorized Attorney or Agent.

9. All moneys and other considerations received by the Trustees in respect of any sales, licenses or other profitable use, exercise or disposition of the patent rights shall be held by them in trust to appropriate and divide the same in the manner hereinafter mentioned that is to say: In the first place to pay and defray all costs charges and expenses incurred by or occasioned to the Trustees in the execution of the trusts hereof or in any wise incidental thereto or arising out of these presents or which with the sanction of the parties hereto of the first second and third parts shall be incurred in obtaining or purchasing Letters Patent or other exclusive rights in respect of any invention similar or analagous to the said inventions or in bringing or defending with the like sanction any litigatory proceedings for the purpose of preventing the infringement of the Patent rights (but not such costs as are mentioned in clause 16 nor any costs which ought to be borne by the parties of the first and second parts) and in the next place to pay the commissions as provided for by clause 12 as and when the same shall from time to time accrue due and subject thereto to divide the remainder as follows, that is to say—as to such parts thereof as represent receipts in respect of any land lines or telegraphs only to divide the same into ten parts and pay two tenths thereof to Smith, Fleming & Co., and eight tenths to the parties of the first and second parts or some of one of them or their Authorized Agent. And as to such parts thereof as represent receipts in respect of submarine or subaqueous lines or telegraphs only to pay the whole thereof to the parties of the first and second parts or some or one of them or their authorized Agent subject however to the provisions hereinafter contained in clause 13.

10. But if any moneys or other considerations shall be received in respect of both land lines and submarine or subaqueous lines then the parties hereto of the first second and third parts shall if possible determine and agree how much of the said price or consideration shall be considered as paid in respect of land lines and how much in respect of subaqueous lines and if they cannot agree the same shall be determined by arbitration and the respective amounts apportioned to land lines and subaqueous lines shall be disposed of in the manner aforesaid

11. No beneficial interest created by these presents in the Patent rights nor the right to participate in such moneys or considerations as aforesaid (save as herein after provided) shall be sold or disposed of by the parties hereto of the first second or third parts (without the consent of all of the others of them) to any third parties unless the person or party desiring to make such sale or disposition shall have previously given notice to all the others of the name of the proposed purchaser or purchasers and shall have offered to all of the others of them to sell and dispose of the share or interest proposed to be sold on the same or not less favorable terms than those proposed to be paid or given therefor and all of the other parties shall have refused or after the lapse of a reasonable time shall have neglected to accept the offer so made any person or persons so purchasing any such beneficial share or interest shall be bound by the arrangements herein made and shall if required by the other parties personally or by Attorney subscribe or execute these presents or some suitable deed or instrument of accession in testimony of his or their being so bound. Provided always that the parties of the first and second parts or either of them may sell a portion of their share or interest to the said George Edward Gourand if they shall think fit to do so.

12. Commissions.

Smith, Fleming & Co., have purchased and shall be entitled to the following payments or commissions.

As regards Submarine or Subaqueous Telegraphs.

Five per cent on all moneys or considerations received in respect of any sale, license or other disposition of the Patent rights or any of them.

In lieu of the commission in any case or transaction Smith, Fleming and Co., may elect to take an interest as hereafter expressed.

As regards Land Telegraphs.

Five percent on one half of all moneys or considerations received in respect of any license under the Patent rights or any of them.

The like in respect of any sale as regards any separate Country or Countries

The like in respect of sale or disposition of any share or interest in the Patent rights as a whole.

If any such sale or disposition as last named be made at such a rate as would if all the shares were sold yield a price exceeding fifty thousand pounds then a further commission of Five percent on the price obtained calculated upon the principle of reckoning such commissions (were all the shares sold) upon one half of the difference between the assumed value of Fifty thousand pounds and the value of all the shares as based upon the price or consideration actually obtained.

In like manner if any sale for any seperate Country or Countries be at such a rate that were the whole interest in the Patents sold a sum exceeding Fifty thousand pounds would be realized then a further commission of Two and a half percent on the price obtained. Such two and a half percent to be on account of the extra Commission payable whenever Fifty thousand pounds has been actually realized but in no case to be returnable.

Whenever Fifty thousand pounds has been actually realized from any source then the commission will be five percent on the whole price obtained instead of on one half the above Commissions shall not be so calculated as to render any part of the price or consideration in any case subject to double commission.

Any difference of opinion as to whether the right to payment of the commissions has accrued shall be a subject for arbitration within the meaning of the clause hereinafter contained.

It is agreed that any sale of any share or interest under the proviso contained in Clause 11 to the said George Edward Gourand shall not be subject to any commission.

13. In lieu of the Commission payable in respect of submarine or subaqueous Telegraphs Smith, Fleming & Co. shall have the option of acquiring the right to a one tenth share or interest in the Patent rights so far as they relate to Submarine or Subaqueous Telegraphs and to the moneys or considerations received in respect of any sales, licenses or other dispositions so far as afore said at a sum equal to one half of the price given for a tenth share by any other person or persons.

Also in lieu of the Commission which may be payable on sales or licenses (so far as regards Submarines or Subaqueous Telegraphs) in any particular case as for instance for any one Country or for any particular line of Submarine Telegraphs Smith, Fleming & Co., shall have a similar option of acquiring a one tenth share or interest in any individual Patent rights for any country or in the moneys or in the benefits resulting from any individual transaction and if such rights or interests shall be so acquired the Trustees shall pay to Smith, Fleming and Co., their proportion of moneys or benefits in respect thereof.

All necessary notices with full particulars shall from time to time be given to Smith, Fleming and Co., so as to put them in posession of all facts necessary for guiding them in exercising or declining such option and a sufficient length of time (in no case less than one month) shall always be afforded to them within which they may exercise such option.

14. In case any of the Patent rights and privileges mentioned in the Schedule hereto shall not immediately as and when granted or obtained be assigned to the Trustees then in the meantime and until such assignment shall be made the person or persons in whom the legal right or interest in such Patent rights and privileges shall be vested shall hold the same subject in all respects to the trusts and provisions of these presents.

15. Without prejudice to the generality of any of the provisions herein contained the said parties hereto of the first and second parts Do and each of them Doth hereby appoint the said John Fleming and I H Puleston and the survivor of them the true and lawful Attorneys and Attorney of them the said parties hereto of the first and second parts and each of them for the purpose of Soliciting, applying for and obtaining all or any Letters Patent Brevets d'Invention or other like instruments or authorities granting exclusive rights and privileges for or in respect of any inventions or any improvements thereon as mentioned in the Schedule hereto. Also for the purpose of assigning unto or vesting in the Trustees or Trustee for the time being of these presents all or any of such Letters Patent Brevets d'Invention or other like authorities as aforesaid with full power to use the names or name of said parties hereto of the first and second parts or any or either of them in any such application as aforesaid or in any other step or proceeding incidental thereto and to present, sign, seal, deliver or otherwise perfect all or any petitions, applications, specifications, assignments assurances or other instruments whatsoever which shall or may be in any way necessary or expedient for the purposes aforesaid. And with liberty to delegate all or any of the powers aforesaid to any other person or persons as may be convenient and any such delegation at pleasure to revoke.

As regards any such improvement or new invention as is mentioned in the Schedule hereto the parties of the first and second parts will always at once and without delay make known the full particulars thereof to the Trustees or Trustee for the time being and also to Smith, Fleming and Company.

16. If at any time any proceeding shall be commenced or taken to repeal any of the Letters Patent mentioned in the said Schedule or any other Letters Patent, Brevets d'Invention or other like privileges to be granted in any other Country (except the United States) in respect thereof or of either of them or if any proceedings shall be commenced against the Trustees or any person or persons to be hereafter licensed be authorized to use the inventions or either of them in any Country (except as aforesaid) in which it shall be alleged that the several inventions or improvements mentioned in the Schedule or any one or more of them are infringements of rights previously granted to other persons all necessary and proper legal or equitable proceedings shall be taken by the Trustees for the purpose of protecting the Letters Patent or other privileges granted or to be granted (except as aforesaid) and all costs charges and expenses of and incidental to the prosecution or defence of any such proceedings shall be borne and defrayed by the parties hereto of the first and second parts or some or one of them. And the Trustees may apply all moneys and other considerations for the time being in their hands which under the provisions of this indenture would be payable to the parties hereto of the first and second parts or some or one of them in and towards the defraying of any such costs, charges and expenses and if the same shall be insufficient the deficiency shall from time to time on the request of the Trustees be made good by the parties hereto of the first and second parts or one of them and the Trustees shall not be required personally to advance any sum of money for the purposes of such proceedings but shall always be kept in funds by the parties hereto of the first and second parts or some or one of them but no deduction on account of for the purpose or the proceedings mentioned in this clause shall be made from the share which under and in accordance with the terms of these presents shall or may become payable to Messieurs Smith, Fleming & Co, The parties of the first and second parts or some or one of them shall indemnify and hold harmless Smith, Fleming & Co., and the Trustees against the claims and demands of all persons posessing Patent or other exclusive rights in priority to any of the patent rights agreed to be assigned or shall procure free Licenses under such prior patent rights to be granted to the said Trustees and in particular shall if necessary procure a free license or licenses granted to the said Trustees in respect of any prior patent rights of or belonging to Mr. George Little.[9]

17. If at any time any person or persons shall infringe any rights already granted or secured or to be granted or secured in any Country (except as aforesaid) which for the time being shall be subject to the trusts of these presents all necessary and proper legal or equitable proceedings shall be taken by the Trustees for the purpose of preventing any such infringement. And the Trustees and apply all moneys and other considerations for the time being in their hands in and towards the defraying of all costs, charges and expenses of and incidental to the prosecution or defence of any such proceeding and if the sum so for the time being in their hands shall be insufficient then the deficiency shall from time to time at the request of the Trustees be made good by the parties hereto of the first second and third parts as nearly as possible in the proportion in which they shall for the time being be entitled to have paid to them the moneys received by the Trustees in respect of Sales of interests in the Patent rights and of Licenses or permissions to use the inventions such proportions to be settled by arbitration in the event of difference and the Trustees shall not be required personally to advance any sum of money for the purpose of bringing or defending any such proceedings but shall be always kept in funds by the parties hereto of the first second and third parts in the proportion aforesaid.

18. It is understood and agreed that no Patent rights of the parties of the first and second parts for the United-States of America are included in the arrangements made by these presents provided always that if any such Patent rights in the United States should control or interfere with the use or application of the inventions and Patent rights which are intended to be subject to the trusts of these presents so as to prevent or render void any arrangement or contract in reference to Submarine Telegraphs which shall touch the shores of the United States then all necessary licenses and authorities under such United States Patents shall be from time to time granted to the said Trustees or Trustee so as to enable them or him to make or conclude any such arrangement or contract. And the said parties hereto of the first and second parts agree that they will only sell or part with their said United States Patents subject to the provision intended to be hereby made for the protection of the said Trustees or Trustee and Smith, Fleming & Co.,

19. Any person or persons may at any time hereafter be appointed as Trustee or Trustees by the parties hereto of the

first second or third parts either immediately or remotely in succession to or by way of substitution for the said John Fleming and I^c H^c Puleston or any subsequent Trustee and any new Trustee so appointed shall in testimony of his acceptance of the appointment sign and subscribe his name to these presents subscribing himself as Trustee and mentioning the date of his Signature or else signing and subscribing his name to an appropriate memorandum declaring his acceptance of the trust and the date of such acceptance and shall thereupon be treated and considered as having acceded to and become party to these presents of the fourth part. And upon any such appointment the number of Trustees may be augmented or reduced and all proper assignments and assurances shall be from time to time made for vesting the Patent rights in the Trustees or Trustee for the time being.

20. The Trustee or Trustees shall always be effectually protected and indemnified from and reimbursed any cost, charges, damages losses and expenses which shall from time to time be incurred or occasioned to him or them in or by reason or in consequence of his or their becoming Trustee or Trustees or exercising or performing his or their duties or power as Trustee or Trustees under or for the purposes of these presents and all such costs, charges, damages, losses, or expenses, shall (save such as are or ought to be borne by the parties of the first and second parts alone) as between the parties of the first second and third parts be borne and defrayed by them as nearly as possible in the same proportion as the cost, charges and expenses mentioned in clause 17.

21. Each of the parties hereto will at the request of the Trustee or Trustees or any other party hereto make, do, execute and concur in all acts deeds and assurances necessary in accordance with the Laws of England or any other Country for vesting in the Trustee or Trustees all the said Letters Patent rights and privileges which in accordance with the terms of these presents should now or hereafter be vested in them or him or which may be in any way necessary or expedient for giving full or more complete and satisfactory effect to each and every of the provisions hereby made or expressed or intended so to be and the costs charges and expenses properly incurred in reference thereto shall be treated as an outgoing and paid as provided by clause 9.

22. All stamped duties annuities or other payments which shall at any time hereafter be necessary for keeping on foot all or any of the Patent rights and privileges (existing or future)

mentioned in the schedule hereto shall be paid by the parties of the first and second parts.

23. All questions hereinbefore declared to be matters for arbitration and all or any other questions, doubts, differences or disputes which shall arise between any of the parties hereto or their representatives on the construction hereof or any clause or thing herein contained or any other matter or thing in any way relating to arising out of or connected with the said Patent rights and privileges or the rights and interests duties and obligations of the several parties hereto by virtue of the provisions hereby made shall be determinable and be determined by arbitration in the manner following, that is to say:—Each party in difference whether consisting of one person or several persons (several persons taking one side of a question being considered for this purpose as a single party) shall by some writing under his or their hand or hands appoint an arbitrator. And if either party shall make default in the appointment of an arbitrator for one month next after he or they shall have been required in writing by the other party to make such appointment the arbitrator to be appointed by such last mentioned party shall be the sole arbitrator. And the said arbitrators (if two shall be appointed) shall by some writing under their hands appoint an umpire and if within ten days next after their appointment or the appointment of the last of them (if they shall be appointed at different times) they shall not appoint an umpire an umpire shall be appointed under the provisions of the Common Law Procedure Act 1854 and the award or awards made in writing by the sole arbitrator or by the arbitrators or by the umpire either alone or jointly with the other arbitrators or either of them as the case may require (such award or awards being consistent with these presents and in no case repealing or offering[10] the same) shall be conclusive upon the parties in difference although they or either of them may have died before the making thereof or although the same may not be final or conclusive as to the whole subject of the dispute or difference and shall be in all respects abided by, observed and performed by them respectively on their respective parts and the submission hereby made shall be made a Rule of Her Majesty's Court of Queens Bench of any other court of competent jurisdiction and the arbitrators and their umpire thereunder shall have and exercise all the powers and authorities given to arbitrators by the said Common Law Procedure Act 1854 and the submission hereby made and the proceedings hereunder or consequent thereon shall be gov-

erned and controled by all the provisions of the said Act which are applicable to submissions to arbitration and proceedings thereunder and consequent thereon.

In witness whereof the parties to these presents have hereunto set their hands and seals the day and year first above written.[11]

The Schedule above referred to:—

Letters Patent under the great seal of the United Kingdom of Great Britain and Ireland bearing date the 10th day of June 1872 No. 1751 granted to Henry Edward Newton in respect of an invention of "Improvements in Mechanism for perforation paper for transmitting telegraphic messages" Like letters Patent bearing date the 27th day of February 1873 No. 735 granted to the said Thomas Alva Edison in respect of an invention of "Improvements in circuits and Instruments for Chemical Telegraphs"—[12]

Provisional Protection under the provisions of the "Patent Law Amendment Act 1852" dated 25th of April 1873 granted to the said Thomas Alva Edison in respect of an invention of "Improvements in Electric Telegraphs".[13]

British Letters Patent for this Invention are now in course of completion.[14]

All existing and future improvements upon the said three Inventions or either of them or comprising any art, process device or mechanical or other contrivance relating to the automatic, or rapid transmission of message by Chemical or Electric Telegraphs which has been already or may be hereafter invented or discovered by the parties hereto of the first and second parts or any or either of them.[15]

All existing and future Foreign Letters Patent Brevets d' Invention or other authorities granting exclusive rights and privileges for the said inventions or any or either of them and also for all or any such improvements existing and future as aforesaid for France, Belgium Holland Spain Portugal Russia Austria Germany Brazil India Canada and other Colonies or Dependencies of the United Kingdom and for all other Countries States or Dominions (except always the United States of America) where Patent rights can or may be obtained and where it shall be deemed expedient or beneficial for the parties hereto to secure or obtain Patent or other exclusive rights.

All the above mentioned Inventions or improvements (existing and future) and the above mentioned letters patent Brevets d' Invention or other exclusive rights (existing and future) are comprised in the arrangements made by these presents and are to be assigned to the Trustees or Trustee.

The Invention or Improvement already made by the said Thomas Alva Edison or now in course of being perfected by him and termed by him "The Automatic Typo Telegraph System"[16] is expressly included in the arrangements made by these presents but Messrs. Smith, Fleming & Co., shall be under the obligation of paying for their share or interest therein (similar or analagous in all respects to their share or interest in the other Patent rights) a further sum in adition to the sum of Ten Thousand Pounds mentioned in the body of these presents in proportion to the amount which the parties of the second part shall themselves pay to the said Thomas Alva Edison for their shares or interest and such proportionate payment to be made by Smith, Fleming & Co., will not it is understood exceed the sum of Six Hundred Pounds.

:(sd) Thomas Alva Edison :(sd) Geo. Harrington
:(sd) Josiah C. Keiff

Signed sealed and delivered by the above named Samuel Edison[17] as to T. A. Edison

TD (transcript), NjWOE, DF (*TAEM* 12:1061). Notarization of signatures and Reiff's 26 November 1879 oath (as to veracity of copy from which this document was transcribed) have been omitted; both are typed transcripts. The original document, held in 1879 by the London firm of Wilson, Bristows, and Carpmael (now Bristows, Cooke, and Carpmael), has not been found. [a]Date from text, form altered. [b]"'A'" centered above, probably added c. 1879. [c]Followed by space for full name.

1. Harrington, president of the Automatic Telegraph Co., held the largest interest in Edison's U.S. patents for automatic telegraphy. See *TAEB* 1:190 n. 1.

2. Should be "Reiff" throughout. Reiff, secretary of Automatic Telegraph, provided major financial support for Edison's automatic telegraph work. See *TAEB* 1:243 n. 7.

3. This was a firm of East India merchants. John Smith of Smith, Fleming & Co. had been among the founding investors on the board of directors of John Pender's Telegraph Construction and Maintenance Co. in 1864 (Smith 1974, 118). Apart from McIlwraith's acquaintance with Lyon Playfair (see Doc. 385), nothing is known of the individuals listed here.

4. Should be "J." throughout.

5. John Henry Puleston (later knighted) had been a newspaper publisher and a banker in the U.S. At this time he was one of three partners in Cooke, McCulloch and Co., the London branch of the American bank Jay Cooke and Co., which was under financial pressure. Wilson 1968 (supp.), s.v. "Puleston, Sir John Henry"; see also Doc. 365 n. 2.

6. Should be "Gouraud" throughout. The patents listed in the sched-

ule below had been asigned to Gouraud on 23 July 1873. He represented the Automatic Telegraph Co. in England. See *TAEB* 1:280 n. 7.

7. For a transcript of the accompanying patent assignment by Edison, Harrington, Reiff, and Gouraud see *TAEM* 12:1086. Gouraud was in New York at this time.

8. Should be "Land".

9. Little held several U.S. patents for automatic telegraphy. See *TAEB* 1:186 n. 2.

10. Should be "altering".

11. Signatures probably appeared at this point on the original.

12. See Doc. 291; and Brit. Pat. 735 (1873).

13. Brit. Pat. 1,508 (1873). See *TAEB* 1:550 n. 1.

14. Edison executed the full specification on 16 September; it was filed in London on 25 October 1873. In addition, on 11 September Edison's provisional specification for British Patent 2,988 (1873) was filed, for "Improvements in Peforated Paper for Telegraphic Purposes, and in Means for Receiving and Transmitting with the Same"; he executed the full specification on 13 February 1874.

15. This was interpreted years later as embracing Edison's quadruplex telegraph, and Fleming and Puleston used their power of attorney (clause 15 of this document) to assign Edison's British patent covering that system (384 [1875]) to themselves as trustees. Edison, Fleming, and Puleston, patent assignment, 29 Jan. 1878, Miller (*TAEM* 28:1056).

16. That is, a roman-letter telegraph system. See Doc. 295.

17. Thomas Edison signed on 2 September 1873; on 3 September, Harrington and Reiff signed before a notary in New York and Samuel Edison attested to his son Thomas's signature at the same time, both on this and the accompanying patent assignment.

GLASS-TUBE CARBON RHEOSTAT Doc. 351

Edison designed a compact, inexpensive carbon rheostat in the summer of 1873 to help simulate long submarine telegraph lines.[1] Although the instrument proved impractical, it led Edison to the realization that the electrical resistance of carbon varies with pressure, a property he exploited in his 1877 carbon button telephone transmitter.[2]

The apparatus consisted of a box frame holding a dozen glass tubes, each tightly filled with powdered graphite through which the current passed. The tubes were arranged so that any number could be connected in series. Binding posts between every two tubes allowed the experimenter to attach condensers (to match the electrostatic capacity of a cable) or other devices.[3]

Although Edison started ten of these instruments, on completing the first one, he "found that the resistance of carbon varied with every noise, jar or sound, and were too unreliable where a definite resistance was required."[4] He never finished the other nine, but the first survived until at least November 1880, when Edison entered it as an exhibit in a telephone patent interference case.[5] That original rheostat apparently no longer exists, but the printed record of the case includes the scale drawings reproduced below.

An ordinary rheostat.

1. At this time rheostats commonly consisted of a box of wire coils (Prescott 1877, 180–82). Edison testified in 1880 that he had discovered the pressure-dependent resistance of carbon "about April 1873"; Charles Batchelor testified that it "was made and used as far back as 1873" (Edison's depositions, TI 1:xix–xx, 46 [*TAEM* 11:19–20, 44]; draft deposition, folder 78-028, DF [*TAEM* 18:896]; Batchelor's deposition, TI 1:234 [*TAEM* 11:94]). Although a patent application of 23 April 1873 (U.S. Pat. 147,311 [Doc. 317]) indicates that Edison had considered using different kinds of rheostats with telegraph cables, there is no evidence that Edison designed cable-simulation experiments before his work in June 1873 with an actual cable at Greenwich, England (Docs. 321–36). An account-book entry of 6 September 1873 indicates completion of "1 Carbon resistance box for Edisons exper. cable" (Cat. 1219:100, Accts. [*TAEM* 21:755]). The only other apparent references to this instrument are in technical notes from the fall or late summer of 1873 (Cat. 299:53, 58, Lab. [*TAEM* 6:116, 120]. There is no indication in Edison's records of any other carbon rheostat apparatus until 1878 (Cat. 996:270–83 and NS-78-003, both Lab. [*TAEM* 3:342–48, 7:779]; Prescott 1879, 138–39; Jenkins and Nier 1984, 193 n. 2).

2. Edison believed that he was the first to discover the variable resistance of carbon, which was also the basis for his later, controversial claim to have invented the microphone. However, various European engineers and scientists had noticed the relationship earlier (see, e.g., Du Moncel 1974, 144).

3. As in most other rheostats, adjustments were made by inserting brass plugs between narrowly separated contacts. For an idea of the kinds of experiments Edison planned, see Docs. 321, 333, and 366; and Cat. 299:46, 53, 68, Lab. (*TAEM* 6:111, 116, 130).

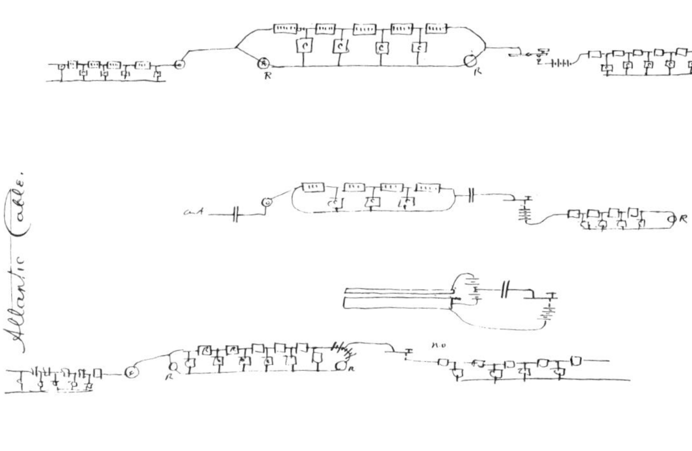

Edison's sketch of an artificial cable set-up, showing resistance boxes, condensers **(c)**, *and simple resistance coils* **(R)**.

4. TI 1:46 (*TAEM* 11:44).
5. Ibid.

–351– [Newark, September 6, 1873[2]]

Experimental Model: Rheostat[1]

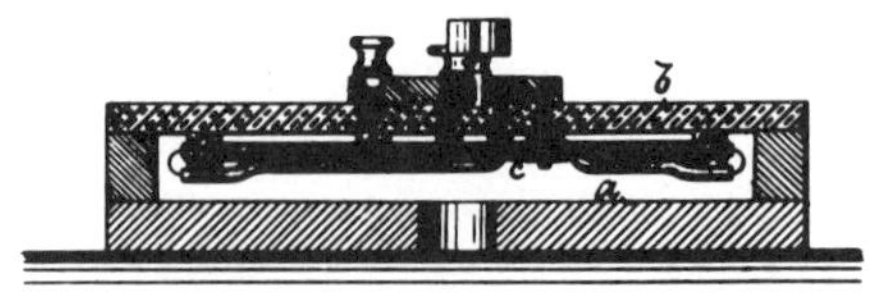

a. brass
b. hard rubber
c. glass tubes containing carbon

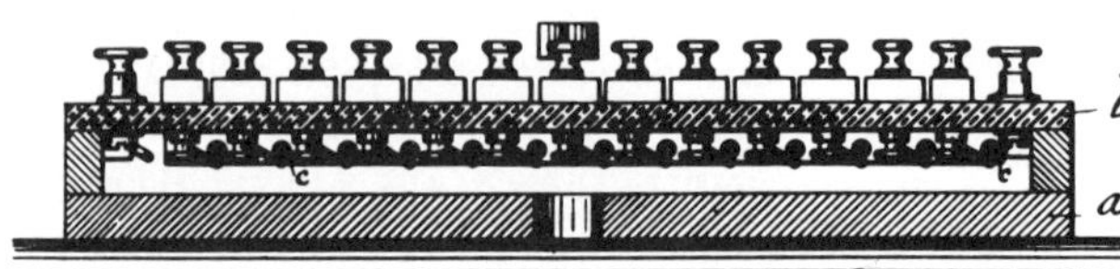

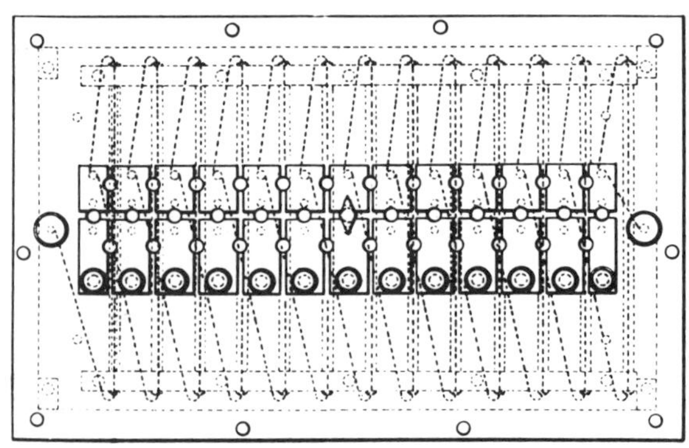

M (historic drawing) (45 cm × 28 cm × 10 cm), NjWOE, TI 2:517 (*TAEM* 11:645). Photolithograph of drawing of exhibited instrument, prepared for printed record.

1. See headnote above.
2. Date taken from an Edison and Murray account book. Cat. 1219:100, Accts. (*TAEM* 21:755).

–352– Newark, Sept 9/73.

R. G. Dun & Co. Credit Report

Edison & Murray[a] Manfrs Telegraph Inst'm'ts

#3700 "J"[1] Both men have been in this bus for some time. "E" was formerly of E & Unger who dissolved & the firm of J T Murray & Co formed composed of Murray Edison & Unger: they did not agree very well & dissolved E & M came together again under above style. They are very reticent abt their affairs but "M" states that they have [-]machinery worth 43m$[2] & that E has RE wor [-]4m$ over encumberance;[3] that their liabilites are abt 3m$ but will be paid on or before the 17th inst as they have 10m$ due them on the way to them from foreign parts. They have sold the Automatic Telegraph System to the Automatic Telegraph Co. for 2mm$[4] one third of which they control. E also owns a Horse Car RR in Port Huron Mich.[5] They have been some what embarassed lately owing to some Englis firm having ordered 11m$ wor of instruments which were made up & abt to be shipped when the order was countermanded & the goods left on their hands[6] but they say they will not lose any thing as they have made arrangements with other parties for the goods. Their principal bus is manufacturing for the Automatic Telegraph Co. 64 Bway & the Gold Stock Telegraph Co 61 Bway N Y City. They keep no a/c[7] as a firm at the Bank but each of them has an a/c at different Banks. they ask no discount[8] & seem to be doing a large bus & to be all right. They are now at work experimenting on a patent battery the object of which is to transmit 100 words per minute over the Atlantic Cable & for which if accomplished the English Government offer 500m$.[9] "E" is very sanguine of gaining this. We find it a difficult matter to make any estimate of their worth except on their machinery & RE. the machinery is of a peculiar kind & only adapted to their line of bus & if closed out at forced sale w'd not probably realize half what they claim it worth. If "E's" statement of July 72 & M's of today are correct it w'd indicate that "E" had paid off the $17500. of Notes due to "U"[10] but of this we have no positive proof. Their way of d'g bus seems seems to us as rather careless & indifferent but we learn oo/ wrong about them. They seem to have paid everything so far with pmptness. We can only give their statement without further comment 2092. 2150

D (abstract), MH-BA, RGD, N.J. 23:206. [a]Followed by "(Prevs see T A Edison 3/2[38])"; "3/238" are the volume and page numbers for related credit reports.

1. The number and initial at the beginning of the report identify the agent sending in the report. The book containing the matched codes and names has been lost. The meaning of the numbers at the end of the entry is unknown.

2. "m" means thousand.

3. Edison had bought a house at 53 Wright St., Newark, in late 1871. See Doc. 206.

4. This may refer to the yearly salary of $2,000 Edison received from the Automatic Telegraph Co. See Doc. 274.

5. In February 1872, Edison had borrowed $3,100 from Josiah Reiff to help his brother, Pitt Edison, with the Port Huron and Gratiot Street Railway. Quad. 70.7, p. 704 (*TAEM* 9:722).

6. About this time the Exchange Telegraph Co. of London stopped buying Edison's instruments and had their printers made in England. Scott 1972, 16.

7. Account.

8. That is, they do not ask a discount of their creditors for paying bills before they are due.

9. See Docs. 368 and 369 for Edison's discussion of his battery experiments. Nothing is known of such an offer by the British government.

10. Edison owed his former partner William Unger this money from the dissolution of their partnership. Some of the notes had been paid off by this time. Doc. 264; receipts and cancelled notes, 72-001, 73-007, DF (*TAEM* 12:663, 665, 667, 670, 672, 674, 1115–16, 1118).

–353–

Memorandum: Experimental Apparatus

[Newark, Summer 1873?[1]]

Thomson mirror astatic high .r.[2]
Standard B A Rheo.[3]
Gaugain tangent Gal[4]
Coulombs balance Gal.[5]
Rhumkoff high r Gal.[6]
Siemens Universal Gal[7]
2 ordinary gal. astatic
1 " Low r"
1 Plate Glass E Mac[8]
1 Holz Large "[9]
Bennetts Electroscope[10]
1 Plate Condenser on slide[11]
1 battery 20 Leyden jars[12]
1 Universal discharger[13]
1 Dry pile 5000 disks[14]
1 Voltometer[15]
1 Small Rhumkoff coil[16]
1 Large "
1 Large Nobili Thermo pile[17]
One Large Electropohsis[18]
" Small d[itt]o
1 Single wire slide Rheo[19]
1 Grame Magneto Mac
1 Small ordinary do
1000 cells vial battery
1 Thomson portable Electrometer[20]
1 Siemens Polzd Relay
2 Large regulating clock

1 Chronograph
1 Varley replenisher[21]
1 Electric Light L Dubocq[22]
1 Proof plane[23]
2 Hughes printer[24]
1 air pump
Pith ball stands.
20 Trays 10[a] gutta .P. batteries.[25]
100 Large Carbon battery Laws[26]
25 Nitro chromic batteries[27]
6. Trays BP.O. battys[28]
50 E Gravity Batteries[29]
1 Ink recorder Siemens[30]
2 Clockworks for chemical paper with Governors.
1 Standard Condsr Equal 5 Knots 1865 Cable.[31]
1 Rheo 1 Million ohms

AX, NjWOE, DF (*TAEM* 12:993). [a]Circled.

1. Edison probably drew up this list of desired experimental apparatus following his 1873 trip to Great Britain, where he saw many of these items for the first time.

2. Galvanometers are instruments for measuring electric currents by movement of a needle or coil in a magnetic field; an astatic form is not affected by the earth's magnetic field. Thomson's mirror galvanometer was used for receiving signals on cable telegraphs. Knight 1876–77, s.v. "Galvanometer."

Gaugain's tangent galvanometer.

3. That is, a rheostat based on the standard ohm established in the 1860s by a committee of the British Association for the Advancement of Science.

4. The standard tangent galvanometer measured current by the tangent of the needle's deflection. Gaugain wound the current-carrying wire on a conical frame and placed the needle at the cone's apex, creating an "apparatus superior to all others for exact measurement." Prescott 1877, 136–39.

5. A Coulomb galvanometer employed Charles Coulomb's torsion balance. Knight 1876–77, s.v. "Galvanometer."

6. A Ruhmkorff galvanometer was designed by Heinrich Ruhmkorff (see *TAEB* 1:82) or employed his induction coil.

7. Probably the sine-tangent galvanometer of Werner Siemens and Johann Halske. Prescott 1877, 143–48.

8. That is, a device for producing static electricity by friction. Atkinson 1890, 721–23.

9. Wilhelm Holtz's machine produced static electricity by induction. Ibid., 726–29.

10. An electroscope detects the presence of an electrical charge; Bennett's, which became the standard because of its sensitivity, employed gold leaf. Ibid., 718–19.

11. Perhaps a variable condenser.

12. The Leyden jar (named after the Dutch town where it was invented) was the first condenser. It incorporated a glass bottle with separate internal and external coatings of metal foil; the internal coating touched a conducting rod passed through an insulated stopper. Atkinson 1890, 741–43.

13. That is, a device for discharging condensers, electroscopes, etc. Ibid., 745–46.

14. A dry pile is a battery in which hundreds or thousands of metallically coated disks of paper or leather replace liquid electrolytes. It produces a high voltage but very little current. Ibid., 789.

15. A voltameter determines current strength by passing the current through dilute or pure sulfuric acid and measuring the volume of gas evolved. Prescott 1877, 157–61.

16. That is, an induction coil.

17. The Nobili pile, devised by Leopoldo Nobili, is a thermoelectric battery consisting of bars of bismuth and antimony soldered together. Atkinson 1890, 956.

18. Perhaps an electrophorus, a device invented by Alessandro Volta for producing static electricity by induction. Ibid., 720.

19. Probably an adjustable rheostat.

20. Sir William Thomson devised a delicate electrometer, an instrument to measure electrical charge. Ibid., 749.

21. A small generator of static electricity used to maintain the charge on an electrometer. Atkinson 1910, 814.

22. Jules Duboscq, Parisian instrument maker, manufactured an arc light. King 1962b, 339.

23. A proof plane (or "carrier") transferred static charges in experiments. Atkinson 1910, 699.

24. See *TAEB* 1:109 n. 5.

25. See *TAEB* 1:607 n. 4.

26. Possibly a battery, employing carbon as one of the electrodes, such as that used with Samuel Laws's gold-reporting telegraph.

27. Probably a battery employing nitric acid with bichromate of potash. This type of battery gave a fairly constant current and was considered useful when a strong current was required. Sprague 1875, 108.

28. The British Post Office telegraph used a modified Daniell battery. Culley 1871, 19.

29. Possibly Edison's battery; see *TAEB* 1:481.

30. The ink-recording telegraph of Werner Siemens and Johann Halske was used extensively in Germany as well as on government telegraph lines in Great Britain and India. Prescott 1877, 489–96.

31. That is, the 1865 Atlantic cable.

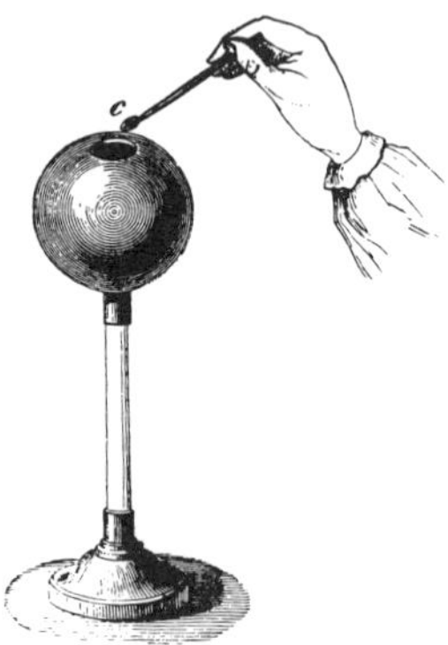

A proof plane was a small copper disc at the end of a glass or shellac rod.

LABORATORY NOTEBOOK Docs. 354–361, 366, and 419

Edison used this notebook extensively during the summer of 1873 to record ideas and information pertaining to automatic, multiple, and cable telegraphy. He made a few further entries during the fall of that year and a final entry on 10 April 1874. Most of the entries from the summer, one from the fall, and the April 1874 entry have been selected. Many of the early entries—including a series of notes taken from John Pepper's *Cyclopedia of Science*—concern chemical experiments related to automatic chemical recorders. Other automatic telegraph experiments detail the use of secondary batteries, magnets, and condensers to prevent tailing. The material on multiple telegraphy includes some of Edison's experiments on his first successful diplex design, work that was essential to his development of the quadruplex telegraph the following year. Many of the entries on automatic and multiple telegraphy show Edison using the same techniques and apparatus in both technologies. The 1873 entries end with cable experiments, reflecting Edison's continuing attempts to apply his automatic telegraph to a British cable.

Although none of the entries from 1873 carries a date, links to Edison's other materials and activities place each one with reasonable accuracy. Following his return from England at the end of June, Edison appears to have perused various books on chemistry for information pertaining to his automatic chemical recorders, and the first entry (Doc. 354) includes some of these notes. Another (Doc. 356) discusses a diplex experiment related to the caveat Edison signed on 2 August (Doc. 348). The last summer entry (Doc. 361) includes circuit designs in which the chemical receiver of an automatic telegraph is placed in a Wheatstone bridge with magnetic (inductive) compensators. The magnets shown in this entry have short cores, but when Edison executed a corresponding caveat on 28 October 1873 (Doc. 375) he specified magnets with long cores, which he devised at the end of September (Doc. 363). The fall entry (Doc. 366) describes cable telegraph experiments that Edison probably performed in early October. The 10 April 1874 entry is the earliest record of the electromotograph phenomenon.

Edison transcribed at least some of these entries from notes on loose pieces of paper, running individual entries together. He sometimes recorded experimental ideas and results as work progressed, and at other times he went back to an entry to indicate the results of proposed experiments. Charles Batchelor indicated the number of hours spent on certain experiments. In one case, a series of experimental strips from tests of different recording solutions for automatic telegraphy were pasted into another laboratory scrapbook.[1]

1. Some of these notes and the experimental strips can be found in a laboratory scrapbook (Cat. 299, Lab. [*TAEM* 6:67–82]). The purchase order for chemicals used in the experiments on automatic recording solutions also is in that scrapbook (*TAEM* 6:86–88).

–354–

Notebook Entry: Chemistry and Automatic Telegraphy[1]

[Newark, Summer 1873]

Try experiment on the delicacy of a frogs leg. it is stated by Pepper[2] that a frogs leg is 56 000 times more sensative than an Electroscope. (condensing)[3] This frogs leg may be arranged to work a second circuit by attaching a circuit breaker to it. It is possible that a live frog may be used, in this case the instrument would be of a permanent character.

~~Try~~Remark[a]

Try if the freed oxygen from a chemical recording pen can be attracted away from the pen and a decomposition partially prevented by a powerful local magnet with pointed poles The object being to attract the oxygen away from the iron pen after the current ceases to prevent an elongation of the mark upon the chemical paper. Thus.

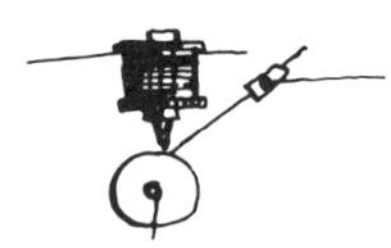

It is stated by Pepper that oxygen is magnetic.[4]

Mix a solution of Pyrogallic acid with potash and make some chem paper. p[latinum] & iron pen. Pepper says freed oxygen gives a dark Brown color.

Try chem paper in solution of Sulphate of manganese.

Schonbeins test[5] for ozone is 1 part Iod Pot in 200 parts distilled water, 10 parts starch thickened by heating, brushed on bibulous paper. Try effects of decomposing with pens.

Pepper says ozone is present in the oxygen tube of a Voltometer, which shows that both are set free Ozone bleaches Black Sulphide of Lead or Plumbic Sulphide, oxydization takes place, and white sulphide of lead formed. This may give good telgh solution.[b]

Powdered Antimony becomes very hot and combines when placed with iodine.[6b]

Palladium absorbs great quantity of hydrogen graham used two coils like a watch spring immersed in water (acidu[lated] prob[ably]) and connected the two ends to a small battery. decomp of water took place the free hydrogen being absorbed by one of the spirals of palladium which quickly expanded, when the current was reversed the first coil contracted and the other expanded[7] It may be that with a proper coil arranged with a lever that the smallest current could be made to work the lever and close a secondary circuit or with mirrow throw a light upon a screen.

Test Schober paper—also Boston.[8]

Procure samples = for chemical decomp = of paper.

X, NjWOE, Lab., Cat. 1171:25 (*TAEM* 6:218). [a]Followed by brace spanning several blank lines. [b]"X" in right margin in unknown hand.

1. See headnote above.

2. John Henry Pepper, lecturer and honorary director at the Royal Polytechnic Institute, London, was a physicist and analytical chemist well known as the author of many popular scientific works (*DNB*, s.v. "Pepper, John Henry"). Here Edison refers to p. 288 in the "Electricity" section of *Cyclopedic Science Simplified* (Pepper 1869a), a work which Edison later recommended to his nephew Charles Edison. The different sections of the book were also published separately and in various combinations.

3. Electroscopes sometimes incorporated condensers in order to detect very weak charges. Culley 1871, 75.

4. This and the following paragraphs dealing with oxygen and ozone are based on the "Chemistry" section of Pepper 1869, 549–50.

5. Christian Schönbein, a German chemist of the early nineteenth century, was noted for his work on ozone (*DSB*, s.v. "Schönbein, Christian Friedrich"). A description of Schönbein's test is found on p. 169 of Fox 1873, a work to which Edison made reference in a 21 February 1874[?] notebook entry (Doc. 402).

6. See Pepper 1869a, 614.

7. The preceding is based on Pepper 1869a, 622, quoting a report on experiments of Thomas Graham, the warden and master of the mint in Great Britain, and William Chandler Roberts. The chemist Graham was best known for his work on diffusion of gases and liquids. Roberts (Roberts-Austen after 1885), a metallurgist and chemist, was first Graham's assistant, then assayer of the mint. *DSB*, s.vv. "Graham, Thomas," and "Roberts-Austen, William Chandler."

8. "Schober paper" is unidentified. "Boston" probably refers to rolls of paper bought from S. D. Warren of Boston. Bill of 21 Jan. 1873, 73-006, DF (*TAEM* 12:1101).

-355-

Notebook Entry: Automatic Telegraphy[1]

[Newark, Summer 1873]

Ascertain the highest practical speed attainable with a short relay operated by perforated paper upon 300 mile line have static capacity of regular with proper amount of leakage to ~~ma~~allow the discharge of relay to have its proper evil effect.[2] the record is given upon Iodized paper the relay repeating into local circuit. Try speed with the relay armature at different distances away also shunted with condenser, magnet etc. diversify. Then at best speed subs~~it~~titute Morse Register and large local and obtain highest speed. The polarize Relay in one direction with large permanent magnet and see if speed is increased or decreased. Then on same circuit substitute 200 ohms Relay R for 100, then 50 ohms and see if decreased induction is ~~b~~of greater consequence than decreased strength. Then substitue a regular polarized relay, with shunt on trans and extra bat or N[egative] 25 P[ositive] 50 paper cutting out P, and obtain highest speed here. ⟨Edison Batchelor[3] 12 [hours] 1873⟩[a]

X, NjWOE, Lab., Cat. 1171:29 (*TAEM* 6:220). [a]Marginalia written by Batchelor.

1. See headnote, p. 46.

2. That is, the inductive discharge of the relay coils, which prolonged the signal.

3. Charles Batchelor. See *TAEB* 1:495 n. 9.

–356–

Notebook Entry: Multiple Telegraphy[1]

[Newark, Summer 1873[2]]

Duplex experiment—Sending two messages in the same direction over the same wire.[3]

The principle is sending reversed currents recorded on a polarized relay for one message and increasing and decreasing the strength of the current to effect another relay. The evils to contend against is to prevent bad effects upon the relay worked by increase and decrease at the moment of reversal. in sending P & N currents through an ordinary magnet or Morse Relay, the cores must entirely lose their magnetism ~~bef~~ due from an N current before they can acquire an opposite polarity due to the P current. at the moment of the polar change ~~there~~ the magnet exerts no force upon the armature hence the spiral spring will draw the armature away from the face of the magnet for an instant, and when the reversals are very rapid the armature lever scarcely comes in contact with the front contact point. To get over th~~o~~e effect of this upon the local circuit containing the sounder, which in this case would vibrate at each reversal and work the sounder, I reverse the ~~useful~~ usual method of connecting the local circuit and make connection with the back point, so than when the lever of the relay is away from the magnet, the local circuit is closed. This closes a repeating sounder, (ie) a sounder whose lever breaks and closes another local circuit ~~in~~ When this sounder is closed it opens the second local circuit, so the same effect is brought about as if the relay lever operated the local circuit and sounder in the ordinary way The reason this secondary plan is used is because when the current upon the main line is supposed to be of full strength and the relay closed the reversals constantly tend to vibrate the lever ~~as~~ the lever flies away from the front contact point but the cores of the relay become re-magnetized and reattract the lever before it touches the back point at a slow speed. consequently the repeating sounder is unaffected and of course the receiving sounder. but if the reversals are sent very rapid the lever of the relay makes a slight contact with the back point This would close the secondary sounder were it not that that sounder being a non-receiving instrument may be adjusted very high consequently the slight contact which the relay lever makes with the back point in the act of vibrating is insufficient to allow the repeatg magnet ~~of~~ to reach it maximum strength which is necessary to overcome the tension of its spring.

~~If~~ I find in actual trial this.

R varied from $500^{\circ 4}$ to 9000° ~~That the~~ relay. (Phelps 126 ohm) ~~with~~ battery reversed by Double springs operated by sounders. [-]Number of cells increased or decreased by a second sounder and contact spring polarized relay 75 ohms. 8 cups always on for working polarized relay, increased by adding by working sounder by 16. That on 500 ohms Perfect signals were received and relay had great amount of margin increased to 1000, signals perfect, margin on adjustment of relay increased. increased to 2 3 5 7 & 9000 ohms Margin on relay increased with increased resistance. decreased extra battery to 8 cups results same decreased battery 8. regular from 8 to 4 cups signals just as perfect. inserted in line two 200 ohm relays. signals on high resistance 3000 & 4000 ohms perfect but on low resistance 500 ohms inductive discharge prolonged the opening at the moment of reversal when inverted currents are sent through ordinary electromagnets the inductive discharge are much more powerful than when a current of one polarity is intermitted hence the inductive ~~a~~effect of the coils of the receiving magnet themselves ~~are~~ seriously effect the adjustment on short circuits, because the current which passes through them is stronger &^{a} the Route for the discharge short, by reason of the low resistance of the circuit. One curious thing observed in this experiment is that with a permanent reversable ~~b~~current sufficient to work the polarized relay that the ordinary relay may be adjusted to give good signals by the addition of but 3 cups to the permanent battery on a R of 8000 ohms. The battery may be increased from 8 cups to 40 without effecting the adjustment of the polarized relay.[b]

Trial in New York unsuccessful on account of probably Relays in line, insufficient battery, Battery on line direct by cross or leakage, & difference in R of Relays used & poor facilities[5]

X, NjWOE, Lab., Cat. 1171:31 (*TAEM* 6:221). [a]Interlined above. [b]Followed by horizontal line after which Edison's handwriting changes.

1. See headnote, p. 46.

2. Edison experimented with this diplex in July and August (testimony of Edison, Charles Batchelor, Uriah Painter, and Norman Miller, Testimony and Exhibits on Behalf of T. A. Edison, *Nicholson v. Edison*, pp. 25, 65–66, 90–92, 97). He successfully demonstrated the system when Thomas Eckert, Albert Chandler, Alfred Brown, Robert Clinch, Reiff, and Painter visited the Ward St. shop on or about 1 October 1873. Testimony of Edison and Reiff, Quad. 70.7, pp. 226, 272–73, 382, 444, 451 (*TAEM* 9:480, 503, 558, 590, 593); and testimony of Edison, Batchelor, Painter, Chandler, and Brown, Testimony and Exhibits on Behalf of T. A. Edison, *Nicholson v. Edison*, pp. 14, 65–66, 89–93, 107–9, and 111–12.

3. This is an experiment based on the plan laid out in Doc. 348.

4. Here "o" means ohms.

5. It is not known where Edison performed this trial, but see Edison's testimony, Quad. 70.7, pp. 227, 265 (*TAEM* 9:480, 499); and Orton's testimony, Quad. 71.1, pp. 123–25, 220 (*TAEM* 10:66–67, 116).

–357–

Notebook Entry: Automatic and Multiple Telegraphy[1]

[Newark, Summer 1873]

It may be that instead of changing the amount of magnetic resistance to cut the writing that it may remain constant and an adjustable rheostat placed in that side of the shunt containing the receiving paper.—no good[a] with a paper having a resistance of several hundred miles[2] I can [not?] conceive how the shortening of the resistance of the magnetic shunt[3] can bring any perceptable quantity of current from the line considering the immense resistance of the paper Therfore I think that decreasing the length of the magnetic shunt does not furnish much more if any counter discharge but only weakens the effects of the static so that it is imperceptable on the paper. If this is the case it is better to use one high resistance magnet only in the shunt and regulate the weakening of the static by an ordinary rheostat in the shunt containing the instrument. This will make the resistance at the receiving station much greater than in the old plan[4] hence less discharge will run out or be attracted to that end the bulk going to earth at the transmitting station.

It is possible that a very fine wire magnet having a resistance as high as 5000 ohms used as a shunt and and a Rheo in the other branch properly adjusted might be able to give a greater proportionate counter charge than the ordinary method.

I ~~e~~think that in wet weather the effacy of the shunt is much less than in dry weather, as the discharge from the magnets in dry weather have but one route practically through the paper while in wet weather there is an extra route for their discharge viz the derivations or leakages near the receiving end thus

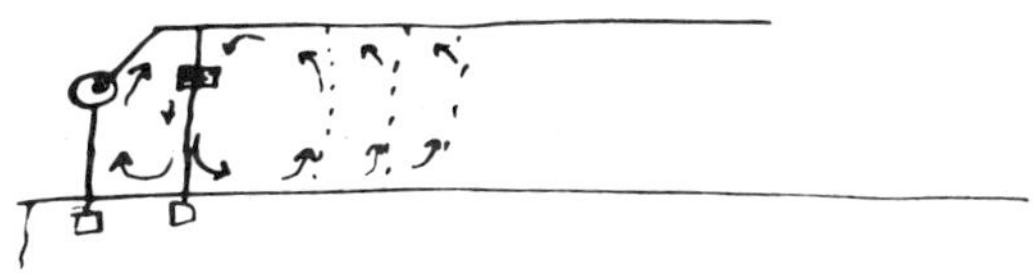

It would probably be a good idea to insert a plain rheo ~~to~~as shown below to ~~we~~ increase R of the extra route of the discharge & thereby deflect it upon the paper & at the same time weaken the main & static charge so that the counter discharge would bear a nearer proportion and be able to overcome— A noticeable change for worse ⟨Edison Batch & Tom[5] All Night 12 Hours⟩[b]

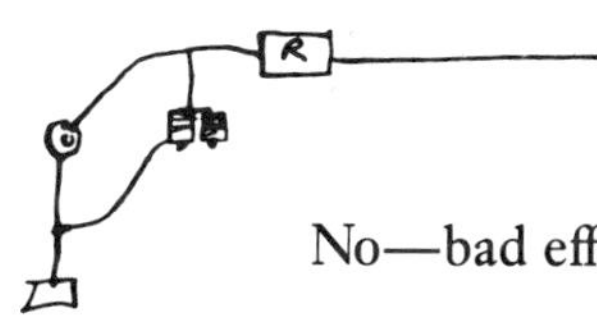

No—bad effect

Try this experiment—The additional R.X. necessary to get effacy second .C. insulates neg charge, exactly as much as it adds & even in excess. So it's failure[c]

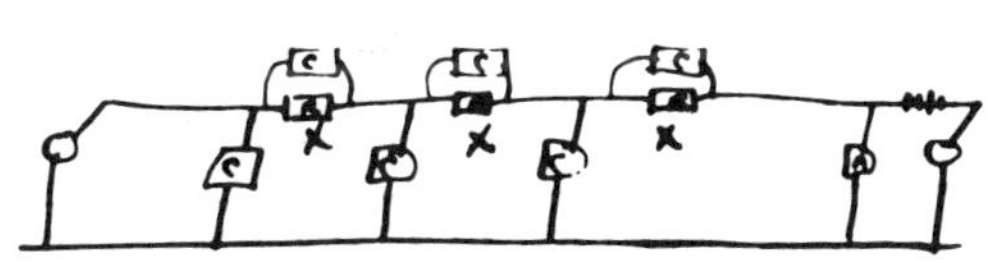

It may take double or thrible the number of condensers to shunt the resistances as the other to kill the static charge, as a portion of their discharge is killed or short-circuited by the R themselves and do not go on the line

Tried extra R brought in counterbalance exactly the good effects of the ~~o~~extra condensers In fact it did more harm than good

Test to ascertain if the discharge from a magnet is lengthened by an increase in the resistance of the circuit in which it discharges Thus[6]

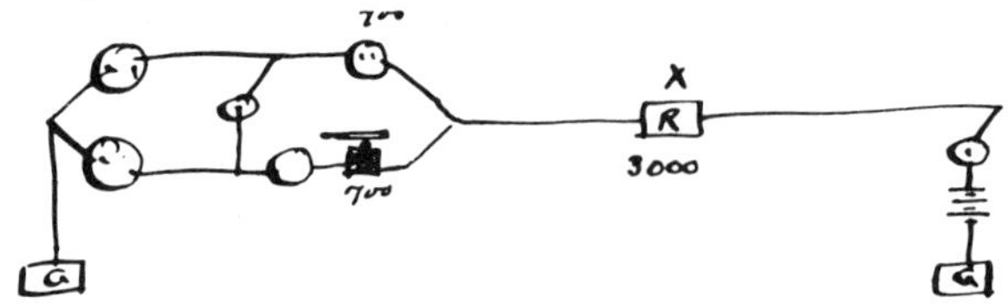

Increase the length of or[d] R of the discharging circuit but insert a galvanometer in main circuit so that the same strength of current is observed. The loss due to the increase of the inductive circuit is made up by decreasing the resistance of the Rheostat X. The same strength of battery will pass through the relay but the R of the discharging circuit will be doubled. Thus

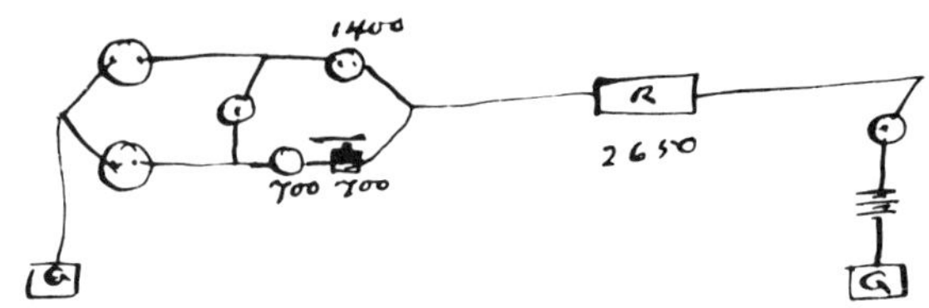

This arrangement should give a weaker, but longer discharge if the discharging time depends upon the exterior R if not then the discharging time unlike a condenser is independent of the the r of the discharging circuit.

If the current is not weakened the tension[7] must be very high.

Remarks[e]

Try sheet lead for secondary batteries,[8] also sheet iron and tin, etc

Try this

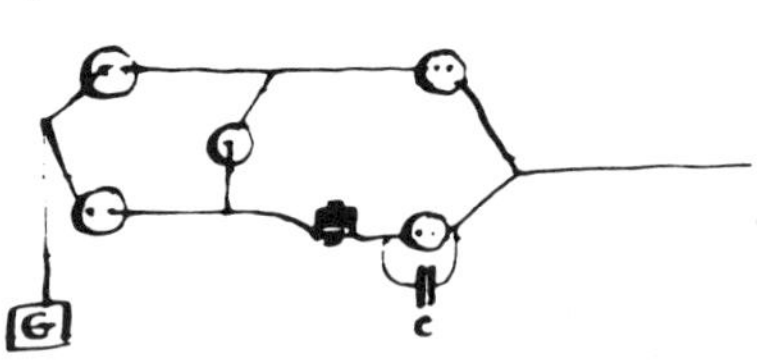

This

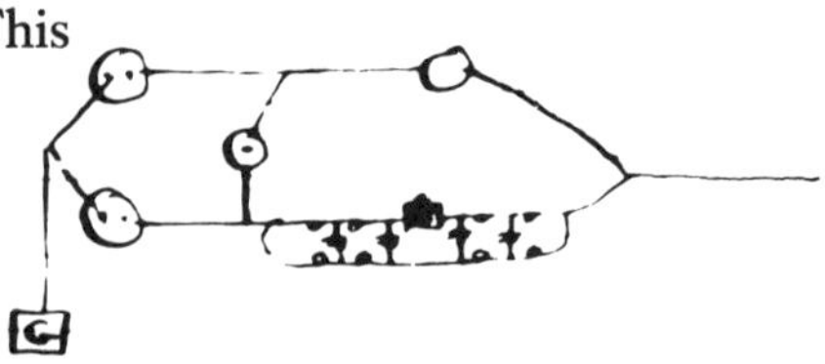

Kill the discharge. ⟨Edison & Batchelor 26 hours⟩[f]

Try this to lengthen the discharge

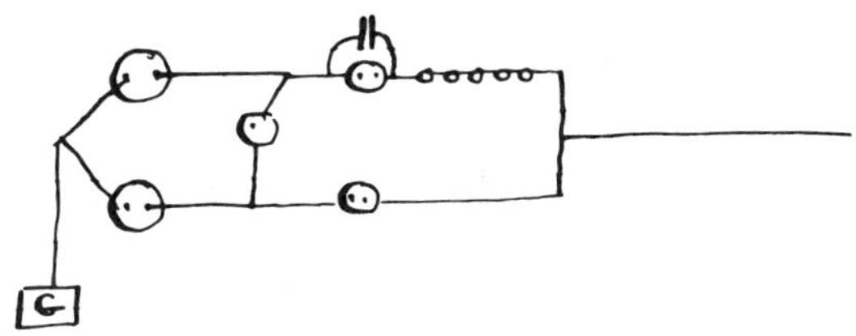

This

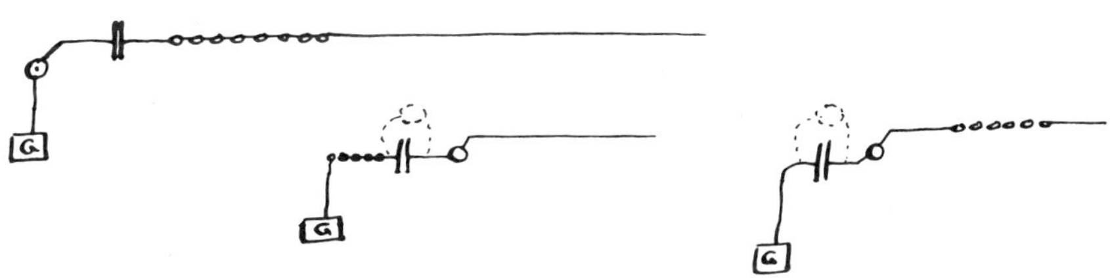

Try the amount of discharge from 6 bottles with water alone, and with various quantities salt. also ~~wit~~

Try this and ascertain if the action is not like a cable

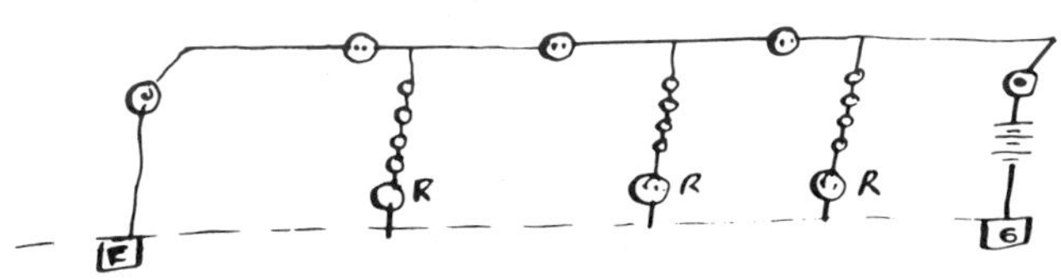

Try this.

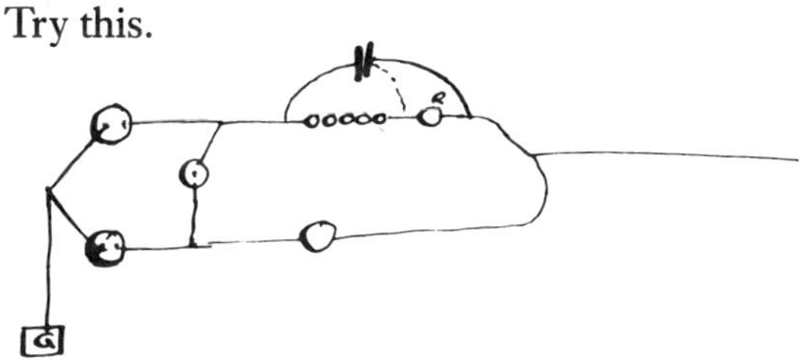

This.— The placing of the secondary battery in the direct circuit branch instead of the shunt alters its action as regards the relay but not the line, hence by placing it in this way, the sticking of the relay may be prevented Thus

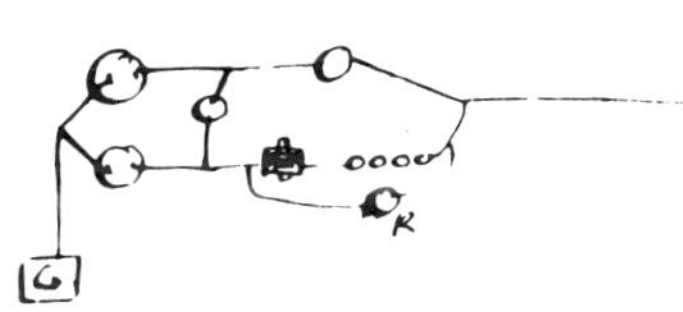

Remarks. both ways work. The cause of sticking is that on the opening of main circuit the secondary sends a longer

charge than the relay and the excess goes out on line opposed to the static chge, but a portion circulates within the shunt and this excess makes the relay stick. This is good thing for line but bad for relays, but as the R of the secondary can be increased, it will not perceptably interfere with the relay. The remedy is to get the secondary pile to send shorter after currents. 7 cells kill the ~~d~~inductive discharge of the heavest relay I have, 1400 ohms. The R at 7 cells well salted is about 600 ohms with only water 5000 ohms. first way made relay stick worse than a plain shunt as the S battery current was added to dischge but placing i~~n~~t in last diagram it works against & you cant stick relay but acts same as to line[g] with relay 1400 ohms dash comes thus — shunt ~~rece~~ sending end, none R[eceiving] end.[h]

Twas the jar of circuit closing lever—made secondary contact.

In my reversal duplex,[9] the effects of the discharge current from the ordinary & polarized magnet interferes with the ordinary relay making it vibrate while the polarized is rendered sluggish, but if the secondares are applied as shown and not too strong & the shunt is considerable it entirely destroys the effect of the secondary discharge from relay and a greater margin is obtained on the ordinary relay while the polarized Relay follows the reversal key promptly = ~~when~~

When a positive current has been on line for any length of time say for a dash the secondary battery becomes highly charged & when the current is reversed this charge acts with it hence this wave will be stronger than the other and it is this that interferes with the ordinary Relay to some extent as this relay depends on changes in strength of current = still I think the secondary will benefit greatly this form of ~~d~~Duplex[10]

Try this

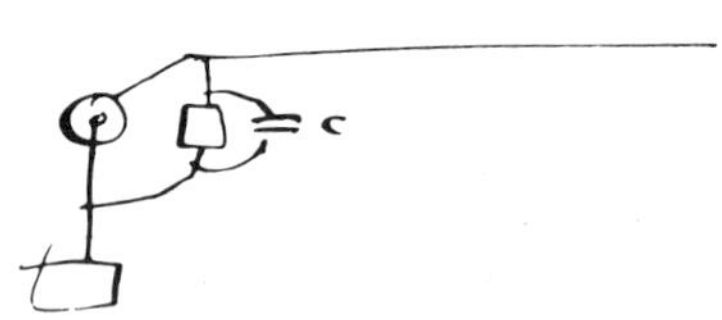

Passing Oxygen down one side of a platina electrode and hydrogen (common gas) down the side of the other platina electrode immersed in water [-]acidulated with sulphuric acid and also again with chloride sodium[11] gave no current perceptable upon a 6 ohm sounder of a delicate ~~21~~126 ohm Relay

Try experiments with battery at sending—mistake at receiving end only Thus

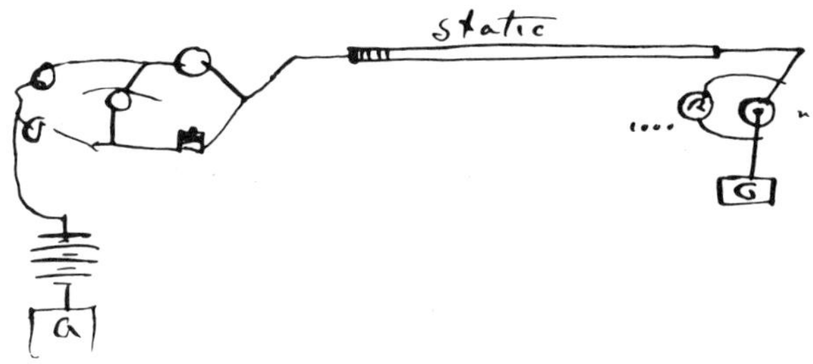

Try a line statically fixed as one side of the bridge[12] to study its nature. ⟨Edison &— Batchelor 16 hours⟩[f]

Test the induction

X, NjWOE, Lab., Cat. 1171:37 (*TAEM* 6:224). [a]"no good" interlined above. [b]Marginalia, written by Batchelor, overwrites sequence of numbers and unfinished drawing on facing page. [c]"& . . . failure" interlined above drawing. [d]Interlined above "of". [e]Followed by several blank lines. [f]Marginalia written by Batchelor. [g]"in . . . line" interlined. [h]Followed by centered horizontal line.

1. See headnote, p. 46.

2. Before agreement on an international standard unit of resistance (the ohm), electricians customarily made measurements in terms of the known resistance of standard wire for a given number of miles.

3. "Magnetic" refers to an electromagnet with high induction rather than high resistance.

4. For examples of earlier circuit arrangements see Docs. 317 and 342.

5. Probably Tom Gartland.

6. The following circuit designs with magnets, rheostats, and secondary batteries arranged in a Wheatstone bridge (see *TAEB* 1:530 n. 17) appear to be the genesis of a caveat Edison drew up in October 1873. See Docs. 361, 363, and 375.

7. Voltage.

8. See Doc. 334.

9. See Docs. 348 and 356.

10. See Doc. 360.

11. Edison drew the accompanying sketch in a preliminary description of this experiment. Cat. 299, Lab. (*TAEM* 6:74).

Edison's design for an experimental battery.

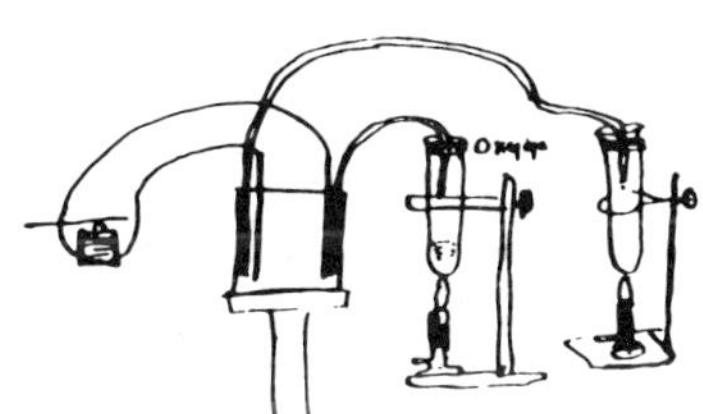

12. That is, an arrangement like the Wheatstone bridge.

–358–

Notebook Entry: Automatic Telegraphy[1]

[Newark, Summer 1873]

Chemical Solution:

To a gill[2] of water add tea spoonfull Nitrate Ammonia. To this add what is held on small knife blade of aurichloride of sodium. The paper is white, the marks with iron pen are blue, but with a tin pen yellow at first but soon become purple (purple of Cassius). The sensitiveness is increased by adding a slight amount (less than of of aurichloride of sodium) Bichloride Mercury[a] There appears after a few hours a continuous mark running through the dots and dashes.—[3] ~~So~~Chl Sodium does not appear to act as well as Nitrate Ammonia

Find out if it is the addition of the corrosove Sublimate (Bichl)[4] that cause the continuous line to appear

When Ferrocyanide solution, with Nitrate Ammonia as given in Culleys book[5] is used the iron pen is polarized and even on very short circuit the dash commences thus

When Sulphuret of Potash is used the polarization of the pen is greatly augumented

Make 3 large secondaries with Callaud jars[6] & pure water to get a shorter after charge.

⟨Edison & Batchelor 12 hours⟩[b]

X, NjWOE, Lab., Cat. 1171:53 (*TAEM* 6:233). [a]"Bichloride Mercury" inserted in margins. [b]Marginalia written by Batchelor.

1. See headnote, p. 46.
2. Four fluid ounces.
3. Edison executed a patent for this solution on 14 October 1873 (U.S. Pat. 160,850).
4. Bichloride of mercury.
5. Culley 1871, 202–3.
6. That is, the jar for a Callaud battery. Developed by A. Callaud of France, this was a simplified version of a Daniell gravity battery (see *TAEB* 1:615 n. 3). King 1962a, 243, 245; Pope 1872, 106–7.

–359–

Notebook Entry: Automatic Telegraphy[1]

[Newark, Summer 1873]

I find that the ~~r~~secondaries should be arranged with the relays Thus and not as a shunt[2]

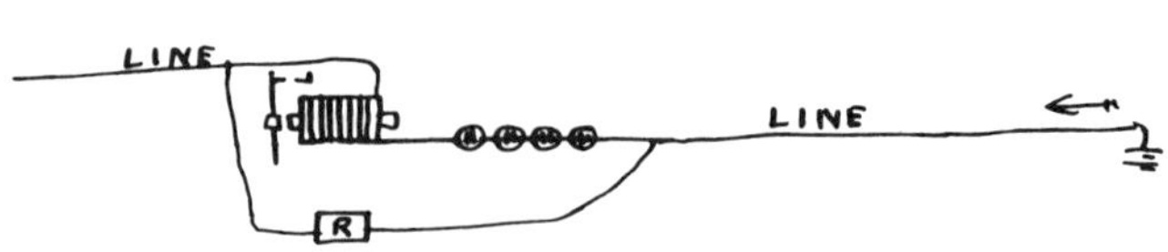

The magnet tends to send a current after the battery is taken off to the left. The secondary battery to the right. The two sources of power keep all portion of the derivation circuit at the same potential hence no current is generated. if too many cups S are used the current from them is greater than that from the discharge of the magnet. ~~hen~~ this would close the relay again but if the number of cups are reduced the balance can be obtained or the resistance of the shunt may be made greater and the balance obtained in that way.

In ~~receiving Th~~ If the cups are arranged as a shunt thus

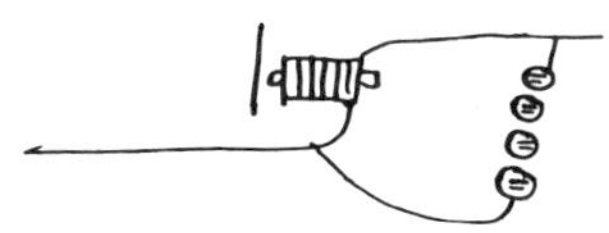

it is obvious that as the magnet generates a current opposing the main current on closing and in the same direction on opening, and The secondary battery ~~sends~~ reverses this order, that by placing this in the shunt, The two electromotive forces within it are working together, although the inductive effect will not go out upon the line. The relay is made to work sluggish owing to the currents both from the magnet and secondary battery circulating in the derivation after the main battery has been disconnected.

In ~~the~~ receiving with a secondary battery is should be arranged with the receiving paper the same as with a relay thus

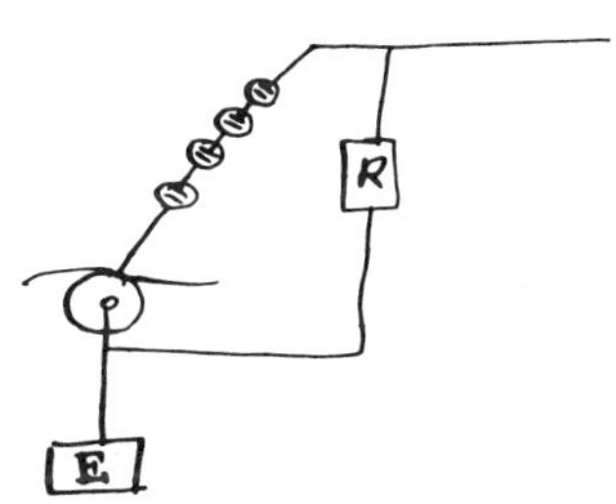

In a trial with Baltimore seven to eight hundred words per minute were received with Relay at Phila shunted. With no Earth shunt at Baltimore, The number of Cups first used were 15 but 200 words could be received. The number of Cups were increased to 35 with the above result. I presume that with 100 Cups almost any speed might be obtained with shunt and probably without shunt at sending station =

X, NjWOE, Lab., Cat. 1171:55 (*TAEM* 6:234).

1. See headnote, p. 46.
2. Cf. first addendum, Doc. 348.

–360–

Notebook Entry: Secondary Batteries and Multiple Telegraphy[1]

[Newark, Summer 1873]

Make a secondary battery of disks of Copper cloth & parafined paper Thus

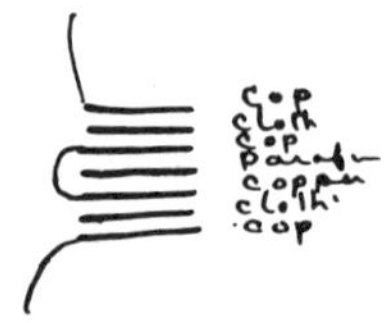

Attach the secondary batteries to the two relays in my ~~D~~Reversal Duplex[2] So see if the destruction of the induction of the magnets does not make less opening of the armature of the ordinary relay.

Yes vast differnce— There is slight difficulty if the Secondary battery is allowed to be discharged. if a long positive has been sent a negative current will flow from the battery when the Main battery is taken off. as it is not taken off but is followed by a negative this in addition to that of the secondary battery makes the negative wave stronger than the preceding positive & thereby interferes with the adjustment of the ordinary, but I think that if the Secondary is proportioned So as to neutralize the discharge from the magnets only no difficulty will be had from variations in strength of the current.

Salt well all the secondaries on hand. Connect for intensity connect them to a fine relay. Then take 6 cups Carbon battery a charge 6 secondaries at time quickly one set after another. see if these intermittant charges are not continuous on relay—

In charging the carbon battery current is thrown into circuit through high R of S battery this kills effect this experiment but the relay shows continuous current less that due from cutting in

A Revolving Rheotomatr[3] might be arranged with several sets of secondary batteries, and a charging battery so that one set could be charged for an instant thrown into circuit and the previous one withdrawn & ~~son~~ so on. Problem. Does 100 cups secondary ~~h~~charged with 10 cups primary have the tension of 100 cups in discharging or only 10 cups. = 3 cups alternated on 6 bottles charges it strongly lasts 1 min—trouble is to keep ckt closed when changing from one charged battery to the other & keep chg batty[a]

X, NjWOE, Lab., Cat. 1171:59 (*TAEM* 6:236). [a]"3 cups . . . batty" inserted between "10 cups. =" and beginning of next entry.

1. See headnote, p. 46.
2. Cf. the first addendum to Doc. 348.

3. Edison probably meant a design like that shown in his patent for "Improvement in Rheotomes or Circuit-Directors" (U.S. Pat. 131,334).

The patent drawing for Edison's revolving rheotome (U.S. Pat. 131,334).

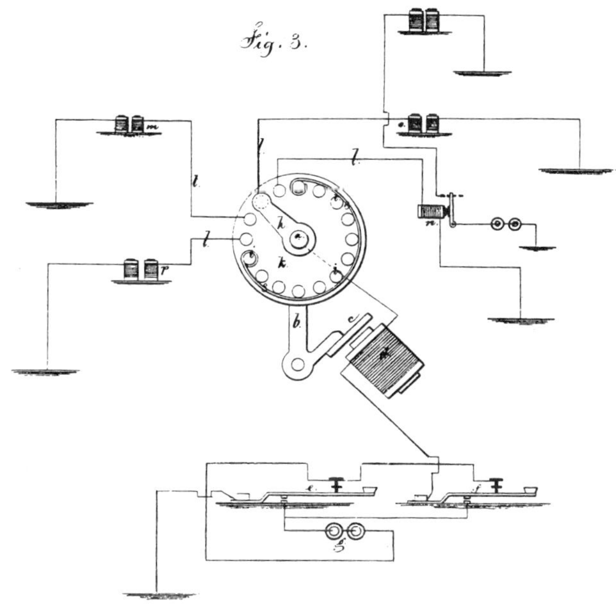

–361–

Notebook Entry: Automatic Telegraphy[1]

[Newark, Summer 1873]

Duplex relay = work over line by induction currents

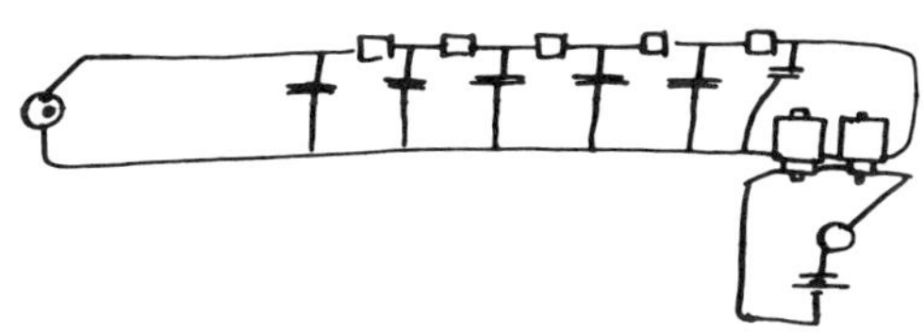

arrange in first place with battery direct with 300 ohm shunt and small relay in, with best magnet shunt at receiving end, and increase R so that it is impossible to get 100 perfect words per minute. then replace shunt by double coil relay and work with inductive currents see if 100 can be obtained with these induced currents—

Wet 10 rolls Boston perm Reg for records arrange as above (previous inductive ex) if induction dont work and replace shunt and use Duplex relay as receiving shunt, use one coil only. Then arrange a box of battery with a Bradley Rheo, with second coil, & unplug till the main line current neutralizes the local current and leaves an excess. The Theory being that the reversal of the polarity of the iron cores generate more induction than without second ckt. Try both with & without second circuit Try without battery by closing coil Thus

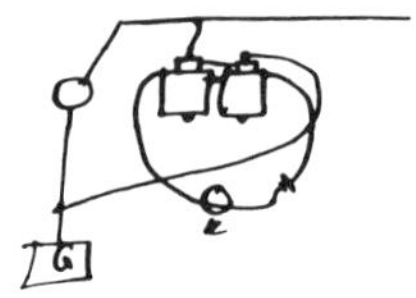

add Condenser thus ☞

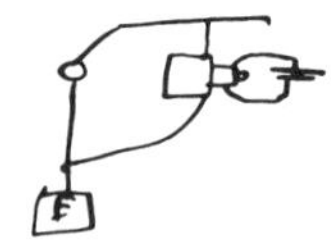

Put Duplex relay in brid[g]e, ~~o~~balance The put a second battery to second coil to as to balance effect of main Then see if there is any induction if so if it is not in the contrary direction to what it would be regular—

Also.

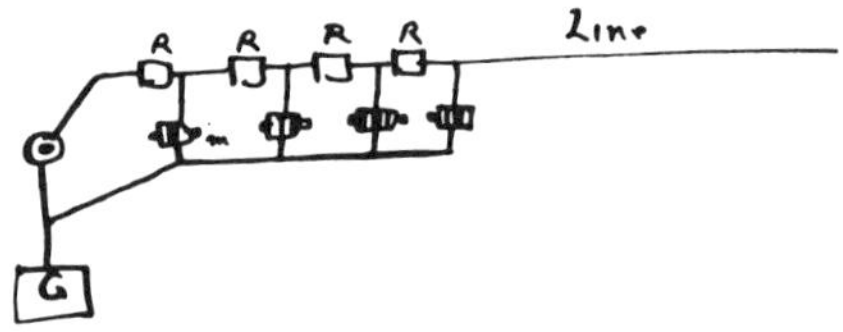

⟨Batchelor Edison—12 [hours]⟩[a]

Try this— ⟨Tried it, OK⟩ Obtaining the inductive effect free from the main current and regulating the latter so it will just make the dashes.[2]

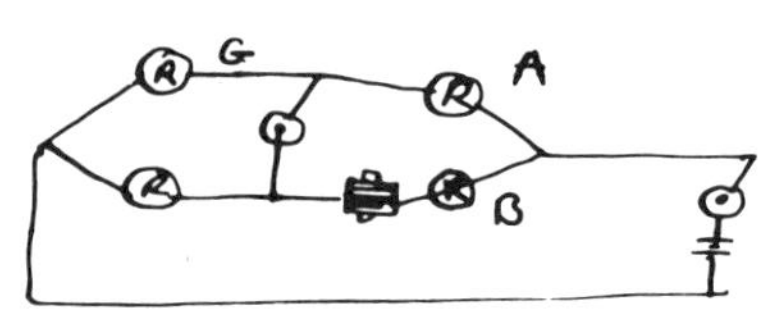

All sides being equal in Resistance no current from the battery passes through the chemical or other instrument in the bridge wire, but the inductive action on the opening and closing of the main is generated within the circuit formed by the two sides ~~of~~ and the bridge The currents are recorded upon the paper. any inductive arrangement acting like a magnet condsr etc[b] may be inserted in these branches and the inductive action observed free from the currents which form it. The difference in resistance of the different apparatus which it is desired to observe is equated for by the Rheosats A.B. so that the same quantity of current is made to pass over the two branches in all cases, an~~d~~ extra magnet inserted at G doubles the effect. If placed in the top branch[3] if placed in the lower branch it neutralizes the effect of the other magnet. This device ~~a~~is very convenient to observe the polarization in liquids with metallic electrodes.

See if the inductive effect of the magnet is affected by the batteries, balanced in the bridge thus

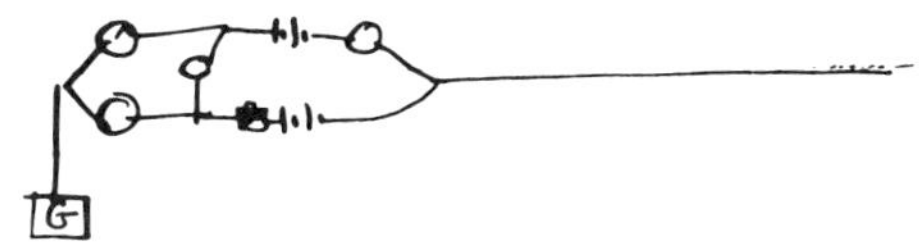

Try this 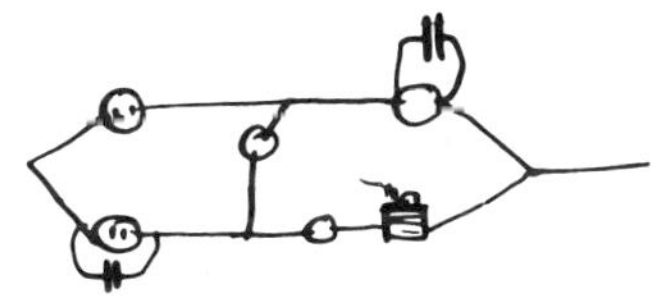

Try with magnet & without, see what condensers will do. Then thus to get longer charge

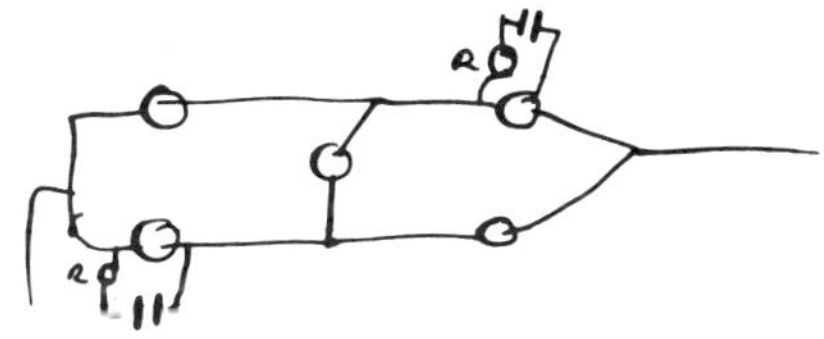

Also 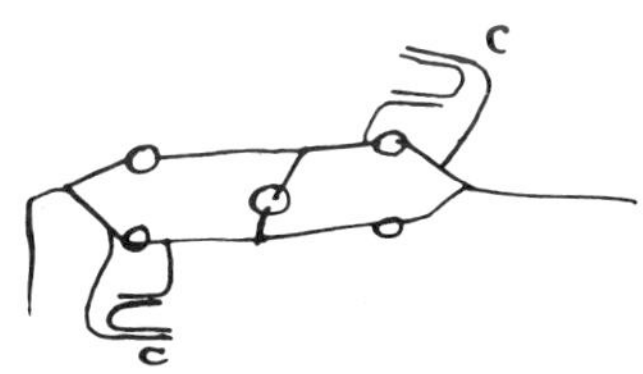

Try this 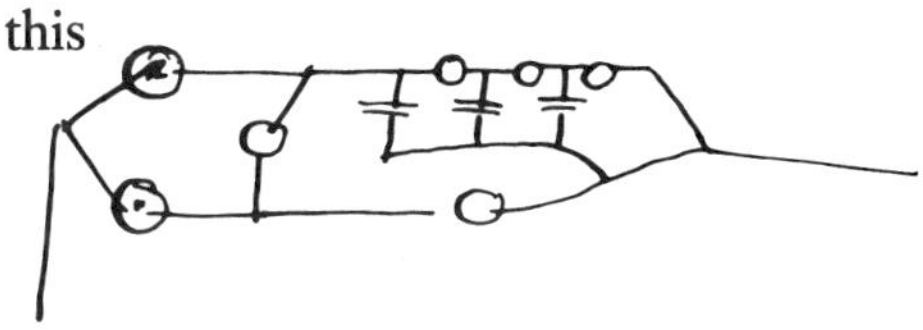

This 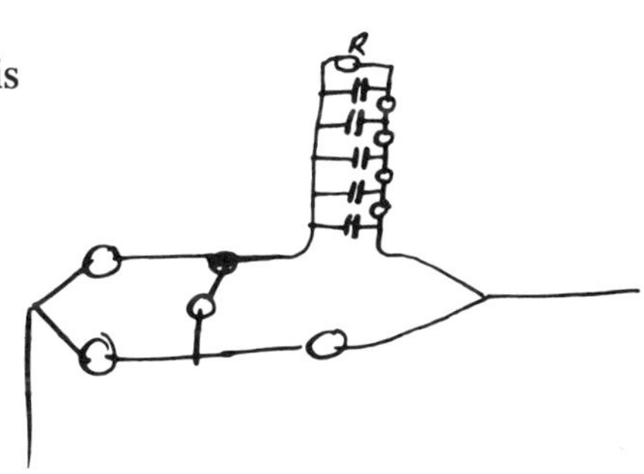

This 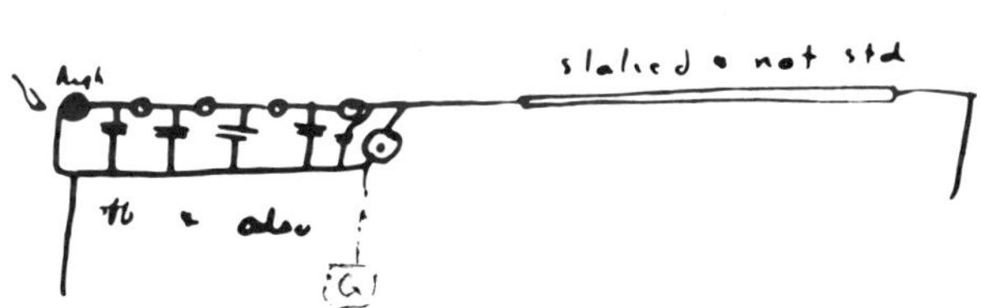

Try if working on well staticed Line if doubling quantity will form dots & dashes thus be careful

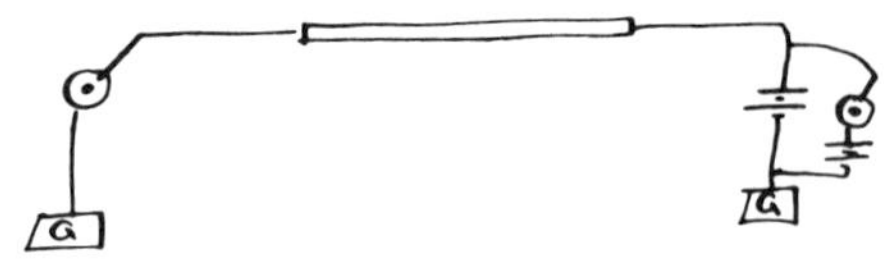

The excessive amount of iodide of potassium used for the regular solution led me to believe that only a certain amount was necessary to give good mark and the balance only increased the Conductivity but I find that by using a small quantity & adding Chl Sodium to give the requsite conductivity that its a failure, & the excess of iodide Pot is necessary probably some chemical might be used that would not redisolve the Iodine & at same time give the paper proper conductivity. ⟨Edison & Batchelor 25 hrs⟩[a]

Try effects on Duplex relay in bridge by closing second coil =Nothing noticable=

Try receiving on well staticd line with battery at receiving end ~~and arrange thus~~

Try and see if a number of bottles with pure water only arranged along resistances after the manner of an artificial Cable does not imitate the Atlantic. C

Thus[c]

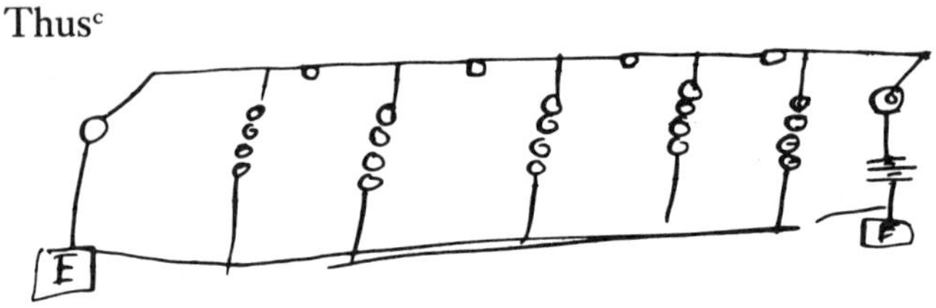

Try this to lengthen the discharge

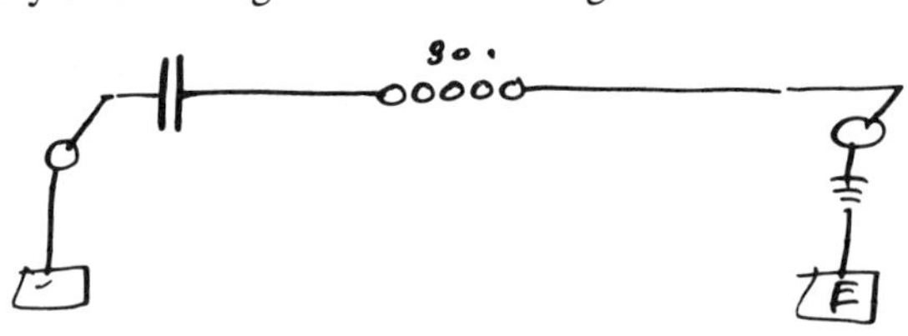

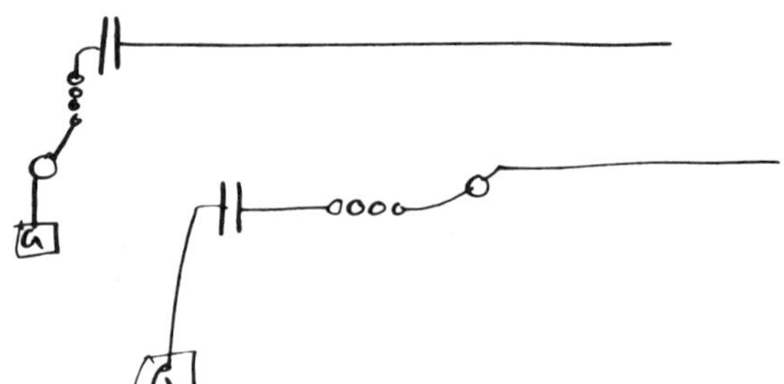

X, NjWOE, Lab., Cat. 1171:63 (*TAEM* 6:238). [a]Marginalia written by Batchelor. [b]"condsr etc" interlined above. [c]Followed by "over" to indicate page turn.

1. See headnote, p. 46.

2. Edison later drew up a caveat (Doc. 375) incorporating variant versions of the following circuit designs, arranging magnets, rheostats, and secondary batteries in a Wheatstone bridge around the receiving instrument. Cf. Docs. 357 and 363.

3. "If . . . top branch" probably should have been cancelled.

-362-

Technical Note: Miscellaneous

[Newark, Summer 1873[1]]

Wet paper galvanometer—wind two bobins 50 feet paper in solution Chl Sodium and wind seperating with oiled silk, thin rubber cloth or gutta Percha foil.

In a galvanic battery try outside, Chloride Sodium and Carbon, in porous pot, solution nitrate silver and carbon, use new carbon & new porous pots—also replace carbons with platina if unsatisfactory Theory being ~~e~~recipitation of the chloride by the nit silver will set up an active current.

Try. same but use cold chl sod outside & boiling hot chl sod inside porous diagram[2]

Try one cell secondary with clear coppers with distilled water and Iod Kali[3]—also replace coppers with platinas or carbons.

Associate with the iron receiving pen a platina pen very close to it so that the oxygen set free at the platina pen shall form the protoxide on the iron. theory being that iron when so associated is more susceptbl of oxidization, also probability that 1/6th of effect is lost by polarization iron pen which will be destroyed by the associated platina pen

Ascertain if anything will combin~~a~~e with protoxide iron to make it soluable in water if so add to Callaud battery to dissolve the yellow sesquioxide[a] on the zinc due to the iron in the zinc. The NJ Zinc works abstract their zinc from a compound containing iron pyrites & consdble pr centage iron remains in their zinc[4]

See if there is a noticeable effect produced in these modifications

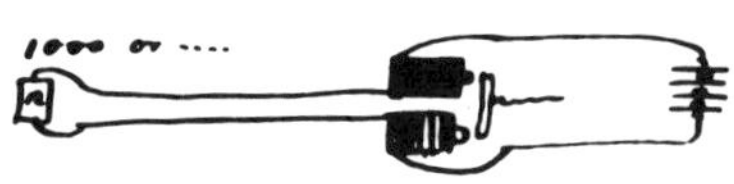

and:

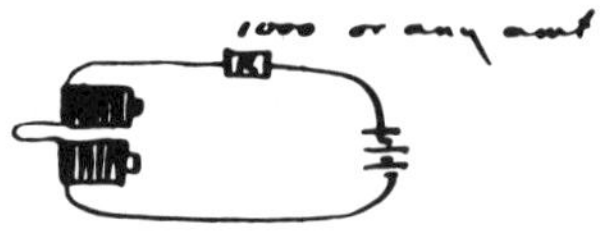

Theory being that in the centre of resistance or bulk of Conductor there is a nil point[5] and that each side from that point the R is P & the other N ~~or~~ & that a battery on a R should if it is strong attract a finely balanced magnetized needle to connection wire at zinc end its N end & at Coke end its P end   Try it=

Try wet paper—sulphuric acid and nut galls. on the passage of the current, protoxide of iron is formed. The sulphuric acid combining therewith forms protosulphate of iron or copper as, the nut galls will then strike an intense black forming common ink—

Make a Zamboni dry pile.[6]

Make a condenser of sheets of copper and sheets of thin stove zinc.

Try electrotypers plumbago rubbed on parafine sheets for condenser

Make trough   one side place long sheet zinc at intervals carbon, try effect.

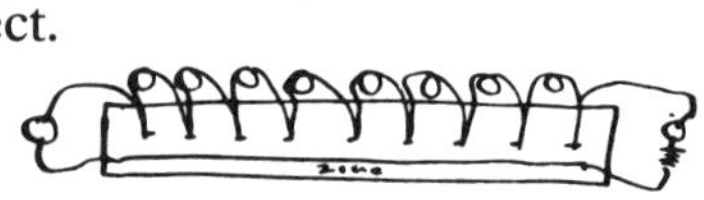

Try 1 Cup then increase, see ~~a~~how many it takes[b] to neutralize= use pure water first then Chl Sodium=

A galvanometer needle hugs the stop pin after the current has ceased. Theory: the needle being thoroughly becomes charged P one end N other end statically & is attracted by pin—try connecting pin with ground.

On a shop line staticd[7] get its R. carefully. Then insert a leak of know R in center and get total .R. I believe: that in a submarine Cable that there is a constant opposing electromotive force in the core against the battery & that they do not get the right R of the core.

Try[a] all the secondaries connected for quantity on a R of 500m[iles]. ⟨× 13 = 1500 + 5000 = 6500⟩[c] 6500 ohms, in the centre 3250. Connect all the secondaries to ground or return and the other end to R through an R of 250 000 ohms. See the Charge due to Secondary Action   Use wet string

Duplex

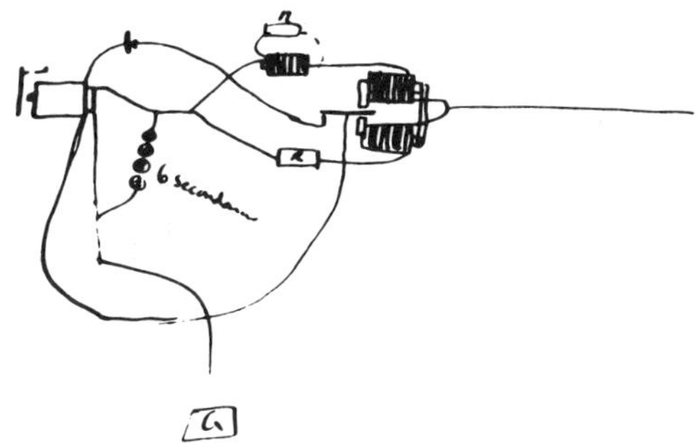

May work without secondaries The lever of the pol[ar-i]z[e]d R[elay] might be made to short ckt the secondary when open.[d]

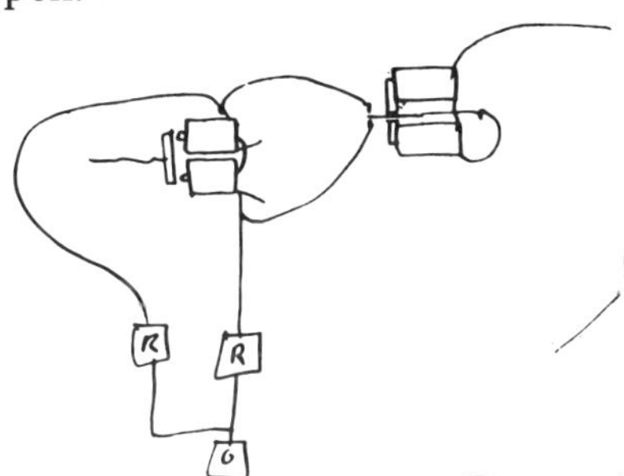

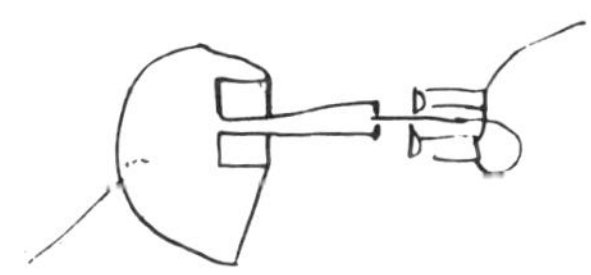

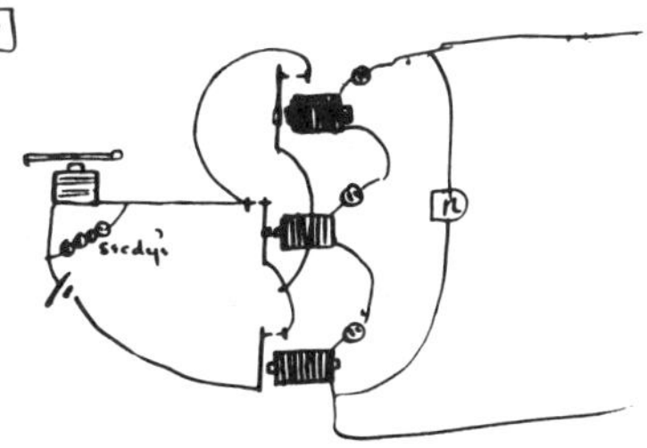

AX, NjWOE, Lab., Cat. 299:64–65, 68–69 (*TAEM* 6:126–27, 130–31). Page numbering and resemblances in style, paper, and ink link these separate scrapbook items. [a]"oxi" interlined above. [b]Starts separated page. [c]Calculation in vertical form, interlined below "500m." [d]Figure begins new page.

1. These miscellaneous notes are typical of Edison's experimental work during the summer of 1873.

2. Not found.

3. That is, potassium iodide.

4. The New Jersey Zinc Co., with headquarters in New York and mines in Ogdensberg and Franklin, N.J., was the leading U.S. manufacturer of zinc. A discussion of the company's ores and extracting processes is found in Ford 1874, 142–43.

5. See Doc. 317.

6. This dry battery, made with tin (or silver) and manganese dioxide, was used primarily in the construction of a very delicate electrometer. Prescott 1877, 83; see Doc. 353 n. 14.

7. Probably a line with a large electrostatic capacity.

[Newark, c. September 27, 1873[1]]

Technical Note: Magnets

Bridge[a] sheet iron Magnetic Condenser

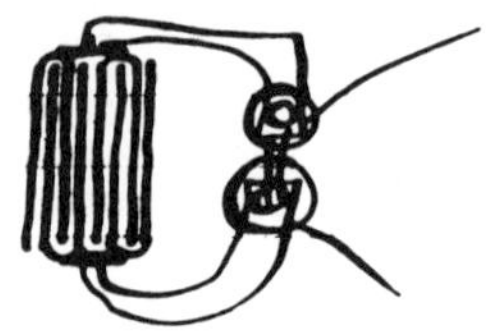

Connect one armature tin foil Condenser to one pole magnet & other armature to the other pole =
Wind a Relay Magnet with long roll tin foil =

10 feet Long

sheet iron plates

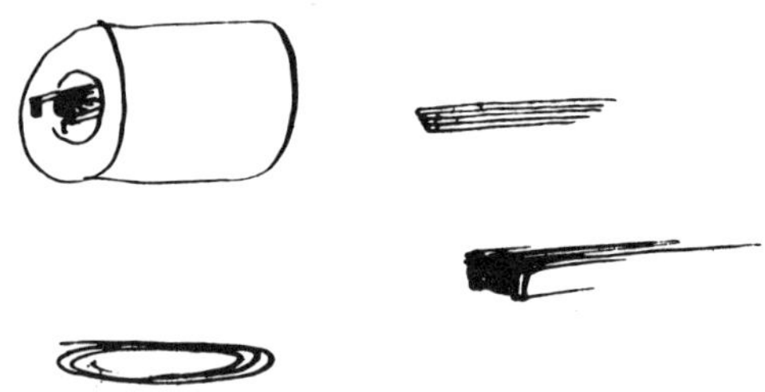

Try following Experiments upon the inductive effect of magnet made as follows in Bridge

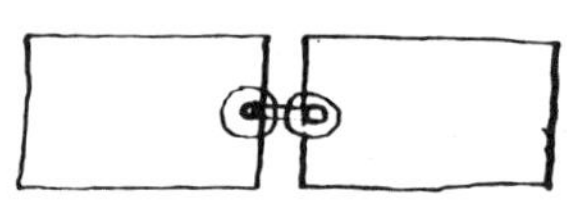

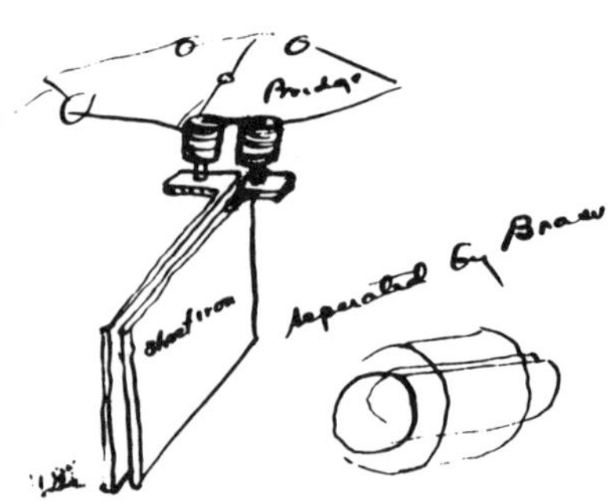

Sheet iron tube 1 foot diameter wound 6 layers 23 wire[2]

10 feet long wound 23 wire 4 layer

AX, NjWOE, Lab., Cat. 299:60 (*TAEM* 6:122). [a]"C" written above the drawing.

1. An account book entry of 27 September 1873 notes "2 spools wound with Tinfoil every layer for Edison." Cat. 1219:112, Accts. (*TAEM* 21:761).

2. Probably no. 23 Brown and Sharp gauge wire, .022571″ diameter. Maver 1892, 514, 562.

–364–

From Ernest Fischer

Philadelphia, Sept 30 1873[a]

Dear Sir:

I have today sent you a copy of my recently obtained Letters Patent,[1] which I hope will duly come to hand. It is my desire to sell the same in state Rights. Now to make it a quick sale I will sell you the State Right for New Jersey, for Eight hundred Dollars or its Equivalent.

Please give the matter your attention & you will find it of great Value & Need throughout the Comml World. Please let me hear from you soon, & Oblige[2] Yours very Respectfully

E. J. Fischer[3]

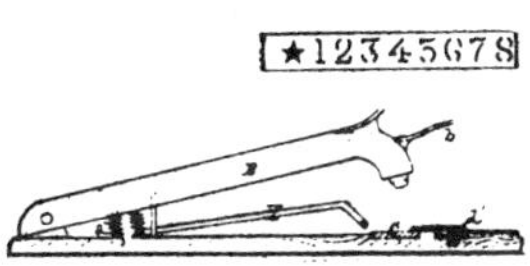

A Patent Office Gazette *illustration of Ernest Fischer's antiforgery punch.*

ALS, NjWOE, Lab., Cat. 299:25 (*TAEM* 6:90). Letterhead of Fischer antiforgery device. [a]"Philadelphia," and "187" preprinted.

1. Fischer's device punched numerals out of paper to prevent forgery. U.S. Pat. 138,559, "Stamp for Cutting Out Figures."

2. No reply has been found.

3. An Ernest J. Fischer is listed in the 1874 Philadelphia city directory as a clerk. Gospill 1874, 477.

–2– October–December 1873

During the last months of 1873 Edison followed the lines of work he had pursued in the summer, spending most of his time in the laboratory and executing few new caveats or patent applications. In extensive chemistry experiments he focused his attention particularly on batteries and automatic telegraphy. Investors who bought the rights to Edison's automatic system in Great Britain extended the tests he had begun in the spring, battling Post Office engineers over the fitness of the system. On 1 October, Edison demonstrated his quadruplex telegraph system, and he continued his work in multiple telegraphy through the fall. These months also marked the beginning of Charles Batchelor's role as Edison's chief experimental assistant.

Edison's increasing chemical sophistication reveals itself throughout his notebooks. Detailed tables record battery tests that extended over weeks and sometimes months as the laboratory crew tackled impediments of polarization and diffusion of electrolytes. Edison's interest in batteries led him to use them in novel circuit designs, such as having extremely weak secondary batteries act as condensers. He even considered sending one of his findings—that electrolytic action affected the diffusion rate of liquids—to the English physicist James Clerk Maxwell, who had recently used diffusion experiments as important evidence for the size of molecules.

Edison prepared only two patent applications and three caveats during this period; all concerned automatic telegraphy and reflected his persistent interest and experimental work in that area.[1] Both patent specifications described chemical solutions, two of the caveats covered roman-letter systems, and the third caveat detailed an automatic circuit. Edison tested many receiving solutions with numerous metal styli, still looking for the combination that would mark permanently and be sensitive, cheap, and easy to handle. Account records show he began experimenting again to develop a three-key perforator for office use.[2] In late December, responding to the continuing priority claims of George Little, Edison assembled several important telegraph promoters to witness what he hoped would be a decisive test between their automatic systems.

In England, meanwhile, the investors who had obtained the rights to Edison's automatic system were struggling to have it installed on Post Office lines. Unfortunately for that enterprise, influential members of the Telegraph Service felt that they could improve the original Bain technology sufficiently to make licensing the Edison system unnecessary—if, indeed, any electrochemical replacement for the Wheatstone ink-recording automatic was wanted at all. As in the spring, successful tests were not enough to ensure governmental acceptance. The investors were also interested in using the automatic on undersea cables and had experiments carried out to that end.

Edison was experimenting with cable technology himself, continuing his summer research and probably reacting to the British interest as well. Few laboratory records remain, but October and November accounts reveal that he built all the components he needed to perform cable tests in his laboratory, such as condensers, resistance boxes, and insulated wire.[3]

Edison also pushed on with his experiments in multiple telegraphy. On 1 October he gave a laboratory demonstration of his quadruplex system to several important officers of Western Union.[4] At various points during the fall he sketched groups of multiple telegraph circuit designs, and late in the year he sent one design to England hoping to promote its commercial application there. The U.S. Patent Office had twice rejected the duplex patent applications Munn & Company had prepared in the spring;[5] Edison now undertook their revision himself, but soon turned the process over to his regular patent attorney, Lemuel Serrell. Further delays in this round of amendments caused problems with later applications.

Edison collaborated with several partners during these months. Little affected so far by the emerging national business depression, he and Murray kept up their manufacturing and continued to rent out shop space.[6] Edison worked with an old friend, telegrapher and instrument maker Jesse Bunnell, to develop a railway signaling device. Charles Batchelor, originally hired as a mechanic, proved to be a careful, nimble-minded experimenter and assumed an increasingly important role in the laboratory. He soon emerged as Edison's principal assistant, shouldering ever-broader responsibilities. Batchelor's role as co-experimenter and even co-inventor typifies the team approach that became characteristic of Edison's laboratory; from this point on it is increasingly difficult to assign ideas (or their experimental exploration) to individuals.

1. Another caveat (45) was executed at the same time, but most of it was prepared in August and it was not filed until 1 September 1875. See Doc. 348.

2. Cat. 1219:100, 116, 119, 133–34, Accts. (*TAEM* 21:755, 763–64, 772–73). For Edison's earlier work see *TAEB* 1:253 n. 5, 258–59, 265 n. 3, 281, 474 n. 3.

3. Cat. 1219:100, 122, 129, 133–34, Accts. (*TAEM* 21:755, 766, 769, 772–73).

4. Edison's testimony, Quad. 70.7, pp. 272–73 (*TAEM* 9:503); testimony of Edison, Charles Batchelor, Uriah Painter, Albert Chandler, and Alfred Brown, Testimony and Exhibits on behalf of Thomas A. Edison, pp. 14, 65–66, 89–90, 107–9, 111–12, *Nicholson v. Edison;* Doc. 356 n. 2.

5. Docs. 304–5, 308–11, and 314–15.

6. For manufacturing records see Cat. 1219:115–47, Accts. (*TAEM* 21:762–79). Both Keene and Prosser continued to rent space; see Chapter 1 introduction, p. 5 n. 7.

–365–

Bill to George Harrington

Newark Oct 1st 1873.

Geo Harrington D[ebto]r ~~G~~To Thomas A. Edison.

One third of money paid by Smith Fleming & Co Leadenhall St London on account of foreign Automatic Patents as per agreement.[1]

Sixteen Thousand six hundred and sixty six dollars $\frac{66\frac{2}{3}}{100}$. gold $16,666. $\frac{66\frac{2}{3}}{100}$

ADDENDUM[b]

[Newark, 1873?]

Memorandum

This money was paid in three checks on Jay Cooke & Co[2] drawn to Edison Harrington and Reiff= The checks were brought over to Newark by Reiff & I endorsed my signature to same which was witnessed by my father, Reiff acting for Harrington. Reiff and Harrington divided the whole of the 50,000. gold & I recd none which is contrary to agreement with said Harrington= The above bill is to be presented when said Harrington returns to the United States He having been repeatedly asked for it ~~wh~~before leaving and always promised to settle, but left without doing so[3]

Thos A Edison

AD (copy?), NjWOE, DF (*TAEM* 12:1119) [a]Followed by horizontal line. [b]Addendum is an ADS.

1. Doc. 350.

2. Jay Cooke & Co. was one of the country's most influential banking houses. Its collapse on 18 September triggered the Panic of 1873 and the subsequent several-year depression (*DAB*, s.v. "Cooke, Jay"; Sobel 1968, 178–79; Fels 1973, 83–112). John Puleston, one of the trustees in the agreement with Smith, Fleming & Co., was a partner in Cooke's London branch.

3. Harrington traveled to Cornwall, England, in the summer of 1873. He also went to Europe in 1875 and remained there for many years. Harrington's testimony, 1:126, 136, Box 17B, *Harrington v. A&P*.

–366–

Notebook Entry: Cable Telegraphy[1]

[Newark, c. October 5, 1873[2]]

Cable Experiments—[3]

Transmitting a .P[ositive]. current both through ground and Cable with two static ab~~a~~utments for grounds.

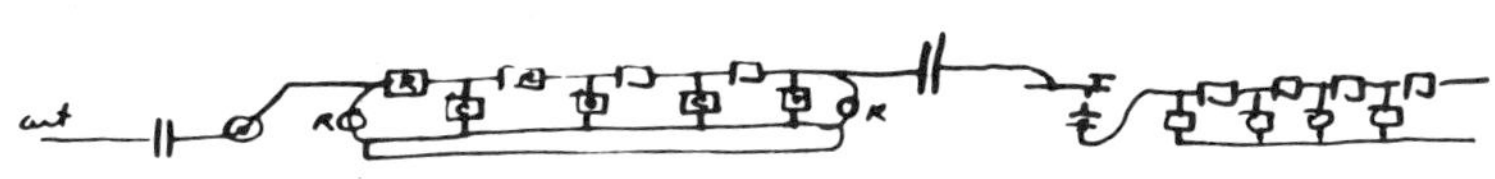

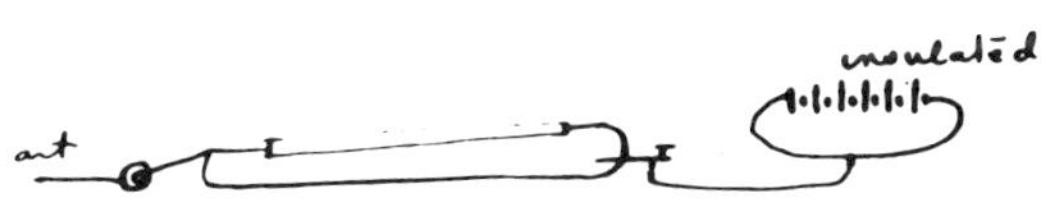

See what speed obtainable on 9000. ohms R with one cup battery Iodide paper. If no mark try to improve Iodide ppr till mark obtained

Try this

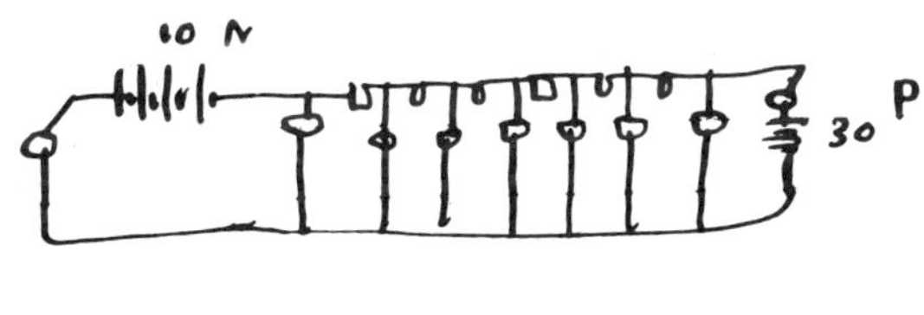

also test if a Voltaic battery[4] put in circuit immediately after the dot has been ~~made~~ sent & the sending end to air[5] will have any effect on stopping the flow from the Condensers. Use same number of elements to kill charge from Condensers as is used to send with. if it does not kill the discharge or have any effect then it is perhaps possible to transmit with Voltaic & static E at the same time, and perhaps a Compensation might be devised upon the difference. in these two forms

It is also possible that it is better to work at sending end with static from an artificial,[6] or to make the mark with Voltaic & send Reversal with static or Vice Versa— Try with 20 P at receiving end and 20 in same direction sending end

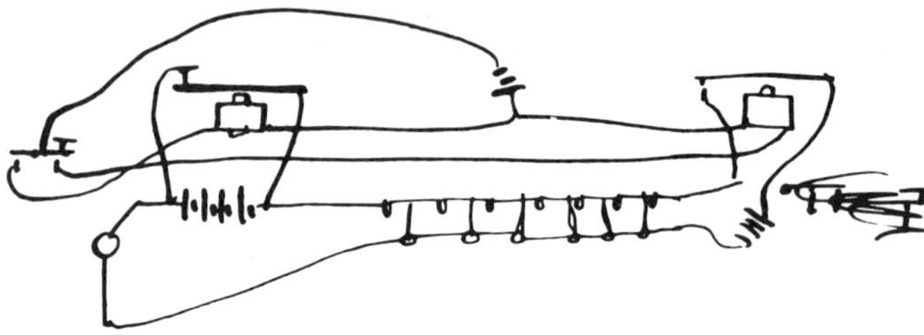

Also reverse send with P & = N the batty at Recg end to oppose static when cut in.

I have noticed that a Morse Relay on a Regular line works much sharper with a battery at both ends than when the battery is at the sending end only. when the latter is used the N —• etc stick and the writing is light which is not the case when 2 batteries are used the sticking acts precisely like the effect of the static charge—. Of course when 2 batteries are used in opening the circuit one battery is cut off but the battery at receiving end remains on & perhaps keeps the line charged or prevents the static from flowing out. ~~Of~~ it however performs some good function

test this—

It is possible when we consider that both batteries form the static that it would perphap be better to have several times the number of cups at the receiving end as at the sending end.

Ascertain if the induction is as great from a magnet with its cores snearly saturated with magnetism as when it is free from magnetism. Use a bridge and a Duplex relay the extra coil having a good battery permanently attached to it the bridge connected to the line free of static. be sure that the magnetism in the Cores due from the local is the same as that induced by the Main battery.

It is possible that 40 cups kept permanently on the cable at

Receiving end the signals could be made better by throwing in & out 10 cups at Receiving end without breaking Circuit than they could by using the whole 50 at sending end.

Attract the static.

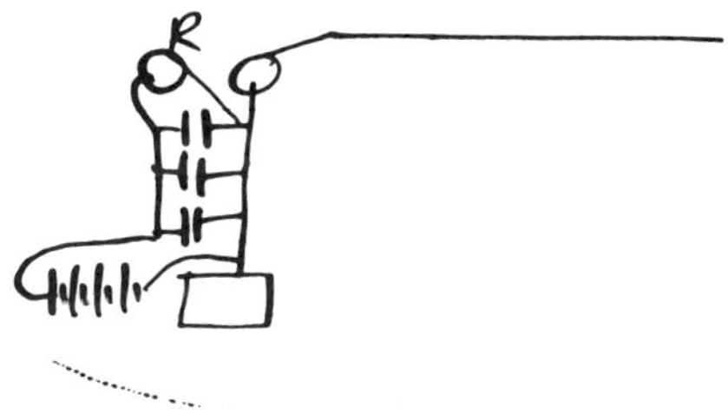

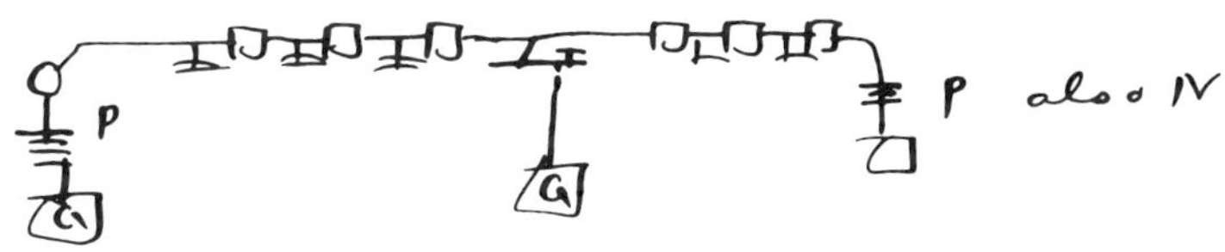

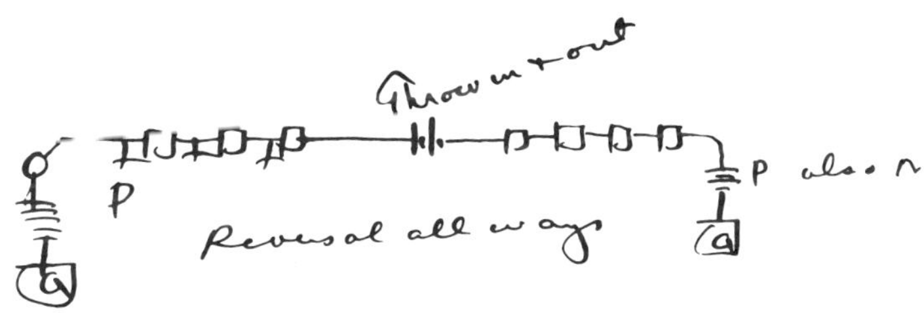

Use the static charge to work the cable

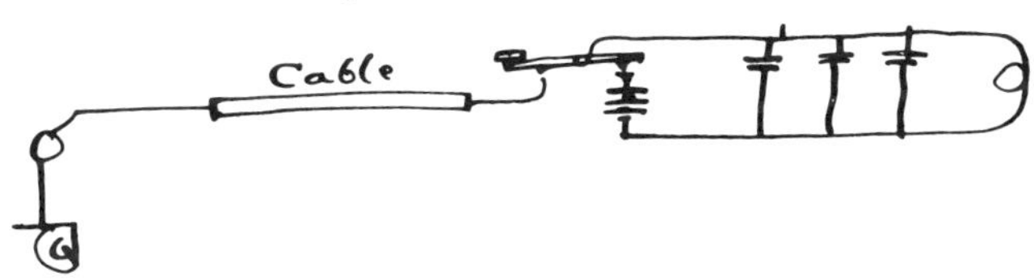

If it is necessary to put regular to ground use this arrangement

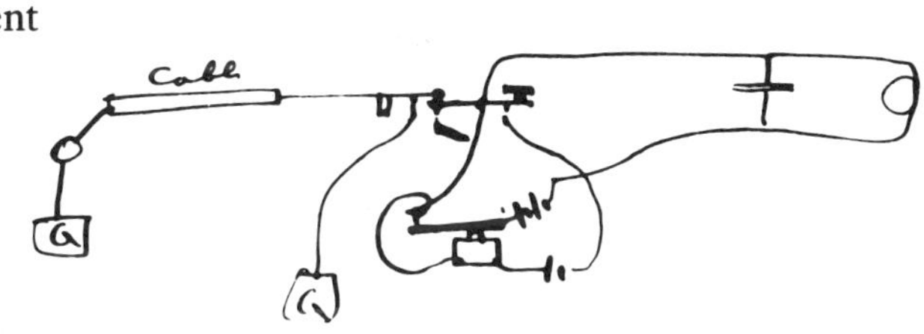

W Siemens etc proved etc—that when a current is sent along through a submerged cable—a quantity of Electy is retained in charge along the whole surface be distributed proportional to the tension at each point— Rep Joint Com Eng p 381[7] Therefore

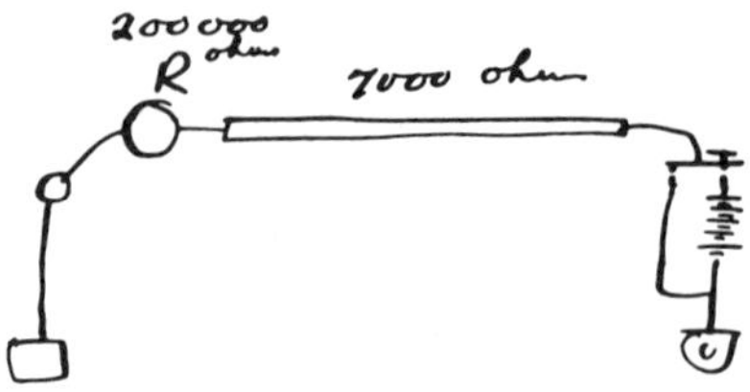

AX, NjWOE, Lab., Cat. 1171:89 (*TAEM* 6:252).

1. See headnote, p. 46.

2. This entry immediately precedes one written about 9 October; see Doc. 374 n. 1.

3. Edison continued to experiment with cable telegraphy throughout the fall (see cable-related account records in Cat. 1219:100, 122, 133, 146, Accts. [*TAEM* 21:755, 766, 771, 779]). It was probably during this time that Edison conceived the idea of a bare-wire Atlantic cable in which the ocean would serve as the battery in combination with two transmitting wires of different metals. NS-Undated-005, Lab. (*TAEM* 8:394–95).

4. That is, a primary battery.

5. That is, the circuit is opened.

6. That is, an artificial line; see Doc. 317.

7. In this paragraph Edison paraphrases William Siemens's result reported in appendix 7 to the "Report of the Joint Committee appointed by the Lords of the Committee of Privy Council for Trade and the Atlantic Telegraph Company to inquire into the construction of Submarine Telegraph Cables; together with the Minutes of Evidence and Appendix" (London: H.M.S.O., 1861).

–367–

From Lemuel Serrell

New York, Octo 6 1873.[a]

Dear Sir

The cups of liquid with copper conductors in them are shown in your patent 141,776 in the branch to earth, and also condensers and batteries. The cups are shown in main line in No 141,773 but not in any that I discover ~~in~~are the shunt cups and resistance introduced as proposed by you when last you were here: the way is clear I think—[1]

The other chemical cases[2] are ready for examination and signature. Yours truly

Lemuel W. Serrell[3]

ALS, NjWOE, Lab., Cat. 299:46 (*TAEM* 6:111). On letterhead of Lemuel Serrell. [a]"New York," and "187" preprinted.

1. The "cups" were simply water (or dilute acid) put directly into the circuit to act as small, weak storage batteries that discharged in the direction opposite to the main current when that current ceased. On 18

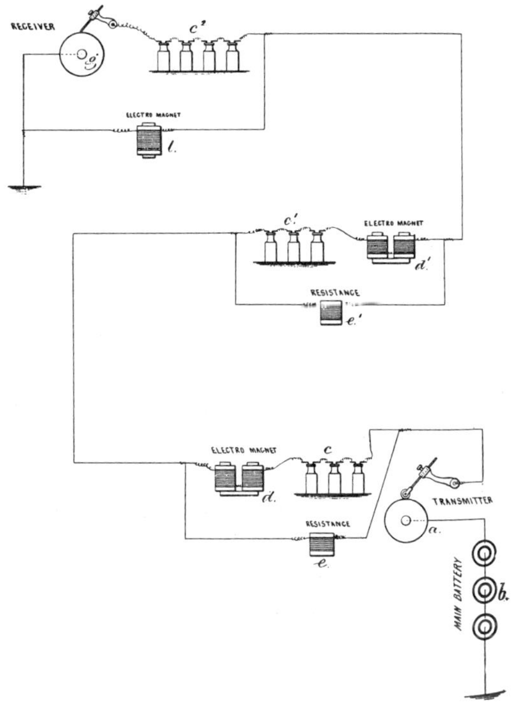

Edison's U.S. Patent 168,243, showing secondary batteries **c**, **c**^{1}, *and* **c**^{2}.

January 1875 Edison executed a patent application for an automatic telegraph circuit containing such cups (U.S. Pat. 168,243).

2. Probably patent applications executed 14 and 29 October 1873 (U.S. Pats. 160,580 and 160,403).

3. Lemuel Serrell was Edison's primary patent attorney from May 1870 until the early 1880s. See *TAEB* 1:196.

–368–

Draft Essay: Batteries

[Newark, c. October 10, 1873[1]]

Proposition.

~~Control the Blue line in a gravity battery (Callaud modification of Daniels) so that the zinc shall at all times be in the zinc solution.~~

To prevent the diffusion of liquids—(sul cop & chl zinc) in a Daniells gravity battery.[2]

~~It has been observed that there is one drawback to the gravity battery which is that the lo[a] sulphate of Copper solution raises[b] gradually raises after the battery has been in use for sometime, and a large quantity of Copper is percipitated on the zinc w~~

When a peice of zinc is thrown into a solution of Sulphate of Copper the sulpuric acid of the sulphate of Copper leaves the Copper and attacks the zinc forming Sulphate of zinc ~~and~~

~~this~~ The metallic Copper is thrown down upon the zinc in a spongy state = and this ~~I~~ will go on until ~~a~~the sul of Copper ~~solution~~ has all been decomposed.

In the gravity battery this same effect is obtained. the Sulphate of Copper raising up towards the zinc soon surrounds it and the chemical as above described takes place which is independent of the Electrical action. the deposition of the spongy Copper upon the zinc leads to local action as well as reducing the power of the battery by continually ~~altering the~~ reducing the difference between the two metals a large amount of Sulphate of Copper ~~is we~~ is decomposed adding nothing to the power of the battery. ~~and locally~~ to obviate ~~som~~ the ~~effect~~ of this deposition, recourse is had to raising the zinc up ~~to the~~ as far away from the sul Cop solution as possible, so that it will be imersed in the sulphate of zinc solution only this answers very well until a complete diffusion of the two liquids takes place but has the disadvantage of increasing the internal resistance =

~~After e In e~~ In experimenting upon a large number of gravity batteries arranged in innumerable forms, I noticed that in two of the number which were arranged like the menotti having saw dust between the elements[c] that I had placed the crystals of blue vitrol below the copper element on the top of which was the saw dust & zinc element, and that the liquid below the element was very blue whists the saw dust was not even colored. It occurred to me that the Electric Current might perform some function and thus by arranging the elements and solutions in a peculiar manner the diffusion of the liquids might be prevented

I ~~the~~st this I ~~obt~~ placed 2 pound of blue vitrol in the bottom of a jar 6 ~~b~~inches in diameter by 9 high a little over the blue vitrol I suspended ~~the~~ the Copper Elements which was a round disk 5 inches in diameter with an ~~smal~~ aperture in the Centre 1½ in diameter over this I suspended the zinc ~~the battery was the plac~~ the poles of the two batteries[3] were then connected together that remained in that way for ~~two~~ several months.

no diffusion of th[e li?][d]quids took place the su[l?][d]phate Copper solutions assumed a very deep blue ~~which~~ up to contact with the Copper Element above the copper element and in contact with it was the sul zinc solution which was a light yellow. The zinc was comparatively clean and no trace of Copper could be discovered.

~~T~~I found subsequently that when the two poles ~~we~~ of the

A gravity battery.

battery were disconnected for a ~~ti~~ short time that the Sulphate of Copper began to raise and mix with the Sul Zinc Solution but it immediately went back under the copper when the poles were connected

~~I also found that the resistance of the battery could be I also found that the action~~ Thus proving that the Electric Current itself prevents the diffusion of the liquids I also found that when the Zinc was ~~pl~~ lowered down within 1/64 of an inch of the Copper Element that the ~~l~~ sulphate of Copper would return back under the Copper element after it had been allowed to rise, in much less time tha~~tn~~ when the zinc was farther ~~wa~~ away from it ~~as~~ therefore the ac[tion?] of this battery is ~~much~~ not only surer when the two elements are very close but the internal resistance is greatly reduced As the liquids in this battery soon difuse it is well ~~to employ as~~ when they are used upon an open circuit to shunt each or all the[e] cups[f] with several feet of small german silver wire.

~~With a large high jar and charged with six or~~ This battery is perfectly constant the galvanometer needlc ~~stood in the same place~~ not varying a degree for months =

AD, NjWOE, Lab., NS-Undated-001 (*TAEM* 8:93). [a]"the lo" overstruck before the rest of the paragraph. [b]Overstruck before the rest of the paragraph. [c]"which . . . elements" interlined above. [d]Document damaged. [e]"or all the" interlined above. [f]"s" interlined below.

1. See Doc. 370 nn. 2–3.

2. Cromwell Varley developed the gravity battery in 1854 as an improvement on the Daniell battery (see *TAEB* 1:615 n. 3). In Varley's design the difference in the specific gravities of the copper sulfate and zinc sulfate solutions keeps them separate (Prescott 1877, 56–61).

3. Edison meant "electrodes."

–369–

Draft to James Parrish

[Newark, c. October 10, 1873[1a]]

Mr. Parish[2]

Please make this suggestion to Mr. Latimar Clark,[3] all of which if he desires may ~~b~~add ~~to~~as my mite to the ~~Te~~Society of telegraph Engineers.[4]

to prevent the diffusion of Liquids in a gravity or a Minotti Battery,[5] ~~discussed by~~ Always place the Sulphate of Copper under the Copper plate instead as heretofore beneath it[6] Grove & Bichromate[7] may be worked in this manner The Sulphate of Copper will never rise above the Copper plate

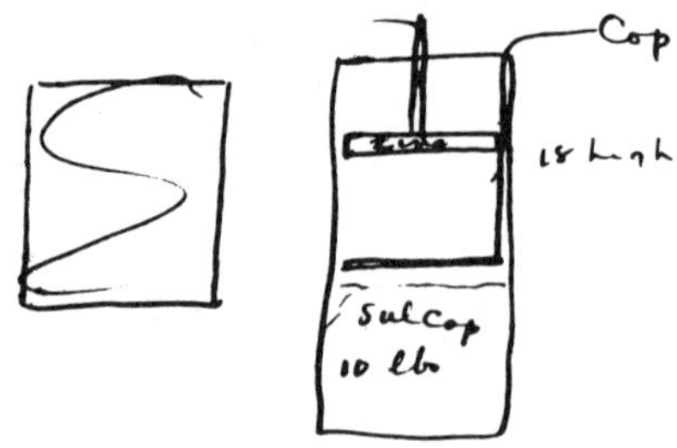

Patented by me 1871.[8]

ADf, NjWOE, Lab., Cat. 297:146 (*TAEM* 5:966). [a]"1" written at top of page.

1. See Doc. 370 nn. 2–3.

2. Little is known of Parrish except that he worked with George Gouraud in promoting Edison's automatic telegraph in England.

3. Josiah Latimer Clarke was a civil engineer and electrical expert best known for his work in submarine telegraphy and in the establishment of standard units of electrical measurement. He was a co-founder of the Society of Telegraph Engineers. *DSB,* s.v. "Clark, Josiah Latimer."

4. The Society of Telegraph Engineers was founded in London in 1871 and later became the Institution of Electrical Engineers. For a history of the institution see Appleyard 1939 and Reader 1987. Edison and Charles Batchelor became members in 1874 (see Doc. 490A).

5. See Doc. 368 n. 2. For the Menotti battery see *TAEB* 1:607 n. 4.

6. Edison meant "above it."

7. A bichromate battery is a modification of a Bunsen battery (see *TAEB* 1:599 n. 3) in which a solution of bichromate of potash and sulphuric acid replaces the nitric acid solution (Prescott 1877, 71–72).

8. U.S. Pat. 142,999. Edison actually executed the application on 31 October 1872; the patent issued on 23 September 1873.

–370–

Technical Note: Chemistry and Automatic Telegraphy[1]

[Newark, c. October 10, 1873[2]]

Send[a] to Clark Maxwell description of prevention of th[e?][b] diffusion of Liquids by electrolytic action.[3]

Try[c] the diffusion of different substances be[t?]ween[b] two platinum disks ~~Try this = try magnets.~~

Molasses—Chromic Acid.
" Chrome Yellow
" Prussian Bl[ue?][b]

Alcohol[d] collored by So[--------]d[b] deep yellow satur[ate?]d solution of Sulphate o[f? --------][b]
Sulphate of copper a[------------------ c?]olored[b] by cochineal.
Protosul Iron and Ca[-----]t[b]
Water and Lamp black
Glycerine & Turpentine;

Reverse connections of battery.

In ascertaining the diffusion first place the solutions in glasses of the same character so that the platinum disks shall fit each one allow them to stand in a quiet place for 24 hours and note the rate of diffusion. Of course The lightest liquid which may be tested with the hydrometer is placed in the glass first and the liquid of a greater weight is poured down a glass tube to the bottom of the glass very slowly and carefully to prevent mechanical diffusion a mark is then made where the two liquids join and a note made as to the sharpness of the line athe glass is then covered w[i?]th[b] a peice of writing paper on which has be[en? ---]bed[b] beeswax and placed over the mout[h of the?][b] gl[ass?][b] and pressed down on the rim Thi[s ---------- e?]vaporation[b] at the end of 24 hou[rs ------- dis?]tances[b] which shows considerable diffu[sion?][b] is [----]n[b] down and the platinum disk placed in it so that the dlower disk shall just come to the mark previously made on the glass. This is connected to the battery the lower to the zinc poinle and the upper platina pole or disk to the copper pole. put aside and note the time it takes to bring back the two Liquids to their original position and so on thwith the whole number note all phenomenon—

The platina disks should be made thus

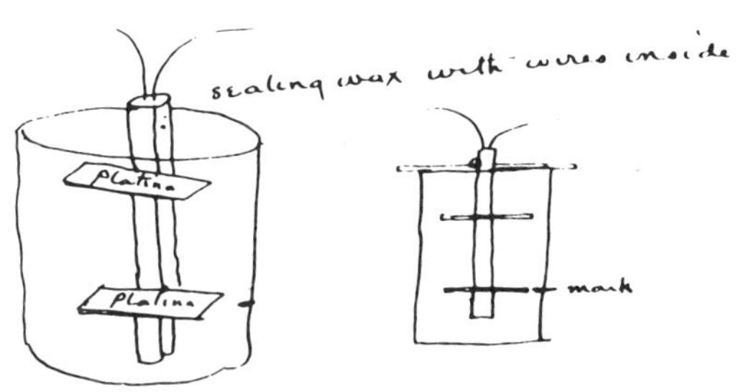

also immerse in two d[-----]cipitates,[b] and connect to battery—[e]

Arrange[f] a test tube w[ith?][b] two platinum strips and connect 4 Callaud[4]—and test all the madetals in Table A, Bloxam under electric influence and ascertain if the reactions are the same.[5] also use very dilute solutions of the metals to be detected and see if the presence of the electric current increases of decreases the delicacy also pass a spark constantly into the solution from the plate glass machine.[e]

Evaporate[f] different metals ~~on~~solutions on clean zinc (polished) by heat and note any specific colors for different metals.

A new reagent ~~is~~for Ammonia is described in Crookes p. 29.[6]

Crookes[f] says that ma[gnes?]ium[7b] has the most powerful electro-m[otive].f[orce]. of a[ll met?]als[b] substitute it for zinc in a small cell [and g?]et[b] deflections. p 37.[8]

When[f] Magnesium is p[ut?][b] in water it liberates pure inodorous hydrogen— hence associate a magnesium pen with the iron pen in reg f[erri]d[9] solution first before the pen in the same path 2nd after the iron pen and note the difference.[10]

A new reagent for Calcium. Common tungstate of sodium. Crookes p 38.

The sulphide of Ammona throws down a black percip with Urinum[11] oxides (ie)[g] protoxide hence a Urinum decomposing pen would form protoxide which in the presence of Sulpht of Amon would give black signals. probably the ferrids & feros[12] would give a colored percip. perhaps more stable than the Sulphide of Urinium which like all other sulphides would in the air bust up and walk on their ear Some description is given of the action of Urinum in Crookes p [55–?]57.[13b] says in testing urinum with ferrocy[ani]de [in?][b] the presense of oxalate Ammonium prevents the [titr?]ation[b] of the red percipitate.

Obtain Chem Pure Zinc by method described in Crooke p 61.

A zinc pen forming a proto. Use ammonia and ferid or ferro, perhaps Chl Sod—The percip is insoluable in Ammonia and is probably white. ~~T~~Were a paper colored black by some organic color used these marks would be quite plain further information see Crooke p. 62.

Get some chromium metal from Beardslee the Nickel-plater[14]

Barium pen. form proto. add chl sod. makes it chl Barium add Sul A. forms percip.[15]

~~I~~To make protoxide of [iron][16b]—Crooke 71.
Sulphocyanide Pot—[iron s?][17b] " 72
Something " 73

To color paper use ie try burnt sugar, Caramel, Roasted Coffee.

Copper pen— manganeese salt. Condctg salt.[18]

Iron pen— Sulphide Am— Copper solution p Crooke 115 ⟨—wont work Sulphide Am Black pen with Copper⟩

Protosalts ar~~k~~e kept from oxidization by placing them with pieces of Camphor in clean paper dry. Crooke 125

Get some Hydroferrocyanic acid[19]

Biscay iron from Spain is the softest known[20]   get some for pens & magnets

Peppers Book of Me[tals][21h] p 385— 391, 393 440 422 453 [46]5[22b] 481

Nickel pen— proto percip by Ammon   470[23h]   ⟨—light Blue very faint—⟩

Get Palladium
    Iridium
    Osmium[24]

Ruthenium pen— Hyposulphite Sodium with Ammonia gives good permanent color in the 1/100,000 part very delicate.

Try and make pens in brass setting of the small peices of metals on hand. Crooke 298.

Protocholoride of Urinum prevents a proto from becoming a per   it also reduces a sesqui to a proto, hence mix some with regular 'fd' sol to prevent the proto going to a per — It is made by formula Crooke 323, 324.— Refer to 347 ~~d~~Crooke[25]

Chl Cal. Carb Sod Hydr[ate][b] pot percip organic bodies— hence in Logwoo[d s?]ol.[b] replace Chl Sod with some Conducting [sal?]t[b] that does not give this effect. See Cr[oo?]ke[b] 348.

Crooke 349. Somthg abt Iodine & Benzol   Try Sol—arsenic

Crookes 350. gives following test for iodine   Water 100 grms starch 1 gramme Net Pot 1 grm boil 5 min add 10 drops Hydrocloric A.

Make some regular paper, 2 lbs to gallon[26]   also same with this test added [in?][b] proper proportion   ascertain the increase in delicacy and permancy   More about it p 351.

P 352. Disulphide Carbon long used in france as a test for iodine

Cyanogen—   p. 363. ⟨good⟩

Get chemical charts.

Antimony pen Chl Sod Antimoniate potash   black percip.

Theorem: ~~S~~Iron pen—sol Sulphocyanide pot. Chl Sod

Nitric A added to convert from proto to a per—chlorine evolved turns it into a chloride   Blood red mark.

Silver pen. Iod Pot. Little nitric a.—yellow mark insoluable

Black paper silver pen—chl sodium—white percip

test the conductivity of continuous jump spark from Electrical Machine for voltaic E. with chemical ~~b~~paper Thus

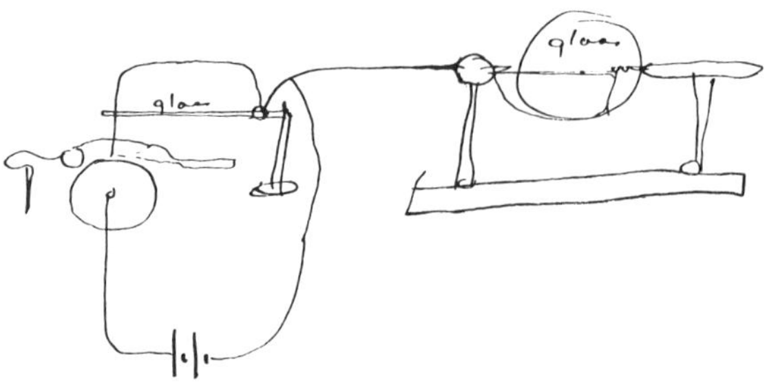

⟨no that wont work⟩

See what effet oil has poured on different chemical solutions

Quickest oxidizer. Mixture Nit[ric] A[cid] & Chlorate Pot. Crooke 424.

Subscribe Boston Journal Chem[27] Chem news,[28] Druggest Circular.[29]

AX, NjWOE, Cat. 298:137, Lab. (*TAEM* 5:270). These eleven pieces of paper are sewn together across their tops and labeled "Volume 1." [a]Preceded by "1" written by Charles Batchelor. [b]Document damaged. [c]Preceded by "2" written by Batchelor and followed by "X" interlined above, possibly in a different hand. [d]Overwrites a ". [e]Followed by centered horizontal line. [f]"X" interlined above, possibly in a different hand. [g]Circled. [h]Interlined above.

1. Some closely related notes and a draft version of this document appear in Cat. 297:92, 93, Lab. (*TAEM* 5:824, 827).

2. An account book entry indicates that various experimental automatic telegraph pens, of the materials Edison uses here, were finished in the shop by 16 September 1873 (Cat. 1219:107, Accts. [*TAEM* 21:758]). The initial portion of these notes was apparently a response in early October to comments by Maxwell. See n. 3.

3. The 25 September issue of *Nature,* which would have reached the United States in about ten days, carried an article by James Clerk Maxwell on "Molecules" that featured experiments on diffusion, including a discussion of electrolysis as diffusion promoted by electromotive force (Maxwell 1873b, 439). However, Edison might have noticed Maxwell's involvement with this subject from his 14 August 1873 *Nature* article on a closely related topic (Maxwell 1873a). Maxwell, often referred to as Clerk Maxwell, was among the most noted of British physicists. His reputation among physical scientists at this time was based as much or more on his work on fluid dynamics, including diffusion, as on electromagnetism (*DSB,* s.v. "Maxwell, James Clerk"). It is possible that Edison was referring to both Maxwell and Josiah Latimer Clark (see Doc. 369); however, only Maxwell's name appears in the draft version of these notes. No correspondence from Edison exists among Maxwell's papers.

4. That is, Callaud batteries.

5. Table A in Bloxam 1869 (pp. 2–3) gives directions for analyzing substances containing various classes of metals.

6. Crookes 1871. William Crookes (later knighted) was a leading chemist, physicist, and science publisher, most noted at this time for his discovery of thallium. His investigations of a possible force of radiation (leading to the radiometer) and of spiritualist claims attracted a great deal of attention. *DSB,* s.v. "Crookes, William."

7. In his draft of these notes Edison wrote "magnesium." Cat. 297:93, Lab. (*TAEM* 5:828).

8. Edison here proposes to test a battery cell using a galvanometer in the standard way by determining how much the galvanometer needle is deflected. Crookes did not say on p. 37 that magnesium gave the greatest e.m.f. of all metals, but he did suggest substituting it for zinc in some apparatus.

9. "Ferrid" is an older form of "ferric," which refers to iron compounds in which iron manifests more than its minimal combining power or valence, as in ferric oxide, Fe_2O_3; cf. n. 12. For Edison's regular ferrid solution see Doc. 410.

10. Beginning with this paragraph Edison reviews various chemical reactions for their potential use on the paper receiving strip of an automatic telegraph system.

11. Uranium.

12. Edison often wrote "ferro" rather than "ferrous" when referring to iron compounds in which iron is at its lowest valence, as in ferrous oxide (FeO).

13. In his draft of these notes Edison cited p. 55 in this connection.

14. Probably George Beardslee of Brooklyn, who had a number of inventions related to electroplating to his credit as well as a military telegraph designed during the Civil War. U.S. Pats. 26,557, 26,558, 129,881, and 149,974; Scheips, 1963, 408–9; Thompson 1958.

15. Cf. Crookes 1871, 64–65.

16. In his draft of these notes Edison wrote "iron." Cat. 297:93, Lab. (*TAEM* 5:831).

17. In his draft of these notes Edison wrote "sulphocyanide of iron etc" (Cat. 297:93, Lab. [*TAEM* 5:831]); both that and sulphocyanide of potassium are mentioned in Crookes 1871, 72.

18. Cf. Crookes 1871, 102.

19. See Bloxam 1869, 86.

20. A Biscayan forge produces malleable iron directly from suitable ore. Knight 1876–7, s.v. "Biscayan Forge."

21. In his draft of these notes Edison wrote "play bk. metals" (Cat. 297:93, Lab. [*TAEM* 5:831]). This refers to Pepper 1869b. Pepper's topics on the pages cited here include oxides of iron and reactions involving iron, bismuth, antimony, cobalt, molybdenum, palladium, and nickel.

22. Edison's list of page numbers in his draft for these notes included p. 465.

23. That is, p. 470 in Pepper 1869b.

24. Chapter 10 in Crookes 1871 is entitled "Gold, Platinum, Palladium, Iridium, Osmium, Rhodium, Ruthenium."

25. The relevance of the material on p. 347 in Crookes 1871 to the rest of this paragraph is unclear.

26. That is, receiving paper for his automatic telegraph with his "regular" chemical treatment. See n. 9.

27. The *Boston Journal of Chemistry* began publication in 1866 as a popular illustrated monthly devoted to progress in zoology, botany, mineralogy, archaeology, and hygiene.

28. Crookes was founder, proprietor, and editor of the *Chemical News and Journal of Physical Science.*

29. This journal began publication in 1857 as the *American Druggist's Circular and Chemical Gazette.* In 1866 "American" was dropped from the title.

–371–

Lemuel Serrell to Commissioner of Patents[1]

New York, Octo 15, 1873.[2]

Sir

At the especial solicitation of Mr T. A. Edison I have consented to attend to the six Duplex Telegraph cases of his that were originally prepared by Mesrs Munn & Co.[3]

If the drawing in each case can be returned I shall be obliged and then the instructions of Mr E. can be taken as to the nature of each invention and entirely new specifications prepared.[4]

Powers of atty. enclosed. Respectfully yours,

Lemuel W. Serrell

ENCLOSURE[a]

New York Oct. 15th 1873

Sir:

I hereby appoint Lemuel W. Serrell, New York, my agent and Attorney, in relation to my application for a patent on Duplex Telegraphs, revoking all previous authority and authorize him, or his substitute, to do whatever may be necessary in the premises, as fully and entirely as I could do if personally present, and to receive the Patent when granted.

Respectfully Yours,

Thomas A Edison

TL (transcript), NjWOE, PS, Abandoned Patent Applications, Case B (*TAEM* 8:599). See Doc. 304 textnote. [a]Enclosure is a TD (transcript), NjWOE, PS, Abandoned Patent Applications, Case C (*TAEM* 8:625); five other transcripts of this, some with minor differences, are included in these abandoned applications.

1. The Commissioner of Patents at this time was Mortimer Leggett. *DAB,* s.v. "Leggett, Mortimer Dormer."

2. During the fortnight before this Edison had begun revising the set of multiple telegraphy patent applications filed the previous April as a result of trials at Western Union's main office, all of which had been twice rejected (Docs. 304–5, 308–11, 314–15). They had been pre-

pared initially, and amended after one rejection, by personnel at Munn & Co. on the basis of notes, drawings, and models supplied by Edison in the rush before his departure for England; Edison later said the results were not understandable (Doc. 303; *TAEB* 1:556). Now, between 4 and 10 October, amendments for the descriptions of five of the designs were sent to the Patent Office. One (Doc. 305) was changed minimally, the others (Docs. 304, 310, 311, 314) extensively (*TAEM* 8:598, 573–77, 698–702, 729–31, 620–24). They were filed with Munn & Co. still as his patent attorney for these cases. Before the Patent Office responded or the full set of applications was amended, Edison changed to the course of action set forth here.

3. Munn & Co. was a major firm of patent attorneys and was the publisher of *Scientific American*. It had actually prepared and submitted eight duplex and diplex telegraphy applications for Edison. Doc. 303; *TAEB* 1:556.

4. See Doc. 372. New specifications were not prepared for any of these cases save the last (cf. Docs. 315 and 555), but aspects of the designs involved were more or less incorporated in other, later applications. See headnote, p. 254, and Edison's testimony, Quad. 70.7, p. 344 (*TAEM* 9:539).

-372-

Zenas Wilber to Lemuel Serrell

Washington,[a] D.C., Octo 18", 1873.

L. W. Serrell Esq

In the matter of your request for return of dwg's in Edisons Appns for Duplex Telhs[1] I am instructed by the Comm'r to say that while resident attorneys are allowed to withdraw dwgs he cannot consent to their withdrawal when they are to be sent to distant points.[2]

The ~~difficulty~~ danger of mutilation and of loss, and the trouble and expense incident to their carriage by mail or express back and forth, are the reasons therefor.

Very Resp &c

Z. T. Wilber Exr[3]

TL (transcript), NjWOE, PS, Abandoned Patent Applications, Case G (*TAEM* 8:733). See Doc. 304 textnote. Original on letterhead of U.S. Patent Office. [a]"Case A." appears above.

1. See Doc. 371.

2. The drawings that accompanied the written descriptions in patent application specifications held a special status; the descriptions could be amended, but the drawings had to remain unchanged to ensure the identity and date of the invention (*TAEB* 1:674). Nevertheless, by the summer of 1874 Serrell had obtained at least copies of the drawings. Serrell's testimony, Quad. 71.1, p. 93 (*TAEM* 10:51).

3. At this time Zenas F. Wilber (d. 1889) was the usual examiner in charge of Electrical Apparatus at the Patent Office. A cousin and ward of Rutherford B. Hayes, Wilber had been major of the Fourth Colored

Infantry. Appointed a clerk in the Patent Office in 1870, he advanced by competitive examination to the post of chief examiner in 1876. Obituary attached to Charles Batchelor to TAE, 27 Aug. 1889, 89-068, GF.

–373–

Caveat: Automatic Telegraphy

New York, October 28, 1873[a]

To all whom it may concern.

Be it known that I, Thomas A. Edison of Newark, in the County of Essex, and State of New Jersey, have invented an Improvement in Transmitting and recording letters or characters, of which the following is a correct description, so far as perfected.

The invention consists of transmitting inverted currents, positive to record one half of the letter and negative the other half through polarized magnets, the levers of which operate ink marking devices.

The following diagram will explain the invention:—[b]

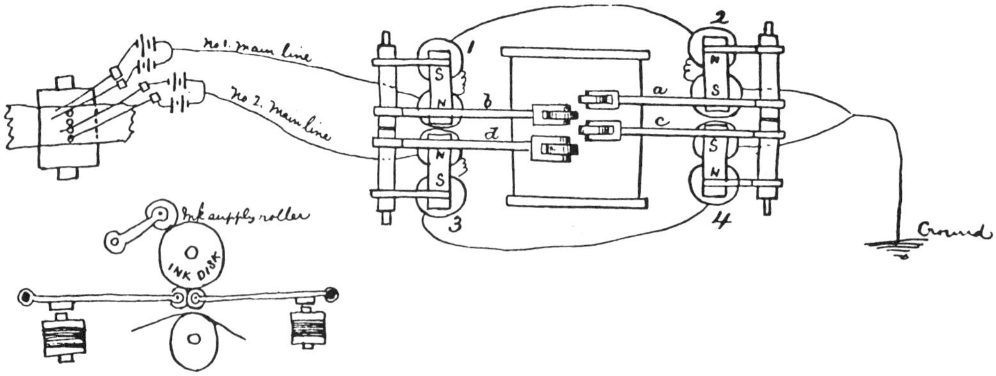

The magnets 1.2. are in one line wire circuit and are polarized so that when a positive current is sent the magnet 1 remains inactive, but the magnet 2. draws the ink marking lever down to the paper and records a mark: if now a negative current is sent over the wire, the magnet 2 is inactive but the magnet 1 draws its lever down a records a mark directly under the first forming the first part of the letter I ⁞. At the same time this was taking place the same effect was produced upon the second wire and two marks were made directly under each other all joined together making the letter I ⁞. The perforated paper has four lines of holes to represent a roman letter and the contact pens are arranged to send the positive and negative currents on both wires.

Each wire is provided with two pens one pen being placed one row of holes in advance of the other so that while one pen is in a hole and sending a positive current the other roller pen is on the paper of the space between and vice versa, thus:—

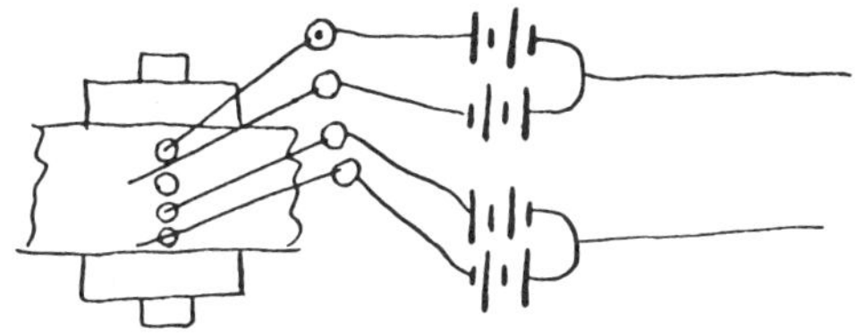

In the receiving apparatus when no currents are being sent the recording arms and rollers are drawn up by a spiral spring and the rollers remain in contact with the ink disk which has a smooth periphery and is inked by a cloth wheel supplied with thickened ink. The disk is rotated by any motor and very rapidly so the most instantaneous touch of one of the recording wheels against it will give the recording wheel a complete revolution and ink its periphery thus always ensuring a supply of ink to the recording rollers.

My claims will probably be.

Four recording levers operated by polarized magnets, two in one circuit and two in another circuit, to record roman letters or other characters substantially as set forth.

The transmission by paper perforated to form the roman or other characters of positive and negative currents at the proper moment over two circuits to operate polarized magnets ink recording levers to form roman or other characters or letters.

Signed by me this 28th day of Oct. 1873.[1]

Thos A Edison

Witnesses. Geo. T. Pinckney[2] Chas H Smith[3]

DS, DNA, RG-241, Edison Caveat 42. Written by Pinckney; petition and oath omitted. [a]Date taken from text, form altered. [b]"$1 50 3 figs" written in an unknown hand at top left of drawing.

1. The docket indicates that the Patent Office received this caveat on 25 July 1874.
2. George Pinckney witnessed many documents in Serrell's office.
3. Charles Smith witnessed many documents in Serrell's office.

–374–

Caveat: Automatic Telegraphy

New York, October 28, 1873[a]

To all whom it may concern.

Specification describing an Improvement in Transmitting and ~~r~~Recording letters or characters, invented by Thomas A. Edison, of Newark in the State of New Jersey.

The object of this invention is to transmit Roman characters over three and two wires.[1]

The invention consists when arranged for two wires, in transmitting one portion of the letter by a positive current and the other part by a negative current: and in arranging the four recording pens & sending pens so that two pens sending the same current shall be one line of dots ahead of the other two pens which send the other or opposite current.[2]

The invention further consists in arranging the perforations or holes in the paper so that but four lines are necessary to transmit a letter thus:—

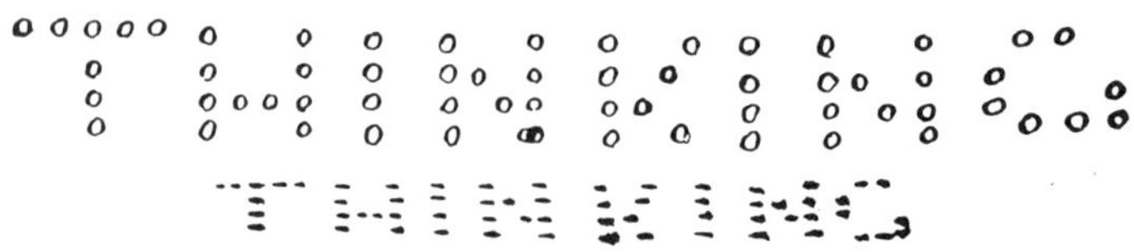

The invention further consists in arranging the two pens of one wire several letters ahead of the two pens of the other wire so as to prevent the dots transmitted over one wire being recorded on the pens of the other wire which would be the case when the chemical paper is very sensitive as will be further on described.

The invention futher consists of devices for running the receiving and transmitting paper through at a uniform speed; and in the arrangement of the circuits, compensating devices and batteries.

In transmitting with two wires I have used letters formed of but three lines of perforation thus:—

and arranged the pens, wires, and batteries, thus,—

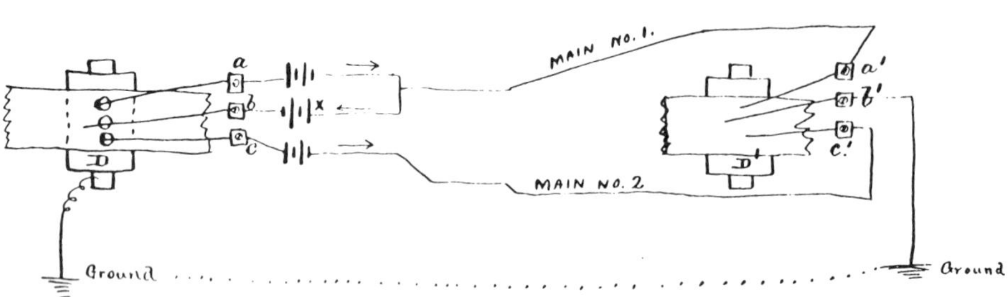

The transmitting rollers a. and c. when they pass through the upper and lower perforation of the letter I $\overset{\circ}{\circ}$ transmit over the two main line wires two currents in the same direction which are recorded by the receiving pens a′.c′: the current from both wires passes up through the pen b′. to the ground but in a contrary direction to what it passes down a′. and c′. hence, no mark is made by the pen b′. Now the transmitting roller b. is on the paper while the rollers a. and c. are in the perforations, being placed ahead a distance equal to the space between each line of holes forming all the letter, so that at no time do the three rollers make contact with the drum at the same time, hence, when the roller b. comes in contact with the drum D the battery X. which transmits a current in the opposite direction to the other two batteries, is sent over the line No. 1. but the current passes by the pen b. and roller D. to the ground and by the latter to the pen b′. recording the mark upon the chemical paper, and then passes up the pen a′. and returns to the battery X by the line No. 1.

From the foregoing it is obvious that two thirds of the letter is made by currents of one polarity sent over both wires and the other third by a current of opposite polarity sent over one wire.

To compensate for the static discharge of the line, I arrange electro-magnets or induction coils, as represented in the following diagram.[b]

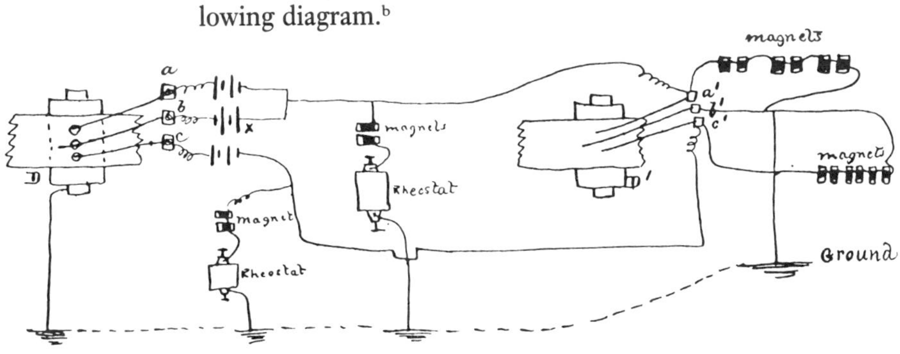

These magnets operate in a manner similar to that described in Letters Patent No. granted to me—[3]

It will be noticed that I use the drum of the transmitter as part of the circuit but the receiving drum forms no part of the circuit and can be made of non-conducting material, which I prefer,[4] the pen b′. being the metallic connection for the current to pass to the ground wire [-]and ground.

The four line letter, formed by four rows of perforations, is preferable to the three line letter, because the letter is much more perfect in form, but in using two wires to make a four line letter, a difficulty is experienced, which is, that the dots which should be recorded by two pens only are faintly recorded by a third pen, owing to the conductivity of the chemical paper and to the four pens being in contact therewith: two of these pens being connected to the ground, they both offer a route to the ground for a current coming over one of the line wires: the current divides, but as one of the pens is connected to the ground farther from the pen connected to t~~o~~he line whereon the current is being sent than the other pen connected to the ground, the current splits, most of it passing through the ground wire pen nearest to the line pen, and the balance through the other pen, the amount being controlled by their distance from each other and the conductivity of the paper.

If one line pen and one ground pen are placed near each other, and the other line pen and ground pen placed near each other but one letter ahead of the other two pens, the resistance of the paper between the two sets of pens will be so great that the extra current recorded will be very slight.

If the[c] ~~s~~distance between the two sets of pens is further increased to several letters, the resistance of the chemical paper will be so great that no extra currents will be recorded.

It is obvious that if the two recording pens of one wire are placed several letters in advance of the two recording pens of the second wire, that the transmitting pens must be arranged in a like manner The following diagram will explain how the current from one wire are recorded upon the pens connected to the second wire:—

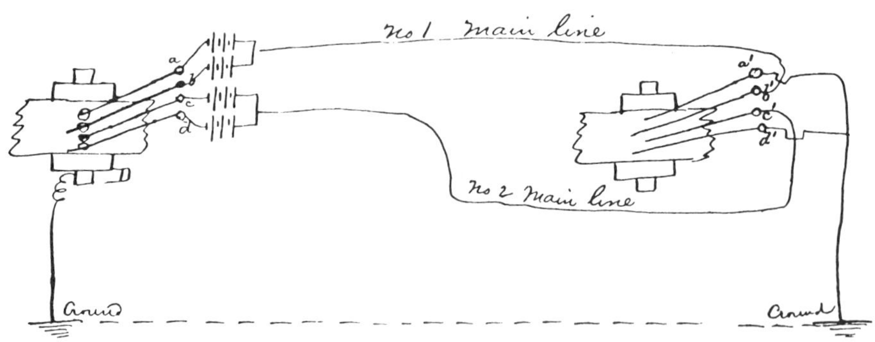

If the letter is such that the roller pen a. only makes contact with the drum D. a negative current will be transmitted, the current passing down the pen a′. over the paper and up through the pen b′. thence over the line No. 1. to battery, but it will be noticed that the pens a′. and d′. are connected together, and instead of all the current passing through the ground and down the pen a′. and up b′. a portion will pass down d′. across the paper and up b′. but as a′. is nearer to b′.

than is d′. the greater portion will pass down a′. owing to the resistance of the paper caused by the extra distance.

In practice I prefer to arrange the pens, drums wires &c. as next illustrated:—

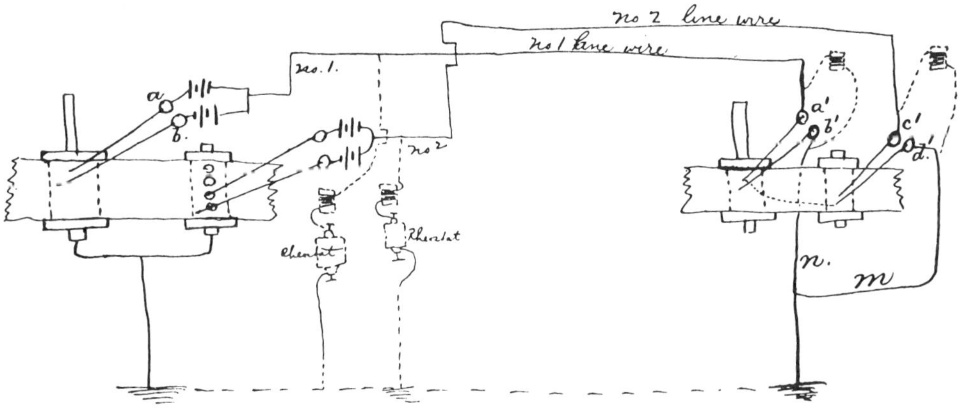

The corresponding pens of one are set several letters ahead of the corresponding pens of the other wire, the object of which is to increase the resistance to the passage of the extra currents,—thus, if a negative current is sent on No. 1. wire, it passes through the ground and up the wires n. and m. thence to the pen a′. as shown by the dotted line and back to the battery. Nearly all of the current will pass through the wire n. and down the pen b′. and across the paper to the pen a′. a space less than 1/32 of an inch, while the current passing up the wire m. and down the pen d′. must traverse the paper to reach a′. a distance one hundred times as great as in the case of b′. hence but one hundredth (1/100) part of the current will pass down d′. which is much less than is necessary to make a mark.

When both the transmitting and receiving machines are running at nearly the same speed (rotated preferably by an electro-motor) the transmitting pen c. first comes in contact with the drum through the second hole from the bottom of the letter.

The letter I thus sends a positive current which is recorded by the pen c′; immediately after, the pen d. makes contact with the drum through the first hole and sends a negative current which is recorded by the pen d′: these recorded dots will be immediately under each other thus : and not thus ·• because one pen being[d] set one line of dots ahead of the other pen, the first pulsation is sent and recorded, then the paper is fed or moved and the second pulsation sent and recorded directly under the first dot. As the paper is drawn along, the pens a. and b. make contact with the drum through the two top perforations in the paper and the two dots are recorded in like manner and immediately over the other two dots, thus, 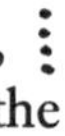 completing the letter I, the pulsations being recorded by the pens a′.b′.; but one half of the letter is formed several inches away from where the letter is completed, the paper being drawn along at a given speed by the transmitter and the receiver, and the two upper dots being made successively when the first recorded portion is immediately under them.

It is obvious that no variation of speed must take place between the time the first part of the letter is formed and the time the second recording pen act, otherwise the letters would not match; thus if the speed did not vary the word man would be recorded like this 

but if there was a slight variation in speed it would be recorded thus 

and if a greater variation, like this 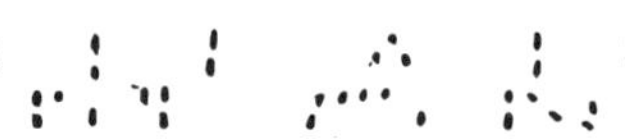

The nearer the two sets of recording pens are to each other, the less effect a variation in speed will have.

The compensating magnets or induction coils are arranged as shown in the diagram last above given.[e]

Instead of placing the pens of one wire several letters distant from each other, the message may be sent from one strip, two roller pens being placed one row of perforations ahead of the other two, and the message received on two strips of chemical paper both running over an insulated drum and separated from each other by a thin insulated partition, or two separate drums can be used.

The top part of all the letters of the message will be recorded upon one slip and the bottom part of all the letters of the message will be recorded on the other strip, the marks being made on the extreme edge of each strip—

For delivery to the public the two strips are matched and pasted together on one strip.

The insulating of one strip from the other is for the purpose of preventing the currents which pass to two of the recording pens going to the other pens and recording false marks.

By using the two strips of paper, any variation in speed will not prevent the letters on the two strips matching when placed together. It has the advantage of being worked with less delicate machinery and being less liable to error than the plan wherein the pens are set several letters ahead of each other, but has the disadvantage of the loss of time necessary to paste the two strips together.

This arrangement is shown in the following diagram:—

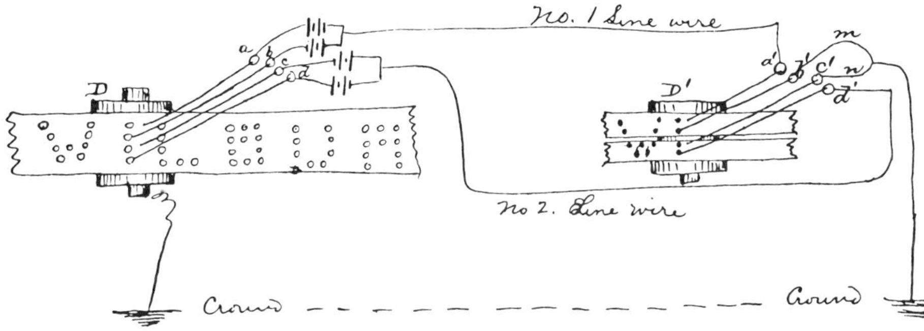

As the two ground line pens are insulated from each other, as regards the chemical paper, it follows that if a negative current is sent on No 1. wire, it would pass through the ground up through m. thence down through b′. across the paper and up the pen a′. back to battery, but no current passes upon[f] the line No. 2. because the current that might pass to the pen c′. by the wire n. cannot take that route to the pen a′. on account of the two strips of paper being insulated from each other, hence the circuit by that route is open, and no mark is made by the pen c′

If three wires are used no difficulty is experienced from the currents of one line acting upon the pens in the other lines, by making one circuit entirely metallic and the other part metallic and part ground: each pen in this arrangement is separately connected.

This arrangement is shown in the following diagram—

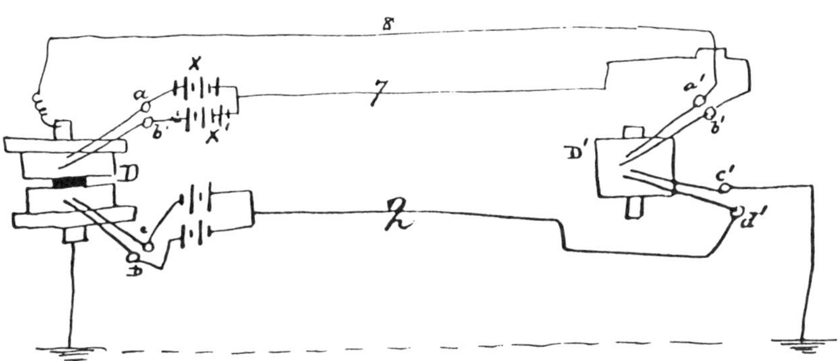

The transmitter drum is divided in two parts, that which the pens a.b. rest upon being insulated from that part whereon the pens c.d. rest.[e]

If the pen a. comes in contact with the drum D. through a

perforation in the paper, the battery X′ is "short circuited" and the battery X. sends a current in a certain direction to record by the pen a′. by passing down a′. across the paper up the ~~a~~pen b′. and back through the other wire 7.[d] If the pen b. comes in contact with the drum, the battery X is "short circuited" and the battery X′ sends a current in the opposite direction passing down the pen b′. across the paper up a′. and back through the other wire 8.

It will be apparent that half of all the letters of a message will be recorded by the pens a′.b′. by currents passing over these two wires 7. & 8. perfectly without any currents from the third wire 2. as the currents in the metallic circuits 7.8. have no tendency to leave that circuit. The other half of all the letters of the message will be recorded by the pens c′.d′. and the third wire 2. as has been previously described

This method is a very practical one and does not depend at all upon the relative speeds of the transmitting and receiving apparatus.

If the receiving instrument should go very fast the letters will be very large and be drawn out, but if it goes very slow, the letters will be small and close together.

For recording Roman or similar letters, I prefer to duplicate the receiving pens, that is, each pen has two marking prongs, this makes the letters compact and large. With single pens the letters appear like this, 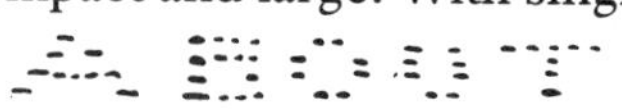

but with the double pens they are similar to this—

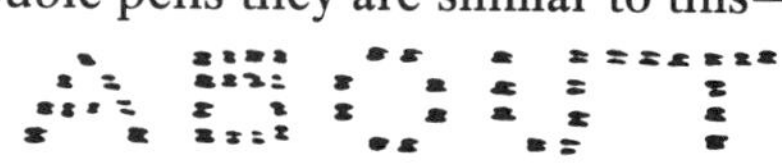

The double pronged pens are shown in the next sketch

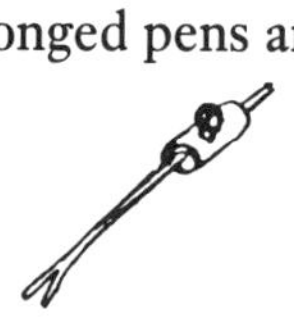

When single-pronged pens are used, they are arranged thus—

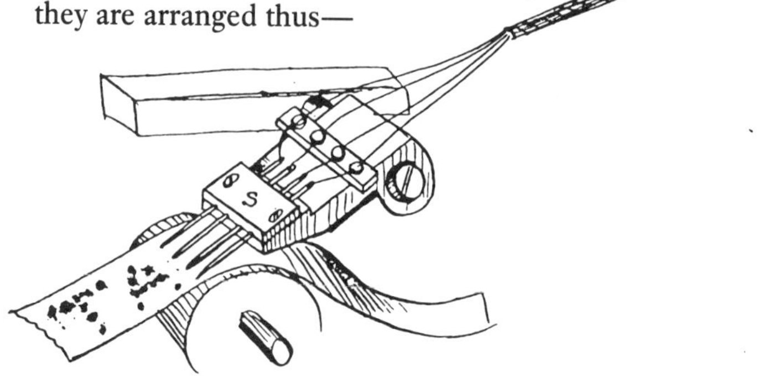

The recording points are common needles clamped together and held rigidly by the plate S. The four needles lie in grooves to prevent their shifting. The wires are connected to the eyes of the needles—

Instead of using perforated paper for transmitting by the modes herein described[g] the letters or characters might be embossed in the paper, and levers with rollers at their ends be employed for opening and closing the circuit, the circuit being closed while the rollers were in the depressions forming the letters and opened while travelling over the surface of the papers

My claims will probably be

First. Transmitting positive and negative currents on one wire to record one half of a Roman or similar letter and positive and negative currents on a second wire to record the other part of the letter, substantially as set forth—

Second Recording Roman or similar letters upon two circuits substantially as set forth—

Third— The method of prepventing the currents of one line recording by the pens of the other line.

Fourth— Setting one of the two pens which are connected to one wire both on the receiving and transmitting paper slightly in advance of the other, the transmitting pens to send Positive and Negative currents and the [-----] receiving[d] pens to record them in proper position—

Fifth— Forming Roman or similar letters of three rows of holes and the manner of sending and receiving the same—

Sixth. Transmitting from a strip perforated or embossed with Roman or similar letters or characters, on two different wires, positive and negative currents for the top two lines and positive and negative currents for the bottom two lines and recording the same upon two different strips of chemical paper near the edges, substantially as set forth.

Seventh. The manner of arranging the batteries, circuits and instruments in the several methods herein described.

Signed by me this 28th day of Oct. AD. 1873.[5]

Thos A Edison

Witnesses Geo. T. Pinckney Chas H. Smith

DS, DNA, RG-241, Edison Caveat 43. Written by Smith; petition and oath omitted. [a]Date taken from text, form altered. [b]"$6" written in an unknown hand at top left of figure. [c]Followed by a series of dots, perhaps to replace an erasure. [d]Interlined above. [e]Followed by a small (terminal?) symbol. [f]"no . . . upon" interlined above. [g]"the . . . described" interlined above.

1. Edison apparently began experiments leading to the two-wire roman-letter automatic described in this caveat during the second week of October. Account records indicate that Edison and Murray billed the Automatic Telegraph Co. on 9 October for work done on "1 Drum etc for Edison's 2 wire Typo" and on 11 October for "fixing transmitter and pen holder for Edison's exp. 2 wire Roman letter arrangement" (Cat. 1219:119–20, Accts. [*TAEM* 21:764–65]). This work probably followed an undated notebook entry in which Edison described the insulated drum used for the receiver (see n. 4) and then wrote: "Use 4 receiving pens for Roman on two wires place 2 sending and 2 Recg pens one letter ahead of the other two" (Cat. 1171:97, Lab. [*TAEM* 6:256]).

2. Edison used reversing currents in automatic telegraphy in 1871. *TAEB* 1:311–12.

3. Probably either U.S. Pat. 135,531 or U.S. Pat. 141,773.

4. In an undated notebook entry Edison indicated how he wanted the drum insulated: "Shellac thickly the iron receiving drum or make a hard rubber one." Cat. 1171:97, Lab. (*TAEM* 6:256).

5. The docket indicates that the Patent Office did not receive this caveat until 25 July 1874.

-375-

Caveat: Automatic Telegraphy

New York, October 28, 1873[a]

Specification describing an Improvement in —Electric Circuits— invented by Thomas A. Edison, of Newark in the State of New Jersey.

This invention consists in arranging the chemical receiving instrument in a "Wheatstone Bridge," and placing the magnetic compensators in the branches.[1]

It also consists in placing secondary batteries in two of the branches and magnets in[b] the other two branches with the receiving instrument in the bridge wire, so that both the magnets and secondary batteries are enabled to send their counter or discharge currents in the proper direction to neutralize the static charge which follows after each signal, and also enables me to adjust the amount of main current passing through the receiving instrument without interfering with the strength of the inductive currents.

The following diagram will explain the action of the main and inductive currents upon the receiving instrument in the bridge.[c]

FIG. 1.

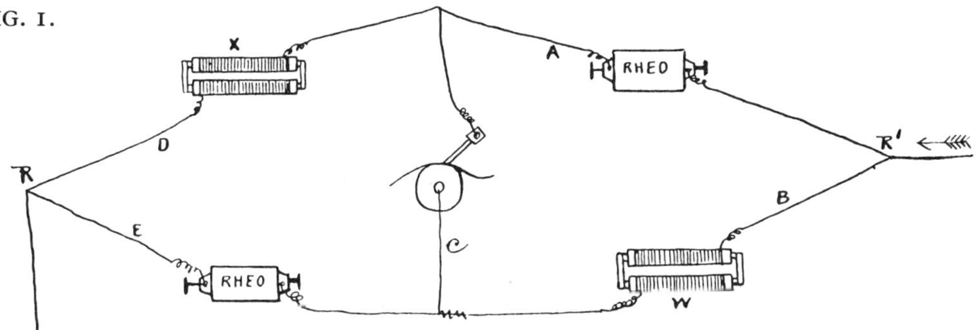

In fig. 1. C. is the bridge wire in which is placed the chemical or other receiving instrument.

In the branches A and E are rheostats and in B and D are long magnets with the keeper permanently attached

When the resistance of A.B.D.E. are equal, no current passes through the receiving instrument. The two large magnets are charged with half of the main current, the other half of the current passing through the branches containing the rheostats. Now at the moment of breaking the main circuit the two magnets discharge their inductive currents within the two circuits formed by E.D.C. and B.A.C. and the direction being from the magnets to the drum thence through the paper up pens and back to each magnet. It is obvious that any discharge current due from the magnet W will circulate within the circuit formed by the wires A. B and .C. and any induction discharge from X will circulate within the circuit formed by the wires E. D and C: in both cases the discharge currents pass through the receiving instrument in the bridge-wire C. in the same direction.

The discharge current from an electro-magnet is in the same direction as the battery current when the charging battery is taken off and in a contrary direction when put on.

Now if the rheostats are so adjusted that the resistance of the wires A and E are less than D and B. a portion of the main current will pass through the receiving instrument and record a mark, the amount being according to the lessened resistance of the routes or branches A.B. & D.E.

If now the main battery is disconnected the static charge would continue to flow out of the line in the same direction and prolong the mark but at the same time that the battery was disconnected, the magnets induced their extra currents into the two bridge circuits which passing up the bridge wire C. in an opposite direction to the main current balanced each other, that is, placed the receiving instrument in an equated position or point of no potential as regards the static and magnetic charges, consequently the mark ended the moment the main battery was disconnected.

When the main battery is put to the line the first part of the mark would be weakened, owing to retardation or electrofication, but the small quantity of current which first appears and passes through the bridge, acts on the electro-magnet which sends an inductive opposed to it: this extra current passing within the bridge circuits passes through the receiving instrument in the same direction as the main current and the inductive current makes up for the current lost by electrofication.

A secondary battery composed of one or more cups containing water and two conducting plates of the same material, and the water charged with some salt to increase its conductivity, transmits charges or secondary currents just the opposite way that an electro-magnet does. The former when a current passes through it sends a current in the same direction, or at least apparently but not really, and when the battery is disconnected, sends a current in the opposite direction.

The electro-magnet just alters this direction and transmits a current opposed to the battery current when it is passed through it and in the same direction when it is disconnected.

It will be noticed that in fig. 1. that there are two branches with ordinary rheostats in them: these substract a greater portion of the main current and furnish no compensation. I take advantage of the opposite qualities of a secondary battery to place one in each branch A.E. as shown in fig. 2.

FIG. 2.

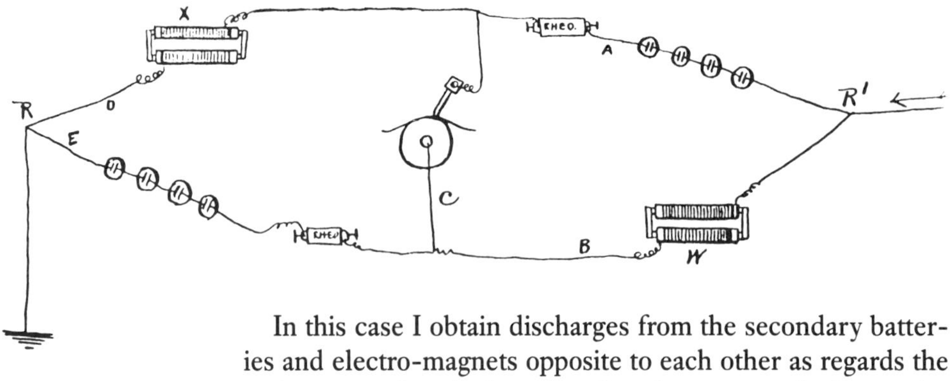

In this case I obtain discharges from the secondary batteries and electro-magnets opposite to each other as regards the main current but in the same direction as regards the receiving instrument in the bridge wire C. so that all the main current which passes from R′. to R. passes through devices which generate compensation or balancing current to counteract the effect of the static charge due from the line.

I prefer in using my magnetic compensation to use but one long magnet with its keeper permanently attached to the poles of the magnet instead of a number of magnets.[2] A long magnet sends a longer and stronger inductive current than a small one or a number of small ones equal in length to the long one. By placing the keeper permanently in contact with the poles, no electricity is transmitted into motion, it being transmitted into induced currents augmenting them nearly double.

I also prefer to wind one layer of fine-~~t~~wire around both prongs of the electro-magnet and connect this to a commutator, and then wind another layer and connect it to the commutator and so on until the magnet is covered with a great number of layers of fine wire.

By means of the commutator I am enabled to throw in and out of circuit as many layers of wire as I desire: Fig 3. explains the connections.[3]

FIG. 3.

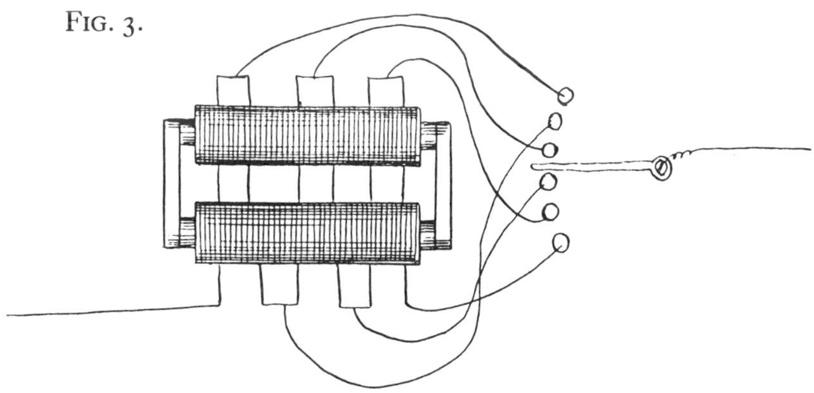

I shall probably claim as my invention,

First. Placing a chemical receiving instrument in the bridge wire of a "~~w~~Wheatstone Bridge," the branches of which contains magnets, or magnets and secondary batteries, substantially as and for the purposes set forth—

Second. The compensating magnet with its keeper permanently attached to it and with its coils arranged with a commutator, substantially as and for the purposes set forth—

Signed by me this 28th day of October 1873.[4]

Thos A Edison

Witnesses, Geo. T. Pinckney Chas H. Smith

DS, DNA, RG-241, Edison Caveat 44. Written by Smith; petition and oath omitted. [a]Date taken from text, form altered. [b]Followed by a series of dots, perhaps to replace an erasure. [c]"$2" written in an unknown hand at top left of figure.

1. See Docs. 357 and 361.

2. See Docs. 363 and 386.

3. Similar drawings are found in Cats. 297:84(3) and 299:67, Lab. (*TAEM* 5:779, 6:129).

4. The docket indicates that the Patent Office did not receive this caveat until 25 July 1874.

-376-

From Lemuel Serrell

N.Y. Oct. 31, 1873

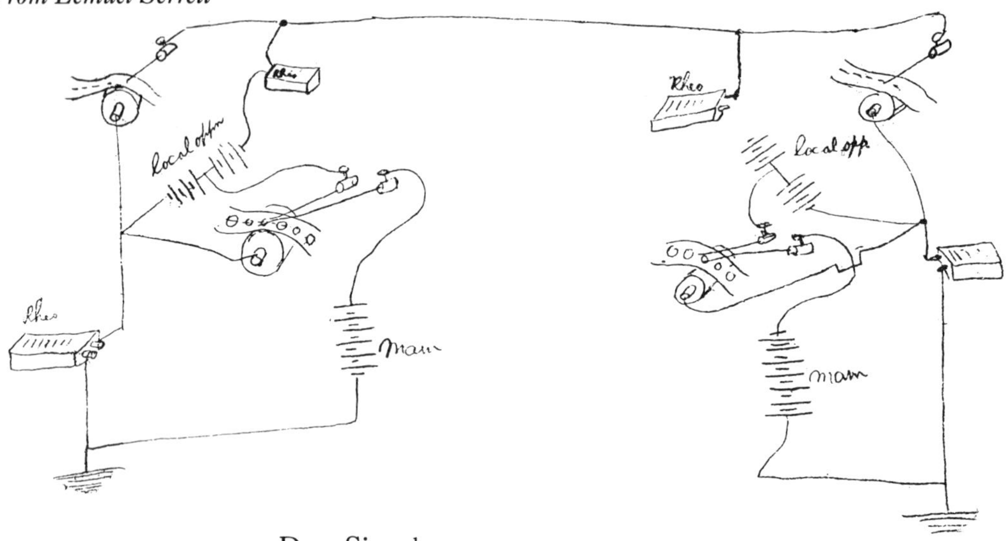

Dear Sir—[1]

This case (69)[2] was filed last March[3] and has remained unacted upon because no model has been furnished. Please have model made and sent to me as soon as convenient.[4] Yours truly

L W. Serrell

ALS, NjWOE, Lab., Cat. 30,094 (*TAEM* 5:325).

1. The figure labels are (clockwise from far left): Rheo, local, opp[ositio]n, Rheo, Rheo, local, opp, main, main.

2. This case-numbering system probably originated in Serrell's office. Edison used it throughout his career.

3. Edison executed the patent application on 7 March 1873 (U.S. Pat. 156,843) while engaged in designing duplex telegraphs under an oral agreement with President Orton of Western Union. However, since this design employed a duplex telegraph circuit for chemical automatic telegraphs, Edison assigned the patent jointly to himself and George Harrington rather than to Western Union. The use of duplex circuits with electrochemical telegraphs was well known (Sabine 1872, 119–23; Pres-

cott 1877, 777–79). Unlike earlier arrangements, Edison's version controlled the balancing current needed in the duplex by electrical rather than mechanical means and employed an automatic transmitter using perforated paper.

4. According to accounting records, Edison constructed and sent this model to Serrell in July (Cat. 1219:80, Accts. [*TAEM* 21:745]). Serrell sent the model to the Patent Office on 3 November. The Patent Office approved the application after Edison amended it in April 1874, describing the specific characteristics of his design more precisely. Pat. App. 156,843; model is Acc. 29.1980.1351, MiDbEI.

–377–

From Alfred Beach

New York, Nov 3d 1873[a]

Dear Mr. Edison

I was much interested in your explanation of your Telegraph improvements, which appear to be important and valuable—[1]

I presume you have taken the steps necessary to secure the same— Yours truly—

~~Munn & Co~~ A E Beach[2b]

ALS, NjWOE, DF (*TAEM* 12:1185). Letterhead of *Scientific American*. [a]"New York," and "187" preprinted. [b]Signature interlined above.

1. Unknown.

2. Alfred Beach, inventor and publisher of *Scientific American*, was also a patent solicitor for Munn & Co. *DAB*, s.v. "Beach, Alfred Ely."

–378–

From Jesse Bunnell

Philadelphia, Nov 15, 1873[a]

T.A.E.

Will be over on tuesday to see you and take at once necessary active steps according to whatever conclusions we jointly arrive at. In todays issue of Popes paper[1] I see issue of patent to him[2] the claims of which seem to indicate that he has barely touched the principles we talked of applying— Since seeing you I have figured out (it seems to me very satisfactorily), all the mechanism simplifications etc which appear to me necessary to give a splendid system and if it proves to be the case as now seems probable, that ~~the~~our combination hasn't yet been thought of, its good enough to go at, hammer & tongs— What I most want to see you about now is: Wording & framing of patent appn[3] & if convenient to try one of your strongest magnetos to see how much strength of an impulse can get through a mile or so of resistance= Also examination of whatever patent[b] data you may have to see ~~if~~ how far[c] we are new in this—

Dont forget the Battery business[4] my partners[5] are much pleased with the idea of taking your agency on a royalty & pushing strong to our regular customers the Penna Reading[6] etc. & advertising it against the Lockwood =[7]

Will be over tuesday— Yrs

Bunnell[8]

Jesse Bunnell, telegraph-instrument inventor and manufacturer.

ALS, NjWOE, DF (*TAEM* 12:988). Letterhead of Partrick, Bunnell & Co. [a]"Philadelphia," and "187" preprinted. [b]Interlined above. [c]"how far" interlined above.

1. *Telegr.*, 15 Nov. 1873. Although Franklin Pope was no longer editor, he continued his association with publisher James Ashley.

2. On p. 282 are the claims from the reissue (no. 5,622) of Pope's U.S. Patent 129,425, "Electric Signalling Apparatus for Railroads."

3. On 5 December 1873, Edison and Bunnell filed a caveat for an "Electric Railway Signal." Neither Edison nor the two men jointly ever received a patent on such a device. Quad. 72.11, p. 47 (*TAEM* 9:188).

4. This may refer to the battery Edison had patented two months earlier (U.S. Pat. 142,999). See Doc. 258.

5. See n. 8.

6. That is, the Pennsylvania and Reading railroads.

7. Robert Lockwood received U.S. Patent 137,556 for a telegraph battery, which was a commercial success by March 1873. "The Lockwood Battery," *Telegr.* 9 (1873): 66.

8. Edison met Jesse H. Bunnell (1843–1899) when they were both telegraph operators in Tennessee during the Civil War. Bunnell formed Partrick, Bunnell & Co. with James Partrick and Franklin Carter in Philadelphia in 1872. Their letterhead billed them as "Practical Telegraphers and Electricians, Manufacturers and Dealers in Telegraph and Electrical Instruments and Supplies."

In 1875 Bunnell joined telegraph manufacturer Tillotson & Co. in New York, and in 1878 he started his own telegraph shop, Bunnell & Co. App. 1.G6; Hardaway 1986; Taltavall 1893, 223.

–379–

To Charles Buell[1]

Newark N.J. ~~Nov~~Dec. 1, 73

Dr Sir

Would be happy to give the Sun recorder[2] a test in my Laboratory at Newark I have every conceivable variety of Electric Apparatus, and any quantity of Chemicals for experimentation. Yours

Thos A Edison

ALS, NjWOE, DF (*TAEM* 12:990).

1. Charles E. Buell (n.d.), inventor and telegraph engineer, began his career in the early 1860s as an operator with the American Telegraph Co. He subsequently worked for several other companies before becoming an accountant and operator for the Car Department of the New York, New Haven and Hartford Railroad Co. in New Haven. He began

experimenting with alarm systems in 1871, filed caveats in 1872, and received his first patent (of many) in 1875. Deposition of Charles Buell, 24 Sept. 1885, pp. 4, 12–13, *Buell v. Martin.*

2. Buell was developing a system of telegraphy that recorded signals photographically. See Doc. 382; and Buell to TAE, 24 Feb. 1874, DF (*TAEM* 13:34).

-380-

From Lemuel Serrell

New York, December 10th 1873[a]

Dear sir

I received today the Spanish Patent granted for your Imp[rovemen]ts. in Printing Telegraphs—[1] Please call in and execute the assignment of this patent to The Commercial Tel. Co. & oblige[2] Yours truly

Lemuel W. Serrell p[er] C H Smith

ALS, NjWOE, DF (*TAEM* 12:1186). Letterhead of Lemuel Serrell. [a]"New York," and "187" preprinted.

1. Unidentified.

2. The Commercial Telegraph Co., incorporated in New York State in 1872, included in its charter a provision for foreign operations. In 1874 the company was acquired by the Manhattan Quotation Telegraph Co. Marchand 1922, 134.

-381-

James Anderson to Frank Scudamore

London Dec 10th 1873[a]

My dear Mr Scudamore[1]

Can you manage to give us a room in which to try the new Addison[2] instruments which have just arrived.[3] They claim to have made decided improvements and Mr. Latimer Clark confirms their statement

I hope you can at an early date give us a Line and a room to make a full and fair experiment and either prove that we have lost our money or obtained the control of an invention worth securing. Yours try

James Anderson[4]

ALS, UKLPO, ATF, Item 85. [a]"London" and "187" preprinted.

1. Frank Scudamore, second secretary in the British Post Office, was responsible for the management of the telegraph system. *DNB*, s.v. "Scudamore, Frank Ives"; Kieve 1973, 128–37.

2. Edison.

3. Accounting entries that may reflect the manufacture and shipment of these instruments and accompanying supplies can be found in Cat. 1214:34 and Cat. 1219:124–25, 131, 134; both Accts. (*TAEM* 21:581, 767, 770, 773).

4. Sir James Anderson (1824–1893) was knighted for his services as

captain of the *Great Eastern,* which laid the first successful Atlantic telegraph cable in 1866. When John Pender created the Eastern Telegraph Co. in 1872, Anderson became the general manager and later managing director (*MED,* s.v. "Anderson, Sir James"; Barty-King 1979, 40). Eastern was the core of Pender's worldwide oceanic telegraph enterprise, the largest of all such networks (Bright 1974 [1898], 119–21; Barty-King 1979, pt. 1 passim). Pender and Anderson, together with John Puleston, in either their personal or corporate capacities, had acquired some interest in Edison's automatic telegraph system through Smith, Fleming & Co. It is not known whether there was a connection between William Nicol of Smith, Fleming and G. C. Nicol, one of Eastern's directors. See Doc. 350; and Josiah Reiff to TAE, 6 Sept. 1877, DF (*TAEM* 14:719).

-382-

From Charles Buell

New Haven, Ct., Dec 12th 1873[a]

Dear Sir

Your favor was duly recd I am engaged as timekeeper, etc, and find it difficult to get away for any length of time, but shall be pleased to accept your kind offer to make experiments at your factory soon as I can arrange to be away

I would say however that I have, in numerous e~~p~~xperiments—succeeded in recording electric signals by Photographing upon both metallic and fiberous surfaces prepared for the same—without the aid of solar rays—the letter or character can be made by shaping sparks by means of stenciled disk or plates or by an arrangement of breaks in the conductors.

the effect is instantaneous, and any desired shape can be had

I have a very convenient process for preparing the paper[1] and a device for indicating the proper letter or character which is quite simple and can be worked either by hand or Automaticaly.

I have not been able to find anything like either the transmitting or recording having been in use.[2]

I hope to perfect the part for transmitting, so as to make it convenient to carry, at an early date. Would not object to Christmas if you find it convenient to make an appointment at that time.

Have you an electric "doubler"?[3] Can you get the use of wires to, say Phila. & return? Very Truly yrs etc

Chas. E. Buell

ALS, NjWOE, DF (*TAEM* 12:991). Letterhead of Car Department, New York, New Haven and Hartford Railroad Co. [a]"New Haven, Ct.," and "187" preprinted.

1. See Buell to TAE, 24 Feb. 1874, DF (*TAEM* 13:34).
2. Buell did not receive a patent on this system.
3. Unknown.

–383–

James Parrish to Richard Culley[1]

London Dec 16th 73

Dear Sir:

Knowing that strong objections exist under ordinary circumstances to the use of a "Grove" battery on your lincs,[2] particularly on the underground portion, also it having been intimated that one would not be permitted without a written permission; I respectfully ask that in our approaching experiments we may be permitted to use either a "Grove" or "Bunsen" under the following circumstances:

First: That the Grove[a] battery be used only for <u>Automatic</u> work using an ordinary one for the Morse or speaking portion:[3]

Second: That the whole, or rather nothing near the whole, strength of the battery, never be permitted to go to line, it being shunted to earth at the transmitting station, thus

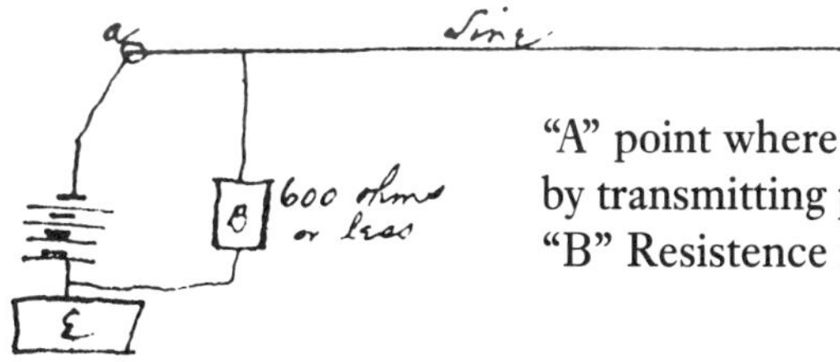

"A" point where contact is broken by transmitting pens.
"B" Resistence in shunt.

ALS (fragment), UKLPO, ATF, Item 103. [a]Interlined above.

1. Richard Culley was chief engineer in the telegraph department of the British Post Office. See *TAEB* 1:34 n. 2.
2. See *TAEB* 1:599 n. 3.
3. In Edison's automatic system, the operators exchanged information ("spoke") about the automatic transmissions by means of a separate Morse line.

–384–

Richard Culley to Frank Scudamore

[London,] 22 Dec 1873

Mr. Scudamore

These further experiments are very satisfactory[1] We certainly need not treat with Edison, who uses a battery which sets ordinary percha[2] covered wire on fire if incautiously applied.

It would seem to be desirable to revert to chemically prepared paper in place of the Wheatstone receiver—on certain circuits[3]

R S Culley

ALS, UKLPO, ATF, Item 102.

1. This letter accompanied David Lumsden's letter of the same date (Item 101, ATF) concerning attempts to operate Bain chemical receivers by employing two currents of opposing polarities (on the Bain automatic telegraph, see *TAEM* 1:65 n. 3). The British telegraph engineers, who objected to the Grove battery required by Edison's automatic telegraph system, sought to develop their own system. David Lumsden to Richard Culley, 15 and 22 Dec. 1873; Culley to Frank Scudamore, 15 Dec. 1873; and Scudamore to Culley, 18 Dec. 1873; all ATF.

2. Gutta percha, a natural latex, was used for electrical insulation.

3. Chemical recording of signals was considerably faster than the ink recorders of the Wheatstone system. The British Post Office experimented with this idea (see Doc. 390) but did not institute the change. For a description of the receiver used in Charles Wheatstone's automatic system, see Preece and Sivewright 1876, 122.

–385–

R. McIlwraith to Lyon Playfair

London E.C. 22nd Decr 1873[a]

My dear Dr Playfair[1]

When I first spoke to you about the new system of telegraphy of which we[2] have taken the agency it never occurred to me that you were likely to have to deal with it officially. I am not sorry at this however, for it will at any rate shew that I am not taking advantage of my personal acquaintance to bring to your notice anything we have taken up since you have received such an important appointment.

My object in now writing you this line is to ask if you can see me for ten minutes any time before Wednesday afternoon when we run down to Yorkshire for a short visit. If you will send me a message to my home, I will call either tomorrow evening or Tuesday morning

The proprietors of the system have kept operators here for a good while on the promise that they would have a room in the new Post office buildings and it was arranged that they were to work a line to Plymouth. Now it appears as if little obstacles are being put in the way of carrying out this arrangement and my desire is to seek your all powerful aid in overcoming these delays and little difficulties—which probably arise from nothing more than the necessities of reference &c &c. Perforators enabling the regular post office telegraph clerks to use the system have been made and I believe you might run them on to this system the very heaviest work you have with immense relief to the office—but that would easily be shewn by commencing to work in the manner arranged, but now temporarily delayed in the way I have mentioned.

Kindly excuse [this big rush?].[b] Yours sincerely

R McIlwraith

ALS, UKLPO, ATF, Item 107. Letterhead of 17 & 18 Leadenhall Street. [a]"London E.C." and "187" preprinted. [b]Possibly "my English".

1. Sir Lyon Playfair, chemist and politician, was postmaster general in 1873. *DNB*, s.v. "Playfair, Sir Lyon."
2. Smith, Fleming & Co.

–386–

Draft to George Gouraud[1]

[Newark, Fall 1873?[2]]

Col Gouraud,

Here is a Duplex, which dispenses with Stearns improvement, (the addition of the condenser,) and replaces it with a more powerful instrument to neutralize the static charge on the receiving instrument at the transmitting station. I have sold it to the Western Union[3] and if you and Mr Latimer Clarke can do anything with it you can have it remembering my "dower" ⅓.—

I will show it as applied to the Am[erica]n System.[4]

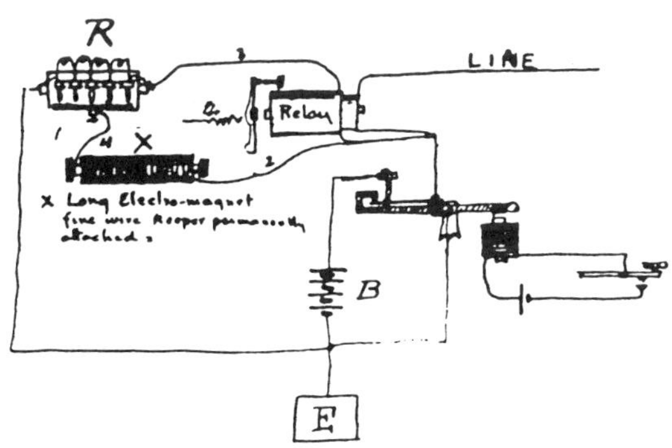

R is a rheostat, X a long magnet wound with[a] fine wire with the keeper permanently attached thus

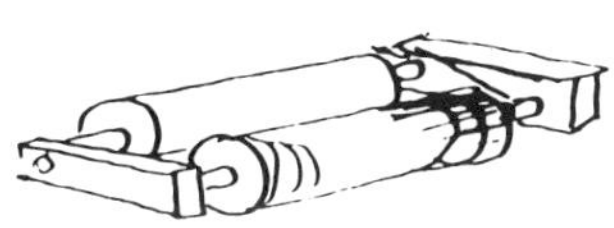

This magnet may be placed anywhere along the rheostat line by the strap on the side of the Rheo to which is connected the wire 4.

~~At the moment of closing The m pulling on the battery .B. the coil inner coil connected to the line receive an impulse~~ When the battery B is connected, and at rest, the ~~current~~ effect of the current is balanced in the double helix relay. The line is statically charged and the long magnet is also charged if now the battery is disconnected the static charge ~~follows~~ passes through the line coil and would throw it out of balance were it not that the long magnet X

ADf (fragment), NjWOE, Lab., Cat. 297:81, 89 (*TAEM* 5:756, 803). Continuity of text and resemblances of style, paper, and ink link these separate scrapbook items. [a]Starts separated page.

1. An earlier, shorter draft appears in Cat. 297:81(5) (*TAEM* 5:762). See also Cat. 297:135(3), 146(4) (*TAEM* 5:930, 969); and NS-Undated-005, Lab. (*TAEM* 8:352).

2. Edison began to use electromagnets with cores longer than normal only in September and October 1873 (p. 47, Docs. 363 and 375). This duplex design is apparently related to other work Edison did in the autumn of 1873 on duplex and cable telegraphy before preparing Doc. 392 (q.v., esp. fig. 8). See Doc. 387; and Cat. 297:62(3), Lab. (*TAEM* 5:650).

3. Nothing is known of such a sale.

4. That is, a closed-circuit Morse system. See *TAEB* 1:10.

-387-

Technical Drawings: Multiple Telegraphy

[Newark, Fall 1873?[1]]

Device 1[2]

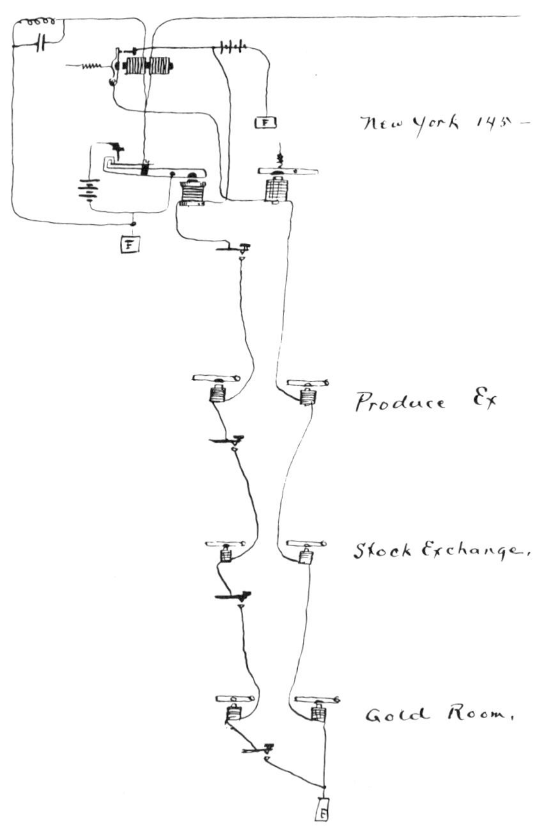

Device 2[3a]

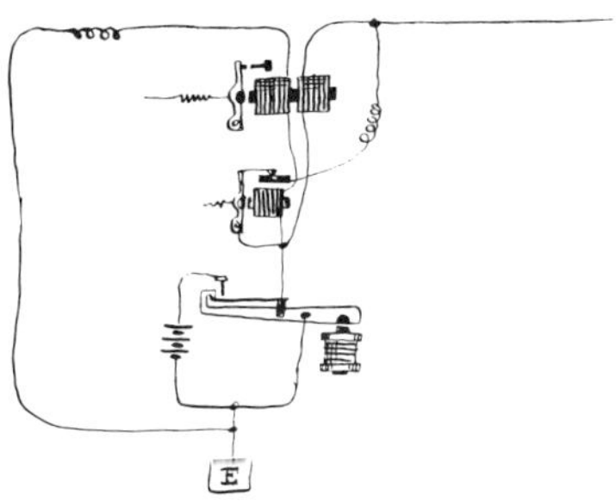

Device 3[4]

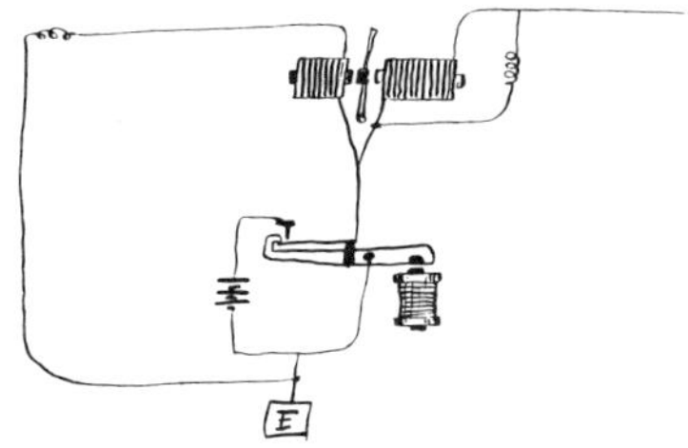

Device 4[5a]

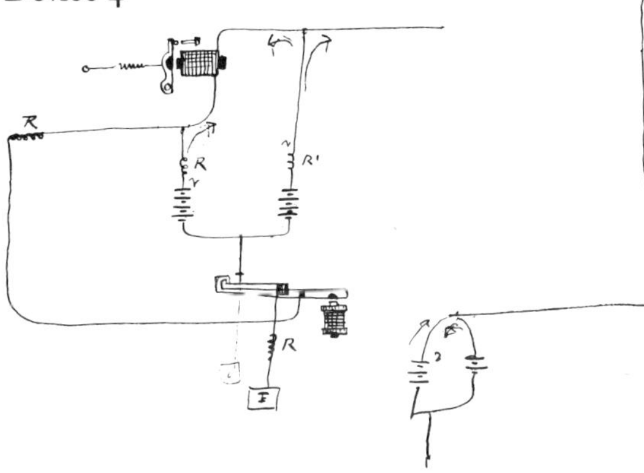

Device 5[6]

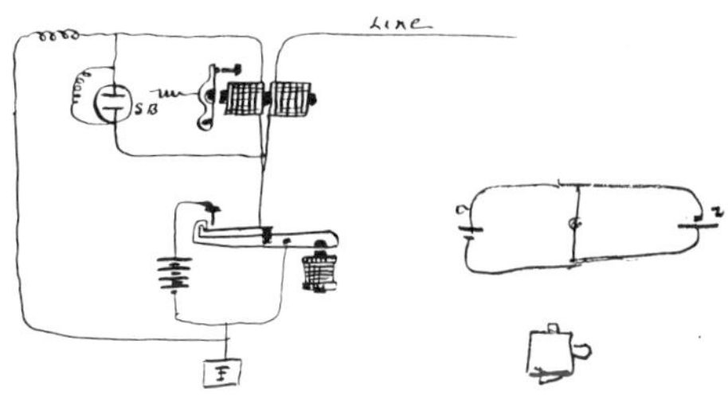

Device 6[7a]

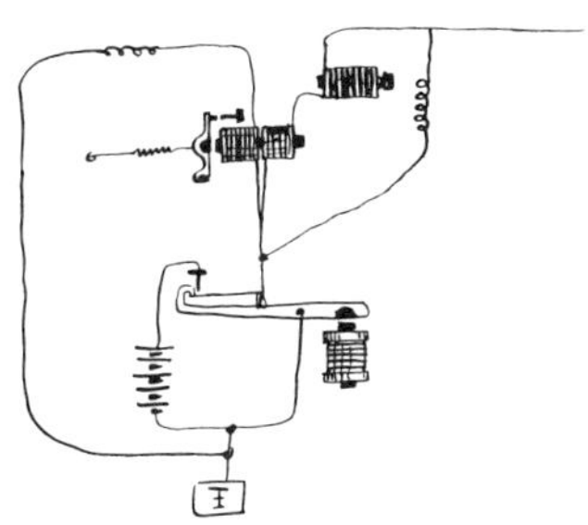

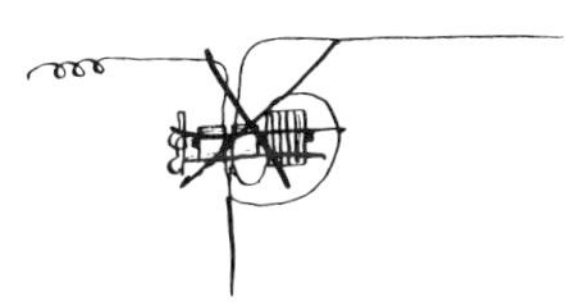

Device 7[8]

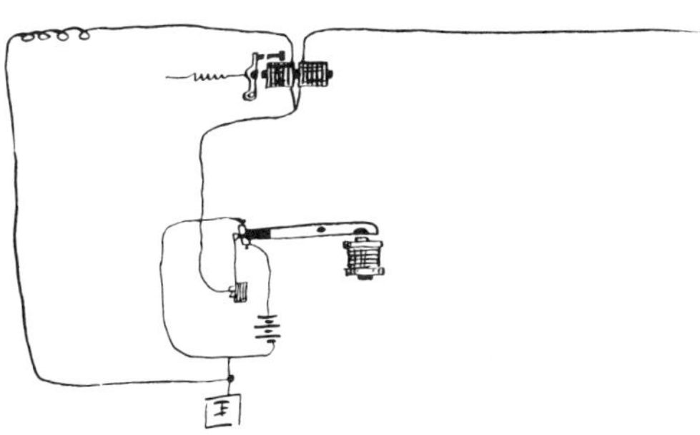

Device 8[9a]

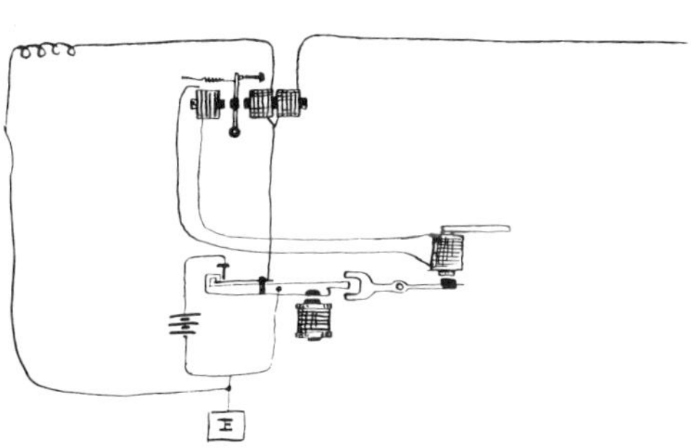

Device 9[10]

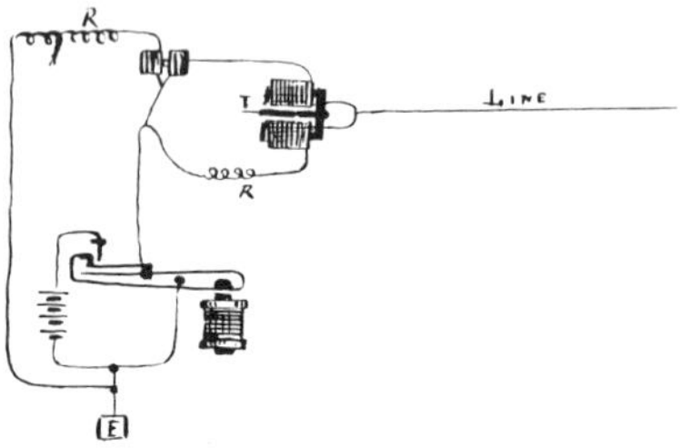

Device 10[11a]

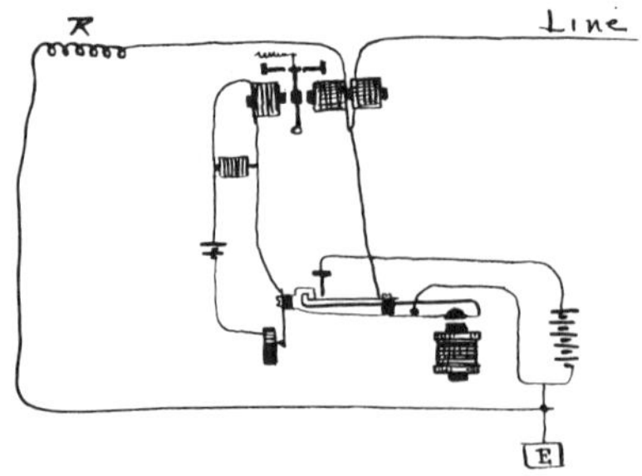

Device 11[12]

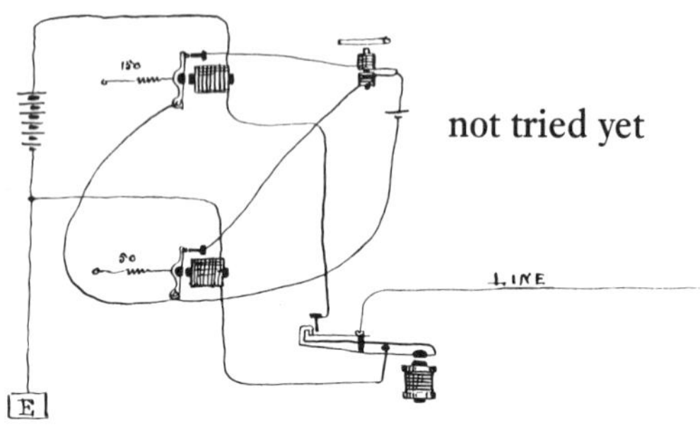

not tried yet

Device 12[a]

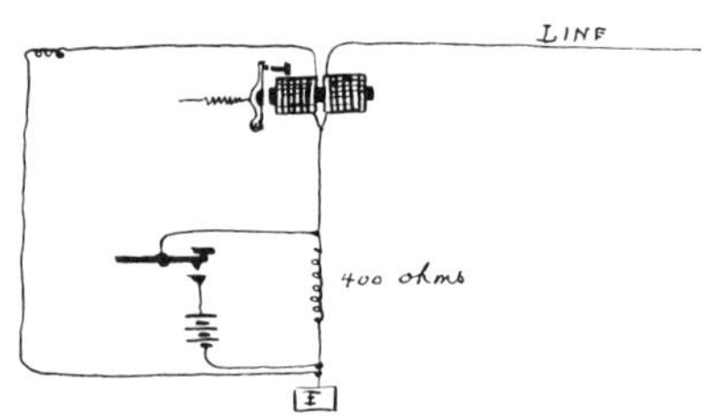

Device 13[13]

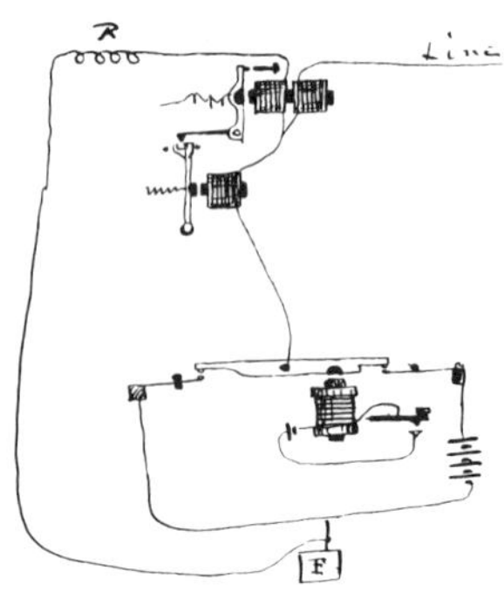

Device 14[14a]

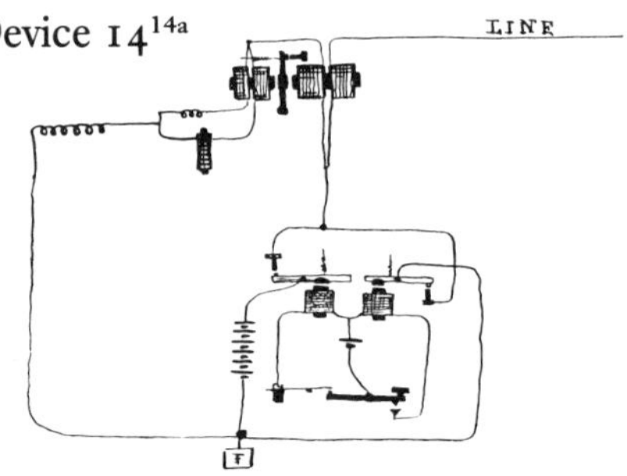

Device 15[15]

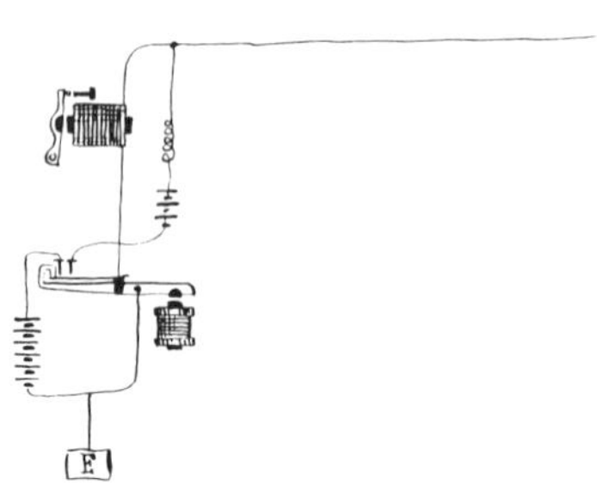

Device 16[16a]

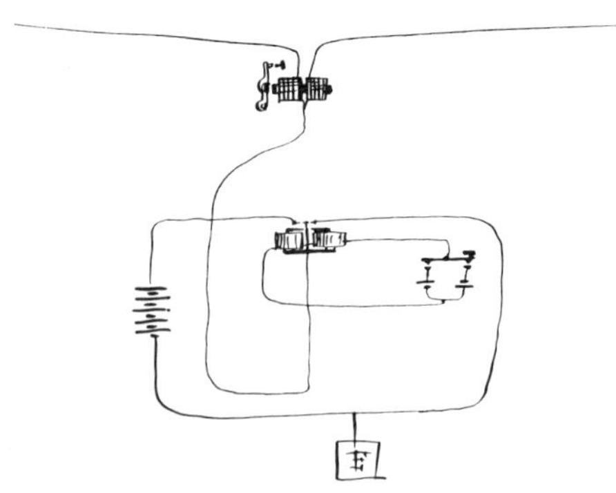

AX, NjWOE, Lab., Cat. 297:136(1), 140(2), 135(2), 140(3), 139(3), 138(4), 138(2), 136(4), 140(4) (*TAEM* 5:931, 949, 929, 950, 945, 942, 940, 934, 951). Resemblances in drawing style, paper, ink, and content, as well as consecutive numbering in Edison's hand, link these separate scrapbook items. [a]Starts separated page.

1. This document is representative of Edison's work in multiple telegraphy in the latter half of 1873 (the design labeled "Device 5" makes use of a secondary battery, an option Edison did not begin to employ until after his return from England). During this time he drew several sets of diagrams for multiple telegraph circuits, experimenting with receiver and transmitter design. This set, one of the most coherent, contains some designs found in other sets and in later caveats, as indicated in the following endnotes.

2. This basic duplex setup (at "New York 145") is arranged so that a key at any of the four stations can transmit, assuming the others are closed at the time. Cf. Doc. 392 (fig. 18); and Cat. 297:72(3), 136(2); NS-Undated-005; all Lab. (*TAEM* 5:710, 932; 8:355, 358).

3. In this design, the armature of the electromagnet in the center is connected to a sliding contact that makes and breaks a local circuit containing a resistance. Since this electromagnet is in the artificial line, it is apparently supposed to affect outgoing signals. Cf. Docs. 442, 453, and 469 n. 3; and Cat. 297:129(2), 152(2), 152(3), 153(1), Lab. (*TAEM* 5:912, 989–91).

4. Here the magnets in the differential relay are of different lengths. Cf. NS-Undated-005, Lab. (*TAEM* 8:208).

5. The two batteries apparently oppose each other at the relay (top center) when the transmitter (bottom center) is closed. Cf. Docs. 285 (no. 13), 309, and 312 (B).

6. Edison has shunted the relay magnet on the artificial line with a secondary battery (**SB**—i.e., a storage battery) instead of a condenser. Cf. Cat. 297:129(2), Lab. (*TAEM* 5:912).

7. Edison has placed an extra magnet in the main line, shunting it and the main-line relay with a resistance. Cf. Cat. 297:153(3), Lab. (*TAEM* 5:993).

8. Here a sliding contact is on the end of the transmitting lever. Cf. device 2.

9. The circuit containing the two extra magnets (one on the main relay and the other to trip the electromechanical device on the transmitter) is clearly incomplete, lacking both a battery and a breaking device. Cf. Doc. 301. See also Cat. 297:31(4, 5), 63(4, 5), 86(3) verso, 102(2), 128(1), 145(2), 152(3), 153(3); and NS-Undated-005; both Lab. (*TAEM* 5:551–52, 660–61, 790, 853, 909, 963, 989, 993; 8:356, 357, 360).

10. The polarized relay (upper right) is in an odd place; with no indication of how it is wound, its function cannot be determined. Cf. Cat. 297:153(3); and Cat. 299:68–69; both Lab. (*TAEM* 5:993; 6:130, 131). See also Doc. 496.

11. Edison again attached a sliding contact to the transmitter. The function of the small magnet (left center) apparently bridging the local circuit is unclear. Cf. devices 2 and 8 of this document; Docs. 283, 308, 392 (fig. 14), and 471; and Cat. 297:31(4, 5), 74(2), 86(3), 139(4),

146(4), 152(3), 153(3), Lab. (*TAEM* 5:551–52, 720, 789, 946, 969, 989, 993).

12. Here Edison drew differentially sensitive relays ("50" and "150"), one for receiving and the other for transmitting. See Docs. 285 (no. 8) and 348 n. 16; cf. NS-Undated-005, Lab. (*TAEM* 8:361).

13. The receiving relay (below the double, differential magnet) cannot be activated unless the armature of the differential magnet is closed by an incoming signal, which pulls the small tooth out of the notch in the receiving armature. Outgoing signals charge the receiving magnet but do not affect the differential relay. Cf. Docs. 301 and 469. See also Cat. 297:86(3); and NS-Undated-005; both Lab. (*TAEM* 5:789; 8:330).

14. Here Edison uses a differential relay, inductance, and resistance as compensation in the artificial line. Cf. Docs. 469 and 513. See also Cat. 297:137(2), 146(3), 152(3); and NS-Undated-005; both Lab. (*TAEM* 5:936, 968, 989; 8:357, 360).

15. Cf. Doc. 285 (no. 4).

16. This transmitter employs a polarized relay. Cf. Docs. 392 (fig. 18) and 473; and NS-Undated-005, Lab. (*TAEM* 8:354).

-388-

Technical Note: Miscellaneous

[Newark, November–December 1873[1]]

Batc[h?][a] ⟨Record all this & proper language⟩

See if you can put a charge from plate of electophorous into a steel needle magnetized to point north and see if the charge will make it point from due north also see how near ~~a su the~~ [-] a substance insulated or not will approach before throwing it out of its track—

See if by any arrangement you can prevent the needle of the silvertown galvanometer from sticking to the points

may be wetting the silk fibre which needle suspended on will do it=

chromic & permanganic acid part with their oxygen with smallest current ~~thus~~ it is a great object with sensitive paper to get the weakest current to liberate the most oxygen so as to produce the most protoxide. Therefore try chromic & permange acids with & without salt or with other substances if it should spontaneously decompose without current a little, but greater with there would be a light continuous mark between dots & dashes but if you run fast 300 words per min with perfd paper the spontaneously made proto w~~l~~ould ~~sh~~not show try it[b]

See if you can arrange a inking pad to ink the embossed marks on Morse Register—

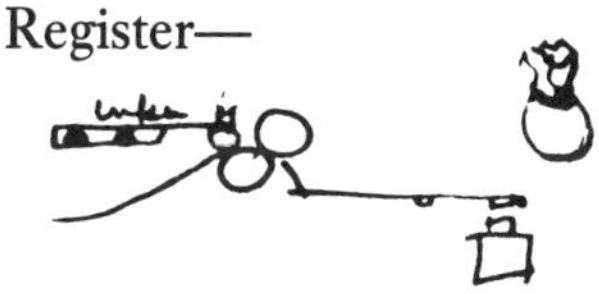

chamois etc

Try conductivity of mercury & zinc amalgam in small hole tube see if its high & constant[b]

Try this

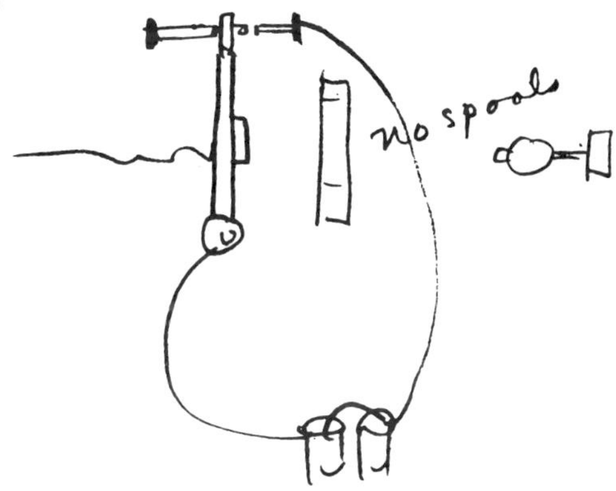

No magnet = see if the mere passage of the current ~~th~~from 2 or 3 good cells or more of Carbon battery through the armature lever & platina points will make them stick I suspect that when ~~no~~the spring is adjusted so as to just pull the lever away from point when no current is on by pushing it up with finger it wont stick but is battery connected if pushed up it will stick, owing to magnetic effect of current on platina points include a sounder in the circuit =[2]

Batchelor

You know that when Sulphuret of potash is used with an iron pen—that the pen becomes polarised thus

sent ——— ——— ———

recorded iron clean

——— —— ——

on closing the circuit there is an intense black mark on opening it ceases on closing again the mark does not commence for some time to get rid of this try a double pen iron & platina thus

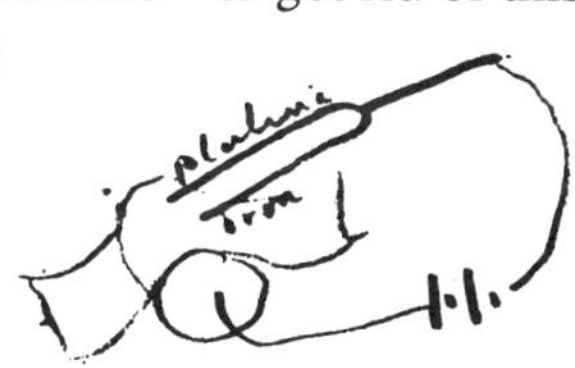

also try Copper Zinc pen with the iron
Batch—

Put of 1~~6~~0 good Carbon & test each one on No 4 of Bradley to get deflection. pick out best one & try & get 9 other that equal it= then connect these 10 to a single cup giving the same deflection—connecting the coke of the 10 cup battery to the coke of the 1 cup battery & the zinc of the 10 cup battery to the zinc of the 1 cup battery then take the deflection on 1 2 3 & 4 of the 10 seperate & the 1 seperate every few hours for 3 days keeping them on closed circuit all the time. also ~~th~~ake the deflection all together= I want to see what action will take place on the single cell—[3b]

Also ~~ins~~[- ----] of connect 5 good with 5 good opposing each other ie[c] the Coke of one 5 with the Coke of the other 5—closed circuit—take deflection on ~~t~~each 5 seperate ~~& try and~~ then connect together & take deflection if there is any current see which battery gives it then change cups so as to get a balance with 1 or 2 degrees on No. 4.[4b]

Put 3 good cups all same deflection on No 4 ~~toge~~ on closed circuit all seperately take deflection on all Coils every few hours wish to know how long a Carbon battery will last on short circuit=[5]

Magnet thus in bridge 

testing for induction—see if the extension of the Cores by putting two extra iron Core pressing up against those of magnet will increase the inductive effect without adding more wire. also try this

1st  2nd 

You can gear up those wheels so that you will have to go quite fast on turning to get a moderate speed on receiver you can get Tom[6] to turn & thereby get an even speed in changing from one experiment to another with the same magnet it would be best to keep Tom turning all the time to keep speed even another thing the pressure of your pen should always be the same probably this would answer

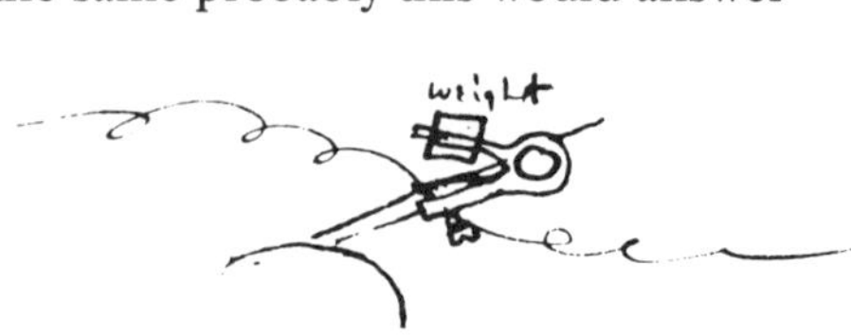

also ~~t~~a whole crock of paper should be wet at once so the paper should have the same thickness take the deflection of battery used on No. 4. coil & always keep it up ~~no~~to that deflection & be sure to get bridge resistance correct so not the faintest shad~~w~~ow of current shows on the paper even in turning slow. it takes a given amount of Resistance to ~~thro~~show current on the next pen after you cant see it on first ½ thus

Try sold Cores thus

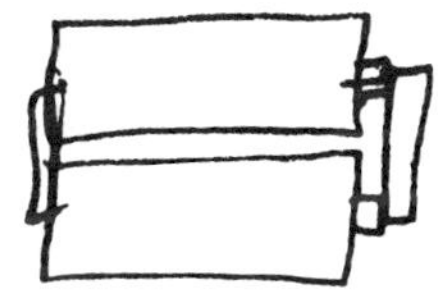

pull out solid Cores insert core of fine iron wires solid—with regular back & armature see which gives most induction solid or wires—

Batch. Try this for induction & strength of attraction

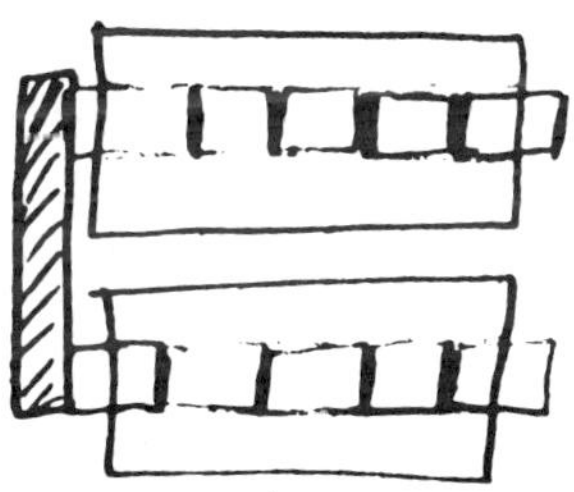

~~see w~~[d] core sawed in length of ¼ inch & seperated by brass & brass screw or by sealing wax. so each part core dont touch but is with a 1/100— Have an Idea that there will be no induction & the attractive strength ½ as great as solid core thrust in same bobbin with same battery if this is so we can make long ordinary magnet thereby getting strength & yet have no induction—

X, NjWOE, Lab., NS-Undated-005 and Cat. 298:138 (*TAEM* 8:232 and 5:281). [a]Document damaged. [b]Followed by centered horizontal line. [c]Circled. [d]Edison apparently wrote "~~see w~~" after making the small drawing of the core and then added the text "core sawed . . . 1/100—".

1. Edison wrote several notes to Batchelor on loose sheets of paper which were subsequently stitched together and numbered consecutively. Batchelor entered most of these (with other experiments copied from Edison's loose notes) in a laboratory notebook and separately numbered each experiment (Cat. 1175:102–14 [*TAEM* 6:156–62]). Batchelor and Edison also indicated the results of some experiments in that notebook. Batchelor dated the first entry in that notebook "Nov 1873." Notes regarding results date from late December and early January.

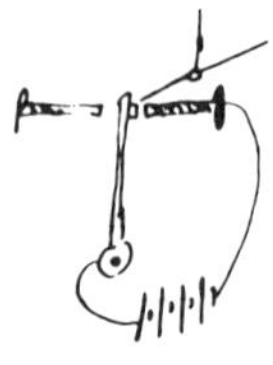

Batchelor's drawing of setup to test the magnetic effect of a current passing through platina points.

2. Batchelor made the following notation on the facing page in the laboratory notebook.

> The stick is very perceptible on 4 cups carbon: polarized realy, 12 ohm sounder. sticking not so great when you substitute resistance for sounder. Platinum is magnetic I tested it by this means I suspended a needle of soft iron by a fibre of silk over the points & every time I closed circuit it drew needle & platina points

Underneath Batchelor's notation, Edison wrote, "Note. Platinum is not magnetic It is the current which passes through it—."

3. The following table of results appeared on the facing page in the laboratory notebook.

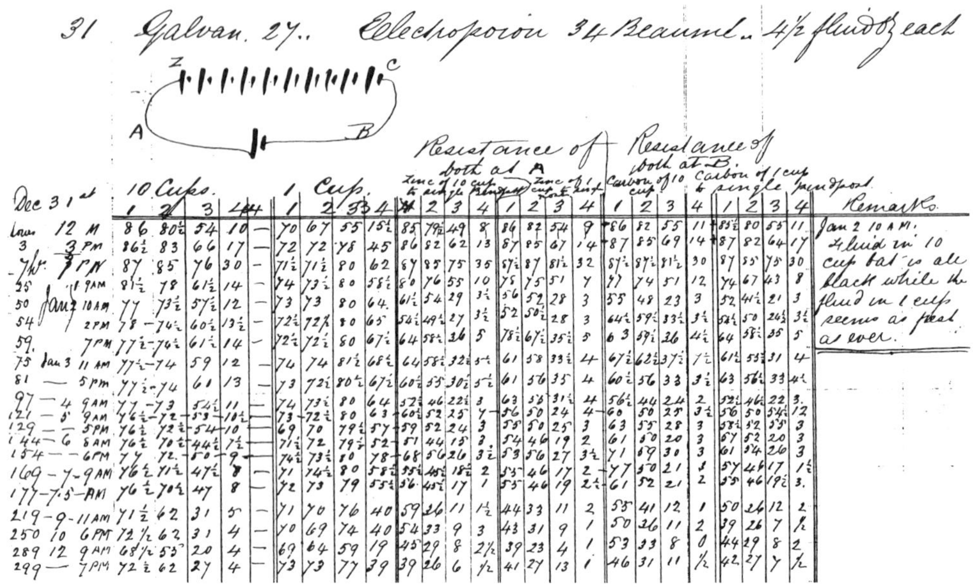

4. The following table of results appeared on the facing page in the laboratory notebook.

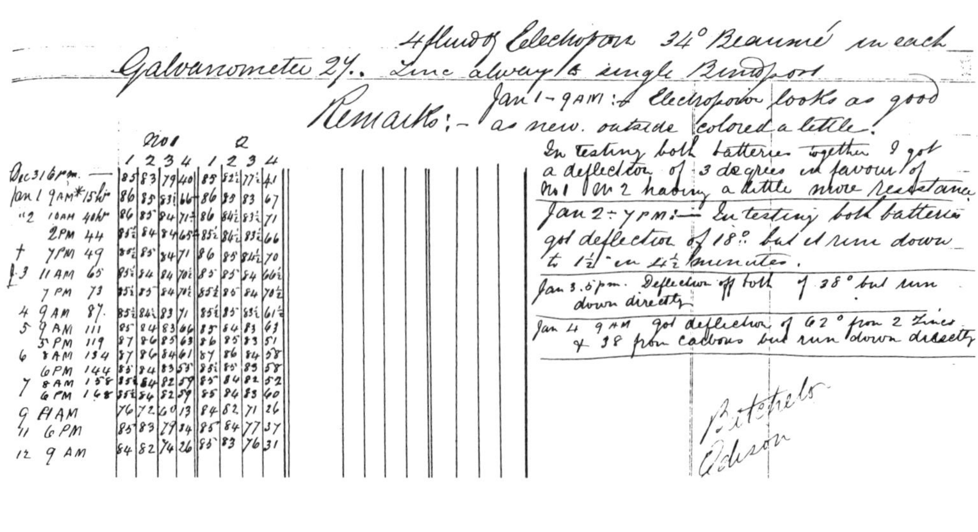

5. The following table of results appeared on the facing page in the laboratory notebook.

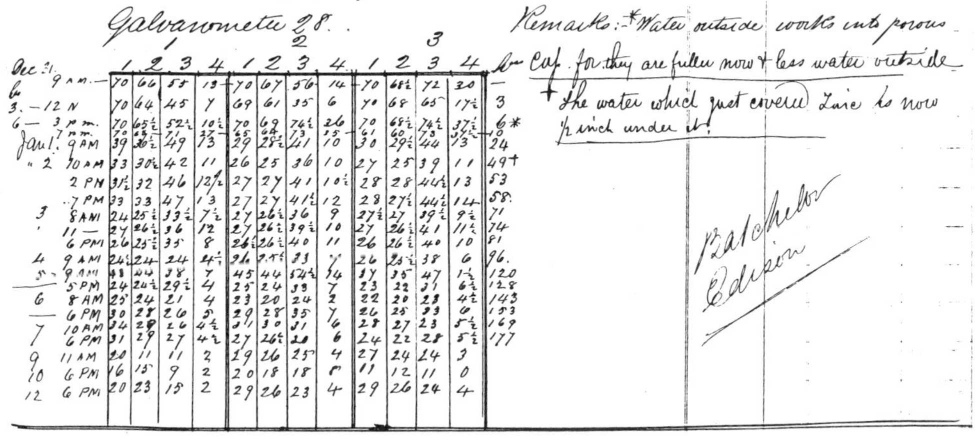

Galvanometer 28.

Dec 31 to	1-1	1-2	1-3	1-4	2-1	2-2	2-3	2-4	3-1	3-2	3-3	3-4	h
9 AM.—	70	66	55	13	70	67	56	14	70	68½	72	30	—
3.—12 N	70	64	45	7	69	61	35	6	70	68	65	17½	3
6— 3 pm.	70	65½	52½	10½	70	69	74½	26	70	68½	74½	37½	6*
7 pm.	70	65	71	27	65	64	73	35	61	60	73	34½	10
Jan 1. 9 AM	39	36½	49	13	29	28½	41	10	30	29½	44	13	24
" 2 10 AM	33	30½	42	11	26	25	36	10	27	25	39	11	49†
2 PM	31½	32	46	12½	27	27	41	10½	28	28	44½	13	53
7 PM	33	33	47	13	27	27	41½	12	28	27½	44½	14	58.
3 8 AM	24	25½	33½	7½	27	26½	36	9	27½	27	39½	9½	71
11—	27	26½	36	12	27	26½	39½	10	27	26½	41	11½	74
6 PM	26	25½	35	8	26½	26½	40	11	26	26½	40	10	81
4 9 AM	24½	24	24	4½	[illegible]	[illegible]	33	[illegible]	26	25½	38	6	96.
5 9 AM	43	44	38	7	45	44	54½	14	37	35	47	1½	120
5 PM	24	24½	29½	4	25	24	33	7	23	22	31	6½	128
6 8 AM	25	24	21	4	23	20	24	2	22	20	23	4½	143
6 PM	30	28	26	5	29	28	35	7	26	25	33	6	153
7 10 AM	34	29	26	4½	31	30	31	6	28	27	23	5½	169
6 PM	31	29	27	4½	27	26½	20	6	24	22	28	5½	177
9 11 AM	20	11	11	2	19	26	25	4	27	24	24	3	
10 6 PM	16	15	9	2	20	18	18	8	11	12	11	0	
12 6 PM	20	23	15	2	29	26	23	4	29	26	24	4	

Remarks:—*Water outside works into porous Cup. for they are fuller now & less water outside

† The water which just covered Zinc is now ½ inch under it!

Batchelor
Edison

6. Probably Thomas Gartland, whose name appears in a later notebook entry (see Doc. 482 n. 1). Gartland's name first appears in the payroll records on 16 August 1872 (PN-72-08-16, Accts. [*TAEM* 22:222]). There is also an 1873 Edison account with Gartland in Cat. 1218:80, Accts. (*TAEM* 21:222).

-389-

Memorandum: Automatic Telegraphy

[Newark, December 1873[1]]

Connect a table in Newark up all above board[2] like Littles diagrams[3] & have W[ashington] send with no shunt use his condsr See what speed if not more than 1050 good[4] ~~invite~~get JCR[eiff] to invite Brown[5] Eckert[6] etc over to witness[7]

AX, NjWOE, Lab., Cat. 297:93(6) (*TAEM* 5:832).

1. See n. 3.

2. That is, with all connecting wires in view.

3. George Little published an article with diagrams describing his automatic system equipped with condensers in the 20 December *Telegrapher* (Little 1873).

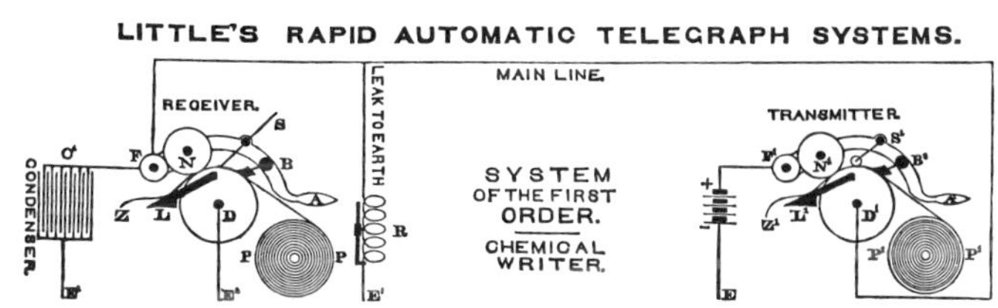

George Little's chemical automatic telegraph with a condenser at the receiving end.

4. That is, 150 words per minute clearly received by the apparatus.

5. Alfred Brown was manager of Western Union's main office in New York. Taltavall 1893, 19–20.

Western Union official Thomas T. Eckert later became president of the Atlantic and Pacific Telegraph Co.

6. Thomas T. Eckert (1825–1910) began his career in 1848 as a telegraph operator, rising to the superintendency of some Pittsburgh-Chicago lines. During the Civil War he was in charge of the military telegraphs at General McClellan's headquarters in Washington, D.C. Following the war he directed the Eastern Division of Western Union.

He later became successively president of the Atlantic and Pacific Telegraph Co., the American Union Telegraph Co., and Western Union. Taltavall 1893, 13–14; *New York Times,* 21 Oct. 1910, 11.

7. Little claimed he was the inventor of the system used by the Automatic Telegraph Co. (see, e.g., Little 1873, 1874a, 1874b, and 1874c). Edison, on the other hand, claimed that the company was using his system, and shortly after the appearance of Little's 1873 article Edison demonstrated the inadequacy of Little's published design to "several gentleman interested with Mr Little," much as he had demonstrated his own diplex and automatic telegraphs to important Western Union personnel in early October (see Chapter 2 introduction). In this demonstration, applying Little's circuit design to a wire upon which Edison's system was receiving 160 words per minute reduced the speed to 75 words per minute. Cat. 297:106, Lab. (*TAEM* 5:859); see also Edison's testimony, Quad. 70.7, p. 289 (*TAEM* 9:511).

-3- January–March 1874

Edison began 1874 as he had ended the previous year, concentrating much of his energy in the laboratory. However, he also turned to the task of producing a manuscript for the book on telegraphy he had begun to plan the previous winter.[1] Although he had apparently done some writing in 1873, he now attacked the project in earnest. Eventually he would compose full or partial drafts of several chapters and fragments of many others.

Edison's principal experimental concern seems to have been the automatic telegraph system. The Automatic Telegraph Company was in business, but the system components needed further modification and refinement. Edison spent time on perforators, repeaters, paper winders, stylus metals, and chemical combinations. He also devoted considerable attention to his roman-letter automatic system, particularly versions that could transmit on only one or two wires.[2]

Edison's cable experiments, closely related to his other automatic system work, continued apace. Using the apparatus he had built in the fall, he tried to develop a system that could overcome the distortion inherent in the electrical properties of cables. The British investors in his automatic system began urging his presence there to assist in their experiments and demonstrations, but apparently Edison's lack of success in the laboratory kept him from returning to England.

District telegraphy, which Edison had briefly investigated in early 1872,[3] occupied him again during these months. In a district telegraph system, transmitting boxes installed in homes and businesses summoned various services—doctors, police, grocers, messengers, and others—through a central

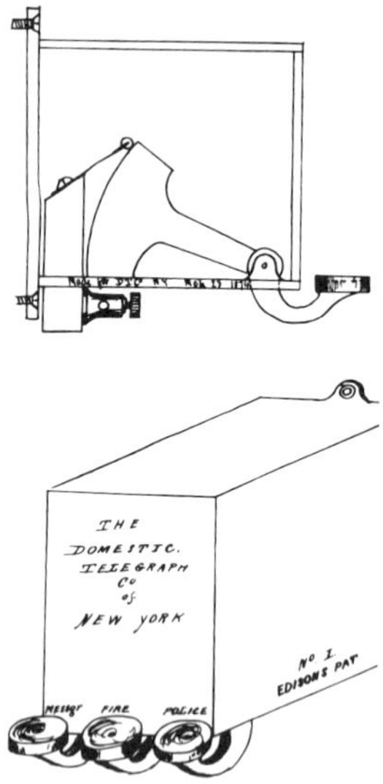

Charles Batchelor's drawings of the domestic telegraph transmitter.

station. The American District Telegraph Company, organized in 1872, was thriving, and in March Edison, Joseph Murray, and two promoters formed a competitor, the Domestic Telegraph Company. Edison's district system derived largely from his automatic telegraph technology, as did the fire alarm system he developed with it. Experimental and account records indicate that during March, Edison and his staff spent more than five hundred hours on these systems. On 2 April he executed a patent application for a new transmitter design. The Domestic Telegraph Co. began to set up its office in New York on 24 March.[4]

The laboratory records contain evidence of other research as well. Batchelor noted an investigation of perpetual motion. At one point Edison billed Automatic Telegraph for work on a "New Copying Machine" that might have been a typewriter.[5] There are also occasional entries of a more mathematical nature (as in Doc. 400).

Edison worked sparingly on the quadruplex during this period. The testimony preserved in litigation records offers conflicting accounts of his actions and intent. According to Western Union's witnesses, Edison kept up a steady, if sparse, series of experiments and tests in the laboratory and on their lines, meeting with both President Orton and George Prescott. According to Edison, he wanted to proceed with the quadruplex, but Western Union made its lines essentially unavailable to him, which precluded any significant tests. At the end of March he wrote to Orton in an attempt to ascertain his standing with the company, but Orton left for Europe before receiving the letter.[6]

The firm of Edison and Murray was busy, primarily manufacturing instruments for Automatic Telegraph and for the Gold and Stock Telegraph Company.[7] It also began taking on small jobs for the Celluloid Manufacturing Company, and was still renting out space and tools on occasion for small jobs by others.[8]

1. See Doc. 293.

2. For his accounts of experimental work for Automatic Telegraph see Cat. 1184:13–30, Accts. (*TAEM* 21:789–98); there is also a bill of 15 February from Rohrbeck and Goebler indicating that Edison bought "358 Chemical Preparation & Alcaloids" (74-001, DF [*TAEM* 13:9]).

3. See Docs. 226 and 228–29.

4. Cat. 1184:31–35 (*TAEM* 21:798–800). Only a few experimental drawings remain of this work, including these drawings by Charles Batchelor of the district transmitter (cf. U.S. Pat. 154,788) and the central station fire alarm bell. Cat. 1307, Batchelor (*TAEM* 90:604, 619).

See also Cat. 30,094 and Cat. 30,100, both Lab. (*TAEM* 5:327, 329–30, 350; 6:405).

5. Cat. 1184:14, 15, Accts. (*TAEM* 21:790–91). For Edison's earlier work on typewriters for the automatic system, see Docs. 132–33, 142, 155–56, 162, 201–3, 231, 234–36, 242, 256.

6. Edison's testimony, Quad. 70.7, pp. 339–41, 389, and Quad. 70.9, p. 91; William Orton's and George Prescott's testimony, Quad. 71.1, pp. 122–25, 130–31, 354 (*TAEM* 9:536–37, 561, 809; 10:66–67, 70, 184); and Edison's testimony, Testimony and Exhibits on Behalf of T. A. Edison, p. 16, *Nicholson v. Edison*, RG-241, MdSuFR.

7. Cat. 1219:148–57; Cat. 1184:12–34; both Accts. (*TAEM* 21:780–84, 789–800).

8. Edison and Murray began manufacturing pins for the Celluloid Manufacturing Co. in March (Cat. 1184:31–32; Cat. 1214:40; both Accts. [*TAEM* 21:798–99; 22:584]). Although Keene the gun inventor disappears from the account records, Prosser continued to rent space (see Chapter 1 introduction, n. 7), and the firm also provided space to a man from the Cottrell Stone Machine Co. (Cat. 1184:21–22, 24, 27, Accts. (*TAEM* 21:793–94, 796).

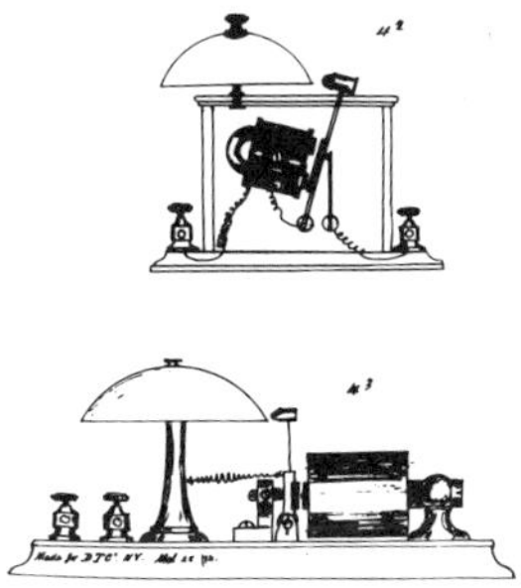

Batchelor's drawings of a fire alarm bell.

-390-

James Anderson to Frank Scudamore

London Jany 5th 1873[1874][a]

private[b] My dear Mr Scudamore

I have only this eveng managed to get through the reports you so kindly lent me to read and I arrive at the following conclusions

Messrs Fischer, Lumsden & Eames[1] give the system the full credit for legibly sending upwards of 500 words per min. and they admit the novelty of the invention

Mr Lumsden refers to the dampness of the paper as an old fault in the Baine's System. this and the fugitive nature of the signals are the only disparaging remarks in his report. he closes his remarks by saying that it is the Baine's System with the important addition of compensating magnets which is new and enables the high speed to be attained—[2]

Mr Eames dwells upon the large staff of transcribing clerks requiresd at all hours on account of the fugitive character of the messages. he also speaks of danger of contact of the paper blurring the marks and of the danger of the paper tearing on account of its tenderness after being moistened.[3]

But our reply to that is—we now use a paper which retains the marks permanently nor is it so damp or tender I believe although the experience in America seems to point to that being no difficulty

Mr Fischer's report is more exhaustive and deals with the practical value of the novelty to your department. he sees the difficulty of obtaining suficiently qualified operators to manage delicate machinery as in the case of the Wheatstone.[4]

I think a little [-]reflection will convince you that the new system only requires an ordinary operator, all the extra supply consisting of perforators and transcribers, and upon those lines where you could use three wires ~~the~~ required for Roman letters fewer transcribers would do.

It appears to me the merits of Eddison system is—that it can do many times more work with the ordinary operators and one wire than any other system yet brought to light.

Mr Fischer alludes to the additional space required We think our system will occupy less space for the same amount of work

Mr F. suggests another trial upon a long circuit where pressure is occassionally felt which cannot be readily overcome such as Dublin—and he thinks the real value of such a system would be found in its use upon costly Lines.

I am very strongly of opinion that this system applied to your French Lines[5] during the busy hours of the day would be well worth the increase of punchers and transcribers and would be invaluable I hope you will upon reflection think so. An extra relay of such operators during the very busy hours is a long way from increasing the whole staff for every circuit.

Mr Culley who at first seems to have doubted the novelty evidently changed his mind as shown in a pencil ~~mark~~ note to his letter of Feb. 27h 1873[6] he concurs in Mr Fischers report but fails to see how it can be profitably employed for a long time to come. but all though the correspondence there is evidence enough that there are times when a rapid method of clearing your wires without a great addition to your highly paid or highly skilled labor would be of great value

I do not know whether it is ungracious on my part to call your attention to the fact that ever since your experts have seen the experiments made last summer their minds and energ~~y~~ies ha~~s~~ve been at work to recusitate the Baines and supply it with some form of compensating addition that will answer the purpose so admirably achieved by the Magnate[7]

I do not know how far Eddisons patents cover all compensating applications but still I venture to think you would not acquire a knowledge of this novelty and come to overtip us by some other appliance ~~h~~which might not have been thought of without seeing these experiments. it is evident that the experiments of last month all point to the discovery of some compensating method and some kind of chemical which will give similar marks to ours.

I hope our prepared paper which costs almost <u>nothing</u> or very little[c] to prepare and provides permanent marks is itself a valuable addition. I do not know whether our new perforator is of much consequence but to send Roman characters with three wires and I believe with two is something worth acquiring So much work can be sent along these wires in so short a time

I share with you however the duty of either Mr Culley—yourself or myself to apprehend seriously any risk from undue battery power, but is there no limit within which you could let us try a fair experiment as compared with your best system using say 200 Daniels cups or as many as Mr Culley would consider safe and see whether in simplicity and speed we have not achieved something worth acquiring

Might it not be that your engineers are infringing upon what Eddison has patented

In any case am I not right in saying that you want speed simplicity and security and that the best systems which can be secured will get your best attention.

I am you will see thinking very freely upon paper to you, but it is private and hope you can yet see your way to give us a fair full trial

I am afraid to push the matter personally upon Mr Culley he may not like my pressuring to ~~say~~ have an opinion about his department, but he is seeking for a compensator, and a chemical and I hope he will not shunt us. we thought we had very high approval for the novelty before we secured it, and we still think it has bona fide merits at least worth a serious trial. we have kept men here ever since summer. Yours very try

James Anderson

ALS, UKLPO, ATF, Item 109. Letterhead of "66 Old Broad Street, E.C." [a]"London" and "187" preprinted; "3" underlined and "?4" added above in another hand. [b]Written at top left. [c]"or very little" interlined above.

1. Henry Fischer was controller (i.e., financial officer) of the Central Office of the British telegraph service in London. J. W. Eames was a staff member in the Central Office assigned to monitor the tests. David Lumsden was an electrician and superintendent of cables for the British Post Office.

2. For Lumsden's report see Doc. 319.

3. For Eames's report see Doc. 318.

4. Fischer to Scudamore, 30 May 1873, Item 72, ATF.

5. British Post Office land lines that carried messages to the British end of cables across the Channel to France. Haigh 1968, 198–99.

6. In this letter to Scudamore (ATF, Item 52) Richard Culley stated he felt that there was nothing relevant in the patent specifications to

which he was referred by George Gouraud. His later pencil note says, "The method of compensation used was not mentioned in the Specns. to which I then referred."

7. See Doc. 384.

–391–

James Parrish to Josiah Reiff

London. 1/17/74

Dear Sir

170 words were done the first night to Plymouth and 214 the second we will probably see 300 words— Sir James Anderson has offered the Lisbon Cable[1] for Experiments— Ask Edison to consider the following

No Cable manager will allow over 10 cells of Daniells—

The present system is a positive and a negative dot the one answering for a dash and the other for a dot— Will a dash in the cable be more difficult to manage than the ~~two~~ dot?

Can ~~p~~reversals be combined with the magnetic shunt to advantage?

They are trying sensitive paper on the Wheatsone instrument[2] and with reversals the full capacity of the instrument, 140 words, has been obtained and they do not know how much more it will do—

Edison ought to be here unless he is necessary to your American organization or unless his cable experiments are nearing completion— I advise you to find out fr[-]om him if there is any chance of his doing anything with cables better than is now done with only 10 Daniell Cells—

We have spent so much time over th~~is~~e ~~questio~~ Automatic that I think the questions I ask ought to be answered by experiment here on the Lisbon cable if there is no other way to find them out. Yrs Ty

James C. Parrish

ALS, NjWOE, DF (*TAEM* 13:81). On letterhead of James Parrish.

1. The Eastern Telegraph Co. cable ran from Porthcurno, Cornwall, to Caravelos, near Lisbon. Haigh 1968, 90–96.

2. That is, the Wheatstone automatic telegraph system.

–392–

Draft Caveat: Multiple Telegraphy[1]

Newark, N.J. Jan. 26, 1874.

Duplex Apparatus. (Caveat)

This invention consists of various devices to improve the working of Duplex apparatus, & various modifications to that end.

When the present Duplex is operated upon a short circuit such as 200 or 300 miles in length & upon which there is but little Static induction, the present arrangement of the Condenser, or Compensating device of the artificial line is sufficient to balance the charge of the line; but when the line is 400 or 500 miles or more in length, the Static charge is greatly increased, & the present method of connecting a single condenser at the extreme ends of the artificial line is insufficient.[2]

Fig 1 shows the present method

Fig 1

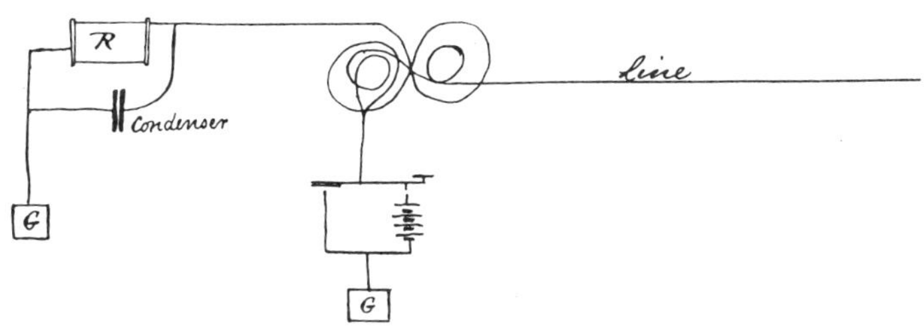

In this case the discharge from the line when long would be recorded thus:[3]—

that from the artificial line due to the condenser thus:—

Hence after the discharge of the artificial line, the discharge from the line continues & alters the adjustment of the receiving magnet. To obviate this I increase the length of the discharge from the artificial line to equal that from the mainline by dividing up the condenser in in several parts & placing it along the resistance coils as in Fig 2:—

Fig 2.

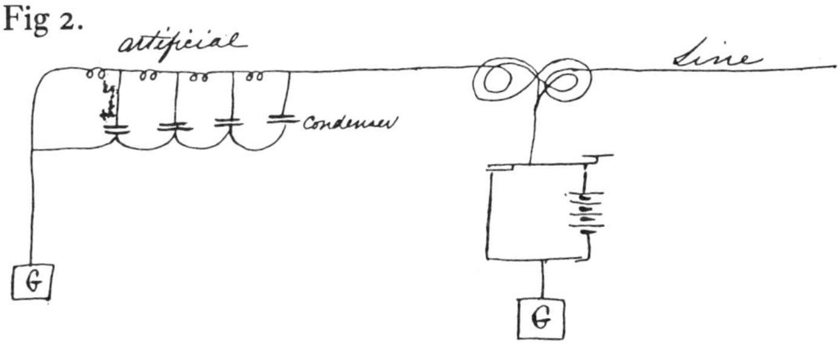

In this case the discharge from one condenser must pass through the other condenser & a multiplication of the discharge takes place, & the condenser may be so subdivided up on the total resistance so as to send a discharge as long as that from the line.[4] This is particularly valuable on submarine cables where the charge & discharge is very long.

When employed on cables I prefer to use the bridge system shown in fig 3:—

Fig 3

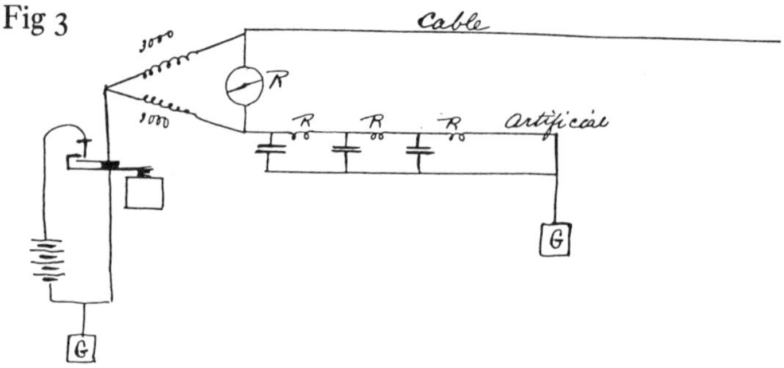

When the resistance of the artificial cable is equal to the resistance of the line or cable, & the bridge is equal, & the condenser & resistances of the artificial cable have been so subdivided as to generate a charge & discharge equal to the cable no effect is produced on the receiving instrument R (which may be of any character) by the outgoing current providing there is no self-induction in Coil G ~~at~~& some in G′ or vice-versa.[5]

When a condenser is used to obviate the earth current on very long cables[6] the arrangement may be modified as in fig 4:—

Fig 4

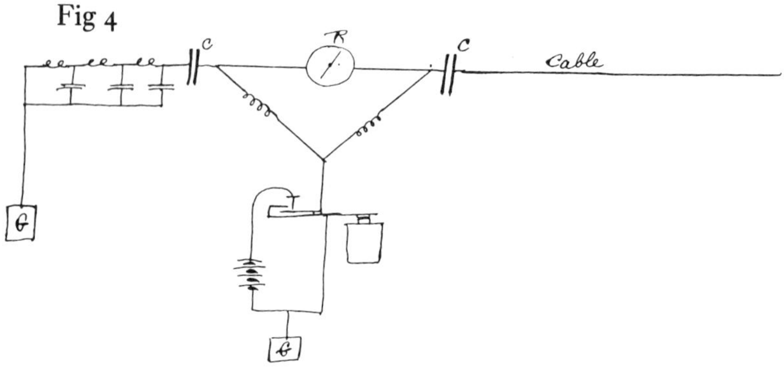

or perhaps it is shown more plainly in

Fig 5

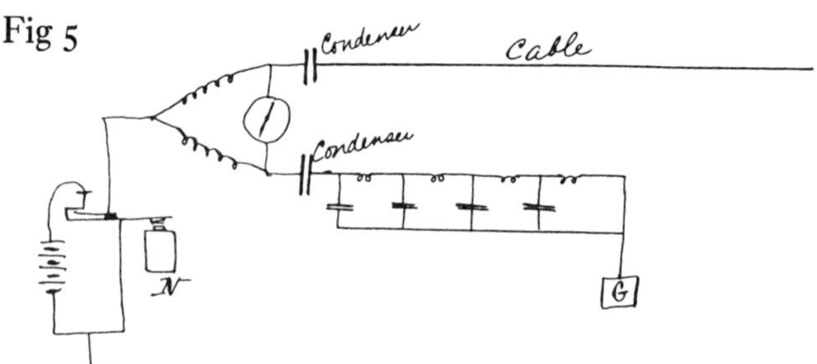

In this case the cable is worked entirely by induction the same as the present French cable.[7] Double current[8] may be used by substituting a reversing apparatus of any kind for the single-sender N. Fig 5.—

Fig 6 shews the reverse I prefer to use[9]

Fig 6

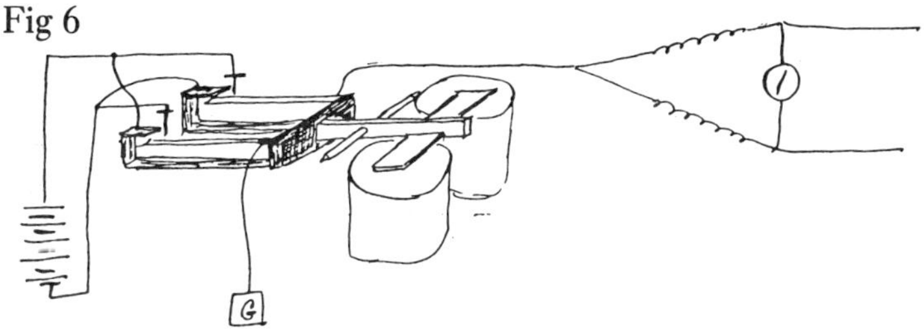

On closing the circuit[10] the two contact springs a a′ come in contact with contact point z & c of the battery sending a current in one direction[11] On opening they come in contact with the point 2 2′ putting the battery on in opposite direction.

The condenser may be dispensed with & a much more effective & economical compensation used instead to wit:—an electromagnet the inductive charge & discharge of which may be made equal to the longest cable.

Fig 7 shows the device. =

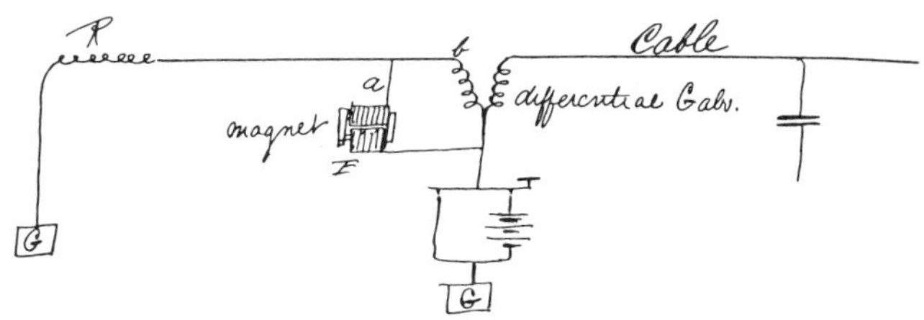

The electromagnet has its armature permanently attached to the core so as to generate as great a discharge as possible.[12] On closing the Key the battery passes through both coils of differential galvanometer, the coils being wound in opposite directions or the current passed through in opposite in opposite directions. If the resistance of cable is equal to artificial line & no static induction were on cable & the magnet E was disconnected, no effect would be seen on needle; but if magnet E be connected the current from battery passing through it would generate an induction charge which would circulate within the closed circuit formed by a & b, & violently deflect the needle for an instant in one direction: if now the cable has an inductive capacity & the magnet was disconnected & battery put in, the static charge would throw the needle violently in the opposite direction, but if the magnet be connected both charges neutralize each others effect on the needle & it remains unaffected

The length & Resistance of magnet can be so proportioned as to generate a charge equal in strength & length to the static charge on the longest cable. The strength of the discharge from the magnet may be regulated by placing it at any point on the resistance scale of the artificial line as in Fig 8

Fig 8.

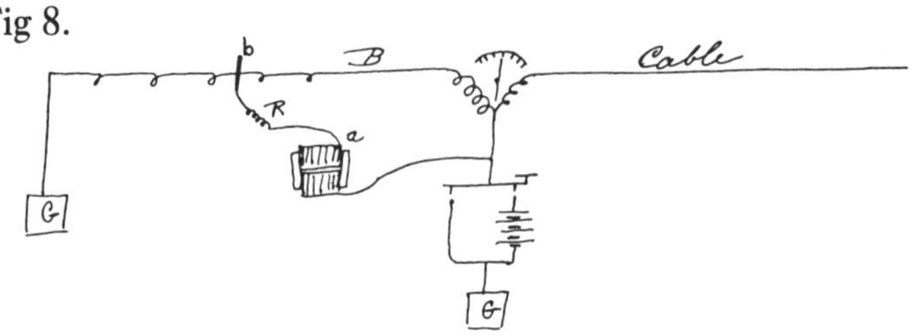

The slide b may be placed at any point in the resistance box, thus increasing the resistance of the discharging circuit aB. The charge & discharge may be also regulated by the adjustable resistance R in the wire & containing the electromagnet, it may also be regulated by making the armature of electromagnet adjustable & receeding from the core to decrease the strength of the induction current, & approaching to strengthen it. By using the electromagnet a much less resistance is necessary in the artificial line.

The electromagnet may be placed in cable & almost the same effect produced thus in figure .9

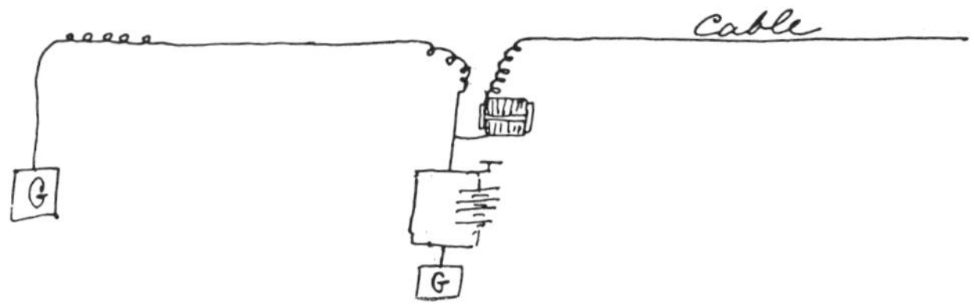

In thus case the induction charge & discharge of the magnet compensates or neutralizes the effect of the static charge directly in the cable itself. When the magnet is to be arranged in a bridge it should be arranged as shown in Fig 10

Fig 10

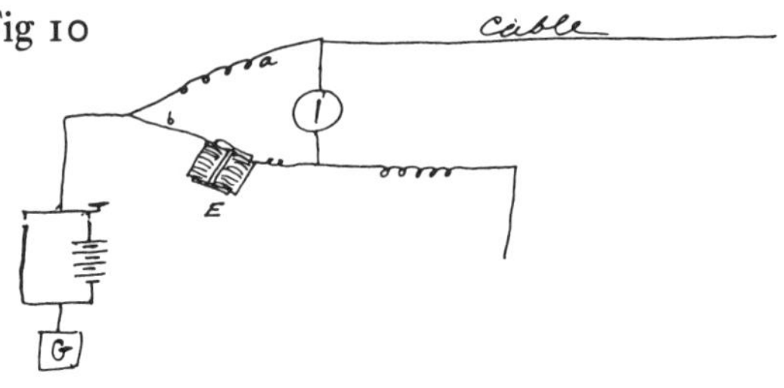

The charge & discharge of electromagnet E circulating in wire a.b. balance the effect of static charge in cable

When the cable is to be worked by induction (ie) with a condenser as is at present used on the French atlantic cable the magnet may be arranged thus:—

Fig 11

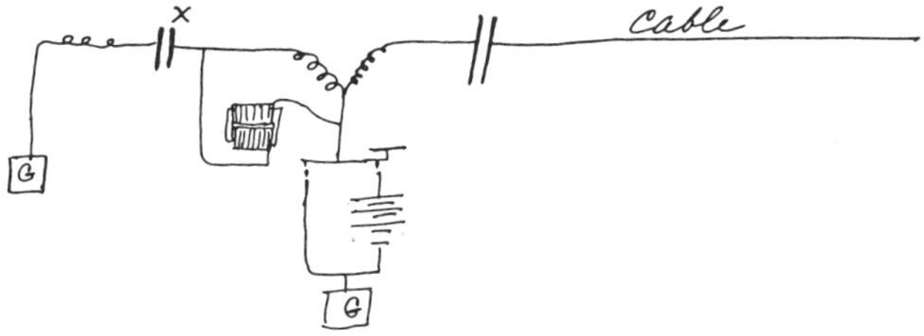

The momentary charge due to the condenser X on the artificial line allows the magnet to send a charge & discharge which acting on the differential galvanometer neutralizes the effects of the extra charge & discharge on the cable:—

Both the magnet & the condenser may be replaced by a secondary battery as in fig 12

Fig 12

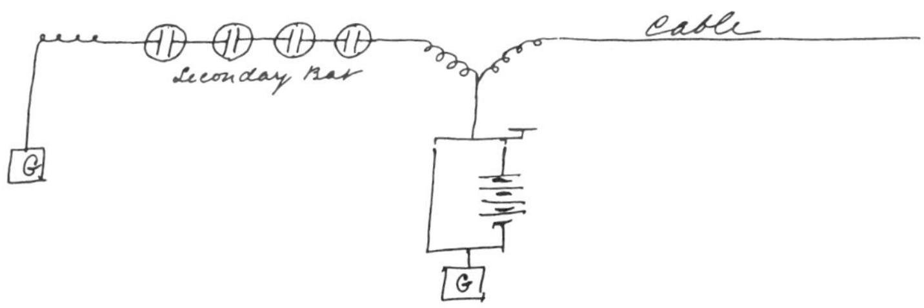

On the closing the circuit (ie) putting on the battery the current on the artificial line is strongest at the first instant of closing which is due to the fact that the secondary battery formed of plates of metal of the same kind immersed in water or chemical solution sets up no opposing electromotive force at first, but very quickly sets up an opposing electromotive force to the battery current When the battery is on permanently the resistance coils of the artificial line should be adjusted so that with the electromotive force the current should be equal to that in the cable. Now on disconnecting battery the secondary batterys send a charge in the same direction as the static discharge on the cable & both acting in the double coils which are wound in the opposite direction neutralize each others effect, & the needle remains unaffected by the outgoing current.—

On land wires considerable trouble is found from the discharge of relay magnets inserted by mistake in the wires working duplex at way stations. The discharge from these relays destroys the balance at both ends of the wire. I remedy this inserting at each end a relay or electromagnet nearly of the same character as the relay which is supposed to be in the line & in that portion of the artificial resistance as will allow of its charge & discharge to balance that inserted in the line

Fig 13 shows this

Fig 13—

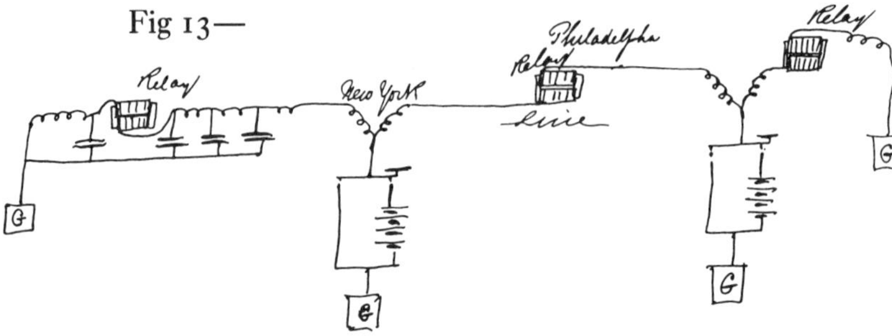

In the instrument at present used in this country but half of the wire upon the magnet is available to give effect to the current from the distant station, the other half being used for the secondary or artificial line & of course as far as the distant current is concerned is of no use. I obviate this to a great extent by combining the double coil or differential principle with the bridge principle as in Fig 14.[13]

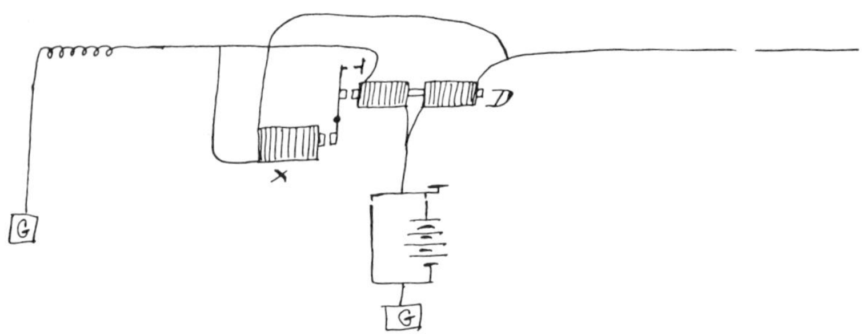

D is a double wound portion of the receiving relay the wire being wound side by side or on two separate bobbins both of the same resistance. If the magnet X is disconnected the outgoing current may be balanced in the differential coils D. If now the magnet X be added It forms a bridge wire between the two circuits & no current passes through but the current from the distant station passes through one coil of D & through the magnet X thus giving double effect to that current without any disadvantage whatever. It also tends to neutralize the self induction of one of the spools of D on the other when the distant current is received

On circuits of ordinary lengths where the static induction is very light the magnet condenser & secondary battery may be dispensed with, & an induction extra magnet used as in Fig 15[14]

Fig 15

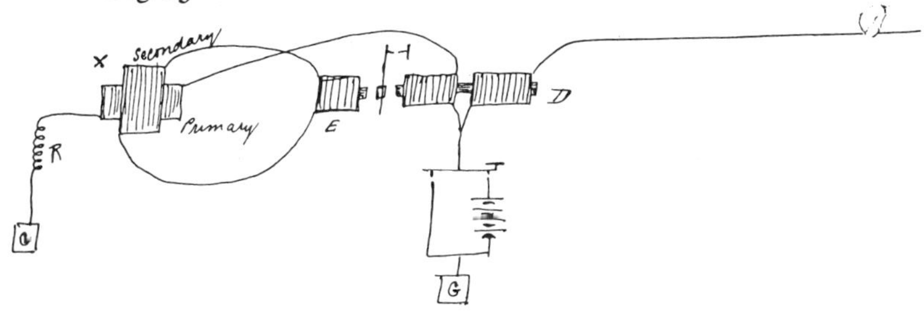

In the equaling circuit is an induction coil X whose primary is in the circuit & whose secondary is connected to an extra retractile magnet E of the duplex relay D. At the moment of putting on the battery the static charge will charge D, but the induction discharge from the induction coil X acting on the magnet E sets up a retractile power equal to the power due to the static charge in D & the effect is balanced.

Considerable trouble is also found with the present contact devises owing to the spark corroding the points & also of the springs getting out of adjustment & mutilating the signals Fig 16 shows a method in which no contact springs are used[15] a common sounder point or Morse Key may be used

Fig 16.

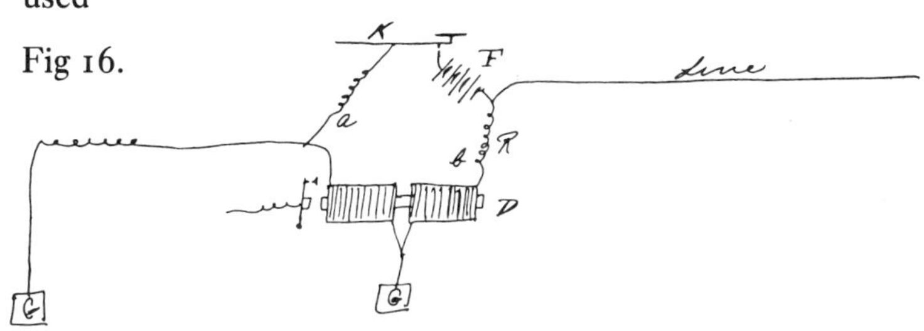

K is the sending Key, F the main battery If the resistance of the two branches containing the spools of D which in this particular instance are wound in the same direction so that the current passes through them in the opposite direction are equal no effect is produced on the instrument by closing the Key K & putting in battery; the current from the battery circulates in 3 closed circuits, first being in circuit a.b., second the artificial line & third on regular line. The battery to be used being a quantity one is capable of supplying several circuits.

The resistance of a.b. is about (1/5) one fifth the resistance of the line & may be even less. To prevent the effect of the static charge a magnet, condenser or secondary battery may be inserted as shown in former figures. It is sometimes required that a way station in a Duplex wire shall receive the signals from one of the ends I attain this object by inserting a short relay[16] at some intermediate station & double the battery at the station from which he is to receive (i.e) I make the strength of the battery twice that of the other end, hence by adjusting for the heavy battery, he will receive the signals due to it but not of the weaker battery. It may be also done by using a differential polarized relay at the way stations & at one of the end stations & using an ordinary relay at the other station. In this case reversed currents are sent from one station with a battery twice the strength of the other battery which sends currents on the line always in the same direction. The heavy battery will actuate all the polarized relays &at the way stations & at one of the end stations, while it will not interfere with the reception of a signal on the ordinary relay at the other station: probably some difficulty will be experienced at the end station having the polarized relay in the reversal of the equative currents figure 17 shows this arrangement.[17]

Fig 187.

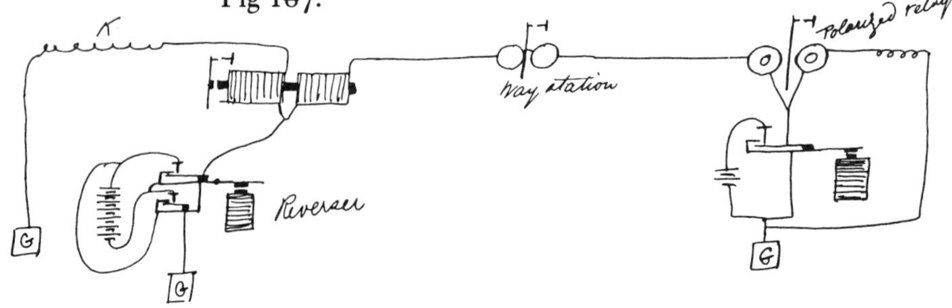

A very important adjunct to Duplex is arranging it at terminal stations with several branch offices so that they can all use the circuit Fig 18 shows a method of obtaining this result without excessive complication[18]

The instruments in the main office are arranged in the usual manner A is the signalling sounder which is placed in a city circuit running to the branch office at each branch office there is a sounder & key for breaking circuit. These Keys, K, K′, K″, have circuit closers

Fig 18

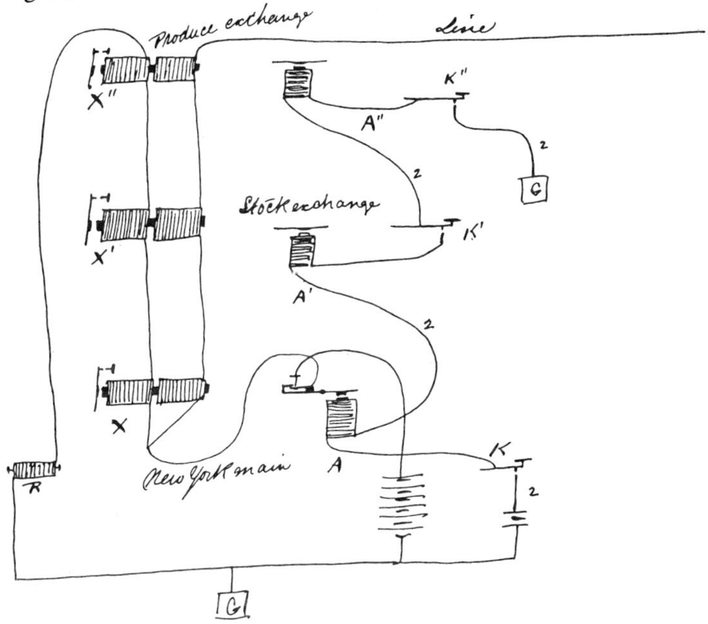

At each branch is a duplex relay through one spool ~~of which~~ or wire of each runs ~~a~~the main line & through the other spools the artificial circuit thence back to main office through the equalling resistance coil R to ground. If all the relays are of equal resistance New York main office operator can equate all the relays of the branches, any branch may signal the distant station by operating the local circuit 2 which actuating the signalling sounder A at the main office connects & disconnects the main battery from the line. The compensators may be attached in th~~u~~e ordinary manner.

It is found in practice that the present duplex apparatus does not double the number of words which may be sent on a wire, for when one operator misses a word he is compelled to sen word back through the other operators the place where he desires a repitition. This of course stops for an instant all four men which would not be the case were two separate wires used, for then when a repitition was required but two operators would be stopped, sometimes when equation is ~~p~~difficult or defective the delay from this cause is considerable Fig 19 shows a plan by which one operator may interrupt his partner without stopping the other 2 operators:—[19]

Fig 19

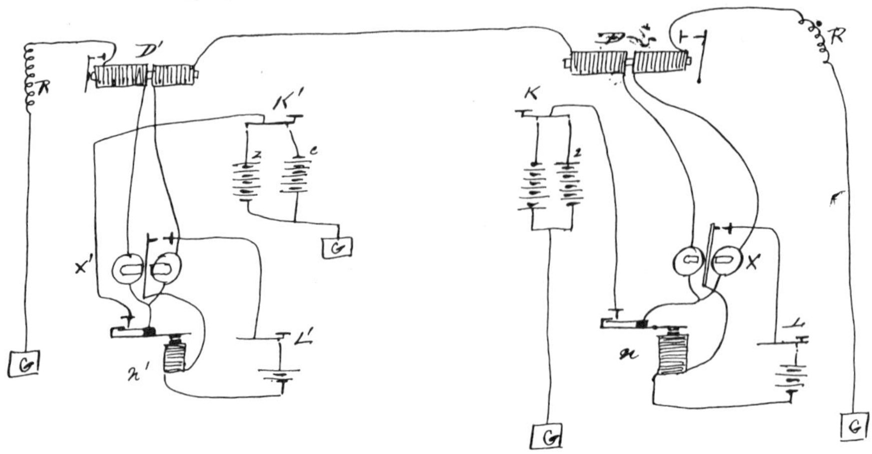

The connections in this method are precisely similar to those ordinarily used with the exception of inserting double coil differential polarized X, X′, at each end of the circuit both of main & equating & an extra reverse battery which may be brought into action by depressing the Keys K, K′,. We will now suppose that D′ is receiving from n, operated by Key L & wishes to interupt he depresses the Key K′ & holds it down: this changes the signalling battery from zinc to copper & the arm of polarized relay X flies over to the other side & opens the circuit in which n is placed. This opening notifies the man at n & L that he must go back one or two words.

This reversal of battery does not affect D which receives its signals from n′L′ as the relay D works as well on a positive battery as a negative The connections being both alake at each station of course D scan stop n′L′ by depressing the key K, ~~& D′~~ & D′ can stop n,L, by depressing the Key K′, no signalling can take place between K′ & LN as a series of change of currents would interupt D, but one single change will not give a perceptible interference as there is a slight break in the continuity of the circuit in the Keys K, K′.

Two signalling sounders might be used as in figure 20.

Fig 20

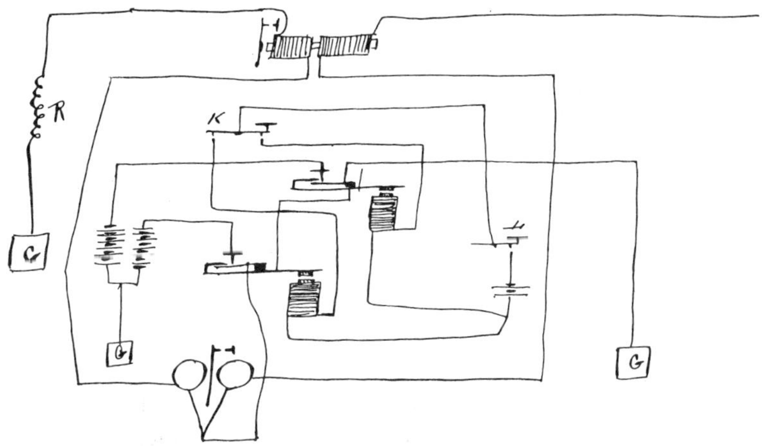

By closing the Key K L would operate either one or the other of the 2 signalling sounders thus reversing the signalling current & interrupt the distant sender

My claims will probably be

1st Subdividing the condensers & resistance coils composing the artificial equating circuit for the purpose set forth.

2nd The arrangement of a subdivided artificial line with a cable or land line worked by induction from condenser—as set forth

3rd The manner of sending reversals in a Duplex apparatus page 11, Fig 6

4th The combination of an inductive electromagnet with a Duplex as in Fig 7 & 8

5 The combination of an electromagnet with with a Duplex apparatus as in Fig 9.

6 The magnet E in the bridge of the Duplex for purpose set forth fig 10

7 The secondary battery arranged as shown in fig 12 for the purposes set forth.

8 Compensating for relays placed in the Duplex circuit at way stations, by the insertion of a similar relay in the artificial circuits Fig 13

9 The extra magnet X arranged with the Duplex relay D for the purpose set forth fix 14.

10 Compensating for the static charge as shown in fig 15

11 The arrangement for connecting & disconnecting etc shown in fig 16.

12 The manner of signalling way stations by an increase of one of the signalling batteries over the other fig 17.

13. Signalling way stations~~from~~ by reversals & receiving from the distant station by a single current.

14 The method shown in fig 18 for working duplex a number of branch offices

15 The manner of "breaking" or correcting errors shown in fig 19.

Copied from Edison's original manuscript, by Chas. Batchelor

Df (copy), NjWOE, PS, Cat. 30,104:7 (*TAEM* 8:855).

1. This draft is almost identical to the final caveat, which Edison executed on 7 August 1874. Edison Caveat 48, RG-241, DNA; a printed transcription is in Testimony and Exhibits on Behalf of T. A. Edison, pp. 163–74h, *Nicholson v. Edison*.

2. This refers to Western Union's new standard duplex methods, based on the approach patented by Joseph Stearns in 1872. See *TAEB* 1:101 n. 3, 521 n. 3.

3. Edison often compared the strength and duration of electrical impulses by the marks they would make if passed through the chemical recorder of his automatic telegraph system.

4. Cf. Cat. 297:131(3); and NS-Undated-005; both Lab. (*TAEM* 5:930; 8:351).

5. It is apparent from the filed caveat that **G** and **G′** here refer to the coils marked "3000" in figure 3.

6. On this use of condensers see Bright 1974, 538–52.

7. Probably the French Atlantic cable, laid in 1869 between France and Nova Scotia. Ibid., 107.

8. That is, currents of opposite polarity or direction.

9. Cf. Cat. 297:87(4), Lab. (*TAEM* 5:795). Finding an optimal reversing transmitter arrangement for circuit designs stemming from Caveat 45 (Doc. 348) concerned Edison for more than a year.

10. That is, the local circuit through the electromagnet at the center.

11. Batchelor left off the labels for this diagram; see the accompanying illustration from the filed caveat. What Batchelor copied as **2** and **2′**

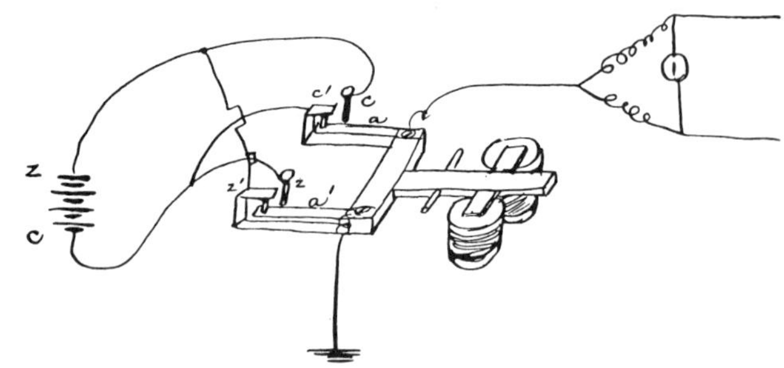

Drawing of a current-reverser from Edison Caveat 48.

in the next sentence should be **z′** and **c′**. For earlier work on this device see Cat. 297:79(2), 87(4), Lab. (*TAEM* 5:746, 795).

12. Cf. Doc. 386.

13. Cf. Cat. 297:137(4), 139(4); and NS-Undated-005; both Lab. (*TAEM* 5:938, 946; 8:352).

14. Cf. Doc. 298; and Cat. 297:74(2), 139(4), 146(3, 4), 152(3), Lab. (*TAEM* 5:720, 946, 968–69, 989).

15. Cf. Cat. 297:137(3), 140(6), Lab. (*TAEM* 5:937, 952).

16. That is, an ordinary neutral relay, but one with shorter coils and cores. The shorter an electromagnet's core, the faster it loses and regains its magnetic field during the reversal of a current. This is the earliest indication that Edison was modifying the coils of ordinary relays in his multiple telegraphy experiments, a modification he incorporated in his successful quadruplex design.

17. Cf. Cat. 297:136(2), Lab. (*TAEM* 5:932).

18. Cf. Cat. 297:72(3), 136(2); and NS-Undated-005; both Lab. (*TAEM* 5:710, 932; 8:355, 358, 362).

19. Cf. Cat. 297:139(1), Lab. (*TAEM* 5:943). See also Doc. 473.

-393-

Edward Johnson to George Harrington[1]

New York, January 28th, 1874.

Sir,—

I respectfully submit the following report of the work done in the demonstration made on Tuesday evening, January 27th, as per your instructions of prior date. The matter selected for the purpose was the President's late message and the Spanish protocol.[2]

STATEMENT:

Matter transmitted,	11,130 words.
Length of Circuit,	281 miles.
Conductors used,	1 wire.

Labor:

New York,	Morse operator,[a]	1	
	Copying operatives,	13	
Washington,	Morse operator,	1	Total, 25
	Perforating operatives,	10	operatives.

Time:

		P. M.	Mins.
Washington,	Perforating commenced,	5.39	
	Perforating completed,	6.24½	45½
New York,	Copying commenced,	5.42	
	Copying completed,	6.48	66
			Total time, 69 minutes.

Cost:

Morse operators,	$100 per month.
Automatic operatives,	40 " "

The characters were perfectly legible and well defined, and were copied with great facility.

The average time during which the perforating operatives were actually at work was forty-five and a half minutes making an average per operative, per minute, of twenty-five words.

The average time of copyists was fifty minutes, making an average per copyist, per minute, of seventeen words.

Unlike the Western Union Co., we had no large corps of operators from which to select our working force, but were compelled to utilize *all,* good, bad and indifferent, which makes it proper to call special attention to the above averages made.

The whole time consumed was sixty-nine minutes, as against the published record of seventy minutes by the Western Union in their late effort.

The average time occupied by Automatic was fifty-five and a half minutes.

The average time occupied by Western Union (as reported) was fifty-nine minutes.

An unfortunate defect in the paper[3] caused much delay in the transmission, otherwise still less time would have been consumed. No attempt, however, was made to attain a high speed of transmission on this occasion, as that point had already been yielded and incontestably proved in the presence of the Hon. Jno. A. Cresswell, Postmaster General, and numerous other gentlemen, including Senators and Representatives in Congress, on the evening of December 11th, 1873, when we transmitted some 12,000 words over our one wire from Washington to New York in twenty-two and a half minutes.

Our operatives were congregated at Washington and New York on Monday, January 26th, and were tested for the first time on the evening of that day. I call attention to this, in anticipation of the charge that the time which has elapsed since the publishing of the message has been improved by our operatives in practising upon it.

With the experience gained in this demonstration, I am confident that in another we could readily dispense with at least two perforators and three copyists, and yet perform a like amount of work. Respectfully,

E. H. JOHNSON, *General Manager.*

PL (transcript), NjWOE, DF (*TAEM* 13:52). Original on letterhead of Automatic Telegraph Co. [a]Several braces omitted.

1. This letter was included as part of a circular reporting on what company officials considered a successful test of the automatic. The circular included a cover letter from George Harrington to U.S. Postmaster General John Creswell describing the test and comparing the results to Western Union's claims for the Morse system time and cost. Also included were signed statements by witnesses to these tests. Harrington to Creswell, 8 Feb. 1874, DF (*TAEM* 13:49–53).

2. President Grant's message to Congress was delivered on 2 December 1873 and was published the following day. The Spanish Protocol was the protocol of a conference held at the State Department between Secretary of State Hamilton Fish and Spanish Minister Plenipotentiary José Polo de Barnaba on 28 November 1873 regarding the seizure of the steamer *Virginius* by a Spanish man-of-war. The text of the protocol was also published on 3 December. *New York Times*, 3 Dec. 1873, 1–2.

3. This probably refers to the chemically treated receiving paper.

–394–

Draft Caveat: Automatic Telegraphy[1]

[Newark,] Feb 1st 1874.

Roman letter Chemical Telegraph. (Caveat.)

The object of this invention is to transmit & receive & record Roman letters, the first of which is done by roman letters perforated in strips of paper of which a patent has already been applied,[2] & the letter recording on chemically prepared paper by electric decomposition the letter so perforated. Several methods have been already described in a previous caveat whereby this may be accomplished on two or more wires[3]

The present invention consists of mechanical devices as well as electrical for operating it upon one wire. The main feature of which is to send half of the roman letter over the wire from the perforated paper with positive & negative currents. One pen both of the transmitter & receiver being set one line of perforations (forming a letter) ahead of the other, & by a director, commutator, or circuit breaker, connected with the receiving instrument throw the line over on another set of receiving pens & record the balance of the letter. The circuit changer is controlled by a sensitive electromagnet provided with a lever which holds & releases an escape ratchet wheel at the beginning of every letter.

Each letter of the perforated paper is formed of small holes; at the commencement of each letter & right below it is a large perforation which has a separate contact roller by itself & when no other roller is through a small hole passes through this large hole & puts a strong current of three times the duration of the current transmitted from the perforation forming the letter. This wave actuates the magnet at the distant station for an instant when its lever is drawn away from the escape wheel allowing the circuit changer to revolve 1 tooth. This circuit changer in revolving one tooth first connects the line & earth wire, first to two recording pens (the last ones) when this has formed half of the letter the circuit changer which is still revolving disconnects the line & earth from back pens & throws it on the first 2 pens until the other half of the letter is formed; at this moment the circuit changer is arrested by the lever of the electromagnet coming in contact with one of the teeth of the escape wheel. It now lays still till another large hole before a letter passes under the extra roller pens at the transmitting station when another wave is sent & another letter is printed Thus by sending an extra wave before each to regulate the recording mechanism & change from one letter to another perfect synchronism is unnecessary & the whole time of the wire is utilized to make the letter or in other words:— Positive followed immediately by a negative forms half of the letter & immediately the wire is thrown over on a new set of receiving pens where the same action takes place, thus keeping the wire constantly at work

The peculiar manner in which the letter must be formed gives its own spacing without ceasing sending currents over the wire to give a space as is usual

Fig 1 shows the arrangement of the contact rollers at the sending station:—

Fig 1

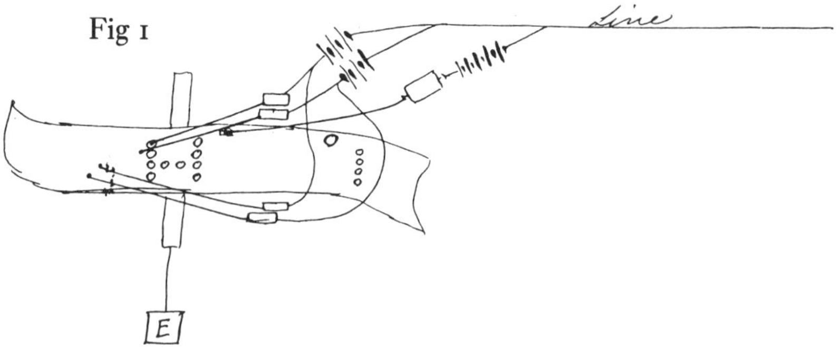

Fig 2

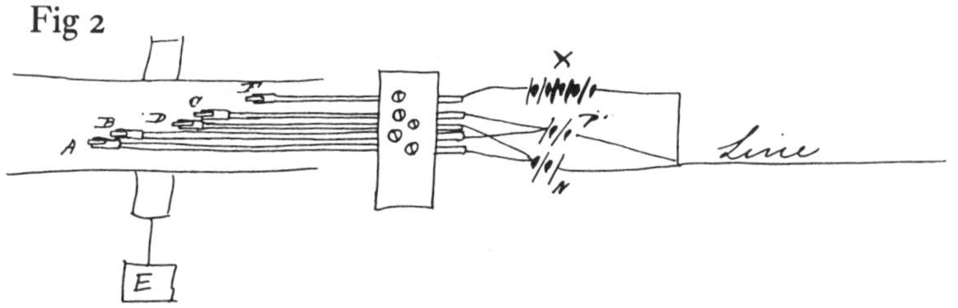

In forming a letter first F comes in contact with the drum through the large hole putting large battery X to line & earth & sending a strong current over the wire which actuating the electromagnet sets the circuit changer in motion (when the circuit changer is arrested the line is not) in connection with either set of pens) As the letter passes along the top passes under C sending a positive to line, then D sending a negative to line & so until the whole of the letter has passed under CD, the lower portion of the letter now comes in contact with .AB. which acts precisely as CD. After the letter has passed under both the circuit changer at the distant end is arrested & another large hole passes under F setting the changer at the distant end going & another letter is recorded

The perforated letters are separated from each other the distance of the room taken up by the largest letter of the alphabet & a little more the extra space being used for the large hole. I do not wish to confine myself to regulating the distant circuit changer at every letter as the large hole may be used only between words providing the transmitting & receiving nearly synchronously.

Neither do I wish to confine myself to changing the line wire from one set of pens to the other set electrically as the pens instead of having the line disconnected from them may be lifted off the paper (ie) when one set is doing work the other set is lifted off the paper & vice-versa. The lifting may be controlled by a magnet actuated by waves sent from the distant or transmitting station, neither do I wish to confine myself to using 2 sets of pens as one pen may be used which may be shifted sideways four or more times the top part of any letter being transmitted first then the pen shifted by the mechanism of the receiving machine & the next portion of the letter sent & recorded then the third & last the fourth when the receiving pen is replaced by the mechanism in its original position ready to make the next letter the time of starting to mark being ~~the~~ controlled by the magnet Fig 3 shows the recording instrument.

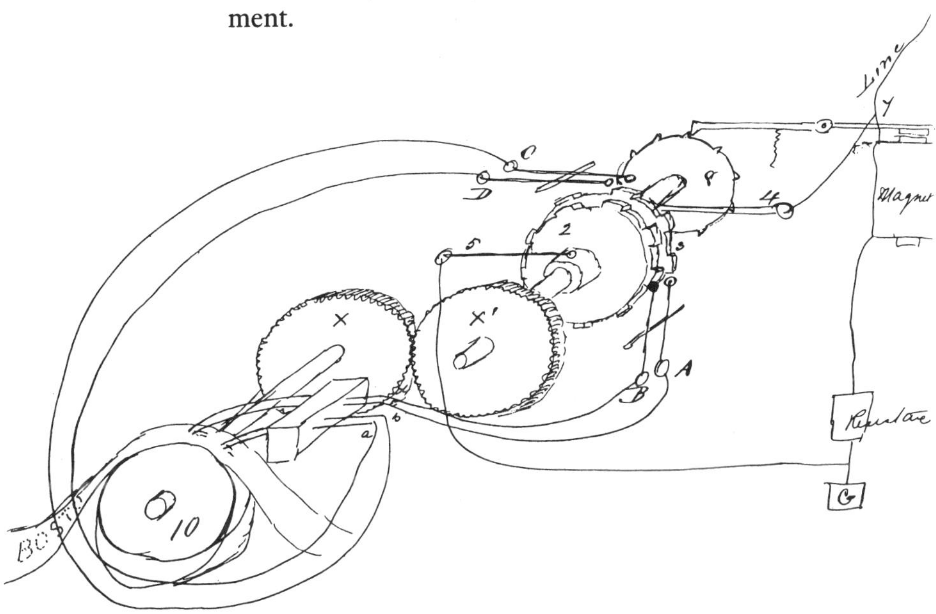

10 is the drum by which the chemically prepared paper is drawn forward, a,b,c,d, are the four recording pens; c,d, are set the length of one letter ahead of a,b, Wires lead from a b to the circuit changing contact rollers D.C. Wires also lead from c d to the contact roller arms A.B. These contact rollers are so positioned regarding the two toothed wheels 2,3, which are insulated from each other that while at rest neither set of rollers have contact with the teeth of either wheel, but while in motion C D first comes in~~motion~~ contact with the two teeth of wheels 2,3, (these wheels are connected one with the line the other with the earth through the points 4 & 5) after they leave the teeth B,A, pass on a tooth; At no time is C D & A B on a tooth at the same time. It will be noticed that the recording pens a.b. are first connected with the line & ground by D C & record half of the letter then the pens c d are connected to the line & earth by points A B recording the other half, when this is done a new tooth on the escape wheel 8 will come in contact with the lever 6 of the magnet & the wheels 2,3, which are carried around by friction only will be arrested while the drum 10 will keep on. At the moment the circuit changing wheels are stopped the magnet receives a new impulse from the large hole in the perforated paper it releases the wheel 2 & 3 & another letter is printed.

1[a] I shall probably claim the arrangement of the transmitting pens as set forth & for purposes specified

2nd The use of an extra perforation in Roman letter perforated strips for regulating the distant recording machinery either after each letter word or number of words

3rd The circuit changer or its equivalent for the purposes set forth

Signed.

Copied from Edison's original manuscript by Chas. Batchelor

Df (copy), NjWOE, PS, Cat. 30,104:41 (*TAEM* 8:871). [a]Written in margin.

1. Edison appears never to have filed this caveat. On 3 February, Charles Batchelor wrote in a notebook, "Having built a machine on the principle set forth by Edison for this extraordinary feat. We proceeded this night to test its qualities." He then drew a picture of the recording mechanism shown in this caveat (fig. 3) and later noted, "E[dison] & B[atchelor] 15 hr". Cf. also the remaining fragment of Edison's draft for this caveat (NS 74-001, Lab. [*TAEM* 7:7–12]) and a technical note of this date (Cat. 30,094, Lab. [*TAEM* 5:356]).

2. U.S. Pat. 151,209.

3. Doc. 374.

–395–

Notebook Entry: Automatic Telegraphy

[Newark,] Feby 5 1874

Manganese & magnesium decompose water at the boiling point, but will not do it cold ie[a] the affinity of .O[xygen]. for H[ydrogen]. is greater when cold than O for Manganese or magnesium, but the affinity of O for H is lessened by heat & O for Mn & Ma correspdly increased.

Other metals will decompose water at red heat Lowest ~~a~~is Zinc then Iron Chr[omium] Co Na Sn Sp[1] Al Pb Cn[2] last Noble which show no tendency to oxidze in air[3] are incapable of removing .O from H & O hence magnesium & Manganese pens (Mem use hot rolls of reg ferrid—)[4]

A fine state of division favors chemical Action hence Calaud round Box Zinc or paper or Gutta box filled with a Zinc & packed granulated Zinc & Copper box filled with Copper wire scraps = ~~g~~Get a porous pot about ½ inch thick to fit 1½ carbon battery Zinc & 1 Carbon to try length time twill last, & to prevent the chemical diffusion of Bichromate from the Carbon pole where it is needed most to give up an atom of O to the .H_2 set free. if Phosphorus exposed in a lump to air dont oxidize so quick but disolve it in a little Bisulphide of Carbon pour on a paper instantly all oxidizes being in a state of fine division hence I dont believe our present form decomposing pen Iron wi~~ll~~th ferrid utilizes all the .O. produced, owing to it being in one mass. hence make a pen 1/64 diameter composed of a large number of no 40 hard drawn iron wires.

X, NjWOE, Lab., Cat. 1175:154 (*TAEM* 6:182). [a]Circled.

1. Possibly antimony (Sb).
2. Possibly cadmium (Cd).
3. That is, the noble metals: gold, silver, platinum, palladium, iridium, rhodium, ruthenium, osmium.
4. Paper treated with Edison's standard iron solution.

–396–

Notebook Entry: Automatic Telegraphy

[Newark,] Feb. 9th 1873[1874].

⟨1874? See page 130⟩[1a] Bully <u>Experiment.</u>[b] by T A Edison assisted by Chas. Batchelor. ⟨12hr⟩[c] Experiment tried on repeating at high rates of speed, by T A Edison Newark NJ

Not being able to [--]repeat more than fifty (50) words per minute through an ordinary relay we constructed one after the following manner:

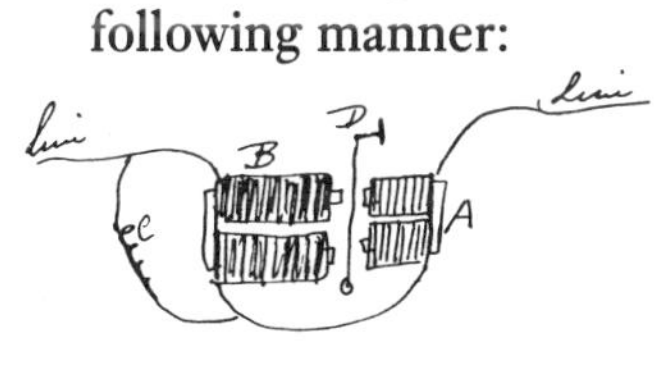

A is (12) twelve ohm magnet
B is a (200) two hundred magnet
C is shunt round B.
D armature lever & Repeat points.

We do away with spring.

When you close key the magnet A charges instantaneously but countercurrent from B prevents it from charging instantaneously & when you open the induction current from B works around the shunt C & prevents the bad effects of B going on line.

The resistance in shunt C should be so regulated that the armature will stay on either side or rather that when the line is kept closed the magnetism is equal on both sides

With this relay we were enabled to repeat & record on chemically prepared paper at the rate of 250 words per minute perfectly;[2] ⟨The sentence we took was "Now is the winter of our discontent made glorious summer by this son of York & all the clouds that lowered upon" with this we got 250 words per minute but counting 5 letters to a word we got 228.⟩[c] if the armature & repeating lever had been lighter we could I have no doubt have [-]doubled the above amount

The connections of app:—

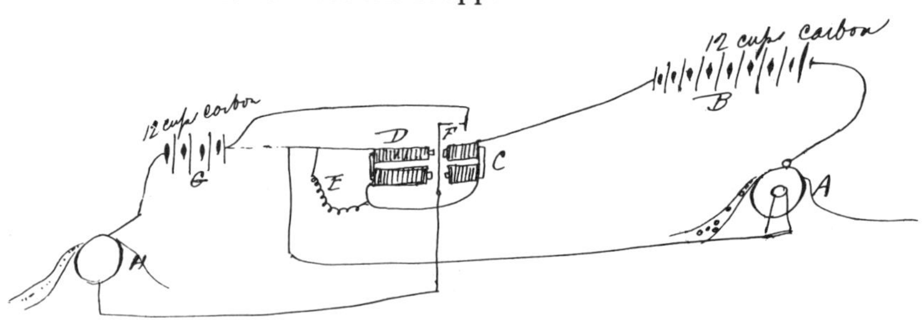

In our experiment the shunt E varied from 10–80 ohms

We found that our rollers at A oxidized very quickly so much so that at that speed they would have to be cleaned every few thousand words, which is a defect which we must remedy

X, NjWOE, Lab., Cat. 1175:163 (*TAEM* 6:186). Written by Batchelor. [a]Marginalia written in an unknown hand. [b]"Bully Experiment" written very large and dark for emphasis. [c]Marginalia written by Batchelor.

1. Edison wrote his original conception for the experiment on p. 130 of this notebook. Cat. 1175:130, Lab. (*TAEM* 6:170).

2. Four pieces of the recorded tape from this experiment and another drawing of the circuit arrangement are found in Cat. 1143:48, Lab. (*TAEM* 27:196).

–397– [Newark,] Feb. 10th 1874.

Notebook Entry: Automatic Telegraphy[1]

2 wire Roman Letters

Experiment by T A Edison ass by Chas Batchelor

The work to be done was to produce Roman letters using only 2 wires.

The principle of this experiment is to make & break connection 8 or 10 times in the length of a dot on each line & never let them both make connection at the same time For the purpose we constructed a machine like this: Fig 1

Fig 1[a]

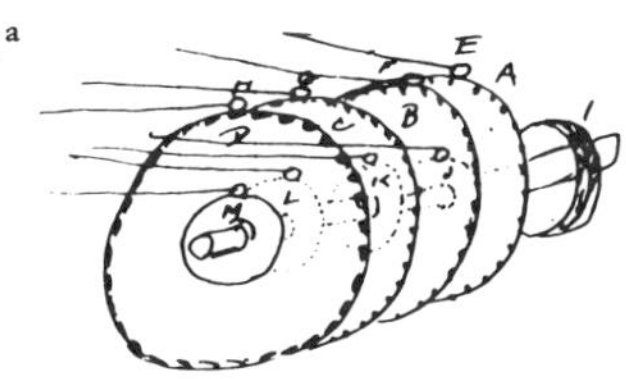

A B C D are four breakwheels with 60 teeth each

E F G H are Breakwheel contact rollers

I is Driving pulley

J K L M are the hubs of breakwheels on which also run contact rollers

This was run from the large grooved wheel of one of our automatic tables.

We used of course reversed currents which only gives us four pens. Our perforations were made from a perforator that was made for 5 wires & consequently is made 5 dots deep thus

Now as we can only use four pens & having no other perforator we were obliged to use these perforations & cut one line of dots which we did ~~in~~thus:

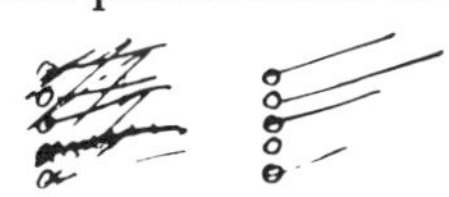

cutting out the fourth line this made our letters come thus:—

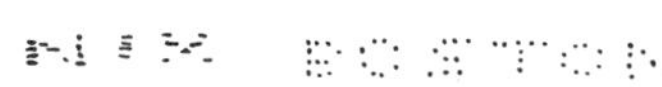

The breakwheels in fig 1 have got to be set so that the contact rollers E F must not make connection at the same time as the rollers G, H. The connections for the whole experimt are shown in Fig 2.

Fig 2

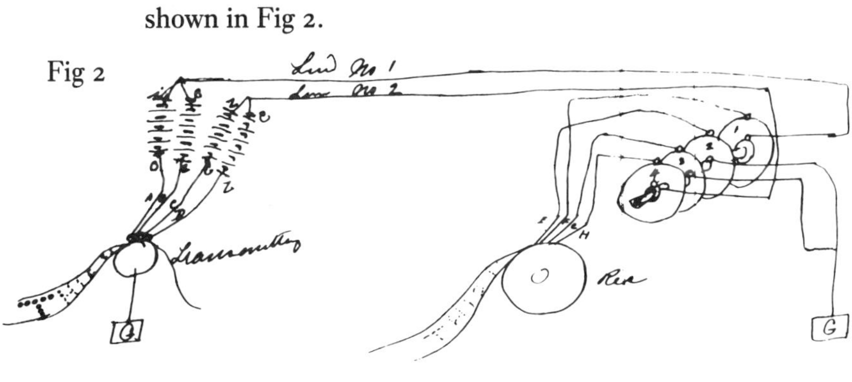

As will be seen by Fig 2 if I close on my pen B at transmitting end it sends positive current to line No 1 which enters breakwheel No 1 at the hub passing through contact roller to pen F where it makes its mark & passes through E & Breakwheel 2 to ground.[b] Now if I close on A at transmitting it sends a negative current to line & positive to ground which caomes up through Breakwheel 2 & marks on E pen etc The same thing is done on Line No 2.

The pens at transmsitting end are set one ahead of the other on both lines thus 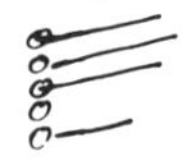

so that one pen is out of a hole when the other is in on each line. The pens at receiving end have to be set the same. If we run at a high speed say 500 to 600 words we do not get it distinct dots are dropped & &others put in owing I think to the 2 rapid reversal of current.

But up to 200 words or under it is all that we could wish the letters all coming out perfect, samples of which can be seen in Scrap book No[2]

X, NjWOE, Lab., Cat. 1175:168 (*TAEM* 6:189). Written by Batchelor. [a]The drawings of two perforated letters "E" appear in left margin. [b]"& passes . . . ground" interlined above.

1. This entry is continued in Doc. 398.

2. Cat. 299:38, Lab. (*TAEM* 6:103), contains undated four-pen specimens "of a Roman letter Telegram Transmitted over two circuits Independent of all synchronism or machinery" on which appear the phrase "Now is the winter of our discontent."

-398-

Notebook Entry: Automatic Telegraphy[1]

[Newark,] Feb 11, 1874

Two Wire Roman letter Chemical Telegraph Experiment by T A Edison ass. by Chas Batchelor[a]

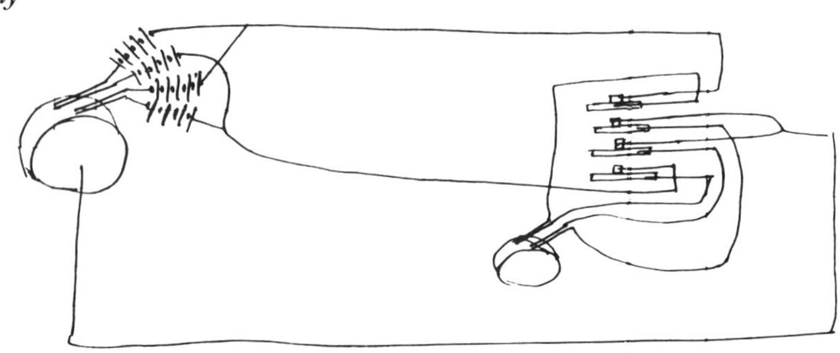

In our last experiment with this subject we noticed that we could gain higher rates of speed if we could do away with such quick reversals of current

The present experiment is for that purpose In this experiment we use reversals but they are of longer duration.

The letter[b] perforations must be made a little wider apart than the length of widest letter.

Then the 3rd & 4th pen must be s~~t~~et the width of the largest letter behind thc others 1s & 2nd thus:—

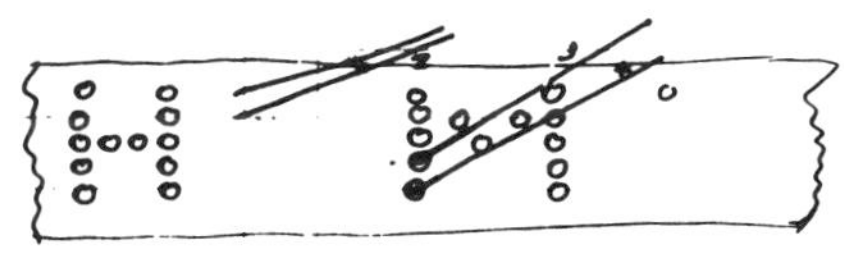

As will [-]be seen the principle of the thing is to send a positive current from 3 & 4 to line No 1 & 2 (Page 169)[2] & record it on the 3 & 4 pens for receiving & after they have got across the letter the pens ~~1 & 2 does the~~ 1 & 2 record their part of letter in same manner.

We can get better results at a higher rate of speed with this, but it necessitates a synchronism, for if the transmitting is sent faster than the receiving the writing will be like this the top half ahead of the bottom 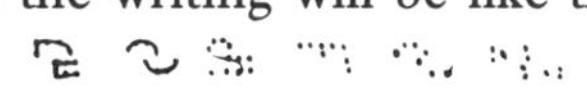
& if transmitting is too slow it will be other way bottom lines ahead of top Very little practice however will enable a person to keep the letters allways straight at the receiving end

X, NjWOE, Lab., Cat. 1175:170 (*TAEM* 6:190). Written by Batchelor. [a]Drawing appears on facing page. [b]Interlined above.

1. This entry is a continuation of Doc. 397.
2. See "Fig 2" in Doc. 397.

–399–

Technical Note: Batteries and Automatic Telegraphy

[Newark,] 17 feby [1874][1]

Batch make a secondary battery of one of the large Blue vitrol battery jars & sheet copper Thus

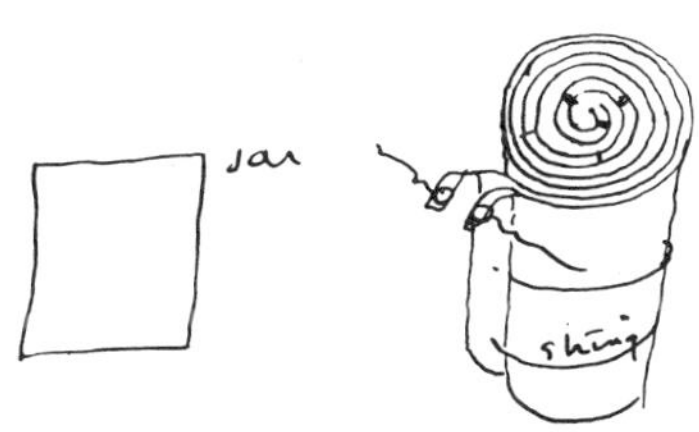

These parallel sheets of copper are nearly as high as the jar about 1½ inch from edge They are coiled together about ¼

inch apart and may be seperated with sticks of wood every 3 inches and then securely bound together with stout marline so that the two plates Cannot possibly touch even with rough handling. I wish to ascertain the amount of charge & discharge this will show ~~as~~in a Bridge as well as the length of discharge. if satisfactory J. C. Reiff authorizes use to make $50 worth which will be a dozen probably to be sent along[a] the line. I expect if these batteries are placed on the line 30 miles apart that they will generate (~~a~~from the addition of the transmitting battery) as much ~~ch~~counter charge as that due to the line that being the case the static charge of the line will be exactly compensated for and any speed on any length of[b] circuit may be obtained—Book it =[2]

Make a Coil of that Gutta percha wire (small) with perfect[-] insulated joints & Coil it on a magnet ie[c] two Large Relay cores so that it will take up the same bulk as if Bradley spools were used want to insert it in the liquid of a battery the current of which passes through the gutta percha wire. also wind the outside of the Bichromate battery jar with 5 or 6 layers of ordinary Relay wire—wish to see if when I close the circuit of the battery there will be induction on the wire—

Charge an insulated secondary battery box with the 8 bottles by connecting the prime conductor on one end of the plate glass machine[3] to one end Battery & other prime conductor to other end & turn for 5 minutes, then with with insulated key throw battery on, Bradley gal[vanometer] ~~g~~also insulated =

AX, NjWOE, Lab., Cat. 1168:32, 36 (*TAEM* 6:337, 342). Continuity of text and resemblances in style, paper, and ink link these separate scrapbook items. [a]Starts separated page. [b]"length of" interlined above. [c]Circled.

1. Batchelor dated this note "17 February 1874" and recorded its substance in Cat. 1175:172, Lab. (*TAEM* 6:191).

2. This is Edison's instruction that this material be entered into a laboratory notebook. On the first page he later noted the date (17 February) and "~~B Batty~~ Book"; on the second, the date and "Book".

3. That is, a static-electricity generator.

-400-

Notebook Entry: Batteries[1]

[Newark, c. February 19, 1874[2]]

Try this

Comparing the electromotive force of 2 elements:[3]

2 elements are connected in same ckt with a Tan[gent].

Galv[anometer]., 1st so that their currents go in same direction & 2nd in contrary directons

Let

Electromotive force be = E E′

The R of elements R R′

Interposed resistance r

Intensities of sum & indif. $I_s\, I_d$

then[4] $I_s = \dfrac{E + E'}{R + R' + r} \quad I_d = \dfrac{E + - E'}{R + R' + r}$

whence[5] $E' - E \cdot \dfrac{I_s - I_d}{I_s - I_d}$

Try[a] small condensers of alternate sheets of:— Copper & tinfoil Tinfoil & Tinfoil

To measure a large ~~f~~EMF with a small one:— use a shunt.

~~Battery E = units.~~ Find resistance of Bat. 1 & 2. Find R of Galvanometer take deflection through about 50 ohms[b]

Insert Battery 2 & change shunt till needle deflects to same degrees. then the equal intensities in galv. are[6]

$$f(\phi) = \frac{E}{(R + r)(1 + \frac{g}{s}) + g} \quad \& \quad f(\phi) = \frac{E}{(R + r')(1 + \frac{g}{s'}) + g}$$

from which $E' = E \dfrac{(R + r)(1 + \frac{g}{s}) + g}{(R + r')(1 + \frac{g}{s'}) + g}$

large battery E′ compared with small one E. shunt is only used in case of the large one & s = 8 may be inserted,[7] in above, by which $E' = \dfrac{R + r + g}{(R + r)(1 + \frac{g}{s'}) + g)}$

X, NjWOE, Lab., Cat. 1175:174 (*TAEM* 6:192). Written by Charles Batchelor. [a]Preceded by centered horizontal line. [b]Followed by centered horizontal line.

1. Although no original has been found, many entries in this notebook were copied by Batchelor from Edison's notes.

2. The preceding and following entries are respectively dated 17 and 21 February 1874.

3. The following discussion is based on Sabine 1867, 317–19. Other notes taken from Sabine on 15 February can be found in NS-74-001, Lab. (*TAEM* 7:32–34; Supp. III).

4. The plus sign in the numerator of the second equation is canceled.

5. The first minus sign in the following formula should be an equal sign.

6. "E" in the second formula below should be "E′".

7. "s = ∞" in Sabine.

–401–

Technical Note: Perpetual Motion[1]

[Newark,] Feby 21. 1874.

Experiment No. 2.

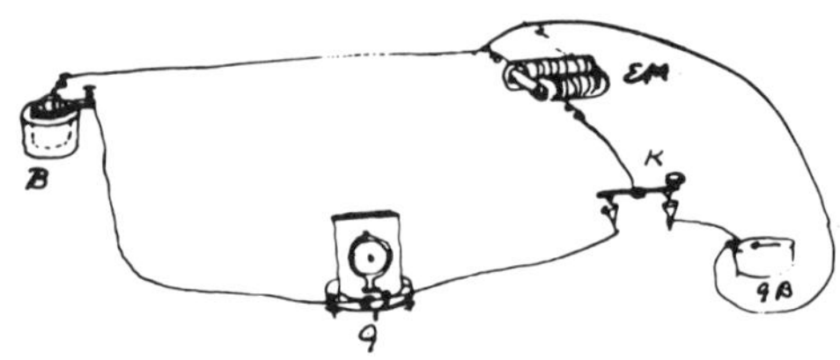

B is a single Daniel cell, G a mirror Galvanometer ~~wit~~ K a key with double contact points EM is the long Electromagnet, GB a Bradley Galv on No 1 Coil. press key down & wait for 5 or 10 seconds then open & close as quick as possible, (very quick) If the Mirror Dont move perceptably, and GB Does then Smash Goes Modern theories.[2] The magnet must be quite long and lots of fine wire on it otherwise you will not be able to work the key quick enough to prevent the Mirror working. The armature on EM should be permanently attached, so as to increase the inductive Effect—

AX, NjWOE, Lab., Cat. 1168:31 (*TAEM* 6:336).

1. Charles Batchelor copied this note into a laboratory notebook under the heading "Perpetual motion or bust the present theories." Cat. 1175:180, Lab. (*TAEM* 6:195).

2. This refers to the first and second laws of thermodynamics (see Ord-Hume 1977, esp. chaps. 2, 7, 9). Cf. Edison's interest in Henry Paine's motor (Doc. 135B; Cat. 1168:27, Lab. [*TAEM* 6:330]).

–402–

Technical Note: Batteries

[Newark, February 21 1874?[1]]

~~N~~Experiment No. 1.

To determine whether any electricity pass from one pole of a battery of 100 elements to the other when seperated—a slight distance.

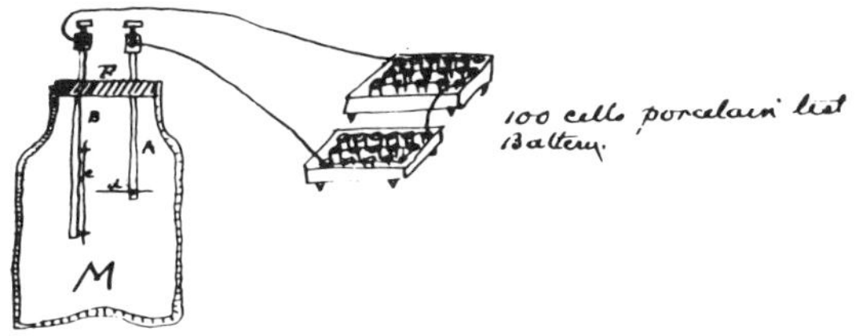

M is a clean dry candy Jar F is a tight fitting ~~w~~Vulcanite or Hard rubber top well secured in the jar with plenty of bees-wax so that not the slightest trace or air can get within or out. A B are two brass rods B has ~~two~~ 4[a] metallic projecting pins on which is secured a strip of chemical paper two inches long moistened with a solution of Iodide of Potassium, And starch. (1 grain starch 2 oz distilled water ¼ oz Iodide potassium.) A is another rod having a platina wire d facing that Iodized Slip c. coming within exactly 1/16 of an inch from it— These two rods project out of the Cover and end in Binding posts. These rods should be as far apart as possible. They are Connected with 3 Trays of the test battery and placed on a table. The tint if any should be noticed and timed as well as the time when the battery is first connected The first tint should be noted and the time taken when it shall have assumed a depth equal to the first tint on the Chromatic Scale shown in Fox's work on ozone[2] Should there be a tint it will prove first, That the particles of air are conductors for Dynamic or Voltaic Electricity of low tension and corollary that induction is the effect of Conduction the fleeting effect being only produced by polarization ie[b] an opposing E.M.F Secondly that Iodide of pot or any other chemical Compound that is decomposible by Electricity are very unreliable tests for ozone—etc.

Book it.

AX, NjWOE, Lab., Cat. 1168:34 (*TAEM* 6:340). [a]Interlined above. [b]Circled.

1. "Experiment No. 2." (Doc. 401) is dated 21 February 1874. Batchelor wrote "Feb. 22 1874" at the top of the first page, probably when he copied this note to Cat. 1175:176–78 (*TAEM* 6:193).

2. Fox 1873. Two chromatic (color) scales appear after the last page of text, the second of which is probably being referred to by Edison; the first is for an ozonometer.

–403–

Notebook Entry: Cable Telegraphy

[Newark,] Feb 22, 1874[1]

Note difference in these two,

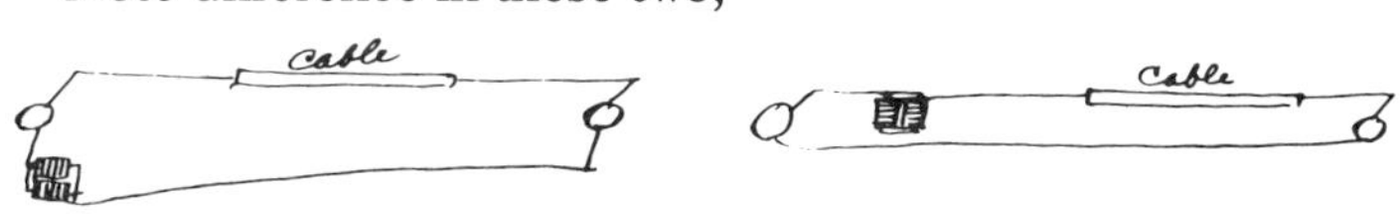

first when ma[-]gnet is placed between instrument & ground & then when it is placed between instrument & cable.

Galvanometer tests or Page 84 Electrical Tables & formulas.[2] Work th~~s~~is & thoroughly understand it

X, NjWOE, Lab., Cat. 1175:182 (*TAEM* 6:196). Written by Charles Batchelor.

1. See also cable telegraph notes from 17 February. NS-74-001, Lab. (*TAEM* 7:35; Supp. III).

2. J. Latimer Clark and Robert Sabine's *Electrical Tables and Formulae* (London: E. & F. N. Spon, 1871) contains a section on p. 84 titled "To measure the resistance of a galvanometer when no second galvanometer is to be obtained," next to which Edison wrote "OK" in his copy (at ENHS).

–404–

From George Harrington

New York 3 M'ch '74

Dear Edison

Will you not send & have an inventory taken— I suppose all movable things have been put in the Counting room & locked up— They could not get at them except they break a lock— I hope within 48 hours to arrange some matters & close the thing up—[1] In the mean time do see to them— very truly,

Harrington

ALS, NjWOE, Scraps., Cat. 1173:30 (*TAEM* 27:355).

1. This marks the end of the American Telegraph Works. Most of the machinery had been transferred to Edison and Murray the previous summer and fall, and the next day Batchelor noted a payment for "Removing machinery at the . . . other shop & pulling down." 73-004, DF (*TAEM* 12:1035–44); Cat. 1184:27, Accts. (*TAEM* 21:796).

–405–

From Jack Wright

[London,] March 9th 74

Friend Edison:

There was not much use in my writing you anything about our first expts between here & Plymouth with the Daniels[1] on the old system. You know what we did with it between here & Lpool & results to Plymouth were the same and of course unsatisfactory.[2] The~~y~~ simple fact that we were allowed only to use the Daniel was enough and Col. G[ouraud]. or Mr. Parrish sent word to that effect at once. We tried our last expt. between here & Plymouth on Friday night with the two batteries as you suggested & have accomplished all that can be desired—got 400 from Plymouth on iron with 95 cells there in each & 500 on iodide.[3] Can send him 100 more than can receive. This is quite up to the mark & stock has gone up accordingly. Our next expts. take place between here & Dublin as soon as we can get there and we expect to do regular business for the first time, so far we have only experimented at night. They wouldn't hear of our doing their business when t'was first suggested. They are softening a bit now & perhaps the old adage may come true "A waiting race is a winning race." The Lord knows we have waited.— Sir James Anderson, one of our present backers, as I suppose you know, has just offered us the use of 3 of his lines (The Indo-European) one from here to Marseilles, one from here to Falmouth & another the route of which I forget.—[4] On these we may do regular business provided our trials between here & Dublin are successful. The one principally spoken of is from here to Marseilles, 800 miles—60 miles from here to the coast, 20 miles cable, rest in France. They ask if we can work this circuit direct? It may not be necessary or possible to work beyond Paris, but in case it should, can we do it? Send me the best suggestions you can in regard to it. I'm sorry that we have not more convenient appliances to work with. The tables getting very shabby & in case we do regular business, want another paper wetter and some more cores. Also would suggest that you get at positively the best iron wire to use. Cant find anything here that beats your needles. Even the same piece of wire varies in quality in the length of a few inches. They say here that even were our system adopted they would never use the crank and if we begin regular business, I'm afraid it will be hard to find a sharp boy to feed instrument. If operator had both his hands free, could do better & more satisfactory work, especially where speed is reduced by cable &c. It is also hard to turn regularly for a long time at slow speed! Think of all these things & a good many others & write me. I'm rushing this as fast as I can to catch tonights mail. Still as homesick as ever. Would have written oftener, but got no encouragement You know the amount of consideration received here by one from the powers that be!— Thats between ourselves tho!— Brown[5] OK at Plymouth yet. Did you get my other letter?—[6]

Give my regards to all the boys. also to Mr Reiff. The "boys" must'nt expect me to write, I'm too busy. I'm married!—

I could show you about now I think. Everybody wants you to come over & I think it would pay well if you did. Everything ripe for you now. Come if you can—be awfully glad to see you. Good Night. Yours.

Jack[7]

ALS, NjWOE, DF (*TAEM* 13:83).

1. That is, Daniell cells.

2. See Docs. 318, 319, and 391.

3. The words "iron" and "iodide" refer to solutions for treating receiving paper.

4. Anderson was associated with the Eastern Telegraph Co., not the Indo-European. The line to Falmouth linked London with the Eastern cables.

5. James Brown.

6. Not found.

7. Jack Wright, an assistant at the Automatic Telegraph Co., had accompanied Edison to England in April 1873 and remained there to continue the tests of Edison's automatic telegraph system. See Doc. 318.

–406–

To Charles Buell

~~M~~Newark NJ Mch 20 74

Friend Buell

Yours recd.[1] Smith & Sawyer[2] propose to record on Cables with a peculiar printing Telegraph.[3] I don't know how they are to do it I think they dont entertain the idea that they can work with it but merely want a charter which is worth something. Yours,

Edison

ALS, NjWOE, DF (*TAEM* 13:35).

1. Buell wrote on 12 March (DF [*TAEM* 13:34]) asking Edison to identify the Smith and Sawyer inventions for facilitating transmission by cables.

2. James G. Smith was general superintendent of the Franklin Telegraph Company, and William E. Sawyer was a newspaperman and electrical inventor. During the 1870s Smith and Sawyer invented jointly in autographic telegraphy, electric motors, and electric lighting. Taltavall 1893, 336; *New York Tribune,* 11 Jan. 1877, Cat. 1240, item 17 (*TAEM* 94:12).

3. This may have been their autographic (facsimile) telegraph (U.S. Pat. 184, 302).

–407–

James Parrish to Josiah Reiff

London March 21/74

Dear Sir

I enclose extracts from[a] a letter from Sir James Anderson received in Paris where I had gone to lay the Automatic before the French Government— Edison ought to be here with sufficient instruments to equip several lines—

There is a desire on the part of Mr Cully to have a self acting machine that will obviate the necessity of turning the handle—[1] Mr Brown tells me that you have machinery of that character in America if so please let Edison bring it with him— If you act promptly I think we can sell the balance of the European interests but unless Edison can come over our friends here will not act with vigor & if we should lose this spring & summer I think it doubtful if another opportunity will be offered for a realisation of money out of Automatic in Europe— Telegraph in receipt when Edison will Sail— Yrs ty

James C. Parrish

ENCLOSURE[b]

[London, 21 March 1874]

Extract from a letter of Sir James Anderson to J. C. Parrish, dated London, March 17/74

"My dear Mr Parrish,

We are on the top of a wave whether to be driven back into the eddy or forward into clear open water depends now upon a very decided push being made—

I am afraid we shall soon fail here and not succeed anywhere unless we are promptly ready with sundry completed sets of instruments complete in all details and fair to look upon— You have established the fact that the thing will work and we are now in the Post office and might hope to keep in if we could take root with all complete— I am ready to employ it upon our Land lines and try it upon our cables and if successful some money would surely be forthcoming but Edison & instruments should now be here if any running is to be made—

I think we have now finessed enough on policy now should be straight hitting & decided action and all our resources brought to a practical focus" We shall not easily resusciate the present enthusiasm or readiness to adopt the system—

Signed James Anderson

ALS, NjWOE, DF (*TAEM* 13:85). [a]"extracts from" interlined above. [b]Enclosure is a D written by Parrish.

1. See Docs. 405 and 418.

BOOK ON TELEGRAPHY AND ELECTRICITY
Docs. 408–409

Edison appears to have begun planning a book on telegraphy and electricity in the winter of 1873, when he first prepared a list of subjects.[1] Most of the work on the book was probably done after Edison's return from England; during the fall and winter of 1873–74, he turned his attention to writings on telegraphy and electricity, and the only dated items related to the book were prepared in February 1874.[2] Drafts and fragments of several chapters have been found,[3] and there are miscellaneous notes in the laboratory notebooks and scrapbooks.[4] The plan of the book can be found in the table of contents (Doc. 408) and the preface (Doc. 409) reproduced below. It is not known why Edison did not complete the book, but material prepared for it may have appeared in some of the articles he wrote during the fall of 1874 as science editor of the *Operator*.[5]

1. See Doc. 293.
2. See Docs. 368 and 410; dated material is in NS-74-001, Lab. (*TAEM* 7:13–24, 27–34; Supp. III).
3. Rough drafts of chapters 3, 10, and 13, as well as fragments of several chapters and notes related to the book, are in NS-74-002, Lab. (*TAEM* 7:42–317). A nearly finished draft of chapter 3 is found in 74-002, Lab. (*TAEM* Supp. III). The drafts of chapter 3 contain significant material in Batchelor's hand. Copies of parts of chapter 10 and fragments of other chapters were made by Joseph Murray.
4. Besides the material in NS-74-002 and the dated items in NS-74-001, other related materials are scattered in Cat. 297 and NS-Undated-005 (both Lab. [*TAEM* 5:424–996; 8:207–504]). In addition, Cat. 30,099:314–20 and Cat. 1175:89–93 (both Lab. [*TAEM* 5:1045–51; 6:149–51]) may be connected with the preparation of the book.
5. See headnote, p. 286.

–408–

Draft Book: Table of Contents[1]

[Newark, Winter 1874?]

Preface[2]

Chap 1. Elementary Movements.
2. Miscellaneous Movements
3 Contact points & Breakwheels.
4 Batteries.
5 Induction of E'magnets & testing
6 Electromagnets
7 Secondary Voltaic Action
8 Translators for Repeataers

9 Static Charge of Tel lines
10 Duplex
11 Chemical Telegraphy.
12 Automatic Chem Tels Dots & Dashes & Roman
13 Printing Telegraphs
14 Miscellaneous Tels
15 Electrical Measurement
16 Miscellaneous Tel. Insts.

Appendix
Notes
Index

Df, NjWOE, Lab., NS-74-002 (*TAEM* Supp. III). Written by Charles Batchelor.

1. See headnote above.
2. Doc. 409.

–409–

Draft Book: Preface[1]

[Newark, Winter 1874?]

The author of the work does not propos to merely reproduce on its pages that which has already been published upon this subject nor to describe elementary laws[a] electricity Nor is it the purpose to put before the public a collection of elaboate and dificult problems ~~on~~ in[a] electricity But the object is to simplify that which appears to be so litle understood upon this subject: The author proposes to do this by presenting to the public in this work a series of experiments as performed by him. Having been for a number of years actively engaged in experimenting ~~on~~ in[a] telegraphy and electricity in general The author also will explain numerous observed phenomena in electricity they having been noticed by him and which is not generaly known.[2]

Df, NjWOE, Lab., NS-74-002 (*TAEM* 7:44). Written in unknown hand. [a]Interlined above.

1. See headnote, p. 158.
2. Edison frequently referred to his experiments in drafts and notes prepared for this book. He may also have intended to include such items as the following fragment concerning his conception of waves of electricity:

> It is probable that a single cell generates the ~~wa~~ its electricity in waves following each other with great rapidity, and that in consequence of static induction even upon bodies at great distance from the cell ~~itself~~ or even the cell itself the time between the waves is bridged over by the static charge Although we are unable as yet to

> isolate as far from all bodies & determine by instrumention the ~~ext~~ existence of these waves an obvious inference of their existence can be drawn from the fact that ~~a current~~ waves intermitted by a telegh key when transmitted through a long conduction close to surrounding bodies that these waves are continuous when made rapidly [NS-Undated-005, Lab. (*TAEM* 8:410)]

See also NS-74-001, Lab. (*TAEM* 7:21–22).

–410–

Draft Essay: Chemistry

[Newark, Winter 1874?[1]]

In Telegraphs and physical Laboratories it is usual to employ for recording currents by chemical decomposition upon prepared paper an iron point the paper being moistened with a solution of Ferrocyanide of Potassium, and I find that this salt is invariably ~~men~~reccomended in all text books referring to the subject= The results are ~~emine~~ invariably unsatisfactory= I have found that no matter what oxidiable[a] metal is employed for the recording point, the protoxide is invarible formed by Electrolysis, and of course to obtain the best results the reagent for that oxide must be used which with an iron point is not that ferrocyanide but the Ferridcyanide of potassium.[2] The best proportions is one ounce of the ~~red salt~~ Ferridcyanide, one pound of ~~the~~ chloride of sodium and one gallon of water the chloride of sodium being used to increase the conductivity of the paper.[3] if ~~more~~ a greater quantity of the Ferridcyanide of potassium is used a prolongation (and if ~~to~~ in great quantity) a weak continuous mark is made upon the paper due to the fact that the Ferridcyanide of Potassium is an ~~oxidizing salt and~~ acid[a] salt and oxidizes the iron point ~~of the~~ independent of the current= but if counter currents be used this effect is not apparent, polarization of the point by Hydrogen takes place & preserves it from oxidation= The paper used must be free from iron and animal size[4] otherwise the ferridcyanide of Potassium will be reduced to the Ferro, coloring the paper blue by combining with the iron of the paper which is invariable a high oxide, and at the same ~~f~~time reducing the quantity of the active reagent= This solution is very economical ~~f~~and sensitive for Telegraphic & experimental[b] purposes ~~and~~ but i~~t~~s unreliable for acurate work insofar as it does not give a true record of the passage of the current, in fact no acurate record can be made unless the coloration is made dependent of the action of Hydrogen, a process which I will presently describe=

As a reagent for the ferric salts of iron the Sulpho Cyanide of Potassium greatly[a] exceeds in delicacy all others and I have endeavored to employ it as a reagent for the oxide, ~~but~~ to 100cc of Distilled water I add ______ grams of Chloride of Calcium & ~~500~~ MG of Sulpho cyanide of Potassium & ______ MG of the Protochloride of iron. = ~~if this~~ no reaction takes place between the iron salt & the sulphocyanide if a strip of paper be moistened in the solution and ~~laid upon~~ passed through the chemical recorder and the battery so connected that oxygen will be evolved on a platina point the proto ~~oxid~~ salt of iron will be raised to a higher state of[c] oxidation and the sulpho ~~cyanide of~~ reagent will combine to form the ~~blu~~ ~~s~~red sulphocyanide of iron. ~~but~~ this coloration is ~~b~~transcient but of great delicacy 2 Bunsen Cells giving readable Morse signals at 20 words per minute through ______ ohms = .

When iron recording points are used they soon become rough and tear the paper ~~and~~ thus requiring them to be smoothed with a file every few minutes Especially if the paper is quite moist and ~~the~~ is moved slowly forward, and on this account I have endeavored to replace the iron point by some other metal which does not become rough by oxidation[d] still retaining the iron salt and their reagents but not very successfully owing to the tendency of the salt to decompose when exposed to the air = If Ferr~~o~~idcyanide of potassium be added to a solution of ~~Protoxide~~Ferrichloride of iron no reaction takes place if now the recording point is made ~~to raise the oxide in~~ of tin which by oxidation does not become rough[e] protoxide of tin will be formed and this being a powerful reducing agent will reduce the Ferric ~~chloride~~ salt[a] to a lower state ~~of~~ of oxidation & this will combine with the ferridcyanide to form Turnbulls blue ~~The trouble with this solution is that the atoms in both solutions change places when exposed to light. The ferric s this paper is turned blue when exposed to the light.~~[f]

Table of delicacies

Tellurium Ruthenium & char[acteristic] phenomenon = Lead & metals that give a mark on H

Claim its due to a Plumburretted Hydrogen etc.

ADf, NjWOE, Lab., NS-Undated-002 (*TAEM* 8:118). [a]Interlined above. [b]"& experimental" interlined above. [c]"state of" interlined above. [d]"which . . . oxidation" interlined above. [e]"which . . . rough" interlined above. [f]In ink to this point; remainder in pencil.

1. Edison might have drafted this piece as part of his book or for separate publication. It appears to predate his intensive investigation of tellurium and could date from late 1873.

2. An 1879 article about Edison's automatic telegraph described this idea in the following manner:

> Previous experimenters had given solutions based only on empirical rules. Mr. Edison found the law, by means of which he was enabled to make many hundreds of working solutions. In all cases it is the proto-salt of the metal which is formed, that is the salt with the least oxygen in its composition. Knowing this, one has only to moisten the paper with a chemical that gives a coloration with this oxide, and called its re-agent. [Fox 1879, 844]

3. This was the standard solution in Edison's automatic telegraph system. TAE to Patrick Delany, 5 Jan. 1888, 88-018, GF.

4. A gelatinous substance made by boiling animal products, size is used as a coating for paper.

–411–

Technical Note: Miscellaneous

[Newark, Winter 1874?]

Try induction with mirror of a Bradley spool of 545 ohms with a needle in it & without

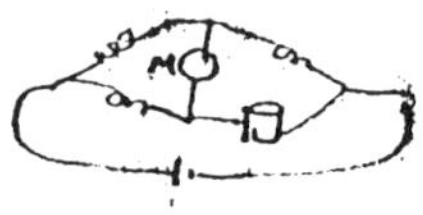

Try Carbon Cup on a secondary battery of clean zinc plates both same size giving no current try with mirror also with 4 tangent=

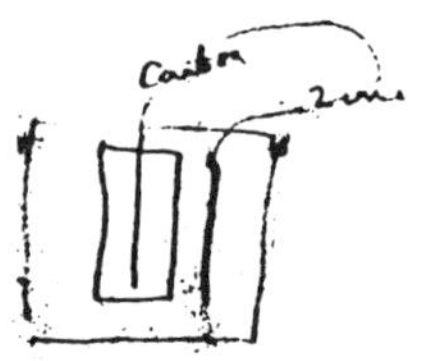

Carbon pole being connected to zinc oxygen is liberated on the Zinc, 1 atom. here it forms protoxide of Zinc & perhaps the Sulphuric A Turns it into Protosulphate of Zinc but this is doubtful. The sul acid merely decreasing R. Now we have protoxide of Zinc, ~~and~~ ie 1 atom O from H_2O, leaving H_2 free. This enter the Nitric Acid cell & substracts one atom of the ~~oxygen~~ 2 atoms of oxygen which go together to form Nitric Acid. This makes water & turn the Nitric Acid into a Nitrous acid. Now the problem is to put in the porous cell a good conducting chemical that will constantly gives off to the free H_2 an atom of oxygen & reabsorb it from the air.

Write to Gouraud & get Clark propose me as foreign member Society telgh Engrs

X, NjWOE, Lab., Cat. 297:97 (*TAEM* 5:843).

-412-

To Charles Buell

Newark, N.J. [March 22, 1874][1]

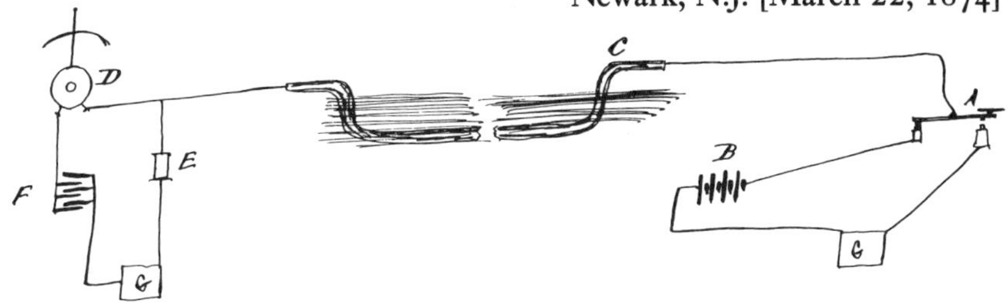

Dr Sir

The above is a sketch of the apparatus used on deep sea cables A is sending key. B is battery C is Cable D is mirror galvanometer E is Resistance coil F is condenser G G is ground plate

As will be seen by the drawing the cable is always kept charged by the key resting on its back point. The condenser F also, In charging which the mirror galvanometer is deflected but returnes to zero as soon as condenser is charged to its full capacity[a]

The signals are sent by pressing down key A & putting cable to earth which causes the electrical tension of cable to fall below that of condenser. a current then flows from condenser through galvanometer (deflecting it) to cable in order to establish an equilibrium of the latter This deflection is the signal

The effects of induction in cables may be obtained on an artificial cable composed of condensers of the proper capacity separated by resistance coils one set of plates of each condenser being placed on same circuit & alternately with resistance coils & the other plates to earth. If you have equal resistance & capacity identical results will be obtained. In experimenting on cables (artificial) in my shop I generally fix them so:—

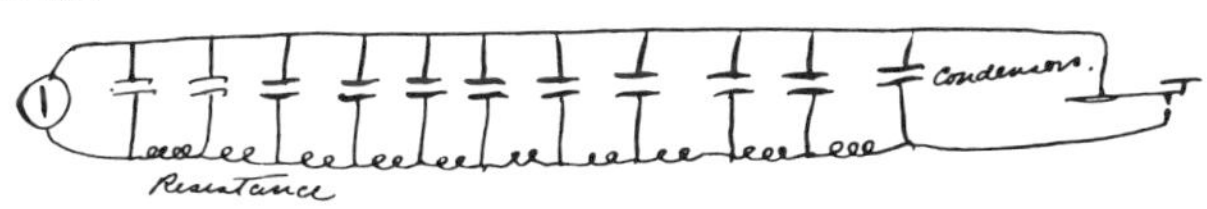

As you wanted an approximate estimate of Atlantic Cable we give you the following—

Length of Cable about 2185 miles
resistance " " " 7584 ohms or about 3½ ohms to 1 mile
Its insulating resistance per mile about 325 megohms, which increased to 2946 after being laid.
Electrostatic capacity 0.3535 Microfarads per knot.[a]

As you are near Yale college you would do me a great favour if you would get from Prof. Silliman[2] information as to where I could get some Tellurium ore & who from, say about 1 barrel. Yours truly

Thos. A Edison per Batchelor

L, NjWOE, DF (*TAEM* 13:36). [a]Followed by centered horizontal line.

1. The date is from the postmarked envelope.

2. Benjamin Silliman, Jr., was professor of chemistry on Yale's medical faculty and editor of the *American Journal of Science and Arts.* Silliman wrote to Batchelor on 18 May 1874 concerning sources of tellurium. *DSB,* s.v. "Silliman, Benjamin, Jr."; Cat. 1143:72, Scraps. (*TAEM* 27:208).

–413–

James Parrish to Josiah Reiff

London 3/23/74

My dear Reiff

Sir James Anderson informs me that experiments brought to a point today have increased the capacity of the ~~Atlantic~~ Lisbon Cable 50% and he hinted to me that if we intended doing anything it was advisable that we did not delay—

I am so well convinced of the energy and purpose of your character that I refrain from writing all I feel in this matter— Not a moment should be lost in dispatching Edison and several sets of machinery Edison first and the machinery to follow if necessary— Wright leaves tomorrow for Dublin and the disposition of the English Post office is very favorable toward us— Mr Watson[1] will leave not before ten days and I have I think made arrangements to see him in the interest of Automatic—

The iron is heating rapidly send Edison at once[a] to help us strike at the right moment Yrs ty

James C. Parrish

ALS, NjWOE, DF (*TAEM* 13:87). [a]"at once" interlined above.

1. Unidentified.

–414–

Caveat: Automatic Telegraphy

[Newark,] March 23, 1874[a]

The following is a description of my newly-invented Solutions for the preparation of Paper for use in Chemical Telegraphs, which is as full, clear and exact as I am at this time able to give.

I employ several solutions, all of which have different characteristics; one solution being cheap and not very sensitive, and suitable for short lines; others being solutions which give no mechanical spreading of the precipitated[b] marks, while others are costly and more delicate.

To a solution of ferrocyanide of potassium (yellow prussiate of potash) I add a few grains of chlorinated lime (common chloride lime) and to this I add a few drops of either sulphuric muriatic or nitric acid which acting on the chlorinated lime sets free the chlorine which again enters into another combination which forms a very delicate test for the protoxide of iron formed by electric decomposition. Nitric acid produces a combination which gives the most sensitiveness, but most any acid will give a mark when combined with the above ingredients. The amount of acid should be just enough to enter into combination with the lime and free all the chlorine, and leave no free acid; a free acid in a chemical solution will oxidize the iron point and a mark would be produced independant of the electric current.

Another solution consists in substituting chloride of calcium for the chloride of lime or more properly oxymuriate of lime, the acids being added until the paper which first assumes a blue color turns yellow, with no free acid. The double acid nitro-muriatic also forms a very sensitive combination

I do not wish to confine myself to the use of any particular chloride, but claim the combination with ferrocyanide of potassium of any chloride, whose chlorine is set free in solution by the addition of an acid.

Another solution consists of two drachms of dilute sulphuric acid to which is added five grains of permangate of potash— these proportions may be greatly altered. This colors the paper red. If the current be passed through a nickel, iron, copper or bismuth point resting on the paper, a white mark is obtained. This is quite sensitive, but the marks are not very permanent.

To a solution of ferrocyanide of potassium and chloride of sodium already patented,[1] I add a small quantity of starch which tends to whiten the paper and the precipitate of the mark combining with the starch greatly augments its size and distinctiveness.

If a little nitrate of silver is added to the above solution without or with the starch, the sensitiveness is increased.

Another solution is formed by adding nitrate of ammonia to a solution of logwood[2] and using a tin pen for the decomposing point This solution is very sensitive, the marks are violet on a yellow background

Several chlorides and nitrates may be substituted for the

nitrate ammonia but the latter gives the best results. Another solution consists in combining perchloride of iron with sulphocyanide of potassium to which may be added chl. sodium to give the solution greater conductiveity: the paper is colorded an intense dark red. If a tin decomposing point be used the passage of the current gives white marks; it is quite sensitive.

Signed by me this 23rd day of March A.D. 1874.

Thos A Edison

Witnesses. Geo. T. Pinckney Chas H. Smith

DS, DNA, RG-241, Edison Caveat 47. Written by Pinckney; petition and oath omitted. [a]Date taken from text, form altered. [b]Interlined above.

1. Edison did not have a patent that combined these chemicals. Three patents executed in the fall of 1873 (U.S. Pats. 160,402, 160,403, and 160,404) mentioned one or the other, but those applications were not filed until 1 June 1874.

2. A Central American and West Indian tree from which a dye is made.

–415–

And Joseph Murray Agreement with Jarvis Edson

[New York?,] March 26, 1874[a]

Whereas, T. A. Edison[b] of Newark, New Jersey, is the inventor of certain Telegraphic Instruments for what is known as District Telegraphing, and being the instruments and method the said Edison is engaged in perfecting at the date hereof and proposes to take out Letters Patent therefor,[1] and[c]

Whereas, J. B. Edson[2] is desirous of organizing a stock company[3] for the purpose of working such patent right, and the said T. A. Edison is desirous of having the said J. B. Edson organize said company for the purpose aforesaid, and[c]

Whereas, J. T. Murray is interested in said inventions, now[c]

This Agreement[d] made this 26th day of March 1874, between Thomas A. Edison and J. T. Murray of Newark, aforesaid, parties of the first part, and Jarvis B. Edson of the City of Brooklyn, party of the second part, In consideration of the premises and oOne dollar lawful money each to the other in hand paid, the receipt whereof is hereby acknowledged:

Witnesseth, that the said parties of the first part hereby covenant, that they will perfect the said inventions and take and secure Letters Patent of the United States of America therefor, so as to fully protect the said inventions thereon which shall be made by the parties of the first part in that branch of telegraphing for one year from date hereof, and the said party of the second part, hereby covenants that he will organize a company under the laws of the sState of New York sunder such style and name as he shall determine, with a capital stock of not exceeding One Million dollars, and with a working capital of at least Twenty Thousand Dollars cash, on or before the 1st day of September 1874, for the purpose of working and vending the said Patent rights throughout the whole United States and that upon the organization of the said company the whole of the capital stock thereof shall be issued to the said parties of the first part,[4] in full payment for the said Patent rights and Letters Patent, whereupon at the same time the said parties of the first part shall forthwith assign and transfer the said Patent rights and Letters Patent and all of them to the said company organized as aforesaid, and the said inventions and Patent rights shall be and become the property of the said company, and the said parties of the first part hereby agree that forthwith and on the delivery to them of the whole amount of said capital stock of said company as aforesaid, they, the said parties of the first part, thereupon in consideration of his services in so organizing said company will transfer to him, the said party of the second part or to such person, or persons as he may designate, Sixty six and Two Thirds percent of the whole of the said capital stock aforesaid to be used by the said party of the second part in such manner as he shall determine in and about the working of said company and as a working capital therefor and the introduction of said Patents and for such other uses and purposes as he shall determine, and the said parties of the first part hereby covenant, that they will transfer to the company all improvments made by them in such departments of Telegraphing as aforesaid within one year from the date hereof[5] and for the consideration aforesaid and all agreements heretofore made are modified accordingly.

Witness our hands the day and year first above written.—

Thos. A. Edison J. T Murrey
Jarvis B. Edson

In Presence of H. G. Batcheller[6]

DS, NjWOE, DF (*TAEM* 13:123). [a]Date taken from text, form altered. [b]"Whereas, T. A. Edison" underlined twice. [c]Horizontal line drawn to right margin. [d]"This Agreement" underlined twice.

1. Edison executed the application for this patent (U.S. Pat. 158,788) on 2 April 1874 and filed it on 11 May. Pat. App. 158,788.

2. Jarvis Borsted Edson (1845–1911), an inventor and former navy engineer, was best known at this time for his steam-pressure recording gauge. *New York Times,* 27 Jan. 1911, 11.

3. The company was initially incorporated as the Domestic Tele-

graph Co. on 4 April 1874 by Edison, Murray, Edson, and Batcheller (see n. 6). On 10 April 1874 they filed a second set of incorporation papers, which differed from the first set by specifying the primary route of the company's wires. Certificates of incorporation, Domestic Telegraph Co., NNYCo. A fragment of what may be a proposed draft of the incorporation papers can be found in Cat. 297:104, Lab. (*TAEM* 5:857). This fragment, in Edison's hand, proposes terms for the establishment of subsidiary companies in cities such as Baltimore and Washington, D.C.

4. On the same day, Edison, Murray, and Edson signed an agreement (DF [*TAEM* 13:129]) dividing a presumed capital stock of one million dollars in the following proportions:

There shall be issued to the treasury of the said company as and for a working capital	3333	shares
To T. A. Edison inventor	2500	"
" Joseph T. Murray "	1600	"
" H. G. Batchellor attorney	300	"
" J. B. Edson	1400	"
" J. B. Edson for distribution	867	"
Total shares	10000	

5. On the same day, Edison, Murray, and Edson executed a supplemental agreement assigning to the company an invention for a fire alarm telegraph that Edison was then completing (DF [*TAEM* 13:127]; Cat. 1214:43–44 [*TAEM* 21:585–86]). Edison did not execute a patent on fire alarms until 9 May 1876 (U.S. Pat. 186,548). He assigned his fire alarm patent to the Domestic Telegraph Co. on 8 March 1876. On 12 October 1875, Edison and Charles Batchelor also assigned to the company U.S. Patent 169,972 (for district signal boxes and chemical receivers). Copy of Edison telegraph patent assignments, pp. 18, 28, WU Coll.

6. Hiland Batcheller was a lawyer with offices at 7 Warren St., New York City. Wilson 1873, 76.

–416–

J. W. Eames to Henry Fischer

[London,] Mar 30 1874

The Controller

The experiments that took place on Saturday and Sunday ~~e~~with Mr. Eddisons instrument did not, as far as I could see, lead to any satisfactory result.

We received no readable marks whatever from Dublin by the automatic apparatus, and therefore the preliminary trial which was arranged to take place at three o'clock yesterday afternoon, did not[a] come off—

It was stated by the American clerk that Dublin could read ~~w~~[-] allright at something like 100 words per minute, and I endeavoured to obtain some confirmation of this report from the clerk in charge at Dublin. He did not however appear to be taking any record of the speed of working, or any particular notice of the character of the marks—

The American clerk informed me that he believed the failure was owing[a] to a wrong arrangement of the batteries in Dublin— The clerk in charge at Dublin however informed me[b] that they had an Engineer in attendance who would do anything that the American clerk required, but it was stated that the battery room at Dublin was locked up and that a rearrangement of the batteries to suit the American clerk, could not therefore be affected.

J W Eames

ALS, UKLPO, ATF, Item 119. On paper headed "*TELEGRAPHS.*" in upper left corner under embossed seal. [a]Written at bottom of page; repeated at top of next. [b]Interlined above.

–417–

Notebook Entry: Cable Telegraphy

[Newark, March 1874?[1]]

Cable
At Western Union[2]

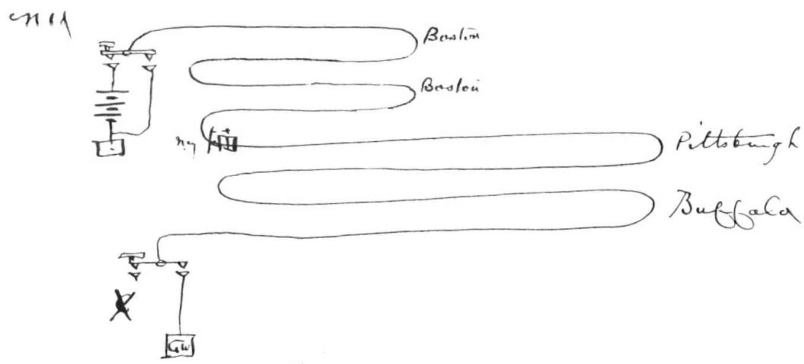

See if Relay closes immediately—whether it closes best with key open at X to earth or insulated =

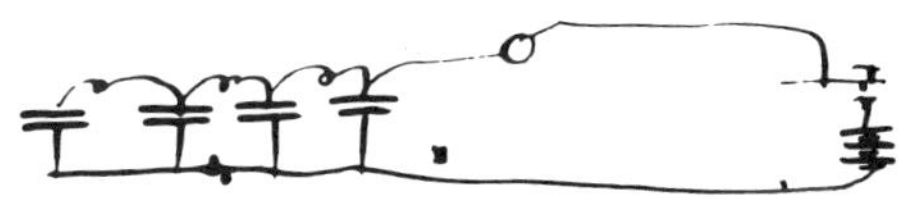

Shop see if there is enough induction to record from my four Condensers

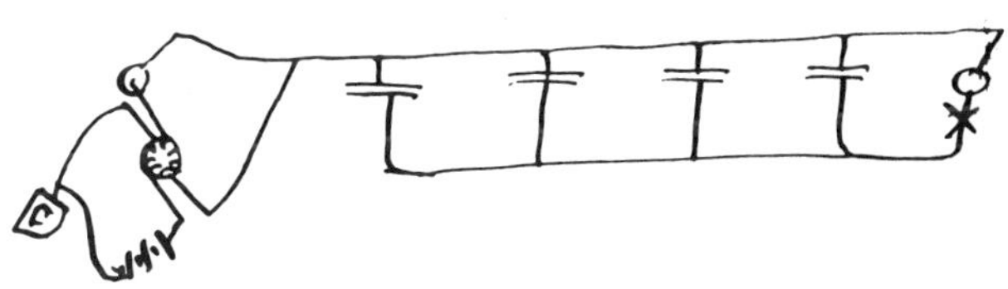

also with battery at X
Put battery on to line & then throw it [to?][a] E through Receiving instrument also with Battery at X

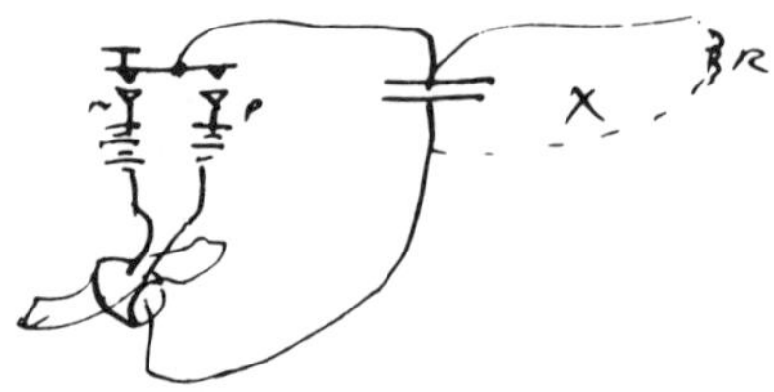

Try what it records. also with X ckt & Resistances.

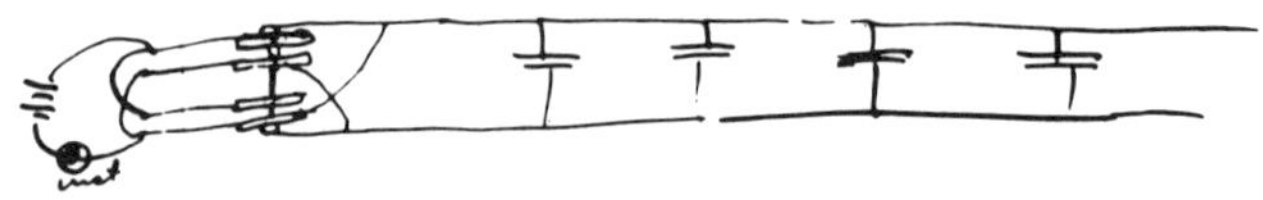

to Reverse Current on Cable but keep it same in instrument—probable that closing distant end will make reversals record which wouldnt if open

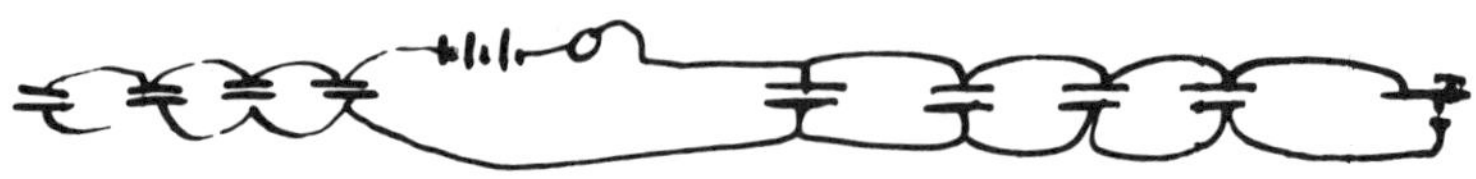

Try

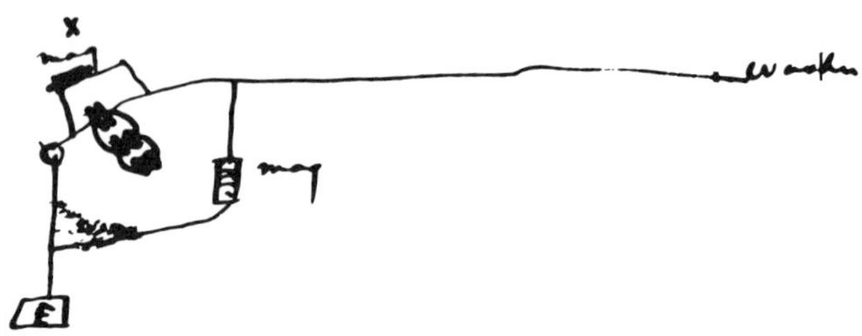

also replace magnet at X by plain Resistance =

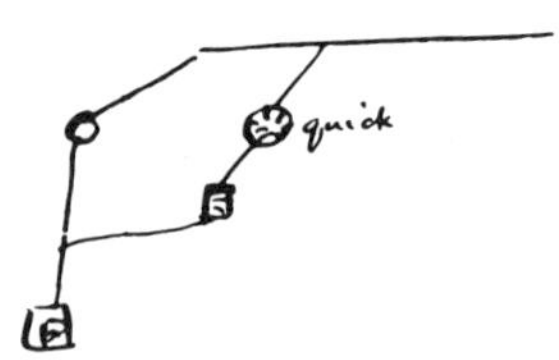

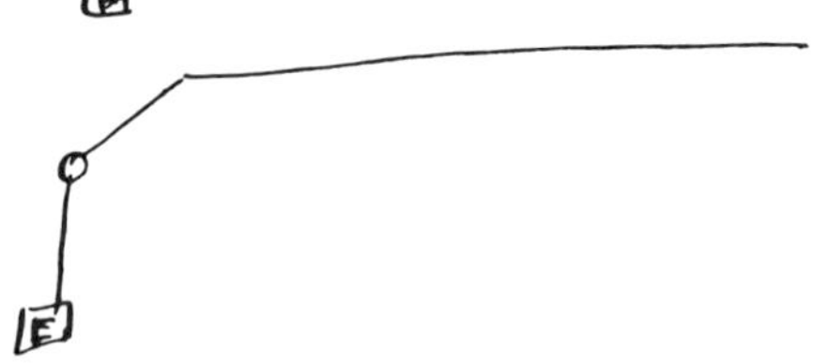

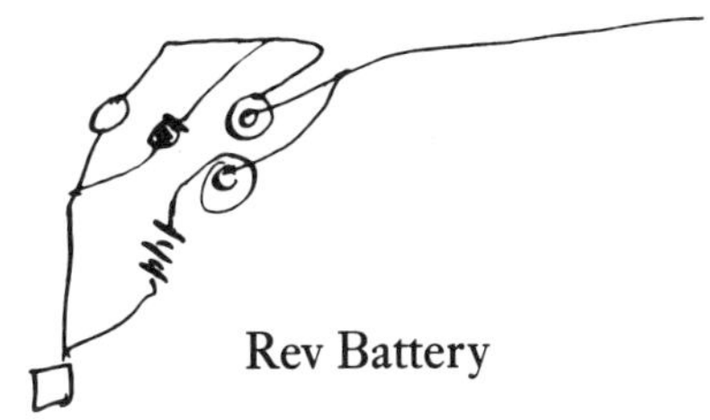

⟨E & B 16⟩[b]

X, NjWOE, Lab., Cat. 1175:186 (*TAEM* 6:198). [a]Obscured by ink blot. [b]Marginalia written by Charles Batchelor.

1. The preceding entry (also on cable telegraphy) is dated 13 March 1874, and the following entry (on automatic telegraph recording solutions) is dated 4 April 1874 (Cat. 1175:184, 190, Lab. [*TAEM* 6:197, 200]).

2. Figure labels are "NY," "Boston," "Boston," "NY," "Pittsburgh," and "Buffalo."

-4- April–June 1874

In the spring of 1874 Edison continued his work on automatic and domestic telegraphy, and he also renewed his efforts on the quadruplex. New endeavors also occupied his time. His ongoing experiments with automatic telegraphy led to the discovery of the electromotograph phenomenon, while experiments with induction coils resulted in the design and marketing of a medical apparatus for giving electric shocks.

During April, Edison was under increasing pressure to return to England in order to solve the technical problems still plaguing his automatic system in its trials on the Post Office lines and, more importantly, on the private undersea-cable lines. British cable investors in Edison's automatic telegraph also apparently expressed a willingness to invest in a proposed new American company to exploit the invention, but commercial failure in Britain threatened this support. Edison's own cable experiments apparently convinced him that such a trip was unlikely to yield success. He did, however, continue experiments in that direction, and even demonstrated a roman-letter, ink-recorder system to the American manager of the French and Anglo-American cable lines.

The projected new American company was the product of George Harrington's and Josiah Reiff's continuing attempts to join the Automatic Telegraph Company with other small telegraph lines in order to offer significant competition to the Western Union Telegraph Company.[1] These negotiations took on new importance as several railroad companies—Erie, Baltimore and Ohio, and Pennsylvania—and the Atlantic and Pacific Telegraph Company expressed interest in such a combination. Jay Gould, who had a financial interest in Atlantic and Pacific, also brought Thomas Eckert, the superintendent of Western Union's Eastern Division, into the negotiations as the potential president of the new firm.[2]

Edison himself seems to have had little direct involvement in these negotiations, focusing instead on experimental work in automatic telegraphy. He sought the right combination of chemical solutions and recording pens for both his automatic telegraph and his new domestic telegraph receiver. Becoming greatly interested in the possibilities of tellurium, Edison wrote to a number of geologists and miners regarding a supply of the metal and filed a patent application for its use as a recording pen for automatic telegraphy. Eventually, both the automatic and the domestic systems would use tellurium recording pens. Edison's experiments with chemicals and recording pens resulted in an intriguing discovery. He determined that certain combinations of chemicals and metals caused a change in friction when the signaling current was passing. After making a few experiments in April with this "new force," Edison turned his attention to other matters and did not experiment again in this direction until August.[3]

Although Edison believed strongly that automatic telegraphy was the more valuable system, he again approached Western Union for support in his efforts to develop multiple telegraphy.[4] In May he wrote Western Union electrician George Prescott and offered him a half-interest in future profits resulting from these experiments in exchange for facilities and personal help. In early June, the agreement of Prescott (and company president William Orton) led to renewed assistance for Edison's duplex experiments. Within a short time, experimental instruments were being made at the company's shop in New York, and Edison began conducting tests over Western Union wires with the assistance of Western Union operators.[5] Edison's work convinced Prescott and Orton of the importance of the diplex he had developed the previous summer.[6] They saw that Edison's diplex, in combination with the Stearns duplex, controlled by Western Union, could make quadruplex telegraphy (two messages in each direction over a single wire) practical, and they pushed Edison to focus his attention in this direction. Although the quadruplex soon came to dominate Edison's efforts in multiple telegraphy, he also tested a variety of duplexes and other diplexes during June. To ensure their control of this work, Orton and Prescott also sought to formalize Prescott's relationship with the inventor.

While Edison and Murray continued to operate their manufacturing shop, their most important customers—Automatic Telegraph and Gold and Stock—ordered few new instruments and required mostly repairs and supplies.[7] The shop did increase its work for the Celluloid Manufacturing Company, producing a large number of comb cutters and a press.[8] Edison continued development work for the Domestic Telegraph Company, which, if successful, would create more business for Edison and Murray.

Edison also conceived the idea of starting a new company to manufacture consumer items—the Scientific Toy Company. He initially designed two items with this market in mind, an electric-shock machine developed during April experiments with induction coils, and a telegraph instrument on which student operators could practice their skills. Except for drawing up a list of items to be manufactured by the Scientific Toy Company (Doc. 428), nothing came of this idea. Instead, Edison and Murray manufactured and sold these two items directly. While both instruments faced many competitors in their respective markets, the one for a telegraph learner's instrument was long familiar to Edison.[9] With the electric-shock machine, which became known as Edison's Inductorium, he was competing in the less familiar market for electromedical devices.

In their attempts to find new markets, Edison and Murray may have been partially driven by financial need. Edison's former partner, William Unger, held a mortgage on their shop as collateral for promissory notes that would be due in July. Although Josiah Reiff promised to relieve Edison of this worry, no money seemed to be forthcoming from Automatic Telegraph, and Edison turned to Western Union.[10]

1. See *TAEB* 1:487 n. 1.

2. See Doc. 522 n. 10 for a summary of these negotiations. See also *Harrington v. A&P*; and Uriah Painter to TAE, 4 Feb. 1877, DF (*TAEM* 14:645).

3. It is possible that some experimental notes are missing, as few laboratory records exist from the period May–August.

4. Throughout this period Edison frequently asserted the superiority of automatic telegraphy over duplex and quadruplex Morse telegraphy. Orton's testimony, Quad. 71.1, p. 119 (*TAEM* 10:64).

5. Prescott's testimony, Quad. 71.1, pp. 329, 335 (*TAEM* 10:171, 174).

6. Doc. 348.

7. Manufacturing records are found in Cat. 1184:35–55, Accts. (*TAEM* 21:800–810).

8. Cat. 1214:44, 46; Cat. 1184:39, 45, 48; both Accts. (*TAEM* 21:586, 587, 802, 805, 807).

9. See Doc. 10 for Edison's earliest interest in such a practice instrument.

10. Orton was out of town at the end of June and Edison asked Prescott for a loan. Prescott in turn asked George Phelps, superintendent of the Western Union factory, to make an inventory of Edison and Murray's Ward St. shop. Prescott's testimony, Quad. 71.1, pp. 349–50 (*TAEM* 10:181–82).

-418-

Draft to George Gouraud

[Newark, c. April 1, 1874[1]]

Col Gouraud.

Culleys note =[2] It is as certain a "the tides that every man who sees the way we send and receive now should suggest some automatic device to replace it, and it is also just as certain that if the same individual would or could examine all the countless clockworks, motors, electric, steam, water etc that have been tried & abandoned and would think on the same for a little while he will come to the conclusion that the hand turning is the best. If Mr Culley will sit down and turn that crank at a ~~rate~~ given rate he will find that he can always turn it at a given rate, then again it dont make any difference at what rate it is turned, whether regular or irregular. I will give some of the reasons why we abandoned clockworks etc for the turning.[3] 1st. Clockworks starting off too quick broke the paper. 2nd. owing to pressure on rollers necessary to ensure good contacts and cut through paper fibres necessitated powerful springs in clockworks.— 3rd sending and receiving ~~at~~ requires different speeds owing to length of perforated characters, and the shortened characters on the chemical paper hence a speed changer = smaller wheel = or two machines. = 4th. the limit of speed is not always the same to all stations, ~~b~~also bad days the operators lower their speeds, ~~whic not sendg~~ ie[a] when N York sends to Washington on a very bad day when scarcely any current reaches him he turns a given speed, but the next instant he sends a message to Phila at a fearful ~~speed~~ rate this allows the wire to be worked up to the limit of its capacity— 5th Clockworks with their devices were constantly getting out of order, while with the wheel and crank out of order is a thing unknown. The advantages of clockworks is that thy look pretty & intensely scientific. Mr Culley appears to have a very poor opinion of his clerks if he call~~is~~ turning a crank anywhere with 50 to 400 revolutions per minute skill.[b] we have two of his operators here turning these cranks. we also find that the operators like the crank and get disgusted with the clockwork as they cannot vary the speeds to suit the condition of the line with the same facility, and requires too much attention. In conclusion I will say that if it is decided to use a clockwork, that I can furnish that article, but I warn you that running dry paper at 150 words per minute is different from wet paper at 300 per minute. As far as I can see the whole tendency over there is to remove all ~~chances~~ devices which give rise to ambition among the clerks on their personal skill, the very thing which makes our American Operators ~~the best~~ so expert ~~in the world. (I wont say world either, because I do not know how tis done in france~~ etc)

Wrights note. =[4] He does well— how long is the wire, Resistance, much buried wire— how is battery arranged, what kind = size, plates, etc. hows your paper wetter. The beauty of your writing depends greatly on the kind of wire you use & the angle & pressure— are those perforators any good. they were invented & made in a hurry— whats their disease— give diagnosis of ther diseases. did you go to Elliot Bros[5] & get those prices. Why dont you answer my letters, any little information in an electrical point of view which you may scoop up in Telegh street thankfully received[6]

—What do you think of clockworks, think there is enough parts in a Wheatstone transmitter to start a watch factory.

Ask Lumsden if he got books I sent him.[7] The self induction I have since found of that Greenwich Cable was more than I ever dreamed of—[8] I was very stupid = Center all right =

Keep the chemical ~~fo~~paper formula to yourself—if it aint to late. —anything wanted = hows Brown tell him write. I want to know ~~i~~from him if an atom of Hydrogen can throw a shadow.

Note of ~~L~~Mr Clarke. =[9] Mr Clark partly mistakes my question What I mean is this. The Copper-Gutta, and sheath, or distant earth plates themselves ~~are n~~ do not act as a battery or give EMFs but that the action of the testing battery upon these elements, makes them set up an opposing electromotive force = which is proportional to the ~~b~~EMF of the battery, and the time it is connected to cable When the battery is disconnected we obtain the static charge and this secondary battery action, (ie chemical action,) which continues for some time after the static charge ~~(although if~~ & wherever there is leakage there must be secondary battery action. Supposing that the conductior of the cable, representing one plate of the secondary battery, is connected to the copper pole of the test battery, the ~~Hydrogen~~ oxygen due to the electrolysis of the leakages is ~~within~~ confined within the gutta, ~~while the and~~ while the hydrogen is on the sheath of distant earth plate. ~~on the~~ now if the battery is disconnected we have all that is ne[--]ccessary for a very peculiar secondary battery charge with ~~m~~a high EMF. This electromotive force will exhaust itself in proportion to the internal R of the battery ie[a] gutta—and the circuiit in which it can circulate, hence with a very high ~~.R.~~ insulation .R. & conductor .R. it will take longer for the secondary battery to discharge than on a circuit of lower resistance. i~~e~~f now we connect the zinc end of the battery to the conductor we obtain oxygen on the sheath and hydrogen inside, in this case we ~~a~~obtain a stronger EMF than with the copper to core. its not a matter of indifference ~~where~~ on which metal portion of the cable the oxygen shall be generated. the ~~amount of~~ EMF of the secondary current (chemical) will be the greatest with zinc to cable. therefore a ~~zin~~ cable tested with a zinc ~~s~~current should ~~show~~ give a higher insulation R than with a copper current. again, the insulation is not increased by what is called absorption. it is the gradual increase in ~~the strength~~ of the EMF of the secondary battery which being opposed to the test[b] battery, tends to create the belief that the ~~incre~~ insulation R ~~i~~has increased whereas its just the same, the error being ~~re~~due[c] to the effect of the gradual increase of the Secondary EMF. Experiment

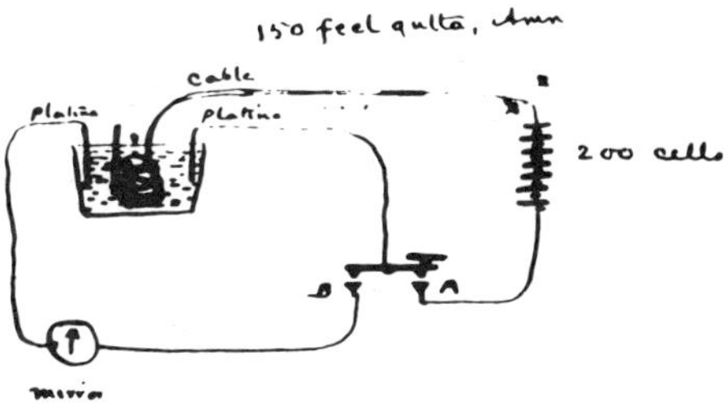

Mirror about[b] 2000 O[hms] Shunted, 100 O Home made

If I press the key down on A I charge the cable. if I open and allow the lever to touch .B. the light is thrown off the scale, but if the platina plates are well washed & give no current themselves it s~~w~~oon returns to zero with the lever on .B. by shunting mirror down I find, after 10 sec electrification 5 deg. limit of[d] throw. 20. sec 7 deg 1 minute 8 to 9 deg. everything insulated = battery = tub = mirror key = etc on vulcanite pillars 3 inches high. Isn't it possible to get the resistance of insulation[e] more accurately by ~~calculating~~ adding the secondary battery action, ~~to the~~ derived from the two platina plates to the ~~curre~~ test current entering cable in given periods of electrification. please explain where & how I have blundered as I feel I have, because what I know about cable phenomenon could be put in a very small book—

Col G. please correct grammatical errors & hand to L. Clark.

Cable— I am not like[b] Micawbar—[10]

Roman O.K. on two wires, 200 per minute from Liverpool to L.

Staff— My proposition hold good stated in former letter,[11] that by a very simple change in the method of using their clerks I can get twice as much work out of them, without any fatigue to the clerks.

Edison

ADfS, NjWOE, DF, 74-006 (*TAEM* 13:88), and Lab., Cat. 297:144 (*TAEM* 5:959). Continuity of text and resemblances in style, paper, and ink link these separate scrapbook items. [a]Circled. [b]Interlined above. [c]Starts separated page. [d]"limit of" interlined above. [e]"of insulation" interlined above.

1. This letter addresses issues raised in Docs. 405 and 407 and precedes Doc. 420, in which James Brown answers Edison's query about whether "an atom of Hydrogen can throw a shadow."

2. See Doc. 407.

3. This statement notwithstanding, Edison was working on a clockwork-driven transmitter-receiver (Doc. 458).

4. See Doc. 405.

5. Edison acquired two copies of an 1873 Elliott Bros. "Catalogue of Electrical Test Instruments" and annotated one of them. *Catalogues, Electrical and Telegraph Instruments,* Library, NjWOE.

6. The main office of the Post Office telegraph. Kieve 1973, 190.

7. Nothing is known of these books.

8. See *TAEB* 1:604.

9. J. Latimer Clarke.

10. Wilkins Micawber, a character in Charles Dickens's novel *David Copperfield,* is noted for his baseless optimism in the face of failure. Edison seems to be despairing of success in his cable experiments.

11. What may be a partial draft of the scheme proposed in Edison's earlier letter can be found in the same scrapbook that contains the last portion of this document. The earlier letter has not been found. Cat. 297:66, Lab. (*TAEM* 5:668).

–419–

Notebook Entry: Electromotograph[1]

[Newark,] April 10, 74

experiments with the new force chem Decomp moving a Lever =

Reversing a Current does not help it a bit with Potassic Hydrate =

Potassic Hydrate in paper rather dry OK— ~~frictio~~ fused Zincic Chloride, increase~~d~~s friction when hydrogen evolved on Lead very strong but probably not so strong as hydrogen reducing friction with Potassic Hydrate.

Silicic Hydrate Acid, increases friction when hydrogen is evolved, probably more noticeable than with Zincic Chl. Oxygen appears to produce no effect in either.

Hydrated Silicic Acid & Potassic Hydrate Hydrogen decreases friction not so strong as without hydrate Silicic A apparently

Sodic Phosphate, H dec fric— not very good

~~Ch~~Ammonic Chloride— very poor if anything on either

Mecuric sulphate— this had little Ammonic Chloride in it[a] increases friction greatly with .O. decreased a little with H

Mercuric Nitric— not so good acts both ways like Mecruic sul

X, NjWOE, Lab., Cat. 1171:118 (*TAEM* 6:265). [a]"— this . . . it" interlined above.

1. This notebook entry is the earliest record of Edison's experiments with the reduction of friction through electrochemical action. Later in the year he termed this phenomenon and the devices through which he explored it "electromotograph." See headnote, p. 46.

–420–

From James Brown

[London,] Apr 12[–15][1]/74

Friend Edison

This morning I received your letter that you wrote Col Gouraud it was forwarded to me.[2] In it you make the request that I shall write you so this effussion I wrote you once before and received no answer so concluded that my letters were not very interesting to you so let that be my excuse for not writing you again I suppose you wish to know the results of Plymouth experiment and the conditions we were forced to work under: The line is 150 miles long and has about 18 miles of underground work in it most of or you might say all at London end. The wire no 8; we~~re~~ were forced to use 100 Daniels cells which of course you know: The principal reason of this is that it was claimed that your fast speed between Liverpool & London was entirely due to Powerful battery power & Iodide paper so they brought us down to that Damed battery that they use which under ordinary circumstances will not pull a setting hen of[f] her nest: Well we went to work Of course on the old plan shunting at the sending end With 100 cells

Danniels and shunt of 1000 Ohms Mr Wright received best results at London being 250 words pr minute on Iron paper We drumed away at this for a long time Mr Wright receiving all the time with only one interval of 15 or 20 minutes when he sent to me in that time I succeeded in receiving 400 Words good on Iron paper having Mr Wright use shunt of 800 Ohms when I done this the wire was taken from me and I had no chance to have any of my Ideas worked upon I could have done better and wrote to that effect but it was no good I never had another chance untill the very last moment which I will tell you of further on: well it went on at this rate untill your letter was received giving plans of working with low battery power and have low resistance transmitting shunt to discharge wire But before this they attempted to work the Wheatstone system over the same wire and to get 250 words pr minute They put up the transmitter alongside of mine at Plymouth set it a going at rate of 120 words pr minute system of permant currents or Double current key Receiving in london on Mr Wrights instrument our Iron paper also It went of[f] gay they were all right They then tied up the fan of transmitter[3] and worked with compensating currents (Culley's) The result was they busted their Transmitter and when they received it appeared like a barn yard with any amount of fowls scratching about: So that was given up as a bad failure which it certainly was After your letter arrived copy was sent to me and I immediately arranged my batteries for it First nights experiments were a decided success we doing 400 & 500 words pr minute with ease So after this passed off they sent me word that we should try once more than for Dublin So at time we went to work Mr Wright still receiving 400 & 500 words Pr minute Then he sent to me ~~I never g~~ I received for a few minutes at rate of 500 pr min when all at once the current ceased to come to me or at least so weak that I could not get a word even at hand speed Wee worked and [-]worked it did not improve but got worse I told Johny he was wrong some where but he said no and that he had galvometer in line and could se his current go out to line all right I told him it made no diference he was not sending me his current: But yes he was so he said He never stopt to think that current of 5 or 10 cells would give him full deflection and still not enough current reach me to work upon There was no help for it it seems so I was sent for to come back to London I did so when the question was put to me why did you not receive at Plymouth I told them

Mr Wright had somthing wrong that it was he had battery wrong Loose connection in it or else his instrument was making earth when he transmitted to me It was no use I could see by their actions that they thought it was my fault So when we got the Dublin wire Mr Wright was sent to Dublin and I retained here First nights experiments English wire 266 miles to cable No 4, 18 miles of underground work in London Cable 66 miles Cable to Dublin 10 miles Dont know gague of wire or if underground from cable to Dublin I commenced to send to Mr Wright with 100 cells of battery arranged like this

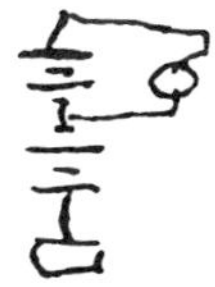

he used Iodide paper and received 120 & 130 words pr minute He then sent to me I could not get a mark at all So that passed of[f] and worked the next evening with same results to him but little better to me as I did get some current but very little That evening passed off when on the third trial I had receiving part of my instrument connected up or in other words the full connections of instrument made now when I came to send to him he could not get any thing so I s[e]t about to see what the matter was I found it The receiving arm when raised and resting on insulation on spring underneath was making Earth by virtue of greace and dirt So when I transmitted I was making Earth through my receiving shunt or magnets So of course Dublin could hardly get any thing Of course when receiving arm was thrown clear back This fault was obviated Now here is where Mr Wright made the mistake and got me into disgrace when he was sending to me while at Plymouth (being the same instrument he used on that occasion) He was sending most of current to Earth at transmitting end I feal all right now since I have found the trouble Well by raising the arm upright or throwing clear back he could read me all right at 120, 130 words on Iodide paper The[n] he sent to me again as usual I could not get any thing from him I told him he was wrong some where and mentioned the fault in my instrument so he raised his receiving arm clear up this helped me some I got better current but no inteligible signals so I told him: Well we tried one or two more evenings with precisely the same results His current varying fearfuly with me I told him trouble was with him but he thought not so it passed on untill last night when we came to work again he received the same as usual from me When he sent to me there was nothing too it so I came to the conclusion that the f[a]ult was certainly with him in his instrument or near him some where (knowing it was not in wire for we had worked two diferent ones) so I had him put in this arrangement: and I had it this way[4]

Result was I got steady mark on paper showing me he was making earth through his instrument or some where near him but he stood to it that he had it all right so I wrote him asking how it was: But this morning I receive a letter from him saying that while I was making connections for you last night I received a shock from an unexpected source and I concluded there was some thing wrong So after we got through I tested the terminals of my instrument and found the insulation gone So you see that it is no wonder I could not get inteligble signals from him up to this time it also explains why he has to use Iodide paper So I hope now I will be exonorated from blame for not receiving I know they thought I did not know enough to do it: And It was my place to go to Dublin So as Mr Parrish and Col Gouraud are both in Paris at present I have writen them the circumstances I tell you Al you have not given me your support since I came here And it has not been all smooth water You have writen Mr Wright as though he was the superior in knoweledge of the subject when you know he is not And he has taken advantage of it I acknoweledge he does it nice in one way for reason he has such good address and every one has helped him I realy have had to stand of[f] on my merits Colonel Gouraud when he came from america would not allow me salry that you promised He only allowing me £6.0.0 pr week Mr Wright £10.0.0 Of course if Mr Wright is worth so much more than me I wont say any thing but they seem to forget that it costs me as much to live & Dress as it does him It is not like being there with you I have said nothing as we have not been buisy and things not looking very bright here But when Colonel Gouraud said that £6.0.0 was what he understood was to be my salry and could not pay any more I said never mind undoubtedly it would be made right there so nothing more has been said (£6.0.0 includes expenses) We work again Monday Tuesday Wednsday when I hope things will go of[f] better Cant you give me the benifit of some of your cable experiments and others since I left it would be very acceptable I think I have done very well since being in England I have studed and worked so I have not gone back any upon what I knew when I left New York I give you some extracts of reports made by Lumsdon upon your experiments between London Liverpool May be you have seen them but I think not It is supposed I know nothing of them so dont get me into trouble

> As compared with Edisons system of single currents Heavy battery power and electro magnetic compensation it appears to me as far as can be judjed from a single trial that our own system of double currents (Reversals) might posibly answer all our requirements: The average speed from Dublin on the Wheatstone is about 60 words pr minute
>
> Found no difficulty in reaching 90 or 100 words perfect using 80 to 100 cells (On chemical paper) With intermittent currents followed by insulation we get good marks at about 80 words pr minute the marks being perfect: Due to the prolongation of the discharge from the cable which was cut off at the proper time by the reversal: But the best marks and the highest speed was obtained by using currents through 10,000 Ohms resistance as compensation between the signals in the Dublin transmitter: Permant currents reduce the speed: There was a great deal of contact on the line the reversals cut it out perfectly: Higher speed might be obtained by increasing the deration of the marking contacts: Bains system has never had a fair trial and the results so far are very satisfactory being an increase of 50% on a circuit having 65 miles of cable in it: Principle drawback is cost of the solution: The marks not being permanent I think is a secondary consideration: What we want is a cheaper solution but equaly sensitive: In these days of high speed it seems probable that we shall have to return to the Bains System again.
>
> Signed Lumsdon[5]

Mr. Scudemore

This as a first experiment will no doubt have great interest to you and will show what the value of Edison's System is not so great to us as may have seemed for we have systems of our own which will compete with it

Signed R S Culley 15 Dec 1873[6]

Mr Culley

The actual speed obtained between glasgow and London was 103 words pr minute being the utmost rate at which the transmitter would run it was by no means the limit: To aberdeen the results were not so good owing to contact & leakage: On the usual Wheatstone Wire gague (no 4) which was working badly we could read well at 80 words pr minute although the contact was so great as to give us perfect reversals from some other Wheatstone Circuit, Showing that the reversals on the double current system cuts of[f] the contact nearly altogether at least up to a certain speed: Through to aberdeen via Edinbourough we did not reach a higher than 60 words pr minute and from same cause contact and leakage So apparently where a Wheatstone would fail we should have a large margin to work upon with a sensitive Bains: The results obtained are I think so far very satisfactory as showing what can be done on the Chemical system with our ordinary double currents and the reversals battery power: The next thing is to set about geting a cheap solution giving if posible a permant mark: I find that a solution of ferrid cyanide of potassium and Sal ammoniac with an Iron stylus is ten times more sensitive then the old bains solution: But it is not satisfactory: However there is no doubt but that we can get an equaly effective solution to that now used by Edison

Dec 22, 1873 Signed D Lumsdon[7]

Mr. Scudemore

These further experiments are very satisfactory we certainly need not treat with Edison who uses a battery which gets ordinary percha covered wire on fire if incautiously applied: It would seem desirable to revert to chemicaly prepared paper in place of the Wheatstone receiver on certain circuits

Signed R S Culley 22 Dec 1873[8]

Pleas dont let out that I have sent those extracts to you as It is a breach of confidence on my part It came from Sir James Anderson I had the pleasure of examining the full reports so

I took those extracts: Posibly you may have seen them before if not It will give you some imformation I delay mailing this untill we have another trial Monday and to give me time to find out about those condensers I will get all the imformation that I posibly can for you: I think by what I have heard say by Smith Fleming & Co. that you would be all right for instruments and appliances to use on cable work if you should come over here You must give me all the imformation you can about the business For my own satisfaction and for the good of the cause wherever I am engaged in it for you I dont think as a pupil you need be ashamed of me at present but I wish to make rapid progress and a word from you would go a long ways in helping me along About our Perforators they have been sick but their disease was not fatal as we found Doctors that was sufficiently skilled to cure them They are liked very much by operators here But allow me to sugest that the motion of key be shortned or reduced and that you had better make them with more space between keys (The base is large enough) as one is very apt unless paying strict attention to strike two keys at once This is the only objection that they find with them with the exception of them making quite a noise even when placed upon rubber undoubtdly you can fix this so it will be all right: I think it would be advisable to pay attention to those things mentioned For as we are competing with them we have to beat them on the main point and do as well on the minor ones O[ur?] auto instruments are not just the thing they always compare the things to a sewing machine of course this does not matter but I think it would be just as easy to get them up in a neat and compact form as it is to have them as they are We want somthing that insulations will not be always bothering and Drum so we will not be continualy rubing it of[f] with emory paper so as to get good results Why not have them fixed with Platina on receiving drum or if it must be of nickel put it on thicker we are bothered with those we have now and besides it looks bad to those that witness the experiments to see us cleaning the drum every five minutes and the receiving paper all blured underneath I dont go in much for Clock work myself with High speeds but those that are in Po Ofs thinks that Clock work~~be~~ would be just the thing You recomended Piano Forte wire for pens: do you get the best results from that wire If so why does it not do just as well here For as yet we cant do as well with it as we do with other wire, even the needles you sent us is better How would it be for you to send us Coils Rheostats

& co so we could put up an artificial arrangement here so as to try diferent things when we had lesiure I think it would help us greatly when we had work to go at It would not be so much like working in the dark You know people here expect when we take hold of a line that we have every thing cut & dryed and that we are to do good work whatever the conditions may be and the conditions to do rapid work over the Lines of England Is the worst that I ever heard of, unless Cable. How are geting along with that roman character auto and are you working it on the Line or excuse me I believe it takes three wires at least to work it practicaly does it not If so how is it accomplished What is the receipt[9] for that Iron Solution we use now. Please let me have it if no objections exist Post ofs People buisy trying to find out somthing equaly sensitive They are experimenting all the time and are trying to work Wheatstone with receiver using chemical paper But Unfortunately for them the use of reversal of currents is limited to low speeds as compared with ours No I was not aware that an atom of Hydrogen would cast a shadow please explain how you discovered this and how is the shadow projected so as you could see it I understand you have lower part of shop fitted up into a labratory and experimental room well I hope I shall have the pleasure of working in it some day in the future I suppose That Mr Batchelor and others have taken my place since leaving I wish them luck Give Batchelor my compliments and tell him I have become quite a cockney know london very well and can find my way about in the fog (Providing it can be cut with a[a] knife) Which is the case occaisionaly but very seldom It is the worst I ever saw

Monday Evening also Tuesday: We have been working Dublin london to day with very poor results indeed at least with me I cant get any thing at all that is readable Contact is somthing fearful over this wire so much so that at this end it is very hard to read Morse I can get Wheatstone just as readily on my relay as I can get Dublin and the worst of it is on the same adjustment that is when I adjust for Dublin so as I can ~~h~~ read him or hear him then I get Wheatstone on the relay too So you can judge what it is to work automatic: By the way I use 130 cells on Morse as well as automatic I dont know if Mr Wright can let me have more battery or not I have asked to try for 150 cells so as it may be posible for me to work over this trouble It seems that when he sends to me that current from some Wheatstone when in opposition to ours cuts the dashes all up into dots I send you specimen[10] When I try to cut under the current is very irregular seeming to commence very faint then gains strength so the last part of mark appears broad and all right then it seems to fall off and I get very faint marks for probably a space of an inch or so when it gets strong again I send specimen Mr Wright sending to me on 110 cells This is the damdest line that I ever saw I cant do any thing with Iodide paper for contacts actualy shows itself when our battery wont make a mark In fact when working with the paper we could read signals evidently from some Wheatstone circuit. I can do this on my relay: of course they are backwards Cant you suggest something so we can get over this: They seem to have a horror of more battery being applied Mr Wright of course gets or feals the cable more than I do but he dont get that contact: he receives about 135 to 140 words pr minute on Iodide paper Battery 130 cells, While I, on any paper cant get key speed We only have 2 hours at a time to work so it dont give us a chance to make many changes in one day and to study the effects The fact of the matter is this we should have instruments and appliances here so as to work privately (Experiment) Then we should know what to do: Where now it takes several days to find out after we have the wire (Why dont you have a shop here if you intend to do much work over here It would pay) Smith Fleming & Co are very anxious to see you here I wish you would come Your presence is neaded very much and I think it is money thrown a way for you to delay much longer as the thing is in such a shape now that it only wants some one here that people have confidence in: Then automatic is a sucess in England and Europe Undoubtedly if we should succeed in doing 200 words pr minute both ways over this Dublin Wire automatic is Hunky boy all over this part of the world so if you have any thing that will do it by not putting on more battery you had better send it immediately They may let us have 150 ~~el~~ cells but I doubt it very much as they hesitated to give me 130 By the way would not a relay in circuit cut up the dashes somthing like sample (One of our relays) You know this is too funny that I can get key speed People here seem to be rather discouraged ~~here~~ because we cant do any better over this wire But I have tryed every thing that I can think of as yet different kinds of Iron for pens at different angles and all that but results remain the same with the exception that different angles of pen gives lighter or heavier marks but not any more inteligble The Line is all right with the exception of this contact as they work Wheatstone beautifully over it so if any faults they exist with us (our instrument connections) One very akward thing here for experimenting we have to depend upon others to make our battery connections through Test box which is out of sight: While at Plymouth I had battery under my own eyes and fixed them as I chose I used to dose them with sulphuric acid well when ever we had an experiment to make I found it made an auful diference especialy in the resistance I tested some of the cells before putting acid in I found them[a] to vary from 3 to 6½ Ohms resistance so you will understand what obstacles we have to contend with The ones that we are using now I dont doubt but what they will average 3½ to 4 Ohms to a cell and we cant help ourselves How would a sulphate of Mercury battery do It has almost the intensity[11] of carbon battery and much less resistance besides being very near as constant as daniels Battery[12] Any way I wish we had somthing better than this They say the resistance of this Dublin Circuit is over 4000 Ohms if that is the case their joints cant be any thing to brag of as it is a No 4 wire to cable 266 miles which should not be over 9 Ohms to mile cable 62 Knots 10 Ohms to Knot and No 8 wire from cable to Dublin 10 miles the whole circuit should not be over 3,300 Ohms resistance and it tests 1000 above this according to what their engineers say Allowences must be made for this of course: they do very bad testing at times: such I have seen

A ½-microfarad condenser from Elliott Bros. of London, now in Edison's reconstructed Menlo Park laboratory at the Edison Institute, Dearborn, Mich.

Wednsday I have found out probable cost of your condensers: Estimate by Elliot Bros £3000.0.0. for 10 of 100 microfarads and 10 of 50 microfarads each they say that they can get them up in 6 months after order is given or they can let you have them by instalment 7 at a time Insulation to be of parifine They say that parifine is an excelent thing to insulate with but in the ordinary condition it cracks after a while on account of crystalization and spoils the insulation but they treat it in a peculiar manner (which they keep to them selves) which prevents this crystalization and preserves the insulation You can judge the cost of those things when I say that they charge £55.0.0 for one of 20 microfarads and £7.0.0 for one of ⅓ microfarad But as they get them up there is no doubt but what they will be first class They say that it is out of the question to insulate them with mica they cant get enough of it posibly it might cost less but is very doubtful I think you can do better within your self They have some of the finest reflecting galvanometers that you ever saw I saw one of them tested to day It gave a deflection with one cell of Danniels through 15,800,100, Ohms of 20°. O.C.[13] and through 1,588,000,000 1°. O.C. It is wonderful I begin to think your are right through a line of 200 miles of glass with 50 cells you could get signals through I send you catalogue of theirs[14] Their workmanship is not to be beat it is the finist that I ever saw Today we had another trial between Dublin London Mr Wright received 120 words pr min on Iron paper good with 130 cells I received 100 words pr minute he sending to me with only 100 cells arrangement to transmit on like this

I get contact fearful so much so that I can get it on my relay as well as Dublins signals I want more battery and must have it to get over this Mr Wright says he will get me all he can Would it not be better for him to send to me with 180 cells old style shunt 1000 or 1500 Ohms than to have 130 in series two batteries as shown above and if he put in two or three magnets in shunt would I not get effect in proportion to magnets put in arrangement above Paper wetter all right nothing ever worked better no fault to find with it only if we have much work to do we shall have to have another Mr Lumsdon not received the books Mr Marsson[15] & Mr Morgan[16] sends compliments and hopes you will succeed in every thing you undertake Mr Lumsdon at present is at sea engaged in some cable work When do you think of coming over Give compliments to Mr Murray & Miss Frazee[17] tell Murray to do that hope they are both well & flourishing Hope you will not be disgusted with this if so it is your own fault If you conssider this worth answering Please address c/o Smith Fleming & Co 18 Leadenhall St E C Compliments to all Friends & co I remain yours Truly

J. W Brown

ALS, NjWOE, DF (*TAEM* 13:90). [a]Interlined above.

1. Brown indicates in the letter that he completed it on Wednesday, which was the fifteenth.

2. Doc. 418.

3. The fan of a Wheatstone automatic transmitter regulated the speed of the clockwork. Finn 1894.

4. In the drawing "z" is zinc and "c" is carbon.

5. This letter from David Lumsden to Richard Culley, dated 15 December 1873, is Item 98, ATF. Brown elided throughout the extract.

6. This letter is Item 99, ATF.

7. This letter is Item 101, ATF.

8. This letter is Doc. 384.

9. Formula for making.

10. Edison's scrapbooks from this period contain many unidentified samples of messages received by automatic telegraph, perhaps including this specimen and the one next mentioned.

11. Voltage.

12. The Daniell cell produced a constant current for several hours. Atkinson 1890, 780–81.

13. Possibly means "off center."

14. See Doc. 418 n. 4.

15. Unidentified.

16. Unidentified.

17. A Miss Frazee is listed in Edison and Murray payroll records between 1872 and 1875 (PN-72-08-16, Cat. 30,106, PN-72-06-14, PN- 75-00-00.2, PN-75-00-00.1, all Accts. [*TAEM* 22:219–368)]. An account book containing a variety of entries from 1873 to 1877 lists her address as 470 1st Avenue, Elizabeth, N.J. (Cat. 1214:516, Accts. [*TAEM* 21:702]). She is probably the Sarah Frazee who married Mary Edison's brother Charles Stillwell (Edison family genealogy prepared by John Deissler for the Charles Edison Fund, EBC).

–421–

From Jacob Edison

Fort Gratiot April 13th 1874

Dear Cousin

I made out the deeds for those two lotts that we were ta~~k~~lking about when you was here and I sent It to you to Execute and I have not heard from you since[1] did you get the deed If so and It suits you please Execute It as soon as you can and send It to me as we are all ready to build and the old Gent is houning me for Rent on the Lot we now Occupy and Pitt[2] dont want me to pay him one cent for he is conducting himself fearfully he told me he was going to marry that thing on Sunday and I am fearful he will do It as he is perfectly Insane and we are all in a great way about it.[3] It is strange how he has turned out he has no regard for himself or any of his family he is now bound to send the hole Edison tribe to Jacksen if he can but I guess he will fail Pit will send a Post Scrip In this letter and If you can oblige me Al I wish you would do It amediately and If not let me know

Jake[4]

PS Pit will write you or has done so today In stead of writing a PS in this letter[5] Jake

ALS, NjWOE, DF (*TAEM* 13:20).

1. Nothing is known of this visit. However, Josiah Reiff later testified that Edison was making trips to Port Huron about every three months during 1873 in regard to financial problems of the Port Huron railway (Quad. 71.1, 49–50 [*TAEM* 10:29–30]). Edison might have visited Port Huron in the winter of 1873 in regard to properties he purchased at that time from his father and brother (Register of Deeds, Liber 49:220, 230, MiPHStCo). The two lots referred to by Jacob may be those purchased from Thomas by Susan Edison in December 1873 (Register of Deeds, Liber 48:324, MiPHStCo).

2. Edison's brother, William Pitt Edison.

3. This is a reference to Mary Sharlow, who had borne Samuel a daughter in December 1873. The marriage did not take place. Conot 1979, 79; Edison family genealogy prepared by John Deissler for the Charles Edison Fund, EBC.

4. Jacob C. Edison and his brother P. M. Edison operated the Edison Brothers general store in Fort Gratiot. Endlich 1981, 102; see also the photographic copy of the advertisement for Jacob Edison's store in the June 1862 issue of Thomas Edison's *Grand Trunk Herald,* printed in Simonds 1940, before p. 65.

5. Not found.

–422–

From Taliaferro Shaffner

New York April 16th 1874

My dear Sir:

I am now ready for American Automatic systems and want your improvements inventions and informations.[1]

When you come to New York again please let me see you and in the mean time let me have your thoughts, and then let me know to what extent you can assist me.

My office hours are from 12 to 2 PM. but I may fail to be there before 1. I can be there at any time to meet you.[2] Your Respectfully

Tal P Shaffner[3]

ALS, NjWOE, Scraps., Cat. 1173:2 (*TAEM* 27:348).

1. Shaffner was revising and expanding his 1859 work, *The Telegraph Manual,* to reissue it as *The Telegraph Monitor.* The latter was never published. Advertisement in *Telegr.* 10 (13 June 1874): v; Reid 1879, 671.

2. Shaffner wrote from 80 Broadway, which also housed the corporate offices of the Automatic Telegraph Co.

3. Taliaferro Shaffner, one of the earliest telegraph promoters in the United States, organized several companies. Besides publishing *The Telegraph Manual,* he edited the *Telegraph Companion* in 1854–55. Shaffner 1859, 841–44; Reid 1879, 670–71.

–423–

Notebook Entry: Induction Coils

[Newark,] Apl 18. 1874

Induction Coils experiments.[1]

1. Wound spool for single current to be made & broke quick of Iron wire. Got shock but not very strong. Size of our relay spool[a]

2.  Wound 3 layers sounder wire next to core & fine iron wire #35[b] over it I felt it very good shock from carbon battery[a]

3 Made coil longer & put 4 layers Sounder wire on & filled up with fine Iron wire spool being about 5′ long, & weighing about 3/4 lb. Made battery out of an old glass jar & put a piece of sheet zinc & small piece of Carbon The induction from this arrangement was very strong

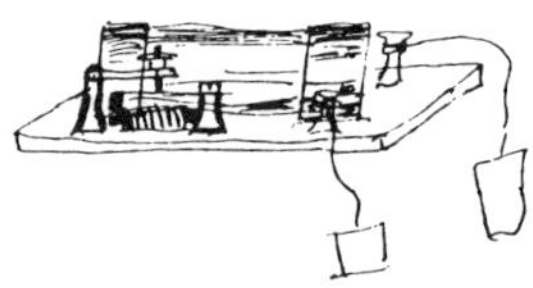

Edison and Batchelor's modified induction-coil design.

X, NjWOE, Lab., Cat. 1170:10 (*TAEM* 6:279). Written by Charles Batchelor. [a]Followed by centered horizonal line. [b]Interlined above.

1. This appears to be the first experiment directed toward the instrument that became known as the inductorium (see Docs. 434 and 435). On 26 April (Cat. 1170:10, Lab. [*TAEM* 6:279]) Edison and Batchelor altered the design of this experimental instrument. On 28 April an order for a model "Shocking Mach" was entered in Cat. 1214:44 (Accts. [*TAEM* 21:586]) and credited to the Scientific Toy Co. (see Doc. 428). Edison later linked the development of his inductorium to the start of his association with James Adams (App. 1.C15, D32).

–424–

Notebook Entry: Electromotograph[1]

[Newark,] Apl 21 1874

Edisons Force

Lead pen Iron base

1 I.K. good =	1[2]
2 Ferridcyanide Pot. =	1 1/4
BiTartrate Potash =	2
Sulphate Potash =	3
Permanganate Potash =	3 1/4
Acid Potassium Tartrate =	3
Nitrate Potash =	3 1/2
Bromide Potassium =	3 1/2
Ferrocyanide Potassium =	2 1/4

Sulphate Iron = ½
Ammonia Citrate of Iron = ⅛
Sulphuret Potash ½
Sulphate of Manganese ¼
Sulphate of Soda = 3
Boracic Acid = 1
Salt Chl. Sodium = 3½
Sulphuric Acid 0
Caustic Potash 3¾
Caustic Pot. & Flour 2~~½~~
HO
⟨E & B 22 hrs⟩[a]

X, NjWOE, Lab., Cat. 1175:192 (*TAEM* 6:201). Written by Charles Batchelor. [a]Marginalia written by Batchelor.

1. This entry is continued in Doc. 425.
2. The meaning of the figures in the second column is unclear.

–425–

Notebook Entry: Electromotograph[1]

[Newark,] Apl 21st [1874]

Night Tried metals as to which was best for testing new force

Platina on Platina
Lead on Platina
Lead on Iron

The base apparently is indifferent as to which metal is used[2]

In testing with carbon to rubber or metallic pad[a] we found that there ~~seemed to be~~ was very little friction at first when line was closed but soon it became very rough which quickly smoothed down as line opened

The reverse of this is the case when zinc is to metallic pad.[b]

With 3 thicknesses of paper on drum it is if anything a little better.[b]

Tried Blue vitriol & Bichromate crystals, Glass & Hard rubber to make it go better by going through it but no go.

Heated metallic rubber no go.

Just got writing through 2 ordinary Water Rheostats 1 large ditto & 1 small ditto with 13 cups carbon bat.

Mirror galv gave deflec 7° through same shunted with 300 ohms (Bradly nix)

200 ohm relay ceased to work when 1 of the reostats was out about 1 inch

Phelp relay 126 ohms refused at about 2″ out of 1 Rheostat.

Tried	wet leather	on	drum	(No	good)
"	Hd Rubber & cloth	"	"	"	"
"	~~Co~~ Wood	"	"	"	"
"	Indiarubber	"	"	"	"
"	Phosphoric Ac.			"	"
"	Card			"	"
"	G[old] & S[tock] paper & [S?]hobart[3c]			"	"
	Dry paper			"	"
	Brown paper			"	"
	porous cup			"	"
	Flour paper			"	"

X, NjWOE, Lab., Cat. 1175:193 (*TAEM* 6:201). Written by Charles Batchelor. [a]"or metallic pad" interlined above. [b]Followed by centered horizontal line. [c]Illegible.

1. This entry is a continuation of Doc. 424.

2. Preceding the first notebook entry recording experiments with the electromotograph phenomenon (Doc. 419) are Edison's notes on an experiment with chemical recording solutions in which he used different metals in his recording pens and in the base of his receiving drum. Cat. 1171:106, Lab. (*TAEM* 6:263).

3. Cf. "Schober paper" in Doc. 354.

-426-

From Jack Wright

Dublin April 28th 74

Friend Tom:

Let the following edify and delight(?) you.!— I have been in Dublin five weeks & have had numerous hacks at the cable and will give you in as condensed a form as possible the results so far.

To start with:—as a rule I receive better than London.! There are however 16 miles of buried wire at that end which seem to have a very bad effect in the way of contact induction etc.[1] In fact until the last 2 or 3 trials Brown could'nt receive a bit, not even at 50 words a minute. This was owing principally to defective instruments, bad contacts, short circiutings etc. He has had them seen to & is getting on better. I have received as high as 150 words very fair writing, but some days I cant come anywhere near that! This may surprise you, but wait a little & perhaps I can explain.

In the first place let me speak of the magnet boxes.—[2] They test horribly— The following is the result of a careful test of the "circular" box which I have here.[3] Total resistance. 436.— Res of each mag beginning at No 1 27. 0. 47. 57. 88.

90. 112. 0. 144. 0. 158. 190. 0. 0. 202. 0. 220. 256. ~~3~~280. 0. 331. 0. 380. 0. 420. 0. 436— You will see that some of coils have no res at all. This does'nt matter so much at the ~~be~~ end of series, but at the beginning it is awfully deceiving— There is very little "Kick" or countercurrent[4] from this circular box. In fact its anything but what it ought to be!!— The other box which you sent me last, the one with the 3 mags in with plain res. tests very well.— The mags as follows.— 13. 47. 0.[a] The last 2 being apparently short-circiuted— The plain coils are very near what you put them at,—viz. 5. 14. 34. 53. 102. 200. 400. 890. 1900 and Total 1947.— All these tests were made with the bridge & a very sensitive defferential galv. and I think may be relied upon.— The "long" receiving boxes have also been in bad condition.[5] Brown has had his entirely renovated & mine is now being looked over & made perfect.— The contacts were simply miserable, being made by the plates being only pressed down on the little bit of wire coming from each mag.— This was probably all right for land lines, but on this fine work, I dont think anything but a soldered contact is to be relied upon. I'm getting as bad as these Englishmen you see old man.— All this however amounts to very little for it can be remedied as soon as discovered I will now tell you about the line & cable itself. You know the cable has 7 conductors in it & is just 60 miles long. The 7 conductors dont seem to do much harm strange to say for I dont feel nearly the contact or induction we did between London & Lpool.— The great difficulty is to clear the cable as it were, when working at or above 100 words.— In this respect the mags at the transmitting end have a wonderful ~~result~~ effect, much greater than I expected, in fact without them we could do nothing when batteries are arranged as you last advi~~e~~ssed i.e. opposing each other.— And instead of using few mags as you told us to do, we find that a number, 14 or even 20 give best results— As we increase mags at transmitting end, so can we decrease mags in shunt at receiving end.— When no mags in at transmitting end it seems as tho' discharge from cable killed the dots & we require mags at that end to kill discharge as it were, & let dots come through. Using battery connected up for quantity with a shunt of 750 to 1000 ohms results very nearly same as with battery divided & opposing:—showing that shunt at transmitting end is[b] wonderful thing if it were only perfect.—[6] Between London & Plymouth your last plan acted admirably, but over this blessed cable it falls in a greater proportion than I should have imagined— If a perfect earth could

only be inserted between each signal at the transmitting end I think results would be greatly improved. Mind I say a perfect earth, no insulation or the slightest interval of time, not even that caused by the thickness of the paper.

On a fine clear dry day the amount of static is perfectly awful & it reduces our speed nearly ⅓— Mind, they only allow us 130 cells in each battery, or 260 altogether. This is the maximum. I think that beyond certain speeds & under certain conditions the current from a Daniells battery acts differently to that from a carbon or grove. This is only an idea. The internal res. of Daniells is too great, it has'nt that heating penetrating property that the others have.— I noticed a peculiar thing today for the first time. I got a mark on the paper without a current. In other words the pen was decomposed without electicity. This is probably a fault in making the paper?— Your suggestions about the position & pressure of the pen hardly apply to this work. The finer the pen we find the less current it requires. God knows we twist them in every position & put pressure on as much as they will bear. The highest speed on the Wheatstone experimentally, chemical paper from London to Dublin is 70 & from D to L 80 wds per min. The sending arrangement is this. They send nothing but dots. The dash is made by the insertion of 10.000 ohms res. immedy after impulse & actually making mark at receiving end with the static itself, which is crowded out at that end as it were by the sudden insertion of such an immense res. at transmitting end.— When using ink recorder they get their dashes as you know by retarding the reversal until requisite end of the dash. This when received on chemical paper looks rather comet-taily as the tailing is used to [-] make dash—

In duplex working, to balance the line they use a condenser of from 15 to 23 microfarads according to circumstances. This will give you an idea of el~~t~~ectro-static capacity of line— Condensers so far have not been perfected & are not as a rule permanent, some of them losing capacity after time..—

You must accept all the above in the spirit its written, I may have made mistakes & foolish errors which you must forgive. However there may be something in it worth knowing— I think it would be a very good thing if you were here personally. Nothing would stand in your way now in the shape of opportunities for experiment. Our people seem to have things now at the point that nothing remains to be done but get good scientific results.

I answer questions in your last[7] as follows— Plymouth wire is 360 miles long. Res. averages 13.5 per mile. Nearly 18 miles uried wire.— The batteries are of the "U.K." form if you remember them. about ½ the size of an old fashioned "local".— The perforators will require some adjusting & perhaps improvment The punches dont cut clean, the feed is slightly irregular and they (the English perforators) complain of the length of stroke. Dont forget also that they will have to be applied to pneumatics—

Clockworks will be almost absolutely necessary when the speed drops to 200 words or under.— The personal equation is too great to do good work at that speed with the present arrangements— Paper wetter red hot.— Nobody knows formula for paper— Salt alone is an old dodge.—So they say!!—

I dont like to say anything against Mr Latimer Clark, but to tell the truth I dont think he knows any more than I do, and thats not much. Your questions to him are "pearls before swine" I'm afraid— Excuse my cheek in saying so. Dont be too fresh about telling these fellows over here anything new you may discover. Their cheek & unscrupulousness is only equalled by their [~~cheek~~?][c] politeness!!—

I enclose sample recd. from London, its a very good one though.[8] On a clear dry day cant do so well, at least not yet. May after little more experience—

Just try & take the trouble to answer this to me personally?— I want to hear from you very much but would rather see you.— Give my love to the boys. If you see Cottrell[9] say I'm still alive & will write soon. enclosing photo. of myself & wife!!—

Respects to Mr Reiff & Mr Harrington also dont forget to remember me to Edwards.—[10] How's things going on with "New Company[11] etc?— Good Night. Yrs as ever

Jack.

ALS, NjWOE, DF (*TAEM* 13:100). [a]Underlined twice. [b]Interlined above. [c]Heavily canceled.

1. The underground lines at London apparently caused interference between wires ("contact") and created greater inductive effects.

2. See *TAEB* 1:608 n. 2.

3. See illustration, *TAEB* 1:563.

4. Inductive discharge.

5. It is not clear what form of rheostat Wright means.

6. Wright and James Brown found that they needed different circuit arrangements for transmitting from the two ends. Brown to TAE, 29 Apr. 1874, DF (*TAEM* 13:104).

Transmitting circuits for the Dublin (left) *and London* (right) *ends of the automatic line.*

7. Doc. 418.
8. Not found. See Doc. 420 n. 10.
9. Probably Herbert Cottrell, a Newark jeweler who developed a stone-cutting band saw and founded the Cottrell Stone Machinery Co. Ford 1874, 78–81.
10. Unidentified.
11. Probably a reference to the Domestic Telegraph Co.

–427–

James Parrish to Josiah Reiff

London 4/30/74

My dear Reiff

After three days trial, ~~hou~~two hours each day, we are compelled to admit that our work would not be good enough for "commercial" purposes— So far as I can learn the cause is insufficient machinery ie a want of magnets with cores of the right size— The meaning of it all is that we want Edison here with as large a variety of machinery as he can bring— Our friends who propose to assist in floating the American Company will be very lukewarm if our European Experiments fail with Edison absent, all winter they have constantly said why dont Edison come and now [-]after obtaining every facility from the P.O. and not succeeding between London And Dublin they say "Ah we were afraid of that, you ought to have Edison here—" Understand that for the financial sucess of your American scheme you must make the European good, the one in which the men who are to aid us in America have put in £10,000 for a fifth interest—[1]

Brown & Wright are attentive but even if Edison can do no more he should come over if you expect our London friends to keep up their interest— There will be trouble if Edison does not come soon— For nearly a year we have been keeping our friends encouraged and using them and now when our combined efforts have opened doors that were closed to us we are without the means to take advantage of the opportunity.—

If you expect any more money from Europe send Edison and let him bring clock work and any combination of machinery that may be wanted Yrs Try

James C. Parrish

I know you cannot send Edison without making sacrifices but I think you are sacrificing all by not sending him—

ALS, NjWOE, DF (*TAEM* 13:106).

1. The European scheme probably refers to the arrangement with Smith, Fleming and Co. (Doc. 350), and the American scheme probably refers to the negotiations to form a new company for automatic telegraphy (see Doc. 521 n. 10).

-428-

Memorandum: Scientific Toy Co. Proposal

[Newark, April 1874?[1]]

List of articles to be manufactured by Scientific Toy Co= with prices.

Sewing Machine for Little Girls	2.00
Magneto-elec-shocking Machine[2]	2.00
Induction coil for "	2.00
Magnetic Engine with battery	1.50
Magnetic Locomotive with track	2.00
Printing Telegraph Machine.	2.00
Printing Press	3.00
Printing case of Types	5.00
Printing Machine	1.00
Microscope	2.00
Galvanometer	1.00.
Gas Making Machine	4.00.
Loom for weaving	4.00
Band Saw.	3.00.
Air Pump	2.00
Spectroscope	3.00
Electric-Bell.	1.00
Lathe.	5.00
Planer	5.00
Drill Press	3.00
Steam Engine	1.00
Caloric[3] "	1.50
Gas Engine	1.50
Kal[ei]descope	.50
Turbine Wheel	1.50
Fire Engine	1.50
Lithograph Press	5.00
Electric ~~Steam~~ Boat	3.00

AD, NjWOE, Lab., Cat. 297:59 (*TAEM* 5:637).

1. This document was probably written about 28 April, when an order was placed for models of a "Shocking Mach[ine]" and a "learners Inst recording" for the Scientific Toy Co. Cat. 1214:44, Accts. (*TAEM* 21:586). It is possible that this document was written after Edison acquired both a spectroscope and a microscope in January 1875 (see Doc. 544 n. 2). However, no other record of this proposed company exists, and Edison and Murray was selling both the inductorium (shocking machine; see Docs. 434 and 435) and the learner's instrument (see Doc. 438) by the summer of 1874.

2. See Doc. 423.

3. A name applied to hot-air engines. Knight 1876–77, s.vv. "Air-Engine," "Caloric-Engine."

–429–

From Josiah Reiff

New York, May ~~9~~11th 1874[a]

My dear old boy

Inclose you 3 important letters just recd—Parrish, Wright & Brown.[1] I opened the two latter & read them casually, but of course there is too much electricity for me—

Wright seems to take to it very intelligently & Brown wades thro it like a worker—

Evidently my boy Edison better take a Sea trip—

When you were here the other day—Harrington was down for first day & his throat was so bad was afraid to talk to you— If you go now you will be in time to do the work at once reinspire our friends & have them ready to advance more money on our new ~~Co~~American Co.[2] if not also a/c foreign patents & out of the first—you shall have $10,000—to take up your mortgages—[3]

Meantime we will have our stock certificates made & if necessary will make some Bonds on[-] our present line, so that in case we dont receive the English money in time, I will borrow the money outside on my own & the Cos securities & protect your property—

You know whether you can rely on me—

My well has never yet gone quite dry—& the interest of Edison is the interest of Reiff—— Wrights letter shews that we have things in good trim— take your wife & baby with you, if you desire—

After reading the letters & judging what is wanted, let me know what you desire to take with you—magnets etc etc— Yrs

J C Reiff

As these letters are all important please let me have them for file—[b]

ALS, NjWOE, DF (*TAEM* 13:115). Letterhead of Kansas Pacific Railway Co., J. C. Reiff, Agent. [a]"New York," and "187" preprinted. [b]"As . . . file—" written in left margin.

1. Probably Docs. 420, 426, and 427.

2. This probably refers to the proposed combination of the Automatic Telegraph Co. with several railroad company telegraph lines. See Doc. 522 n. 10.

3. Edison owed his former partner William Unger $10,000, secured by a mortgage on the Ward St. shop. See Doc. 452 n. 3.

–430–

From Josiah Reiff

[New York, c. May 15, 1874[1]]

Now my dear Boy

Stand by us—

I stand by my letter I showed you the other day—

In addition to the fact that we will have money here before your Mtge comes due, I will say that I have some RRd & other stock, which ~~n~~if necessary I will pledge to pay Unger so that if you go, your property shall be protected & the Mtge cancelled.[2]

Our future can only be affected now by failure of Dublin line, in delaying our money arrangements. You can do that work the[a] first day after you land & the name of Edison in Europe & America will be greater than Morse or Thompson.[3] Let me send the cable & give you money to fit out.

J C R

ALS, NjWOE, DF (*TAEM* 13:114). [a]Remainder of letter written in left margin.

1. This letter was written on the back of a letter from James Parrish to Reiff of 2 May (DF [*TAEM* 13:113]), and the transatlantic crossing took at least ten days. Parrish's note said: "You were yesterday telegraphed asking when Edison would leave. all that has been said urging his departure I now confirm—".

2. See Doc. 429 n. 3.

3. Sir William Thomson (later Lord Kelvin) was the most prominent British physicist of the time and had been knighted for his telegraph engineering work in connection with the successful establishment of Atlantic cables in 1866. *TAEB* 1:335 n. 2.

–431–

Technical Note: Cable Telegraphy

[Newark,] May 16, 1874.[a]

Experiments por Socitié du Cable Transatlantique Français[b] P~~e~~ar Mrs Edison & Batchelor[1]

No. 1.[2]

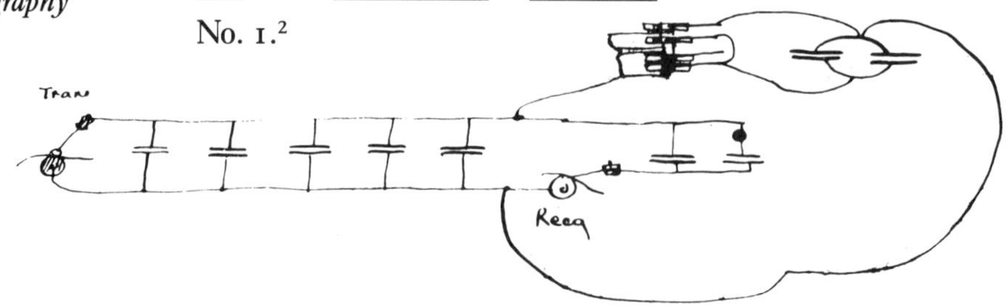

X is of same Capacity as now used on end viz 1 Bunster & [-]Red[3]

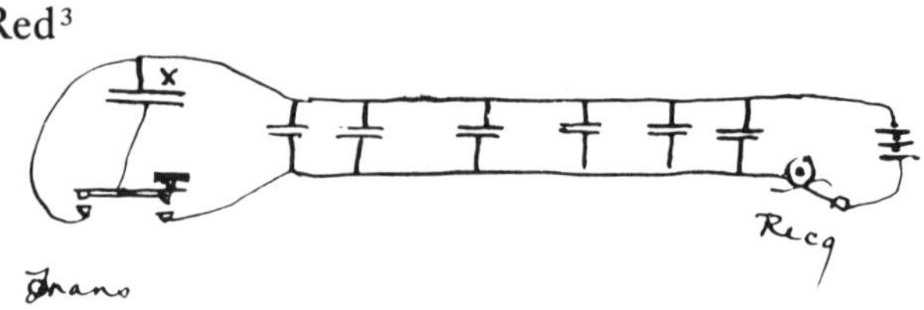

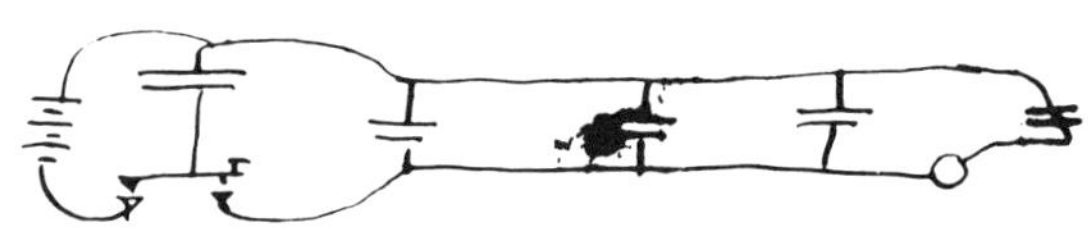

⟨Ink Recorder⟩

X, NjWOE, Lab., NS-74-001 (*TAEM* 7:36). [a]Followed by centered horizontal line. [b]"Experiments . . . Français" underlined twice.

1. Although the Société was a French company established in 1869 to lay an Atlantic cable, the cable itself and most of the funding were provided by British cable interests. In 1873 the Société merged with the Anglo-American Telegraph Co. Bright 1974, 107.

2. This drawing appears to show a roman-letter receiving instrument. Edison had prepared a demonstration of his roman-letter automatic system on four and two wires for George Ward, American manager of the French and Anglo-American lines on 16 April. In that demonstration he probably used an ink recorder such as that proposed in his caveat of 28 October 1873 (Doc. 373). The preparation for the demonstration is recorded in Cat. 1184:39, Accts. (*TAEM* 21:802). For Ward see Reid 1886, 526.

3. Unidentified.

–432–

To George Prescott[1]

NEWARK, N.J., *May* 19, '74.

DEAR SIR:

You probably think it strange that I have done nothing with duplex. The fact is, Mr. Orton's sudden disappearance took the bottom out of my boat, and I can do nothing without his or your coöperation.[2] As I have a great number of duplex combinations, which may be of value to the company in the future, I dislike to lose all the time I have given towards perfecting them.

I make this proposition—

That you give me facilities and personal help to test them and then take the patents out in our joint names,[3] and then present them to the company for purchase on their merits alone; profits, if any, to be divided equally.[4] Yours truly,

THOS. A. EDISON

PL, NjWOE, Quad. 71.2, p. 13 (*TAEM* 10:228). Original on letterhead of Western Union Telegraph Co., George Prescott, Electrician.

1. George B. Prescott was the chief electrician for Western Union and the author of a standard work on telegraphy. *TAEB* 1:258 n. 4.

2. William Orton, president of Western Union, left on a European trip on 28 March 1874, returning on 24 May. Edison had written him on this subject just before that departure but received no reply. Prescott was gone, on a trip to Florida, from 18 April to 4 May. Orton's and Prescott's testimony, Quad. 71.1, pp. 128, 390 (*TAEM* 10:69, 202); App. 3, Edison's Affidavit, p. 810.

3. Patents could be jointly issued in the names of co-inventors or they could be issued to one inventor but assigned from the start to more than one person. Edison later claimed that he had intended here the latter kind of partnership with Prescott, much like his partnership with George Harrington. Edison's "Answer," Quad. 72.14, p. 7 (*TAEM* 9:208); App. 1.G15–16; App. 3, Edison's Affidavit, p. 810.

4. Apparently Prescott initially agreed in general to this proposition in a telegram to Edison on 1 June 1874. However, that telegram, along with various other documents, was lost while in the possession of Grosvenor Lowrey, one of Western Union's lawyers, before the Quadruplex Case came to trial (Prescott's and Phelps's testimony, Quad. 71.1, pp. 329, 309 [*TAEM* 10:171, 161]). As a result of Edison's proposition, during the late spring and summer of 1874 he and Prescott prepared patent applications on devices already tested, tested variations of generally understood apparatus, tested new designs, and extensively developed one line of commercially practical equipment. Prescott and Edison also each met with Orton before that work got well under way (Edison's testimony, Quad. 70.7, pp. 267, 342–43 [*TAEM* 9:500, 538]); Orton's testimony, Quad. 71.1, pp. 132–35 [*TAEM* 10:72–73]). This agreement was later formalized (see Doc. 451).

-433-

Memorandum to Charles Batchelor

[Newark, May 19, 1874[1]]

Batch= Kinney of the Advertiser[2] wants a set of telegraph instruments for a line from Advertiser Office to his house in Broad St Think our Students Ink recorder would suit—[3]

When you have any write him & ~~te~~ say— I think we have some telegh instruments that would suit your proposed line from ur ofs to ur house Call and see their price $24. per pair—w~~e~~ill take it in advertising[a]

Write to Messrs Powers & Weightman Mafctg Chemists Phila[4] for price of Metallic Tellurium & if they have none to be kind nuf to inform you where its ore may be obtained in quantity write to Prof Cooke Harvard College[5] [-]ask Tellurium =

Write to the ~~M~~Edi~~s~~tor Mining Gazette[6] dont know where pub— Where you can procure Tellurium ore in quantity. Sig "C. B." Chas Batchelor

Where is that ~~d~~little round district[7]

AD, NjWOE, DF (*TAEM* 13:39). [a]Followed by short horizontal line at left margin.

1. At the top of the page Edison wrote and then crossed out: "Newark May 19. 74 Mr Kinney Dear Sir. I think I have an".

2. Thomas Kinney was editor and proprietor of the *Newark Daily Advertiser*.

3. See Doc. 438.

4. The firm headed by Thomas Powers and William Weightman was a major manufacturer of chemicals. An 1859 survey described it as the largest chemical manufacturing firm in the United States and one of the largest in the world. Freedley 1859, 207.

5. Josiah Cooke was professor of chemistry and mineralogy at Harvard College from 1850 to 1894. His most important work was on the classification of elements by atomic weights. *BDAS*, s.v. "Cooke, Josiah Parsons."

6. Probably the *Mining Gazette* of Houghton, Mich.

7. Edison is probably referring to a district telegraph transmitter, possibly the one he assigned to the District Telegraph Co. of New York in 1872. *TAEB* 1:415–27.

–434–

Draft Essay: Inductorium

[Newark, c. May 20, 1874][1]

Edisons Inductorum

~~This coil has been designed especially for family use. The in~~ The principle of the induction coil[2] was first discovered by the celebrated Experimental philosper Faraday in ______[a] and the first coils were constructed by Rhumokoff an ingenious French ~~in~~mechanician and by Prof Page an American in ______.[3a] ~~Since~~ But little improvement in their form i has been made since that time with the exception that machines of greater power have been constructed. ~~In~~ the ~~p~~Royal Polytechnic Institition of London has an immense induction Coil ~~some th~~ the wire bobbin of which is about ______ feet in length and ______ f~~oo~~eet in diameter, which gives an ~~sp~~ electric spark of 2 feet in length.

In the Stevens Institute of Technology at Hoboken is another very powerful induction Coil Constructed for that institution by Ritchie of Boston Mass.[4] which is exceeding[ly][b] powerful considering its si~~n~~ze when compared to ~~the~~ its English prototype in the polytechnic Insitution Prof Morton[5] has suceeding in making the spark[c] penetrat~~ing~~e through a sold block of glass 4 inches square ~~in th~~ the course of the spark being beautifully shown by the fracture in the line of the path. The induction coil in the hands of the hands of the sciencists has proved & is proving[d] a powerful instrument of investigation. = ~~But~~ It was not long after the introduction of the induction coil before it was discovered that the transmission of the electricity generated by it ~~had a power~~ through the different parts of the human body ~~had a powerful effect.~~ had a ~~powerful~~[e] tendency to overcome certain obstinate diseases ~~which~~ when all other means had failed, for instance chronic Rheumatism. ~~in the fo~~ Latr years a special branch of medical practioners have ~~be~~ sprung up whose rely soley upon Inductive Electricity as a ~~means~~ remedy and many Valuable facts have be observed many of which have been[f]

power of the Coil is greatly lessened. ~~The effort has until now proved unsuccessful The lowest priced coil now sold~~ and the want of a clean powerful & economical battery which should be of a lasting character. The effect has until now proved unsuccessful

The ~~lowest priced coil having~~ lowest price which a coil of like power & workmanship can be purchased in this county is twenty five dollars [a]nd[b] even at this price, owing [to th?]e[b] facts mentioned above very little profit acrues to their makers.

The cheapest coil sold cost twelve dollars but ~~the~~ the ~~d~~current from it is comparatively feeble.

This induction coil is wound with many turns of a four strand fine Insulated[e] iron wire, ~~instead of copper wire used in the in all previous coils, the price of which is much less. it is of the poorest quality it~~ in place of a single insulated[e] wire of copper. ~~there the~~ By using Iron wire the whole coil becomes magnetic and sets up an extra inductive current which is not the case with the coil wound with copper The~~r~~ res

The coils are then boiled for several hours in specially prepared parafine to thoroughly insulate each convolution from its neighbor and prevent loss of power.

The workmanship on this apparatus is fully equal to that

given by the best philosophical instrument makers, and unlike that of most cheap coils now sold, which are made without regard to beauty or lasting qualities. The battery accompanying this coil is a Bunsen Bichromate of potash battery complite, and the same as used by the telegraph Co with the exception that the zinc of th~~e~~is battery is chemically pure and will last twice as long as the ordernay zinc universally used besides giving a stronger current and evolves no hydrogen. It is the most powerful battery known and it may be used for other purposes than in connection with the coil— When connected to ~~e~~a small electromagnet it will cause it to sustain several pounds of iron in contact with the core   Among the uses to which the battery ~~m~~can be put to may be mentioned, electropolating with silver and copper etc, working ~~tele~~ magnets telegraph instruments, ~~s~~for short lines of telegraph for ringing electric bells, for decomposing water into Oxygen & Hydrogen gas. twenty cells will produce an intense electric Light between ~~Lead pencil point~~ carbon points & t~~wo~~hree cells will heat fine platina wire red hot= and innumerable other uses. The power of the coil may be greatly increased beyond human endurance by adding two to four cells instead of one, although it is seldom that a person can stand one half of the current generated from the one cell

The following are the specific diseases which it is known to cure although it is maintained by ~~many t~~ many works on medical electricity that it produces good effects in many diseases not mentioned   first of all & for which it is the best and only remedy is Rh~~a~~uematism.

add here diseases[g]

ADDENDUM[h]

[Newark, Spring 1874]

It may be made to act as a burglar alarm by attaching ~~a~~ two[e] plates of tin one the window sill and connecting the halfs of door knobs and allowing the coil to vibrate all night= or by connecting a wire so that when a window is raised or a door opened it will close battery ckt & the handles being connected to a cat she will give an unearthly & diabolical yell & wake all up

Patent applid for[6]

By pulling the regu~~g~~lating rod just out of the end of the coil a knife may be magnetized by pressing it on the tin part of the rod & passing it back & forward ½ an inch—

ADf (fragment), NjWOE, Lab., NS-74-002 (*TAEM* 7:298). [a]"in _______" interlined above. [b]Document damaged. [c]"making the spark" interlined above. [d]"& is proving" interlined above. [e]Interlined above. [f]Page 4 of manuscript ends here; page 5 is missing. [g]Followed by two vertical lines. [h]Addendum is an AD, NjWOE, Lab., NS-74-002 (*TAEM* 7:308). Each page is headed "note" at top center.

1. See headnote below.
2. See *TAEB* 1:59 n. 2.
3. See *TAEB* 1:82 n. 10.
4. See *TAEB* 1:82 n. 11.
5. Henry Morton (1836–1902), a professor of chemistry, had been appointed in 1870 as the first president of the Stevens Institute of Technology in Hoboken, N.J. He served in that capacity until his death. *BDAS*, s.v. "Morton, Henry."
6. No application has been found.

INDUCTORIUM Doc. 435

Edison's inductorium was one of a number of electromedical devices available that employed induction coils for giving electric shocks.[1] Edison began experiments leading to the inductorium in April 1874 (Doc. 423) and was selling some version of the instrument by 22 May.[2] The first reference to an instrument called an inductorium appears in a letter dated 8 June (Doc. 440). Instructions for maintaining the batteries, which were attached to the inside lid of the inductorium box, were printed about 20 May, and advertising circulars were printed on 16 June.[3] Edison and Murray initially sold the instrument for $4.50, but had raised the price to $6.00 by the beginning of June.[4] The higher price appeared in their circulars, which described the inductorium as "a specific cure for rheumatism, and as an inexhaustible fount of amusement."[5] By early July more than one hundred instruments had been sold. While a few were sold directly to individuals, most were sold through agents—Edward Johnson in Philadelphia; James Adams, A. G. Stolbrand, and Partrick, Bunnell & Co. in New York City; and Austin and Howe in Utica, N.Y.—who were able to purchase them at $4.50.[6] At the end of July, Edison and Murray began to advertise the inductorium in more than three hundred newspapers, and many more instruments were sold directly to consumers.[7]

1. On the use of electricity in medicine in the nineteenth century see Beard and Rockwell 1871 and Greenway 1987.
2. PN-74-05-22, Accts. (*TAEM* 22:111).
3. Cat. 1184:60, Accts. (*TAEM* 21:813). Copies of the instructions

A circular for the inductorium.

EDISON'S INDUCTORIUM.

This is an exceedingly powerful induction coil, designed expressly for medical and family use. It is constructed upon a principle recently discovered by its inventor, whereby most extraordinary effects are produced, without a corresponding increase in the size of coil, or battery power, as has heretofore been necessary. Hence we are enabled to furnish the public with an apparatus at one-third the price asked by other makers for coils of equal power.

The workmanship is of a very superior style, and of a solid and substantial character.

The battery which accompanies the coil is the same as used by the telegraph companies.

An almost infinite number of experiments may be tried, by manipulating the electrodes. This instrument should be in every family as a specific cure for rheumatism, and as an inexhaustable fount of amusement.

The second cut shows the manner of connecting the coil with the battery, and electrodes. The latter when applied to different parts of the body for medical purposes, should have moistened sponges placed in the hollow part of each. The handles, which are held by the persons applying the current, should be wrapped with dry paper to prevent its passing through his own body.

Directions for charging the battery accompany each box.

PRICE of each complete apparatus, SIX DOLLARS, sent C. O. D. by express to any part of the U. S. Liberal discount to Agents.

Address.

EDISON & MURRAY, 10 & 12 Ward St., Newark, N. J.

Manufacturers of Recording Telegraph Instruments for learners and private use, Telegraph Supplies, Batteries, Mirror Galvanometers, Resistance Coils, Condensers, Philosophical and Experimental Apparatus, etc.

can be found in Cat. 1178 and Cat. 1143:53, both Scraps. (*TAEM* 27:141, 199). The scrapbooks contain copies of a one-page circular illustrating the inductorium (Cat. 1178 and Cat. 1143:72, both Scraps. [*TAEM* 27:137, 207]).

Edison's draft of this circular and the draft of a second circular (in an unknown hand), which describes in some detail the instrument's medical uses, are in 74-001, DF (*TAEM* Supp. III).

4. PN-74-05-22, Accts. (*TAEM* 22:111–12).

5. See Doc. 434 for a description of the inductorium's uses.

6. Nearly half were sold through Johnson. PN-74-05-22 and Cat. 1184:76–79, both Accts. (*TAEM* 22:111–15, 821–22).

7. Bill of E. W. Foster, 28 July 1874, DF (*TAEM* 13:167). See also clippings of classified advertisements, Cats. 1178 and 1143; and E. W. Foster to Edison and Murray, 22 July 1874, Cat. 1173:40; all Scraps. (*TAEM* 27:141, 172, 173, 206, 374).

–435–

Newark, [c. May 20, 1874]

Production Model: Inductorium[1]

M (10 cm × 15 cm × 20 cm), NjHum.

1. See headnote above.

–436–

From Josiah Reiff

New York, May 26 1874[a]

My dear Edison—

I sometimes fear you feel dissatisfied, because things move slowly, & you see no return for your labor & &.ct—

Now you know I am your staunch friend & hence I never will complain of you or your intimates impatience, but shall go along as heretofore to work out for the general interest as rapidly as possible—

Of course I have not yet recd a cent, or none of my freinds but all has gone & continues to go into old automatic & I believe it & a great deal more will come out, but I have made up a little statement to show that th we have not been idle—

I find I have paid Murray a/c shop since Sept last $8,646 & T.A.E. on his personal account $7,987.0 $16,733.[b]

You get it in driblets & it dont seem to[o] correct but when added up it aggregates pretty round figures. Therefore only be patient.[c]

I send you these figures simply & only to show that I propose to stand by this thing & see us all through— Better destroy this after reading— Yours

J C R[d]

ALS, NjWOE, DF (*TAEM* 13:55). Letterhead of Automatic Telegraph Co. [a]"New York," and "187" preprinted. [b]Preceded by brace indicating summation of the two accounts. [c]This paragraph written in left margin. [d]Last paragraph and initials written in right margin.

–437–

Memorandum to Charles Batchelor[1]

[Newark, May 1874?[2]]

Write to R Faulkner[3] 40 Endell St Long Acre London Catalogue.[a]

Statham, 111.A.1. Strand London Chemists. Catalogue, illustrated 3 stamps.[4a]

Dr Caplatzi[5] Science Depot Chenies st 209 Tottenham court road, Catalogue.

Stanley,[6] 13 Railway Approach London Bridge .SE 1 stamp—Catalogue.

Write to J Lawrence Smith Esq[7] Louisville Ky to obtain Tellurium O~~u~~res

Write to Delaney[8] Washn Automatic Ofs If he will be so kind as to obtain & Send by Mail This book at Govt printing office Mines & Mining in states & Territories west of Rocky Mountains being 5th Annual Report (1872) of R W Raymond US Com Mining Statistics, also to see[b] what works of like character[c] he can obtain free provided we want them.[a]

AD, NjWOE, DF (*TAEM* 13:6). [a]This paragraph canceled with a diagonal line. [b]"to see" interlined above. [c]"of like character" interlined above.

1. Batchelor was acting as Edison's secretary at this time.

2. The date is estimated on the basis of dated correspondence about tellurium found in Cats. 1143 and 1144 (both Scraps. [*TAEM* 27:208, 210, 213–14, 346–47]).

3. Forty Endell St. was the address of a William Faulkner, MRCS, surgeon and registrar. *Post Office London Directory* 1874.

4. William Edward Statham was an operative chemist (*Post Office London Directory* 1874; for the meaning of "operative chemist" see Bud and Roberts 1984, 54–55). His advertisement is in Cat. 1143:86 (*TAEM* 27:217).

5. Dr. James Caplatzi, a science lecturer, also manufactured and sold scientific and photographic apparatus; surgical, surveying, and musical instruments; and much else. *Post Office London Directory* 1874.

6. William Ford Stanley, an optician, manufactured surveying, photographic, electrical, and other instruments. *Post Office London Directory* 1874.

7. John Lawrence Smith, the first American to study chemistry under Justus Liebig at Giessen, and president of the American Association for the Advancement of Science in 1872, was noted for his work in analytic chemistry and his method for the separation of alkali metals from silicates. He was also coauthor of an important series of articles in the *American Journal of Science* (1853–55) on the "Reexamination of American Minerals." *BDAS*, s.v. "Smith, John Lawrence."

8. Patrick Delany, manager of the Washington, D.C., office of the Automatic Telegraph Co. (Delany's testimony, 3:2, Box 17B, *Harrington v. A&P*; see also *TAEB* 1:619 n. 9). On 5 June, Delany wrote to say that he would attend to Edison's request as soon as possible (Cat. 1173:21, Scraps. [*TAEM* 27:349]).

–438–

Draft Circular: Student Telegraph Instrument

[Newark, May 1874?[1]]

EDISONS INK RECORDER.
For Amateurs, Students, and Short Lines.

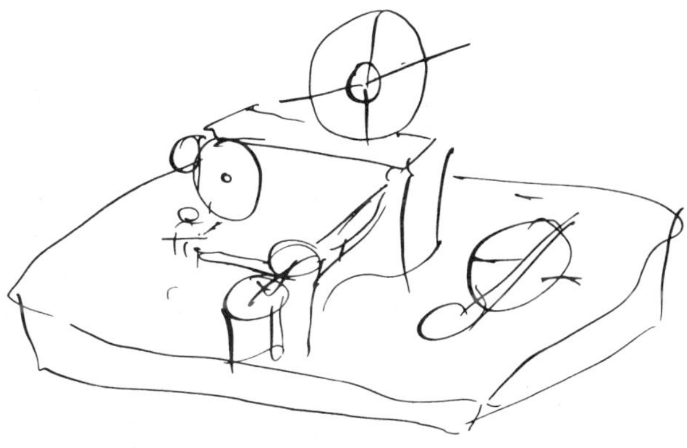

(Patent Applied for)[2]

~~This apparatus is is precisely similar~~ to

This apparatus has been especially devised to supply a want long felt by students of telegraphy to wit an instrument ~~which should~~ to[a] record the message ~~upon paper~~[b] so that the student may be enabled to learn the proper method of forming the letters by the key by referrance to the printed slip. ~~All Amateur Apparatus unti~~ and[a] also to allow of immediatie use as a means of Communication without the necess~~ity~~ary ~~of~~ tediousl~~y~~ process of[c] learning to read by sound. the fact that the message can be read by sound and at the same time is recorded in ink facilitates wonderfully the aqusition of Sound Reading.[3]

It supplies another want among those who do not intend to ~~p~~become professional telegraphers but[d] merely require an apparatus to communicate between one house another or any office and a factory or between houses of several persons for ~~a~~recreation & amusement ~~and who do not intend to become professional telegraphers or can spen give the time and attention necessary to learn Sound Reading~~ it may be connected ~~by~~to a relay o~~r~~f ~~a~~lines of any length worked like a Morse register[4e]

This apparatus is highly finished and is better made than ~~7/10~~ a majority[f] of the ~~tele~~ best telegraph instruments as for price we defy any ~~telegraph~~ maker ~~in this Country~~ to even Manufacture it at the price which ~~we sell it~~— now[a] ~~it~~ is sold—

Price of ~~Complete~~ each[a] Instrument $1~~0~~2.00.
~~R~~Including Roll of paper Ink etc
~~Pri 2 instruments 22.00.~~
Bunsen[a] B~~att~~ichromate Battery. 75 per cell
Price of two machines with battery sufficient
to work a line 1/~~4~~2 mile long. $25.00.
Instructions accompany each instrument[5]
Liberal[g] discount to agents

Edison & Murray
10 & 12 Ward St

Mfrs e[6g]

AD, NjWOE, DF, 74-004 (*TAEM* 13:42; Supp. III). [a]Interlined above. [b]"~~upon paper~~" interlined above. [c]"process of" interlined above. [d]"do . . . but" interlined above. [e]"it . . . register" interlined above. [f]"a majority" interlined above. [g]Underlined twice. [h]Followed by three broken horizontal lines.

1. It is not known when Edison drafted this circular. A printed version appears in an Edison and Murray circular that also featured the inductorium, as well as listing products manufactured and sold by the firm

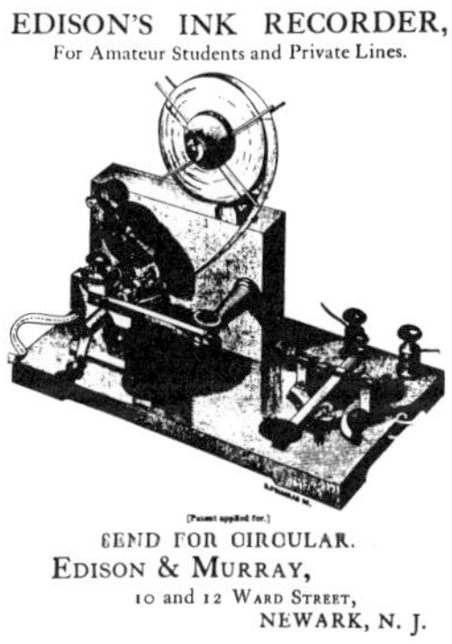

The first published advertisement for Edison's student instrument.

(Cat. 1178, Scraps. [*TAEM* Supp. III]). The first reference to this instrument is a 28 April order book entry for a "learners Inst recording Mod[el]" (Cat. 1214:44, Accts. [*TAEM* 21:586]). In mid-May, Edison suggested a possible customer for the instrument in a memorandum to Charles Batchelor (Doc. 433). The first published advertisement for the instrument appeared in the 1 July 1874 issue of the *Operator* (p. 8). The first recorded sale occurred on 5 August 1874, and eight more sales of the instrument and supplies for it were recorded over the next two months (PN-74-05-22, Accts. [*TAEM* 21:116–20]). Other sales in the New York City area may have preceded this. Edwin Eagles to Edison, 22 July 1874, Cat. 1173:42, Scraps. (*TAEM* 27:377); Cat. 1184:67, Accts. (*TAEM* 21:816).

2. Edison did not receive a patent for this device.

3. For a comparison of Edison's devices with those of other inventors and manufacturers, see advertisements in the *Telegrapher* and the *Journal of the Telegraph*.

4. See *TAEB* 1:10.

5. What appears to be an undated draft of instructions for this instrument can be found in Cat. 30,099:305–14, Lab. (*TAEM* 5:1051–60).

6. A separate, draft business card lists the variety of items manufactured by Edison and Murray that would be listed here. 74-008, DF (*TAEM* Supp. III).

A draft business card for Edison and Murray.

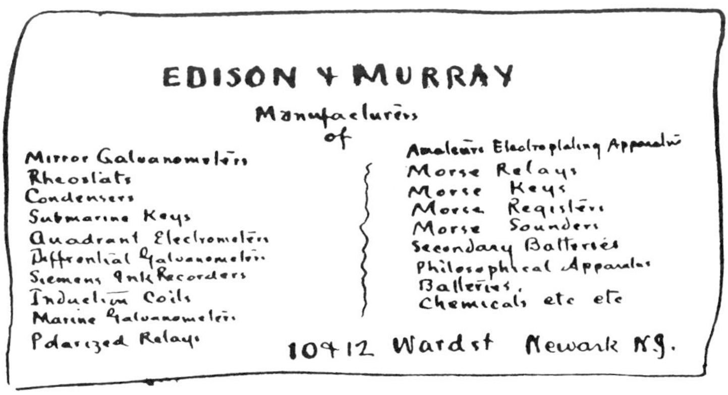

–439–

Patent Application: Automatic Telegraphy

New York, June 1, 1874[1a]

1[b] To all whom it may concern.

Be it known that I, Thomas A. Edison, of Newark in the county of Essex and State of New Jersey, have invented an improvement in Chemical Telegraphs, of which the following is a specification.

2 The object of this invention is to record dots and dashes upon chemically prepared paper by electric decomposition for telegraphing.

3 The invention consists in forming the decomposing stylus or pen of tellurium and wetting the paper with water or

any liquid conductor;[2] I prefer water to which has been added a quantity of hydrate lime or potash.

4 If the stylus tipped with tellurium is connected to the zinc end of a battery, and the drum or plate upon which the moistened paper rests be connected to the copper end of the battery, hydrogen is evolved at the point where the tellurium touches the paper. The hydrogen combines with the telluriam to form hydrotelluric acid which has a red color, but is instantly decomposed in contact with the air and forms a black pigment which is permanent.[3]

5 I have found that this reaction is more delicate than that of iodide of potassium, which is the most delicate known in chemistry.

6 This decomposition of tellurium has the peculiar characteristic over all other metals that it produces the mark with hydrogen and also that the mark at its first formation is not a precipitate like the ferro cyanide of iron when an iron pen and ferro cyanide of potassium is used, which has a tendency to adhere to the stylus after the current has ceased and produce attenuation of the signals, but the permanent color is formed by the action of the air after the decomposition has taken place, hence the signals will be much sharper.[4]

I claim as my invention,

First. I claim the method of[c] recording telegraph or other signals by the[d] electro-decomposition upon moistened paper or other material ~~by~~ of[d] a stylus of tellurium.

Second. I claim a paper moistened with hydrates of lime, potash or the cyanides of potassium ~~when used~~ for use[e] with a tellurium stylus for the purpose set forth.

Signed by me this 1st day of June A.D. 1874.

Thos A Edison

Witnesses Geo T. Pinckney Chas H Smith

DS, MdSuFR, RG-241, Pat. App. 166,859. Written by Pinckney; petition and oath omitted. Interlineations probably by Edison. [a]Place taken from oath; date taken from text, form altered. [b]Section numbers written in margin in an unknown hand. [c]"the method of" interlined above. [d]Interlined above. [e]"for use" interlined above.

1. The application was filed on 25 July; the patent issued on 17 August. This and other Edison applications for automatic telegraph recording solutions were referred by electrical examiner Zenas Wilber to chemical examiner Prof. B. F. Hedrick. Pat. App. 166,859.

2. An 1879 *Scribner's Magazine* article on Edison's automatic system stated that after many months of research, Edison had found that "the metal tellurium used on paper moistened with salt water" provided the best recordings (Fox 1879, 845). Nevertheless, Edison did continue to

use iron pens and to experiment with tellurium after filing this patent application (Davis and Rae 1876, 37; Cat. 1172:38, 167, 171, Lab. [*TAEM* 3:96, 122, 124]). Tellurium apparently also became standard for the pens in the receiver of his domestic telegraph (James Browne to TAE, 6 Apr. 1892, 92-046, GF).

The *Scribner's* article indicated that the "paper found to give the clearest and sharpest signals was thick, well- washed bibulous cotton-paper, moistened with water, in which a little salt had been dissolved,— the purpose of the salt being to make the moist paper a better conductor of the electricity, thereby augmenting the amount of decomposition."

3. The *Scribner's* article reported that the "hydrogen set free by the current forms with the tellurium a yellow compound," suggesting that a different chemical solution was later used. The article goes on to say, "A curious scientific measurement occurs in connection with this. . . . If the paper be passed very rapidly under the tellurium a yellow mark can be seen near the point of contact, which in a short distance turns black. If this distance is measured and the speed of the paper be known, one can easily reckon how much time it requires for telluretted hydrogen to oxidize. In the small space of an inch, the inventor more than once succeeded in recording the Lord's prayer in telegraphic characters, all of which were perfectly distinct under the microscope." Fox 1879, 845.

4. According to the *Scribner's* article, "Tellurium has the property of cutting the signals off very sharply. This serves as another device for overcoming the static discharge from the line." Ibid.

-440-

From Jesse Bunnell

Philadelphia, June 8. 1874[a]

Tom.[1]

Came back here Saturday P.M. Will have copy of those papers over there last of this week for your assistance in revising making up claims etc.[2] Have you received an order for six inductoriums[3] from us here? If not please enter it now & send them on to us here— Carter[4] is at 22 Dey St[5] this week attending to business there. Spose he has your case of goods by this time— Please do not mention in NY anything bout the Eagles transaction—[6] Dont want them to tumble until we get those papers in in proper shape—

I am coming to you with a proposition to take some of your jim cracks[7] two or three at a time and run them into the market in good style soon as we get started in N.Y. Dont you think twould pay both of us— Will also want you to make up [-]— Tell you all about it when I see you— Yours

Bunnell

ALS, NjWOE, Scraps., Cat. 1173:35 (*TAEM* 27:360). Letterhead of Partrick, Bunnell & Co. [a]"Philadelphia," and "187" preprinted.

1. Bunnell wrote "Taed.—" and the Morse code for "ison" before the salutation.

2. See Doc. 378 n. 3.

3. An induction coil intended for medical use. Edison sold his for $6.00 as "a specific cure for rheumatism, and as an inexhaustible fount of amusement" (Cat. 1178, Scraps. [*TAEM* 27:137]). See Docs. 434 and 435.

4. Franklin Carter. See Doc. 378 n. 8.

5. Partrick, Bunnell & Co. had just announced the opening of a branch office at this address. *Telegr.* 10 (1874): 138.

6. The Eagles battery, which had a lead outer vessel, had for some time been licensed exclusively to F. L. Pope & Co. Although Pope continued to advertise the battery, the license had been transferred to Partrick, Bunnell & Co. by the early fall of 1874. "The Eagles Battery," *Supplement to the Operator,* 1 Oct. 1874, 1; "A Card," *Operator,* 15 Dec. 1874, 6; Pope's advertisement, ibid., 8.

7. A gimcrack is a trinket; here, it may be a reference to Edison's inductorium.

–441–

Equipment Specification: Multiple Telegraphy

[Newark, June 10, 1874?[1]]

Relay with a back point, to have the spools made half the length as marked on rubber covers.

Two new ones shortened and with back points (the front point is insulated with ivory.[2]

~~All to have~~ The first relay to have 100 ohms resistance. The two new ones to have about 130 to 150 ohms each.[3]

Two Duplex Sounders altered to D~~u~~ouble points like model.[4]

Three ordinary sounders made thus[5]

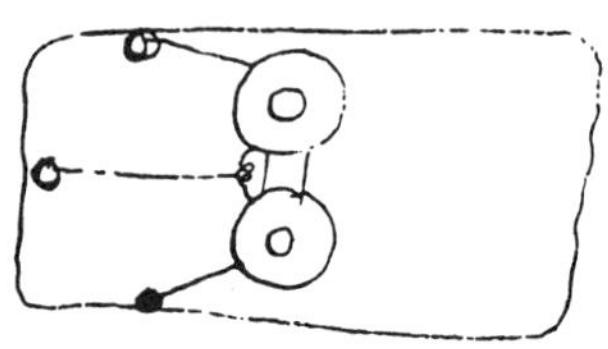

an extra binding ~~point~~ post being added with a connection to the wire connecting the two spools together.[a]

One ordinary ~~s~~Phelps sounder with the magnets only one inch in length.

Get from Gold & Stock two polarized Relays (new) which we used on Pope & Edison Private Line instruments.[6]

AD (photographic transcript), NjWOE, Quad. 71.2, following p. 14 (*TAEM* 10:230). Another copy follows p. 146 (*TAEM* 10:314), and there is a copy in Testimony and Exhibits on Behalf of T. A. Edison, *Nicholson v. Edison.* [a]Followed by stamp of Western Union shop to note delivery, dated 25 June 1874.

1. This is the first of several orders from Edison and George Prescott for equipment to be made for testing. On 10 June, Prescott sent this document to George Phelps at Western Union's factory with instructions to make the apparatus "as soon as possible" (Quad. 71.2, following p. 14 [*TAEM* 10:233]). Some or all of the apparatus was apparently delivered, along with other instruments requested by Edison, on 25 and 27 June. (Phelps's testimony, Quad. 71.1, pp. 307–9 [*TAEM* 10:160–61]).

2. See Doc. 392 n. 16.

3. Edison wanted coils that would be sensitive to signals of different strengths.

4. Not found.

5. These could be used in an alternative "bug trap," as explained in connection with fig. 4 in Edison's British Patent 384 (1875). Batchelor (*TAEM* 92:41, 46).

6. Prescott wrote in his cover letter to Phelps, "I understand the G.&S. are not using the polarized relays mentioned" (Quad. 71.2, following p. 14 [*TAEM* 10:233]). For the "Pope & Edison Private Line instrument" see Doc. 107.

-442-

Equipment Specification: Multiple Telegraphy

[Newark, June 12, 1874?][1]

Spec one

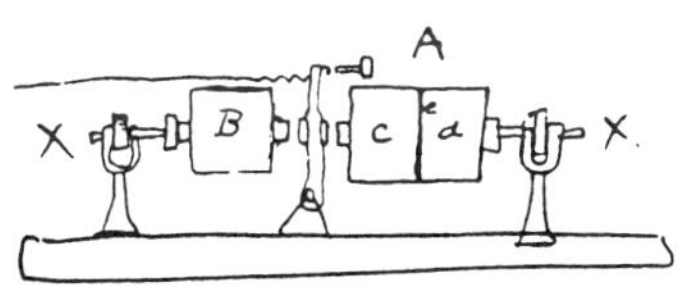

The magnet "A" is the same Length and size as ordinary plain Relay, divided in two compartments "c" and "d" by a thin~~e~~ rubber head "e". Each compartment to have 130 ohms or thereabouts 15 ohms either way will not matter, the ~~eo~~other retracting magnet "B" to be of same ~~s~~diameter as "A" or if more convenient to get adjusting spring in it may be smaller, to be inch and half in length and with 100 ohms Resistance or thereabouts. Both magnets to have the new improved magnet adjuster X X[2]

The armatures to be flat scant 1/16 of inch thick and placed on each sid~~o~~e of lever

wires to be connected thus[3]

The magnet A is to be one inch longer than a Duplex Relay. ~~t~~To be divided into two compartments c & d seperated by a thin rubber head 'e' each compartment to be wound with 200 ohms or thereabouts B is an angle peice rigid to the base to which is permanently secured an armature "a". The magnet is to be made to approach or re[--]cede from the fixed armature by the adjusting nut. Diamater of magnet same as ordinary morse relay[4]

Spec 2

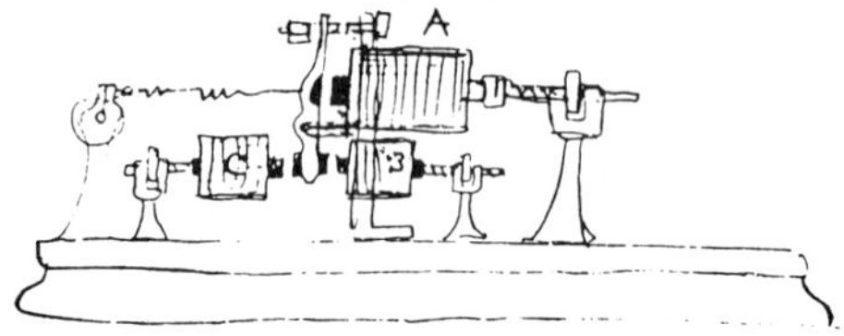

The magnet A same length as ordinary morse relay same diameter about 150 ohms

Magnet B inch and quarter long, same diameter as A or it may if inconvenient be made smaller wound with 100 ohms.

Magnet .C. same as .B. but wound with 12 ohms resistance[5]

Design may be altered if inconvenient.

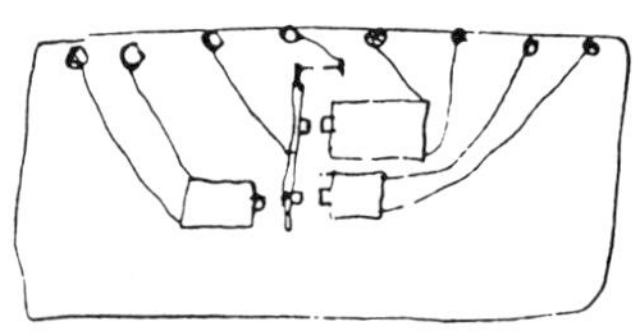

Spec 3.

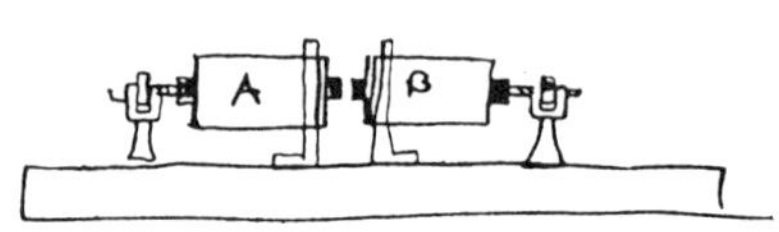

A & B are magnets of the same length as the Duplex Relay, same diameter each having 400 to 450 ohms each.

By means of the adjusting nuts the magnets core[a] are to touch each other and to be seperated ¼ of an inch or more.[6]

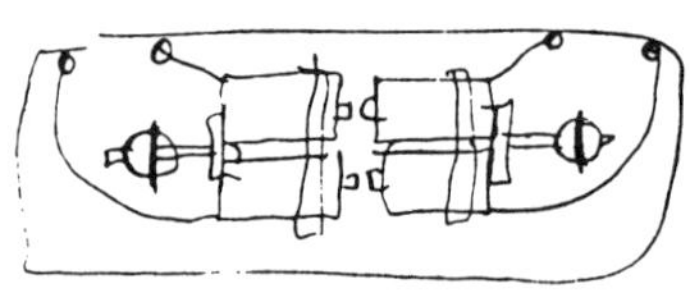

AD (photographic transcript), NjWOE, Quad. 71.1, following p. 146 (*TAEM* 10:302). Also found in Testimony and Exhibits on Behalf of T. A. Edison, *Nicholson v. Edison*. [a]Interlined above.

1. On 12 June, George Prescott sent this document to George Phelps at the Western Union factory with instructions to make the instruments and send them to Prescott when ready. Phelps sent the instruments and

returned the specification to Prescott on 14 July. Quad. 71.2, following p. 146 (*TAEM* 10:301); Phelps's and Prescott's testimony, Quad. 71.1, pp. 308, 336 (*TAEM* 10:161, 175).

2. These adjusted the distance of the magnets from the armature.

3. Cf. Docs. 387 (devices 8 and 10) and 392 (fig. 15); and Cat. 297:114(2), Lab. (*TAEM* 5:870).

4. Cf. Docs. 392 (fig. 7), 471, and 532 (fig. 3).

5. Like the first design above, this device, depending on the polarity of the connections, employs cooperating and opposing electromagnets to affect the armature (cf. Docs. 449, 453, and the illustration in Doc. 469, n. 3). The functional difference here is that no magnets share a common core, so there would be little inductive interaction. The role of such relays depends on the way they are wired in a circuit (cf., e.g., Cat. 297:119[1] verso, Lab. [*TAEM* 5:881]; and Doc. 449). This type of compound relay may have been associated with later, successful quadruplex designs. See Docs. 485, 513, and the illustration in Doc. 348 n. 9.

6. Cf. Docs. 308 and 532 (figs. 5 and 6) and fig. 17 in Edison's British Patent 384 (1875).

–443–

Jarvis Edson to Edison and Murray

[New York,] 13th June '74

D Sirs

I have already written to Utica[1] that I would exhibit our system[2] to them during the middle of next week and consequently hope you will have the Recorder you have under way finished up in time and thouroghly tried as this one here has some objections which I believe you are going to provide against in future— Amongst them the armature lever throwing both ways so as to prevent the reel from making more than one revolution at a time even if the sender does not proper let go the key or pull. A much larger amount of paper could be put on at a time I think if the "stripper"[3] was pivoted much lower down—[a] Orange[4] appears to be working along all right and I think we shall get it— Have no paper to show the instrmnt to two of the important men who live out there & do business near by here so hope you will get me in a supply soon— The two rolls Tom[5] brought in were dry & I think the solution was not exactly right but it may have been on acct of the dryness of the paper— How about the deliquesent solution? I have got big bug after Washington— Dont forget I want some paper, & am all out, and am looking every day for a telegram from that Capitalist from Providence[6] saying when he will be here to examine into it— Yours truly

J. B. Edson

ALS, NjWOE, Scraps., Cat. 1173:126 (*TAEM* 27:422). [a]Underlined twice.

1. See Doc. 447.
2. Fire alarm; see Doc. 415 n. 5.
3. Unidentified.
4. Orange is a suburb of Newark, N.J.
5. Probably Tom Gartland.
6. Unidentified.

–444–

Equipment Specification: Multiple Telegraphy

[Newark, June 15, 1874?[1]]

A

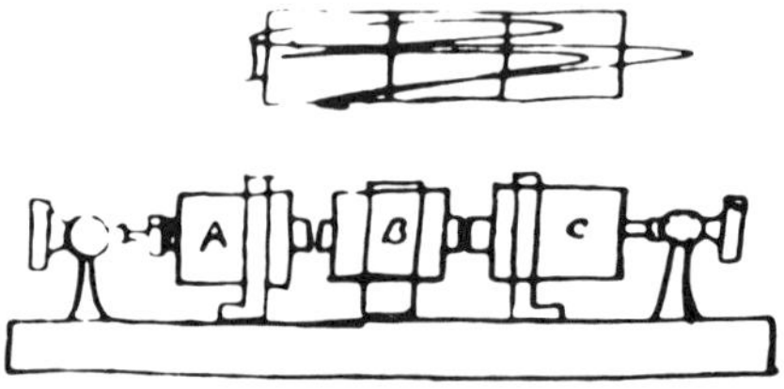

A B & C are three magnets C is the same length & diameter as a Duplex Relay and wound with the same Resistance ~~as~~ ie[a] about 1000 ohms, in the regular ordinary way.

B is a magnet, inch & ¾ in length and same diamater as C wound with 150 ohms. A is a magnet same length & diamater as .C. wound with 1000.

A & C are adjustable B is rigid.

the ends of the Cores are to be perfectly flat so that the three magnets may be adjusted towards each other to form a single Core[2]

The connections are to be made thus

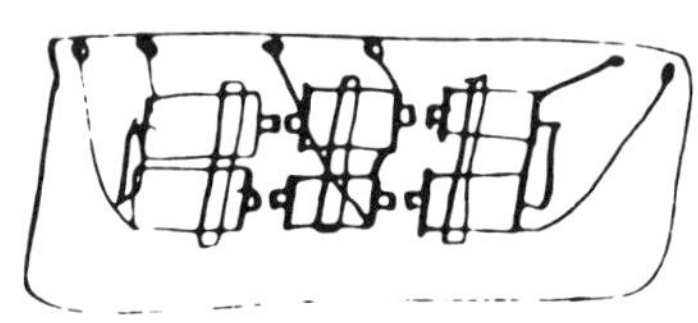

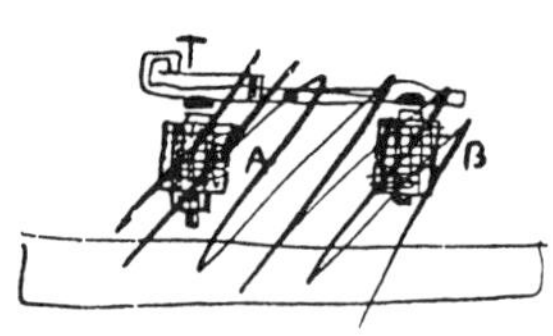

X

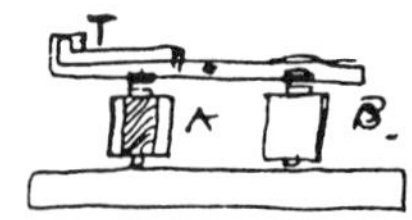

~~AB~~. X is ordinary Duplex ~~magnet~~ Transmitter with exception of an extra magnet .A. which is wound with double wire.[3] first 4 layers of wire is to be No 23, and to go to two binding posts. over this and insalated from it is wound fine wire about 100 ohms Resistance The Extra armature is to be flat so that one may be enabled to pack it so as to give a required play to the lever, hence the magnet A should be a little shorter than

B. There should be[b] 1/16 of inch play from A to B of the lever.

One of those Keys used in Connection with the shunted Relay.

An ordinary Phelps' Sounder with a front point thus

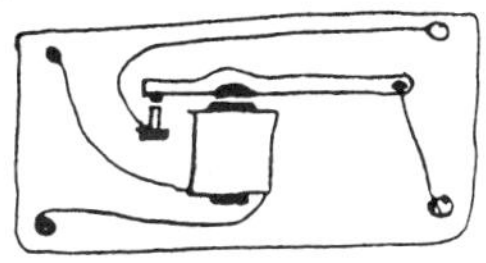

An ordinary Phelps Sounder with a back point thus

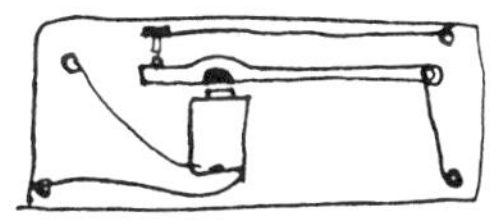

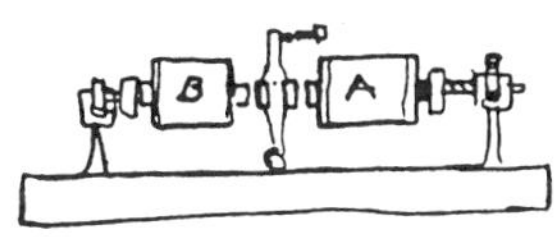

A. and B are two magnets equal, same diameter as ordinary morse relay but only one inch and three quarters in length.[4] Each magnet is to be wound with ~~2~~300. ohms. Binding posts as in figure two

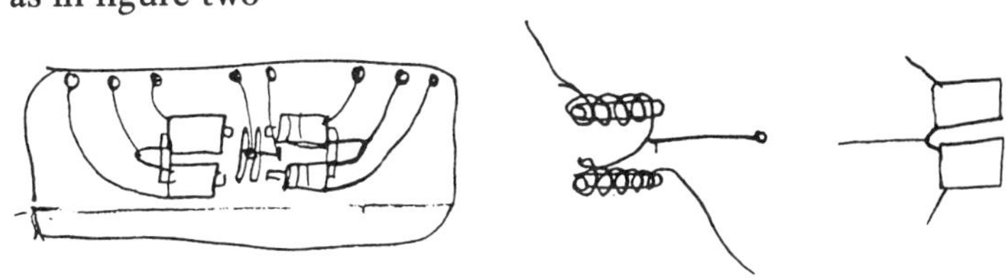

wound just like ordinary relays only an Extra wire connected to cross wire connecting spools together, which runs to an extra binding post

AD (photographic transcript), NjWOE, Quad. 71.2, following p. 146 (*TAEM* 10:338). There is another copy in Testimony and Exhibits on Behalf of T. A. Edison, *Nicholson v. Edison.* [a]Circled. [b]Preceded by what may be two pen rests.

1. On 15 June, George Prescott sent this document to George Phelps at the Western Union factory with instructions to send the instruments to him when ready (Quad. 71.2, following p. 146 [*TAEM* 10:337]). These instruments were not ready until 24 July; the cost was $105 (Phelps's testimony, Quad. 71.1, p. 308 [*TAEM* 10:161]).

2. Edison used a similar device in a later patent application (Doc. 467) and caveat (Doc. 532).

3. Edison used a related device in Doc. 531 (fig. 3).

4. Edison had occasionally used magnets of this form in previous duplex work; cf. Docs. 297 (K, L), 308, and 312 (D).

–445–

Draft Agreement with George Prescott

[New York?,][1] June 21 1874

Memorandum of an agreement[2]

I hereby agree with Geo B Prescott Electrician of the Western Union telegraph to go on and perfect ~~my~~ the several magnetic Duplex Apparatus ~~B~~for sending two messages in opposite directions, and those which send two messages in the same direction~~s~~, also those which send two messages in one direction & two messages in the opposite direction at the same time on one wire[a] ~~and~~ [~~but~~?][b] the patents of which are to be taken out in our joint names, and make assignments if such assignments are necessary[c] & execute all proper papers securing to him one undivided one half of said inventions,[3] with the understanding that all profits derived either directly or indirectly are to be divided equally, also that ~~no~~the price at which the inventions ~~w~~are to be sold shall be satisfactory to myself ~~Th~~Also that when the proceeds from these inventions are divided I am to receive Eleven hundred and twenty five dollars Extra which were expended previous to this agreement on models patent fees etc.

~~The~~In Consideration of this agreement Mr Prescott pays me one dollar the receipt of which is hereby acknowleged

Thos A. Edison

ADfS (photographic transcript), NjWOE, Quad. 71.2, following p. 146 (*TAEM* 10:343). Letterhead of Western Union Telegraph Co., George Prescott, Electrician. [a]"~~B~~for sending . . . wire" is written at bottom of page and signed "Edison", with a line indicating insertion here. [b]Heavily canceled. [c]"if . . . necessary" interlined above.

1. This was probably drafted by Edison at the Western Union building during the period in which he conducted tests with Prescott. Prescott's testimony, Quad. 71.1, pp. 318–19, 328 (*TAEM* 10:166, 171); App. 3, Edison's Affidavit, p. 810.
2. See Docs. 451 and 466.
3. See Docs. 432 n. 3 and 451 n. 2; and App. 1.G15–16.

–446–

Draft Caveat: Multiple Telegraphy

[Newark, Spring 1874?[1]]

Quadruplex.[2]

The object of this invention is to transmit four 4[a] messages over one wire at the same time two in one direction and two in the other direction.

The invention consists in transmitting positive and negative currents to effect a polarized relay and increasing and decreasing the same to effect an ordinary relay or magnet, substantially as described in a previous caveat filed by me relating to this invention.[3]

The invention further consists in providing the polarized and common relays with double coils and arranged an artificial circuit at each end of the line so that the effects of the outgoing Battery shall or can be balanced upon the two relays at the receiving stations.[4]

The invention further consists in arranging electro magnets shunted arounds the extra coil of both relays to nullify the effect of the static charge upon these relays.[5]

The invention further consists in shunting the Line coils of the polarized Relays to prevent their self induction being thrown on the line wire.[6]

It further consists in the various combinations ~~singly~~ and as a whole—

ADf, NjWOE, Lab., Cat. 297:24(3), 24(4) (*TAEM* 5:527, 528). [a]Circled.

1. Edison's reference to his caveat (Doc. 348; see n. 3) but not his patent (Doc. 472), as well as the particulars of the description, suggest that he wrote this draft around the time he began collaborating with Prescott.

2. This is the earliest use of this term by Edison. He apparently never filed a comprehensive caveat for his quadruplex design; he never received a corresponding U.S. patent. However, he received comprehensive foreign quadruplex patents (e.g., Brit. Pat. 384 [1875]; Canadian patent application, 74-008, DF [*TAEM* 13:337]), and he filed an application for a U.S. patent that seemed "to cover the entire Quad. system" (Case 121, executed 30 June 1876 and later abandoned; "Abstract of Edison's Abandoned Applications," p. 10, PS [*TAEM* 8:537]; a drawing from the application appears at *TAEM* 8:739).

3. Edison refers to Doc. 348, which he had executed on 28 October 1873 but which actually had not yet been filed.

4. The accompanying illustration, showing doubled coils on the receiving relays, probably dates from about this time.

A quadruplex design with doubled receiving coils (the two upper assemblies).

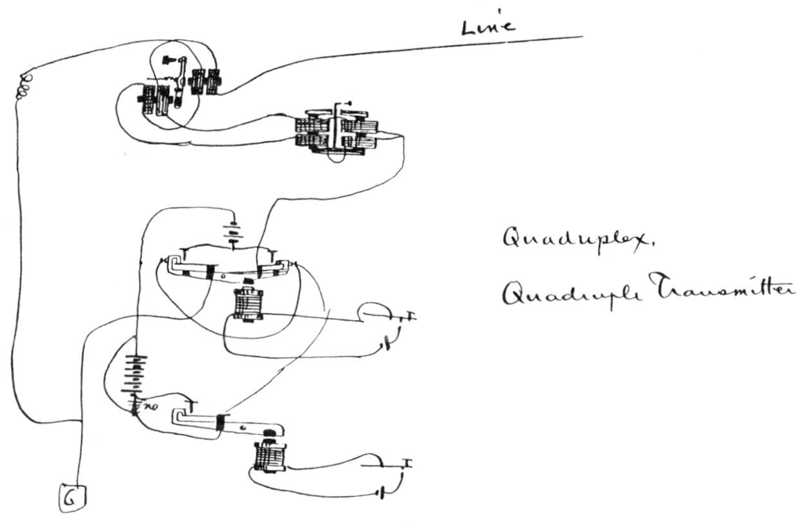

5. Cf. Doc. 532 (fig. 5).
6. Cf. Doc. 533 (fig. 1).

–447–

From Howe & Austin

Utica, N.Y. June 30th 1874.[a]

Dear Sir.

~~Since~~ The fire comrs have appointed a committee to examine the alarm and if they deem it advisable, to, make necessary arrangments for introducing the City Alarm =[1] I think there is no doubt whatever but the work will be ordered at once— The President of Fire Board asked me if a guarantee would be given that no trouble should be given the city in regard to the patents owned and operated by others— Please write by return mail what position to take in relation to this— Send register paper first express— ~~I~~ Howe will be in Newark for part next week Yours

Howe & Austin[2]

⟨Answered Batch⟩[b]

L, NjWOE, Scraps., Cat. 1173:40 (*TAEM* 27:371). Letterhead of the Walter Electric Burglar Alarm Co. [a]"Utica, N.Y." and "187" preprinted. [b]Marginalia written by Batchelor.

1. No action had been taken by 27 July, although a favorable decision seemed imminent. However, the city's next budget was presented before Howe & Austin submitted an estimate, and then Howe & Austin dissolved, leaving Howe in charge of the business. Nothing further is known of this enterprise. A. S. Howe to TAE, 27 July and 11 Aug. 1874, Cat. 1173:44, 76, Scraps. (*TAEM* 27:382, 396).

2. Western Union operators A. S. Howe and W. A. Austin were general agents for the Walter Electric Burglar Alarm Co. in Utica. Cat. 1143:22, Scraps. (*TAEM* 27:182).

–448–

R. G. Dun & Co. Credit Report

Newark, June 30/74

Edison & Murray[a] Manfrs Telegraph Inst'm'ts

#3700[1] Prev repts give all that we can get relative to their m[ea]ns & these are mainly locked up in machy & telegraph instrumts of their manfr. One thing is certain that they hve done quite a lge bus in manfg machines for the Gold & Stock Telegraph Co & ought to hve mde considl money. "E" is considered remarkably inventive & to genly produce some thing pecuniarly val. Has sevl patents now of promising char. What the firm is wor is we believe beyond anyones ability to est, not excluding E & M themselves, the firm hvg no Bank a/c that we can learn of. We believe that ordinary eng[agemen]ts will be pd & hear of no complaints in town in anyway.

D (abstract), MH-BA, RGD, N.J. 23:206. [a]Followed by "(Prevs see T A Edison 3/2[38])".

1. See Doc. 352 n. 1.

–5– July–September 1874

The summer of 1874 opened with Western Union making plans for a demonstration of the new quadruplex that would lead to the first public announcement in the *New York Times* on 10 July. At the same time, Edison and Prescott reached an agreement making Prescott co-owner of the related patent applications, most of which Edison executed on 19 August and filed on 1 September. These applications were intended to protect the quadruplex invention and give Edison and Prescott something to formally sell to Western Union. Ownership of these patent rights would later become the subject of extensive litigation.[1]

Edison continued to refine the quadruplex for Western Union. Gerritt Smith, chief operator at Western Union's main New York office, was assigned to Prescott's office in early Septembe to assist in these experiments. By the end of the month they had successfully tested the quadruplex on a line between New York and Boston. Smith, who later played an important role during the installation of the quadruplex on Western Union's lines, subsequently received credit for improvements that appear to have been derived from Edison's experiments. Another inventor, Henry Nicholson, laid a claim to the quadruplex as early as August 1874.[2]

The potential value of Edison's quadruplex and his partnership with Prescott caused concern at the Automatic Telegraph Company. Company president George Harrington, whose own partnership agreement with Edison[3] became the basis of his claim to ownership of the quadruplex, sought to limit Edison's involvement with Western Union. He and others connected with Automatic Telegraph raised money to relieve Edison of his financial worries stemming from the

$10,000 due William Unger early in July.[4] They also continued their negotiations with the railroad companies and Jay Gould's Atlantic and Pacific Telegraph Company to form a new combination that would use Edison's automatic telegraph to compete against Western Union.[5]

The *New York Times* announcement of the quadruplex brought Edison to the attention of a larger public. Although he was already well known in telegraph circles, his reputation was not entirely a positive one. His former partner James Ashley, the editor of the *Telegrapher*, now launched attacks on Edison's inventions and his character.[6] At the time these attacks began, Edison was associated with a new journal, the *Operator*, which came to his defense. As science editor of the *Operator*, Edison also began to contribute original articles, some of which may have been adapted from his proposed book on telegraphy and electricity, which he had apparently abandoned.

Edison gained further public exposure when *Scientific American*, the leading inventor's magazine, published his letter detailing his recent discovery—the electromotograph. This article detailed Edison's August experiments, during which time he conceived his name for the phenomenon and designed a telegraph relay to employ it. He applied for a patent for this relay, which did not need an electromagnet and thus held the potential for getting around the Page patent, which Western Union controlled.[7]

Besides quadruplex telegraphy and the electromotograph, Edison's experimental work focused on automatic and domestic telegraphy. He persisted in his attempts to improve automatic telegraph system components and to develop a roman-letter automatic system. He briefly turned his attention to attempts to duplex the Wheatstone ink-recording automatic system. Edison also continued his experiments with tellurium recording pens, and he even expressed interest in developing a process for reducing telluric ores.[8] In mid-September, Edison's extensive chemical investigations led him to employ Robert Spice, professor of chemistry and natural philosophy at Brooklyn High School, to provide private instruction in chemistry. During the summer other individuals also began to take part in Edison's experiments. James Adams, an agent for Edison's inductorium, began working as an experimental assistant in August. Edison's nephew Charley, who visited briefly in early August, also acted as a laboratory assistant. Both Adams and Charley would later play important roles in Edison's laboratory.

Edison's experimental work for the Domestic Telegraph

Company focused on developing a new receiving instrument, which was demonstrated at Orange, N.J., at the end of July.[9] Experimental accounts show that Edison was also developing a private-line printer for Domestic Telegraph.[10] Although the domestic system remained experimental, Edison and Murray did prepare instruments for a demonstration of the company's fire alarm system in Utica in early July.[11]

Edison and Murray continued its regular manufacturing and repair work for Automatic Telegraph and the Gold and Stock Telegraph Company.[12] The firm also continued to manufacture and sell the inductorium, advertising it in more than three hundred newspapers. Additional income came from manufacturing contracts for the Celluloid Manufacturing Company, Partrick, Bunnell & Company, and the Recording Steam Gauge Company.[13]

1. See Chapter 7 introduction and App. 3.

2. Nicholson 1874.

3. Doc. 109.

4. These financial arrangements would ultimately lead to other litigation. See Doc. 516.

5. Reiff's testimony, Quad. 70.7, pp. 446, 698–702, 708 (*TAEM* 9:591, 719–21, 724); Orton's testimony, Quad. 71.1, pp. 144–45, 155, 160–62 (*TAEM* 10:78, 83, 86–87).

6. Ashley's animus against Edison derived from their earlier partnership with Franklin Pope. See *TAEB* 1:225–26 for Edison's falling out with Ashley and Pope.

7. For the Page patent see *TAEB* 1:448 n. 2. Matthew Josephson (1959, 117–19) relates the filing of Edison's electromotograph relay patent (U.S. Pat. 158,787) to the suit brought by Western Union against Jay Gould's Atlantic and Pacific Telegraph Co. for infringement of the Page patent. That suit was not filed until the following summer (Bill of complaint, filed 11 Aug. 1875, *Priscilla Page and Western Union v. Atlantic and Pacific Telegraph Co.*, Equity Case 7-287, Box 355, RG-21, NjBaFAR). He also mistakenly claims that Edison installed the instrument on Atlantic and Pacific lines in the fall of 1874, but if this ever did occur, it would not have been until 1875. Josephson ascribes too great a value to the Edison patent in claiming that it "saved the Gould telegraph empire from extinction" (p. 118). Indeed, there is no evidence that, as Josephson claims, Gould paid $10,000 for the electromotograph relay. Edison did not sell the patent rights until 1880, and he sold them to Western Union. Although Edison was aware of the electromotograph's value in evading the Page patent, he focused his attention on its use as an automatic telegraph repeater. See Reiff's testimony regarding Edison and the Page patent in Quad. 70.7, pp. 713–15 (*TAEM* 9:726–27).

8. On 3 August, Edison had an apparatus made for "moulding & working" tellurium (Cat. 1214:74, Accts. [*TAEM* 21:601]). Nothing is known about Edison's ideas regarding a reducing process (see Doc. 480 n. 3 for references to his proposal).

9. There are undated drawings of domestic receivers which might represent this work. See, in particular, Cat. 30,095, Lab. (*TAEM* 6:365–528 passim). A notebook entry from August indicates that Edison also experimented with chemical recording solutions for the receiver (Cat. 1172:171, Lab. [*TAEM* 3:124]). Preparations for the demonstration at Orange are listed in an account entry of 21 July (Cat. 1184:62, Accts. [*TAEM* 21:814]).

10. Cat. 1184:63 and Cat. 1223:6, both Accts. (*TAEM* 21:814; 22:134); Docs. 545 and 651.

11. See Doc. 447; Howe and Austin to TAE, 27 July 1874; Cat. 1173:44, Scraps. (*TAEM* 27:382); and Cat. 1184:58, Accts. (*TAEM* 21:812).

12. Manufacturing records are found in Cat. 1184:56–81, Accts. (*TAEM* 21:811–23).

13. Jarvis Edson of the Domestic Telegraph Co. was vice-president of the Recording Steam Gauge Co. Partrick, Bunnell & Co. to TAE, 31 July and 3 Aug. 1874, and M. B. Edson to TAE, 24 Sept. 1874, Cat. 1173:43–44, 140, all Scraps. (*TAEM* 27:379, 383, 444). See also Cat. 1214:44, 46, 48, and Cat. 1184:78, 81, both Accts. (*TAEM* 21:586–88, 822–23).

QUADRUPLEX TELEGRAPH CIRCUIT DESIGN
Doc. 449

The drawing of the receiving portion of a bridge quadruplex circuit in Doc. 449 describes a patent model to be constructed at George Phelps's Western Union shop. The design was to be included in Edison's 9 July agreement with Prescott.[1] However, Edison did not include Prescott's name when he finally executed and filed the covering patent application (Case 112) in December 1874.[2]

This circuit represents a partial solution to the major quadruplex problem—namely, how to prevent the neutral relay's response to reversals of polarity. Although on short lines the "bug trap" (circuit **iopq** in the illustration below)[3] adequately by-passed the neutral relay problem, on long lines distortion lengthened the gap between positive and negative signals, rendering the bug trap inadequate. Here Edison replaced the retracting spring on the neutral relay with magnet **u** operated by polarized relay **t** in the main circuit; when the current reversed, the magnet momentarily discharged and the neutral relay armature did not move.[4] The other novel feature of this plan is the variable rheostat/condenser arrangement **b aaaa**, which allowed the operators to balance the artificial line more easily than before. The small magnets on the ends of the neutral relay magnet cores are a design Edison patented in April 1873.[5] Connected in a local circuit with a variable resistance, they could work with or against the primary magnets to compensate for varying current strength on the main circuit.

1. The agreement (Doc. 451) mentions "twelve . . . specifications now in the hands of George M. Phelps." See also Prescott to Commissioner of Patents, 2 and 7 June 1875, Quad. 72.2 (*TAEM* 9:30, 32); and Edison's testimony, Quad. 70.7, p. 355 (*TAEM* 9:539).

2. U.S. Patent 207,724 (Case 112), which details the working of the circuit, was executed on 14 December and filed on 28 December 1874.

3. For the bug trap, see Doc. 348 n. 16. The two transmitters in this diagram are at the lower left. Galvanometer **v** indicated when adjustment of rheostats **f** and **f'** balanced the two sides of the bridge.

Edison's U.S. Patent 207,724 covered this early quadruplex circuit design.

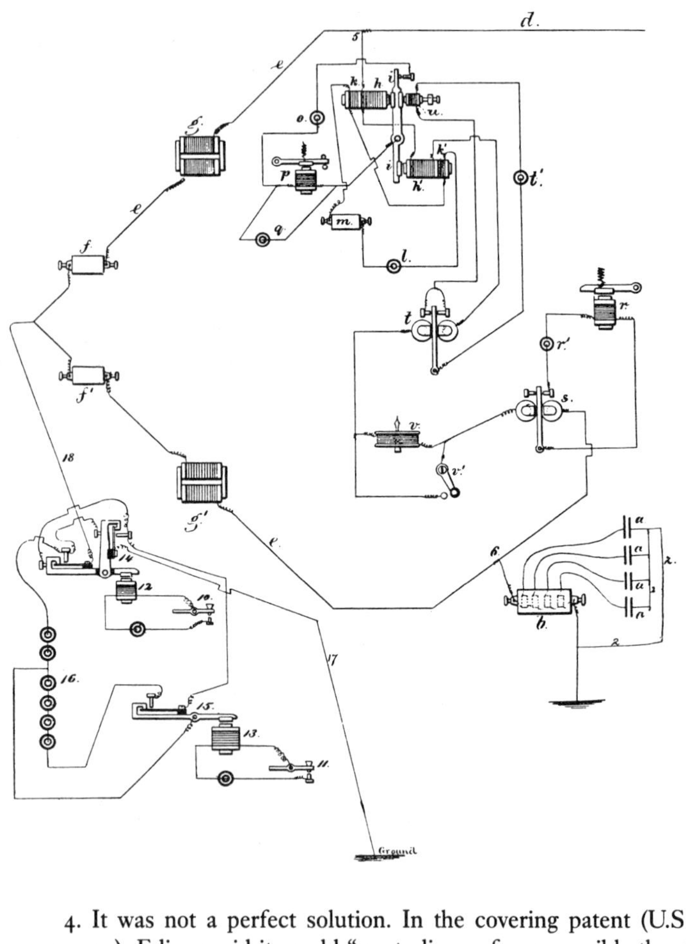

4. It was not a perfect solution. In the covering patent (U.S. Pat. 207,724), Edison said it would "neutralize as far as possible the risk of a false movement" of the armature. He continued to experiment with similar relays (Docs. 442, 453, 477, 485). Other arrangements were adopted in practice; see Doc. 515 n. 2.

5. U.S. Pat. 150,846.

–449– [Newark, c. July 1, 1874]

Patent Model Drawing: Multiple Telegraphy[1]

Model

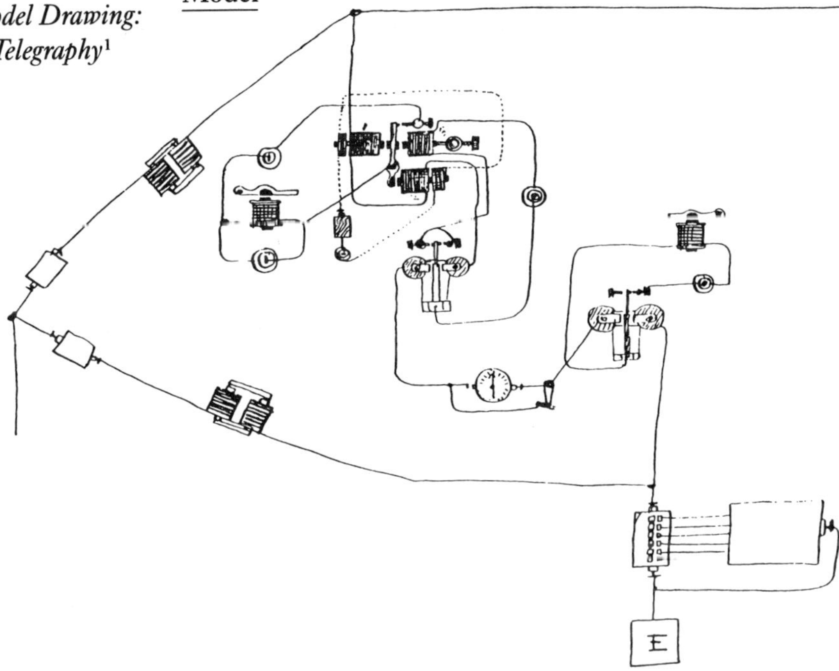

AX (photographic transcript), NjWOE, Quad. 71.2, following p. 146 (*TAEM* 10:336). Another copy is in Defendant's Exhibit 43, *Nicholson v. Edison.*

1. See headnote above.

–450– [Newark, c. July 1, 1874[1]]

Patent Model Specification: Multiple Telegraphy

Model. A.[2]

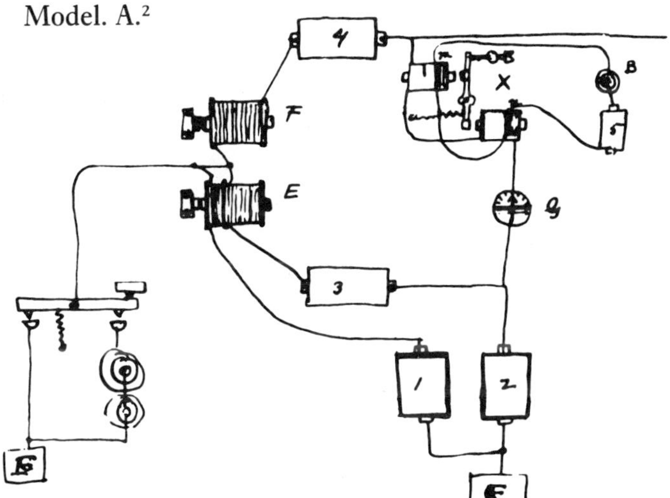

1 2 3 4, 5 are Rheostats. G is a small galvanometer like a pocket compass with two or three turns of 23 wire wound around it. ~~X is a double spiral relay with the local point at the back~~ F is a magnet adjusted up to a fixed armature wound with fine wire. E is another magnet adjusted up to a fixed armature, but wound with two coils one of fine & one of very fine The latter is the smallest & connects to Rheostat .1.

X is a double relay. Each spool is wound with two Coils n & m are 23 wire the balance fine wire 5 is a Rheostat B a battery all connected in circuit with the coils m & .n.

AX (photographic transcript), NjWOE, Quad. 71.2, following p. 146 (*TAEM* 10:333). Another copy is in Defendant's Exhibit 43, *Nicholson v. Edison.*

1. See Doc. 449 headnote, n. 1. This specification was sent by George Prescott to George Phelps on 13 July. Quad. 71.2, following p. 146 (*TAEM* 10:335); George Phelps Jr.'s affidavit, Pat. App. 168,385.

2. This model was for Edison's patent application Case 111, which he did not execute and file until 18 and 26 January 1875 (relay **X** is covered in his U.S. Patent 150,846). This case was to have been included in Edison's agreement with Prescott; its issuance as U.S. Patent 168,385 sparked intense objections by Prescott since he was not assigned the half due him under that agreement (see Docs. 569 n. 5 and 577 n. 10).

Edison designed this circuit to offset the static discharge of the line; to that end he inserted electromagnets **E** (with double coils) and **F** in the bridge and added the extra line with rheostat 1. The operation is described in the patent.

–451–

Agreement with George Prescott

New York, July 9, 1874[a]

This memorandum of an agreement made the ninth day of July, 1874, by and been[1] Thomas A. Edison and George B. Prescott, witnesseth:

Whereas the said Edison and Prescott are the joint inventors[2] of certain improvements in telegraphic apparatus described as magnetic duplex apparatus, being the invention and improvements more particularly described hereafter, for which invention they are about to apply for letters-patent of the United States, to be issued to them jointly.[3]

And whereas it is desired by both parties to enter into certain engagements with each other as to their respective interests in such patent, and in the use and benefit thereof.

Now, in consideration of one dollar to each of said parties by the other paid before the sealing and delivery hereof, and the receipt of which is hereby by each of them acknowledged, it is covenanted and agreed by each of said parties with the other as follows:

1st. That the improvements and invention of which they are the joint inventors and in respect to which this agreement is made are all those inventions for making multiple transmission of magnetic signals for use in telegraphy which are described in twelve several specifications, now in the hands of George M. Phelps, for the purpose of making models of the machinery whereby such invention can be operated,[4] and being all the inventions of said parties, whereby, at the same time and on the same wire, one message may be sent in one direction and one message in the opposite direction, or two messages at once in the same direction, or at the same time and on the same wire two messages may be sent in one direction and two messages in the opposite direction.

2d. That both of them shall have an equal, undivided interest in all future improvements of either of said inventions which may be made by either of them, and that, if it be necessary in order to secure such interests to either, the other shall make an assignment and transfer of such interest to him in due form sufficient to vest such interest in him, and to entitle it to be recorded in the United States Patent Office.

3d. That both of the parties shall have an equal undivided interest and be joint grantees of all letters-patent of the United States or any foreign countries which may be granted for all or any of said inventions, or of any future improvements thereof, and of all extensions and reissues of any such letters-patent.

4th. That whereas Edison has heretofore expended $1,125 for models and patent fees, the benefit of which he contributes to the common interest, and waives reimbursement of that sum, or any part of it, Prescott hereby agrees to pay solely and without contribution from Edison all the future expense and cost of specification, drawings, models, Patent-Office fees, and patent solicitors' and agents' fees, and all other charges incident to the procuring of letters-patent for any of said inventions.

5th. That neither of said parties will sell, assign, or otherwise dispose of the whole or any part of his interest in said inventions or letters-patent therefor, or any of them, without the written consent thereto first obtained of the other party.[5]

6th. That neither of said parties will himself manufacture, use, or sell, nor grant licenses, or the right in any way to any other party to manufacture, use, or sell any of the said inventions, or any improvements thereof, or any machine embodying or article containing any of said inventions or improvements, or protected by any of said letters-patent, without the written consent first obtained of the other party.

7th. No sale of any of the said inventions and no license or right to make or use the same in any way shall be made or given except at a price to which both parties agree, and all net profits shall be equally divided between the parties hereto.

8th. The covenants and provisos of this agreement binding either of the parties hereto shall also bind his executors, administrators, and assigns.

In witness whereof the said parties have hereunto set their hands and seals the day and year first above written.

THOMAS A. EDISON, [L.S.] GEORGE B. PRESCOTT. [L.S.]
Sealed and delivered in presence of—R. H. ROCHESTER.[6]

PD (transcript), NjWOE, Quad. 72.11, p. 40 (*TAEM* 9:185). Notarization omitted. There are three other versions of this document in the Quadruplex Case, two of which contain insignificant variations, primarily in punctuation. [a]Place taken from notarization; date taken from text, form altered.

1. Probably "between" in the original, which was prepared by Western Union's lawyers (the firm of Porter, Lowrey, Soren, and Stone) after Prescott took Doc. 445 to them. App. 3, Edison's Affidavit, p. 810.

2. Edison testified that this shift from simply giving Prescott a partnership in the patent rights (as earlier proposed in Docs. 432 and 445) to labeling Prescott as a joint inventor was made at Prescott's insistence, though Prescott invented no part of the devices in question. Quad. 70.7, pp. 390–91, 405–8 (*TAEM* 9:562, 569–71). See also Edison's "Answer," Quad. 72.14, p. 7 (*TAEM* 9:208); App. 1.G15–16; App. 3, Edison's Affidavit, p. 810; and Dyer, Martin, and Meadowcroft 1929, 1:157.

3. Edison executed seven applications and two caveats—all in his own name—on 19 August 1874. See headnote, p. 254.

4. The specifications included Docs. 449, 450, 467–73, the illustration ("Model No. 3") on p. 265, and U.S. Patent 209,241 (see the illustration on p. 348) (Edison's testimony, Quad. 70.7, pp. 327, 343–44 [*TAEM* 9:530, 538–39]). Phelps had eight models ready on 10 July, another on 31 July, and another on 7 October (Phelps's testimony, Quad. 71.1, p. 308 [*TAEM* 10:161]).

5. Edison apparently had this clause inserted to protect the interests of his partners in automatic telegraphy. Josiah Reiff's testimony, Quad. 70.7, pp. 449–50 (*TAEM* 9:592–93); Lemuel Serrell's and Prescott's testimony, Quad. 71.1, pp. 82, 84, 341 (*TAEM* 10:46–47, 177); App. 3, Edison's Affidavit, p. 810.

6. Roswell Rochester had been the treasurer of Western Union since 1871. Reid 1886, 687–88.

–452–

From George Harrington

[New York,] Migdnight July 9th/74

My dear Edison

I returned this P.M.[1] Have not seen Reiff but having learned what was going on have been all this Evening investigating & therefore beg of you to see me before you sign any more papers, take any money or go to any other place.[2] Come to 80 Broadway. I am in hopes that I can relieve you.[3] At this moment adverse actions will cause a loss of 100 000$[4] Truly

Harrington

ALS, NjWOE, Scraps., Cat. 1173:48 (*TAEM* 27:386). A transcription of a copy of this letter is in Quad. 70.7, p. 419 (*TAEM* 9:577).

1. Harrington had been out of town for a few days. His son Chase hand-delivered this note to Edison at Edison's house on Wright St. Chase Harrington's testimony, Quad. 70.7, pp. 417–36, 439 (*TAEM* 9:576–87); Joseph Murray's testimony, Quad. 71.1, p. 297 (*TAEM* 10:155).

2. This refers to Edison's joining with George Prescott and Western Union to develop multiple telegraphy. Edison earlier had told Josiah Reiff about the prospective agreement (Doc. 451) and had shown it to him right after signing it. Edison's and Reiff's testimony, Quad. 70.7, pp. 391, 452 (*TAEM* 9:562, 595).

3. Edison owed William Unger $10,000 as of 3 July to meet the terms of the dissolution of their former partnership; failure to pay meant forfeiture of the Ward St. shop (see Doc. 264). Edison turned to his associates at both Western Union and the Automatic Telegraph Co. He offered the shop as collateral, but neither Prescott, Orton, nor the Western Union Co. provided a loan, though Prescott had George Phelps make an inventory of the equipment and premises, and Orton arranged for a $3,000 advance on Edison's Gold and Stock account. Reiff raised $1,000 and got Unger to allow a short delay, but Harrington and the other investors in the Automatic Telegraph Co. were still unable to raise the rest of the needed money. They finally sold some land bonds held by William Seyfert, increasing Seyfert's share in Automatic while providing $6,600 to Unger on Edison's account, canceling Unger's lien on the shop. In this connection Edison signed some promissory notes that generated considerable trouble in later years (see Doc. 516). Reiff's testimony, Quad. 70.7, pp. 448–53, 706 (*TAEM* 9:592–95, 723); Edison's testimony, Quad. 70.9, p. 90 (*TAEM* 9:809); Orton's testimony, Quad. 71.1, pp. 148–51, 191–97 (*TAEM* 10:80–81, 101–4); depositions of Harrington and Reiff, *Seyfert v. Edison,* Lit. (*TAEM* 46:410, 412); PN-75-01-05, Accts. (*TAEM* 20:29). See also App. 1.D157, G15.

4. Negotiations were in progress to include Automatic in a combination of telegraph companies to challenge Western Union, or alternatively to sell Automatic to Western Union. See Doc. 522 n. 10.

–453– [Newark, July 13, 1874?[1]]

Equipment Specification: Multiple Telegraphy

Quadruplex instruments[2]

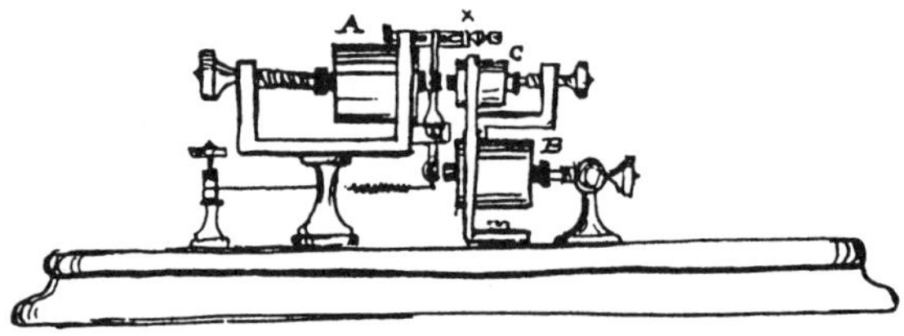

A and B are two magnets one inch in length, they may have a diameter the same as ordinary Morse relay spools, or Duplex relay spools. The resistance of each magnet is to be 200 ohms C is another magnet, one inch long, whose diameter may be small, resistance 50 ohms. The top armature is to be double Thus, The weight of the two not to exceed that of a single "half round" armature, which may be used with the bottom magnet.[a]

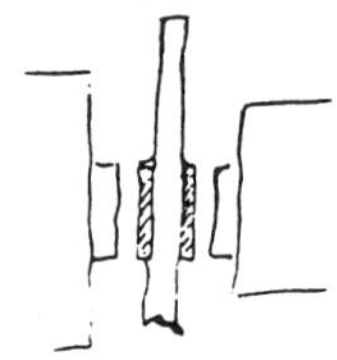

The platina is to be "hard drawn."[3] The [-]local platina point is at the back X, with ivory in the point of the front screw The cores of A B and .C. are to be ¼ of an inch in diameter ~~and free as possible from permanent ma~~ well annealed, if this is inconvenient the ordinary Morse size will answer nearly as well. The cores are to be slotted from end to end, wi~~t~~dth of slot about 1/32—depth, to the centre. Thus:—

A single layer of 23 wire is to be wound on each spool of A. and .B. after the fine wire has been wound on them These layers of wire must be well insulated from the fine wire

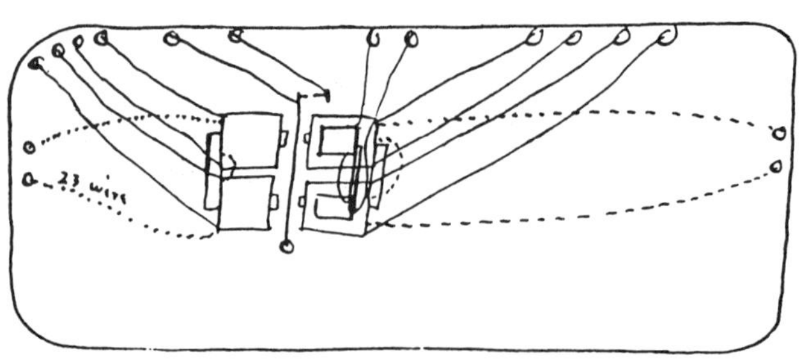

16 Binding posts The next diagram may make it clearer.

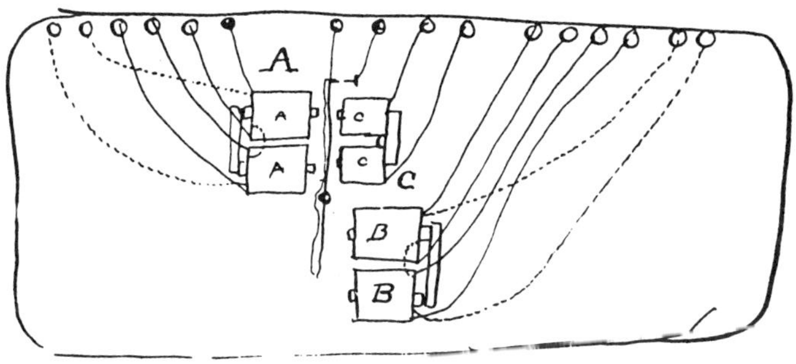

The dotted lines represent the 23 wire, use ordinary Phelps binding posts.

Four of these relays will be required.

~~F~~Six 6[b] [-]sounders with spools one inch long altered as before thus:—

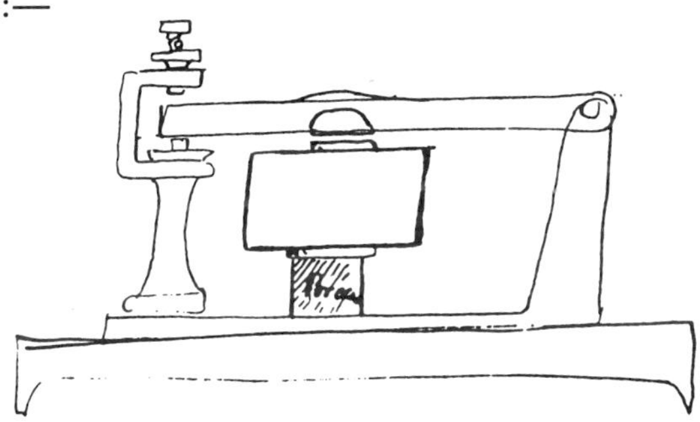

Th~~e~~ree polarized relays, same style as before to be obtained from Gold and Stock.[4] Their tongues to be put in same as in Siemens Relay.[5] The brass projection from the tongue to be a full 1/16 thick. The magnets to be inch and half long and wound with two sets of coils[6] Thus

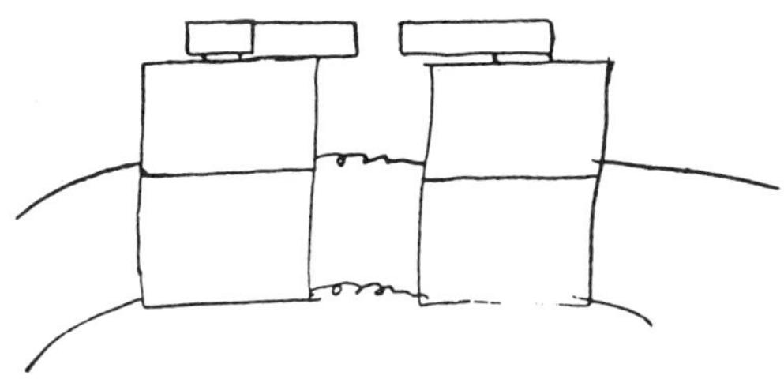

if it is not convenient to put on rubber covers they may be dispensed with. connections thus

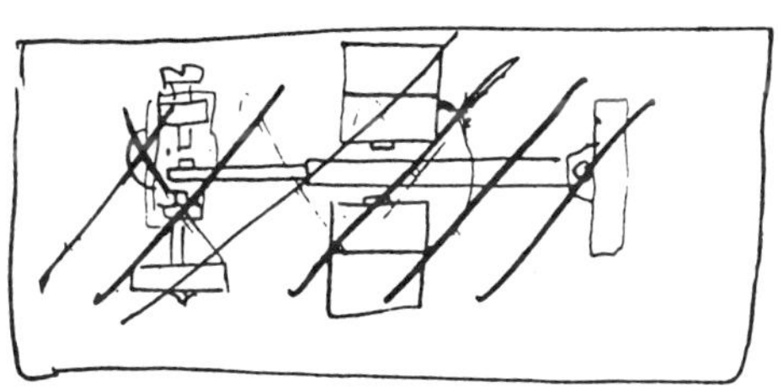

Thus

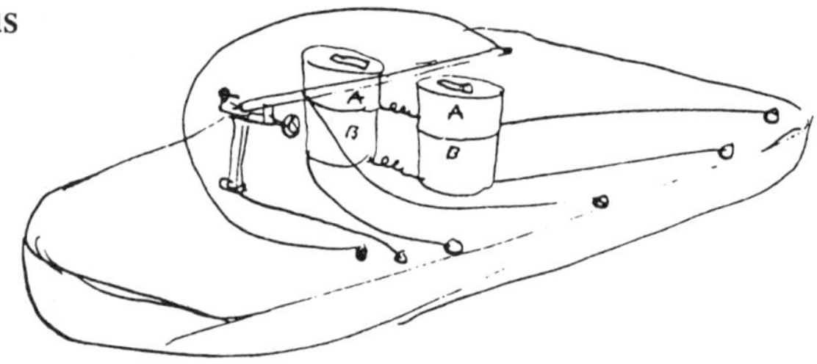

The two returned relays are to have their tongues connecte[d][c] to the angle peice same as ~~s~~Siemens, [-]the brass projection ~~con~~ from the tongue to be 1/16.—it spring[s][c] now= The resistance of A and .A. are to be about 150 ohms together[d] and B .B. the same. The returned polarized Relays are to be rewound and made 150 ohms resistance, with a single layer of 23 wire wound over the fine wire and running to seperate binding posts.

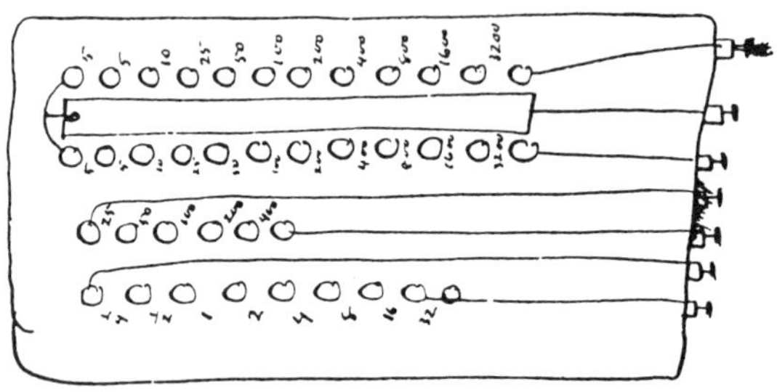

Four such Rheostats as this.

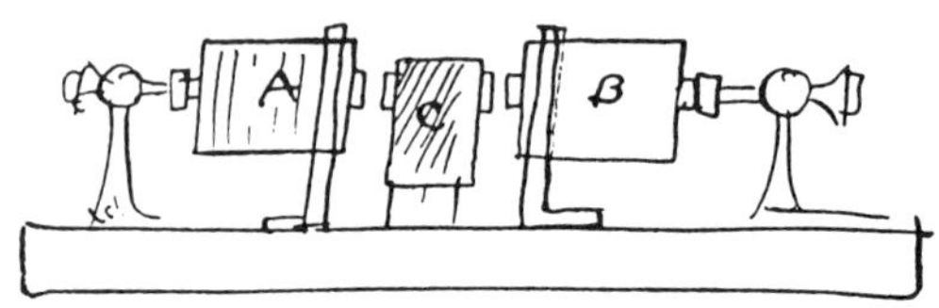

A & B are two magnets of the same length as ordinary Morse wound with 200 ohms each, and so arranged that they adjust up to fixed armatures in a peice of brass C.

The Connections are made thus[7]

I should like to have this arranged as compact as possible, which might perhaps be done this way

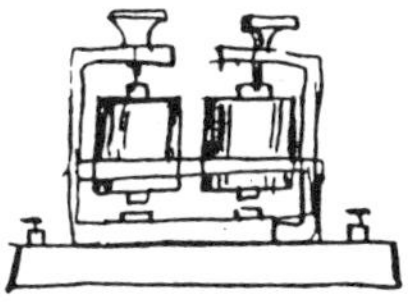

I have very little room for it on the table.

AX (photographic transcript), NjWOE, Quad. 71.2, following p. 146 (*TAEM* 10:306). Another copy is in Defendant's Exhibit 43, *Nicholson v. Edison.* [a]"The weight . . . magnet," interlined above the drawing. [b]Circled and interlined above. [c]Page trimmed. [d]Interlined above.

1. George Prescott sent this specification to George Phelps on 13 July, requesting the instruments as "soon as possible." Phelps had them ready on 29 August. Quad. 71.2, following p. 146 (*TAEM* 10:313).

2. Doc. 449 contains a similar relay without the retracting spring; cf. also Doc. 442 and the relay illustrated in Doc. 469 n. 3.

3. In a fragmentary earlier version, Edison wrote, "The platina used on previous instruments is so soft that it creates too much oxide" (Cat. 297:104[1], 116, 117[1], Lab. [*TAEM* 5:856, 875, 876]; the quote is on p. 117[1]). See also Doc. 489.

4. See Doc. 441 n. 6.

5. That is, a particular style of polarized relay; see *TAEB* 1:38–39.

6. Cf. Doc. 513.

7. Cf. Doc. 532 (figs. 3 and 4).

-454-

Article in the Operator

New York, July 15, 1874.[1]

Mr. Thomas A. Edison, of Newark, N.J., has for some time past been engaged in perfecting a system of "Quadruplex Telegraphy," or sending two sets of messages and receiving two sets at the same time over a single wire.[2] His apparatus being in order for experimental work, Messrs. Orton, Prescott, Hunter[3] & Brown[4] of the W.U. Teleg. Co., and Mr. Grant,[5] of the Montreal Teleg. Co., met in the electrician's office on Wednesday, 8th inst., where they witnessed the successful working of the Quadruplex over a circuit of about two hundred miles in length, two wires having been looped to Philadelphia and back for the purpose.

Messrs. Bogart, Phillips,[6] Boileau, Mixer, Cook, Fullum, Gramzow and Kennedy of the night force took seats at the instruments, and for an hour exchanged newspaper specials at a rapid rate, all the instruments working nicely and promptly.[7] Some changes in the instruments have suggested themselves to Mr. Edison,[8] and as soon as these are made, in a few days, the experiment will be repeated on a longer circuit, it being proposed to use then a wire to Boston and back.[9]

PD, *Operator,* 15 July 1874, 5.

1. By this date the quadruplex had already generated considerable publicity, beginning on 10 July 1874 with a *New York Times* article, "Telegraphy. New Inventions in the Science of Electrical Transmission" (p. 2), which was widely noted and reprinted. This account contrasts both in content and tone with that article, which praised the invention as a technical marvel and a valuable business asset while barely mentioning Edison. This more modest assessment accords with Edison's view; at the time, he still regarded the quadruplex as of little worth compared to the automatic. He was closely associated with the *Operator* and undoubtedly had much to do with the writing of this article.

2. Edison based the system used for these trials on the diplex design of Caveat 45 (Doc. 348), using the differential mode of duplexing the diplex (Testimony and Exhibits on Behalf of Thomas A. Edison, p. 33, *Nicholson v. Edison*). Thus, at least the receiving instruments were much like those in the illustration accompanying Doc. 446. No full, detailed description of the system exists.

3. William Hunter oversaw the purchasing of supplies for Western Union. Reid 1878, 485–86.

4. Alfred Brown managed Western Union's New York Office. Taltavall 1893, 19–20.

5. Angus Grant was an electrician for the Montreal Telegraph Co. Reid 1879, 335.

6. Walter Phillips, later general manager of the Associated Press, wrote many humorous pieces and sketches connected with telegraphy under the pen name "John Oakum." He also provided his own recollection of the quadruplex tests. Phillips 1897, 184–85; Taltavall 1893, 336–38.

7. Edison later said that he had told the operators to carry on normally even if troubles developed on the line or in the instruments. See App. 1.G15.

8. Although this report says the apparatus worked well, Edison later testified to the further development needed "to effect quadruple transmission commercially." He stated: "In the early state of the art, with poor instruments, the two messages [one way] were not transmitted so smoothly as when afterwards good instruments were made. When it was attempted to put two of these devices together by duplexing the same, then the defects of bad instruments were increased, great complication ensued, and the working current was decreased," and he elaborated that "the art being new" meant that "the margin for adjustment was very small, hence endeavors were made by testing a variety of instruments to increase the margin for adjustment, both for equating the relays and other points where adjustments were required to meet the various conditions of line as to static discharge and leakage during wet weather." The differential duplex arrangement accentuated this as it was hard to get "two relays so near alike that if you balanced the outgoing current upon one relay the other would be the same; it was generally out when the second one was balanced; then the first one would be out." This led Edison to try the bridge system, which relied less on perfectly matched coils. Testimony and Exhibits on Behalf of Thomas A. Edison, pp. 44, 40, 33, *Nicholson v. Edison.*

9. Another test, using a looped line to Albany, took place on 9 Sep-

tember. The New York–Boston tests did not occur until the end of that month and did not use a looped line. Tests on a wire as long as a New York–Boston loop were not made until December. See Docs. 494 and 515.

–455–

John Clark to Edison and Murray

Cairo Ritchie Co W ~~W~~Va July 20/74

Messrs Edison & Murray

Your favor of 13 inst[1] just received and I send enclosed a rough sketch of the main feature in my improvement in the Electro Magnetic Engine

The figure is intended to represent a double contracting Electro Magnet, composed of 34 sections (17 in each leg of the magnet) besides the curved end pieces. When the magnet is extended the sections open leaving an interval of 1/18 of an inch This gives when closed a stroke of one inch. One layer of insulated copper wire is wound on these sections after the manner of electro magnets, but on each alternate section the wire is wound in the opposite direction. By a peculiar arrangement of the Breakpiece the current is sent in one direction and the magnet contracts with great force. Then the current is reversed on each alternate section and the magnet opens its length by the repelling force. The sections are of soft iron, hollow, slightly convex and concave where they touch when closed. The covering wire is not broken at the junction of the sections but is looped out so that the movement of 1/18 inch does not injure it. Another feature in this arrangement and which I have never observed in any other, is that when the machine is in operation, there is no surplus material or dead weight—every grain of copper wire or of soft iron being continually in active operation. I have here purposely avoided giving the details that you might better understand the main principle on which the machine operates, and I feel convinced that it will develop the full mechanical power of a given battery, and if the day should ever arrive when we could obtain electricity cheap enough, it would supersede steam as a motive power. At present however I see no reason why it could not be applied successfully to sewing machines and other machines where safety cleanliness &c would be more of an object than cost of running. From your experience in electrical pursuits you should be able to form a good opinion of the merits of my plan, and if you think proper I will send a sketch of the machine in detail.[2] Yours respectfully

John Clark

P.S. I have resided in West Virginia since the war but was born and raised in Paterson NJ. My grandfather started the first cotton mill in that place and my father started what is now called the Danforth Locomotive Works

ALS, NjWOE, Scraps., Cat. 1173:41 (*TAEM* 27:375).

1. Clark wrote Edison and Unger on 6 July offering to sell a half interest in his invention. Cat. 1173:40, Scraps. (*TAEM* 27:373).

2. Clark sent Edison and Murray a sketch of the device and further explanation of its working on 13 August (Cat. 1173:93, Scraps. [*TAEM* 27:401]). Nothing more is known of this motor.

–456–

From Willis Engle

Indianapolis, July 24 1874[a]

Friend "Al."

I suppose that you will be able to call to remembrance the undersigned, as an old school fellow and playmate; and the times we used to have in the "cave" and the many wonderful things we planned but did not accomplish in our boyhood days. You, I am glad to see, have succeeded better since you have come to man's estate, and have greatly improved upon old Ben's chaining-lightning-performance May success still crown your efforts in future work. I have see short newspaper accounts of you late invention which enable one wire to carry four messages, but would be glad to see a fuller account. Your telegraph printing instrument I have not seen, but am likely to need something of the kind, as we intend to construct a telegraph line from our office to our yard, about one mile in length. Will you give me some particulars as to its workings &c and what a couple of them will cost us. We just started in business in February last and are doing very well for new beginners. I worked, previous to that time, 9 years in the Local office of the "Vandalia Line" at this point as Cashier, my oldest brother, George being the Agent.[1] Father and mother, my sister and two brothers, George & Frank (the latter married) are now living here. Father has charge of a church here, while my brothers are both still in the employ of the "Vandalia" Where are your folks now? I suppose some of them are still in Port Huron Let me hear from you as soon as convenient,

and let me know about the instruments Your Old Friend
Willis D Engle[2]

ALS, NjWOE, Scraps., Cat. 1173:101 (*TAEM* 27:412). Letterhead of Engle & Drew, Coal Dealers. [a]"Indianapolis," and "187" preprinted.

1. This was part of the Pennsylvania Railroad system running between Indianapolis and St. Louis. "General Map of the Pennsylvania Railroad and its Connections," in *Report of the Investigating Committee of the Pennsylvania Railroad Company* (Philadelphia: Allen, Lane & Scott's Printing House, 1874).

2. On Willis Engle and his family see *TAEB* 1:27 n. 1, 28 n. 6.

-457-

Edison and Murray Account

[Newark,] Friday July 24th 1874

Geo. Harrington Dr

To[a] 9 Patent office Models—	~~20~~
/ 2 Wire Roman letter[1]	20
/ 1 Wire Roman letter[2]	20
/ Repeating Relay (Automatic)[3]	20
/ Improved Wheatstone[4]	20
/ Winding & unwinding model[5]	20
/ Wheatstone Duplex[6]	45
/ Wheatstone Duplex	20
/ Aut. with air condenser[7]	20
/ Aut short magnet =[8]	20
	$205

D, NjWOE, Accts., Cat. 1184:64 (*TAEM* 21:815). Written by Charles Batchelor. Three other entries appear under this date; each is separated by a centered horizontal line. [a]Checkmark in left margin.

1. Edison executed the patent application on 7 August 1874 (U.S. Pat. 172,305).

2. Edison executed the patent application on 7 August 1874 (U.S. Pat. 173,718).

3. This may be related to an undated, draft patent application (Doc. 524).

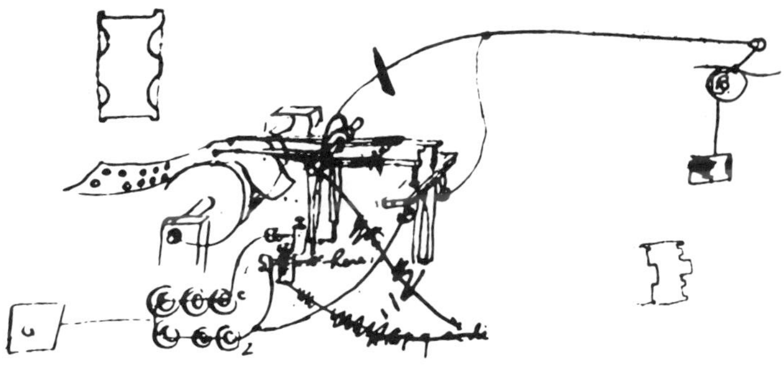

Edison's drawing for a patent model for his "improved Wheatstone" automatic telegraph.

Edison's device for winding and unwinding rolls of automatic telegraph paper.

4. That is, the Wheatstone automatic telegraph system. Edison very likely conducted experiments with the Wheatstone instruments in the Western Union electrician's office, which had been acquired by George Prescott and Thomas Eckert during their 1873 trip to England (Reid 1886, 735). Although these experiments probably took place on Western Union lines, the patents would have been assigned to Harrington.

No patent application has been found and the nature of the proposed improvement remains unclear. The sketch on the preceding page shows this model (Cat. 1168, Lab. [*TAEM* 6:325]). Edison would typically draw a sketch for a model, which would be made by workmen in the shop. He or Batchelor frequently put a price on the model sketch, but at least in the case of the Wheatstone models listed here, they did not necessarily match the amount charged in the account books.

5. Edison did not execute the patent until 11 February 1875 (U.S. Pat. 171,273). He drew the following sketch of the model (Cat. 1168, Lab. [*TAEM* 6:323]); the model is at the Edison Institute, Dearborn, Mich. (Acc. 29.1980.1358). This invention was apparently used in offices employing his automatic telegraph system; a number of the contact-rollers seen at the right of the sketch can be found at ENHS.

Edison's drawing of the patent model for his apparatus to wind and unwind the paper tape for his automatic telegraph (U.S. Pat. 171,273).

6. No patent applications have been found and the nature of the proposed improvements remains unclear. There are two sketches that may be the Wheatstone duplex designs listed here (although the prices on the

Edison's drawing for a patent model of his duplex circuit for the Wheatstone automatic.

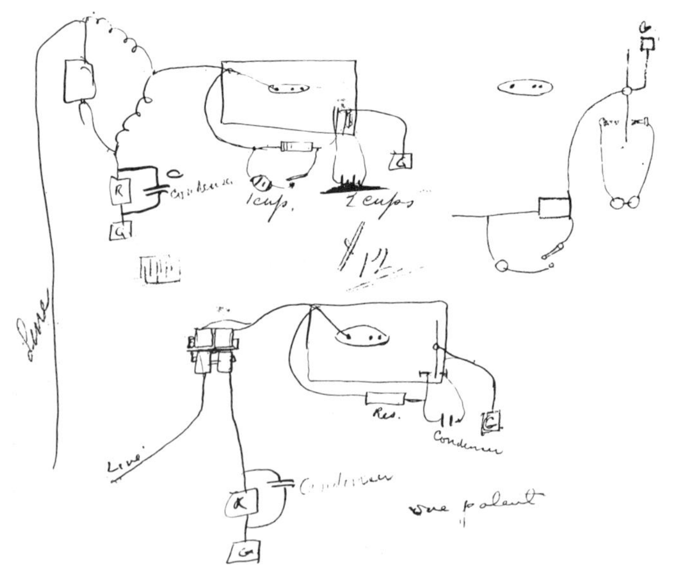

Another Edison drawing for a patent model of a Wheatstone duplex circuit.

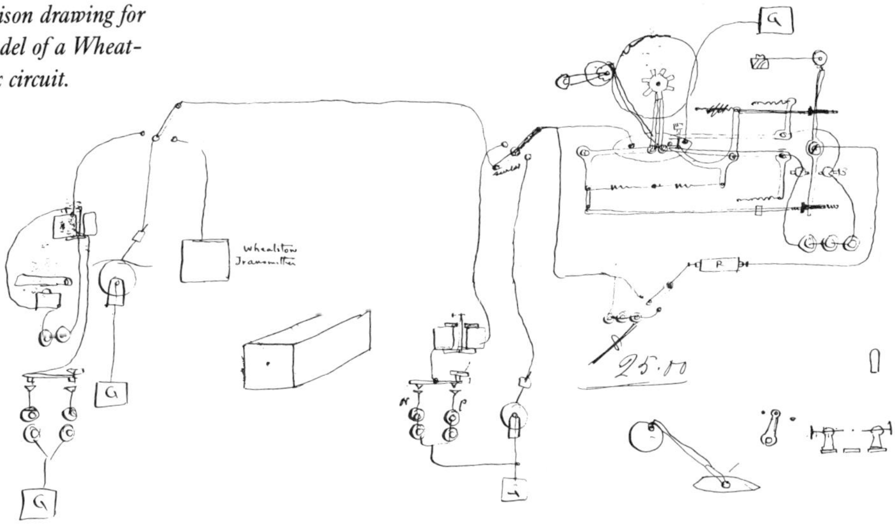

sketches do not match those in this account entry). On the sketch priced at twelve dollars, Edison wrote, "Wheatstone Duplex will explain." Cats. 1168, 30,100, both Lab. (*TAEM* 6:324; Supp. III).

7. Edison did not execute the patent application until 19 January 1875 (U.S. Pat. 195,752).

8. No patent application has been found. A patent executed by Edison on 18 January 1875 (U.S. Pat. 195,751) does specify the use of relays "not to exceed one inch in length" for shunting the induction current on lines employing his automatic telegraph. Although this relay design was not included in the claims, Edison noted that "I have found that by decreasing the length of the cores and coils, the induction is decreased to such an extent" that the relays do not stick, as often occurred in common relays with longer coils.

AUTOMATIC TELEGRAPH TRANSMITTER-RECEIVER Doc. 458

In July 1874 Edison created the model for his first patented transmitter-receiver for automatic telegraphy.[1] Although the Automatic Telegraph Company had been using a combined instrument for some time, that instrument was operated by a hand-crank.[2] In his new design Edison used clockwork to rotate the transmitting and receiving rollers, which were placed on the same shaft and engaged by turning a switch.

In this design Edison made the transmitting roller larger than the receiving roller. Since the perforated paper would pass over the transmitter more rapidly than the chemical recording paper would pass over the receiver, the marks recorded at the receiving end would "be contracted proportion-

A manually powered transmitter-receiver for automatic telegraphy. The transmitter is on the left, the receiver on the right.

ately" and the instrument would use less chemical paper.[3] To help electrically balance the line, Edison incorporated two switches (on the base of the machine, one on either side of the switch used to engage the rollers) that controlled variable resistances "under arrangements of circuits adapted to automatic telegraphy."[4]

Edison had been working on a clockwork-driven transmitter-receiver since the end of March 1874.[5] The impetus for it likely came from England, where tests were continuing on his automatic system. Letters from his agents in England often noted that the British Post Office engineers favored clockworks over hand-cranks. Edison criticized this enthusiasm for clockworks, although he stated that "if it is decided to use a clockwork, . . . I can furnish that article."[6] The use of a clockwork in this transmitter-reciever may also have influenced (or perhaps been influenced by) his simultaneous work on a domestic telegraph receiver.[7] Little evidence remains of Edison's experiments with clockwork mechanisms for automatic telegraphy,[8] and the wooden patent model depicted below is the only known example of this design.

1. Cat. 1184:64 (*TAEM* 21:815). Edison executed the covering patent application on 18 January 1875 (Pat. App. 168,242).

2. It is not clear when Edison designed his standard hand-cranked transmitter-receiver for automatic telegraphy. In his early work on automatic telegraph transmitters and receivers, Edison designed separate

instruments and operated them by both hand-cranks and clockworks. He soon abandoned the clockwork but continued to employ separate instruments. By the time he began working on this clockwork design in March 1874, account records indicate that a combined transmitter-receiver was being employed by the Automatic Telegraph Co. (Cat. 1214:41, Accts. [*TAEM* 21:584]). The transmitter-receiver used by Automatic Telegraph may have been similar to the production model shown on p. 246, which could also be the new design used on Atlantic and Pacific lines in April 1875 and discussed in Docs. 559 and 560.

3. U.S. Pat. 168,242.

4. Ibid.

5. A shop order book indicates that Automatic Telegraph wanted "2 Transmitters & Rec together to run with clockwork." Cat. 1214:43, Accts. (*TAEM* 21:585).

6. Doc. 418; see also Docs. 420 and 481.

7. See Doc. 615.

8. See Doc. 481. See also Cats. 30,094 and 30,095, both Lab. (*TAEM* 5:321, 361; 6:398); and Cat. 1307, Batchelor (*TAEM* 90:628).

-458- [Newark, c. July 24, 1874]

Patent Model: Automatic Telegraphy[1]

M (26 cm × 17 cm × 14 cm), MiDbEI(H), Acc. 29.1980.1352.

1. See headnote above. Made largely of wood, this model has some brass parts.

–459–

Notebook Entry: Electromotograph and Chemical Experiments

[Newark,] Aug. 3 1874[a]

Experimental Philosophy

1.

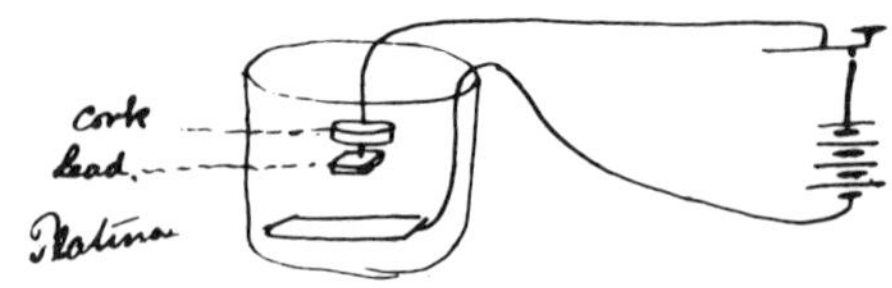

See if H[ydrogen] evolved on the Lead will decrease its specific gravity   Solution Caustic Potash.

1[b]   12 cups of battery used Bunsen   Cork raises slowly but is entirely due to mechanical means so much H collecting under the lead as to considerably lessen its weight. Acts slow.

⟨Batchelor   20 hr   different exp.⟩[c]

Batchelor.

2.

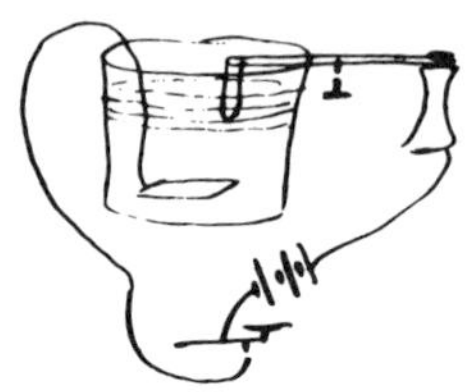

See if it dips when circuit is open. ⟨Emg⟩   ⟨Exp.   Made for investigation of Emg action⟩[1c]

X, NjWOE, Lab., Cat. 1172:40 (*TAEM* 3:97). Written by Batchelor. [a]Date also appears with signature. [b]This paragraph, marginalia, and signature written on facing page of notebook. [c]Marginalia written by Batchelor.

1. Batchelor's notations identifying these experiments as "Emg" mark the first appearance of the standard laboratory abbreviation for electromotograph. They were probably not made before 7 August. Account records show that on 1 August, Edison had taken a model of his "new force" apparatus to Lemuel Serrell, and only on 7 August did the term "electromotograph" appear in notebooks and account records. Cat. 1184:67, 69, Accts. (*TAEM* 21:816, 821); Doc. 462.

–460–

Notebook Entry: Chemical Experiments

[Newark,] Aug 4 1874

Sulphide of Iron (FeS.)

Made this from bar Iron heated to a white heat & pressed against a bar of sulpher over a water bath into which it drops after combination with sulpher   It is brittle & has a look slightly bronzed

Batchelor

Hydrosulphuric Acid ~~& H~~

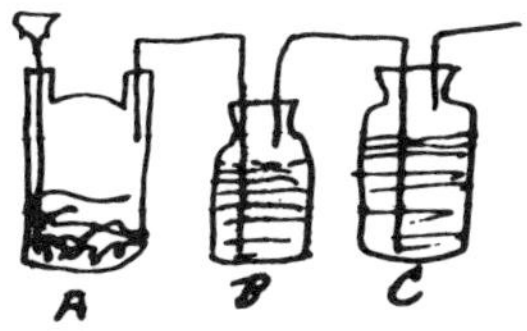

Made this in the following way A contains Sulphide Iron (FeS) & dilute Sulphuric Acid ($HOSO_3$) Strong chemical action in which Sulphuretted Hydrogen (HS) is given off which passes through the bottle B which only has water in it to wash the gas & from that into the bottle C filled with distilled water which is impregnated with Sulphuretted Hydrogen (HS) The gas after leaving bottle C would smell obnoxious, & in order to prevent this we pass it into still another bottle containing Ammonia Hydrate (NH_4OH) which combines to form Sulphide of Ammonium (NH_4S). The chemical change which takes place in first bottle is shown by the following:—

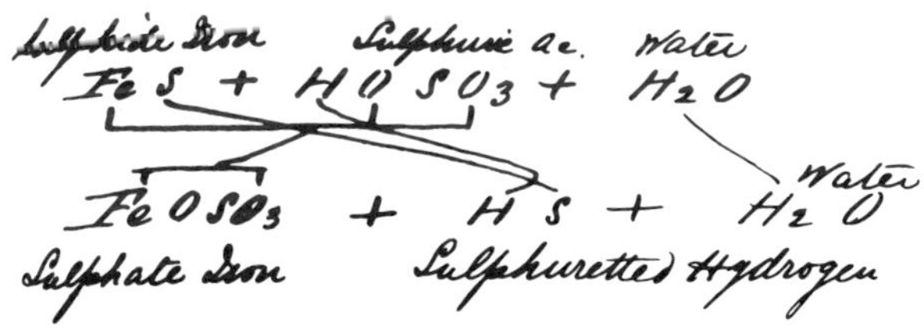

Batchelor

X, NjWOE, Lab., Cat. 1172:200 (*TAEM*) 3:130). Written by Batchelor.

-461-

Notebook Entry: Automatic Telegraphy and Electromotograph

[Newark,] April[August][1] 5 74

⟨Automatic and Emg⟩[a]

Qinine Sulph. HO. ~~N~~Chl Sod. Hyd Ammon.

Marks with tin pen on Hydrogen side.[2] Platina shows Edsons N[ew] F[orce] good on both. Tin on Hydrogen side shows well. The tin pen marks black but its not so very sensitive. Lead pen is most sensitive. N.F. Shows splendid on this solution. Extra good I think on account of the lead being reduced & giving a mark I think this is necessary for that purpose that the lead [-] should be reduced to get sensitiveness, though platina shows it on short ckt it wont on 10 000 ohms & 6 cells Carbon while lead is fearful strong.

Aniline,[3] Excess lime water, boiled for some time & then

filtered hot, Yellow solution. No marks but Lead shows extraordinary strength N.F. better than Qunine above. I find on trying Potash that it is as good probably morse sensitive but above appears to act sharpest and evenest, paper smoother. I added salt to last, but no marks obtained.

⟨Edison Batchelor Adams[4] Fox[5] Charley[6] 12⟩[a]

X, NjWOE, Lab., Cat. 1172:159 (*TAEM* 3:118). [a]Marginalia written by Batchelor.

1. The preceding page is dated 5 August, and the following page (Doc. 462) is dated 7 August.

2. That is, attached to the battery pole that will evolve hydrogen gas from an electrolyte (the cathode). See Doc. 439.

3. Aniline, a cheap extract from coal tar, attracted great interest following the discovery in the 1850s and 1860s that it was a source for many valuable artificial dyes. This colorless, poisonous, oily liquid became the base for a whole new rapidly growing and extremely lucrative industry, first in Britain and France, then in Germany and Switzerland. The initial discovery in this line by English chemist William Perkin came during an attempt to synthesize quinine (Haber 1958, 80–83). The preceding page in this notebook concerns aniline solutions (Cat. 1172:158 [*TAEM* 3:118]).

4. Little is known of James Adams, who became one of Edison's principal assistants in his electric-pen, acoustic telegraph, and telephone experiments. Edison later related that Adams was a sailor who came to Menlo Park looking for a job, and that he gave him small induction coils to sell to his fellow sailors (App. 1.B7, C15, C17, D262, D380, D382, E11). A card pasted in one of Edison's scrapbooks (Cat. 1143:94 [*TAEM* 27:222]) and hand-dated 1874 indicates that Adams was an agent for Edison's inductorium and for the Walter Electric Burglar Alarm Co. He may also have been an agent for other electrical equipment manufactured by Edison and Murray. His office was at 510 Warren St. in Brooklyn.

5. Nothing is known of Alexander P. Fox except for the following item from the spring of 1875 found in one of Edison's scrapbooks: "It is reported that a young man named Fox, in the employ of Edison Murray, telegraph instrument maker at Nos. 10 & 12 Ward street, has decamped, taking with him between $400 and $500 in money belonging to the firm." Cat. 1178, Scraps. (*TAEM* 27:146). See also *Operator,* 1 June 1875, 9; and *Telegr.* 11 (1875): 102.

6. Charles P. Edison (1860–1879), son of Edison's brother, William Pitt Edison.

–462–

Notebook Entry: Automatic Telegraphy and Electromotograph[1]

[Newark,] Aug 7th 1874 4. AM.

Pyrogallic Acid with salt & lead pen very sensitive on Hydrogen to pen. White paper & Blackish brown mark. Paper seems to be less sensitive as it is kept. All pens mark on Hydrogen side except silver.

Excess of Chl Na & Pyrogallic acid seems to be best[2]

Quite sensitive even when small amt of Pyrogallic Ac is used

Chl Zinc kills it entirely

NH_4 added to Salt solution no good Seems to be exceedingly sensitive to 'new force'

Platinum marks on this but not so sensitive as lead which seems to be the best.[a]

Platinum also gives on O[xygen] side[3] a not very sensitive pale yellow mark[a]

Changed the Instrument here Platinum to Iron-drum[b]

Pyrogallic Acid & Citric Acid no mark at all on any pen

Pyrogallic Acid & Nitro Acid & Salt lead marks on small resistance not very good

Gallic Acids mark something like the same as Pyrogallic[a]

These last it seems may be due to the metallic drum on which they were tried as we changed it from Platina to Iron.

witness[c] Batchelor Edison

⟨See Page 162[4]⟩[d]

X, NjWOE, Lab., Cat. 1172:160 (*TAEM* 3:119). Written by Batchelor. [a]Followed by centered horizontal line. [b]"to Iron-" interlined above. [c]Apparently added later. [d]Marginalia written by Batchelor.

1. This entry is continued in Doc. 463.

2. One of Edison's scrapbooks contains a test strip for automatic telegraphy labeled "Pyrogallic Acid Chl Na Aug 7 1874." Cat. 1143:128 (*TAEM* 27:242).

3. That is, the anode; see Doc. 461 n. 2.

4. Doc. 464.

–463–

Notebook Entry: Electromotograph[1]

[Newark,] Aug 7th 1874 4 AM

New Force[a] Electromotograph

Alcohol	no good
Alcohol + H_2O	" "
Sulphide Potassium	" "
Ferrid Sol. appears to be very good	

Repeated & recorded with Electromotograph (300) three hundred words per minute 10 000 ohms & 6 cells of Carbon with ordinary Ferrid paper & that rather dry. with Lead to a point.

Also tried winding several layers of paper round the drum but impractible. Put a piece of paper under lead & kept it there constantly. Impractible. Shunted Instrument with battery tried both ways apparently of no use. The new machine with double spiral spring adjustments is a success much better than the other one gone to P[atent]O[ffice].[2]

We tried if the mere passage of a current produced the force by making a bow of the lead but it did not.

Phosphorus Water No good.[b]

32 Continuous hours work

Batchelor witness[c] Edison

Adams Charley[d]

X, NjWOE, Lab., Cat. 1172:161 (*TAEM* 3:119). Written by Batchelor. [a]Underlined twice. [b]This sentence written by Edison. [c]Apparently added later. [d]"Adams Charley" written by Batchelor.

1. This document is a continuation of Doc. 462 and is continued in Doc. 464.

2. On 7 August 1874, Edison executed a patent for the employment of his electromotograph as a telegraph instrument. He would have supplied Lemuel Serrell's office some time earlier with a model from which to prepare drawings (see Cat. 1222:130, Accts. [*TAEM* 21:405]). Serrell filed the application with the Patent Office six days later (U.S. Pat. 158,787).

In the accompanying patent drawings, moistened paper passes over drum **a** and under the "vibrator" **c** at the end of arm **d** (which is pivoted at **f**). When key **y** is closed, a current passes through the vibrator onto

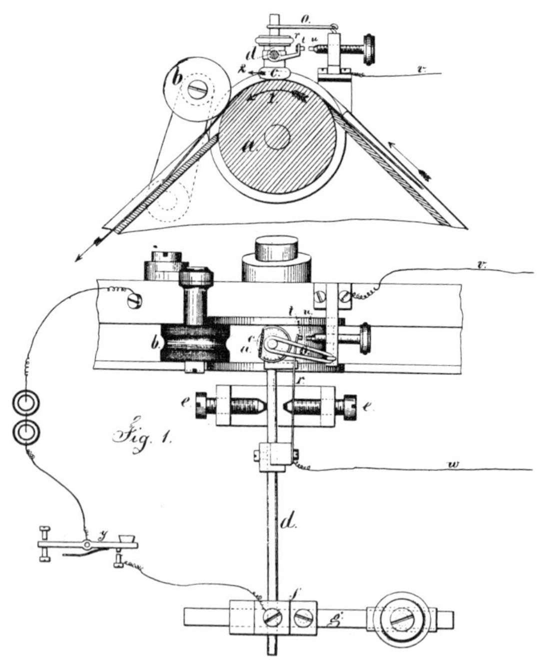

The patent drawings of the electromotograph (U.S. Pat. 158,787).

the moving surface below it, and the resultant lessening of friction allows rubber band **o** to pull lever **d** back, closing local (or relay) circuit **vw** through contact points **t** and **u**.

–464–

Notebook Entry: Automatic Telegraphy and Electromotograph[1]

[Newark,] Aug 7th 1874 6 pm

Relative Delicacy

The Pyrogallics & Gallics give very sharp writing but they lack delicacy they will make very nice salts for short circuit work & are not as yet of sufficient delicacy for long circuits but probably can be made so by further experiment. Excess of salt as yet seems to be the best for them

Solubility of Gallic although slight appears to give as good results as the more soluble Pyrogallic

Oxalic acid added to Reg Ferrid in[a] which Prussian blue is said to be insoluble did not increase the delicacy of reaction nor alter the characterics of marks but somewhat whitened the paper.

Nitrate Mercury & Amm. Nitrate (chlorides form pcpt.[2]) gives a delicate solution in pen on O with small normal tail it is ~~g~~a little more delicate than Pyrogallics but not of a very permanent character Could some compound be p in the paper in which the mark is insoluble it would be a valuable addition

When combined with Aniline red or Magenta which was previously oxydized by Nitros acid & discoloured greatly increases the sensitiveness of the paper it is then more sensitive than regular Ferrid with little normal tailing (Tin pen on O) as yet we have not succeeded in making it very permanent More delicate than all these solutions is Aniline red or common small crystallized Magenta the proper name of which is perhaps Hydrochlorate of Rosaniline This readily dissolves in water forming a deep red or carmine with extraordinary coloring prop[er]t[ie]s By adding to the solution in water a certain quantity of the red fuming Nitros Ac. the aniline becomes oxidized & made nearly colorless.

In this solution is added a small quantity of vegetable Ac[3] the action of which I do not clearly understand of Tartarac Citric Acetic & Oxalic acids the latter of which is the most sensitive & gives the best writing

Tin pen is used on O

The action of which is that the O forms a proto oxide ~~or perhaps~~ which is one of the best known reducing agents & that portion of the discolored oxidized aniline in the immediate vicinity of pen is reduced or deoxidized to its original color which is red at first but by contact with air it becomes dark violet The marks are very permanent The sensitiveness nearly equals if not quite Iodine. one defect of this sol is that the marks are so faint at first that they can scarcely be seen in fact at first they are not at all visible This defect we shall probably be able to get rid of[b]

A combination of Pyrogallic with Gallic & Chl. Sodium gave no better results

Edison

X, NjWOE, Lab., Cat. 1172:162 (*TAEM* 3:120). Written by Charles Batchelor. [a]Interlined above. [b]"able . . . of" overwrites a horizontal line and is followed by another horizontal line.

1. This entry is a continuation of Doc. 463. Edison continued to experiment with different solutions and pens for automatic telegraphy and the electromotograph phenomenon during the next two weeks. Cat. 1172:164–72, Lab. (*TAEM* 3:121–25).

2. Precipitate.

3. Vegetable acids are organic acids derived from plants. *OED*, s.v. "Vegetable."

–465–

Notebook Entry: Electromagnetism

[Newark,] Aug 1~~2~~3 1874

Spark. Induction Current:—

1. When you put one end of Secondary[1] into fine brass filings & then dip the other wire into them & steadily lift it away from them a thin thread of filings follows & sticks to the wire. This thread we drew out to 2 inches. It seems as if the filings were soldered together for the time with Iron filings I got about ¼ inch if the filings were lighter we might get more Powdered Carbon will not do it at all it rather repels if anything

X, NjWOE, Lab., Cat. 1172:170 (*TAEM* 3:124). Written by Charles Batchelor.

1. The secondary circuit of an induction coil.

PARTNERSHIP AGREEMENT AND CASES 94–100: THE QUADRUPLEX CASE PATENT APPLICATIONS Docs. 466–473

On 19 August 1874 Edison executed a set of seven patent applications involving multiple telegraphy designs; they were filed at the Patent Office, along with two caveats prepared earlier, on 1 September 1874.[1] These designs—which came to be known as Cases 94–100[2]—evolved from Edison's work of the previous eighteen months and were essentially completed before his 19 May 1874 partnership proposal to George Prescott. Nevertheless, Edison named Prescott as co-inventor when he first submitted the specifications to Lemuel Serrell. While preparing the applications, Serrell recognized their close relationship to some previous applications of Edison's.[3] He met with Edison and Prescott to discuss the matter, which led to the replacement of Doc. 451 by Doc. 466 and the adoption of wording identifying Edison as the sole inventor.[4] Ownership of patent rights to these designs remained the basis of the agreement with Prescott, however, and became the object of extensive litigation starting in 1875.[5]

All the applications except Case 97 (Doc. 470) eventually led to patents, in most cases after modification, although Case 99 (Doc. 472) was not allowed until 1892.

1. Models to accompany the drawings and specifications had been prepared by early July (Phelps's testimony, Quad. 71.1, p. 308 [*TAEM* 10:161]). The caveats, numbered 45 and 48, were drafted much earlier (Docs. 348 and 392).

2. The case numbers were assigned by Lemuel Serrell.

3. Docs. 304–5, 308–11, and 314–15.

4. Serrell's testimony, Quad. 71.1, pp. 54–55, 78–86 (*TAEM* 10:32, 44–48).

5. See Chapter 7 introduction and App. 3.

–466–

Agreement with George Prescott[1]

[New York,] August 19, 1874[2a]

Articles of agreement made and entered into in this nineteenth day of August, A.D. 1874, by and between Thomas A. Edison, of Newark, in the State of New Jersey, and George B. Prescott, of the City and State of New York,[3] witnesseth:

Whereas, said Edison has invented certain improvements in duplex telegraphs, for which he has executed, or is about to execute, applications for letters patent of the United States, and such applications are numbered 94, 95, 96, 97, 98, 99 and 100, and are dated August 19, 1874, and said Prescott is entitled to an equal interest in the same and others hereafter mentioned.[4]

Therefore, in consideration of the premises and the sum of one dollar in hand paid, the receipt whereof is hereby acknowledged, the said Edison has sold and assigned, and does hereby set over and convey unto the said George B. Prescott one undivided half part of the right, title and interest of every character, in, to, under, and connected with each and all the aforementioned inventions and letters patent on the same when granted, and authorizes and requests the Commissioner of Patents to issue the said letters patent to Thomas A. Edison and George B. Prescott, as the assignees of said Edison, for the use and behoof of themselves and their legal representatives. And whereas the said Edison has also invented other improvements in duplex telegraphs, the descriptions of which have been lodged with George M. Phelps for the purpose of models being constructed, it is hereby agreed that such inventions are included in this present[b] agreement, and that when the applications for patents are made,[5] the patents to be granted in accordance herewith, and that the said Edison shall sign the required papers therefor.

This transfer is made on the following terms and conditions, which are hereby made part of the consideration in the premises:

1. That both of the parties shall have an equal undivided interest in all letters patent of the United States, or of any foreign countries, which may be granted for all or any of said inventions, or of any future improvements thereon made by either party, and of all extensions and reissues of any such letters patent.

2. That whereas Edison has heretofore expended $1,125 for models and patent fees, the benefit of which he contributes to the common interest, and waives reimbursement of that sum, or any part of it, Prescott hereby agrees to pay solely and without contribution from Edison all the future expenses and cost of specification, drawings, models, Patent Office fees, and patent solicitors' and agents' fees, and all other charges incident to the procuring of letters patent for any of said inventions.

3. That neither of said parties will sell, assign or otherwise dispose of the whole or any part of his interest in said inventions or letters-patent therefor, or any of them, without the written consent thereto first obtained of the other party.

4. That neither of said parties will himself manufacture, use or sell, nor grant licenses, nor the right in any way to any other party to manufacture, use or sell any of the said inventions, or any improvements thereof, or any machine embodying, or article containing, any of said inventions or improve-

ments, or protected by any of said letters patent, without the written consent first obtained of the other party.

5. No sale of any of the said inventions, and no license or right to make or use the same in any way shall be made or given, except at a price to which both parties agree in writing, and all net profits shall be equally divided between the parties hereto.

6. The covenants and provisions of this agreement, binding either of the parties hereto, shall also bind his executors, administrators and assigns.

In witness whereof, the said parties have hereunto set their hands and seals, the day and year first above written.

THOMAS A. EDISON, [L.S.] GEORGE B. PRESCOTT, [L.S.]
Witnesses: HARALD SERRELL,[6] LEMUEL W. SERRELL.

PD (transcript), NjWOE, Quad. TLC.3, p. 200 (*TAEM* 10:921). There are eight different versions of this document in the Quadruplex Case; the variations are trivial except where noted. [a]Date taken from text, form altered. [b]Some versions have "patent".

1. See headnote above.

2. Lemuel Serrell, who witnessed this agreement, had his office in New York. He testified that the terms of this contract were settled on 19 August but that the document was not signed until 23 or 24 August. Quad. 71.1, p. 87 (*TAEM* 10:48).

3. See Docs. 432, 445, and 451.

4. See Docs. 432 n. 3 and 451 n. 2.

5. See Docs. 449 and 450. These applications (Cases 111–13) were not filed until the following winter, and only the third listed Prescott as an assignee, although he claimed a half interest in each. Pat. App. 168,385, 207,724, 209,241; Prescott to Commissioner of Patents, 7 June 1875, Quad. 72.2, pp. 9, 11 (*TAEM* 9:35–36); Prescott to Commissioner of Patents, 18 Dec. 1875, Quad. 72.3, p. 7 (*TAEM* 9:42).

6. Harold Serrell, Lemuel Serrell's son, was a clerk in Lemuel's office.

–467–

Patent Application: Multiple Telegraphy[1]

New York, August 19, 1874[a]

94[b]

1 To all whom it may concern. Be it known that I, Thomas A. Edison, of Newark, in the County of Essex and State of New Jersey, have invented an Improvement in Duplex Telegraphs, of which the following is a specification.

2 I make use of a compound induction coil through which the currents pass, and those from the sending station are balanced, but the current from the distant station is operative.[2]

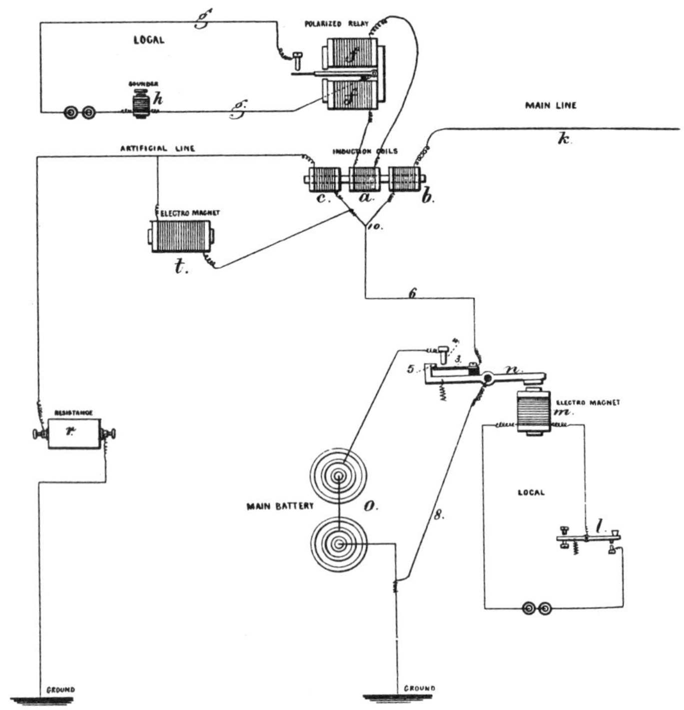

Inventor Thomas A. Edison per Lemuel W. Serrell atty
Witnesses. Chas H. Smith Harold Serrell[c]

3 The helix a. surrounds the central part of the core that passes through the electro magnets b. and c. hence a secondary or induced current is set up in the helix a only when there is an excess of current in one of the helices b. or c. because if the current acting in b. is equal to that acting in c. and the helices are properly wound the magnetizing actions of the helices on the core will neutralize each other and there will not be any secondary or induced current in a. but when the current in one helix is greater than that in the other, the core will be magnetized and a secondary current set up in the helix a.

4 I avail of this feature of the compound differential induction coil to operate a duplex telegraph instrument by causing the current at the sending station where this compound differential instrument is placed to divide and act equally in both b. and c. but when the current from the distant station increases the energy of the helix b. then the induction or secondary current set up in the helix a.[3] magnetizes the core sufficiently to set up a current in a.

5 The current in a. operates in the polarized magnet .f. to

open and close the local circuit g. in which is placed the receiving or sounder instrument h.

6 When the pulsation passing along the line k. from the distant station, ceases ~~there is a reactionary current producing~~ the core of the helix b de-magnetizes and in so doing sets up[d] a second induced current in a. of opposite polarity to the first and that acting in the polarized magnet f. instantly throws the contact point of the armature the other way and opens the local circuit.

7 These operations in the compound differential induction coil, being borne in mind, it now becomes necessary to explain the manner of sending through such coils without producing any action on the helix a.

8 The key l. in the local circuit to the magnet m. operates the lever n. that contains an insulated spring closer 3. acting against the circuit point 4. and the hook end 5. of the lever n. so that when the key l. is closed, the lever n. moves the spring 3. into contact with 4. closing the circuit from the battery o. through 4. 3. and the wire 6. to the helices b. and c. and at the same time breaking the contact of 3. and 5. and hence cutting out the ground wire 8. from the lever n. but the moment the lever n. returns to its normal position by the demagnetizing of m. the spring 3. closes the circuit at 5. just before separating from 4. hence there is always a metallic circuit complete for the pulsation coming from the distant station whether the circuit of the sending battery o. is opened or closed.

9 In order to balance the action of the sending current that divides at 10. and passes through b. and c. I introduce in connection with the helix c. an artificial line equal in resistance and conditions to the line k. hence compelling an equal current to pass through b. and c. To effect this the resistance r. is placed in the ground connection from c. which resistance should be adjustable, so that the rheostat or resistance r. equals the line, and in order to set up in c. a counter magnetism equal to that set up in b. by the static from the line, I make use of the electro magnet t. placed in a shunt that passes around c.

10 By this construction of compound differential induction coil and the arrangement of the connections the inductive effects of pulsations from the sending instrument are balanced and neutralized while the pulsations from the distant station operate the receiving instrument.

I claim as my invention

1st The compound differential induction coils a. b. c. in

combination with the polarized relay f. and the circuit connections substantially as set forth.

2nd The artificial line composed of the rheostat r. and magnet t. and ground connection in combination with the compound induction coil and line connections substantially as set forth.

Signed by me this 19th day of August 1874

Thos. A. Edison.

Witnesses Chas H. Smith Geo. T. Pinckney

DS and PD, MdSuFR, RG-241, Pat. App. 178,221. Written by Pinckney; petition and oath omitted. [a]Place taken from petition; date taken from text, form altered. [b]Case number and section numbers (in margin) written in an unknown hand. [c]Drawing and accompanying signatures from printed patent. [d]"the core . . . up" interlined above.

1. See headnote, p. 254.

2. This is Case 94. It evolved from Doc. 310 (Case F), differing only in the arrangements of the batteries and locations of extra compensating coils. However, the design is distinctive enough that Prescott later included it in his published review of duplex telegraphy, where he stated that it had worked well in tests (Prescott 1877, 823–24). Although the Patent Office had rejected both the original Case F application in April 1873 and its modification the following May, the rejection notice had indicated that changes in wording would make some aspects of the design patentable (Report of Zenas Wilber, in Commissioner of Patents to TAE, 10 June 1873, Abandoned Patent Applications, PS [*TAEM* 8:696–97]). Case 94 was at first rejected because of a conflict with an amended version of Case F that had been filed in October 1873 (Report of Zenas Wilber, in Commissioner of Patents to TAE, 31 Dec. 1874, Pat. App. 178,221). Eventually this application was approved without amendment as U.S. Pat. 178,221. The drawing for the patent model appears as "Model No 4.," Quad. 71.2, following p. 146 (*TAEM* 10:325).

3. This probably should read: ". . . secondary current is set up since the helix b. . . ."

–468–

Patent Application: Multiple Telegraphy[1]

New York, August 19, 1874[a]

95[b]

1 To all whom it may concern.

Be it known that I, Thomas A. Edison, of Newark, in the County of Essex, and State of New Jersey, have invented an Improvement in Duplex Telegraphs, of which the following is a specification.[2]

2 A balanced battery is used for transmitting when the balance is disturbed. An electro magnet is used through which both the received and transmitted pulsations pass, and the[c] connections are made so that the action of the current sent is

balanced, while that coming from the distant station is operative to work a balanced relay and local circuit or sounder.[3]

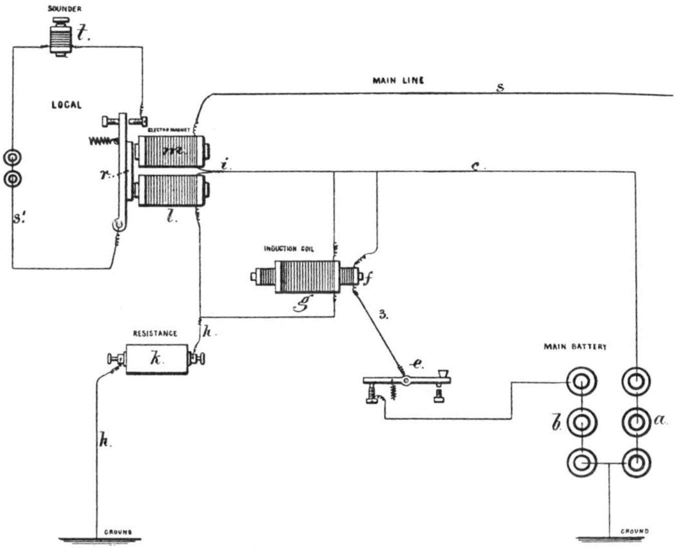

Inventor Thomas A. Edison per Lemuel W. Serrell atty
Witnesses Chas H. Smith Harold Serrell[d]

3 In the accompanying diagram drawing the battery a.b. is connected at one end to the line wire c. at the other end to the closed key e. and in the middle to the ground.

4 The connection 3. from the key e. to the line c. passes through the induction coil and core f. and there is a second induction coil g. around the coil f. that is in a shunt between the line connection c. and the branch h. to the ground in which branch h. there is a resistance k. that is adjustable.

5 The line-connection c. bifeurcates at i. passing one way through the helix l. to the ground-branch h. and the other way through the helix m. to the main line s.. The electro-magnet l.m, being wound in the usual way, will not respond when the connection is made in the middle, because the current passing from c. goes one way through one helix and the other way through the other, polarizing the cores so that the armature r. is not attracted. It is important that the resistance of the branch h. and rheostat k. should be about the same as that of the main line s. so as to cause the current to divide equally at i.

6 It will now be understood that any current from the distant station passing through the magnet m. in the usual direc-

tion will cause the armature to respond, whether there is any current passing through the helies or not from the sending station, and this electro magnet m. and armature act as a relay to operate the local circuit s'. and ~~repeater~~ sounder t.

7 The main battery, it will be seen, is in a local circuit when the key e. is closed, hence if both sides are equal there is no current passing upon the main line, but when the key e. is open, the local circuit is broken and the portion a. of the battery sends the pulsation through l. and m. and upon the line s. to the distant station where the pulsation passing through m. operates the local and sounder or receives the portion through l. returning to a. through the branch h. and ground.

8 When the key e. is closed the induction coil f is charged, and it discharges when the key e. is opened; the helix g. that has been charged by induction also discharges and sets up in l. a current that equals that resulting from the static charge of the line, and the reverse currents are produced in the induction coils as the circuit is closed at e, thus such induction coils serve to neutralise or balance the effect of the static charge, and prevent any false pulsation on the main line, resulting from the return static charge acting in m.

9 It will be apparent that the closing of the key e. and the connecting of the battery b. with the line tends to set up in the line and to earth, currents of opposite polarity to those resulting from the battery a. because the positive of the one and the negative of the other are to the ground and line respectively, and this local circuit (c.a.3.e.b.) serves to maintain an unbroken connection that offers but little resistance to the pulsation from the distant instrument passing to the earth, and the resistance is nearly uniform to the current received, whether there is a current that is being sent or not.

I claim as my invention

The battery a.b. in a local circuit connected to the line, in which is a circuit breaker, a finger key in combination with the magnet l.m. branch h. and resistance k. and the induction coils f. g. the parts operating substantially as set forth.

Signed by me this 19th day of August, A.D. 1874.

Thos A Edison

Witnesses Chas H. Smith Geo T Pinckney

DS and PD, MdSuFR, RG-241, Pat. App. 178,222. Written by Pinckney; petition and oath omitted. [a]Place taken from petition; date taken from text, form altered. [b]Case number and section numbers (in margin) written in another hand. [c]Interlined above. [d]Drawing and accompanying signatures from printed patent.

1. See headnote, p. 254.

2. This is Case 95. It resulted in U.S. Pat. 178,222.

3. Cf. Docs. 285 (no. 6), 286 (no. 19), 298, 312 (E, F), and 392 (fig. 15). The drawing for the patent model appears as "Model No 1.," Quad. 71.2, following p. 146 (*TAEM* 10:318–20).

–469–

Patent Application: Multiple Telegraphy[1]

New York, August 19, 1874[a]

To all whom it may concern.

1[b] Be it known that I, Thomas A. Edison of Newark, Essex County, New Jersey have invented an Improvement in Duplex Telegraphs, of which the following is a specification.[2]

2 The transmitting battery is connected with the line by a lever that simultaneously breaks the earth connection so as not to interrupt the continuity of the circuit; the current sent divides and operates equally in two helices; in one helix there is a sliding core that moves with the armature of the other heliex and this latter[c] responds to the pulsation from the distant instrument and closes a local circuit to a sounder or other receiving instrument, and there is a mechanical device that serves to compensate the attraction in one of the magnets that is due to the reverse action of the static discharge, thereby causing the forces to be accurately balanced.[3]

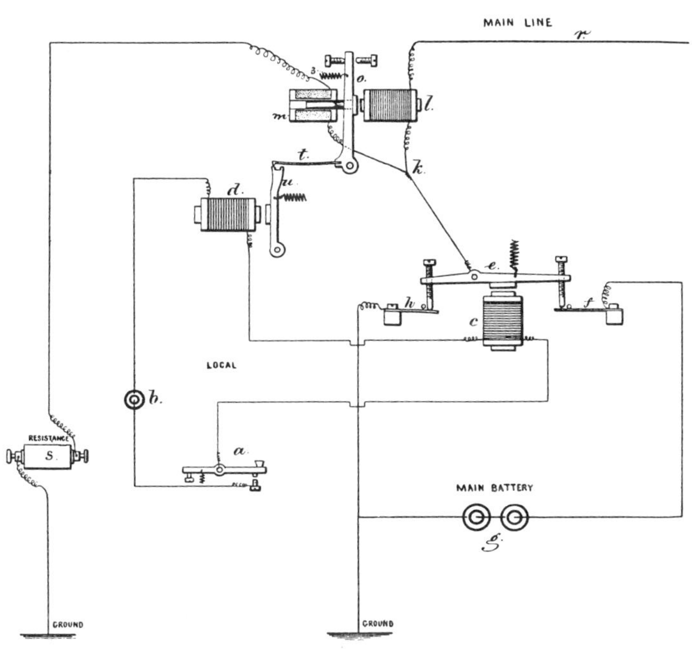

Inventor. Thomas A. Edison per Lemuel W. Serrell atty

Witnesses Chas H. Smith, Harold Serrell[d]

3 In the diagram of the drawing the transmitting key a. opens and closes the local circuit of battery b. to the helices of the electro magnets, c. and d.

4 The lever e. actuated by the armature of c. closes the line connection f. from the battery g. just before breaking the earth circuit of the line at h. so that there is always a path for the pulsation from the distant station.

5 The circuit from e. bifurcates at k. and a portion of the pulsation sent passes through l. and upon the line r. and an equal portion of the pulsation passes through m. to the artificial line composed of the rheostat s. and ground connection, said rheostat being adjusted to equal the resistance of the line.

6 The armature lever o. is inoperative by the current sent, because the action of the two magnets on the same is balanced, the attractive forces of such magnets being equal, and I make the core of the magnet m. to slide in the helix and attach it to the armature lever o. so that the cores of l. and m. may be equally energized and not influenced by the current sent, whether the armature is near the core of l. or drawn back therefrom by the spring 3.

7 It will now be understood that the lever o. will not respond when the pulsation sent passes through l.m. but that the armature lever o. will respond to the pulsation from the distant station because the same only acts in l. and by this lever o. a local circuit and sounder or other receiver is operated.

8 The pulsation sent from g. upon the line r. is increased by the static charge and upon cessation of the pulsation static charge reacts and these operations might disturb the balance between l. and m. I therefore employ the mechanical compensator formed of the spring t. and notched armature lever u. to the magnet el. hence when the current is closed at a. and the armatures of e. and u. are attracted the motion of the lever u. brings the notch at the end of such lever across beneath the v projection on the spring rt. thereby the force of the spring rt. is relieved and then again bent, and when relieved, the spring 3[c] exerts its full power, and when the spring rt. is bent its force lessens the power of the spring 3. hence this spring 3. is able to resist the increase of magnetism due to the static condition of the line, when the circuit is closed and to compensate for the reaction of the static charge as the circuit at a is broken thus leaving the magnet l. entirely uninfluenced by the current sent and capable of the most delicate adjustment by the spring 3. for receiving from the distant station.

I claim as my invention

First. The magnets l. m. through which the current sent passes to the line and artificial line, in combination with the armature lever o. and sliding core, substantially as specified.

Second. The electro magnets c. d. in the local circuit containing the circuit breaker a. in combination with the circuit closing lever e. notched armature lever u. spring t. and armature lever o. substantially as set forth.

Third. The mechanical compensation by spring t. and notched lever u. for neutralizing the effect of[e] the static charge in a duplex telegraph substantially as set forth.[4]

~~Thos A. Edison~~

Signed by ~~us~~me this 19th day of August A.D. 1874.

Thos A. Edison

Witnesses. Chas H Smith Geo. T. Pinckney

DS and PD, MdSuFR, RG-241, Pat. App. 178,223. Written by Pinckney; petition and oath omitted. [a]Place taken from petition; date taken from text, form altered. [b]Section numbers written in margin in an unknown hand. [c]Interlined above. [d]Drawing and accompanying signatures from printed patent. [e]"neutralizing . . . of" interlined above.

1. See headnote, p. 254.

2. This is Case 96. It is directly related to Docs. 285 (duplex no. 5) and 387 (device 13). The drawing for the patent model appears as "Model No 7.," Quad. 71.2, following p. 146 (*TAEM* 10:329–30).

3. This design is representative of Edison's electromechanical experiments in multiple telegraphy. Cf. Docs. 301, 387 (devices 2, 7, and 8), and 488; and the accompanying patent model drawings, which appear as "Model No. 3," Quad. 71.2, following p. 146 (*TAEM* 10:323–24). The figure labels on the detail sketch are "Brass wheel," "Rubber," and

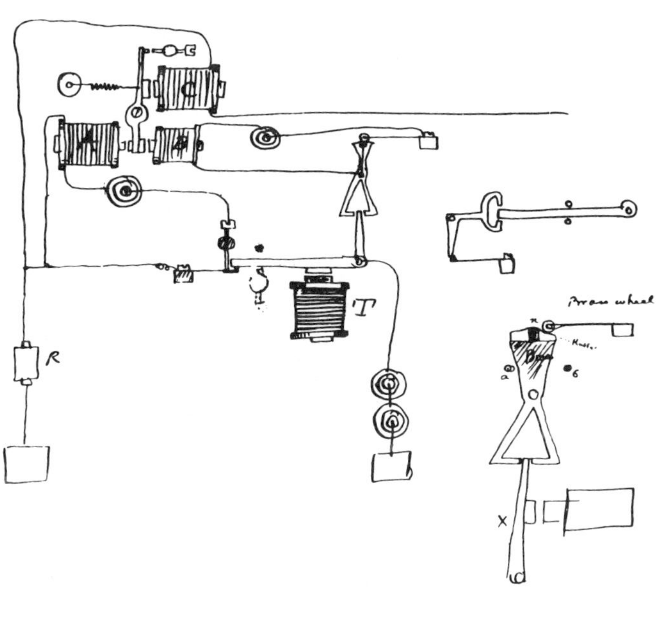

In the electromechanical device at the center of this circuit (enlarged in detail), the roller makes and breaks contact across the brass piece **n** *as the lever swings from side to side.*

"Brass"; pin "a" is to the left, pin "b" to the right. The text accompanying the sketch says:

> When the sounder lever suddenly hits the anchor peice either way it throws it against the pins a b by momentum   The contact roller passing over the brass portion n closes the circuit in its passage   The Relay magnets A.B.C. . . . A is wound with 23, B with 23 & C with fine No 30   R is a rheostat.

4. The patent examiner declined to approve this application absent further explanation, which Edison provided in the form of amendments on 20 January 1875. After a delay due to litigation, this resulted in U.S. Patent 178,223.

–470–

Patent Application: Multiple Telegraphy[1]

[New York,] August 19, 1874[a]

To[b] *all whom it may concern:*

Be it known that I, Thomas A. Edison, of Newark, in the County of Essex and State of New-Jersey, have invented an improvement in duplex telegraphs, of which the following is a specification:[2]

Two batteries are employed that are balanced by rheostats, and the static charge, compensated by an electro magnet, and the same poles of the batteries are connected at opposite sides of a receiving or relay magnet, hence the batteries neutralize each other in the constantly closed circuit.

A shunt wire, with a key or sounder between the centre of the combined batteries and the earth circuit, serves to destroy the balance of the batteries, allowing the one-half to act on the line, but the receiving magnet or relay is at a neutral point, where action and reaction are equal, except for the current from the distant station; to this it responds with delicacy.

In the diagram *a.* and *b.* are the batteries, connected in opposition to each other at *c.*, and provided with adjustable rheostats at *d. e. l.*, and one part of the circuit passes to the line *f.*, and the other to the earth circuit at *g.;* the line *f.* and earth *g.* are connected with the electro magnet *h.*, the armature lever of which opens and closes a local circuit to a sounder or receiving instrument.

The rheostat *r.* is of resistance equal to the line, and in the connection *s.* from *c.* to *g.* is the finger key or lever *k.*

Where a circuit closing lever is used it may be the lever of an electro magnet *l.*, or sounder, operated by a local circuit and key *m.*

When the connection *s.* is open, the currents of the batteries *a. b.* simply neutralize each other, the rheostats *b.* and *e.* allowing only equal currents to reach *h.*, which currents, being

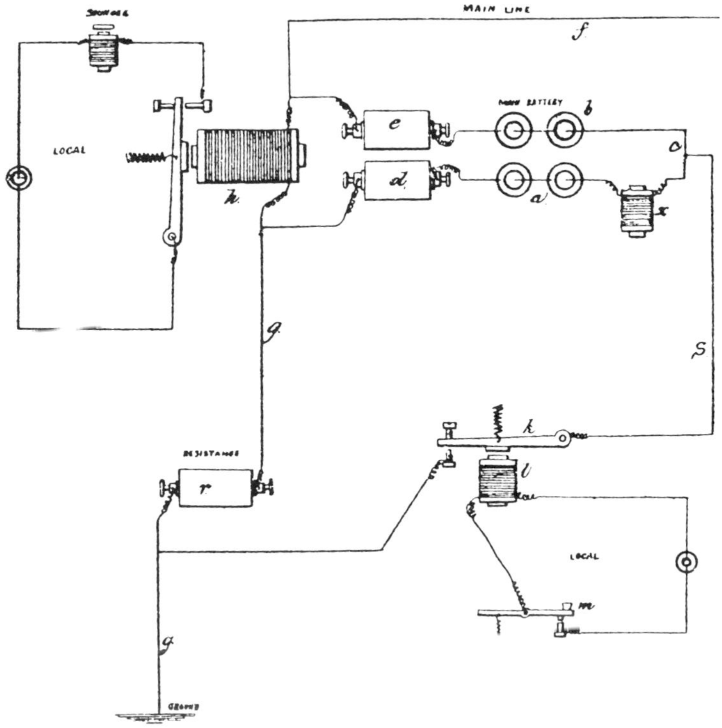

Inventor: Thomas A. Edison. per Lemuel W. Serrell. atty
Witnesses: Chas H Smith Harold Serrell.[c]

of the same polarity, and opposed to each other, produce no effect in *h.*, but leave the magnet *h.* free to respond to the pulsations coming over the line from the distant station.

When the lever or key *k.* is closed, the battery balance is destroyed through *c.s.k.g.r.d.*, and the current from *b.* passes most through *e. f.* to the distant station, returning by ground *k. s.* and *c.*, and the current from *a.* passes mostly through the artificial line *g. r.*, equalling the line *f.*, but the relay *h.* is placed at a point where the action of the currents is balanced, as the portion passing from *a.* through the same to the line is balanced by that from *b.* through the artificial line, hence neutralizing each other.

When the lever *k.* opens, the forces of the batteries *a. b.* are brought again into opposition, but as there is a static discharge from the line which acts in *h.*, I make use of the electro magnet *p.*,[3] which has been charged when the batteries are active or unequally set up, producing an induced current which balances the static charge of the line, and neutralizes the effect of the same in *h.*, thus leaving the magnet *h.* entirely uninfluenced, except by the pulsation from the distant station.

I claim as my invention the batteries *a.* and *b.*, connected in opposition to each other in the circuit containing the rheostats *d. e.*, magnet [-]x.[d] and receiving magnet *h.*, in combination with the line circuit *f.* through *h.*, the artificial line *gr.*, and the shunt, *s.*, containing the key or lever *k.*, the parts being arranged for operation substantially as set forth.[4]

Signed by me, this nineteenth day of August, A.D. 1874.

THOMAS A. EDISON.

Witnesses: CHAS. H. SMITH, GEO. T. PINCKNEY.

PD (transcript) and PD (photographic transcript), NjWOE, Quad. 70.6, p. 63 and facing p. 64 (*TAEM* 9:364, 365). [a]Date taken from text, form altered. [b]Preceded by "CASE NO. 97. APPLICATION AND OATH, AS IN CASE NO. 94. SPECIFICATION." [c]Drawing and signatures on separate page. [d]Correction written in an unknown hand.

1. See headnote, p. 254.

2. This is Case 97. It is related to Docs. 285 (duplex no. 13) and 309, but it incorporates an electromagnet instead of a condenser in the artificial line. The drawing for the patent model appears as "Model No. 5," Quad. 71.2, following p. 146 (*TAEM* 10:326).

3. Should be "*x.*"

4. No patent resulted from this application. However, Edison had not yet abandoned it as late as 13 November 1878, as shown by a disclaimer of that date in Pat. App. 168,385.

–471–

Patent Application: Multiple Telegraphy[1]

New York, August 19, 1874[a]

1[b] To all whom it may concern.—

Be it known, that I, Thomas A. Edison of Newark in the County of Essex and State of New Jersey, have invented an Improvement in Duplex Telegraphs of which the following is a specification.[2]

2 The main battery is in two parts connected in reverse in the line circuit; in the same cuircuit is the receiving instrument or relay, to which is connected a local ciricuit and sounder: One half of the battery is short circuited at the sending station by the depression of the key; simultanioulsly with this depression a local circuit is closed and a reverse current sent from a battery through the receiving magnet and in that circuit is a rheostat that is adjusted so that the effect of the battery on the line is neutralized in the receiving relay at the sending end but the battery at the distant end is free to act at the receiving station. At the same time the effects from the static charge in the line are neutralized in the receiving instrument by induction.

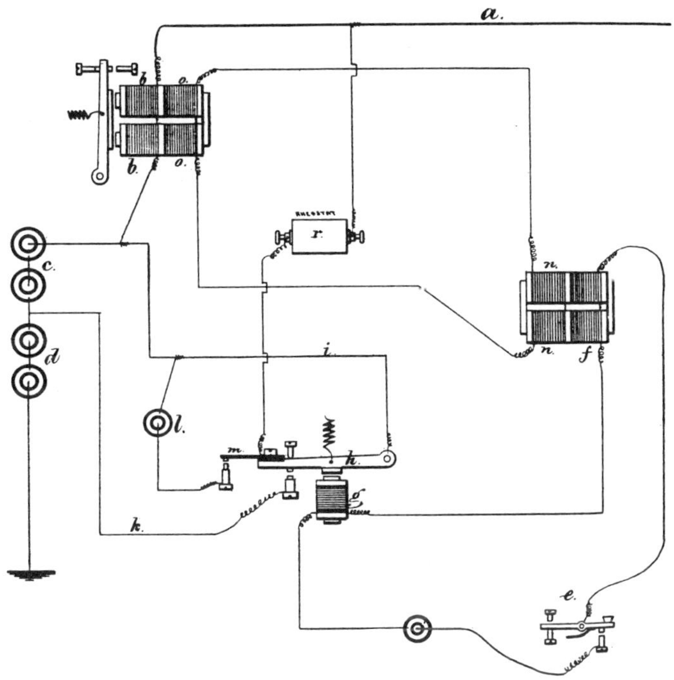

Inventor Thomas A. Edison per L. W. Serrell atty
Witnesses Chas H Smith, Harold Serrell[c]

3 In the diagram a. is the line passing through the electro magnet b. that operates a relay or sounder. The connection from b. is through the batteries c.d. or local circuit i.l.k. to the earth.

4 The batteries c. and ~~a~~d. oppose each other and being equal are not operative in b.

5 The key e. is in a local circuit in which are the helices f. g and when the key is closed the electro magnet g attracts the armature and lever h. closing the circuit i.k. to the centre of the batteries cd. hence short circuiting c. and allowing d. to act in the line.

6 In order to compensate the action of the battery d. in b. the local battery l. is used and the local circuit from l. through the insulated spring m. and rheostat to the line a. is closed simultaneou~~l~~sly with the short circuiting of c. and this rheostat r. is adjusted so that the action of l. in b. equals the action of d. in b. and being in reverse the forces are neutralized.

7 The helices f and g being charged and discharged simultaneou~~l~~sly there is an inductive current set up in the core of f and the helix n. and that gives a secondary charge to the

helix o. that surrounds the core of b. and hence when e. is closed, the secondary effect in n. neutralizes the static effect as the line is charged, and as the circuit at the key e. is broken a reverse induction current is set up in [-]n. ~~m~~neutralizing the discharge of the static charge of the line: the helices being wound so as to produce this reverse and neutralizing effect in the core of the electro magnet b by the helix n.

I claim as my invention.

First.—The local battery l and re~~h~~eostat r in a shunt around the receiving magnet b in combination with the batteries c.d. shunt i. and lever h. substantially as and for the purposes set forth.

Second.—The magnets g and f and induction coils ~~u~~n. o. in combination with the magnet b. batteries cd and shunt circuits substantially as set forth.[3]

Signed by me this 19th day of August A.D. 1874.

Thos A Edison

Witnesses Geo. T. Pinckney Chas H. Smith

DS and PD, MdSuFR, RG-241, Pat. App. 180,858. Written by Pinckney; petition and oath omitted. [a]Place taken from petition; date taken from text, form altered. [b]Section numbers written in margin in another hand. [c]Drawing and accompanying signatures from printed patent.

1. See headnote, p. 254.

2. This is Case 98. It derives from Doc. 308. The drawing for the patent model appears as "Model No. 2," Quad 71.2, following p. 146 (*TAEM* 10:321–22).

3. The patent examiner objected only to the wording of the first claim, which Edison amended in a letter of 21 January 1875. After a delay due to litigation, this resulted in U.S. Patent 180,856.

–472–

Patent Application: Multiple Telegraphy[1]

New York, August 19, 1874[a]

99[b]

1 To all whom it may concern.

Be it known that I, Thomas A. Edison, of Newark in the County of Essex and State of New Jersey, have invented an Improvement in Duplex Telegraphs, of which the following is a specification.[2]

2 The object of this invention is to enable two operators to simultaneously send over one wire, in one direction, by reversal of a battery current in one instance and increasing and decreasing the strength of the current in the other instance, and the connections are so arranged that the party at the re-

ceiving station can signal to the sender to repeat in case of inaccuracy.

3 By duplicating the parts herein described, four transmitting operators and four receiving operators can work simultaneously over one wire, two of each being at each end.[3]

Fig: 1

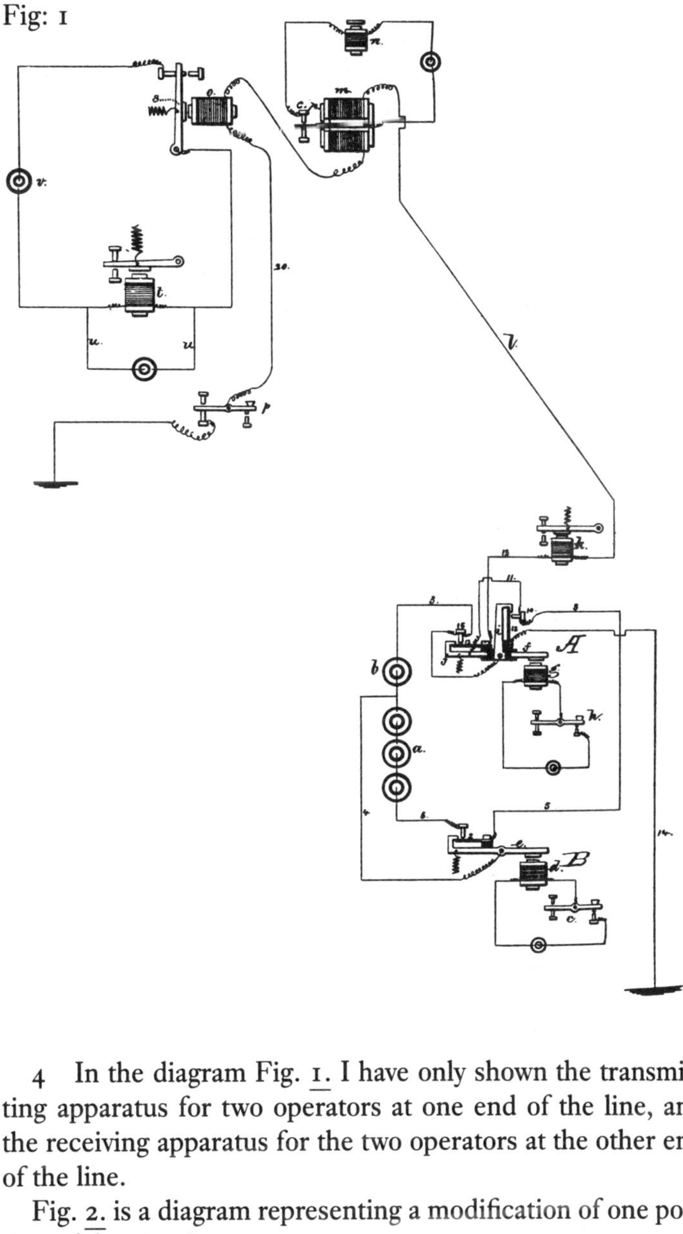

4 In the diagram Fig. 1. I have only shown the transmitting apparatus for two operators at one end of the line, and the receiving apparatus for the two operators at the other end of the line.

Fig. 2. is a diagram representing a modification of one portion of the circuits.

5 The battery a.b. is composed of unequal elements, sup-

pose a. to represent seventy five cells, and b. twenty five cells.

6 The key c. is in a local circuit that operates the magnet d. and circuit changing lever e. so that when the circuit c. is open, the battery b. is out of circuit and inoperative, but the continuity of the metallic circuit is maintained through 4.e.~~3.~~ and 5. to the reversing key f. next described, and when the key c. is closed, the entire battery a.b. is connected through the insulated spring .2. and wires 3. 5. and 6. to said key f. and hence the operator of c. controls the amount of battery connected to the line and this change of power operates at the receiving station as hereafter set forth.

7 The key f. is made for reversing the battery connections alternately whether the whole or only a portion of the battery is thrown upon the line by the operator of c. and this key f. is operated by the electro-magnet g. local circuit and key h.

8 When the key h. is open as shown, the current from 5. passes by 10.11.12. and 13. out through the tell tale magnet k. to the line l. and the return is through ground to 14. spring 15. lever-arm i. 16. and 3.

9 When the key f. is attracted by g. the circuit connections are reversed so that the current from 5. passes by 10.15. and 14. to ground, thence returning through line l. 13.12.16. and 3. to battery.

10 By bearing in mind that one operator at c. makes his signals by changing the amount of battery power and that without breaking the metallic circuit, and that the operator at h. signals by reversing the circuit regardless of the battery power and without breaking the metallic connections, hence both parties can work without hindrance from the other, it now becomes necessary to show what instrument at the receiving station are employed, the one to respond by the change of battery force regardless of polarity and the other to respond by change of polarity only, regardless of battery force.

11 The polarized relay magnet m. is in the line l. and is made with a permanent or magnetized armature, hence it will only respond when the circuit is reversed, and when it does respond by the reversal of the circuit by the operator of h. the armature lever works a local circuit and sounder n.

12 The electro-magnet o. is also in the main line circuit that passes by 20. and key p. to the earth; the normal condition of this key p. is closed, but whenever either of the parties receiving, perceives any inaccuracy or interruption of the message, then the parties sending are notified by either re-

ceiver breaking the circuit at the key p. and the tell-tale magnet k. no longer responding to the pulsations of the keys h. or c. indicates to the senders that the circuit is broken.

13 In order to produce a receiving instrument that is operative only by the excess of the current, the spring of the armature s. is tightened to the proper degree of tension so that the magnet o. will not respond except when the closing of the key c. brings the entire battery a.b. into action.

14 If a reversal of current takes place when the entire force of battery is on the magnet o. there occurs a false movement of the armature s. To prevent an incorrect record or sound the local circuit and receiving instrument are made as next described.

15 The local circuit and battery u. is constantly operative in the receiving magnet or sounder t. and hence the armature of t. only responds when an opposing current is sent through t. to neutralize the current of u. and de-magnetize the core: This is effected by battery v. and circuit that is closed by the lever s. on the back movement, hence, when s. is attracted by o. the circuit of v. being broken, allows u. to act unbalanced and give the sound, but should s. fly back on reversing the current, it will be so instantly re-attracted by o. as to prevent the circuit of v. being closed long enough to neutralize the action of u. hence the signals given by the operator at c. will be correctly responded to by the receiver t.

Fig: 2

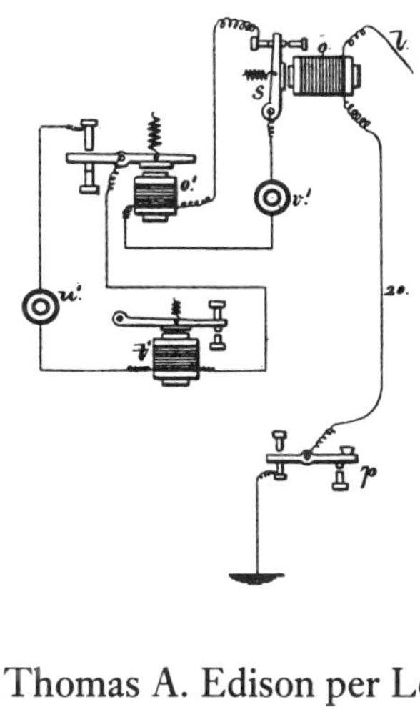

Inventor. Thomas A. Edison per Lemuel W. Serrell atty
Witnesses. Chas H Smith Harold Serrell[c]

16 The diagram fig. 2. shows a modification of the receiving circuit to prevent a false indication at the receiver.

The armature s. operates the circuit of the battery v′. to the electro magnet o′. and the armature of o′. closes and opens the circuit of u′. to the receiving instrument or sounder t′. and as the magnet o′. is rather sluggish in movement its armature does not have time to move and break the circuit u′. before the armature again flies back should the circuit connections have been reversed while the key c. is closed.

I claim as my invention,

First. Transmitting two distinct messages over one wire in the same direction and at the same time, one operating by reversal of the battery current and the other by increasing or decreasing the current from the battery.

Second. In a duplex telegraph, the polarized magnet and an ordinary magnet in the same circuit, when arranged substantially as specified, so that one responds to change of polarity of the current and the other to an increase and decrease of the current substantially as set forth.

Third. The arrangement of two keys e. f. battery a.b. and reversing circuit connections substantially as set forth, to bring into action the whole or a portion of the battery and to reverse the polarity of the current passing on the line without entirely interrupting at any time the metallic circuit.

Fourth. The reversing key f., made with the arm i. springs 12. and 15. and circuit closing points 10. and 16. substantially as set forth.

Fifth. The receiving instrument or sounder in the constant circuit from the battery u. in combination with the balancing local battery v. circuit and circuit closing armature s. and electro magnet o. substantially as set forth.

Sixth. In the duplex telegraph arranged for sending two separate messages simultaneously in the same direction and from the same end, the circuit breaking key p′. at the receiving end and the tell-tale magnet k. in the line at the transmitting end for the purpose set forth. Signed by me this 19th day of August A.D 1874.[4]

Thos A Edison

Witnesses[d] Chas H Smith Geo. T. Pinckney

DS and PD, MdSuFR, RG-241, Pat. App. 480,567. Written by Pinckney; petition and oath omitted. [a]Place taken from petition; date taken from text, form altered. [b]Case number and section numbers (in margin) written in another hand. [c]Both figures are from the printed patent on a single page, signed and witnessed at the bottom. [d]Followed by a brace including both names.

1. See headnote, p. 254.

2. This is Case 99. Though closely modeled on Doc. 348, this design reconfigures the transmitters and specifies the simpler of two bug traps as the preferred option. The September trials on the Boston line (and subsequent tests) employed the more complex bug trap ("Fig: 2" in this document; see illustration, p. 311). The drawing for the patent model appears as "Model No. 8," Quad. 71.2, following p. 146 (*TAEM* 10:331–32).

3. This "duplication" actually implies either (1) doubling the coils of the receiving relays and wiring them into appropriate artificial lines for a differential duplex arrangement, as in the illustration accompanying Doc. 446 or the circuit labeled "Quadruplex: Condsr at recg end . . ." in Doc. 485; or (2) arranging the lines for a bridge type of duplex, as in Doc. 449 or the illustration accompanying Doc. 477. Compensation for static discharge would be needed as well.

4. The Patent Office declared this application to be in interference with a 14 October 1874 application of Henry Nicholson. Lengthy proceedings ensued; after totally rewriting the specification, Edison finally received U.S. Patent 480,567 in 1892. *Nicholson v. Edison;* Pat. App. 480,567.

–473–

Patent Application: Multiple Telegraphy[1]

New York, August 19, 1874[a]

1[b] To all whom it may concern.

Be it known, that I, Thomas A. Edison, of Newark, in the State of New Jersey, have invented an Improvement in Duplex Telegraphs, of which the following is a specification.[2]

2 Where two persons are sending and two receiving the entire line is sometimes deranged by a signal from one of the receivers to repeat. My invention is made to allow either party that is receiving to interrupt the person sending to him so that he is thereby warned to repeat, and that without interfering with the other message that is being sent or received.

3 In the diagram drawing a. is the line, b. the receiving relay instrument, d. is a helix around the same core as b. and this helix is in a circuit passing to the artificial line and rheostat e. and this rheostat is to be adjusted to equal the line, so that the pulsation from the sending station acting in reverse in the helices b. and d. produces no magnetizing effect in the core, but the pulsation from the distant station passing along a. acts in b. unbalanced and either produces the sound by the armature magnet or else works a local circuit and sounder.—

4 The pulsation received from the distant station passes by 3. f. and 4. to the earth connection.

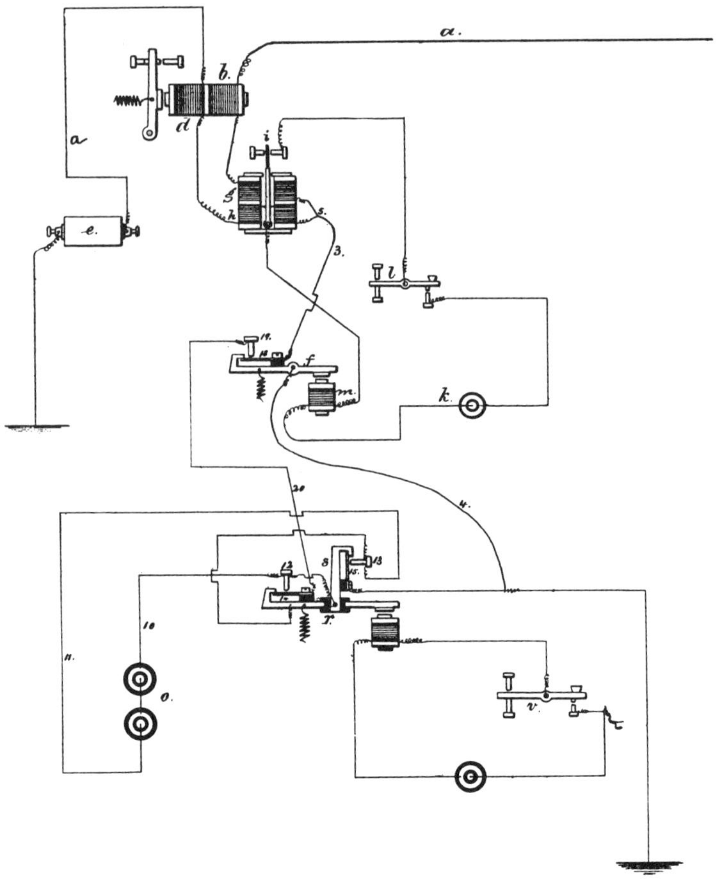

Inventor. Thomas A Edison per L. W Serrell atty
Witnesses. Chas H Smith Harold Serrell[c]

5 The lever f. is operated to give the pulsations that are sent upon the line and these pulsations divide at 5. passing equally through the helices g. and h. thence through the helices b. and d. aforesaid.

6 The helices g. and h. are of a differential polarized relay, the tongue or polarized armature lever i. of which forms a circuit breaker in the local circuit k.l.m. and in this circuit the key l. acts to open or close the circuit and by the electro magnet m. operate the lever f and send the pulsations upon the line.

7 The main battery o. is connected by wires 10. 11. to closing points 12. 13. and the reversing lever r. has springs 14. 15. and an insulated arm s. When the key l. is operated the lever f. closes 18. and 19. and with the lever r. in the position

shown, the circuit is closed by 10.12.s.15. to the ground returning through e.a.d.b.g.h.3. spring 18. of f. screw 19. wire 20. to 14.r.13. and 11.[3] By these connections the pulsations are sent to the distant station and received there in a similar instrument. It will be apparent that at the sending station the current divides at 5. and passing through both g. and h. the polarized armature will be retained in the normal position with the circuit of k.l.m. closed regardless of whether the current sent is reversed or not, but the current from the distant station only acts in g. hence the polarized tongue is changed by the reversal of the battery connection at the distant station so that if the receiver closes the key v. he does not interfere with the party sending by the key l. from his own station, but the battery connection being reversed the differential relay at the distant station is changed by the change of polarity in g. and the party sending the message is warned to repeat by the fact that his own magnet m. does not work because[d] the polarized tongue ~~breaking~~ has broken the circuit of the magnet m. and it will not respond to his key l. It is to be remarked that the receiving operator only closes his key v. sufficiently long to give the signal, and then releases it so that the parts resume their normal condition to allow of the message being again sent to him.

8 The [-]reversal of the polarity at the sending station does not change the balanced condition of b.d. nor make any difference in the action at the distant station because that current sent is operative in b. at the receiving station whether of one polarity or the other.

I claim as my invention

The differential polarized relay magnet g.h. and local circuit containing the key l. and magnet m. in combination with the reversing key r. and circuit connections substantially as set forth.

Signed by me this 19th day of August A.D. 1874[4]

Thos A Edison

Witnesses Chas H Smith Geo. T. Pinckney.

DS and PD, MdSuFR, RG-241, Pat. App. 207,723. Written by Pinckney: petition and oath omitted. [a]Place taken from petition; date taken from text, form altered. [b]Section numbers written in margin in an unknown hand. [c]Drawing and accompanying signatures from printed patent. [d]Interlined above.

1. See headnote, p. 254.

2. This is Case 100. It is based on the design shown in Docs. 392 (fig. 19) and 387 (device 16). The drawing for the patent model appears

as "Model No 6.," Quad. 71.2, following p. 146 (*TAEM* 10:327–28).

3. Edison amended this, replacing "returning through . . . 11." with "there dividing, part returning through e d h to 3, part going to distant station thence by a.b.g to 3, whence the united currents pass by 3.18.19.20.14.r.13 and 11. to the battery." Pat. App. 207,723.

4. After several amendments, this resulted in U.S. Patent 207,723.

–474–

Caveat: Automatic Telegraphy

New York, August 19, 1874[a]

To all whom it may concern.

Be it known that I, Thomas A. Edison, of Newark in the State of New Jersey, have invented an Improvement in Circuits for Automatic Telegraphs, and the following is declared to be a full and correct description of the same so far as perfected.

The object of this invention is to obviate the effects of static induction on telegraph wires so as to obtain high speeds.

The invention consists in arranging inductive devices at the transmitting station so as to increase the strength of the first part of each signal and weaken the last part.

The diagram fig. 1. illustrates my invention.

FIG. 1.

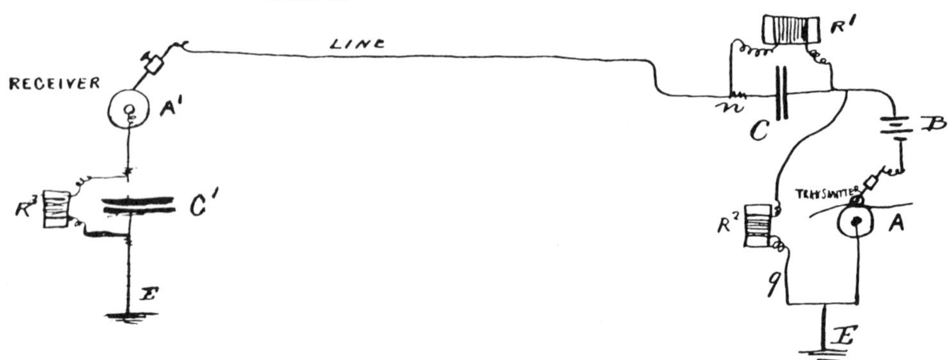

A. is the transmitting instrument. B is the battery. The wire 9. and resistance or rheostat coil R^2. forms a shunt to earth to preserve the continuity of the line when the battery is disconnected from the line by the interposition of the perforated paper.

C is a condenser, preferably an air condenser,[1] which may or may not be shunted by the resistance coil R^1. according to the condition of the line.

A^1. is the chemical recording instrument. C^1. an air condenser shunted with the resistance coil R^3.

The operation is as follows:—

At the moment that the battery B. is connected to the line and condenser C. by the contact roller passing through a perforation in the paper and coming in contact with the metallic drum, the induction current of the condenser C. which flows in the same direction as the battery current on closing it, is added to that portion of the primary current which passes to the line through the resistance R^1. This excess of current is but momentary and ceases when the condenser is fully charged.

Were the line free from static induction when the battery is connected the dots and dashes would be recorded upon the chemically prepared paper, thus—

and the first part of each signal being heavier than the other is due to the extra current sent into the line from the condenser C. each time the battery is connected.

At the moment when the battery is disconnected, the condenser sends into the line a current which circulates in the opposite direction to the battery, and were this current recorded it would show thus:—

Now if the line has considerable electrostatic capacity (i.e.) capacity for giving a static current when the battery is connected the record will be different from the above at the receiving station but like it at the sending end, only greatly magnified.

Were no condenser used at the sending end, the record would be thus:—

As it should be 

As recorded 

The reason why the first portion of each signal is weakened is because the effect of the static induction when the battery is first placed on the line, is to set up a current in the opposite direction to it and prevent it from showing its full strength; this effect is but momentary and the dash gradually increases in depth of coloration due to the full action of the battery. If, now, the latter be dis-connected from the line, a second static current is set in motion which circulates within the line and acts upon the chemical paper in the same manner as the primal current and prevents the sudden cessation of the chemical decomposition, hence, it is evident that the effect of the static end action is to weaken the first part of a recorded signal and add to the other part.

Now by referring to the record made by the condenser C. upon a line having no electro-static capacity, it will be seen that the first part of the signal sent over the wire is the strongest while at the other end a current of an opposite polarity is seen Now the effect of adding the condenser C. to a line having a certain electrostatic capacity is to exactly balance these opposite conditions. The extra current from the condenser added to that part weakened by the charge of the line, makes the record commence abrubtly and the second current from C. passing off the line in a contrary direction to the static current which prolongs the end of the signal, exactly balance each other and the signal abru~~b~~ptly ceases and the record would be thus:—

The object of the condenser C^1. and resistance R^3 is to increase this compensating charge or current should the electro static capacity of the line allow of setting up currents greater than those set up by the condenser C.

I will mention that should the condenser C^1. be found in practice to set up a counteracting current, the current from the condenser C. might be short circuited after each signal by running a wire from n. to another contact roller connecting with the drum through an extra set of holes placed just opposite the space between the regular signal holes, or the pen so arranged in line of the latter holes that it would only pass through them when the signalling rollers were on the bridge or space between.[2]

My claim will probably be

To the condenser at the sending station applied in the line and arranged with reference to the batteries and transmitting instrument in the manner specified for increasing the first part of each pulsation and weakening the last part so as to render the marks at the receiving station sharp and distinct.

Signed by me this Nineteenth day of August A.D. 1874

Thos. A. Edison

Witnesses Chas H Smith Geo. T. Pinckney

DS, DNA, RG-241, Edison Caveat 50. Written by Smith; petition and oath omitted. [a]Place taken from oath; date taken from text, form altered.

1. A condenser in which air serves as the dielectric or internal insulating material.

2. Cf. Docs. 320, 332–33, and 344.

-475-

Article in the Telegraphic Journal

Newark, N.J., U.S.A. Aug. 28, 1874[a]

ON A NEW FORM OF RELAY.

BY T. A. EDISON[b]

I HAVE taken advantage of the induced currents set up within the coils of an electro-magnet at the moment of closing and opening the circuit to construct a telegraphic relay, the lever of which is worked or controlled by them.

The accompanying diagram will serve to illustrate the principle—

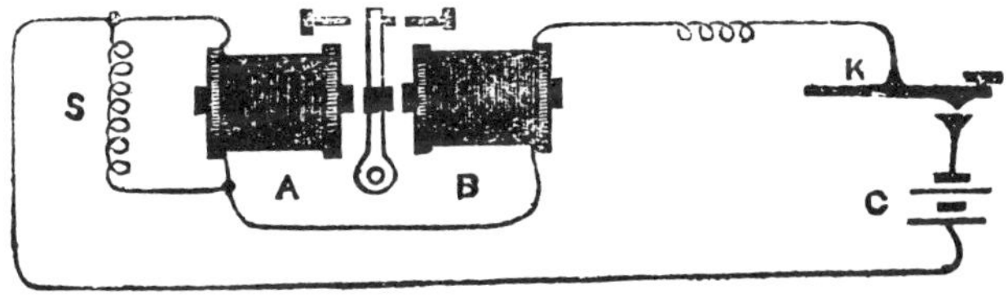

A and B are two ordinary electro-magnets, of equal diameter, length, and resistance. Both are included in the same circuit, together with a battery, C, and key, K.

The magnet A is permanently shunted with a resistance, S, twice as great as that of the magnet,

The armature or tongue is so adjusted that it will remain in contact with either point when the circuit is permanently closed, the same as the tongue of a Siemens's relay when no current passes through it. It should also remain where placed when the circuit is interrupted. The operation is as follows:—Supposing that the lever is in contact with the point on the left, and the circuit be closed, the iron cores of the magnet B acquire their magnetism almost instantly, while the induced current from A, being provided with a current of low resistance formed by the shunt S and magnet sets up an opposing electro-motive force to the primary current, and delays the magnetisation of the cores of A, and allows B to attract the armature. If the current is now interrupted, B instantly loses its magnetism, while the secondary current from A circulating within the closed circuit in the same direction as the primary current, prolongs the time of demagnetisation, and this secondary or after magnetism attracts the armature to the left.

It is obvious that the greater the resistance of the circuit in which A and B are placed, the greater will be the difference in the induced currents set up by A and B. When the resistance of the current is small the induced current from B acts only upon closing the circuit, and reduces the differences, but by increasing the resistance this current is proportionately weakened, while that from A remains constant.

When properly constructed this relay is one of great delicacy.

The speed by which signals can be relayed into a secondary circuit is fully 18 per cent greater than with a Siemens relay.

A greater difference in the charging and discharging times of the two magnets may be obtained by increasing the length of A.

PD, *Teleg. J. and Elec. Rev.* 2 (1874): 319. [a]Place and date not those of publication. [b]Followed by centered rule.

–476–

To the Editor of the Scientific American

Newark, N.J., August 1874.[1a]

To the Editor of the Scientific American:[2]

In my new system of telegraphy, it would seem that power was obtained or that electricity had been passed into a new mode of motion, as with magnetism; but this is only apparent, not real, if I understand it right.

The electricity, acting by electrolysis, changes the nature of the surface of the paper, either by depriving it of some constituent, or the hydrogen, in conjunction with the metal and paper, form substitution compounds, the surfaces of which are smoother than the paper in its natural state, in the manner that the surface of rough paper is made smooth by dipping it into sulphuric acid. The strangest thing connected with this phenomenon, however, is this:

In trying to ascertain what caused the lever to move, whether it was by reducing the lead by hydrogen to a finely divided powder that acted as a lubricant, or whether the nature of the surface of the lead were changed by the absorption of hydrogen, like palladium, or whether the effect were due to the effort of the gases to escape from under the lever: I was led away from these notions by finding that platinum, with sulphate of quinine, will likewise show the movement. It then struck me that the nature of the paper was changed by the electrolysis. To test this, I had a long message received over the Automatic Telegraph wire from Washington (this wire runs in my laboratory at Newark), and recording the same on ordinary chemically prepared paper. The speed with which the message was sent from Washington was about 800 words per minute, and the colorations forming the dots and dashes were rather faint. I then passed the strip into the electromotograph (I use this name for the want of a better one), the colorations being in a direct line with the lead point. On rotation of the drum, and when no coloration was under the lead point the lever was carried forward by the normal friction of the paper. But the moment a coloration passed under it, the lead point slid upon the paper as upon ice, the friction was greatly reduced, and the lever moved in an opposite direction to the rotating drum.

In this experiment, no battery was connected to the instrument. This proves that electrolysis produces a change in the nature of the paper.

I afterwards fou[n]d that, if a tin pen were used to receive the message from Washington, although no marks were seen, the paper appearing unchanged, yet, on passing the paper through the instrument, the movement of the lever was more marked than before Receiving the message with a lead pen did not give so good results, although lead is the best when used, standing at the head of the twelve metals tried. The next is thallium. On paper moistened with aqueous solution of pyrogallic acid, tin is as good as thallium. Of all the solutions yet tested, potassic hydrate has been found to give the most marked results. The second best is sulphate of quinine. Third, rosaniline oxidized and discolored by nitrous acid.

A peculiarity of the quinine solution is that platinum shows an action, and shows it when either oxygen or hydrogen is evolved on its surface. With hydrogen the friction is lessened, as with all other metals; but with oxygen the friction is increased. This is so with all the metals subject to oxidation; but it appeared strange, at first, that it would show with a metal upon which the nascent gases had no effect.

With a lead point and a solution of the disinfectant known as bromo-chloralum, the evolution of hydrogen increases the friction of the paper enormously.

Silver seldom shows a movement with any solution; and when it does, it is very weak.

Sulphuric acid shows least movement with any metal.

It appears to be a matter of indifference as to the character of the metal used for the drum, which acts as one of the decomposing electrodes. Considering that the lever will close a secondary circuit under the great pressure used upon the lever, its sensitiveness to electricity is wonderful. With a delicately constructed machine, moved by clockwork, which I have nearly finished, I have succeeded in obtaining a movement of the lever, sufficient to close the local circuit with a current (through one million ohms, equal to 100,000 miles of telegraph wire), which was insufficient to discolor paper moistened with potassic iodide, or move an ordinary galvanometer needle. Messages may be read from the sound of the lever, when the most delicate telegraph magnet shows no current.

The uses of this instrument are many; in fact, it gives an entire new system of telegraphy.[3]

As no secondary currents are generated, as with an electromagnet, to prevent the instant magnetization or demagnetization of the iron cores, and electrolysis being instantaneous, it is obvious that the lever will respond to signals transmitted with great rapidity. I have succeeded in transferring signals from one circuit to another at the rate of 650 words per minute; hence it may be used to repeat the rapid signals of the automatic telegraph into secondary circuits.[4]

By attaching an ink wheel to the extremity of the lever, opposite a continuous strip of paper moved by clockwork, messages transmitted at a speed of several hundred words per minute may be recorded in ink. By attaching a local circuit to the repeating points, and adding thereto a sounder, it may be used as a Morse relay to work long lines of telegraph.[5]

T. A. EDISON.

PD, *Sci. Am.* (n.s.) 31 (1874): 145. [a]Date and place not those of publication.

1. This appeared in the 5 September 1874 issue of *Scientific American.* It was reprinted in a number of other journals, and parts of it also appeared in Sprague 1875, 360–61. John Sprague later indicated that he had thought it "best to use Mr. Edison's own description, not withstanding the somewhat peculiar and confused manner in which it was written." He did, however, give Edison the benefit of the doubt, suggesting that "in fact, I suspect it has passed through the hands of two or three compositors before it reached me" ("Edison's Motograph," *Operator,* 1 May 1875, 9). A *Scientific American* reader also had difficulty understanding Edison's description and wrote to him requesting a clearer description and a diagram (John Pratt to TAE, 31 Aug. 1874, Cat. 1173:129, Scraps. [*TAEM* 27:427]).

2. In an introduction preceding the salutation, the editor called Edison "well known as a telegraph engineer of the highest ability, and the inventor of a larger number of electrical devices, probably, than any other person living."

3. Among other potential uses, Edison saw the electromotograph as an "aid in the solution of the problem of fast working through long submarine cables, as it requires but small battery power and is remarkably sensitive to feeble currents, and can be used to record delicate signals without electro-magnets." Davis and Rae 1876, 39.

4. Designing an adequate repeater for the automatic telegraph was a continuing concern of Edison's, particularly with regard to dropping messages at way stations and transferring messages to branch lines. Edison later claimed that he had "succeeded in applying [the electromotograph] as a repeater, and transmitted 1,400 words from one circuit to

another in *one minute*, which requires at least 400 full and perfect movements of the lever *each second*" (ibid.). There are no indications, however, that Edison actually used such an automatic repeater commercially.

5. Edison executed a patent application for such a relay in August 1874 (U.S. Pat. 158,787).

–477– [Newark, August 1874?[1]]

Equipment Specification: Multiple Telegraphy

armature & lever very light

The Rubber Cores should fit so that the uprights will not be moved when they are drawn in & out perhaps it would be a good idea to secure the two upright D & E together by two hard rubber studs so they couldnt spring.

One[2]

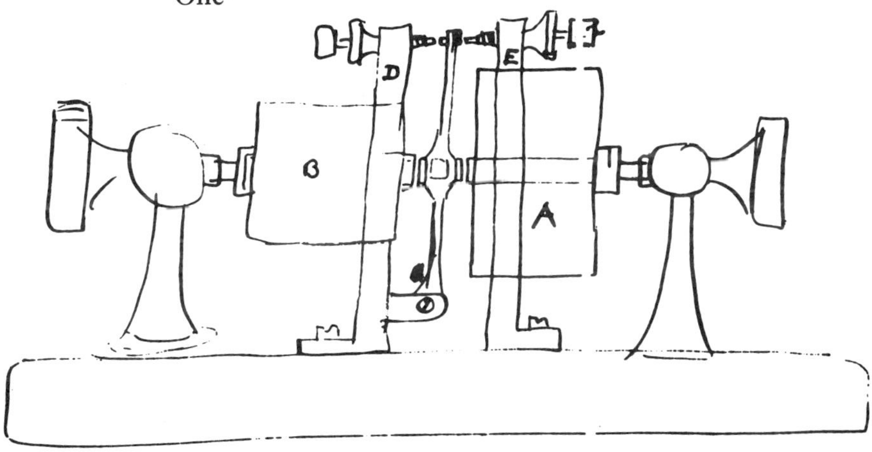

Hard platina points & Cores well annealed.

B 30 ohms

A. 400 ohms, or thereabts same size and length. Cores same size as shn and slotted thus[3]

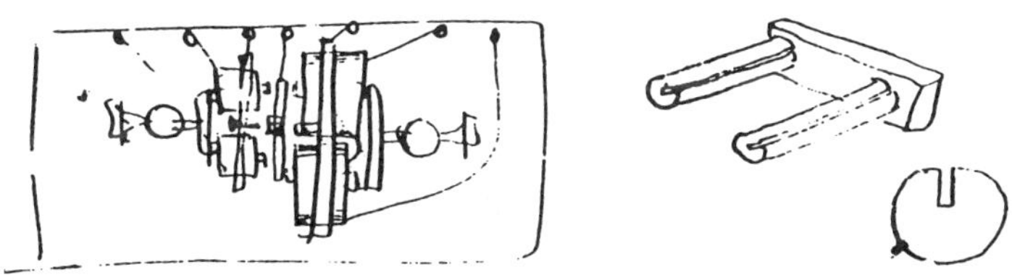

Cores of B need not be slotted

AX (photographic transcript), NjWOE, Quad. 71.2, following p. 146 (*TAEM* 10:316). Other copies appear in Quad. 71.2, following p. 14 (*TAEM* 10:233), and in Defendant's Exhibit 43, *Nicholson v. Edison.*

1. Phelps generally completed his work for Edison and Prescott in a few weeks, and this instrument was finished by 25 September. Docs. 442

n. 1 and 444 n. 1; Phelps to Prescott, 25 Sept. 1874, Quad. 71.2, following p. 146 (*TAEM* 10:317).

2. The coils with the shorter cores (**A**) would be the neutral relay in a quadruplex circuit; shorter cores lost and regained a magnetic field more quickly than long ones, reducing the period of no magnetism. Edison testified that this device was for experimental work during the testing period, when he tried "several different relays" (Quad. 70.7, p. 346 [*TAEM* 9:540]). For indications of how it might have fit into a circuit see the accompanying illustration and Doc. 485. Neutral relays with short cores later became standard on the quadruplex. Maver 1892, 214 (fig. 171).

Quadruplex circuit incorporating Edison's short-core relay.

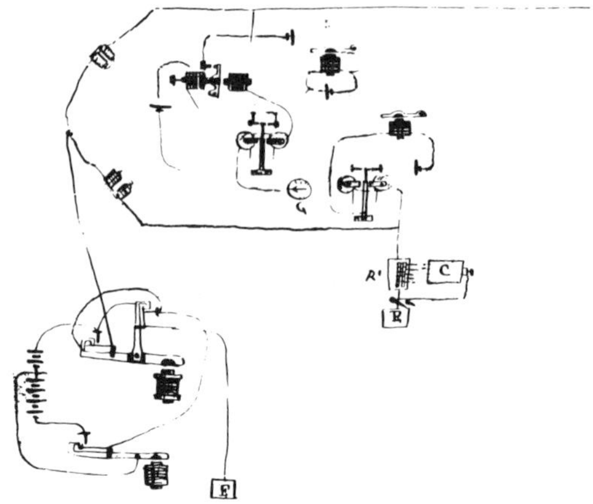

3. Edison claimed that slotted cores released the armature faster. U.S. Pat. 138,869; Doc. 120 n. 2.

EDISON AND THE *OPERATOR*

Docs. 478, 479, 489, 495, and 505

Edison began his association with the *Operator* during the summer of 1874.[1] The 1 July issue contained the first advertisement for his student telegraph instrument[2] as well as an advertisement for the Automatic Telegraph Company. By August, Edison had agreed to serve as the journal's science editor (Doc. 478), and the 15 August issue announced that the journal had "secured the services of a prominent electrician who will furnish us with short and pithy articles of a scientific character. . . . These articles will be worded in such a manner that they will be easily comprehended by the veriest tyro in the business" (p. 4). Edison's name was never mentioned, and his subsequent contributions were unsigned; nonetheless, it is possible to attribute authorship of many articles to him because draft versions of some items exist and because others are derived from Edison's experiments and telegraph designs. His first articles appeared on 1 September, when he began a three-part series on duplex telegraphy that discussed both the historical development of the art and his own experiments (Docs. 479, 495, and 505).[3] That issue also contained two shorter pieces describing his early practice instrument (Doc. 10) and an observation based on his experience with relays. One of the many subsequent miscellaneous items by Edison—on the use of platina points in telegraph instruments—is somewhat longer and more detailed than most and has been selected as a sample (Doc. 489). Edison continued to write for the *Operator* through the end of the year, primarily contributing short pieces. It is likely that Edison also prepared some articles that merely reprint information found in other journals or report designs of other experimenters.[4] This kind of secondary reporting continued throughout 1875, but the extent of Edison's involvement with the journal is unclear. See Docs. 578 and 632.

1. The *Operator* published its first issue on 1 March 1874. Edison later said he had started the telegraph periodical that was taken over by W. J. Johnston and turned into *Electrical World*; although the *Operator* was that journal, the documents included here contradict Edison's recollection of his initial involvement. App. 1.B78, D371; untitled editorial, *Operator* 13 (1882): 710.

2. See Doc. 438.

3. Drafts of the duplex articles are found in NS-74-002, Lab. (*TAEM* 7:156–209). Miscellaneous pages from other drafts appear in Cat. 297:103, 120, 127, Lab. (*TAEM* 5:855, 883–84, 906).

4. Edison's scrapbooks contain copies of printed illustrations, including some of those appearing in the *Operator* during this period. For example, there is one of an aerial screw designed by a M. De Fonvielle that appeared in the 1 December 1874 issue (p. 1) (Cat. 1178, Scraps. [*TAEM* 27:153]). Edison also prepared material that was never published, as evidenced by some illustrations for the *Supplement to the Operator* (Cat. 1143:6, Scraps. [*TAEM* 27:172]).

–478–

Draft Masthead for the Operator[1]

[Newark, August 1874?]

THE OPERATOR[2]

The Journal of the Telegraph Operators

Published in the City of New York on the 1st and 15th of each month, and sent postpaid to any part of the US on the following terms:—

Three Months	35	cents
Six "	70	"
One Year	1.30	

On and after the 1st of Sept 1874 this journal will devote a large portion of its space to the publication of original scientific articles, written in a[a] clear and simple style without technical phrases or algebraic formula, and accompanied with numerous illustrations. Send two stamps for sample number

Address all communication to:—

Allen & Cahill[3]
WU Telegraph 145 Broadway NY

ADf, NjWOE, DF, 74-004 (*TAEM* 13:45). [a]Interlined above.

1. See headnote above.

2. This masthead revision was never used. A likely earlier draft appears in Cat. 297:119, Lab. (*TAEM* 5:880).

3. Tom Allen and R. J. Cahill were editors and proprietors of the *Operator.* Though it shared Western Union's address, the journal apparently had no official connection with that corporation.

–479–

Article in the Operator[1]

New York, September 1, 1874[a]

DUPLEX TELEGRAPHY[b]

It is our aim in the following article, which is to be continued from time to time, to clearly describe, by the aid of numerous diagrams, the principles of Duplex and Quadruplex telegraphy; using neither technical phrases nor algebraic equations, which of themselves require explanation and might not be understood.

The first attempt at sending two messages over the same wire at the same time in opposite directions, was made by Dr. Gintl, an Austrian telegraph superintendent.

The main problem in that class of Duplex by which two messages are sent over the same wire in opposite directions, is to send a current from one station by some means into the line through its own receiving relay, without causing the cores to become magnetized or affecting it in any way whatever; but to magnetize the cores of the distant relay, untrammeled by the current sent through or around it at its own station.

This principle may be clearly illustrated if we consider that electricity travels over wires in the same manner as water travels through pipes. To work a Duplex by the transmission of water through a circuit formed of water pipes, we might arrange it as in Fig. 1.

Fig. 1

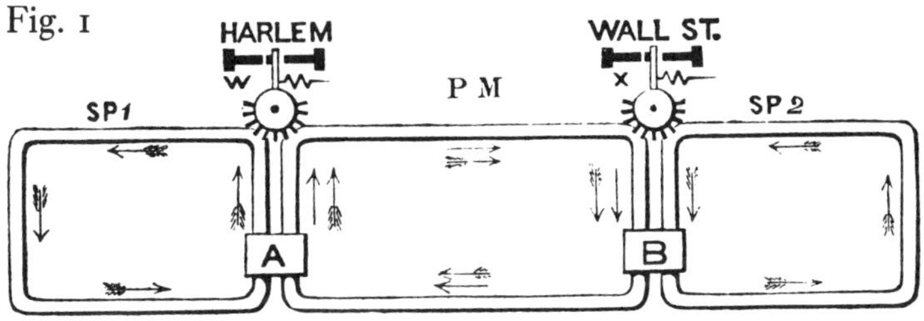

We will suppose that this water pipe circuit is extended from Wall street to Harlem, and that A and B were engines and reservoirs, by which a current of water of any desired amount might be made to circulate in the direction indicated by the arrows.

W and X are two water wheels provided with levers, which strike between fixed points; spiral springs are secured to these levers, to keep them against the proper point when no water passes. The object of the secondary water pipe circuits, SP 1 and SP 2, will be explained as we proceed further.

Now suppose that a stream of water was thrown into the main pipe by A, that would but half fill them and traverses them in the direction shown by the unfeathered arrows. The action of this passing stream of water would be to slightly move both the wheels and take their levers from one fixed point to the other; when the water is stopped the levers are drawn back by the springs; so far the signal has been received at both the home and distant stations.

We will describe now how the wheel X is left to respond, and the wheel W prevented from doing so. You will notice, by referring to the diagram, that the water circuit SP 1 is so arranged upon the opposite side of the wheel, that the stream which is injected into SP 1 from A tends to turn the wheel in a direction opposite to that injected into MP; hence, if the amount and speed of the water in both the main and secondary pipes are equal, it is obvious that the wheel W will not move; as long as these balances are maintained any amount of water, traveling at any speed, may be sent from A into both circuits without moving the wheel W. At the distant station the water moves the wheel, because none of it can circulate in the secondary circuit—SP 2.

We will now suppose that a stream of water from A is circulating in SP 1 and the main circuit, and the lever of the wheel X is moved to the left hand point, W remaining unaffected as shown before. If now the apparatus at B is so arranged that the water may be injected into the main pipe and secondary circuit in the same direction as that injected by A, the effect of the water injected in main circuit, MP, on the wheel X, is exactly balanced by that which circulates in the circuit SP 2; but it does not prevent the wheel X from being affected by water from A.

The water injected at B into MP destroys the balance at W, between MP and SP 1, and the excess throws the lever over to the left hand point. It will be seen that it is not necessary that the water should flow in opposite directions to obtain these signals.

In the electrical duplex the secondary circuits are replaced by the Rheostat or resistance box, in which are coils of fine wire so arranged that an amount of fine wire may be inserted, sufficient to retard or set up a resistance to the passage of the current equal to that offered by the line. The water wheel is replaced by a relay having double coils of wire, so that the current passing over both the line and secondary circuits in the same direction, may pass through the two coils of the relay in opposite directions, just as in the case of the water wheel, the effect is obtained by allowing the stream of water to circulate in the pipes on the opposite sides of the wheel. The current of electricity, passing through one of the coils of the relay tends to make one of the iron cores take a north magnetism, while the other takes a south; but the effect of the current passing through the other coil is just the opposite, it tending to make a north magnetism in the same core as the other coil is endeavoring to make a south; and as the two magnetisms cannot exist in the same core, at the same time, none is formed. Thus we see that the current passing through the relay does not affect it, if the conditions of the main line are equal to those of the artificial line formed by the rheostat.

(To be continued.)[2]

PD, *Operator,* 1 Sept. 1874, 1. [a]Place and date of publication. [b]Followed by centered rule.

1. See headnote, p. 286.
2. See Doc. 495.

–480–

From Thomas Page

Toledo, [Ohio,] Sept 2d 74.

Friend Al

To day I send by American Express Co. "charges paid" a small box of Chlorides, Sulphurets & Galena ores. The Chlorides are from "The Fuller" Galena from "Relief" & the copper-stained Chlorides from The Miller Mine—all owned & operated by our Co.[1]

We are now raising 20 tons pr day of Chlorides from the Fuller. The Chlorides sent you to day are not as rich as we are now raising, as they were from surface of ground Should it be necessary for more ore, you may telegraph Wm B Welles our Supt. at Salt Lake City for a sack or two, to be sent you by Express.—using my name to the telegram instead of yours.[2]

In reply to your questions as to Salt, Iron, Sulphur, Copper &c. There is an endless amount of them tributary to Salt Lake City

Soon as you are through testing, I will look for a report from you, and I hope your plan will result in greater yield than we got by selling to smelters[3]

Do you think of going to Colorado this Month? If so why not get two passes out to Salt Lake & return, and take me along with you for ballast

Hoping to hear from you soon, and with best respects to your Wife & Sister I am my dear Sir Yours Respectfully

T. S. Page[4]

ALS, NjWOE, Scraps., Cat. 1173:105 (*TAEM* 27:415).

1. This was the Toledo-Utah Silver Mining and Smelting Co. (J. M. Haskell to TAE, 30 Oct. 1874, Cat. 1173:157, Scraps. [*TAEM* 27:471]). Edison had apparently already received a shipment of ores. In a laboratory note of 26 August 1874 Charles Batchelor indicates that he "Received from T. A. E. for Analysis Some Ore from Relief mine Utah (Black,) sulphurets of silver Some ore from Fuller mine Salt Lake (Yellow) Chloride of Silver" (Cat. 1172:173, Lab. [*TAEM* 3:125]).

2. More ore was sent from Utah at the end of October. (J. M. Haskell to TAE, 30 Oct. 1874, Cat. 1173:157, Scraps. [*TAEM* 27:471]).

3. Although Edison does appear to have considered developing a system for reducing these ores (Page to TAE, 5 Nov. 1874, Cat. 1173:158, Scraps. [*TAEM* 27:473]), his main interest in them was as a source of tellurium, which he had first expressed an interest in obtaining the previous May (Docs. 433 and 437) for use in the receiving pens of his automatic and domestic telegraphs (Doc. 439).

4. Thomas Page's brother Homer and Edison's sister Marion had been married since 1849.

–481–

Notebook Entry: Miscellaneous

[Newark,] Sep 5th 1874

Method of reversing a current through a relay without opening the points.[1]

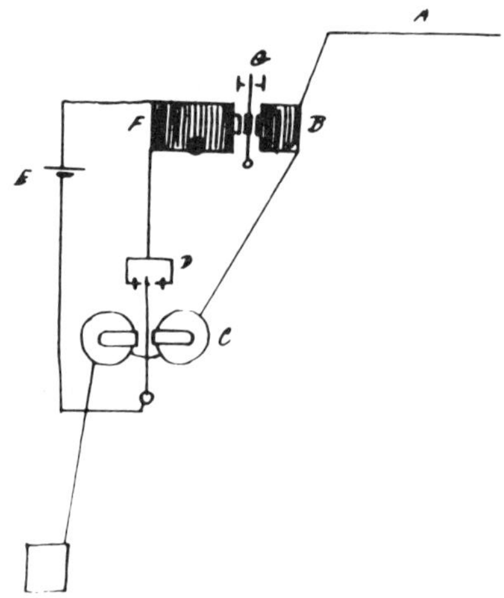

A is line passing through B & C to ground. C is polarized magnet which whether the line is charged or not, closes the local circuit for magnet F which acts as a spring for armature

A positive current now comes over at A drawing armature G to magnet B. & also lever to C. Reverse it & the polarized magnet working quicker than the ordinary open local & prevents F from pulling armature[a]

Speed Regulator for clockwork.[2]

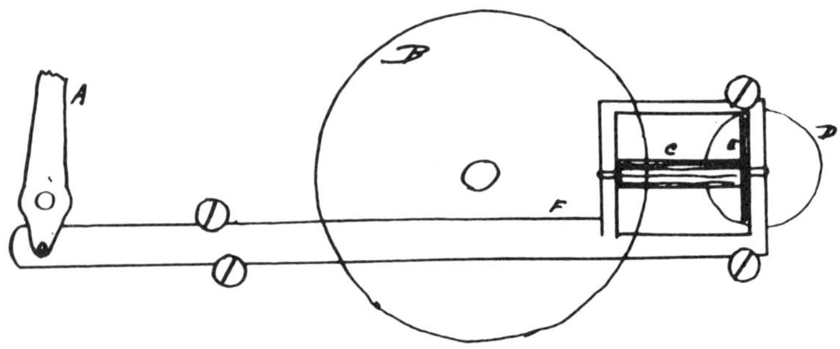

B is a crown wheel on the end of the 3 or 4 shaft & runs in the pinion C on which is disc E Disk E rolls on D disc which is fastened to fan shaft.

Whole carriage F is moveable & when moved to the left so that Disc E is nearest the edge of D it allows the pinion C & of course the clock work to move quicker.

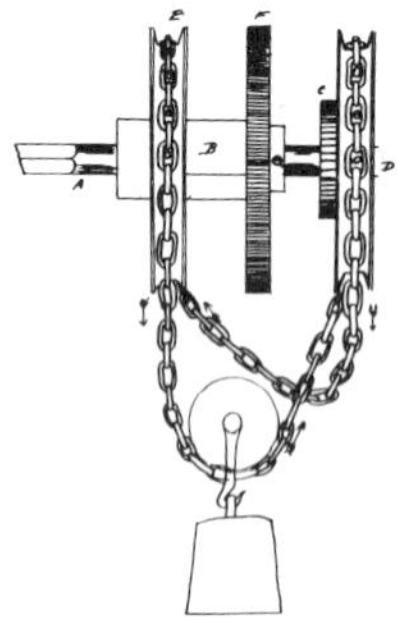

Charles Batchelor's clock-work-winding mechanism.

X, NjWOE, Batchelor, Cat. 1307:11 (*TAEM* 90:612). Written by Charles Batchelor. [a]Followed by centered horizontal line.

1. Cf. Doc. 472.

2. In addition to the clockwork in Edison's automatic telegraph instrument (Doc. 458), his patent application for the electromotograph suggested its use (U.S. Pat. 158,787). Two days before, Batchelor had

drawn and described "a method of winding a weight up without stopping the instrument that it drives." Batchelor, Cat. 1307:9 (*TAEM* 90:610).

–482–

Notebook Entry: Automatic Telegraphy

[Newark,] Sep 7th 1874

Made experiment of Roman letter with one wire with the two instruments & O.K. we also found out that the extra dots could be blotted out by Caustic Potash[1] The connections for it are thus

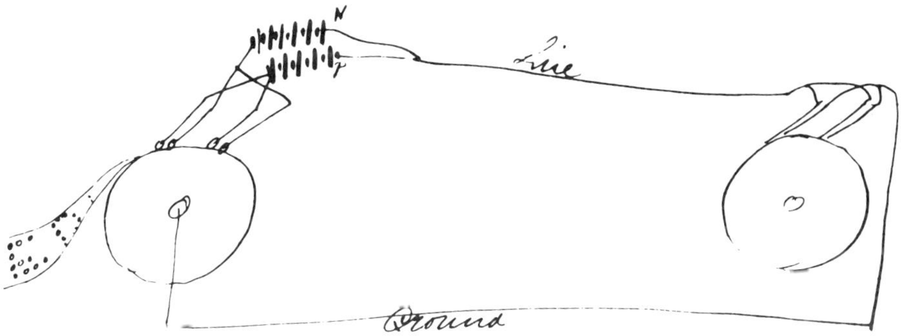

The two rollers that transmitt the bottom of the letter are set more than a full letter behind the two top ones in order to get off the letter before the other ones get on. Two batteries are used you send a positive & neg current on the 3 & 4 pen which are recorded on all f[-]our & the same on 1. & 2 which record on all four ~~& the way they can dbe distinguished~~ thus: in the word 'this' when you send on 3 & 4 it commences on T & makes the two bottom dots but the two top pens make also two dots but which are ahead of the other two so:—(: :) now when you send on 1 & 2 3 & 4 also marks but behind but the two top marks meet the bottom ones in the right place & the consequ[-]ence is there is the same amount of dots between the letters as the letters themselves

in the next letters the extra dots lap over these so in the word 'this'

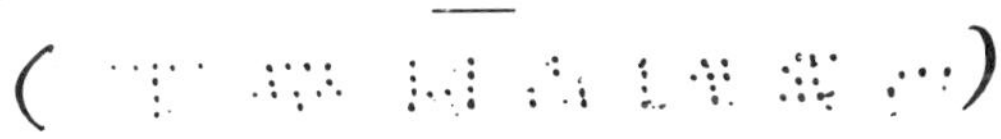

Now blot out the extra dots & you have the word (THIS) If you take the strip when they are still damp they are very easily erased by caustic Potash weak. This of course is with ordinary ferrid sol ⟨Cost of Insts 35.00 E & B. 12h⟩[a]

Chas Batchelor T. A. Edison[b]

X, NjWOE, Lab., Cat. 1172:174 (*TAEM* 3:126). Written by Batchelor. [a]Marginalia written by Batchelor. [b]Names written by Batchelor.

1. The following page in this notebook contains a record of Tom Gartland's ten hours of "Experiments to blot out the marks made by Plat on Pyrogallic and salt. And Logwood and salt." Cat. 1172:176, Lab. (*TAEM* 3:127).

–483–

From Charles Edison

Port Huron, Mich. Sept 9 1874.[a]

Uncle ~~a~~Al.

You told father you wanted me to come down as soon as I could.[b] father wants me to go and I want to go to but mother dont want me to go, but if you have got any thing for me to do I will come [-]Right down as soon as I can get Ready answer as soon as you can.

C. P. Edison

I have got a new kind of a telegraph machine I want to make as soon as I go to Newark one of my own Invention[1] C P. Edison

ALS, NjWOE, Scraps., Cat. 1173:130 (*TAEM* 27:431). Letterhead of Port Huron and Gratiot Street Railway Co. [a]"Port Huron, Mich." and "187" preprinted. [b]"as I could" interlined above.

1. Charley, who had been at the laboratory in August, returned to Newark by the end of January 1875, when he began a notebook that includes several telegraph novelties (PN-75-00-00.2, Accts. [*TAEM* 22:342]; Cat. 1169, Lab. [*TAEM* 6:528–602]). After about six months in Newark he returned to Port Huron, where he devised and installed the town's first fire alarm telegraph system. He also opened a telegraph school in Port Huron called the "Electrical Institute for Women" (Mitts 1970).

–484–

William Orton to Thomas Eckert

N. Y. Sep 10/74

Respectfully referred to Gen T. T. Eckert. The employment by the Electrician[1] of a competent assistant to aid in conducting the very important experiments now going on,[2] was suggested by me. I am unwilling, however, to direct the detail of one of your men except through you and with your consent.[3] Unless some reason occurs to you why Mr Smith[4] should not be appointed to the position named (in which case you will return this paper with a statement of such reasons) you will please direct him to report to Mr Prescott.

Wm. Orton President

Gerritt Smith, telegrapher and inventor.

LS, NjWAT, Box 1048.

1. George Prescott.
2. See Doc. 494.
3. When Eckert resigned from Western Union on 14 January 1875 he reminded Orton that their acrimonious relationship dated from May 1873. *New York Tribune*, 16 Jan. 1872, 2.
4. Gerritt Smith (1838–1915), who began his telegraph career in 1853, served as an assistant electrician at Western Union from 1874 until 1879, when he became engineer-in-chief of the American Rapid Telegraph Co. He returned to Western Union in 1882 as electrician of circuits, in which position he served until his retirement in 1907. Smith is credited with several telegraph inventions, including improvements in duplex and quadruplex telegraphy. Smith 1914; *NCAB* 15:387. See also Doc. 512 nn. 3 and 5; and Doc. 513 nn. 2 and 4.

At the time of this letter, Smith was chief operator at Western Union's main office in New York. Eckert responded the same day and agreed to Orton's request, transferring Smith from the operating department to the post of assistant electrician under George Prescott to work on the quadruplex experiments. Eckert to Orton, Box 1048, NjWAT; *J. Teleg.* 7 (1874): 296.

–485–

Notebook Entry: Multiple Telegraphy

[Newark,] Entered[1] Sept 10, 74[a]

best arrangement for static[2]

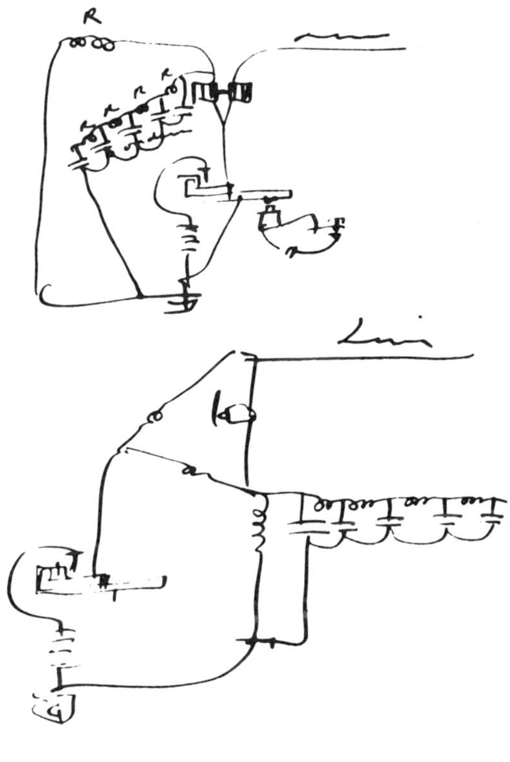

Quadruplex: Condsr at recg end shown arranged in differential[3]

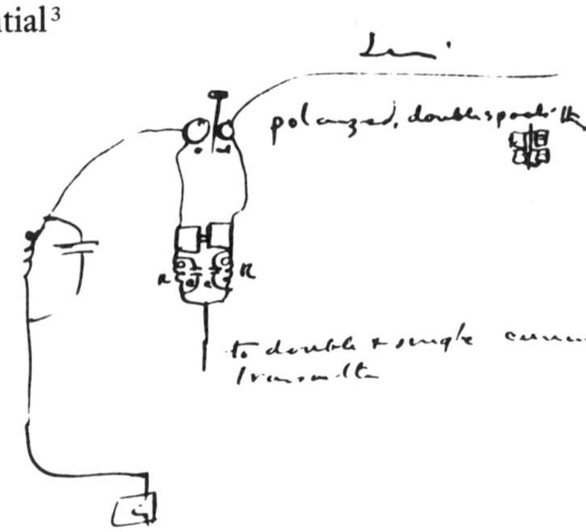

Caveated[4]

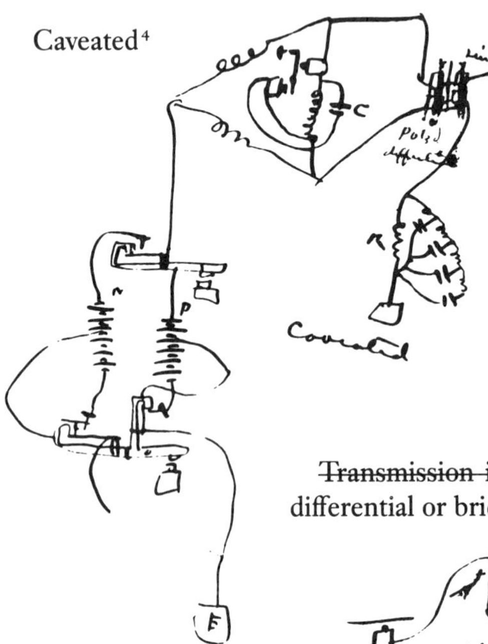

Duplex[5]

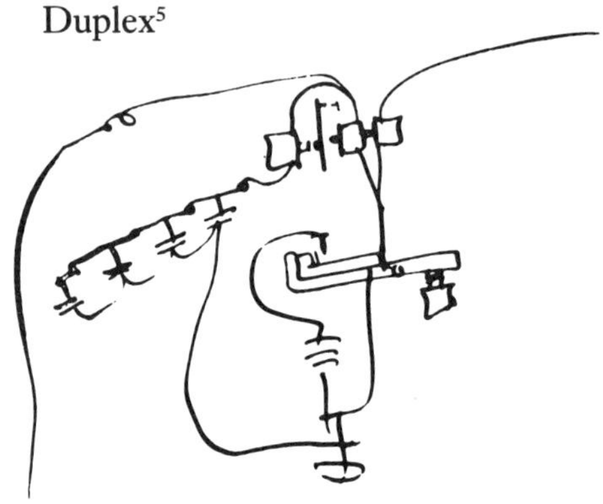

~~Transmission in same direction~~ Duplex: which is neither differential or bridge[6]

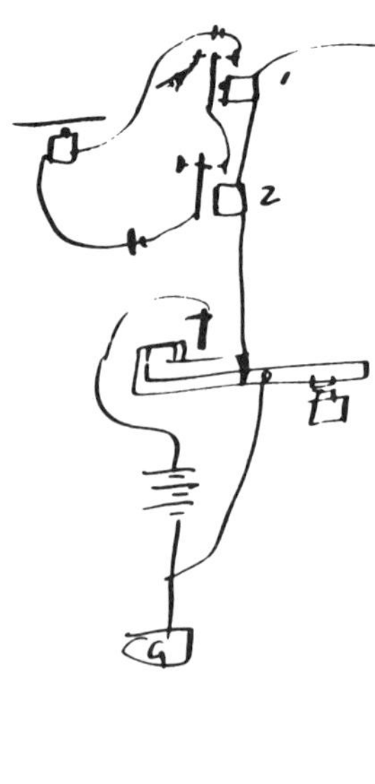

~~& work~~ dont look as if it worked

Problems Reversing with single battery Morse key[7]

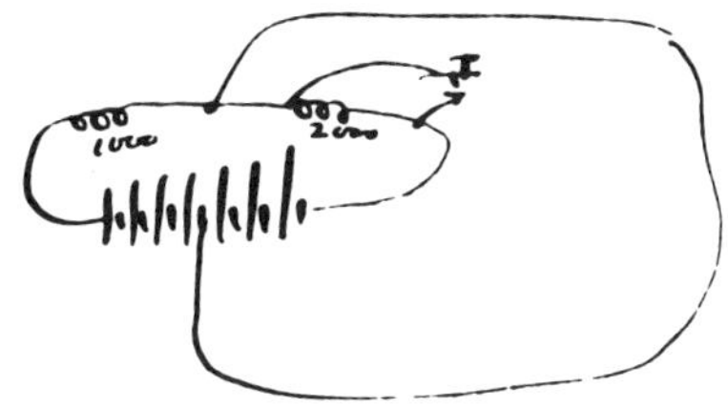

also

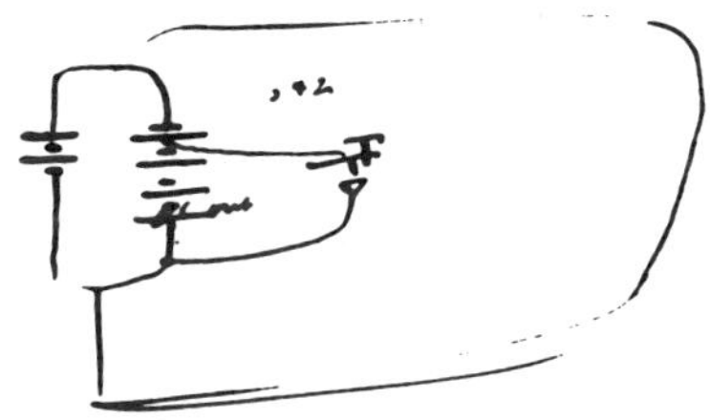

Repeater[8]

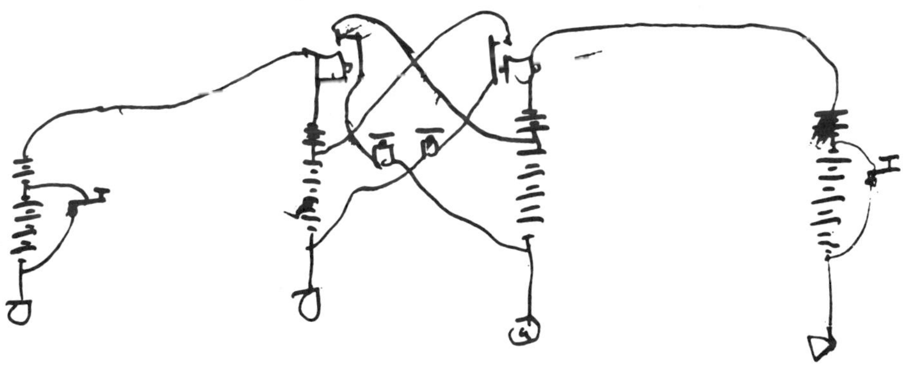

Transmission in the same direction[9]

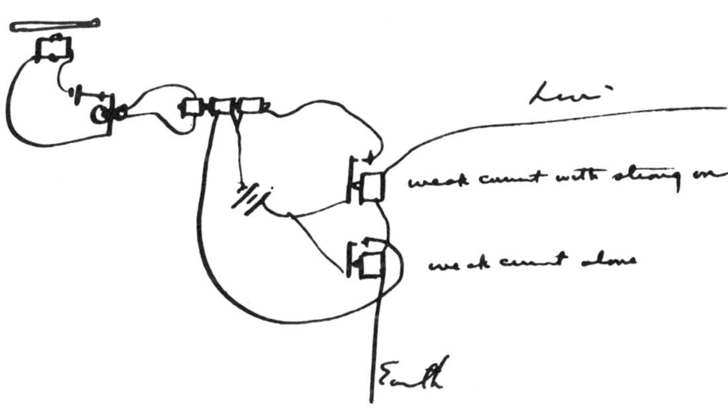

Thos A Edison

X, NjWOE, Lab., Cat. 1176:57 (*TAEM* 6:56). [a]Several drawings also bear this date.

1. These designs are all directly or indirectly related to Edison's development of a commercially practical quadruplex. The design, construction, connection, and extensive testing of alternative components and arrangements—together with his other projects—kept Edison very busy during the summer and fall, as evidenced by his 2 September telegram from New York City to Batchelor at the Ward Street shop: "Inform wife will not be home tonight. Come over with pyrogallic solution for domestic before eleven o'clock. Call at Western Union for me. Edison" (Cat. 1173:128, Scraps. [*TAEM* 27:426]). The day before recording the designs in this entry, Edison tested one version of the quadruplex on a circuit to Albany and back ("Advancement," *Operator,* 15 Sept. 1874, 4). The content of this document may date from an earlier set of loose pages; at irregular intervals Edison and Batchelor copied into the notebooks various notes and drawings originally done on loose sheets and scraps of paper. For work probably preliminary to this see Cat. 297:125(1, 3), 128(3), 144(2); and Cat. 1176:55; both Lab. (*TAEM* 5:896, 898, 911, 960; 6:55). Cf. Cat. 297:84(3), Lab. (*TAEM* 5:779), which may indicate, together with fig. 3 in Doc. 375, that some designs here and in Docs. 512 and 533 date from as early as October 1873.

2. Edison found that balancing the line with an artificial circuit to cope with "static discharge" or "kick" in duplex circuits became more problematic on long lines or when a set of diplex instruments was being duplexed to make a quadruplex (Edison's testimony, Quad. 70.7, p. 279 [*TAEM* 9:506]). The two drawings shown here are equivalents—the first is a differential circuit and the second is a bridge. They are quite similar to the designs in figures 1 and 5 in a caveat executed on 4 December 1874 (Edison Caveat 54), where some account of their detailed operation is provided. Though they are shown here only in the simple duplex forms, the point of interest—the structure and placement of the artificial line and connections—would be unaffected by adding a second, polarized, receiving relay and a second transmitter. They thus apply as much to quadruplex as to duplex telegraphy. The figure label in each case is "Line."

3. Figure labels are "Line," "polarized, double spools thus," and "to double & single current Transmitter." This is the differential counterpart of the design in the accompanying illustration; each circuit has a condenser, shunted by a resistance coil, connected directly to the neutral relay receiver as well as to the one in the artificial line. (For some account of how it could operate, see Edison's discussion of figure 2 in Doc. 513.) This may have been related to the general problem of compensating for static discharge and varying conditions on the line as well as to the prob-

One of Edison's bridge duplex designs, an equivalent to the differential design shown in Doc. 485.

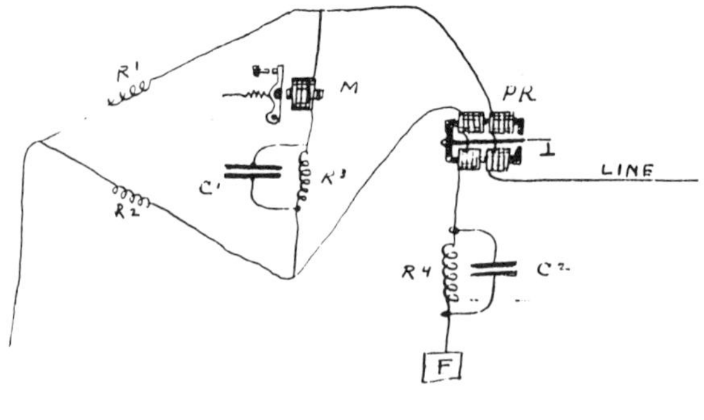

lem addressed in the next circuit design in this document (see n. 4). This arrangement, in the bridge form, may have been the one used in the successful December tests of the quadruplex on the line from New York through Buffalo to Chicago. See Doc. 515.

4. Figure labels are "Line" and "Polzd differential." Among other things, this circuit was designed to handle the problem tackled in Doc. 449 (see headnote, p. 229). In the receiving instruments here, the shunted condenser in the bridge would send a momentary current through an extra electromagnet as the signal current dropped to zero; that extra magnet coil would hold the armature lever long enough to allow a bug trap to do the rest. The accompanying unfinished illustration, most likely from midsummer 1874, suggests other work in the same direction, as does a drawing on the back of the same sheet (Cat. 297:128[1], Lab. [*TAEM* 5:910]; cf. Docs. 315 and 392 [fig. 14]). Figure labels in the illustration are "Differential Polarized Relay," "R Box," "Resistance Box," "Condenser," "Line," "Battery Cup," and "Condenser." Western Union used this form of compensation—having a shunted condenser activate an extra magnet coil on the neutral relay—

Another (incomplete) quadruplex terminal circuit.

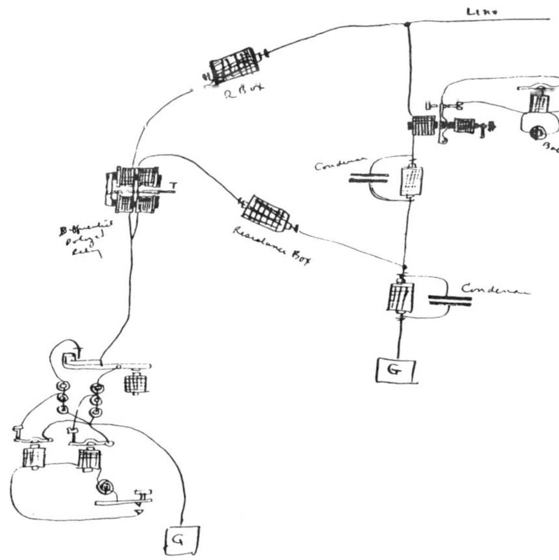

extensively starting in 1875, attributing its development to Gerritt Smith (Pope 1875, 337–39; see Doc. 484). Doc. 513 is the caveat covering this part of the design; Doc. 512 is the caveat covering the transmitting arrangement. See also n. 7.

5. This design is similar to the first two drawings in this document and to figures 3 and 4 of Edison Caveat 54, where some description of its operation can be found.

6. This is part of an extensive range of similar designs, mostly undated, in which Edison uses coils of different sensitivity in series. Examples date at least as far back as February 1873 (Doc. 285, duplex no. 8) and forward to January 1875 (Doc. 533, fig. 2), and include the illustration accompanying Doc. 348 n. 16 as well as the last drawing in this document.

7. In his basic quadruplex design, Edison used polarity reversals for

one transmitter at each end. He and others considered many designs while seeking a practical way of reversing polarity for a telegraph signal. See Docs. 506 and 512 n. 5; cf. Doc. 294 (penultimate drawing) and Cat. 297:64(3), Lab. (*TAEM* 5:665).

8. Repeaters were needed for the practical working of long quadruplex lines. See headnote, p. 347.

9. Figure labels are "Line," "weak current with strong on," "weak current alone," and "Earth." This is the first of a number of different diplexes (and thus potential quadruplexes) that Edison considered over several months. The operation of this incomplete design—which, like that in Doc. 467, uses a triple magnet—is not clear. It shows only one receiver, for "weak current" signals; other receiving apparatus presumably would receive only "strong current" signals. The drawings on Cat. 297:144(2), Lab. (*TAEM* 5:960), suggest a relation to the circuits in Doc. 533 as well.

–486–

From Josiah Reiff

New York, Sept 11th 1874[a]

My dear Edison—

I did not expect to get the amt promised ($100) until to-morrow, hence can only send you $25 by Bearer.

Sorry to hear you are unwell. I have nearly completed my plan for the signal etc[1] & when all is arranged will submit same to you before proceeding further—May be ready for you Monday—

The old gent (Mr H[arrington]) may be here tomorrow.

Smiles will wreath your lips & enthusiasm fill your Brain very soon— Keep quiet, but our night of suspense is over.

We will shake the foundation of things— Stearns[2] is entirely too unimportant to cause you a moments thought— The money & satisfaction is from another direction— I want you to understand that in all I or we do—Batchelor the faithful & Murray the persistent will be duly cared for— The old wheel horse T. A. E & I will pull together & I shall try & keep up my side— Yrs

J C Reiff—

ALS, NjWOE, Scraps., Cat. 1173:133 (*TAEM* 27:435). Letterhead of Automatic Telegraph Co. [a]"New York," and "187" preprinted.

1. Unidentified.

2. Joseph Stearns; Reiff is referring to multiple telegraphy as competition for the automatic.

–487–

From John Elmore

New York Sept. 12th 1874

Dear Sir

When I last saw you I think you mentioned that you had entertained the idea of employing a Chemist.

Allow me to introduce to your notice the bearer Mr. Robert Spice[1] with whom I have been acquainted the last two or three years and whose varied experience could not fail to be of great advantage. Respectfully Yours

John N Elmore[2]

ALS, NjWOE, Scraps., Cat. 1173:133 (*TAEM* 27:434).

1. Robert Spice (b. 1845) was professor of chemistry and natural philosophy at the Brooklyn High School. He first met Edison on 14 September and gave him private instruction in chemistry between 21 September and 24 October 1874. Spice's testimony, TI 1:298–99 (*TAEM* 11:126).

2. John Elmore sold chemicals and chemical apparatus in New York and was one of Edison's chemical suppliers. Wilson 1874, 380; Elmore's bill to TAE, 12 Aug. 1874, 74-001 (*TAEM* 13:12).

–488–

Notebook Entry: Multiple Telegraphy

[Newark,] Sept 12 1874

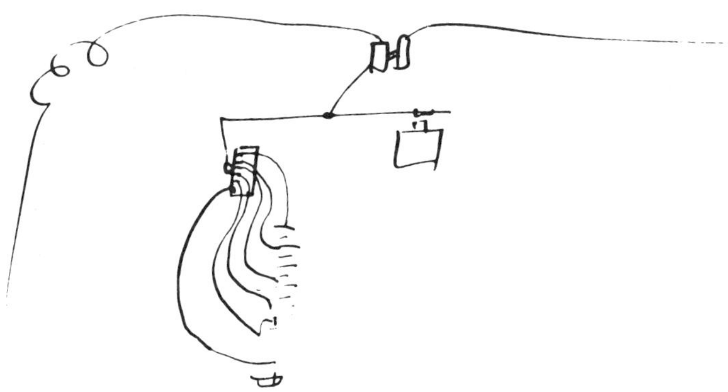

put on battery little at time to break up static not let it all come at once[1]

Edison

X, NjWOE, Lab., Cat. 1176:59 (*TAEM* 6:58).

1. As the magnet under the right end of the lever at center pulled down that end of the lever, the rolling contact attached to the other end included more cells of the battery in the circuit. Cf. Doc. 285 (duplex no. 5) and Cat. 298:48(2), Lab. (*TAEM* 5:97).

Article in the Supplement to the Operator[1]

New York, September 15, 1874[a]

Platina Points.[b]

This metal, which plays an important part in telegraphic instrumentation, was discovered in 1741, in the silver bearing sands of the river Pinto, in South America. It is generally found combined with several very rare metals, among which are iridium and osmium; the former, could it be obtained in sufficient quantities and be worked with facility, is superior to platina for circuit contacts: the latter, when in the form of an acid, is stated by St. Clair Deville, the French chemist, to be the most poisonous substance known. In a paragraph, which is now going the rounds of the papers, this chemist is said to have stated before the French Academy, that the vapor from the one thousandth part of a grain[2] of osmic acid, liberated in one hundred cubic feet of air is sufficient when inhaled to produce death, and that twenty-five pounds of it was sufficient to kill every person in the world, which is rather a *Frenchy* statement if it takes one thousandth of a grain to kill one person.

Platina is mostly obtained in the Ural Mountains in Russia, and the total production of the world does not exceed two tons per year, the major portion of which is used in chemical utensils and in the manufacture of platinic salts, such as sulphate and chloride of platinum, which are used by analytical chemists in quantitive analysis.

Platina is the heaviest metal known, it is harder than copper but softer than iron, and is found to excel in tenacity all other metals except copper and iron. Wollaston, by placing a lump of platina inside a silver wire and drawing them together and afterward removing the silver by an acid that did not attack the platina, succeded in making nearly a mile of wire 1/30,000 of an inch in diameter, which is considerably less than the finest spider web.

The platina sold in this country is all imported, and the prices are subject to a variation. On August 1st the price was eight dollars per ounce. It is fortunate, on account of its cost, that it is not necessary to use any great amount of this metal in telegraphy, though it hardly seems excessive, when we consider that one ounce will make about 250 relay points. The writer of this article having had considerable experience with this metal, has found that platina, as purchased from the dealers, is entirely too soft, and makes very bad points,[3] *i.e.*, points which are easily oxidized by the spark, and can never be made to close a local circuit of low resistance perfectly, except the relay lever is moved with considerable force. The other defects of soft platinum are that the oxide is pressed into it, and does not clear itself, and that the oxide increases the spark to such an extent, that the relay lever does not entirely open the circuit. Should the relay points be adjusted close, and there is considerable spark, the sounder is worked merely by increasing and decreasing the resistance of the local circuit, the spark preserving its continuity. The resistance offered to the passage of the current by the spark on relay points is sometimes as low as thirty ohms, therefore the sounder is operated by merely changing from a three ohm circuit to a thirty ohm circuit. It is obvious that the sounder will work sluggish owing to a portion of the current still remaining, after the circuit is supposed to be opened.

The same effect is observed on keys, especially on those which are provided with large flat platinum wire, when thoroughly cleaned they may be worked close, but in the course of an hour they fail to open the circuit, a bright spark is noticed (due to the incandescence of the finely divided oxide), and more play must be given, or the points cleaned. By using these large flat contacts no clearance is given the oxide; in other words, the oxide is formed on the faces of the large flat points, and hammered artistically into them, without any method being provided for its removal. If the points were made small, and the lower one beveled or rounded, the oxide would have a chance to be crowded away from the point of contact. If the lower point be rounded, and the upper one left flat, the negative pole of the battery must be connected to the latter, otherwise a cup will be formed; the point connected to the negative pole becomes pointed by accumulation of oxide, and that connected with the positive, cupped.

There exists an astonishing difference between contact points made of soft platina and that which is hard drawn. The slightest pressure on the latter will close a low resistance current perfectly, the spark will be reduced more than four-fifths, and if the points are properly shaped the oxide will free itself, it being almost impossible to hammer it in.

For the reasons explained above, two flat plates or a point working against a flat plate, should never be used either upon relays or kegs.[4] The points should never be filed, for it does not matter how fine the file is, it will leave innumerable points which are easily oxidized. They should be scraped with a sharp knife and then thoroughly burned which prevents the formation of these points.

PD, *Operator,* 15 Sept. 1874, 2. [a]Place and date of publication. [b]Followed by centered rule.

1. See headnote, p. 286. This article is preceded by a reprint of Doc. 476 and an article on units of electrical measurement.
2. 1 grain = 0.0648 grams.
3. See Doc. 453 n. 3.
4. Should be "keys."

–490–

From Robert Spice

Brooklyn, Sep: 16. 74

Dr Sir,

To save unnecessary loss of time when I come on Monday, please note the following things we shall require.

Sulphydrate of Sodium sol:)

Apparatus all ready to make and pass H_2S through solutions

Also please have about two gallons of Distilled water in White glass stoppered bottle.

I am arranging & writing in tabular form a system of Analysis wch from experience I know to be efficient & clear.[1]

I have the honour to be Yours obedient Ser[van]t

Robert Spice

ALS, NjWOE, Scraps., Cat. 1173:136 (*TAEM* 27:440).

1. Not found.

–490A–

From George Preece

London, 25 September 1874

My dear Sir:

I am much obliged for your letter of the 1st and the extract from the Scientific American, which I shall insert in our Journal.[1] The Electromotograph seems a wonderful thing and I should be very glad of a drawing.[2]

If you would kindly write me a short and interesting account of your quadruplex accompanied by a drawing I shall be very glad to give it insertion.[3] We are always glad to receive original communications.

I trust to hear of you soon joining the Society which is steadily increasing.[4] Yours faithfully

Geo E. Preece[5]

ALS, NjWOE, DF (*TAEM* 13:776).

1. *J. Soc. Tel. Eng.* 3 (1874); the extract (Doc. 476) appeared on pp. 161–63.

2. See Doc. 552.

3. No such account appeared.

4. David Brooks nominated Edison and Batchelor as foreign members of the Society on 30 October 1874. They were elected on 9 December 1874. Membership applications of Edison and Batchelor, Institute of Electrical Engineers Archives, London; *J. Soc. Tel. Eng.* 3 (1874): 485.

5. George Preece, a British telegraph engineer, was Secretary of the Society of Telegraph Engineers in London. He was also the brother of William Preece, later chief engineer of the Post Office telegraph in Great Britain. Baker 1976, 95–96, 107.

–491–

George Harrington to Lemuel Serrell

NEW YORK, [September 26,] 187[4][1]

Dear Sir:

If you can find it convenient this evening will you please look over the Harrington & Edison contract[2] and see if it does not cover the *Duplex* and *Quadruplex.* I also leave the power of attorney[3] and will call Monday to see you. Respectfully,

GEO. HARRINGTON, Per C. B. HARRINGTON.[4]

PD, NjWOE, Quad. 71.2, p. 128 (*TAEM* 10:293). Original on letterhead of Lemuel Serrell.

1. Serrell testified to the date of this note. Quad. 71.1, p. 62 (*TAEM* 10:36).

2. Doc. 109.

3. Doc. 155.

4. Chase Harrington.

EDISON AND THE *TELEGRAPHER* Doc. 492

This editorial is representative of the many vituperative articles and editorials James Ashley published in the *Telegrapher.* The attacks began on 18 July 1874,[1] in the wake of Edison's announcement of the quadruplex, and continued for several years. Ashley's animosity toward Edison had been simmering for at least three years prior to this, apparently rooted in the circumstances surrounding Gold and Stock's purchase of the American Printing Telegraph Company. This acquisition had prompted Ashley and Frank Pope to sever their connections with the printing telegraph company and terminate their business association with Edison.[2]

Ashley focused on several issues in his campaign to undermine Edison's professional reputation and personal integrity.

The first was Edison's claim that the quadruplex represented a new process of multiple transmission. According to Ashley, a number of other telegraph inventors had already devised far more efficient systems of multiple telegraphy.[3] The quadruplex bore "a strong resemblance to Falstaff's lunch—'a pennyworth of bread to a most intolerable deal of sack,'" he quipped.[4]

Ashley also disputed assertions that made Edison the primary inventor of a practical automatic telegraph system, printing numerous letters which gave that honor to either George Little or Marshall Lefferts.[5] "So far as we can prevent it, no charlatan or unprincipled appropriator of other's property and rights in connection with electrical and telegraphic inventions shall meet with success," Ashley promised his readers.[6]

Through the fall of 1874 Ashley's attacks became increasingly shrill. "We hate and despise the pretended inventors who are forever making new and startling discoveries—which are for the most part, either stolen from others or dug out of books where they have lain fallow, because they were not of sufficient importance or practicality to be worth bringing forward," he wrote in October. Readers of the *Telegrapher* knew the identity of one of these fellows, he continued. "We can conceive of no meanness or moral degradation deeper than that of such a person. Any one having such characteristics should be ignominously kicked out of any respectable telegraph office whenever he sets his foot in it, or dares to bring any of his spurious wares seeking for a market."[7]

When Western Union and the Atlantic and Pacific entered their protracted struggle over ownership of the quadruplex patents, Ashley made his view of Edison's role in the affair clear by dubbing Edison the "professor of duplicity and quadruplicity."[8] "Every person who has had dealings heretofore with Edison knows that he is utterly unreliable," Ashley declared, "that it is a well known characteristic of his to systematically and deliberately 'go back' on every person who has ever endeavored to aid or cooperate in his inventions or business."[9] Similar sentiments are found in almost every issue of the newspaper between January and May 1875.[10]

1. "The Dutch have Taken Holland!" *Telegr.* 10 (1874): 169.

2. See *TAEB* 1:225–26.

3. For example, Henry Nicholson claimed his "triple relay" performed the same function as the quadruplex. Nicholson 1874.

4. "More Startling Inventions for Rapid Telegraphy," *Telegr.* 10

(1874): 172. See also "The Western Union Company Caving In," "The Wheatstone Automatic and the Western Union Telegraph Company," and "Telegraphic Inventions and Inventors and Western Union Officials," ibid., 188, 191, 196.

5. "Automatic Telegraphy," "Automatic Telegraphic Inventions," and "Automatic, Duplex and Quadruplex," ibid., 177, 178, 195. Little himself engaged in a public denigration of Edison's contribution to automatic telegraphy through the pages of the *Telegrapher*. His attack on Edison also spread to the *Scientific American* (see Doc. 518) and the *New York Tribune* (see Little's letter to the editor, reprinted in *Telegr.* 10 [1874]: 178–79).

6. "Telegraph Inventions and Inventors and The Telegrapher," *Telegr.* 10 (1874): 179. See also "Telegraphic Inventions—Old and New," ibid., 181. Daniel Craig, in a letter to Marshall Lefferts from about this time (Thursday, n.d. [c. 1874], Lefferts), also disparaged Edison's originality as an inventor:

> Edison has not one patent upon any material point, which cannot be clearly proven to be a mere evasion of other peoples' inventions—& I venture to assure you that Edison has very little ability as an originator—his ability is only as an adapter of other people's ideas. He is simply a plausible plasterer, and your confidence in him & in his inventions is misplaced. I say this after careful investigation & with not one unkind feeling towards Edison. His inventive reputation would break down a strong case, if he sustained it, at any point between Boston and Washington, & he can & will, if need be, be proven to have perjured himself in several glaring cases.
>
> . . . it is just equal to Edison's genius to steal your ideas & magnet-coil resistance, & patent it under a new name & perjure himself. That exactly, is Edison's gift of genius, & he has not other but lieing.

7. "The Policy, Creed and Practice of The Telegrapher," *Telegr.* 10 (1874): 238.

8. Ashley first called Edison the "professor" on 6 February 1875 ("The Telegraphic Situation," ibid., 11 [1875]: 34).

9. "The Quadruplex Controversy," ibid., 136.

10. See, for example, "The Title to the Quadruplex Patents" and "The Contest for the Title to the Edison Quadruplex Patents," ibid., 76, 124.

–492–

Editorial in the Telegrapher[1]

New York, September 26, 1874.

The "Organ" Business.

It has been generally supposed that the *Journal of the Telegraph* was the official organ of the Western Union Telegraph Company. We have shared in this belief, and have occasionally referred to it as such.[2] Of late, however, its character has been changed, and it has now become the personal organ of TOM EDISON and Mr. GEORGE R. PRESCOTT, the latter occupying the position of electrician to the company. It is EDISON and

PRESCOTT and PRESCOTT and EDISON, *ad nauseum.* It was always heavy and dull, as became an "official organ," of which, as its circulation is mainly to the officers and employes of the company gratuitously, and as those to whom it was sent were under no painful obligation to read it, there was no reasonable cause for complaint, and it served a useful purpose in circulating the orders, tariff rates, etc., of the company.

As a personal organ, however, run in the interest of the two parties named, to advance their purposes and manufacture fame, reputation and fortune for them exclusively, it is quite a different affair, and is open to criticism. The electrical idiosyncracies of EDISON and PRESCOTT (since they have entered into partnership) are fully set forth in the columns; and if much puffing could render them valuable they would be the most remarkable scientific geniuses of the present or any former age. Unfortunately for them, outside the limited "ring" of which they are the principal members, they are not regarded as the bright and shining lights which they evidently consider themselves. However effectually they may succeed in convincing the President of the company of their preeminent abilities and attainments, electricians and practical telegraphers refuse to accord them the high position in electrical science and practical telegraphy to which they aspire. In fact, they are regarded rather as pretenders than as actual electricians and scientists.

Not content with one "organ," however, we notice that they have adopted a nondescript sheet called *The Operator,* of which a few of our readers in this vicinity may have heard, as a sort of sub-organ or tender upon the larger and more pretentious sheet. Recent issues of this publication have contained such fulsome and absurd slaverings of EDISON as to make that person appear even more ridiculous than he otherwise would.[3] It is generally believed that these are the productions of his own pen, as it is not reasonable to suppose that any other person would venture to expose even him to such ridicule as must be caused by the stuff that appears in its columns. The "organ" business is truly being run into the ground.[4]

PD, *Telegr.* 7 (1874): 233.

1. See headnote above.

2. Western Union began publishing the newspaper as a medium for executive orders and information to all its offices in December 1867. James D. Reid served as its first editor. "Journal of the Telegraph," *J. Telegr.* 1 (2 Dec. 1867): 4; *TAEB* 1:12.

3. The 15 September 1874 issue of the *Operator* described a test of

Edison's quadruplex at Western Union headquarters, noting, "It is not singular that the teeming brain and untiring nerve of this great genius should bring forth a telegraphic combination that will startle the electric world." The article concluded with a word about Edison's critics: "There are already a number of persons ready to decry this great invention, and, from their *Little*-ness of intellect, venture opinions that are as worthless as their weak, vacillating minds are capable of maturing." "Advancement," *Operator* 4 (15 Sept. 1874): 4.

4. This issue also included a front page reprint of an article from the *Telegraphic Journal* of London in which the originality of Edison's quadruplex was questioned, and an editorial reaffirming the newspaper's intention to reveal "appropriators of the ideas and inventions of others—charlatons and would be impostors . . . in their true character and colors," which immediately preceded an editorial challenging Prescott's criticisms of the automatic telegraph system. Arthur R. Granville, "Duplex and Quadruple Telegraphy," "To the Friends of the Telegrapher," "The Edison and Prescott Organ and Fast Systems of Telegraphy," *Telegr.* 7 (1874): 229–30, 232.

–493–

W. S. Williams to Charles Batchelor

[New York,] Sept 26 1874

To[a] Batch

Tell Edison to look at the telegrapher Sept 26th[1] Read it Well see if he thinks best to answer it. and if so to give us some points "Solid ones" in ~~wariting~~ Writing. if he thinks it Worth while and gives us some points to ans his part of it we'll take ~~advance~~ advise of plymouth church & "Give Em Hell"[2] Allen[3] & I will Work on our part tomorrow and then put whole thing together. if he thinks better not answer it why we will drop his part & may possibly answer his our part. They are a party of Damned Slanderous Damned rascals and ought to be pushed to the front at least I think so. Dam em tell him to give it to us With his cuts[4] Monday so we can straighten up things in time

Wms.[5]

ALS, NjWOE, DF (*TAEM* 13:41). Message forms of Western Union Telegraph Co. [a]"187" and "To" preprinted.

1. The 26 September *Telegrapher* contained an editorial that attacked the duplex and quadruplex in defense of the automatic system ("The Edison and Prescott Organ and Fast Systems of Telegraphy," p. 232) and another that viciously attacked Edison, Prescott, and the *Operator* (Doc. 492). On 15 October the *Operator* replied in "This Is Not an Organ" (p. 4); Edison was not mentioned.

2. The Reverend Henry Ward Beecher, noted for his powerful oratory, preached at the Plymouth Congregational Church in Brooklyn, N.Y. *DAB*, s.v. "Beecher, Henry Ward."

3. Tom Allen, co-publisher of the *Operator.*

4. Edison supplied several drawings that were published in a supplement to the 1 October *Operator.*

5. W. S. Williams, a Western Union telegraph operator, was editor and publisher of the *Operator*. See *Operator* masthead and issue of 1 March 1875, p. 6.

–494–

To George Prescott

[New York,] *September* 30, 1874.[1]

G. B. PRESCOTT.

Started to equate duplex[2] about ten o'clock P.M.: No. 60 wire, eight gauge to Boston.[3] Both me and Smith[4] bothered in equating. Soon found that 60 was crossed;[5] changed it for No. 61, eight gauge; waited till twelve o'clock to get men; started for an hour's trial at 12.03 P.M.; messages taken from ten days' files; no picking.

Operators at New York—Fallam, De Graw, Cook, Bennett. Operators at Boston—D. Sanford, Davin, Colson, Wood.

Weather in N.Y.: No rain; very high wind.

" " Boston: N. East storm; all day fine; drizzling rain at 12:03.

Boston sent us 142 in one hour.[6]

N.Y. sent Boston 125 " "

Worked well both sides.[7]

Found, after got through, that I was out of balance 380 ohms,[8] as the wind had dried insulators on my end, while it still drizzled in Boston. Total breaks, thirty-eight.[9] Condensers old and leaky.

EDISON.[10]

PD, NjWOE, Quad. 71.2, p. 14 (*TAEM* 10:229). Original on message form of Western Union Telegraph Co.

1. This is Edison's earliest progress report on the first trial of the quadruplex on a New York–Boston line. Edison testified in 1880 that the sketch shown on the next page "represents correctly the connections used in the apparatus between New York and Boston" (Testimony and Exhibits on Behalf of Thomas A. Edison, p. 32, *Nicholson v. Edison*; see also Doc. 513 n. 4). Western Union set up the equipment for these tests in the regular operating rooms in both cities, not just in the electrician's experimental rooms in New York as in earlier trials (Docs. 454 and 485 n. 1).

2. For the Boston trials, Edison and Prescott principally tested a quadruplex circuit that used a bridge rather than a differential arrangement for its duplex aspect, though at some point they also tested a differential system (Prescott 1877, 835; on the bridge duplex, see Doc. 285 n. 17). Like most duplex systems it had to be balanced or equated, setting the resistance and capacity of the artificial lines equal to the values

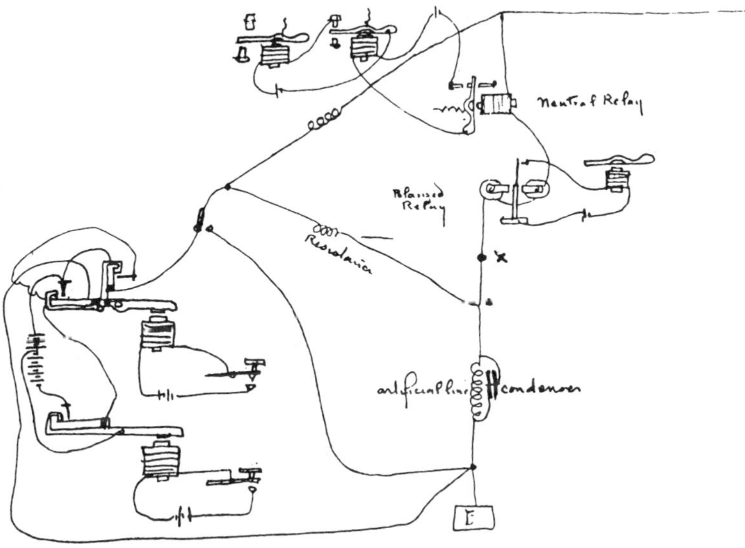

The circuit design Edison used on the New York–Boston quadruplex trials. The dot and **x** *show where Edison made additions for the Chicago trials (cf. Prescott 1877, 839).*

for the actual line. Maver (1892, 231–40) and Prescott (1877, 860–62), describe how to balance a quadruplex circuit. See also Doc. 454 n. 8.

3. Eight-gauge iron wire (in the Birmingham gauge system) was the common size for main-line American telegraphy; its diameter was a bit over 4 millimeters (Prescott 1877, 969). The quadruplex would not work well on long circuits with the smaller wire that might serve a simple Morse telegraph. This line was 240 miles long (ibid., 835).

4. Gerritt Smith was in charge at the Boston end. Smith's testimony, Quad. 71.1, p. 298 (*TAEM* 10:156); Reid 1886, 676.

5. Most telegraph lines were bare wire. If one sagged and touched another (crossed), the electrical contact disrupted the circuits and signals.

6. That is, 142 complete messages.

7. The two different kinds of signals and apparatus in the quadruplex were referred to as "sides," one working with reversals of current polarity, the other with increasing and decreasing current strength.

8. In comparison, the total resistance of the line wire would have been about 3,000 ohms, and that of a set of receiving relays about 600 ohms. Maver 1892, 554, 556.

9. A "break" was the interruption of a sender by a receiver, commonly to ask the sender to repeat a portion of a message.

10. A day or two later Edison reported: "Found quadruplex working when I came; Brown balanced; works O.K.; doing all the biz.; works better than ever; found a bad connection last night." Edison also told Prescott that one particular sounder—the "Giant sounder" made by Partrick, Bunnell & Co.—was "a great assistance" for the quadruplex and should be employed throughout (Edison to Prescott, c. 30 Sept. 1874, Quad. 71.2, p. 15 [*TAEM* 10:234]; see *Operator,* 15 Dec. 1874, 8). Western Union used the quadruplex for normal telegraphic business on the Boston route starting on 2 October; in his annual report, President Orton made much of this accomplishment (*J. Teleg.* 7 [1874]: 297, 312; *Operator,* 15 Oct. 1874, 5; Western Union 1874, 13–15).

–6– October–December 1874

During the last three months of 1874 Edison needed money. His shop expenses were high,[1] and the financial depression that had gripped the nation for more than a year now seriously affected the business of telegraph companies and their suppliers, Edison and Murray among them. Edison had to sell his house. He not only received considerably less than he had paid for it, but he also lost all of his equity. He, Mary, and Marion had to move into an apartment in a commercial area of downtown Newark. The technical side of his life was brighter. His work was gaining wider attention among scientists and engineers, and he expected financial settlements for his quadruplex system and for the sale of the Automatic Telegraph Company to the Atlantic and Pacific Telegraph Company.

In the fall, Edison demonstrated the commercial viability of his quadruplex telegraph. Though by no means his only project, it was the center of his activity. On 2 October a quadruplex circuit began operation on a New York–Boston wire; by early December, Edison had eight operators communicating between New York and Chicago, with a repeater in the circuit at Buffalo. The modifications behind this advance resulted from extensive experimentation, both in the laboratory and on Western Union's lines.[2] The New York–Chicago operation satisfied the conditions set by Western Union President Orton, and Edison expected a considerable settlement for his achievement.

Edison was also further developing his district telegraph system; a keyboard-operated, chemically recording, private-line printer to accompany it; and a new fire alarm system. In

December he contracted to have the district system introduced in Canada, and the Domestic Telegraph Company began to establish a circuit in New York.[3] Edison explored the possibility of beginning a stock-quotation enterprise in Brazil.[4] With Murray, he continued to manufacture his own inductorium and student telegraph instrument. They also sought and received orders from within and without the telegraphic community, ranging from instruments for the vendor Partrick, Bunnell & Company to quadruplex apparatus for Western Union to work for the Celluloid Manufacturing Company of Newark, Jarvis Edson's Recording Steam Gauge Company, and a New York electrotherapeutics entrepreneur.[5]

While devoting his laboratory time largely to refining his quadruplex and district telegraph systems, Edison continued his quest for new forces of nature, for the first time mentioning an antecedent, the "odic force" postulated some decades earlier by the German scientist Karl Reichenbach.

During this period Edison's name spread beyond the telegraph community to a wider public. His electromotograph brought him to the attention of American scientists for the first time. At the request of George Barker, professor of physics at the University of Pennsylvania, Edison exhibited an electromotograph device and discussed its principle before the National Academy of Sciences in Philadelphia. The general public could read Orton's appreciative comments on the quadruplex in newspaper accounts of Western Union's annual report, which received wide notice because in 1874 the company paid its first dividend in years. Within the American telegraph community, James Ashley's *Telegrapher* continued its contumely, while the *Operator,* for which Edison still wrote as science editor, persisted in its praise. Overseas, at the 2 December *conversazione* of the Society of Telegraph Engineers, Edison's electromotograph "became the attraction of the evening";[6] one week later Edison and Charles Batchelor were elected to the society.

President Orton's deferral of an agreement for the quadruplex precipitated a crisis in Edison's relations with Western Union. Despite Edison's desperate need for money, Orton left New York for the Midwest at the end of December without committing the company to Edison's terms. At the same time, in what might otherwise have been an unrelated development, Jay Gould's Atlantic and Pacific Telegraph Company agreed to take over the struggling Automatic Telegraph Company. This merger culminated year-long negotiations by Automatic to form a larger network. As a part of the deal, Atlantic and Pacific acquired Edison's automatic telegraph patents and his services as electrician. On the evening of 30 December, having just signed the merger agreement, Gould, Thomas Eckert, and Albert Chandler visited Edison's laboratory to examine the technology they had purchased and to meet its maker.[7] Eckert and Chandler brought Gould to discuss the quadruplex as well, knowing that Edison and Western Union had not come to terms.[8] Within a week, with Orton still absent, Gould would purchase Edison's interest in his quadruplex patent applications.

1. App. 3, Edison's Affidavit, p. 812–13.

2. Edison's quadruplex had well over one hundred possible permutations of current strength, current direction, and mechanical positions, and he had to anticipate several current changes during any single signal. Theoretical consideration could not guarantee success; the quadruplex required extensive bench and line testing. "Years later Mr. Edison remarked that the problems in incandescent electric lighting were relatively simple compared to the complex and tantalizing ones that came up during the period that he was laboring with the solution of quadruplex telegraphy" (Jehl 1937, 1:74). William Maver's description admirably illustrates the system's complexity:

> For example, it is quite demonstrable that during the making of a dash on the neutral relay at one station the distant pole-changer may reverse its battery several times; the home pole-changer may do likewise, and the home transmitter may increase and decrease the electromotive force of the home battery, repeatedly. At the same time, and, of course, as a result of the foregoing actions, the home neutral relay may have had its magnetism reversed several times, and the *signal* will have been made, partly by the home battery, partly by the distant and home batteries combined; partly with current on the main line, partly without; partly by the main line "static" current, and partly by the condenser current, and yet on a well adjusted circuit it will have been heard on the quadruplex sounder as clearly as any dash on an ordinary "city line" sounder. [Maver 1892, 198]

3. Cat. 1214:51, Accts. (*TAEM* 21:589).

4. Fernando Rohe to Edison and Murray, 4 and 17 Nov. 1874, Cat. 1173:158, 169, Scraps. (*TAEM* 27:472, 501); Cat. 1214:43 and Cat. 1184:99, both Accts. (*TAEM* 21:588, 832).

5. Cat. 1214:43, 51, 204, 208; Cat. 1184:95, 98–99; both Accts. (*TAEM* 21:588–89, 643, 645, 830, 832); see also Docs. 501 and 517 n. 1.

6. Sprague 1875, 362; "Edison's Motograph," *Operator,* 1 May 1875, 9.

7. Eckert and Chandler, although still nominally in the employ of Western Union, were in fact preparing to join Atlantic and Pacific. See Doc. 528.

8. App. 1.D102, D264, G15; Edison's testimony, Quad. 70.7, pp. 289–90, 294 (*TAEM* 9:511–12, 514).

–495–

Article in the Operator[1]

New York, October 1, 1874[a]

DUPLEX—NO. II.[b]

There is an extra phenomenon in the electrical duplex which does not occur in the water device. This is electrostatic induction, the effect of which is to destroy the balance of currents in the coils of the differential relay.

In sending a dash, for instance, through the relay to the distant station, the static inductive current is added to that of the battery current, thereby increasing the strength of the first portion of the signal within the main line helice, while the artificial line, being free from these currents, is not thereby increased. Hence the cores of the differential relay will become magnetic by this excess of current, and the lever will be attracted for an instant only, or "kick," as the operators say.

On disconnecting the battery a static current flows from the line in a contrary direction to the first current, and this again produces a movement of the relay lever.

Now, it follows that if we could by any means create a static induction current upon the artificial line, equal to that of the main line, at the moment the battery is connected and disconnected from both circuits, that the two currents would balance each other in the differential relay.

Mr. Stearns was the first to bring about this necessary result, and without which previously devised apparatus were impracticable on lines of ordinary length. Mr. Stearns effected it by adding a condenser to the artificial line.

A condenser consists of a number of sheets of tin foil or other metallic surfaces, connected to the artificial line, and facing an equal number of plates connected to the earth. These plates are separated from each other by sheets of insulating material, usually paper dipped in boiling paraffin.

If the main line is very short, as between New York and Philadelphia, the strength of the static induction currents are insufficient to produce a movement of the relay lever and the condenser is unnecessary; but when the line is long, as between New York and Buffalo, the static induction is sufficient to prevent all signalling until the condenser is added to the artificial line, and then it is requisite that a large number of sheets are used, otherwise it will not set up currents of sufficient strength to maintain the balance.

Another essential device to obtain perfect signaling is the addition of a small resistance coil. R2 (fig. 2) in the earth wire.

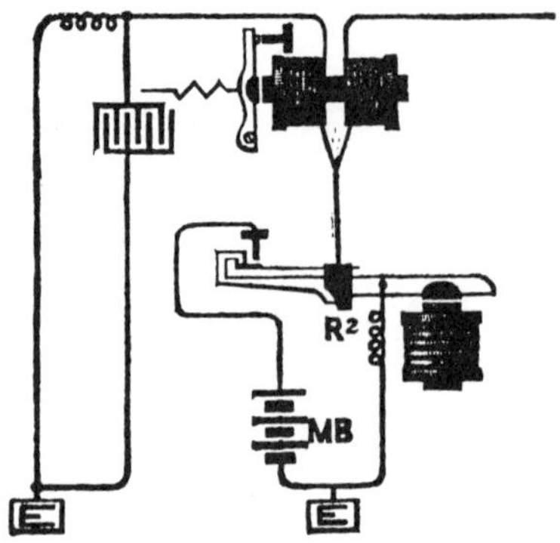

This coil has a resistance equal to that of the battery, so that if either be made to form part of the main line the resistance will be constant. It also serves to reduce the spark which occurs when the battery is for an instant "short-circuited" by the signaling point and spring. Were no coil inserted in the earth wire the resistance of the main line would be increased when the battery was inserted, and decreased when put to earth. This variation in the resistance of the line would vary the strength of the received signals. If the battery was inserted the distant signals would be "light," and when disconnected they would be heavy; and when the received current was exceptionably weak it would prevent signaling altogether, if the battery resistanec was considerable.

The spark, if a grove battery is used, would quickly destroy the contact points on the sounder lever. This device is also due to Mr. Stearns.

With these exceptions the electrical and water duplex are alike.

One peculiarity of the duplex is, that it may be operated when the signaling batteries are connected to the line in such a manner that they oppose each other, *i.e.,* when the positive pole of each battery is connected to line. When this takes place, no current passes over, but the current circulating within the artificial lines keeps the relays closed.

The water duplex diagram, slightly changed, will show this action clearly:

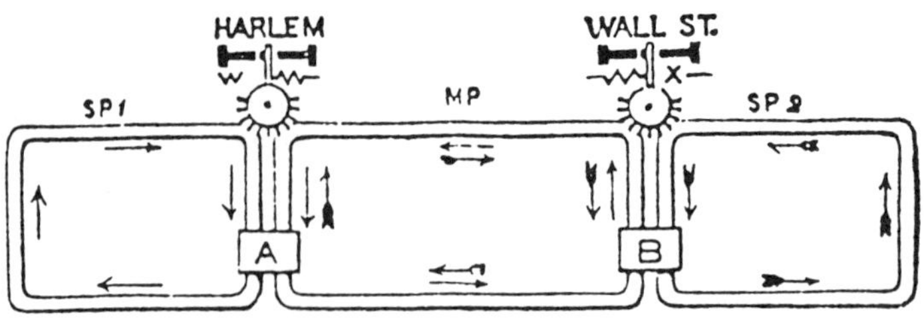

When no water passes from the reservoir A through the main or secondary pipes, the water from B passing through M P, throws the lever of W to the left, and the local circuit is closed, so to speak; but, as explained before, the wheel X remains unaffected. If, now, the water is made to pass from A into the main and secondary pipes, in the direction indicated by the unfeathered arrows, it will act, or we will suppose it to act, against the other stream passing in a contrary direction, in such a manner that the water will become still or cease to flow. The effect of this would be to allow the lever of W to be drawn to the right by the spring, were it not for the fact that now instead of the lever being held by the water from B the water in the secondary pipe, SP 1, performs this operation. If the water ceases from A, the balance between the two streams in the main pipe is destroyed; the water ceases in SP I, but the lever will still be held against the point on the left by the water from B. It will be seen that when the two streams oppose each other, the wheels and levers are held to their forward points by the water in the equating pipes, while if only one stream is passing they are held by the water in the main pipe.

If the direction of the water in the main pipe is the same, then all the signalling is done by these currents only and independent of the equating circuits.

The arrows shown in figure 3 will show this action in the electrical duplex.

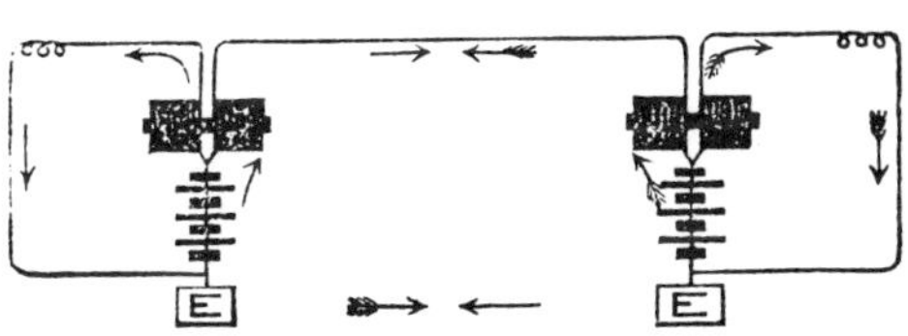

When both stations connect the battery to line they oppose each other, and no current passes over it but the relays close, nevertheless; for the current in the coils on the artificial lines, not being balanced by an equal amount flowing through the main line coil, closes the relay; when the batteries have their dissimilar poles connected to line, a current flows and the relay is closed only by the main line current.

It has been found in practice, that if a relay be inserted in a wire worked duplex, the balance is destroyed.

The reason for this is, that the passage of the signalling current through the coils of the relay sets up secondary magnetic inductive currents (these currents are distinct from static induction currents), the effect of which is to weaken the main current on closing and sending a current in the same direction after the battery is disconnected.

This effect, which is absent from the artificial line, destroys the equation. If a relay of the same character be inserted there the balance is preserved.

In fact, if all the evil effect which take place on the main line could be produced upon the artificial, the equation would be maintained and perfect signaling would always be obtained.

There are many effects produced on the main circuit which are not produced upon the artificial, among which may be mentioned.

1st. Static Induction.

2d. Magnetic Induction, as from relays.

3d. The induction from other wires.

4th. The phenomenon of secondary battery action.

5th. Leakage.

6th. A permanent current from the other wires.

7th. Variability of leakage, caused by opening other wires which acts as earth connections.

8th. Variability of the electrostatic capacity by opening and closing wires in close proximity.

9th. Momentary currents from other wires by leakage across the arms.

10th. The variability of leakage by increasing and deceasing the strength of the signaling currents.

11th. Lightning.

12th. Auroral currents.

13th. The difference between the electrostatic capacity of the wire at noon and midnght.

There are other causes which prevent perfect signaling to be found in the various instruments, among which may be mentioned the following:

1st. Permanent magnetism in the cores of the relay.

2d. Any difference in the capability of the two spools of the differential to produce magnetism with same strength of current, which necessitates either an increase or decrease in the resistance of the artificial line and a consequent increase or decrease of the electrostatic capacity of the condenser.

3d. The secondary currents from the receiving relay passing within a circuit formed by the line and combined leakage.

4th. The momentary decrease of the resistance of the line when the battery is short-circuited by the contact points.

5th. The short-circuiting of the battery from which two or more duplex wires are attached.

6th. The induction currents thrown into the equating coil by the passage of the distant signaling current through the main line coil.

7th. Leakage between the two sets of condenser plates.

8th. Induction from the resistance coils when not wound double.

With one or two exceptions, which are equated for by the condenser, all of the phenomenal effects which take place on the line, as well as the instrumental defects, still combine together in the present duplex to prevent perfect signaling.

As we proceed with this article we will describe various duplex apparatus in which many of these defects are provided for.

Having partially set forth the principle of modern duplex we will go back a little and describe the earlier forms and endeavor to explain why they were inoperative.

The first apparatus devised was that of Dr. Gintl, one end of which is shown in figure 4.[2]

C is the differential relay provided with two spools. The spool A is of large wire and is placed in a local circuit. B is in the main circuit and has fine wire. K is a double-pointed key, its lever, when resting upon the back point C, connects the main line to the earth, and when on the front point D, connects it to the battery.

The spring E, in connection with the point F, serves to close the equating circuit at the same instant the battery is connected to line R, as a resistance coil for adjusting the strength of the equating battery EB.

It is obvious that this device is inoperative for several reasons, one of which is that the continuity of the main circuit is broken when either key passes from the back to the front point.

Were one of the levers permanently connected to D and the other operated, the effect of the slight interruption of the current would not be perceptible upon the relay, providing that the tension of its adjusting spring was not great and that very little movement was given the key lever, but when both keys are operated these interruptions are doubled and the signals are mutilated.

If to this defect be added the greater effect of static induction, signaling becomes impossible. Even if there were no interruption in the continuity of the circuit, or any static induction, a perfect equation would be impossible with the large and small wire coils, owing to the difference in their capability to set up secondary currents and the difference in the resistance of the circuits upon which they would circulate.

Gintl himself found that this device was inoperative and adopted a chemical recording instrument in place of the relay, from which he says better results were obtained; but we surmise very imperfect signals, if any, were ever recorded over a line.

(To be continued.)[3]

PD, *Operator,* 1 Oct. 1874, 1. [a]Place and date of publication. [b]Followed by centered rule.

1. See headnote, p. 286.
2. The drawing was left out of this issue but was published as part of Doc. 505.
3. See Doc. 505.

–496–

Article in the Telegraphic Journal[1]

Newark, N.J., U.S.A., October 16, 1874.[a]

ON A NEW METHOD OF WORKING POLARISED RELAYS.

BY T. A. EDISON.[b]

BY means of the arrangement shown in the accompanying diagram any number of polarised relays may be worked in closed or permanent circuits.

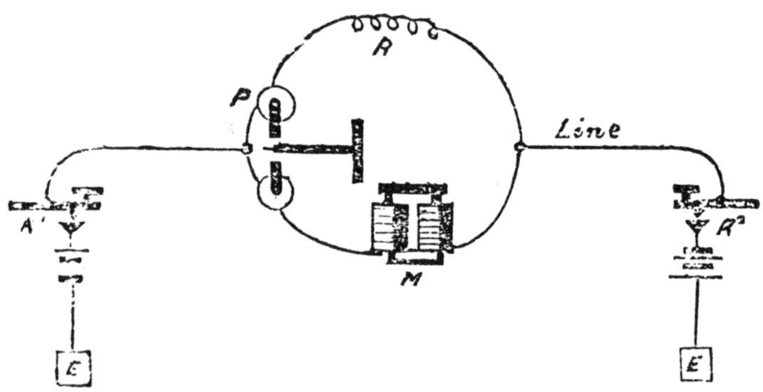

P is an ordinary Siemens polarised relay. The main-line wire is attached to that connecting the two coils together. M is an electro-magnet, having a resistance equal to that of the relay P. R is a resistance equal to M.

When both keys are closed, and the current is made permanent, the tongue of the polarised relay will remain on either side, as the main-line current passes through the two coils in opposite directions, and produce no magnetism. But the electro-magnet M not being so balanced, becomes magnetic.

If the circuit be interrupted, by opening either key, the secondary inductive current—from the latter circulating within the closed circuit of low resistance, formed by P, R, and M—acts upon the two spools of the polarised relay, and the tongue is thrown to one side. If the circuit is now closed, an inductive current from M passes through the circuit in the opposite direction, and the tongue is thrown to the other side. The armature of M is permanently attached to its poles to increase the strength of the inductive currents.

When once the main current is balanced within the coils of the relay P, by the coils R, no further adjustment is ever required.

PD, *Teleg. J. and Elec. Rev.* 2 (1874): 361. [a]Place and date not those of publication. [b]Followed by centered rule.

1. Two partial drafts of this article are in Cat. 297:134, Lab. (*TAEM* 5:925–27).

–497–

From Jarvis Edson

NY. Oct ~~7~~26th 74

Friend Edison

I met Dr Beard[1] 53 West 33d St N.Y. City today and called his attention to your Inductorium

He asked what you claimed for it & said that he would be pleased to do anything for you he could. Should particularly like to have you either come in person or send some competent one with an instrument that you might both exchange views— I have known him a long while as he is ~~devoting &~~ writing (publishing) specially on the ~~na~~Treatment of Disease by Electricity & devoting himself to original research in the matter I commend him to you— He could also place many of your instruments among his patients & a certificate from him would be of great value—as he is an acknowlegdged authority on such matters I believe— Yours Truly

J. B. Edson.

ALS, NjWOE, Scraps., Cat. 1173:156 (*TAEM* 27:466).

1. George Miller Beard (1839–1883) was an American physician noted for his researches on neurology and electrotherapeutics. Rosenberg 1976, 98–108; *DAB*, s.v. "Beard, George Miller."

[Newark,] Nov 2nd 1874[a]

Notebook Entry: Multiple Telegraphy

Duplex[1] leaks around X 1000

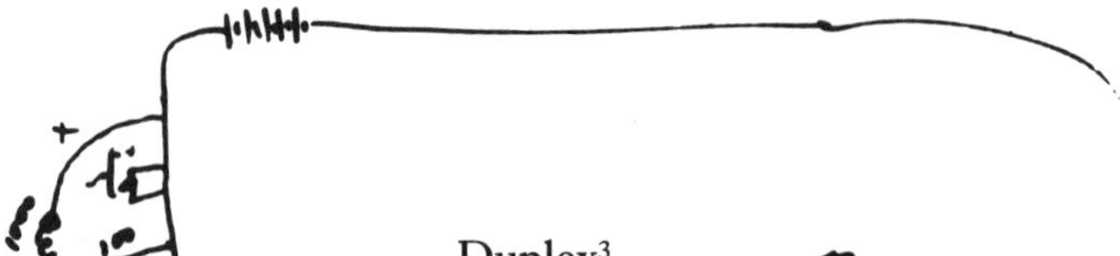

Duplex[3]

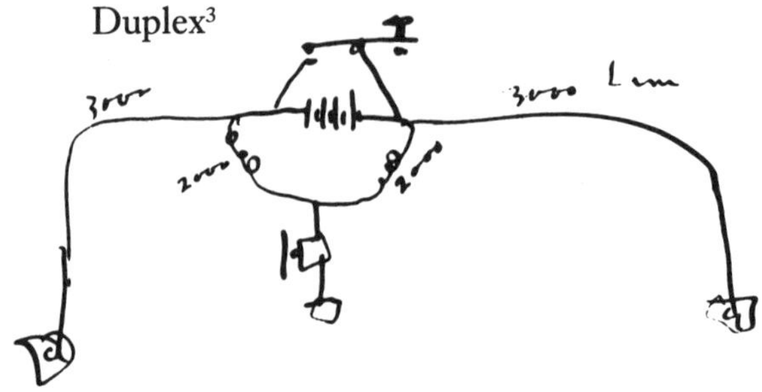

Duplex[2]

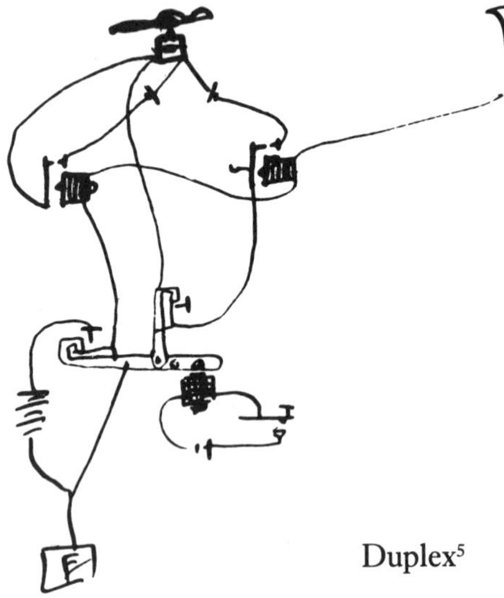

Duplex[4]

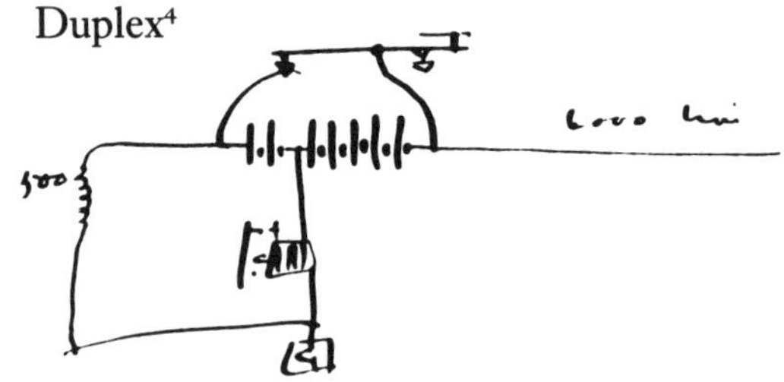

Duplex[5]

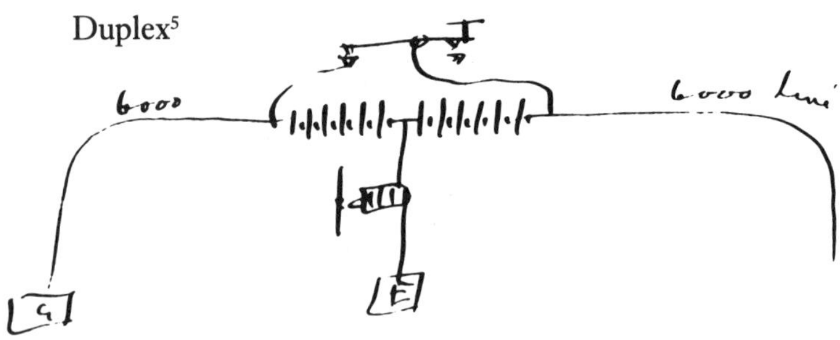

Duplex[6]

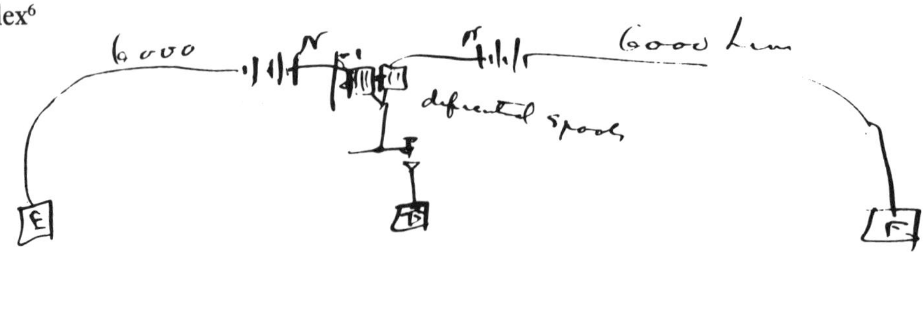

Duplex

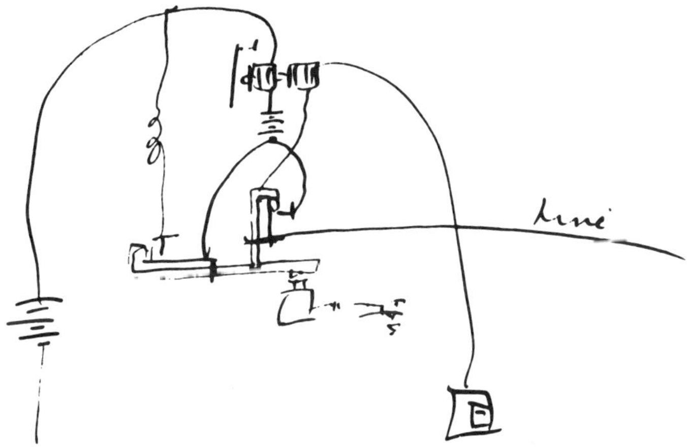

Duplex[7]

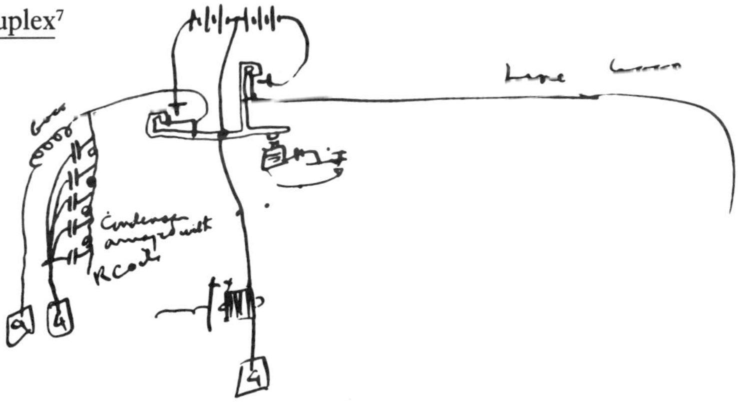

Duplex[8]

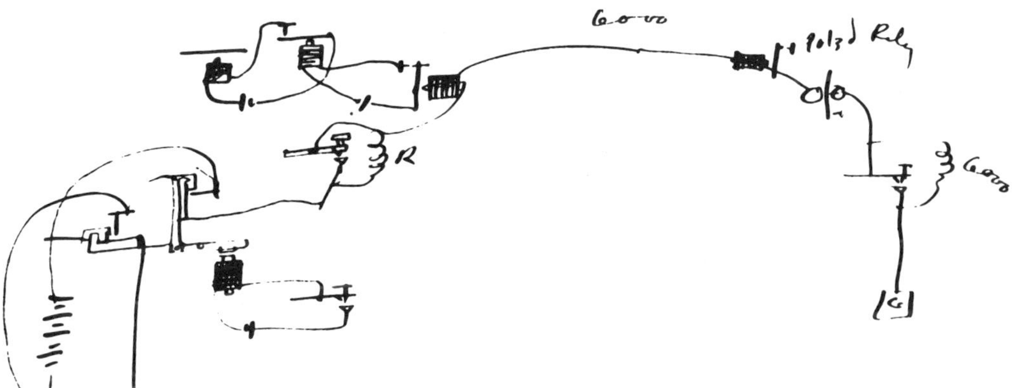

X, NjWOE, Lab., Cat. 1176:60 (*TAEM* 6:59). [a]Each drawing bears this date in some form.

1. This and the other designs in the entry are alternative approaches to the basic problem in duplex telegraphy—the interference of outgoing signals with the reception of incoming signals. These sketches were preliminary to the preparation of Caveat 53 one month later, where further explanation of the operation of such circuits can be found (see headnote, p. 347). As is usual with Edison's duplex work, related but undated documents also exist. Edison intended this design—in which the two batteries neutralize each other until the key is pressed—to be used "upon lines where there is but little static charge." See the discussion of figure 4 in Edison Caveat 53.

2. Though not labeled in the drawing, this design probably has the left-hand relay set to work only with the stronger currents produced when the keys at both ends are closed, while the right-hand relay is operated by weaker signals. Cf. Docs. 285 (duplex no. 8), 387 (device 11), and 485 (last drawing); and U.S. Pat. 147,917.

3. Roughly comparable designs are Doc. 392 (fig. 16) and the loose sketch of 5 February 1874, NS-74-001, Lab. (*TAEM* 7:25); see also Cat. 1175:160, Lab. (*TAEM* 6:185).

4. This design seeks "to economize battery" by using a small battery and low resistance in the artificial line (the loop on the left). See figure 2, Edison Caveat 53.

5. In this basic design, both batteries are short-circuited until the key is pressed. At that point the currents cancel each other in the receiving relay (center). The artificial line, with resistance equal to the main line, is on the left. See figure 1, Edison Caveat 53.

6. Cf. figure 2 in Edison Caveat 54; figure 4 in Caveat 53 also bears some resemblance to this design.

7. The "condensers arranged with R coils" allow this circuit to compensate on lines of high electrostatic capacity. See figure 3 in Edison Caveat 53.

8. This arrangement for sending two messages in one direction and one in the other is a cross between Edison's designs in Docs. 315 and 348.

–499–

Technical Note: New Force

[Newark,] Nov. 2nd 1874[a]

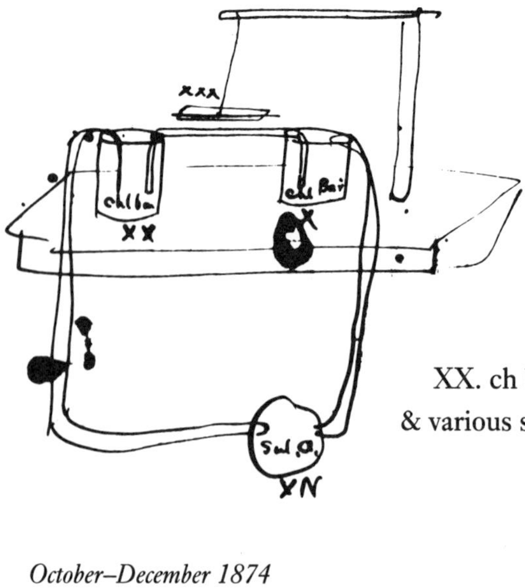

XX. ch bar XX also H_2O
& various salts XXX ch Bar

also H_2O in XX & X. Cl Ba in XX & SO_3 in XN. Chemical affinity ought to draw XXX from right angle to paralell.

⟨Tried this   No good   E & B⟩[b]

Also

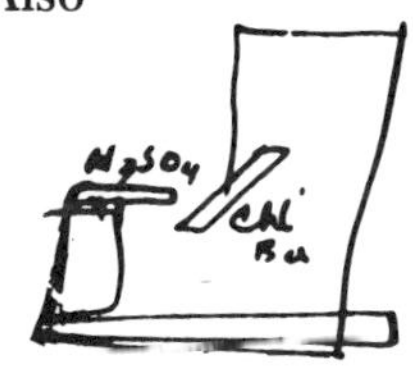

drop the [-]SO_3 on paper see if Chl Ba is attracted.

Also[c] try this way.

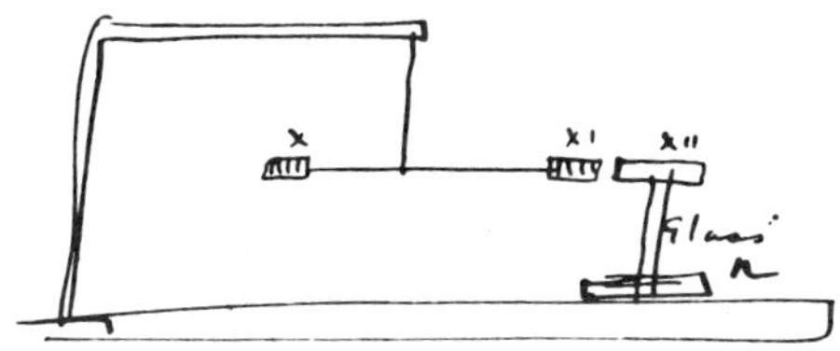

X~~X'~~ X' are peices of paper moistened with the solution and secured to peices of glass rod, also to metallic rod ~~&~~ N is a standard with glass rod, on top is X″ a peice of paper moistened with another solution.

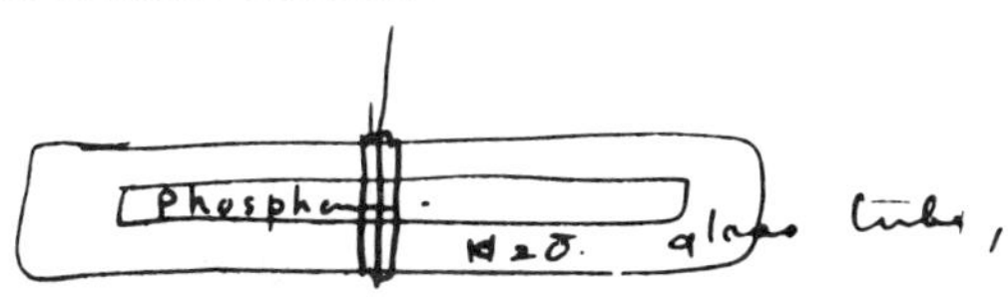

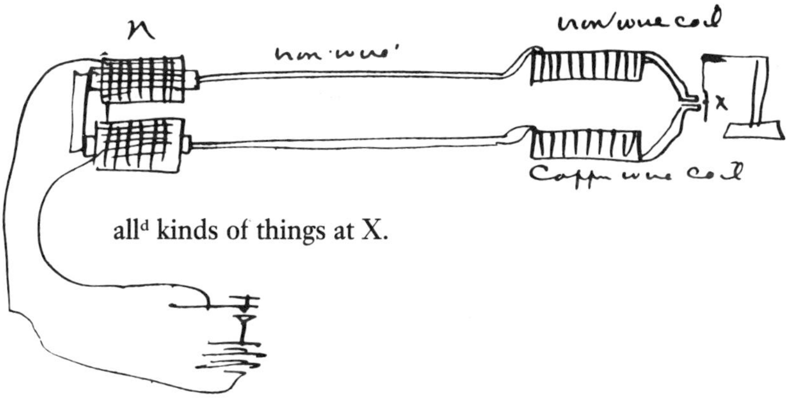

all[d] kinds of things at X.

N Batches Copper disk quantity magnet here with quantity magnet—

dispense with copper coil & iron coil used large iron wire 50 feet long both ends also do. copper & iron also wet string & iron wire, string moistened with ~~as~~various chem sols; also copper wire & wet string also close these various conductors of a odic m[-]agnetism[1] together when X is attracted =

when current is on break contact of iron or copper from cores of magnet also[e] send reverse currents.

Try at .X tubes with various chems in—

Also connect the two helices of electromagnet ~~w~~opposite so they oppose & produce no magnetism and try all over again— Thus[2]

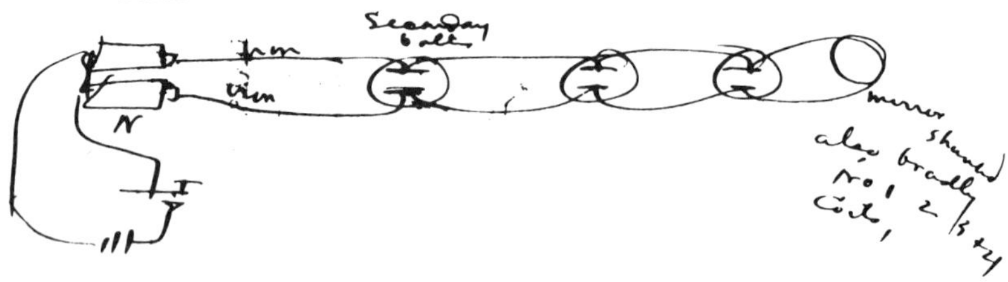

Cores of N connected to opposite & in regular way also used 1 iron 1 copper also 2 copper wire, also wet string, organic conductor ~~a~~dry & wet with various chems.

strip of paper[3] wet in various subs that decompose in air & by batty see if pyrogallic a & Iod K decomp in magnetic field also connecting mag to oppose

shellac[f]
glass
Tubes with hydrocarbons
Tubes with metal inside with acid attracting them—
Salt crystalizing—[g]

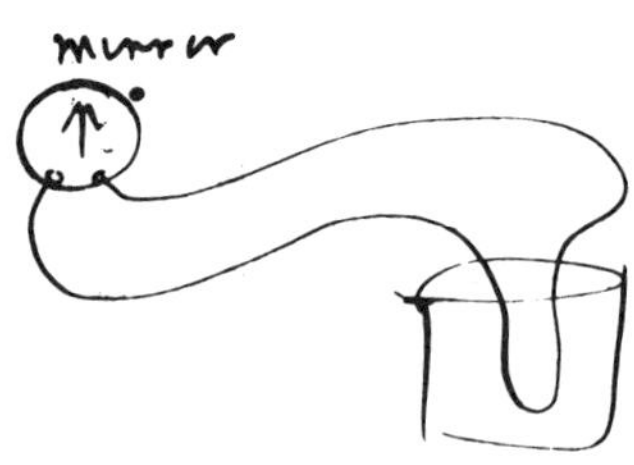

Various chemicals with loops of various wires, percipitates formed & crystalizing also acids to attack metals immersed, also thus

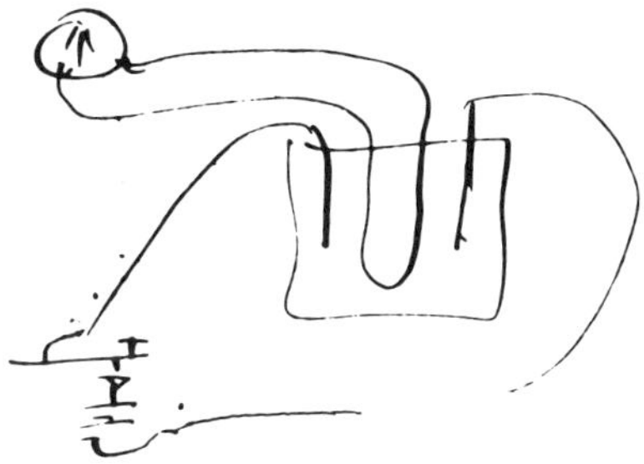

⟨Exp to find some new phenomena⟩[b]

X (fragment), NjWOE, Lab., NS-74-001 (*TAEM* 7:37). [a]Date written by Charles Batchelor. [b]Written by Batchelor next to drawing. [c]Page numbered "4" begins here. [d]Page numbered "5" begins with drawing; dated by Batchelor. [e]Page numbered "6" begins here. [f]Preceded by vertical line; page numbered "7" begins here, dated by Batchelor. [g]Followed by horizontal line across page.

1. "Odic force" (or just "od") was the name given by Karl Reichenbach to a force he thought he discovered in the 1840s. Reichenbach, a German/Austrian industrialist, chemist, and natural philosopher, was the first to isolate paraffin and creosote. He thought od was a "dynamic" magnetism and regarded Faraday's discovery of diamagnetism as closely related to his own researches, but he also saw od as a physical power that would explain various phenomena associated with spiritualism. He attempted to specify through many experiments the ways in which od operated and he set out detailed summaries of his evidence in order to prove that od existed and was a force distinct from ordinary magnetism, electricity, etc. These results were never accepted as valid by the general scientific community. Edison evidently knew of Reichenbach's work and was interested enough to undertake investigations related to aspects of it; more than a year later he cited Reichenbach by name as a partial precursor when announcing what he regarded as his own discovery of the new "etheric" force (Doc. 678). An undated clipping, apparently from the *English Mechanic*, reporting comments (by "Sigma" and "A. C. G.") on experiments by a Dr. Stone with Reichenbach's supposed discovery is in Edison's Cat. 1143, Scraps. (*TAEM* 27:237). Reichenbach's works (with English translations after 1850) had an international circulation, and the term "odic" became common in spiritualist circles and in the general culture. *DSB*, s.v. "Reichenbach, Karl (or Carl) Ludwig."

2. Figure labels left to right are "N," "iron," "Secondary batts," "mirror shunted also bradly No 1 2 3 & 4 coils."

3. This refers to horizontal line at right of drawing.

-500-

From George Barker

Philadelphia, Nov. 3d 1874[a]

My Dear Sir:—

It occurred to me while examining the remarkable little instrument of yours—the motorgraph—at the Franklin Institute Exhibition[1] this evening, that perhaps you would be willing to come on and show it to the highest scientific body in this country, the National Academy of Science,[2] of which Professor Henry,[3] is President, now in session in this city. Mr. Partrick was good enough to say he would send you a message from me and I wrote one.[4] I send this letter to explain matters more fully.

The first notice I saw of your curious discovery was in the London Journal of the Telegraph.[5] I at once became very much interested in it, as a scientific fact of very important bearing in many directions. And I have had it in mind ever since to write and ask you to prepare me an article for this Journal[6] describing it and if possible, illustrating it. This request I have now to make. If you really cannot do it, can you not place me in possession of the facts & material so that I can make up a paper on it.

I am very anxious too, that the Franklin Institute should recognize the merit of this discovery and properly reward it. As a member of the Board of Managers I shall be very glad to do what I can toward this end.[7] So that if you can make it convenient to come on either tomorrow or Thursday, there are several matters we can profitably talk over.

I am the Professor of Physics in the University of Pennsylvania (Locust & 34th Sts.) and shall be glad to see you there when you come; or at my house, 408 S. 41st St.

Trusting that you may be able to run down here for a day and see us, I remain Truly yours

George F. Barker.[8]

ALS, NjWOE, Lab., Cat. 297:132 (*TAEM* 5:920). Letterhead of the *Journal of the Franklin Institute.* [a]"Philadelphia," and "187" preprinted.

1. The Franklin Institute, originally founded as a mechanics' institute in 1824, was now a nationally prominent center for science and technology (Sinclair 1974). The exhibition was the Twenty-seventh Exhibition of American Manufactures; the official report of the exhibition makes no mention of Edison (Franklin Institute 1874). However, a comment in the *Operator* of 15 November 1874 ("Dots and Dashes," p. 4) read: "The display of telegraphic and philosophical apparatus at the Fair of the Franklin Institute in Philadelphia is very complete. Mr. Edison, by request of the managers, has sent on his electromotograph for exhibition."

2. The federal government created the NAS during the Civil War as a private advisory board for military technology. After the war it became largely an honorary society. Dupree 1957, 141–48; Bruce 1987, 301–5, 315–17.

3. Joseph Henry, director of the Smithsonian Institution, was one of

the elder statesmen of American science. His major research concerned electromagnetism. *DSB,* s.v. "Henry, Joseph."

4. Barker to Edison, 4 Nov. 1874, Cat. 297:310, Lab. (*TAEM* 5:920).

5. The *Telegraphic Journal and Electrical Review* (2 [1874]: 321–22) reprinted Edison's letter to the *Scientific American* (Doc. 476), including some of the introductory editorial material.

6. The *Journal of the Franklin Institute,* which Barker edited. No such article appeared.

7. Apparently no action was taken by the Institute in connection with the electromotograph.

8. George Frederick Barker (1835–1910) had become professor of physics at the university in 1873. His interest in the motograph marked the beginning of a long association with Edison. *DAB,* s.v. "Barker, George Frederick."

–501–

From Louis Drescher

New York, Nov 4 1874[a]

Mr Edison

I am about to contract a machine[1] to b[e?][b] made in large numbers and [be?]sides[b] this another ammount [o?]f[b] work—continuously & permanent—at an average of $1000.– pr month.

I have heard of your ability etc. and if convinient, please call here at your earliest convinience if you should think it to mutual benefit Respectfully

Louis Drescher[2]

ALS, NjWOE, Scraps., Cat. 1173:162 (*TAEM* 27:485). Letterhead of the Drescher Electric Manufacturing Co. [a]"New York," and "187" preprinted. [b]Document damaged.

1. A magnetoelectric generator, possibly one designed by Drescher (U.S. Pat. 168,560). Order book entries indicate that Drescher ordered one magneto on 12 November and fifty on 24 November 1874; the same book contains an account entry for this work. (Cat. 1214:50, 200–201, Accts. [*TAEM* 21:589, 641]). Drescher also asked Edison to manufacture electrometers. Drescher to TAE, 11 Nov., 17 Nov., and 14 Dec. 1874, Cat. 1173:167, 170, and following p. 199, Scraps. (*TAEM* 27:493, 503, 534).

2. Dr. Louis Drescher, electrical manufacturer and inventor, was vice-president of the Drescher Electrical Manufacturing Co., 154 Fourth Ave., New York (H. Wilson 1876, 352). He manufactured electromedical apparatus, including items similar to Edison's inductorium ("Dr. Drescher's Illustrated Catalogue and Price List of Electro-Therapeutic Apparatus," Cat. 1143, Scraps. [*TAEM* 27:236], cover only on microfilm).

–502–

Edward Johnson to George Barker

Hall of Exhibition[1] [Philadelphia,] 10[11]/5/74[2]

Dear Sir

I am informed that you have invited Mr Thomas A Edison to appear before your convention with his Electromotograph I am acting for Mr Edison & on his behalf thank you cordially for the Invitation & if I may have 2 minutes audience with you, will arrange to have Edison meet you at your option as to time & place

A reply left at the Telegraph stand will quickly reach me With respect I am yours

E. H. Johnson Genl Manager Automatic Telegh. Co.

ALS, Barker.

1. See Doc. 500 n. 1.

2. Johnson apparently misdated this, as it was a response to the invitation from Barker to Edison (Doc. 500).

–503–

To the Editor of the Scientific American

Newark, N.J. November 7, 1874[a]

Cable Telegraphy.

To the Editor of the Scientific American:

In your last issue you print a paper read before the British Association by W. K. Winter on an improvement in cable telegraphy.[1]

Allow me to state that the principle shown was invented by myself and patented both in England and this country some three years ago.[2] It is used by the Automatic Telegraph Company of New York.

THOMAS A. EDISON.

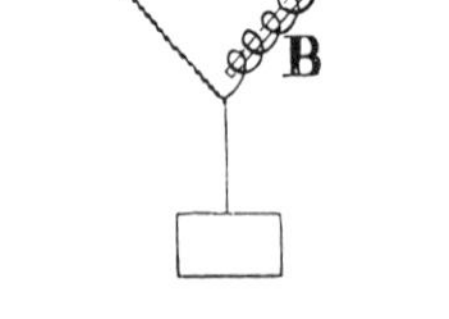

In G. K. Winter's cable telegraph circuit, coils **A** *and* **B** *were actually the primary and secondary of an induction coil.*

PL, *Sci. Am.* (n.s.) 31 (1874): 292. [a]Place not that of publication.

1. Winter's paper was reprinted in the 24 October issue of *Scientific American* as "Improvement in Cable Telegraphy" ([n.s.] 31 [1874]: 258).

2. It is unclear what Edison's reference could mean. The principle at issue was the use of a modified Wheatstone bridge in which the coil in one branch was wrapped around that in another branch, thus developing inductive effects. No Edison patent application is known that embodies this idea, though various related circuits are scattered through Edison's loose notes and sketches and his notebooks. Circuits using induction coils in automatic telegraph receiving arrangements are found in Edison's U.S. Patent 150,848, figs. 3 and 4 (these also appear in his British Patent 735 [1873] as figs. 5 and 6). Doc. 375 shows automatic receivers in Wheatstone bridges, and the automatic receivers in practice were arranged to "produce a current balance as in the bridge of a Wheatstone balance" (Davis and Rae 1876, 38). However, these hardly employ a principle similar to that described by Winter.

–504–

To Samuel Edison

Newark NJ Nov 8 74

Dear Father,

Yours received,[1] send on your deeds and I will sign them =[2] We have moved from Wright street down to 65 Bank street ~~n~~only two blocks from the corner of Broad and Market.[3] Everybody well, my little one is very healthy & can talk.[4] Wish you would come here this winter and keep the Laboratory. Am paying a man $12 per week to do it now only to clean apparatus etc. You could live with me or anywhere you pleased, will give you $20 per week— Would like ~~l~~Little Charlie to come also.[5] Everything is approaching a climax and I expect to get my money from Automatic & Western Union within a ~~m~~six week.[6] Write. Your Son

T. A. E

I have made a new discovery in electricity[7] which I have been exhibiting to the American Academy of Sciences[8] in session at the University of pennsylvania at Philadelphia. They all agreed that it was original and important discovery[9] E

ALS, MiDbEI, EP&RI.

1. Not found.

2. One of the deeds in question is probably for a lot in Fort Gratiot sold by Samuel to Thomas for $400 on 16 November. Register of Deeds, Liber 50:501, MiPHStCo.

3. Edison sold the Wright St. house to Mary Chandler, but the deed was not signed until 1 December 1874; it was certified on 18 February and recorded on 4 March 1875 (Deed Records, NjNECo, D18:175–77). Edison nominally sold it for $4,000, which was $1,500 less than the price of the house when he purchased it three years earlier. In fact, he received no actual payment since the entire $4,000 was accounted for by Mrs. Chandler's relieving Edison of his outstanding mortgage and loan obligations (Cat. 1185:8–9, Accts. [*TAEM* 22:566]; see Doc. 206). One of Edison's machinists, Frederick Ott, remembered the 65 Bank St. residence as a little flat over a drugstore (clipping of 17 Oct. 1931, *Newark* [*Evening News?*], "Ott, Frederick," Pioneers Bio.); the 1875 *Newark City Directory* lists the Newark City Pharmacy at 67 Bank St.

4. Edison's daughter Marion was nearly twenty-one months old.

5. Samuel did agree to come and take care of the laboratory. Both he and Charley (Edison's nephew) were apparently at the shop by the end of January. PN-75-00-00.2, Accts. (*TAEM* 22:342–43).

6. This refers to the expected closing of negotiations regarding a new company to run the automatic (see Doc. 522 n. 10) and to expected payments from Western Union for the quadruplex (see Docs. 517 and 519).

7. That is, his electromotograph.

8. That is, the National Academy of Science.

9. See Docs. 500 and 502. A report in the 15 November 1874 issue of the *Operator* ("Dots and Dashes," p. 4) claimed that those at the NAS meeting had concluded that the electromotograph principle was "an en-

tirely original discovery in physics, and capable of numerous and valuable applications."

-505-

Article in the Operator[1]

New York, November 15, 1874[a]

DUPLEX—No. III.[b]

Having partially set forth the principle of the modern duplex, we will go back a little and describe earlier forms, and endeavor to explain why they were practically inoperative.*

The first duplex, if it deserves to be so called, was that of Dr. Gintl, the connections at one station being shown in fig. 4.

Fig. 4.

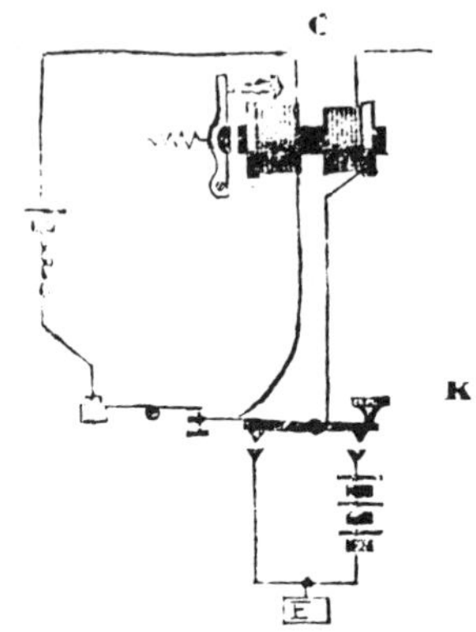

C is the differential relay, provided with two coils. The spool on the right[2] is of large wire, and is placed in a local circuit.

The coil on the right is placed in the main line, and is of fine wire. K is a double-pointed key; its lever, when resting on the back point, connects the line to earth, and when on the front point, connects the battery.

Upon the extreme end of the key lever is a spring, which, in connection with the point above it, serves to complete the local circuit, containing a compensating battery and resistance coil for adjusting the strength of the current passing through the coil on the left, so as to obtain a perfect balance between it and the line current passing through the other coil. It is obvious that this device is inoperative for several reasons, one of which is that the continuity of the circuit is interrupted every time the lever of the key passes from the back to the front point.

Were one of the levers permanently depressed while the one at the distant station was worked, the effect of the slight interruption of the current would scarcely be perceptible upon the relay, provided that the tension of the adjusting spring was not too great, and that very little movement was given to the key lever; but when both keys are operated at the same time, these interruptions are doubled and the signals are mutilated.

If to this defect be added the greater one of static induction, signaling becomes impossible. Even if there were no break in the continuity of the circuit by the key levers, or any static induction, a perfect equation could not be obtained with the large and small wire, owing to the difference in their capability of settling up secondary induction currents, as well as the difference in strength caused by passing through circuits having a great difference in resistance.

Gintl himself found that this device was inoperative and adopted a chemical recording instrument in place of the relay from which he says better results were obtained; but we surmise very imperfect signals, if any, were ever received over a line.

The next method was that devised both by Frischen and Siemens-Holske,[3] of Berlin, independently of each other, and is a great improvement over that devised by Gintl, if a device which is operative can be called an improvement over one which is inoperative.

It can be worked upon short lines, and was in fact worked some years ago in Holland, but abandoned probably on account of the peculiar manner in which telegraphy is done in that country.

There are one or two defects which are insufficient to prevent it working on short lines, but preventing signaling altogether upon longer ones.

This method is shown in fig. 5.

Fig. 5.

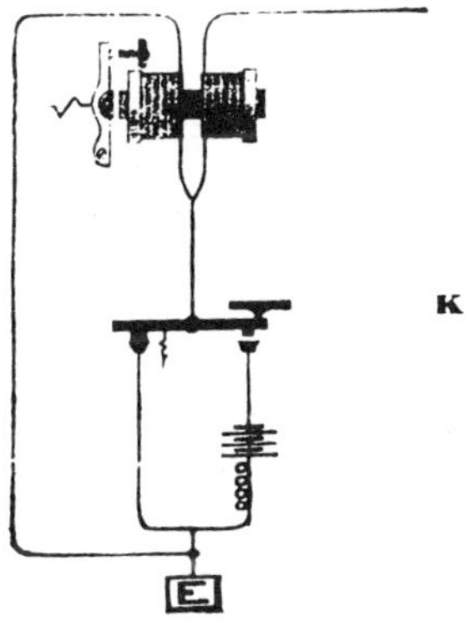

A is the receiving relay provided with two coils wound in opposite directions. K is the transmitting key provided with a front and back contact point; to the former is connected the main battery and to the latter is connected the earth wire.

R is resistance, by which the resistance of the circuit in which it is a part, can be made equal to that of the line; when this is done the battery may be connected and pass through both coils without producing any magnetism in the cores of the relay, providing the line is so short that the static charge is insufficient to effect the balance. The operation is as follows: Suppose the lever of K rests on the back point and connects the line to earth, a current from the distant station coming over the line, passes through the right hand coil to the key lever, thence through the back point to the earth and distant station. If the lever of K be connected to the front point, the only change that takes place in the direction of the distant current is, that it now passes through the main battery to the earth instead of the simple earth wire. So the current from the distant station will operate the relay whether the key lever is in contact with the back or front points, but it will be noticed that in operating the key K, there is a short interval of time when the lever does not touch either point, this would apparently break or greatly weaken the effect of the distant current upon the receiving relay as it must in that position pass through a resistance equal to the line itself. But although the resistance is at once doubled the amount of magnetism produced by the distant current in the cores of the relay is precisely the same, because the current passes through both coils not in an opposite direction like that from the main battery but in a direction which increases the magnetism. This somewhat parodoxical result might perhaps be better explained by stating that if a relay of 500 ohms be inserted in a line of 5,000 ohms resistance, its cores will sustain say one pound—if the resistance of the circuit be increased to 10,000 ohms and that of the relay to 1,000 ohms, its cores will lift practically the same; so we see that although the resistance of the circuit in which the relays are placed is at times doubled in the act of signaling it does not effect the adjustment of the relay.

When both keys are operated the writing cannot be said to be perfectly even, the defect being due to the change in the polarity of the iron cores of the relays.

Suppose that the key K be depressed and the battery connected to the line, if the resistance of the artificial line be equal to the main line, the relay will remain unaffected. If now the key at the distant station be held in such a position that it touches neither the front or back points, the resistance of the line will be doubled, while that of the artificial remaining the same the balance will be destroyed and the relay will close. This would be just what is wanted, as the key is never depressed without it is desired that the relay should close, but it must be remembered that the polarity of the cores of a relay cannot be changed without the armatures flying away from the point for an instant. When the key at the distant station is depressed the moment it leaves the back point the balance is destroyed, as above, and the relay armature is attracted by the excess of current in the artificial line. Now the moment the distant key lever touches the front point the balance is regained and the battery is put to line, but instead of holding the distant relay closed as it should to give a perfect signal, its effect is to give the relay as strong a magnetism as that produced by the destruction of the balance but of the opposite character; hence there is a period in this change of polarity when there is no magnetism, and the relay armature opens for an instant and then closes.

However, the trouble from this defect is very small when compared with the lack of means for compensating or balancing the static change[4] of the line. Even were the condenser added to the resistance coil as is done by Mr. Stearns, the balance could not be obtained, because the resistance and consequently the capacity of the line is constantly varied by the working of the key at the distant station. It is true that when the lever of the key touches neither point, the static current of the line and that of the condenser would partially balance each other (providing the distant key was quiet), but a greater portion of the charge would pass to the distant station and tend to prolong the signal, *i.e.*, produce "sticky" writing.

Notwithstanding these defects, it must be considered the first duplex telegraph, Gentl's device being entirely unworthy of being used as a means of detracting from the honor due to the ingenious inventors of this device.

(To be continued.)[5]

*We republish this portion of our last article, the diagram which should have accompanied it being left out by mistake.[6]

PD, *Operator,* 15 Nov. 1874, 1. [a]Place and date of publication. [b]Followed by centered rule.

1. See headnote, p. 286.
2. Should be "left."
3. Carl Frischen, Werner Siemens, and Johann Halske.
4. Should be "charge."
5. There is no continuation of the series on duplex telegraphy. The

15 December 1874 issue of the *Operator* appears to be the last issue to which Edison contributed "scientific" material on a regular basis. He was subsequently asked by the editor to contribute other material (Doc. 630).

6. See Doc. 495.

–506–

To the Editor of the Operator

NEWARK, N.J., NOV. 25, 1874.[a]

To the Editor of the Operator:

I submit the following problem to your readers for solution:

"Transmit alternately positive and negative currents within a closed circuit from a battery all the poles of which are connected in the ordinary manner using an ordinary Morse key, to which no extra point or appliance whatever is to be added. No device other than the battery key and connecting wires is to be used." I inclose two diagrams, both solving the problem in a different manner.[1]

THOS. A. EDISON.

PD, *Operator,* 1 Dec. 1874, 3. [a]Place and date not those of publication.

1. This same problem was published in the 15 January 1875 issue of the *Telegraphic Journal* (3:28). The drawings were not published in either journal. Edison's draft is in a laboratory scrapbook, but no drawings accompany it (Cat. 297:124, Lab. [*TAEM* 5:894]). Drawings of designs related to this problem appear in Doc. 485 and Cat. 297:64(3), Lab. (*TAEM* 5:665).

No answer appeared in the *Operator*, but one did appear in the *Telegraphic Journal* on 1 May 1875 (3:102–3). This was written by the British mathematical physicist Oliver Heaviside, who described the problem as one of a "highly interesting nature." After submitting his own solutions, Heaviside noted that "Mr. Edison's own solutions would also be very acceptable," suggesting that he had seen Edison's drawings.

–507–

Technical Note: Multiple Telegraphy

[Newark,] Nov 25 1874

Quad.[1]

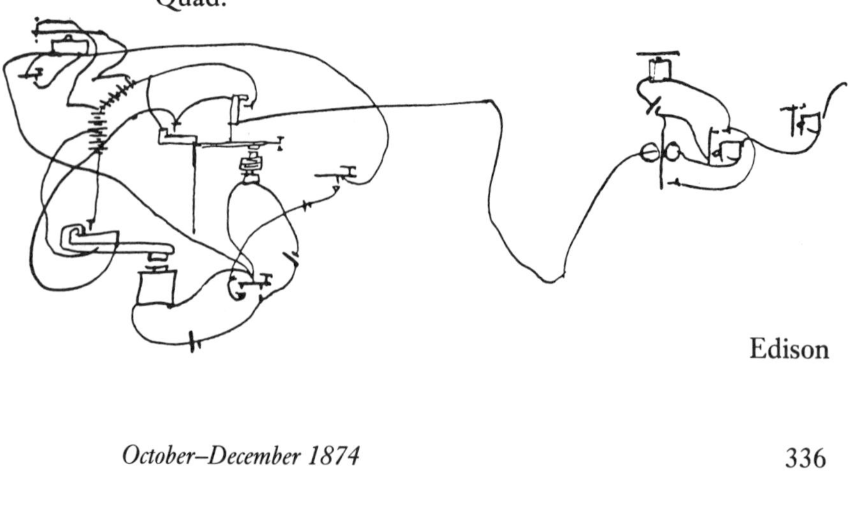

Edison

when heavy batty off reversals work plzd relay when heavy battery [on?] reversals prevented from being sent[2] & extra heavy battery used instead which wkg an extra magnet keeps local closed which plzd tongue goes over & wld open[3]

Quad principle[a]

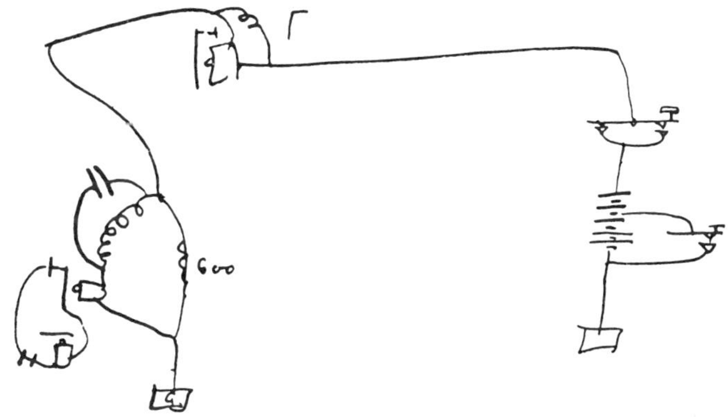

Quad principle[a]

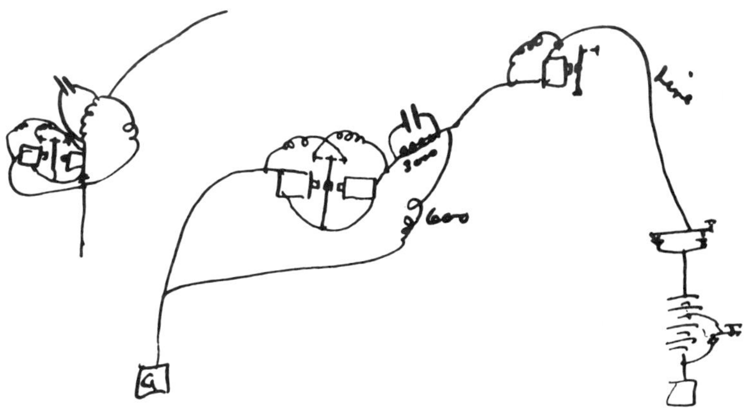

worked[4]

AXS, NjWOE, Lab., NS-74-001 (*TAEM* 7:41). Resemblances in style, paper, ink, and topic link these separate items. [a]This design is on a separate piece of paper.

1. Each of these sketches shows a quadruplex unlike Edison's basic design, which he was then modifying for tests on the Chicago line.

2. Probably should be "received."

3. The precise mode of operation of these alternative designs is not clear, and they may be incomplete in important ways. Aspects of their functions may be understood by comparing them with Docs. 533, 663, and 716, and with Edison Caveats 58 and 72. Cat. 297:128(1) (Lab. [*TAEM* 5:909]) shows a related design that probably predates these.

4. It is not known whether Edison tested this on an actual line or only in the shop. See also Doc. 477.

–508–

Section in William Ford's Industrial Interests of Newark, N.J.

[Newark, October–November 1874[1]]

TELEGRAPH INSTRUMENT MANUFACTORY.[a]

EDISON & MURRAY, 10 and 12 Ward-street. Within the last few years Newark has been the location of important developments in the science of telegraphy and the manufacture of electrical apparatus. This branch of industry was first attempted in Newark during 1869, in a small factory at 15 N.J. Railroad Avenue.[2] Soon after a removal was made to the commodious factory at the location first named above. The present firm is composed of Messrs. Thomas A. Edison and Joseph T. Murray. The senior member, Mr. Edison, is the well known telegraph engineer, and his several inventions form important contributions to the science of telegraphy.

In 1870 he invented his Gold and Stock reporting telegraph instrument, no[w] used so extensively in Wall-street for telegraphing from the Stock Exchange the sales of gold and stock to some 900 brokers' offices, etc., in New York.[3] The peculiarity of this instrument is that the despatches are received and recorded automatically. It is in use in other large cities of the Union, as well as in the London and Continental Stock Exchanges. Thirty-six hundred of these machines have been made within the last three years, many of which have been exported to Europe, where their use is constantly increasing. Messrs. Edison & Murray are the sole manufacturers of this instrument and to facilitate and cheapen its production they have introduced valuable labor saving machinery and ingeniously devised tools. In this way the parts of the machine are made "interchangeable," as in a Remington gun.[4]

The use of the magnetic telegraph, as a means of transmitting intelligence, increased so rapidly during recent years that, in the large cities of this country and also in Europe, great difficulty has been experienced in providing sufficient wires and operators for the telegraphic business of metropolitan cities. The cause of this difficulty has been that by the Morse system the transmission of a single message employed a single wire entirely. To remedy these defects and avoid the expense of erecting new wires, Mr. Thomas A. Edison has perfected inventions whose value as a means of facilitating the transmission of intelligence by electricity can scarcely be overestimated,

The minor invention which Mr. Edison has contributed to the science of telegraphy is known as the Quadruplex Telegraph, now in use by the Western Union Telegraph Company. By this device four messages are sent over one wire in various directions, at the same time, without interference with each other. It is readily seen that by this method the working capacity of every telegraph wire in the country can be instantly quadrupled.

The most important result, however, of Mr. Edison's genius and inventive skill, is known as the American Automatic Telegraph System. Many electricians have previously attempted to telegraph automatically. It remained for Mr. Edison to perfect a system, and by his method the capacity of a single wire for the transmission of messages may be almost indefinately increased, this showing its great value at a glance, as by its use the great expense for the erection of new wires is entirely obviated, not to speak of numerous other advantages. The salient features of the Automatic System are as follows: Preparatory to transmission, the message is prepared by special machines, with which round holes, representing the telegraphic characters, and hence, the letters and words of a message, are punched in a continuous strip of paper. The strip of paper containing the several groups of holes is then taken to the transmitting machine, which consists of a revolving drum upon which rests a metallic point. This point is connected to the line, while the drum is connected to the battery and thence with the earth.

If the end of the strip of paper be put in the machine it passes through it like a flash. The point referred to rests on the passing slip of paper, and the latter being a non-conductor the circuit is thereby broken; but, the moment a hole occurs, the point, falling through, comes in contact with the drum, thus forming a connection with the battery, when, the circuit being completed a wave of electricity passes over the wire and records itself at the receiving office as a dot on a continuous strip of paper moistened with a chemical solution, and the effect of the electricity in passing through it is to make or form "ink." As there are large and small holes in the perforated paper, long and short marks are made upon the moistened paper thus forming the dots and dashes of the Morse alphabet. As the speed with which signals can be transmitted is practically unlimited it follows that by employing many perforating operators, 20 to 30 times as much intelligence can be transmitted as on a Morse wire or as much as thirty wires by the Morse system can accomplish. The Automatic telegraph is now successfully working between New York and Washington and other lines are soon to be built in this country, and it is destined to become a powerful competitor against the Western Union Company. Four sets of the Automatic system are in use by the British Post Office between London and Dublin and upon the submarine telegraph from Falmouth, England, to Vigo, Spain.[5]

The Automatic apparatus is manufactured by Messrs. Edison & Murray. It is highly creditable to the city of Newark that such grand results should be achieved within her limits..

Messrs. Edison & Murray also manufacture mirror and marine galvanometers, inductive coils, secondary batteries and similar apparatus. This is the largest manufactory of its kind in the country. The production for 1873 amounted in value to $200,000 with a working force of 70 hands, the weekly pay roll being $750.[6]

PD, Ford 1874, 231. [a]Followed by centered rule.

1. Ford's book was published in 1874. As an account record shows that Joseph Murray purchased a copy on 27 November 1874, it is likely that the book was not brought out until October or November. Cat. 1184:102, Accts. (*TAEM* 21:834).

2. This was the location of the first shop of Edison and Unger. *TAEB* 1:156 n. 1.

3. Ford is referring to Edison's universal stock printer (Doc. 195).

4. See *TAEB* 1:347.

5. These were experiments only. See the related documents in this volume.

6. Edison and Murray was certainly one of the country's largest telegraph manufacturing firms at this time. Only Western Union's New York shop (run by George Phelps), the Western Electric Telegraph Co. in Chicago (formed by the consolidation of Gray and Barton and the Western Union Ottawa shops in 1872), and L. G. Tillotson & Co. rivaled it in size. Edison and Murray payroll records indicate that during 1873 the size of the labor force varied between 45 and 60, although day laborers might have brought the total to 70 at times. No equivalent records exist for Western Union or Western Electric, but what information does exist indicates that these shops also were quite extensive. Pope 1871; "Western Electric Manufacturing Company," *J. Teleg.* 5 (1872): 215; Smith 1985, 39–40; Western Electric 1876; Fischer and Preece 1877, 415–29; "The Machinery of the Telegraph," *J. Teleg.* 2 (Supp. 15 Mar. 1869): 1; "Proposed Enlargement of L. G. Tillotson & Co.'s Manufactory," *Telegr.* 8 (1872): 188.

–509–

William Orton to Joseph Stearns[1]

New York, 2 Dec 1874

I have received your letters[2] of November 10th and make haste to reply.

I have not forgotten the points you make touching our understanding on the subject of multiple transmission and I shall be ready to perform all the stipulations I have made whenever you are ready on your part.

It is true that we have been working successfully the invention of Messrs Prescott and Edison between New York and Boston for nearly two months, and yesterday we opened a Quadruplex Circuit between New York and Buffalo which worked a No. 8 wire even better than the Boston circuit.[3] This, however, was due to the greater perfection of the apparatus resulting from the experience acquired since the Boston circuit was opened.[4] Apparatus has been forwarded to Chicago, and we expect to try an experiment on a circuit between that city and New York by the end of this week.[5] The Chicago circuit was my crucial test with the Duplex, and when I found we could work it successfully, I then decided it could be worked anywhere. The difficulties, although numerous, and some them serious, have all been overcome and we can now work Duplex between all the principal cities of the country. When we can do the same with the Quadruplex, I shall feel that we have added greatly to the value of all telegraph property.

The circumstances under which the Quadruplex comes to us are these: About a year ago I became apprehensive that processes for working Duplex would be devised which would successfully evade your patents, and also that your attorneys had not done their work as thoroughly and efficiently in respect to shutting out competitors, as might have been done. I, therefore, sent for Edison, and after several conferences and much discussion with him, I employed him to invent as many processes as possible for doing all or any part of the work covered by your patents.[6] The object was to anticipate other inventors in new modes and also to patent as many combinations as possible. The object was not to glorify Edison, but to protect the Western Union in the use of what they had purchased from you. It became necessary for Edison to use Mr. Prescott and his facilities, and soon after they entered into an agreement to take patents for new inventions relating to multiple transmission in their joint names. It was while conducting experiments thus inaugurated that the process for sending two messages in the same direction at the same time was discovered and also that the Duplex could be applied to it. I have not settled definitely the question of compensation to Messrs Prescott and Edison for two reasons, first, we do not yet know what the Quadruplex is capable of doing on a long circuit, and second, I am not clear as to what they have invented. Patents have not been issued to them, because as I understand them they desire that none should be issued for the present, in order to conceal the details as long as possible.[7]

"Trouble on the 'Quad,'" published with the 15 December 1874 Supplement to the Operator.

I did not move too soon in respect to entrenching the Duplex as it is pretty evident that several entirely new patents will be issued which do not infringe yours in the least. We expect to control all of these except one and that is a patent issued to D'Infreville[8] which ought not to have been issued,[9] and which I am confident, is an interference with yours.

I have regretted your absence from the country during the last year very greatly. And I am confident that if you had returned six months ago, and abandoned your inventions to all the Governments of Continental Europe you would be better off pecuniarily a year or two hence.[10]

I should comprehend your motives better in holding back your multiplex system if the reasons you give were sound. Among them you speak of the difficulties in the way of introducing and working the Duplex. That you have had difficulties abroad I can readily understand. But there have been no difficulties here. Moreover, we have been for five years constantly threatened with annihilation by the Automatic system of Little and Craig and Edison and Harrington & Co. Within a fortnight after my return from Europe the market price of Western Union Stock fell off four per cent, because of an announcement that the Automatic Company had formed an alliance with the Erie, Pennsylvania and Baltimore and Ohio Railroads.[11] The first use I made of the Quadruplex was to outblow the Automatic people and my success was complete. Now, in utility it realizes all we promised concerning it, it renders Automatic unnecessary. They have not made alliances with anybody, but they fell out among themselves, and Craig has just brought suit against his associates, and filed a complaint which covers more than thirty printed pages.[12] You will see, thereforc, that whatever may be the facts in other countries, we could not afford to turn away from so valuable an auxiliary as was offered us in the Quadruplex.

The International Ocean Telegraph Company has no occasion to use the Duplex and is not likely to have any for a long time to come.[13] We have two perfect cables between Key West and Havana and expect to lay a second between Key West and Punta Rassa within a few months. The object of the latter is to diminish the risk of losing communications and not for the purpose of increasing facilities. There is not business enough now to keep one cable busy one quarter of the time. When we have two therefore, we have provided for all the increase of business we are likely to get for several years to come.

I shall be pleased to receive any proposition you have to submit concerning the use of your inventions for multiple transmission on the lines of the Western Union Company.

Wm. Orton

P.S. I was pained to hear some time ago that your health was giving way. I sincerely hope that you will be able to continue your work, but trust you will permit nothing to induce you to go beyond your strength.

LS (letterpress copy), NjWU, LBO 14:34–38. The original was unavailable for final proofreading prior to publication.

1. Joseph Stearns invented the first commercially successful American duplex, which Western Union owned. *TAEB* 1:101 n. 3.

2. Not found.

3. No. 8 was the typical gauge of wire on Western Union lines; the value of the quadruplex would have been less had it worked well only on extra-large, heavy wires.

4. See Doc. 515 n. 2.

5. See Doc. 515 n. 1.

6. See *TAEB* 1:513–15.

7. Patent examiner Zenas Wilber, who would evaluate the applications, later stated that he had not acted quickly because he was given the impression by Lemuel Serrell when Cases 94–100 were filed that they were only part of a set of related applications, as the relevant assignment

indicated. After Caveats 51–54 and another case (111) were filed during December, Wilber took up the applications. Statement of Examiner Wilber, Quad. 72.5, p. 1 (*TAEM* 9:64); see also Docs. 450 and 466, and headnote, p. 347.

8. Georges D'Infreville, an important French-born telegraph inventor and engineer, devised a duplex that the Atlantic and Pacific Telegraph Co. used to evade the Stearns patents. He later succeeded Edison as electrician for the Atlantic and Pacific Telegraph Co. and was frequently an expert witness in telegraph litigation. Israel 1989, 225; Reid 1886, 679.

9. This is either U.S. Patent 139,302 or 147,487.

10. In 1873 Stearns traveled to Europe in an attempt to interest European governments in his duplex. Butrica 1986, 112–13; George Prescott to William Orton, *J. Teleg.* 6 (1873): 169; "Adoption of Stearns's Duplex by the British Government," ibid., 329; editorial note, ibid., 392.

11. See Doc. 522 n. 10.

12. Complaint of Craig and Brown, Quad. 70.8, pp. 35–71 (*TAEM* 9:781–99).

13. Western Union acquired International Ocean in the spring of 1873. Reid 1879, 440.

–510–

Agreement with Robert Clinch and Thomas Robinson

[New York?,] December 3, 1874[a]

Memorandum of Agreement made and entered into this third day of December, A.D., one thousand eight hundred and seventy four, by and between Thomas A. Edison, inventor and proprietor of certain instruments and machinery, for telegraph purposes, of Newark, N.J., U.S., and Robert T. Clinch[1] and Thomas M. Robinson,[2] of St. John, N.B., Dominion of Canada.[b]

The[c] said Edison agrees to furnish, without cost to the said Clinch and Robinson, the necessary instruments of each kind and description proposed to be patented, to be used as models, for the purpose of procuring letters patent from the Dominion Government.[b]

The[c] said Edison agrees to furnish, without cost to the said Clinch and Robinson, instruments to be put in circuit for exhibition for the purpose of facilitating the organization of a Stock Company, and securing subscriptions thereto, namely, one recorder, and two transmitting boxes, with relay bell, and appliances necessary to put them in operation.[3b]

The[c] said Clinch and Robinson agree to procure and pay for patents in the Dominion of Canada, and to organize a Joint Stock Company in the City of St. John, New Brunswick, for the purpose of bringing the said inventions into public use.[4b]

The[c] said Clinch and Robinson agree to organize and establish a like Company, subject to the same conditions as govern the Company at St. John, New Brunswick, and to the terms of this Agreement, in such other cities and towns in the Dominion of Canada as it may be found practicable.

It[c] is mutually agreed that thirty per cent. of the capital stock of each and every Company which may be organized and established under this Agreement, shall be issued to the said Edison, inventor, without liability to assessment.[b]

The[c] said Edison further agrees that he will furnish instruments for the use of the said Clinch and Robinson at a price not to exceed one dollar and sixty five cents, United States currency, for each transmitting box, and sixty five dollars, United States currency, for each recorder, with such reduction as may be found practicable in view of actual cost, and that the said Clinch and Robinson shall be entitled to the full use and benefit of any and all improvements in instruments and machinery applicable to what is known as "District," "Fire Alarm" and "Signal" telegraphy, together with the full use and benefit of his key-board attachment, for transmission in connection with chemical recorder,—and improvements theron,—which may be made by the said Edison during two years following the date hereof.[b]

In[c] witness whereof the parties hereto have affixed their hands and seals the day and year first above written.

Thomas. A. Edison Thos M Robinson

R T Clinch by his attorney T M Robinson

In presence of Albert B Chandler[5]

DS, NjWOE, Miller (*TAEM* 28:975). [a]Date taken from text, form altered. [b]Followed by line to fill in space. [c]Preceded by line to fill in space.

1. Robert Clinch was a director of the New Brunswick Electric Telegraph Co., which leased its lines to Western Union, and manager of all Western Union lines in New Brunswick and Nova Scotia. Reid 1879, 342–43.

2. Unidentified.

3. An order book entry from the same date indicates that instruments and supplies were sent to Robinson and Clinch at St. John's, New Brunswick, Canada. Further instruments and supplies were sent in January and March 1875. Cat. 1214:51; Cat. 1220:88, 93; Cat. 1223:6, 19; all Accts. (*TAEM* 21:589; 22:42, 44, 134, 140).

4. An account book entry indicates that Clinch and Robinson were charged $38 for a Canadian Patent Office model on 24 December 1874 (Cat. 1184:113, Accts. [*TAEM* 21:839]). The model, ordered on 10 December, appears to be the same as the American model found at MiDbEI(H) (see Doc. 545; Cat. 1214:52 [*TAEM* 21:590]).

5. Albert B. Chandler (b. 1840) began his career in telegraphy in the mid-1850s and met Thomas Eckert in 1858. Entering the military tel-

egraph service in 1863, he soon moved to management. In 1866 he joined Western Union as Eckert's assistant, and he followed Eckert to Atlantic and Pacific in 1875. He briefly headed an arc lighting firm (Fuller Electrical Co.) in the early 1880s before becoming president of the Postal Telegraph Co., Western Union's chief competitor. Taltavall 1893, 161–62; *NCAB* 3:171.

-511-

Agreement with Thomas Eckert, Albert Chandler, Robert Clinch, and Thomas Robinson

[New York?,] December 3, 1874[a]

In consideration of services rendered, and to be rendered by Thomas T. Eckert and Albert B Chandler of New York City, New York, in furtherance of the purposes indicated in the Agreement this day entered into, between Thomas A. Edison, Robert T. Clinch and Thomas M. Robinson[1] it is hereby agreed that ten per cent of the capital stock belonging to the said Edison, and ten per cent of the capital stock belonging to the said Clinch and Robinson, twenty per cent in all, shall be issued, without liability to assessment, to the said Eckert and Chandler, and the said Eckert and Chandler, on their part, agree, to act, without compensation, as Agents in New York City, for the transaction of any business pertaining to the aforesaid agreement, and to do whatever lies, or may lie in their power, to further and promote the interests of the several parties concerned therein.—

In witness whereof, the parties hereto have set their hands and seals, this third day of December, A.D. one thousand eight hundred and seventy four.

Thomas .A. Edison R T Clinch by his Atty T M Robinson
Thos M Robinson Thos T. Eckert
Albert B Chandler

In presence of Wm H. Eckert[2]

DS, NjWOE, Miller (*TAEM*28:978). [a]Date taken from text, form altered.

1. Doc. 510.

2. An order book entry of the same date lists two private-line chemical transmitters and recorders to be shipped to someone named Eckert in Cincinnati; the Cincinnati city directory lists a William H. Eckert as living at 137 Broadway. Cat. 1214:51, Accts. (*TAEM* 21:589); *Williams' Cincinnati Directory* 1874, 293.

QUADRUPLEX TELEGRAPH CAVEATS
Docs. 512–513 and 531–534

After working on the quadruplex through the late summer and fall of 1874, modifying and supplementing his basic design and devising alternative methods, Edison finished preparing ten caveats.[1] He executed four (numbered 51–54) on 4 December 1874 and six (55–60) on 18 January 1875.[2] Six of these caveats are included in this volume.[3] They include explicit discussion of difficulties Edison encountered in tests of quadruplex circuits, in regular operation of the New York–Boston quadruplex arrangement, and in his efforts to find an arrangement that would work well on the much longer line to Chicago (which had a repeater at Buffalo). Edison applied some specific features of these designs in commercial operation; others apparently were never used. Only one of these caveats led to a successful U.S. patent application.[4] That one, number 56, described the repeater used on the New York–Chicago quadruplex line in December.[5]

1. Edison's procedure in preparing caveats (and patent applications) was not uniform. Sometimes he wrote out complete drafts, but other descriptions supplied to his patent attorney were less comprehensive, resulting in confused or inaccurate texts and drawings (Serrell's testimony, Quad. 71.1, p. 73 [*TAEM* 10:41]; see, for example, Docs. 348 and 467 n. 3). During this period Edison also executed two related patent applications, Cases 111 (18 January 1875) and 112 (14 December 1874) (see Docs. 449 and 450).

2. Edison filed the first group on 9 December 1874, and the others on 27 and 29 January 1875.

3. Not printed here are Caveats 53, 54, 56, and 58. They dealt respectively with duplex transmission, compensation for static discharge, a quadruplex repeater, and an alternative diplex system in which one signal consists of changes of current strength and the other simply of complete breaks in the circuit.

4. U.S. Pat. 209,241. However, some points were included years later in his U.S. Patents 452,913 and 512,872. Edison also included much of this caveated material in some foreign patents; for example, his British Patent 384 (1875) and his Canadian patent application of 12 April 1875 (DF [*TAEM* 13:376]). Features of these designs also appear in patents obtained by Gerritt Smith, the Western Union electrician assisting Edison on the tests. See Doc. 512 nn. 3 and 5, Doc. 513 nn. 2 and 4, and Doc. 717.

5. A quadruplex repeater consisted of the terminal equipment for two quadruplex lines, wired so the receivers on one side operated the transmitters of the other. Edison's repeater diagram therefore reveals the circuitry used at New York and Chicago; it is faulty at some points and should be compared with the formal drawing for the patent. Since a quadruplex (or duplex) circuit was designed to work in both directions

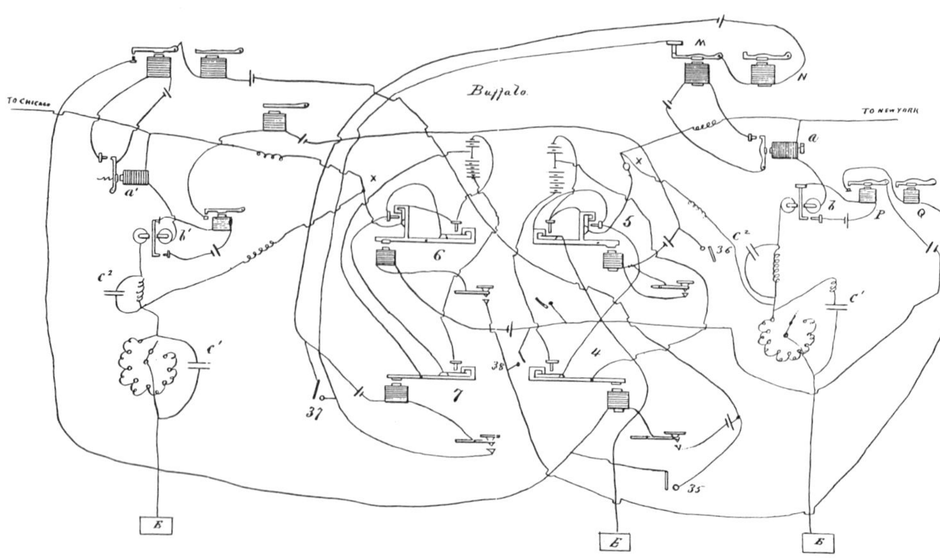

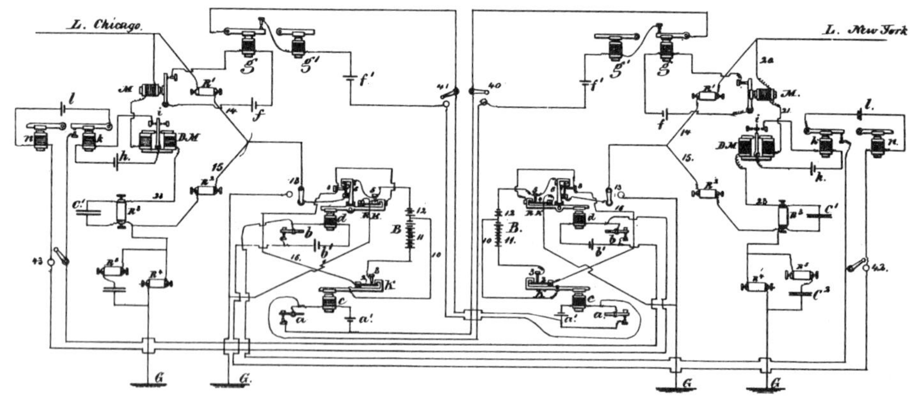

Drawings of Edison's quadruplex repeater from his Caveat 56 and U.S. Patent 209,241.

at once, this repeater did not need the switches necessary to reverse the signal direction on ordinary, single Morse lines. Edison went so far as to describe the circuit as one "which although it appears complicated is very simple," and George Prescott called it "very simple in principle," described it in two lines, and deemed it unworthy of an explanatory drawing. Brit. Pat. 384 (1875), p. 19; Prescott 1877, 862.

See also Doc. 515 n. 4. Note that the execution of caveats, like patent applications, often significantly postdated the conception of a design; Caveat 56 was signed on 18 January 1875.

Caveat: Multiple Telegraphy[1]

New York, December 4, 1874[a]

To all whom it may concern.

Be it known that I, Thomas A. Edison of Newark in the county of Essex and State of New Jersey have invented an Improvement in Quadruplex Telegraphs of which the following is a specification

The object of this invention is to transmit four messages over one wire at the same time without interference with each other.

The invention consists in the combination of two polarized relays within the main circuit one of which will respond to reversed currents and the other to increased and decreased strength of such currents.[2]

Fig. 1.

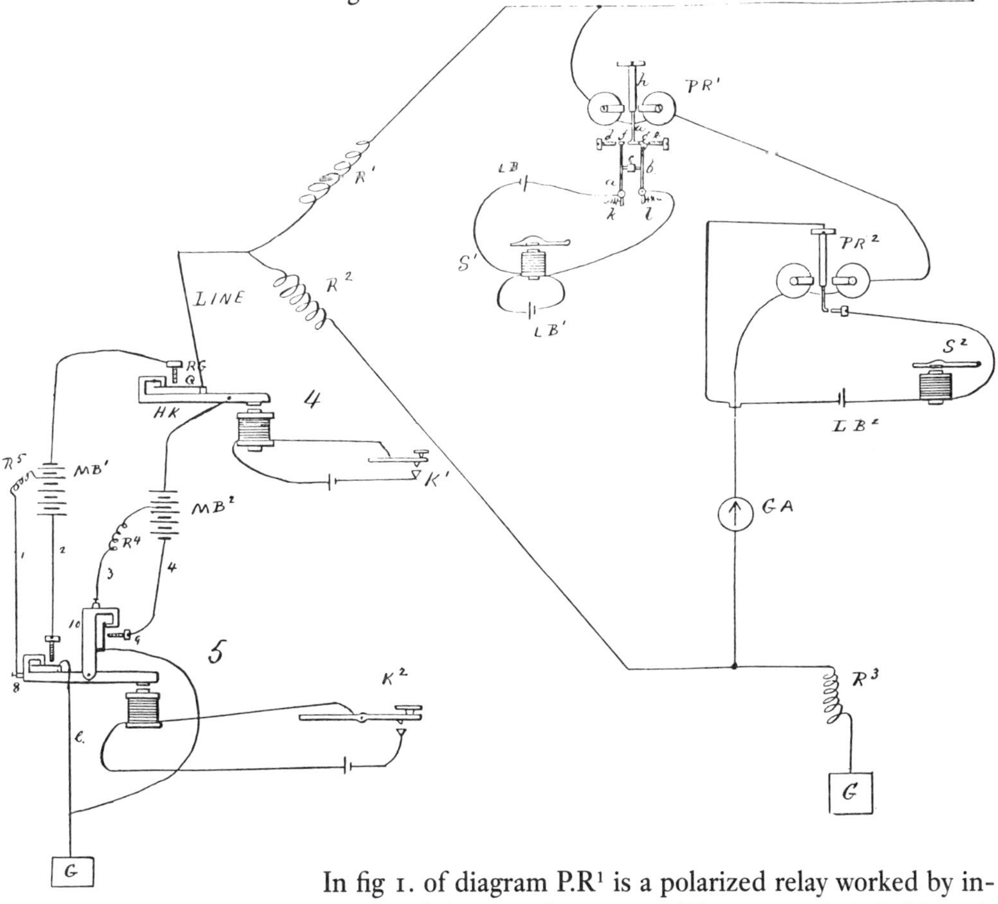

In fig 1. of diagram P.R[1] is a polarized relay worked by increased and decreased currents.[3] The tongue h, is held equidistant from the points d, e, by the two arms a, b, which arms are held against c, by their spiral springs k. l. These arms have platina points g. f, and the tongue h, when the weaker battery which works the other polarized relay P.R² is sent, remains centered by the arms a. b, because the power of the magnet is insufficient to overcome the spiral springs k. l. When thus centered the platina point of the tongue is in contact with both the platina points on the levers a. b, and the local circuit of the sounder is closed, but the sounder itself is not closed because it is placed in a bridge wire or a balance of batteries. The current from LB, passes through the sounder in one direction is neutralized by the current from L.B² passing through it in the opposite direction. Supposing the double lever sounder at the distant station is closed the main battery is increased several times in strength: this increased current coming over the line and passing through the polarized relay P.R is sufficient to overcome the springs k. l, and the tongue goes to the right or left according to the polarity of the current, say to the left; in doing so the local circuit is broken between the tongue and lever b, allowing the battery LB² to close the sounder: The tongue being to the left and the sounder closed and the operator at the distant station should commence signalling to the relay P.R² by reversing the battery, then the tongue h, would fly from one side to the other at every reversal but the local circuit will not be closed for just as the tongue h, comes in contact with the lever b, it leaves a, thus keeping the local open and the sounder closed, so it is obvious that although the effect of the reversals is to constantly work the tongue of P.R² it will produce no effect on the sounder which is essential for obtaining perfect signals. If the double pointed sounder at the distant station is opened the battery will be decreased still leaving enough for the transmission of all the signals from reversals received on P.R² but insufficient to hold the tongue h, to either side, hence it flies to the center by action of the adjusting spring K L and closes the local circuit and opens the sounder.

The polarized relay P.R² will work without change of adjustment in the weak or strong battery.

4. and 5, are the two transmitting sounders;[4] 4, is the sounder for sending reversals.

MB¹ and MB² are the two main batteries say of 100 cells each. MB¹ has zinc to line while MB² has copper to line.

The main line is in contact with the spring Q of 4: this spring when the local circuit working 4, is open is in contact with the lever HK, thence to the battery MB² and if the local which works 5, is open, it passes through 25 cells of MB² thence through a resistance coil R equal to the resistance of the remaining 75 cells, thence by wire 3, to the lever 10 to the spring and thence by wire 5 to earth: thus 25 cells copper to line will pass over the circuit to the distant station and throw the tongue of the polarized relay PR^2 to the left and open the local circuit. If now the key K^1 be closed the current from the zinc of MB^1 passes to line, thus from line to spring Q thence through point RG, to the battery MB^1 thence through 25 cells to the resistance R^5 via wire 1, to the arm 8, through the spring to earth putting 25 cells zinc to the line which passing to the distant station throws the lever of the polarized relay PR^2 to the right closing the local circuit and sounder; thus it will be seen by manipulating the key K^1 the sounder 4 sends positive and negative currents over the line from two separate batteries.

Now while these reversals are being sent if the key K^2 be closed then the springs connected to earth will come in contact with the points 7 and 9 separating them from 8 and 10 and throwing the extra 75 cells of each battery in circuit so that the sounder 4 transmitts reversals from 100 cells on each battery instead of twenty five. This increase of current passing over the line is sufficient to produce a magnetism in the cores of the polarized relay PR^1 that will overcome the springs k. l, and the sounder will close in the manner before described. The object of the resistances R^4 and R^5 is to keep the resistance of the main line always the same for when the seventy five cells having say 75 ohms resistance are out the resistance of the coils R^4 and R^5 equal it.

At each end of the line is a Wheatstone bridge: the two sides are formed by the coils R^1 and R^2 and the receiving instruments are placed in the bridge wire. If the resistance of R^1 and R^2 are equal and that of R^3 made equal to the line the reversals and increased and decreased currents from 4 and 5, may be sent to the distant station to actuate similar relays as PR^1 and PR^2, without at all effecting those shown in diagram. Thus I am enabled to send two messages in one direction and receive two from the opposite direction. I do not wish to confine myself to placing these two relays in the bridge wire of a Wheatstone paralleleogram for it may be done as in figure 2, both relays being made with differential coils and worked on what is known as the differential methods.[5]

Fig. 2

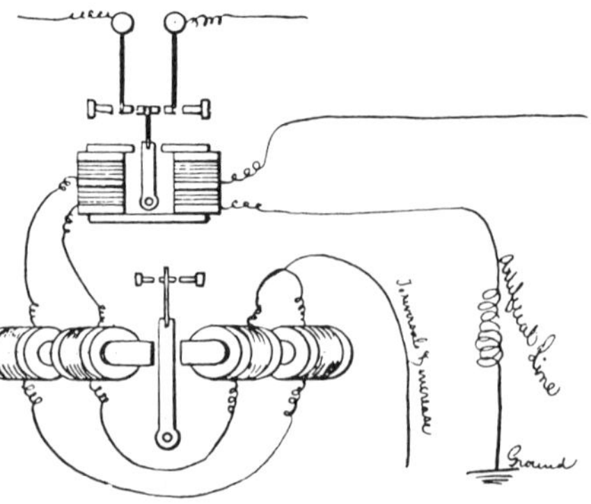

Inventor Thos. A. Edison per L. W. Serrell atty
Witnesses. Chas H. Smith Harold Serrell[b]

My claims will probably be.

First. The combination of two polarized relays in the same circuit one of which has its tongue centered as described or in an equidistant equivalent manner for the purposes set forth.

Second. A polarized relay having a centered tongue arranged with a sounder and local so that the sounder will not be perceptibly affected when such relay tongue is moved by reversal of the current substantially as set forth.

Third. The combination of a polarized relay with a centered tongue in a circuit with a polarized relay not having its tongue centered and operating the latter with currents so weak that they will not throw the tongue of the former out of center.

Fourth. The transmitters 4, and 5, or equivalent devices in combination with two main batteries MB^1 and MB^2 all arranged and operated substantially as described.

Signed by me this 4 day of December A.D. 1874.

Thos. A. Edison.

Witnesses Geo. T. Pinckney Chas H Smith

DS, DNA, RG-241, Edison Caveat 51. Written by Pinckney; petition and oath omitted. [a]Place taken from petition; date taken from text, form altered. [b]Both figures were together on one page, signed and witnessed at the bottom.

1. See headnote above.

2. This is Caveat 51. The transmitter described in this caveat can be found in Doc. 485. Edison here substituted a second polarized relay for the neutral relay in his basic quadruplex receiver design (in bridge or differential arrangement). Edison probably began sketching such circuits as early as the summer of 1874 and possibly much earlier. See Cat. 297:37(2), 37(2) verso, 81(3) verso, 84(3), 85(3), Lab. (*TAEM* 5:570, 571, 760, 779, 784).

A polarized relay with movable contacts for the quadruplex.

3. After Edison's estrangement from Western Union at the end of the month, Gerritt Smith and George Prescott continued working on the quadruplex to improve its operation on the company's wires (Orton to James Gamble, 11 Nov. 1875 and 6 Feb. 1876, LBO 15:154–61, 16:148–67). Smith, who received a salary increase for his efforts (Orton to Prescott, 19 Jan. 1875, LBO 15:124), developed several alternative quadruplex apparatus and circuits which were said to greatly improve its operations ("The Quadruplex Controversy," *Telegr.* 11 [1875]: 136; Orton to Stager, 12 Nov. 1875, LBO 15:167–70; Reid 1886, 676–77; Doc. 513 n. 4). He patented several (U.S. Pats. 185,588, 185,589, and 189,276). Elements of these patented designs are quite similar to or identical with designs found in Edison's sketches and caveats from about the time Smith worked with him (e.g., Docs. 485 and 513).

Polarized relays with two movable contacts were used in the quadruplex system commonly operated by Western Union for some years before the company returned to the use of short-core neutral relays (Prescott 1877, 851–53, 858–60; Maver 1892, 212–16; Edison's testimony, Testimony and Exhibits on Behalf of T. A. Edison, p. 33, *Nicholson v. Edison*). While the rights to Edison's quadruplex patents were being challenged in litigation, Prescott credited Smith with this relay design and its use in quadruplex circuits, along with several other designs. He dropped the attribution later when Western Union's title to Edison's designs was secure; without naming an inventor, he included them in a chapter titled: "Edison's Quadruplex Telegraph" (Prescott 1877, 815–19, 842–52; idem 1879, 309–32; idem 1885, 848–66; cf. Doc. 484). Edison did not promptly patent this in the United States but did so elsewhere; see, for example, figure 20 in British Patent 384 (1875). This became an issue in connection with similar devices in Smith's application (executed 3 December 1875) that led to U.S. Patent 185,589 (see Doc. 717). These relays are part of Edison's U.S. Patents 452,913 and 512,872. Smith's U.S. Patent 189,276, executed 9 February 1877, shows relays of this sort in various circuits (cf. his U.S. Patent 185,588, executed 23 December 1875). See also n. 5 below; and NS-Undated-001 and NS-Undated-005, both Lab. (*TAEM* 8:46, 436, 444).

It is unclear whether Smith's designs represent independent invention, his reduction to practice of Edison's general ideas, or his appropriation of Edison's work. It should be noted that Prescott regarded himself as half-owner of Edison's designs from this period, including this caveat and Docs. 513 and 531–34, and at some point arranged to receive a half-interest in Smith's patents as well. Though caveats were confidential documents, both Prescott and Smith knew the contents of Edison's caveats. Prescott to Commissioner of Patents, 2 and 7 June 1875, Quad. 72.2 (*TAEM* 9:30, 32); Gerritt Smith, Pat. App. 189,276. See also Doc. 513 nn. 2 and 4.

4. The alternative transmitter shown here is but one of many Edison had been considering for over a year (see Docs. 348, 387 [device 14], and 392 [fig. 6], and illustrations accompanying Docs. 446 and 485; see also Cat. 297:23[1], 79[2], 81[3], 87[4], 140[1] [*TAEM* 5:522, 746, 759, 795, 948]). The goals sought included absence of sparking and corrosion, nearly instantaneous action, distinctness of signals, and perfect circuit continuity. The quadruplexes tried on the Boston and Chicago lines apparently included tests of some different transmitter designs (see il-

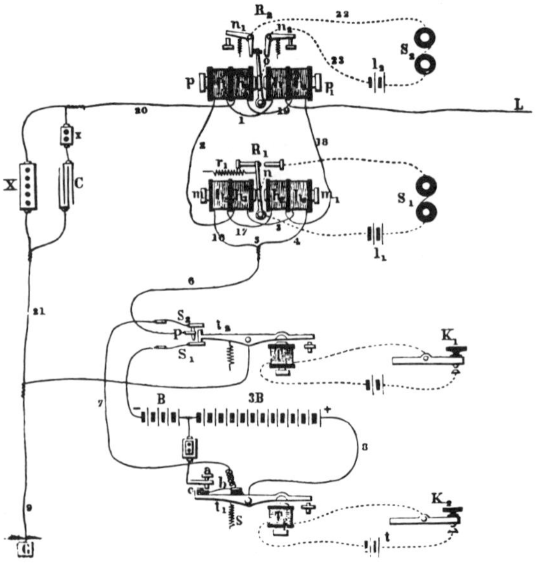

Western Union's standard quadruplex circuit.

lustrations with Doc. 515). Edison included still other transmitter designs in other caveats prepared during this period.

5. In 1876 Prescott reported that this plan had become the standard form of quadruplex for Western Union and ascribed the design to Gerritt Smith with some unspecified contributions from himself, which attributions he later dropped. Smith did not include this exact circuit arrangement in his patents, although a very similar one is in his U.S. Patent 185,589 (Prescott 1877, 849–63; idem 1879, 309–32). See Doc. 485.

–513–

Caveat: Multiple Telegraphy[1]

New York, December 4, 1874[a]

To all whom it may concern.

Be it known that I, Thomas A. Edison of Newark in the county of Essex and State of New Jersey have invented an Improvement in Telegraphing, of which the following is a specification.

The object of this invention is to send and receive four messages at the same time over one wire in various directions.

The invention consists in placing an unpolarized relay in the bridge wire of a Whetstone paralleogram and then passing the main and artificial line through the double coils of a differential polarized relay and a combination of the bridge and differential systems.

The invention further consists in placing the unpolarized relay in a derivation from the bridge wire and including in this derivation a large amount of resistance R^3. This resistance I shunt with a condenser if considerable capacity.

The invention further consists in placing an extra magnet C′ on the unpolarized relay in such a manner that it will assist the main line magnet and including this magnet in the condenser circuit, so that when the condenser discharges or charges the current must pass through this magnet C′ and its cores acquire a momentary magnetism.

The invention further consists in the method of increasing or decreasing the strength of the two main batteries used in signalling by shortcircuiting a portion of each.

I will now describe the working of the apparatus with reference to the drawing.[2]

Fig. 1.

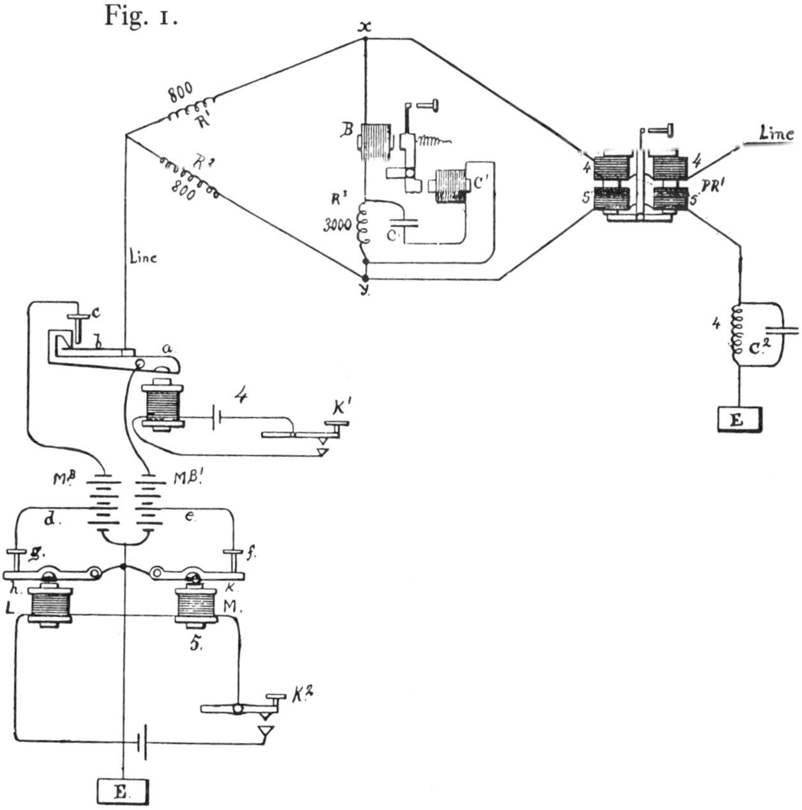

Supposing that no messages are being sent from 4, or 5, but that two are being received on the unpolarized relay B, and the polarized relay P.R′ from apparatus precisely similar to 4 and 5, at the distant station, we shall consider 4 and 5, to be the distant transmitting devices.[3]

When the key K^1 is open at the distant station, the spring

6, to which the line is connected is in contact with the lever a, which connects to the zinc pole of the main battery M.B′, the course of the circuit is thence through 25 cells and then down the wire e, to the point f, through the lever h, to earth; the lever h, and wire e, serves to short circuit 75 cells of the battery and prevents the current from thence going out on the line hence a zinc current from 25 cells will pass over the line; this in passing through the main line helices 4, of the polarized relay P.R′ throws the tongue to the left and opens the local circuit it then passes to the junction x a portion passing through R^1 to the transmitting device to earth while another portion passes through the bridge wire containing the unpolarized relay B to the point Y, thence through R^2 to earth through transmitting devices; owing to the resistance of R^4 scarcely any passes to earth by this route.

Now the small amount of current although on account of its superior delicacy is abundently sufficient to actuate the polarized relay P.R′, has scarcely any effect on the unpolarized magnet B, owing partly to the great resistance of the branch on which it is placed; secondly to the small amount of current which passes through the bridge wire even with the R^3, and thirdly owing to the want of delicacy of ordinary electro magnets, hence the transmission of alternate waves of positive and negative electricity from 25 cells does not affect B; if now the key K^1 at the distant station be depressed the line is disconnected from MB′ or zinc battery of 25 cells and put in contact with one main battery M. B. or copper battery of 25 cells. This current passing over the line through 4.4, of the polarized relay P.R′ throws the lever to the right and closes the local circuit, and this current does not affect B, as before described. Now if the key K^2 be closed at the distant station the short circuit of the extra 75 cells of each battery is removed and each battery is made 100 cells: this excess of current does not change the adjustment of PR′ because that relay does not depend on strength of current for working but merely on its direction but the excess of current passing through the bridge wire containing B, is of sufficient strength to cause the magnet to attract its armature: this current in passing through the resistance R^3 allows a charge or charges the condenser C, (ie) a short momentary wave of electricity passes into the condenser.

Now as this wave must pass through the extra magnet C′, it follows that the electricity acting upon the relay is double at the first instant to what it is after the circuit is permanently closed and if the key K[2] at the distant station is now closed and both batteries are of full strength, 100 cells each, the relay B, will remain closed. If the operator on 4, now reverses the current this reversal necessitates the change in the magnetism of the cores of B, from say N. P to P N, hence there is an interval where there is no magnetism and the tendency of the armature is to be drawn away from the cores by the spiral spring, but just at the instant that this occurs, the condenser discharges a short wave and this short wave passing through the magnet C′ gives it a short sharp magnetism; this magnetism supplies for an instant the force which is last in B, in its change in polarity in the cores, hence the lever remains attracted. The main object of the extra magnet, condenser etc, is to bridge over this short period of "no magnetism" of B, brought about by the reversals.[4]

When the line is very long and the electrostatic very high, this interval of no magnetism is greatly increased. As the first portion of Zinc current following after a copper current is greatly weakened by being neutralized by the static current of the line, hence the change of magnetism in the cores of B, would be effected slowly and signalling[b] would be impossible were it not that the first part of the weakened reversed current of the condenser which current passes within the circuit formed by the relay B, and resistance R[1] and R[2], thus making the signal even; again this current acting in a second manner within the magnet C′, places a magnetism in it at that instant when there is none in the magnet B, hence with this powerful compensator I am enabled to work lines of high electrostatic capacity. Having described the manner of receiving two messages from the distant station, I will now describe the method of transmitting two messages in the opposite direction without affecting B, or PR′.

R[1] and R[2] are the two parallel wires of a Wheatstone bridge, of equal resistance. 5.5, are extra coils on the relay P.R.′ R[4] is the resistance to form the artificial line of equal resistance to the line. When this is done no current passes through the bridge wire containing B, and the current on the line passing through 4.4, is neutralized by an equal amount passing over the artificial line through 5.5.

The relay PR′ might be inserted in the bridge wire, but I prefer the method shown as the discharge from the condenser C, appears to affect it and it has not so much margin or working current.

Fig. 2.

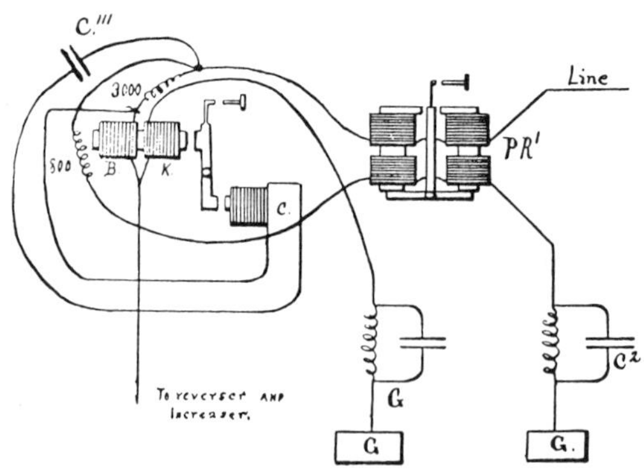

Inventor Thos. A. Edison per L. W. Serrell
Witnesses, Chas H. Smith Harold Serrell[c]

I do not wish to confine myself to any particular arrangement of the relay B, R³, condenser C, and relay PR′ in various circuits by which the outgoing currents may be balanced and made to have no effect upon them as it may be done as shown in figure 2, in which PR′ is arranged as in fig. 1, but B, instead of being in a bridge wire is provided with two coils wound in opposite directions; one, B, is placed in the main line circuit and the other K in the extra artificial line, where the resistance G, a shunt of 800 ohms is placed around the relay B, and the 3000 ohms in the relay derivations is shunted with a condenser precisely as in figure 1. The reason why I employ an extra artificial line is that were both relays placed on the same artificial it would be difficult to obtain a proper balance.

My claims will probably be.

First. The double relay B.C, in combination with a condenser and resistance arranged substantially as described and for the purposes specified.

Second. The combination of a double balancing device consisting of the bridge and differential systems.

Third. The combination of the polarized relay with the unpolarized relay, in a quadruplex telegraph when operated substantially as described.

Fourth. The method of increasing and decreasing the strength of the two signalling batteries and devices for reversing the same.

Signed by me this Fourth day of December A.D. 1874.

Thos. A. Edison.

Witnesses Geo. T. Pinckney Chas H Smith

DS, DNA, RG-241, Edison Caveat 52. Written by Pinckney; petition and oath omitted. [a]Place taken from petition; date taken from text, form altered. [b]Both figures are on a separate page, signed and witnessed at the bottom.

1. See headnote, p. 347.

2. This is Caveat 52. The receiver described in this caveat can be found in Doc. 485. George Prescott described a quadruplex circuit with receivers arranged this way but using two polarized relays as in Doc. 512; at first he ascribed its invention to Gerritt Smith and himself (Prescott 1877, 849–51). See Doc. 512 n. 3.

Prescott claimed that this quadruplex circuit design, similar to that in Edison Caveat 52, was invented by Gerritt Smith and himself.

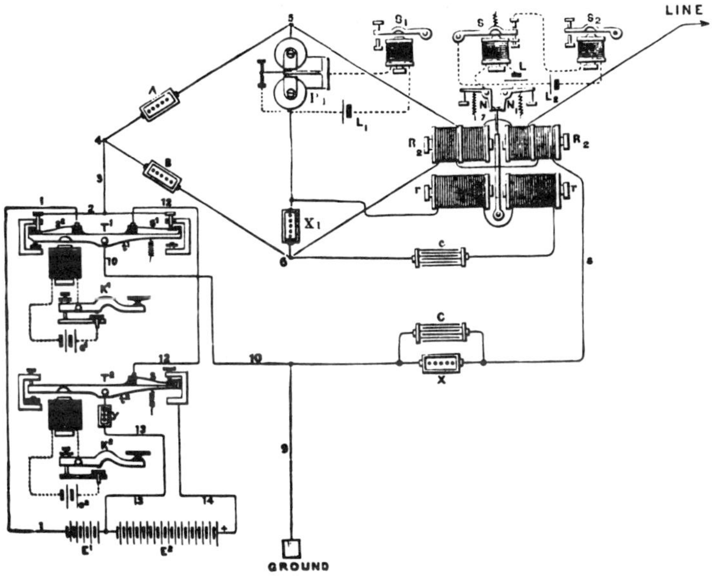

3. Cf. Doc. 387 (device 14) and the illustration accompanying Doc. 485 n. 4. On alternative quadruplex transmitters see Docs. 512 n. 4 and 485 n. 7.

4. An extra coil on a receiving relay, activated by a shunted condenser, later became an important part of quadruplex circuits but was credited to Gerritt Smith (Maver 1892, 203–7, 212–16; see Doc. 512 n. 3). Though designed by Edison at least as early as September 1874 (Doc. 485; see also Cat. 297:128[1], Lab. [*TAEM* 5:909]), it apparently was applied in practice only sometime during 1875. A report of 11 May 1875 in the *Telegrapher* credited Smith with having just introduced a "new double coil, short core relay" that much improved the performance of the quadruplex ("Origin of the Underground Telegraph . . . ," *Telegr.* 11 [1875]: 117). In the *Operator,* Edison responded that he had devised and used "the double deck relay lately put in use on the Quadruplex" in his "first experiments" ("Domestic Notes," *Operator,* 1 June 1875, 9; cf. Docs. 442 and 453).

Prescott and Pope represented the arrangement shown here to be the initial form of the quadruplex (Prescott to Commissioner of Patents, 7 June 1875, Quad. 72.2 [*TAEM* 9:32]; Prescott 1877, 839–41), although it was clearly not used on either the Boston or the Chicago line in 1874 (see Docs. 494 and 515). Pope's drawing for Prescott's 7 June 1875 letter to the Patent Office was first published in the 15 November 1875 *Journal of the Telegraph* (Pope 1875) and reprinted immediately in the *Operator* (cf. the illustration in Doc. 348 n. 9). This condenser-relay combination is shown and described in Gerritt Smith's U.S. Patent 189,276 (fig. 10), but it is not included in any of the claims granted in that patent.

–514–

Notebook Entry: Thermostat

[Newark,] Dec. 6th 1874

Hard Rubber Thermostat.

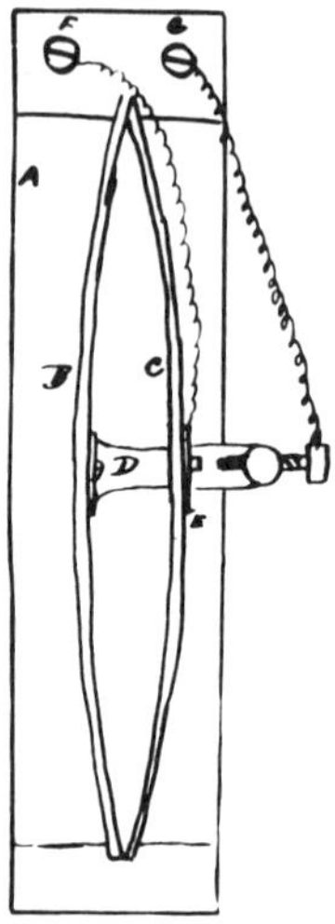

About Nov 8th Edison suggested to me that as Hard Rubber had such great expansive properties we should make a thermostat of it & try it. Made one like drawing & it worked well.

X, NjWOE, Lab., Cat. 1307:12 (*TAEM* 90:613). Written by Charles Batchelor.

–515–

To William Orton

[New York, c. December 7, 1874[1]]

Our duplex[2] worked from 7.30 P.M. last night till 11.02 P.M. without a single hitch.[3] Best yet. After 11.02 worked double to Chicago, double to Buffalo, and Buffalo double to Chicago on regular business just the same.[4] After the line men on Erie[5] get through moving the wire we use on[to] new set of poles we will work right along. They crossed wire to-day about every twenty minutes. I need 10, 9, 8, 7, 6, 5, 4, 3 or 2,000 dollars—any one you would like to advance.[6] Yours,

EDISON.[a]

PD, NjWOE, Quad. 71.2, p. 17 (*TAEM* 10:235). [a]"3*" appears at bottom center of document.

1. This is the only extant report by Edison on the trials of the quadruplex on the line to Chicago, which Orton regarded as the crucial test of its commercial value (see Doc. 509). Edison and Orton testified that Edison gave this note to Orton at the latter's office (Orton thought it had been written in his anteroom) a few days before the date of Doc. 517 (10 December) (Quad. 70.7, pp. 283, 421; Quad. 71.1, pp. 137–38, 239 [*TAEM* 9:508, 578; 10:74–75, 126]). The New York–Chicago tests (with a repeater at Buffalo) began on 1 December with a successful trial on the line from New York to Buffalo. The first Chicago-Buffalo tests, begun the next day, had only limited success, but regular operation was achieved on 4 December, when trials lasted until midnight (Doc. 509; Orton to Anson Stager, 2 and 26 Dec. 1874, LBO 14:39–40, 97–100; "Quadruplex Telegraph," *J. Teleg.* 7 [1874]: 376–77; "Telegraphs and Telegraphers in Chicago.—The Quadruplex, etc.," *Telegr.* 10 [1874]: 296; "Successful Working of the Quadruplex.—Good Time on President's Message.—Reduction of Salaries, etc.," ibid., 303). On 9 December, Orton officially reported to Western Union's directors that the quadruplex "is now working" on the line to Chicago ("The Western Union Telegraph Company.—Its Business and Prospects," ibid., 301).

2. That is, the quadruplex. Edison testified that the details of the circuit and of the apparatus had to be changed from the pattern used on the Boston line; "this perfecting of the details and making the adjustment of the parts" continued through the tests, particularly coping with the much more troublesome static discharge on the four-times-longer line (Serrell's question and Edison's testimony, Testimony and Exhibits on Behalf of Thomas A. Edison, pp. 40–41, 44–45, *Nicholson v. Edison*). Edison stated that a key modification for this was adding a shunted condenser in the bridge wire between the connection to the artificial line and the receiving instruments, a design he covered in his Caveat 60 and probably had experimented with already in the summer of 1874 (ibid., pp. 32, 45; Doc. 534; see also the illustrations accompanying Doc. 485).

No drawing of the exact equipment or circuit employed in these tests survives, but the illustrations on the following page provide a close approximation, as indicated by Edison's drawing of the repeater used (see illustration, p. 348). The first diagram (made by M. L. Lawson, a night operator in Chicago) appeared in a paper by telegrapher Francis Jones, a participant in the tests ("Duplex and Quadruplex on the Wires and at the American Electrical Society's Meeting . . . ," *Telegr.* 11 [1875]: 62). Jones presented his paper to the newly formed American Electrical Society in Chicago on 17 February 1875, by which time the quadruplex worked regularly on the New York–Chicago and Chicago-Cincinnati lines (Jones 1875; also published in *Teleg. J. and Elec. Rev.*, 1 July 1875). The other diagram, possibly reflecting a later circuit, was published in the October 1875 *Journal of the Franklin Institute* by S. M. Plush, an inventor who claimed to have carefully inspected the quadruplex as installed in Western Union's offices. Despite visual differences, the drawings substantially agree in electrical details. Plush's drawing includes repeating sounders for the polarized relays as well as for the neutral relays, which Edison covered in a caveat (Doc. 534) and included in the quadruplex repeater design but which were not shown in other representations of commercial quadruplex circuits. Plush's drawing also has the

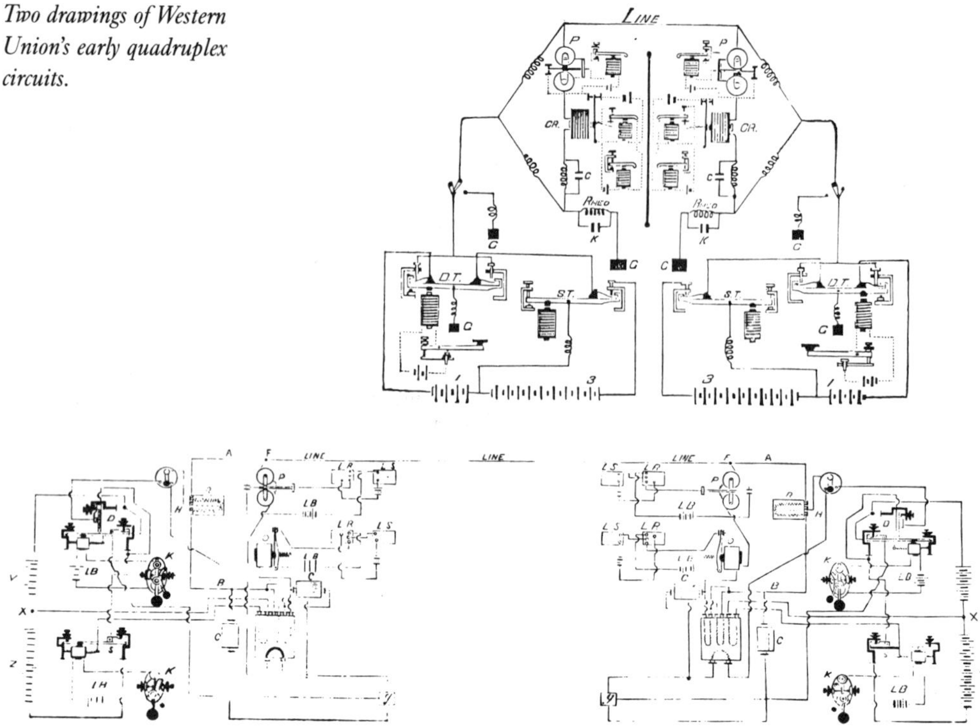

Two drawings of Western Union's early quadruplex circuits.

artificial lines' condensers connected at points that are probably incorrect. The shape of the armature levers for the pole-reversing transmitters differs; Plush's matches that found in Edison's repeater design and various other drawings (including Doc. 348), while Jones's (which he calls "the Prescott and Edison key system") matches different Edison drawings (e.g., the illustration on p. 223), the depiction in Pope 1875 and Prescott 1877 (p. 839, first included in Prescott to Commissioner of Patents, 7 June 1875, Quad. 72.2, facing p. 8 [*TAEM* 9:34]), and the only other early drawing of a quadruplex circuit—that is, Charles Davis's from as early as May 1875, which accompanies Doc. 348 n. 9 (Davis and Rae 1876; "Origins of the Underground Telegraph Bill . . . ," *Telegr.* 11 [1875]: 117). Plush's drawing, however, includes more of the physical layout, not just the electrical circuit, showing how resistance coils can be packed together (in **R** and the large rectangles between the condensers **C**; see Cat. 297:119[1] verso, Lab. [*TAEM* 5:881]; and the rheostat in Doc. 453).

3. These first successful long-distance quadruplex trials required the attention of first-class operators and electricians at each end, including Edison, Prescott, Gerritt Smith, Charles Summers, and Francis Jones. *J. Teleg.* 7 (1874): 371; Taltavall 1893, 166. On Summers see *TAEB* 1:18, 32 n. 8; on Jones see Taltavall 1893, 165–67.

4. Wherever there was a repeater in a quadruplex circuit (Buffalo in this case), and the circuit's entire capacity was not taken up by traffic between the end points, the apparatus could be switched to serve as one duplex circuit from each end to the repeater office along with a through

duplex circuit over the whole length of the line. This flexibility was cited as one of the best features of Edison's quadruplex. Preece and Sivewright 1891, 200; articles from the *Buffalo Courier* and the *San Francisco Bulletin* reprinted in *J. Teleg.* 7 (1874): 371 and 8 (1875): 21.

5. The Erie Railroad.

6. Orton offered $5,000, to which Edison assented; this led to the transaction recorded in Doc. 517 (Quad. 70.7, pp. 283, 421; Quad. 71.1, p. 138 [*TAEM* 9:508, 578; 10:75]). Edison said he needed funds to keep his and Murray's workshop and to pay their workmen. He possibly was also already being asked for payment in connection with Doc. 516 (App. 3, Edison's Affidavit, pp. 813–14; Edison's testimony and Grosvenor Lowrey's question to Josiah Reiff, Quad. 70.7, pp. 280, 451 [*TAEM* 9:507, 593]; see also App. 1.D157, G15, and Doc. 530). Business was still depressed in the wake of the Panic of 1873, Edison had lost his house, and after two months of commercial quadruplex operation he had received nothing for his work (Doc. 504). Orton claimed that during the fall he had brought up the issue of a quadruplex sale but that neither he nor Edison had moved to settle the matter (Orton's testimony, Quad. 71.1, p. 137 [*TAEM* 10:74]; see also Doc. 509). Edison denied Western Union's view that this request for an advance (or Doc. 517) constituted an implicit acknowledgment by Edison of a pre-existing agreement obligating him to sell the quadruplex design only to that company (Quad. 70.7, pp. 256, 280–84, 329; Quad. 71.1, pp. 117, 132–33, 137–38 [*TAEM* 9:495, 507–9, 531; 10:63, 72, 74–75]; App. 3, Western Union Bill of Complaint, p. 798).

–516–

Promissory Note

NEW YORK, Dec. 9, 1874.[1]

$3351.70.

Three months after date I promise to pay to the order of Geo. Harrington, three thousand three hundred and fifty-one 70/100, at office of Messrs. E. D. Randolph & Co., in New York. Value received.[2]

Due March 9, 12, '75.

THOMAS A. EDISON.

Endorsed—Geo. Harrington.

PD (transcript), NjWOE, Lit., *Seyfert v. Edison* (*TAEM* 46:408).

1. This note is a renewal of a July note described in n. 2. Josiah Reiff's testimony, p. 9, *Seyfert v. Edison*, Lit. (*TAEM* 46:413).

2. As part of an effort to pay off Edison's obligation to his former partner William Unger, investors in the Automatic Telegraph Co. sold some land bonds owned by William Seyfert (see Doc. 452 n. 3). Edison signed one or more promissory notes for Harrington in connection with those bonds, which notes Harrington then endorsed for Seyfert's personal use. The notes were to be returned at such time as Automatic Telegraph was sold or otherwise began to provide substantial returns. However, one note was never withdrawn, passing instead into the hands of William Seyfert's wife, Lucy, and serving as the basis of a significant lawsuit against Edison. *Seyfert v. Edison*, Lit. (*TAEM* 46:407–17).

–517–

Receipt to Western Union

N.Y., Dec. 10, 1874

Whereas, Thomas A. Edison and George B. Prescott are the inventors of certain improvements in telegraphy, relating to duplex and quadruplex telegraphing, for which letters patent of the United States have been applied for by said inventors; and

Whereas, said Edison and Prescott have agreed to assign all their right, title and interest in and to said inventions and letters patent to the Western Union Telegraph Company, provided the terms of payment for such assignment and transfer shall be satisfactorily adjusted between the said parties and the said telegraph company:

I, the said Thomas A. Edison, hereby acknowledge the receipt of five thousand dollars to me in hand paid in part payment for my interest in the said assignment and transfer.[1]

Witness my hand and seal, this tenth day of December, 1874.

Witness: A. R. Brewer.[2] Thos. A. Edison. (L.S.)

PD (transcript), NjWOE, Quad. TLC.3, p. 207 (*TAEM* 10:924). Notarization omitted. There are eight other versions of this document in the Quadruplex Case, containing insignificant differences.

1. Edison later claimed that Orton intended this amount to cover the operating expenses of Edison's shop, particularly in light of the $15,000 order Orton placed on 17 December for twenty sets of quadruplex instruments (74-011, DF [*TAEM* 13:186]; Edison's testimony, Quad. 70.7, pp. 282–83 [*TAEM* 9:508]; Edison's statement, Quad. TLC.2, p. 135 [*TAEM* 10:887]). See also Cat. 1214:52, Accts. (*TAEM* 21:590), for the account record for this order. Jay Gould agreed to take the instruments in his 4 January 1875 contract with Edison; it is not clear that Edison ever finished making them (Doc. 526; receipt of 7 Jan. 1875, Complainants' Exhibits 6.2, 1:285, Box 17B, *Harrington v. A&P*).

2. Brewer was assistant to George Mumford, vice-president and secretary of Western Union. After Mumford's death in July 1875, Brewer became company secretary. Reid 1879, 562.

–518–

To the Editor of the Scientific American

Newark, N.J. December 12, 1874[a]

Cable Telegraphy.

To the Editor of the Scientific American.

Mr. Little's assertion, in your number for November 21,[1] that Mr. Winter's improvement in cable telegraphy consists in working a galvanometer by an induction coil having primary and secondary wires, is incorrect, as a reference to the diagram and description printed in a previous number of the Scientific American will show.[2]

T. A. Edison.

PL, *Sci. Am.* (n.s.) 31 (1874): 372. [a]Place not that of publication.

1. George Little, "To the Editor of the Scientific American," *Sci. Am.* (n.s.) 31 (1874): 324. Little also responded to Edison's earlier letter to the *Scientific American* (Doc. 503) in the pages of the *Telegrapher* (10 [1874]: 273).

2. See Doc. 503. Winter's plan does involve an induction coil and a galvanometer. However, in common usage (including Edison's), working a device "by an induction coil" meant that the device was operated simply by the current induced in the secondary coil of an induction apparatus and was not otherwise affected at all by the current in the primary coil. Winter's design involves complex combination effects instead of such simple separation.

-519-

And George Prescott to William Orton

NEW YORK, December 16, '74.

D'r sir,—

Your company has now over 25,000 miles of wire, which can now be profitably "Quadruplexed." [1]

Considering these 25,000 miles to be already Duplexed, the Quadruplex will create 50,000 miles additional.

For all our patents,[2] and efforts in protecting the company in the monopoly of the same during their life, we will take 1/20 of the average cost of maintenance of 50,000 miles of wire for 17 years, one-third down and the balance in yearly payments during the above mentioned period. "Half of such payments to cease the moment any other person shall invent and put into practical operation a quadruplex (not infringing our patents), upon a circuit of 400 miles in length." [3] Yours, etc.

(Signed) THOS. A. EDISON.,[4] GEO. B. PRESCOTT.

PL (transcript), NjWOE, Quad. 70.2, p. 13 (*TAEM* 9:315). There are many other versions of this document in the Quadruplex Case, and another in DF, containing insignificant differences.

1. Western Union actually had over 175,000 miles of wire on more than 71,500 miles of line, but considerations such as insufficient wire size and lack of traffic made quadruplexing much of it unproductive. Western Union 1874, 6.

2. No patents existed as yet for Edison's quadruplex design or related work (see Doc. 509 n. 7). Two days earlier Edison had finally executed an application for Case 112, which he had been working on since the summer (Doc. 449), but two more weeks passed before he filed it.

3. Western Union did not accept this proposition. Orton instead asked Edison to reduce it to actual figures, which Edison did, arriving at a sum of over $350,000. Quad. 71.2, pp. 141–43 (*TAEM* 10:76–77); "Original calculation in Quadruplex," Cat. 30,096, Scraps. (*TAEM* 27:597); Edison's testimony, Quad. 70.7, pp. 285–86 (*TAEM* 9:509–10).

4. The previous day Orton had told one of his officers,

I have hardly seen Mr Prescott for three or four days, but Edison is almost wild over the Quadruplex, and promises me still more startling developments within a few days. Although he still professes to believe in the utility and economy of Automatic, he admits the Quadruplex promises to surpass it. [Orton to Anson Stager, 15 Dec. 1874, LBO 14:74]

-520-

Receipt from William Unger

Newark Dec. 24th 1874

Received of Thomas A. Edison ~~by~~ check on Merchants National Bank for[a] Seven Hundred Dollars—which amount if paid is to be taken as part payment of the amount due in a certain promissory note dated July 28th 1874 payable in four months and drawn by said Edison to the order of Wm A. Unger for one thousand dollars— This receipt is in duplicate—same being enclosed in said note

Wm A. Unger per R. Byington atty—

D, NjWOE, DF (*TAEM* 13:5). [a]Interlined below.

-521-

Memorandum to William Orton

[New York, c. December 25, 1874[1]]

Two Propositions:[2]

1st. We will take twenty-five thousand down and twenty-five thousand in six months for all patents,[3] and a royalty on quadruplex of $166 per year for each circuit created.

2d. We will take twenty-five thousand down for all patents, and a royalty of $233 per year for each circuit created.[4]

PD (transcript), NjWOE, Quad. 70.4, p. 18 (*TAEM* 9:334). There are five other versions of this document in the Quadruplex Case, three of which contain insignificant differences.

1. Orton later stated that Edison brought this to him on 30 December or a few days before (Quad. 71.2, pp. 143, 252–53 [*TAEM* 10:77, 133]). Edison was asked if it dated from 31 December; he said that it could have been as early as three or four days after Doc. 519 (16 December), to which this was an amendment (Quad. 70.7, pp. 286, 357 [*TAEM* 9:510, 540]). Note also Doc. 535.

2. According to Orton, he told Edison that the differences between these propositions and what he was prepared to offer were insubstantial and that after he returned from a trip to Chicago they would draw up a contract (Quad. 70.7, 143, 249–51 [*TAEM* 10:71, 131–32). Edison testified that he could not remember Orton's making such a statement and that these two propositions were among several he gave Orton intended only as talking points in the negotiations (Quad. 71.2, pp. 348–50 [*TAEM* 9:541–42]). Orton subsequently sent Edison and Prescott a letter accepting the second proposition (Doc. 535). This became the basis of Western Union's claims that it had bought the invention.

3. No patents had been granted yet for any of Edison's relevant applications. See Doc. 509 n. 7.

4. Although Edison and Orton disagreed over whether the royalty should be for the life of any covering patents (seventeen years) or for only ten years, either of these propositions would have given Edison and Prescott substantial royalties (Orton testimony, Quad. 70.7, p. 250 [*TAEM* 10:132]). There were 60 quadruplex circuits on Western Union lines by 1878 and 160 by 1884 (Prescott 1879, 309; idem 1885, 843).

–522–

Jay Gould Agreement with Josiah Reiff and John McManus

N.Y. Decr 30/74[1]

It is hereby understood that the undersigned will heartily cooperate in concluding an alliance between the A&P Tel Co & the Automatic System on the general basis following[2]

A&P to increase her Capital to	$15,000,000[3]
Automatic interests to receive[4]	4,000,000
To remain in Treasy	1 000 000
	5,000,000

The 14,000 shares A&P now in the Cos Treasy to be distributed to the A&P stockholders as a dividend. Automatic System covering Patents, Contracts, etc etc to be turned over to A&P Tel. Co.[5] Management to be mutual and subject approval of Mr Jay Gould & Col Thos. A Scott[6]

Genl T. T. Eckert to be Presdt[7]

T A Edison to be Electrician[8]

D H Craig to organize the news Deptmt.[9]

The Automatic are to conclude the pending Contracts with Erie, PRR, & B&O & turn them over to A&P The A&P Tel Co to assume the liabilities under said contracts[10a] Automatic to have representation on Executive Committee[11]

Jay Gould[12]

Josiah C Reiff
John McManus[13]

DS (photographic transcript), NjBaFAR, *Harrington v. A&P,* Box 17B, Complainant's Exhibit No. 2, 3:98. [a]"under said contracts" interlined above.

1. Gould, Thomas Eckert, and Albert Chandler visited Edison's laboratory on the evening of 30 December. See Chapter 6 introduction, p. 314.

2. Reiff's version of the drafting of this document and of the prior negotiations can be found in his 1893 memorandum for the directors of the Western Union Telegraph Co. enclosed in a letter to Edison dated 27 November 1893. 93-046, GF.

3. The board of trustees of the Atlantic and Pacific Telegraph Co. voted to increase the capital stock of the company from $10 million to

$15 million on 27 April 1875; the stockholders approved it on 6 May. Exhibits, Extracts from Atlantic and Pacific Minutes, 1:391–94, Box 17B, *Harrington v. A&P*.

4. Stockholders in the Automatic Telegraph Co. were to receive a total of 31,800 shares of stock in Atlantic and Pacific. The proposed division of shares is found in George Harrington to Jay Gould, 16 Apr. 1875 (Exhibit 5), Bill of Complaint, 1:53, Box 17B, *Harrington v. A&P*.

5. The transfer occurred in April 1875. See Doc. 561; and deed of Automatic Telegraph Co. to Atlantic and Pacific, 10 Apr. 1875, and letters of Harrington to Gould, 16 Apr. 1875 (Exhibits 5–7), Bill of Complaint, 1:53–58, Box 17B, *Harrington v. A&P*.

6. Thomas Scott was president of the Pennsylvania Railroad Co.

7. Thomas Eckert was elected president of Atlantic and Pacific on 14 January 1875. Gould and Eckert first discussed his leaving Western Union and taking over Atlantic and Pacific in April 1874. Eckert's report to stockholders of Atlantic and Pacific, 28 Apr. 1875, 2:178; Eckert's testimony, 2:222–23; both Box 17B, *Harrington v. A&P*.

8. See Docs. 529, 539, and 585.

9. This provision was never carried out. By November 1874 Craig, who still claimed that he owned two-thirds of the Automatic Telegraph Co. system, was also claiming that he controlled "a new automatic which is far better" (Albert Chandler to George Harrington, 5 Nov. 1874 [Exhibit 14], 2:210, Box 17B, *Harrington v. A&P*). This was the Foote-Randall automatic, which Craig later offered to Western Union and on the basis of which he formed the American Rapid Telegraph Co. in 1879 (Israel 1989, 214–15).

10. Negotiations between Automatic Telegraph and the Erie, Pennsylvania, and Baltimore and Ohio railroad companies began sometime in the spring of 1874. Automatic Telegraph hoped to organize a new telegraph company by uniting the telegraph lines on the railroad companies' trunk routes into a system employing automatic telegraph equipment in order to offer formidable competition to Western Union. These negotiations were apparently related to earlier attempts on the part of Automatic Telegraph to combine efforts with the Pacific and Atlantic, Franklin, and Southern and Atlantic telegraph companies (see *TAEB*1:487 n. 1). Attempts were also made to interest Thomas Eckert, the Atlantic and Pacific Telegraph Co., and the newly established Direct United States Cable Co. in this opposition company. Negotiations with the three railroad companies and with Eckert continued up to the time of this agreement. A proposed agreement was reached and approved in its general outlines by some of the parties in late November and early December 1874, but it was never concluded, and George Harrington was still negotiating with the railroad companies on 1 January 1875. In January, Atlantic and Pacific took over negotiations with the railroad companies, concluding agreements for rights-of-way on lines of the Pennsylvania and Baltimore and Ohio systems. Exhibit A, 1:140–41; Complainant's Exhibit 1, entered 1 May 1905, 2:297–300; Complainant's Exhibit 14, 2:210–11; Complainant's Exhibits 9–12, 16–19, entered 6 May 1905, 3:104–16; Complainant's Exhibits 9 and 13, entered 11 Apr. 1905, 2:196, 203, 209; Baltimore and Ohio Correspondence Exhibits, 1:417–37; Respondent's Exhibit 8, 2:88–92; George Harrington's testimony, 1:120, 123, 134–35; Albert Chandler's testimony, 2:15–

16, 26–27, 33–43; Josiah Reiff's testimony, 3:41–42; all Box 17B, *Harrington v. A&P.*

11. This provision was never carried out.

12. Jay Gould (1836–1892), the financier whose dealings had made him a public symbol of business corruption, controlled a number of important railroad lines and was seeking to gain a major foothold in the telegraph industry. Gould's attack on Western Union was part of his longstanding conflict with railroad baron Cornelius Vanderbilt, who also controlled the telegraph company. By 1881, using the telegraph lines of his own railroad companies as the basis for opposition telegraph companies, Gould forced a merger with Western Union that gave him control of the telegraph giant. *DAB,* s.v. "Gould, Jay"; Klein 1986.

13. See *TAEB* 1:280 n. 3; and App. 1.G15, G21.

–523–

Prose Poem

[Newark, 1874?[1]]

A yellow oasis in hell =
premeditated stupidity = A phrenological[2] idol.
The sombre dream of the grey-eyed Corsican[3]
A Brain so small that an animacuale went to view it with a compound Microscope The wrestling of shadows, a square chunk of carrion with two green eyes held by threads of gossamer which floats at midnight in bleak old rural graveyards.
Three million miles beyond the limits of the universe where the anglels dare not go
There flies forever from nihil to nihil the foulest demon of the Cosmos

AD, NjWOE, Lab., Cat. 297:15 (*TAEM* 5:501).

1. This appears to be Edison's private response to the personal attacks on him which appeared in the *Telegrapher* in 1874 and 1875 (see headnote, p. 305). Other examples of Edison's fanciful vituperation can be found in Cat. 297:82 (Lab. [*TAEM* 5:767]) and in a technical note on cable telegraphy dated 17 February 1874 (Vol. 9:1, Lab. [*TAEM* Supp. III]).

2. Phrenology was a psychological and physiological theory in which character and personality were related to the shape of the skull. Davies 1971; Young 1970.

3. Probably a reference to Napoleon.

–524–

Draft Patent Application: Automatic Telegraphy

[Newark, 1874?[1]]

The effect of placing electro-magnets in the main line of a chemical telegraph is to neutralize a portion of the signal by sending a counter current and when the battery current is broken the relays or magnets send a current in the same direction and prolong the signal beyond that prolongation due to the discharge of the Line.[2] I have taken advantage of this ~~an~~ evil by placing these magnets in a short circuit or shunt where the conditions are reversed and when the signal commenses the ~~w~~current set up is against the current coming down the shunt but with the one going down the paper the consequence of which is to augment the the blackness of the mark and when the battery is taken off the relays discharge a counter current Neutralizing the current due to the discharge of the Line thus leaving a perfect character upon the paper. I ~~also~~ now place relays or magnets upon the line which are the opposite of the ordinary relay or magnet which causes a prolongation of the current, by putting an extra coil of wire upon the magnet and magnetizing it by a constant local battery, and so connecting the coils that when the main line current enters the relay the magnetism induced by the local battery is neutralized the consequence of which is that a current due to the demagnetization of the iron Core will be sent upon the Line in the same direction as the main current thus helping it instead of neutralizing as an ordinary Relay or magnet would and when the main current is broken the iron Core becomes instantly magnetized again by the Local battery and the act of magnetization would send a current upon the main Line in the opposite direction thus neutralizing the currents due to the discharge of the Line. Thus the ordinary relay magnetizes the Core when the current is closed, but in this relay the magnetizism which is already in the core by reason of the Local battery is neutralized the consequence of which is to reverse the direction of the inductive current and use them for discharging or neutralizing the current due to the discharge of the Line =

I place a number of these relays along the Line to neutralize the discharge current at every part of the Line at the same instant.[3]

These magnets may be provided with armature levers & adjusting springs the same as the ordinary Morse Relay and used as a receiving instrument.

The adjustment of the local battery may be done by a Rheostat described in a late previous patent =[4]

The Local battery & Coil may be dispensed with and an adjustable permanent magnet used

I claim one or margnets as described placed in a chemical telegraph line for the purpose set forth =

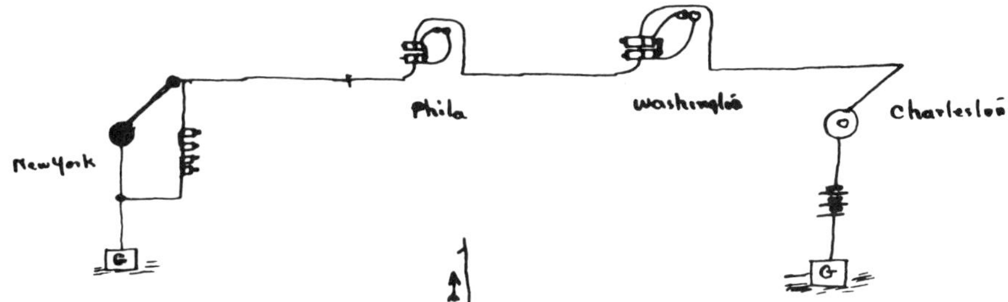

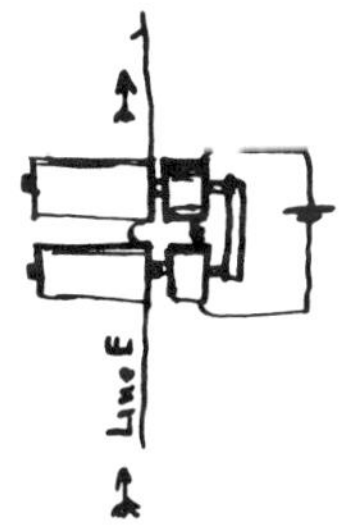

ADf, NjWOE, Lab., Cat. 297:8, 12(10) (*TAEM* 5:464, 481). Content and resemblances in style, paper, and ink link these separate scrapbook items.

1. The patent mentioned below (see n. 4) issued on 12 May 1874.

2. Edison continued to experiment with combinations and forms of relays and shunts to minimize signal distortion on automatic circuits. See, for example, U.S. Pat. 195,751.

3. Edison had experimented with related designs as much as two years earlier but had not then employed devices at intermediate points on the line. Patrick Delany, "Self-Induction as a Remedy for Static Discharge," *Elec. Eng.* 12 (1893), enclosed in Joseph Wetzler to TAE, 31 Oct. 1893, 93-003, GF.

4. Probably U.S. Pat. 150,846.

–7– January–March 1875

As a result of his new arrangements with Jay Gould and the Atlantic and Pacific Telegraph Company, Edison began 1875 by severing his ties to Western Union and focusing his attention on automatic telegraphy. He immediately began to enjoy substantial financial benefits. Edison had essentially finished his technical refinement of the quadruplex telegraph system in late 1874; now the venue of quadruplex developments moved from the Newark workshop-laboratory and the Western Union operating offices to the national scene: the U.S. Patent Office, company boardroom, halls of Congress, and courtroom. All of the multiple-telegraph caveats and patent applications Edison filed during this period described his earlier work, and one amended application submitted in March was simply a legal maneuver, an attempt to reestablish his nominal ownership of the quadruplex.

Two major parties now battled over Edison's designs.[1] Jay Gould, having contracted to purchase George Harrington's interest in Edison's automatic telegraph system, also acquired from Harrington whatever rights he had to the quadruplex.[2] To bolster his claim to the quadruplex, on 4 January, Gould agreed to pay Edison $30,000 for whatever rights Edison held in his own prospective quadruplex patents, even though Edison's possession of any legitimate interest was arguable. When President William Orton of Western Union returned from the Midwest to New York in mid-January, he tried to counter Gould's claims by accepting terms Edison had suggested in December.[3] George Prescott, Edison's partner under the agreement of 19 August 1874,[4] agreed to Orton's offer, but Edison was no longer interested in selling to Western Union.

This caricature of Western Union president William Orton was captioned "The Great Telegraph Juggernaut: How the Telegraph Monopoly Treats its Victims."

In addition to selling his rights to Gould, Edison asked the Patent Office to reassign all his quadruplex patent applications from Prescott and himself to Harrington and himself, which would effectively assign them to Gould. The Commissioner of Patents personally investigated the case and decided to follow the original assignment to Edison and Prescott. In late March lawyers for Harrington and Gould petitioned the Secretary of the Interior to examine the situation. Western Union had opened another front in late January by filing a court petition seeking to force Edison and Prescott to sell the quadruplex patent rights to Western Union.[5]

The legal struggle for control of the quadruplex patents rapidly escalated, with Western Union and Jay Gould waging the contest at the highest levels. They mustered the most

This cartoon shows "The Wall Street Orpheus" (Jay Gould) "playing" various corporations—The Atlantic and Pacific Telegraph Co.; the New York Tribune*; the Erie, Union Pacific, and Panama railroads; and the Pacific Mail Steamship Co. The cymbals on his knees, labeled "New York" and "Yokohama," probably represent the Occidental and Oriental Steamship Co.*

powerful advocates possible in Washington, D.C., at a time when influence-peddling and corruption were common. Western Union hired Senator Roscoe Conkling; Jay Gould used Congressman Benjamin Butler. Moreover, at the outset of the dispute the Secretary of the Interior was Columbus Delano, Gould's sometime ally, but in October 1875 Conkling's old cohort Zachary Chandler assumed that post.[6] Although the case moved beyond Edison's direct influence, he remained an important participant, supplying documents and testimony as needed.

As a result of the quadruplex patent dispute, the Patent Office stopped all action on related Edison patents. Edison pulled back as well and did not execute any new patent applications between 24 February 1875 and 7 March 1876. He instead filed long caveats to protect his ideas.

Another result of the dispute was an increase in the severity of James Ashley's attacks in the *Telegrapher,* where Edison's move to Atlantic and Pacific was described as one more example of his duplicity (and "quadruplicity"). These attacks seem not to have affected Edison's reputation overseas, however. In March he published an article in the British *Journal of the Society of Telegraph Engineers,* and the society's secretary urged him to send more. At the end of that month, a Prussian telegraph official visited the Newark shop to examine several of Edison's telegraph inventions.[7] Within the telegraph community, the New Jersey inventor was gaining increased international recognition.

Having cut himself loose from Western Union, Edison affiliated closely with Gould's Atlantic and Pacific Telegraph Company. He began working as the company electrician, although he was not formally appointed to the position until June. His principal task was overseeing the installation of automatic telegraph equipment on the Atlantic and Pacific lines. By the end of January the New York–Boston and New York–Washington lines used automatic technology to send their regular traffic.[8] Edison visited the intermediate offices on the New York–Boston line in late January, assessing their staffs and instruments. Equipping the offices called for much new apparatus, and as a result the Edison and Murray shop found itself with all the work it could do.

With a new patron for automatic telegraphy, Edison returned to developing a roman-letter system. He successfully tested his four-wire apparatus on Atlantic and Pacific's New York–Philadelphia line at the end of January.[9] Even the advantage of having a message arrive ready to deliver did not offset the disadvantage of requiring four wires, however, and Edison concentrated on devising circuits that used one or two wires to send the four signals needed to form letters.

Work on the district telegraph continued as well. The Domestic Telegraph Company established itself as a going concern, erecting lines in New York and Newark and soliciting customers at fifty cents per month.[10] In February, Edison and Charles Batchelor jointly executed a patent for a district system. At some point Batchelor became the electrician for the company (Edison was nominally the consulting electrician) and assumed responsibility for development work.[11]

Edison put the money from Gould to immediate use. He traveled to Port Huron in early January and gave money to his brother Pitt for the street railway business. Edison bought scientific books and stocked his laboratory with new equipment, including a microscope, spectroscope, thermopile, and air pump. He used his cash to tide Edison and Murray over between payments by Atlantic and Pacific and other customers, and he loaned money to his associate E. H. Johnson in Newark.[12]

The money also made a difference in Edison's family life. Sometime in January, Edison's father and nephew Charley came to Newark, where they assisted in the shop and laboratory. Mary Edison, relieved of the financial pressures of the previous year, spent substantial sums of money during the winter. Edison also paid up bills and gave money to members of Mary's family.[13] On Edison's birthday Mary gave a masquerade party at the family's Bank Street apartment. Sometime in March or early April the family moved again, this time to a rented house at 89 South Orange Avenue, in a residential area about one mile west of the Ward Street shop.[14]

1. E. Baker Welch (on the basis of an 1869 patent assignment [Doc. 61]) and George Little (on the basis of an agreement with George Harrington [Doc. 199]) also advanced their claims to Edison's quadruplex, and telegraph inventor Henry Nicholson had filed an application in October that potentially interfered with Edison's Case 99 (Doc. 472; see *Nicholson v. Edison*).

2. Harrington came to New York at the beginning of January to discuss the agreement of 30 December (Doc. 522) with Gould. According to Harrington,

> During the interview Mr. Gould referred to Edison's quadruplex inventions, and to the complications of title; and I at once, and of my own volition, said I would give him a title thereto. . . . This offer by

me was made voluntarily and promptly, and without discussion; nor was it accompanied by any demand for pay or compensation. [Harrington's testimony, 1:122, Box 17B, *Harrington v. A&P*]

Much of Harrington's claim to the quadruplex hinged on the definition of "fast" telegraphy, a point at issue in the Quadruplex Case. See Doc. 491.

3. Doc. 521.

4. Doc. 466.

5. See App. 3, Western Union Bill of Complaint.

6. Conkling and Butler have been characterized by historian Allan Nevins as "grotesques, . . . more incredible than characters in Dickens" (Nevins 1957, 573). On the other political connections, see ibid., 459–60, 474, 496, 512, 591.

7. "A Visit to Edison's Laboratory," *Operator,* 1 Apr. 1875, 7.

8. "The Telegraphic Situation," *Telegrapher* 11 (1875): 34.

9. Cat. 1307:16, Batchelor (*TAEM* 90:617). According to the *Operator* (15 Feb. 1875, 3), Edison sent twenty-two messages in forty-nine seconds (the *Operator* incorrectly dated the test 31 December).

10. See Doc. 553; and "Domestic Telegraph System," *Newark Daily Advertiser,* 15 Jan. 1875, 2. In a diary entry for 26 May that may refer to the company's Newark operation, Batchelor noted, "Domestic started & took about $3." Cat. 1307:47 (*TAEM* 90:632).

11. Cat. 1144, Scraps. (*TAEM* 27:256). For examples of Batchelor's work on the domestic see Cat. 1307:15, 49, Batchelor (*TAEM* 90:616, 634).

12. Cat. 1185:39; PN-75-01-05; both Accts. (*TAEM* 22:579; 20:28–36).

13. PN-75-01-05, Accts. (*TAEM* 20:31–32).

14. Edison gave a newspaper interview on 16 April at the new address. "A Quadruplex Telegraph," *Newark Evening Courier,* Cat. 1144, Scraps. (*TAEM* 27:278).

–525–

Notebook Entry: Automatic Telegraphy and Electromotograph

[Newark,] January 2nd 1875—

Experiments to obtain a good solution upon which to receive Tellurium marks also to ascertain if on this good solution Tellurium marks can be wiped out by Electricity or some solution put on the marks Also to find a platina solution delicate, also to obtain a cheap delicate solution on other metals

Hypochlorite Calcium wipes out Tellurium marks Completely

Oxide Bismuth & NaCl Platina on H.[1] Te good Platina wipes out Te splendid on .O.[2]

Bichloride Hg. NaCl. Platina on H.

Chromate K & NaCl. Platina on O blots Te Te good but leaves yellow mark after blotting

Ca~~ust~~rbonate Manganeese. NaCl. Tin on H. Lead on H

Clpb—P pen Hypo'ite soda marks good on H. go for this

Caustic Baryta P on H faint Mark.

Bromide of Pot platina on O wipes out Te. leaves slight yellow mark—

Nitrate Lead Platina on both. H best marks when you first touch "Musical Sound on drawing along"

Hypoite Soda. ClPb & Cl.K. & Caustic K.

Carb[onate] Am[monium]. NaCl. Lead on H

Molybdate Soda Tin pen marks on .O.

Te marks splendid on Cl.K. best

Stanate Soda Te Splendid

Nitrate Silver P on O good

Carbonate Am Tin pen slight mark may be good.

Ch Manganeese Platina Pen Marks on .O. good Te bad.

Phenomenon Bisulphate Soda & Sul Zinc Te fumes and spreads, hurts eyes & garlic smell.

Cl Calcium Te marks splendid. P on O wipes mark out

Carb Cadmium & NaCl. Te good. P on O wipes it out good.

Molybdate Am—NaCl. Lead on H. Te good

Phospate Soda & NaCl Lead splendid delicate as Te. dont show so black next morning though

Urate Am—NaCl. Te no good Lead splendid on H black in morning

Sul[phate] Lime NaCl. Te good P on O blots good

Bisulphate K. Te fair P on O blots fair

Ch Magnesium & NaCl. Te Splendid P on O blots Splendid Lead gives splendid mark on H not very black in morning

Phosphate ~~Manganese~~ Magnesia Lead on H Splendid—Not very black in morning

Phosphate Am—NaCl. Te good P on O wipes out slowly

Carb Baryta NaCl. Te good P on O wipes out quick

Phosphate Soda ~~Te~~NaCl~~o~~ Te not very good P marks on H good. Pb splendid on H not extra black in morning

Proto Tin Te Hardly marx.

Bisulphite Soda Te good

Oxide Strontia Te good P on O blots.

Nitrate Lime & NaCl. Te good P blots it out on .O. Extra black in morning

Bi~~s~~sSulphate K NaCl. Very Phenomenal Te good spreads fearful. P on O wipes beautiful; leaves white mark where Te spreads thinly this white mark delicate if could be done cheap twould give good P solution

X, NjWOE, Lab., Cat. 1174:153 (*TAEM* 3:43).

1. That is, attached to the cathode.

2. That is, attached to the anode.

–526–

Agreement with Jay Gould

[New York,][1] January 4, 1875[a]

Articles of agreemt made and entered into this 4th day of Janry 187~~4~~5 Betwn Thos A. Edison of Newark in the state of New Jersey ~~T~~and Jay Gould of the city of New York Witness:

Wheras said Edison has invented certain Imprvnts in Duplex and Quadruplex[b] Telegraphs for which he has Executed or is about to execute applications for Letters Patent of the United States and such applications are numbered 94, 95, 96, 97, 98, 99 & 100 and all other numbers refered to below[c] And are dated Augt 19, 1874[2]

Therfor in Considation of Thirty Thousand Dollars to be paid by the said Gould in cash & the other considration herafter named[3] the said Edisn herbey sells assigns & transfers all his interst right, title claim & demand in & to the aforesaid patents[4] and all renewals thereof being at least[d] one equal undivided half part of said patents.[5] The said Edison for himslf his executes admtratrs & assigns herby agrees to execute full & absolute convey of the said patents including all patents or imprvtns made or to be made[e] by said Edison relating to Duplex or Quadruplex Telegraphy (~~A~~to be hereafter numbered)—[6]

Said Gould agrees in addition to the aforesaid Thirty Thousand Dollars to pay him Edison[f] one half of all sums realized by him for the sale of rights outside of the United States. The contract for Quadruplex Intmts[7] now being made by said Edison ~~an~~ is[f] to be taken by said Gould.[8] In witness wherof said parties have hernto set their hands & seals[9] the day & year above written.

Thos. A. Edison. Jay Gould

ADDENDUM[g]

[New York,] January 4, 1875[a]

Recvd Jny 4, 187~~4~~5 Ten Thousand Dollars in part paymt of the Considatn of Thirty Thousand Dollars herein mentioned

$10,000 Thos. A. Edison

DS, NjWOE, Miller (*TAEM* 28:979). Written by Jay Gould. [a]Date taken from text, form altered. [b]"and Quadruplex" interlined above. [c]"and all . . . below" interlined above. [d]"at least" interlined above. [e]"or . . . made" interlined above. [f]Interlined above. [g]Addendum is a DS, written by Jay Gould.

1. A few days after Gould and Eckert witnessed a quadruplex demonstration at Edison's shop in Newark, Eckert took Edison to Gould's residence on Fifth Avenue in New York, where they settled this sale. Edison's testimony, Quad. 70.7, pp. 293–94, 351 (*TAEM* 9:513–14, 542), App. 1.G15.

2. The numbered cases of that date are Docs. 467–73; no other numbers were "refered to below," but both Edison and Gould understood this document to refer to the designs that were Cases 111–13, though at that point only the application for Case 112 had been filed. See Docs. 449 and 450; and headnote, p. 347 n. 5.

3. The other consideration is not specified later in the document; cash actually constituted less than half of what Edison received. Checks dated 4 and 7 January 1875 accounted for $13,500 of the amount due, and at some other time in "the early part of 1875" Edison received $16,500 worth of Union Pacific Sinking Fund bonds. Another $7,170, paid by Gould to Edison on 7 January, was for a different purpose, probably that referred to in n. 8 below. Testimony of Giovanni Morosini (Gould's bookkeeper and personal secretary), Quad. 70.7, pp. 488–92 (*TAEM* 9:614–16); see also Edison's testimony, Quad. 70.7, p. 351 (*TAEM* 9:542); and PN-75-01-05, Accts. (*TAEM* 20:28).

4. At about this time Edison probably learned that Cases 94–100 and 112 had all been examined in the Patent Office on 31 December 1874 and that none had passed. Examiner Zenas Wilber rejected Cases 94, 96–98, 100, and 112. Case 95, although not rejected, was not separated from the set for independent issue. However, Case 98 faced only minor objections, and the rejection of Case 100 rested on some unanswered questions. Cases 94 and 97 were rejected because of similarities to Edison's earlier Cases F and E (Docs. 310 and 309). Case 99, the crux of the quadruplex, was suspended in consideration of a probable interference (with Henry Nicholson's 14 October 1874 application); such a response was more positive than a rejection, in the sense that an interference could be declared only on inventions deemed to be patentable. Statement of Examiner Wilber, Quad. 72.5, pp. 1–2 (*TAEM* 9:64–65); Wilber's letters are in Pat. App. 178,221–178,223, 180,858, 207,723–207,724, and 480,567.

5. What share of the outcome of his quadruplex work Edison had a right to and could freely dispose of was a central issue in the Quadruplex Case. According to Edison, he did not assert that his minimum interest at this point was one-half the total. Rather, he told Gould the history of the claims that Prescott, Western Union, and Harrington had upon the invention; stated that he was unsure whether he owned anything or had any rights involved; and insisted that he could and would sell only whatever interest he might still have with an explicit awareness that this might be nothing. Gould still agreed to the purchase. Edison's testimony, Quad. 70.7, pp. 293, 351 (*TAEM* 9:513, 542); see also App. 1.G15 and App. 3.

6. Beyond the cases mentioned in n. 2, this refers to the designs included in Edison Caveats 51–60 but not yet in patent applications, and probably to that of Edison Caveat 67 as well. See headnote, p. 347; and Doc. 592.

7. Instruments.

8. Orton had given Edison an order for twenty sets of quadruplex equipment, for which Edison had expected to get about $15,000 and on which some work had been done and expenses incurred. See Doc. 517 n. 1; and n. 3 above.

9. Although signed, the document bears neither seals nor indications thereof.

–527–

From Jay Gould

New York, Jan 6 1875

To[a] Thos. A. Eddison

Please call and see me tomorrow morning at 10 or 11 oclock[1]

Jay Gould

L (telegram), NjWOE, Scraps., Cat. 30,096:64 (*TAEM* 27:598). Message form of Atlantic and Pacific Telegraph Co. [a]"187" and "To" preprinted.

1. See Doc. 528.

–528–

To Jay Gould

[Newark or New York, January 7, 1875?][1]

Gen Eckert to take hold on the 15th of January.[2a]

On the night of the 14th Transfer to be made ready for next morning.[3]

Same night an illustrated article is to be handed ~~tr~~o Tribune[4] for publication next morning.[5a]

Advertisement to be inserted next day in Washn Balto Phila NY & Boston papers.[6]

Then keep the thing boiling in all the papers.[7]

ADDENDUM[b]

[New York, January 7, 1875?][8]

All right—go ahead

Jay Gould

AL, NjWOE, DF (*TAEM* 13:422). [a]Followed by dash in center of page. [b]Addendum is an ALS, in bottom margin of page.

1. This postdates Doc. 526, probably postdates Doc. 527, and antedates 14 January by several days. Edison met Gould on 7 January and then went to Port Huron, Mich., where he stayed at least through the

thirteenth of the month, so most likely Edison handed this to Gould at their meeting on the seventh. Testimony of Giovanni Morosini, Quad. 70.7, p. 489 (*TAEM* 9:614); George Prescott to TAE, 13 Jan. 1875, Cat. 30,096:63 (*TAEM* 27:597); Doc. 530.

2. Thomas Eckert was elected president of the Atlantic and Pacific Telegraph Co. on 14 January 1875, having submitted his resignation from the position of General Superintendent of Western Union's Eastern Division on the ninth. Reid 1879, 580; see n. 5 below.

3. Presumably this refers to the Automatic Telegraph Co.

4. A loan from Jay Gould allowed Whitelaw Reid, former assistant to Horace Greeley, to buy control of the *New York Daily Tribune* in 1872; Gould's interests subsequently received generally favorable attention in the *Tribune*. Klein 1986, 135.

5. No such article appeared in the *Tribune*. However, an indication of its proposed content is given in the short report that the *Tribune* did publish on Friday, 15 January, under the heading "ACTIVITY IN FINANCIAL CIRCLES" in its "LOCAL MISCELLANY" column. That item, dealing with Eckert's election as director and president of the Atlantic and Pacific Telegraph Co., included Eckert's letter of resignation from Western Union and William Orton's response. The story also contained the following passage:

> Thomas Edison of Newark, N. J., will probably be chosen electrician. It is stated that Mr. Edison has invented a machine which will transmit 2,200 words a minute, requiring only the services of one operator and a girl to cut off the messages when received. The messages are received on a strip of paper similar to that used in the printing-telegraph, with the exception that the slip is prepared by solution in chemicals, and the electric spark striking the paper discolors so as to form the requisite letters. Machines for the equipment of the Atlantic and Pacific lines between Washington and Boston have been ordered, and the experiment of telegraphing with them between those cities will probably be made on Monday.

Apparently no such test of an Edison roman-letter automatic telegraph took place until the end of the month, and then only on a shorter line. See Chapter 7 introduction, p. 374.

An undated, unillustrated draft of the larger article by Robert Browne of Newark survives among Edison's papers. Robert Browne, untitled essay, 75-011, DF (*TAEM* 13:464).

6. Gould and the others planned to have the expanded Atlantic and Pacific Telegraph Co., now including the former Automatic Telegraph Co.'s system, challenge Western Union in a rate war. However, they did not announce or institute reductions until mid-February, when the cost of many messages dropped from $.40 to $.25. "The Telegraphic Situation," *Telegr.* 11 (1875): 40; Reid 1879, 584.

7. Short articles about Eckert's shift between companies appeared fairly widely, but additional items in general newspapers were often brief and comparatively cool. George Little responded critically to the brief mention of an Edison high-speed, roman-letter system in the 15 January *Tribune* article (see n. 5 above; and Little's letter to the editor, *Tribune*, 16 Jan. 1875, p. 2). Items with a more forceful tone followed familiar themes, attacking Western Union as an oppressive monopoly and fore-

casting that new technology could revolutionize telegraphy and make it cheap (e.g., the letter to the editor from "A Foreign Banker" [*Tribune,* 16 Jan. 1875, p. 10], which also ran in the *New York Daily Times* that day [p. 7]). Other developments sustained interest in the changes, particularly a congressional investigation of Western Union and an anti–Western Union bill introduced by Benjamin Butler in the House of Representatives and Stephen Dorsey in the Senate. The later contest over Edison's patent applications generated further interest. Many newspaper clippings of stories from as far afield as Chicago and San Francisco are in Edison's scrapbooks (Cat. 1143 and Cat. 1144, both Scraps. [*TAEM* 27:207, 233–47, 278–79, 285]). See also Docs. 554, 569, and 577; and Lindley 1975, 246–47, 264.

8. See n. 1.

–529–

From Jay Gould

[New York,] Jny 8, 187~~4~~5

Dr Sir:

Yours with "Pints" received. I will attend to this being properly signed & sent you—[1] Please confer with Genl E. as to when they shold go in the offices—as the time must conform to his general views— I only saw him a moment last night In hste

J G

ALS, NjWOE, Scraps., Cat. 297:64 (*TAEM* 27:598).

1. The referent here is unknown.

–530–

Account

[Port Huron?,] Jany 11th [1875][1]

Jany	11th	Paid Carleton for St RR Stock 93 shares[2]	78[-]94.00 ~~8000.00~~
"	11	Wm P Edison for trouble on St RR[3]	1000.00
"	11	Samuel Edison loan	120.00
		" " "	50.00
"	11	Paid to Wm P Edison to take up a	
	11	note against St RR held by H Howard[4]	519.00
Jany	11	Paid to Wm P Edison balance check from Carleton.	106.00
		Spent on Pt Huron trip.	200.00

AD, NjWOE, Accts., PN-75-01-05 (*TAEM* 20:28).

1. The account book containing this entry dates from 1875.

2. J. D. Carleton was made president of the Port Huron and Gratiot Street Railway after he loaned Pitt Edison money to gain a controlling interest in the company. Jenks n.d., 7; see also Doc. 175.

3. This probably refers to costs incurred as a result of suits involving the Port Huron and Gratiot Street Railway in 1874. Jenks n.d., 11–13.

4. Henry Howard, president of the First National Bank of Port Huron, also owned a large lumber mill. *History of St. Clair County* 1883, 575; Jenks 1912, 366–68, 428.

–531–

Caveat: Multiple Telegraphy[1]

New York, January 18th, 1875[a]

To all whom it may concern.

Be it known that I, Thomas A. Edison of Newark in the county of Essex and State of New Jersey, have invented an Improvement in Quadruplex Telegraphs of which the following is a specification.

The object of this invention is to transmit four messages over one wire in various directions at the same time. The invention consists in a peculiar polarized relay for recording the signals sent by reversing the current. The method of attaching and working the local circuit on such polarized relays. The modifications necessary with such local circuits and transmitters in repeating signals from one circuit into another. The arrangement of a secondary battery with the common back point relay upon which is received the signals sent by increase and decrease of the strength of the battery, such secondary battery compensating by a secondary current the effect of the static charge of the line at the receiving station, also the arrangement of a condenser in place of such secondary battery when the line is very long and of high electrostatic capacity.[2]

Fig. 1.

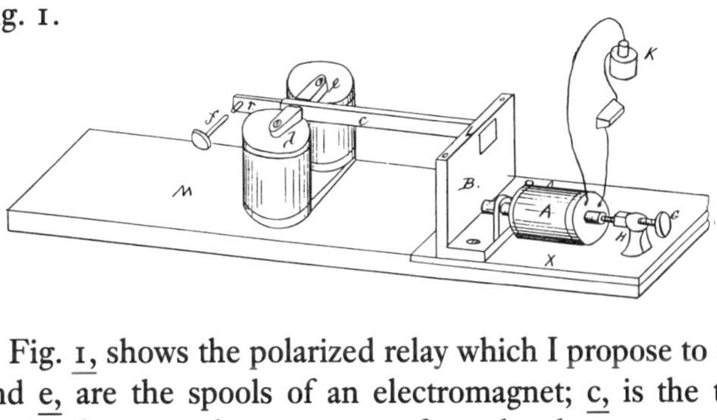

Fig. 1, shows the polarized relay which I propose to use. d, and e, are the spools of an electromagnet; c, is the tongue playing between the two prongs from the electro magnet; r, is a flat brass elongation of the tongue c, provided with a platina contact point, which coming in contact with the point f, closes a local circuit. The tongue c, is pivotted in the soft iron piece B, in the manner usual in polarized relays. This iron piece is secured to a brass plate X. A, is an electromagnet kept constantly charged by the current from a local battery k, circulating within its helices. By means of H & G, the iron cores of this electro magnet can be made to approach or recede from the iron piece B. When the iron core of A, touches B, it imparts say a maximum north polarity to it, this in its turn imparts the same polarity to the tongue c, Hence if the polarity of the iron core of the electromagnet d,e, is say North for d, and South for e, the tongue will be attracted by e, and d, will be neutral or repel; if now the current be reversed in d,e, the polarity of its cores will be exactly reversed and the tongue c, will be attracted by d, and e, will remain neutral or repel c. When as in my Qudruplex apparatus reversed currents are sent over the wire and through d, and e, of variable strength a very strong permanent magnetism or magnetic constant is employed, irregular signalling is the result.

I have found as has already been described in a previous Caveat[3] relating to my Quadruplex apparatus that if the magnetic constant of permanent directive magnetism is weaker than the magnetism produced by the passage of the weakest signalling current through d, and e, that the signals will be even. Hence to enable me to use the polarized relay on circuits of variable length and resistance I must provide some means whereby the strength of the permanent magnetism can be increased or decreased to suit the conditions of the line. By using the arrangement shown in Fig. 1. I am enabled by adjusting the electromagnet A, to or from the iron piece B, so that any degree of magnetism can be given to the tongue c. Instead of adjusting the magnet A, to and from the tongue and piece B, a resistance coil might be inserted within the local circuit in which A, is a part and adjust the strength of the current by increasing or decreasing the resistance, or the tongue c, or even the magnet d,e, might be surrounded by a coil of wire connected in a circuit with a battery and adjustable resistance coil and accomplish the same purpose, but I prefer to use the arrangement shown in Fig. 1.[4]

Fig. 2.

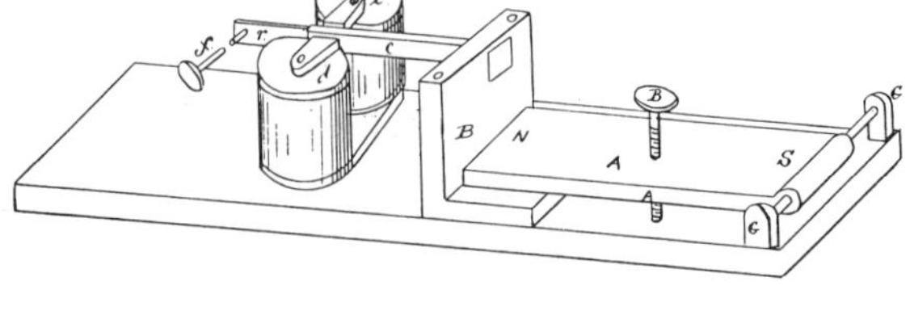

In figure 2, is shown a method of adjusting the amount of permanent magnetism by means of a permanently magnetized steel bar A, which can be made to approach or recede from an iron piece B, to which the tongue c, is pivotted, by means of the adjusting screw B.

This steel bar A, is pivotted upon the shaft G,G, and it might be made to approach and recede from the electromagnet d,e, and give the same result. I prefer to use the plan shown but am experimenting to demonstrate the best method of effecting the adjustment of the magnetism of the tongue either by an electromagnet and battery or a simple steel magnet.

Fig. 3.

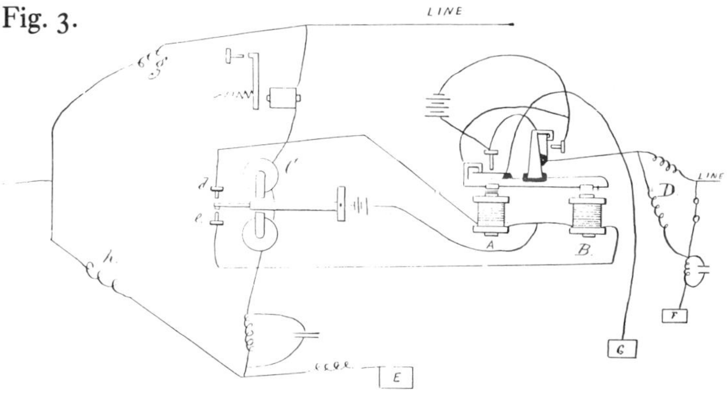

Another part of my invention consists in arranging the local circuit, sounders etc. at the end receiving station and at repeating stations to overcome certain difficulties which I have encountered in operating upon long lines of telegraph. The connection for the transmitting sounder sending reversals into one circuit with the receiving polarized relay upon another circuit, both at the repeating station, is shown in figure 3. C, is the polarized relay provided with two contact points d, e, which are connected in circuits to the double electro magnets A B, of the transmitter: that of A is closed when the tongue touches d, and, of B, is closed when the tongue touches e. This enables me to make the transmitting sounder self adjusting,[5] dispensing with the retractile spring, by which I gain two important advantages, first of which is that less attention is required of the attendant at the repeating station and the length of the wave which passes to the distant station is considerably lengthened by reason that no time is lost in the magnet B, overcoming the inertia or tension of the retractile spring and again in the fact that when the wave is sent to the distant station by the closing of B, by contact of the tongue c, on point e, it does not cease until the tongue has left e, and touched d.

The polarized relay C, is placed say in the Chicago wire at Buffalo and works the transmitter A.B, connected to the New York wire D, at the same station.[6]

Another method of increasing the length of the waves which produce the signals is shown in Fig. 4.

Fig. 4.

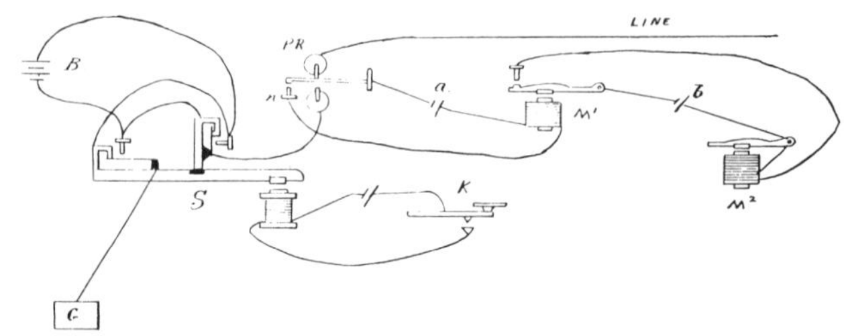

S, is the double pointed reversing transmitter operated by an electro magnet, local circuit key K, etc, all of which has already been described. P.R, is the polarized relay which is supposed to be at the distant station: the current of the battery B, passes through the various contact points of S, in such a direction and through P.R, that when the key K, is closed, the tongue of the polarized relay will fly away from the point n, thus opening the circuit of M^1: the lever of M^1, in its turn closes the local circuit of M^2 and the lever of M^2 is attracted or "closed". When the key K, is opened the direction of the current through P.R. is reversed, the tongue comes in contact with n, closing M^1 which opens M^2 thus the lever of the sounder M^2 follows the lever of the key K. In consequence of the inductive effect of the various magnets used which is apparent when the circuits are closed and not when opened, the distant signal coming through a repeater has heretofore been very "light" or short. My arrangement remedies this defect as the closing of the signalling sounders take place when the polarized relays open.

I thus by reversing the usual method of transmitting the reversed current get rid of the unequal charging time of the various magnets used and signal by the opening of said magnets.

Another part of my invention consists in the method of compensating for the effect of the static charge, both for the incoming and outgoing current. The arrangement of circuits &c are shown in Fig. 5.[7]

Fig. 5.

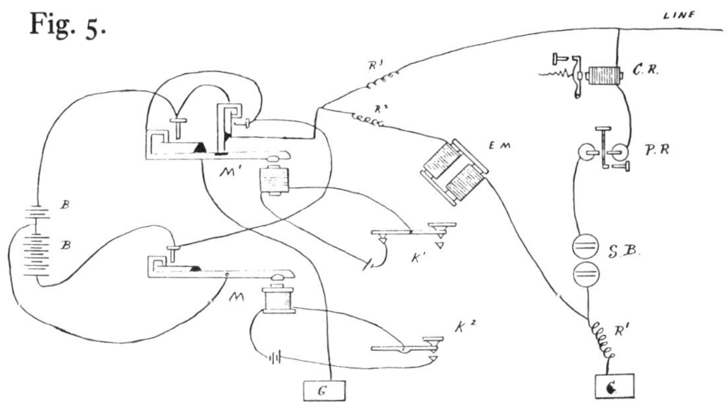

Inventor Thomas A. Edison L. W. Serrell atty
Witnesses Chas H. Smith Geo. T. Pinckney[b]

S.B, are secondary batteries composed of sheets of similar metal immersed in water or other liquid with or without chemicals to increase the strength of the secondary current. These secondary batteries are connected in the bridge wire in which is included the two receiving instruments C.R, and P.R, the first being a common relay and the other a polarized relay, though I do not wish to confine myself to the use of any particular relay or relays with the secondary battery.

The action of this battery is as follows. Supposing the current is permanent on the line and through the bridge wire containing the relays and secondary battery, then S,B, becomes charged. If now the current is interrupted at the distant station the static charge which would still continue to pass to earth through P.R & C.R. and make them stick after the distant current is disconnected, is neutralized by the charge from the secondary battery passing in an opposite direction to that of the static charge through the circuit formed by R^2 R^3 and wire containing secondary battery and relays.

I will mention here that the secondary battery may be used with any form of Duplex or Quadruplex apparatus worked by bridge differential or otherwise if the receiving instruments or that portion of the receiving instrument on the line be included in the same circuit as the secondary battery and the whole shunted or arranged in such a manner that a route is provided for the after current from the secondary battery to circulate on.

E.M, is an electro magnet placed in one side of the bridge and is used for setting up an inductive current upon the passage of the current from B.B, through it which inductive current is used for the purpose of balancing the effect of the static induction current of the line.[8]

Another part of the invention consists in working the double current transmitter M[1] by a back pointed key K[1] so as to obtain a longer wave to form the distant signals than if the front point was used.

I shall probably claim.

First. The combination of a polarized relay with an adjustable magnetic constant of permanent magnetism, substantially as set forth, with a telegraphic circuit within which reversed currents of variable strength are transmitted for the purpose set forth. (See Figs 1 & 2.)

Second. The combination of a polarized relay C. (Fig. 3) having double points with a receiving or retransmitting sounder or relay having an electromagnetic retractor substantially as and for the purpose set forth.

Third. The combination of the receiving relay in a duplex or quadruplex telegraph with a secondary battery composed of plates of similar conducting substance and a conducting liquid when the whole are shunted for the purpose set forth.

Fourth. The electro magnet E.M, (Fig. 5) combined with the relays C.R. P.R and secondary battery S.B. as and for the purposes set forth.

Fifth. The combination of the magnet M[1] M[2] and local circuits, batteries etc. with the polarized relay P.R[c] all operated substantially as set forth and for the purposes specified.

Signed by me this 18th day of January A.D. 1875.

Thos A Edison

Witnesses Geo. T. Pinckney Chas H Smith

DS, DNA, RG-241, Edison Caveat 55. Written by Pinckney; petition and oath omitted. [a]Place taken from petition; date taken from text, form altered. [b]All figures are on two separate pages, both of which are signed and witnessed at the bottom. [c]"with . . . P.R" interlined above.

1. See headnote, p. 347.

2. This is Caveat 55. A condenser appears in the appropriate place in the circuit in figure 3, but otherwise this feature is neither explained nor claimed here. It is covered in Caveat 60 (Doc. 534).

3. Doc. 534.

4. The designs shown in the first two figures, as well as the variations mentioned here, are closely related to Edison's U.S. Patent 134,868 (executed on 8 May 1872); see also Doc. 285 (duplexes no. 9–11, 17).

5. See *TAEB* 1:30.

6. Edison may have experimented with this and the preceding method, but they were not part of the repeater used on the Chicago line as shown in Edison Caveat 56. See illustration, p. 348.

7. Edison here drew upon some of his earliest experiments with his

basic quadruplex design. See Docs. 357 and 360; and Cat. 1171:97–98, Lab. (*TAEM* 6:256).

8. This is based directly on Doc. 470; cf. Doc. 450 and Edison's British Patent 384 (1875).

–532–

Caveat: Multiple Telegraphy[1]

New York, January 18th, 1875[a]

To all whom it may concern.

Be it known that I, Thomas A. Edison, of Newark, in the State of New Jersey, have invented an Improvement in Duplex or Multiplex Telegraph, of which the following is a specification.

The object of this invention consists in dispensing with one of the signalling relays and accomplishing the reception of the two different messages in one relay, and also in the arrangement in circuit of such relay and in other devices necessary to effect quadruplex transmissions.[2]

The invention consists of the combination of a soft iron armature with a polarized relay The soft iron armature is operated by increase and decrease of the current and the polarized tongue is effected by a reversal of the current, this relay is shown in fig. 1.

Fig. 1.

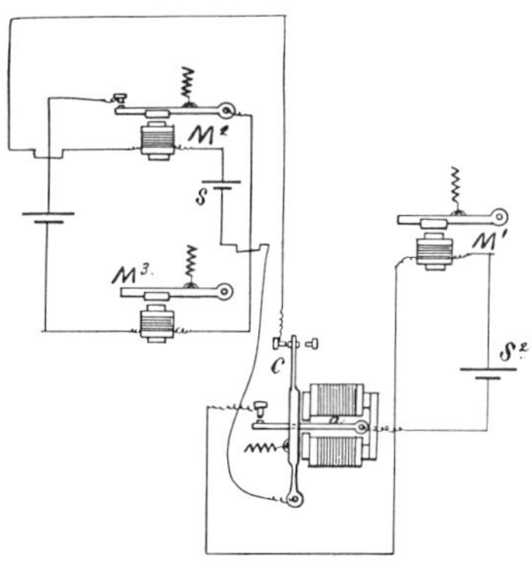

C. is the common iron armature and responds to the increase and decrease in the strength of the current. It is provided with a back point and repeating sounder M² and receiving sounder M³ the action of which has been described in another Caveat of mine, No 55.[3]

a. is a polarized tongue the end of which plays between projections on the end of the electro magnet and responds to the reversals of the current; one object gained in combining the common relay with the polarized relay as in this form, is the lessening in the resistance as two separate coils are unnecessary.

This relay may be placed in the bridge wire of the balance, and operated as in fig. 2. or it may be arranged in the bridge wire as in fig. 3.

Fig. 2.

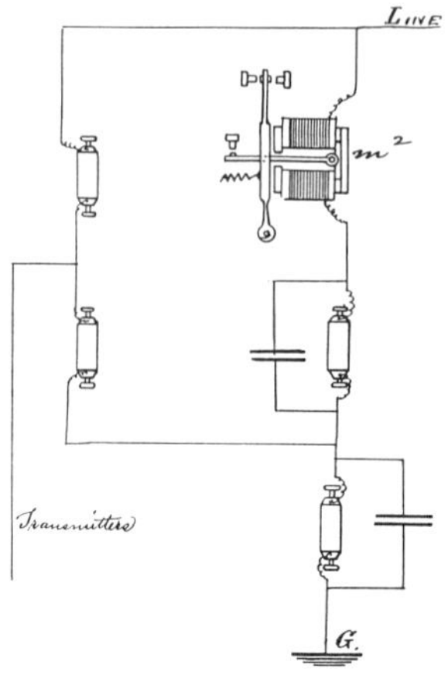

Fig. 3.

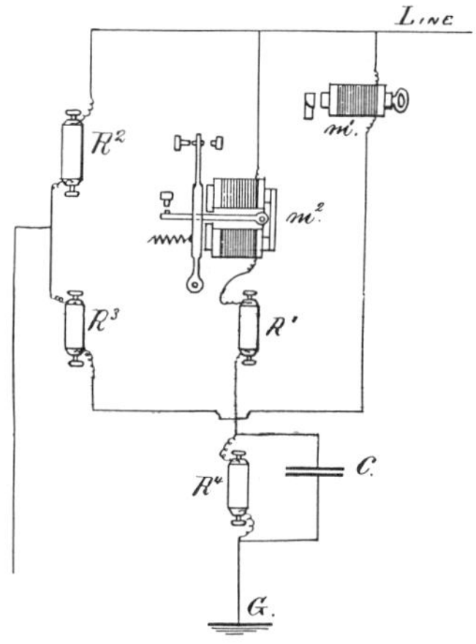

M¹ is a long fine wire electromagnet so arranged that it may be made to approach or recede from a fixed armature, so as to increase or decrease its inductive capacity. The resistance is say 600 ohms. It forms one bridge wire.[4]

M² is the double relay placed in a paralel bridge wire with a resistance coil R¹. of 2000 or 3000 ohms.

The object of the inductive electromagnet M¹ is to compensate for the effect of the static charge of the line upon the received signals.

On a very long line, the first portion of every current is entirely absorbed in either chaarging the line or in neutralizing a return charge consequently at the moment of reversal of the currents at the distant station there is quite a period or interval when no current passes through the relay M^2.

This does not effect the polarized tongue as that will stay on either side according to the last current sent but with the common iron armature the case is different for as that has a retractile spring it follows that if no current is passing through M^2 and the cores are unmagnetized that its lever will open and give a false signal. In another Caveat, I have described a resistance in the bridge shunted with a condenser, the effect of the condenser is to send a current through the relay at that particular point where no current passes through the relay from the distant station and thus hold it closed.[5]

The magnet M^1 performs the same function; supposing the current from the distant station is passing both through M^1 and M^2 the iron armature of M^2 is attracted and the cores of M^1 are magnetic,: ~~a~~if now the current is reversed; the reversed current, owing to the effect of the static charge of the line, does not operate instantly, but in the meantime an inductive current from M^1 is set up which circulating in a circuit formed of the two paralel bridge wires magnetizes the cores of M^2 until the arrival, to maximum, of the distant current and thus prevents the cores of M^2 becoming entirely de-magnetized and the iron armature will not be drawn away from the magnet by its retractile springs.

I do not wish to confine myself to the use of M^2 with the inductive magnet M^1 as a common relay and polarized relay may be included in place of M^2.

Fig. 4.

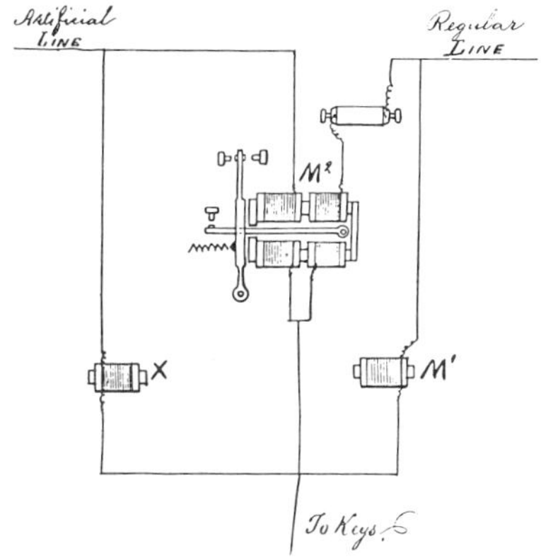

The magnet M^2 might be provided with double coils and worked upon the differential plan as in fig. 4. M^1 is the inductive magnet, which operates precisely as in the bridge wire shown in fig. 3. X is an extra induction magnet which balances M^1 when the outgoing currents are passing through the differential coils but it may be dispensed with and the capacity of the condenser increased two separate relays; to wit, a common relay and polarized relay each having double coils might replace the single combination relay without ~~e~~affecting the result.[6]

This compensation of the effect of the static currents upon the received signals may also be effected by arranging the induction magnet as in fig. 5.[7]

Fig. 5.

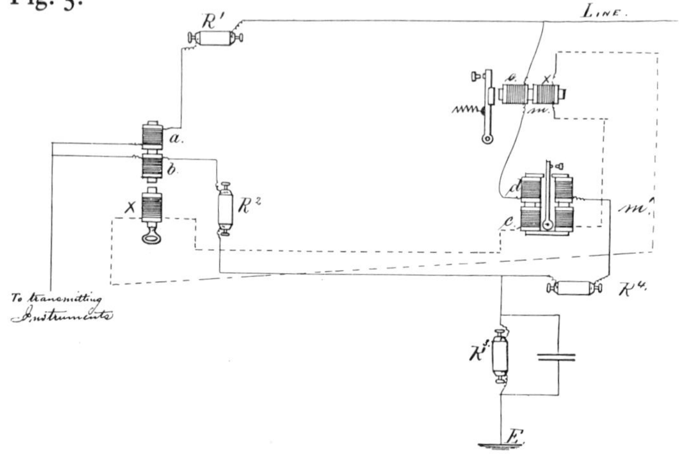

M and M^1 are the common relay and polarized relay, both in the bridge. a. and b. are two differential coils in one iron core, a. is in one prong of the bridge, and b. in the other.

The passage of the outgoing current when the apparatus is properly balanced produces no magnetism in the iron cores, but the most of the distant current passing through R^1 and a. (an equal portion would pass through R^2 and b. were it not for the resistance R^4.) makes the core magnetic. In the act of magnetizing and de-magnetizing, an induction current is set up in X; this current passing through the extra helices on M and M^1 compensates for the effect of the static charge, as has been described. A better arrangement would be to arrange it as in fig 6. then the resistance R^4 could be dispensed with and more margin of working current obtained in M. and M^1.

Fig. 6.

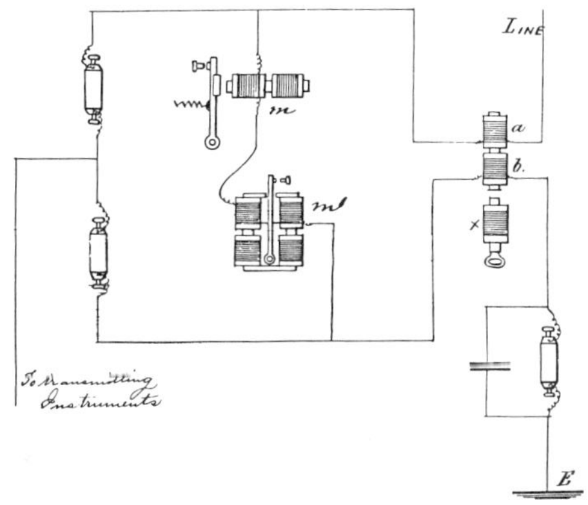

Inventor Thos A. Edison Lemuel W. Serrell atty.
Witnesses, Chas A. Smith Geo. T. Pinckney[b]

In this arrangement the effect of the outgoing current is perfectly balanced while the total amount of received current passes through a. only.

My claims will probably be:—

First. The combination of a polarized and unpolarized armature with an electro magnet for double or quadruplex transmissions.

Second The combination with the receiving instrument or instruments of an inductive magnet or magnets arranged substantially in the various manners shown for the purpose set forth.

Signed by me this 18th day of January AD 1875.

Thos A Edison

Witnesses. Geo. T. Pinckney Chas H Smith

DS, DNA, RG-241, Edison Caveat 57. Written by Pinckney; petition and oath omitted. [a]Place taken from petition; date taken from text, form altered. [b]All figures are on two separate pages, both of which are signed and witnessed at the bottom.

1. See headnote, p. 347.

2. This is Caveat 57. Edison here formalized various designs he had been exploring since the spring and summer of 1874. In his U.S. Patent 189,276, Gerritt Smith refers to the double-armature relay of this caveat. Knowledge of this relay, which was neither used in practice nor described in any publication, indicates Smith's familiarity with Edison's work in this period.

3. Should be "45" (Doc. 348).

4. The drawing errs on this point; the wires that cross between $\mathbf{R}^1$ and $\mathbf{R}^4$ must connect for there to be any bridge wire in this circuit. An earlier sketch of this arrangement, probably from the summer of 1874, is in Cat. 297:128(1), Lab. (*TAEM* 5:896).

5. Doc. 534.
6. Cf. Doc. 446.
7. Cf. Doc. 471.

–533–

Caveat: Multiple Telegraphy[1]

New York, January 18th, 1875[a]

To all whom it may concern.

Be it known that I, Thomas A. Edison, of Newark, in the County of Essex, and State of New Jersey, have invented an Improvement in Quadruplex Telegraph, of which the following is a specification.[2]

The object of this invention is to transmit two messages over one wire in the same direction and at the same time.[3]

Fig. 1.

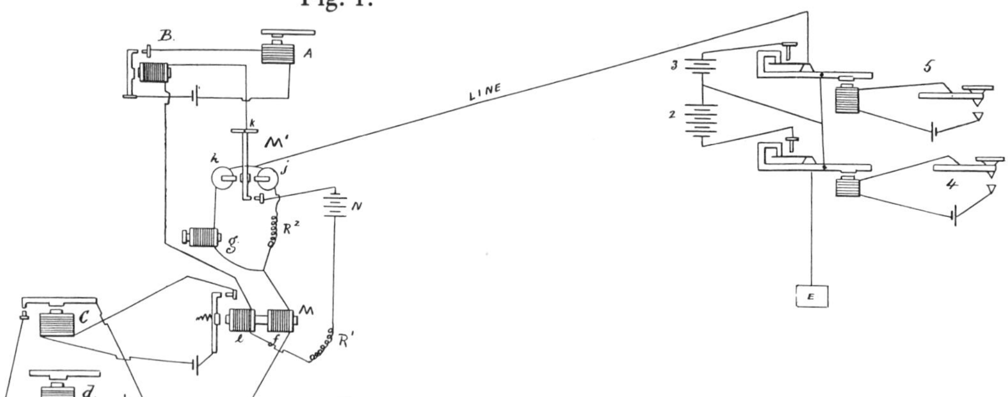

The invention consists in the combination of a polarized relay or common relay with a differential common relay and transmitting increased and decreased currents to effect the transmission of the two messages.[4] M^1 is a polarized relay having the line wire attached to the cross wire connecting the two spools together. The other ends are connected one to a resistance coil R^2 and the other to an ordinary electromagnet g.

The resistance of g. is equal to R^2. Now when the current is made to circulate permanently through the two coils of the polarized relay and through g and R^2. its effect is balanced in the polarized relay M^1. as it passes in opposite direction through its two coils but the magnet g. is charged. Now if the current be interrupted or suddenly decreased in strength the iron cores of g. set up an induced current in the helices surrounding it and this current circulating within the circuit formed of g.R^2. and the coils of the polarized relay passes in the proper direction through M^1 to make its cores magnetic and the tongue is thrown to one side; if now the circuit be closed or the strength of the current increased the magnet g. will in the act of chanrging set up another induction current which now circulates in the opposite direction and throws the tongue of the polarized relay in the opposite direction.

The tongue is to be adjusted so that the weak current from battery 3. at the distant station will not produce sufficient induced current by its passage through g. to effect the tongue of M[1], but only when the battery 2 is put to line, hence the polarized relay will respond only to currents transmitted from a strong battery 2.

M. is a double coil relay, which works by the weak current from 3.

The main current passes through the coil f in one direction and if the transmitter 4. is quiet, workes the lever and the message is received: I use a repeating sounder C. to repeat the signals to the receiving sounder D. for the reason that in signaling with battery 2. the lever of M. receives a forward jerk and would give a false signal, were it not for the repeating sounder which owing to slugishness and having to be fully charged eetc, does not repeat the false signal on d.

I will now describe the method by which I prevent the battery 2 from effecting M.

Suppose the battery 3. is disconnected and the battery 2. is put to line, the polarized relay responds, and the lever of M would also, were it not for the fact that owing to the rapidity with which polarized tongues move a second circuit is closed which contains a resistance coil and a battery adjusted in such a manner that a current passes through the second coil e. of M. equal to that passing through f. from the battery 2. hence the effect of the strong current from that battery is balanced in M[1].

Within this equating circuit is placed a relay B. which operates a receiving magnet[5] from which the signals of the polarized relay are copied.

It is obvious that the current from the battery 2. will pass through coil f. on magnet M. before the equating circuit can be closed, and the effect is to give the lever of the relay a quick forward movement, but the effect of this is neutralized by employing the repeating sounder before described.

I do not wish to confine myself to any particular form of apparatus or arrangement of the same to carry out theis principle as a number of methods are obvious, as in figures 2. and 3.

Fig. 2.

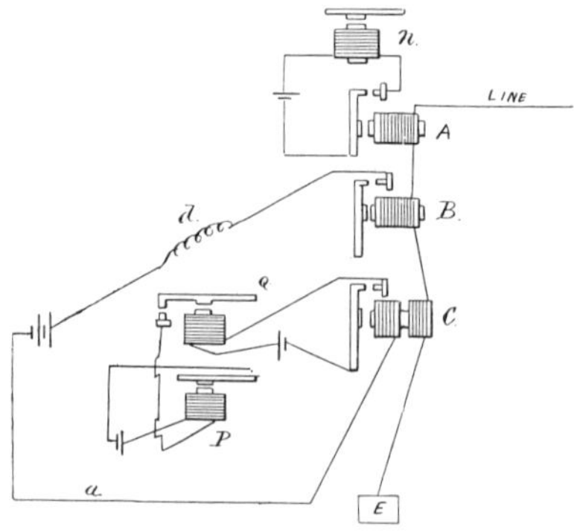

In figure 2. the polarized relay is dispensed with, and two separate relays used, both of which respond to the heavy battery while C respond~~i~~s only to the weak battery.

A. and B. are both adjusted high. One message is received upon A. by the sounder N. while the other is received upon C. and sounder P. B. is used to close the balancing circuit to balance the effect of the large battery within the coils of C.

The magnet A. might be taken out of the main line with and inserted in the balanced circuit, and produce the same effect. The next method is shown in fig. 3.

Fig. 3

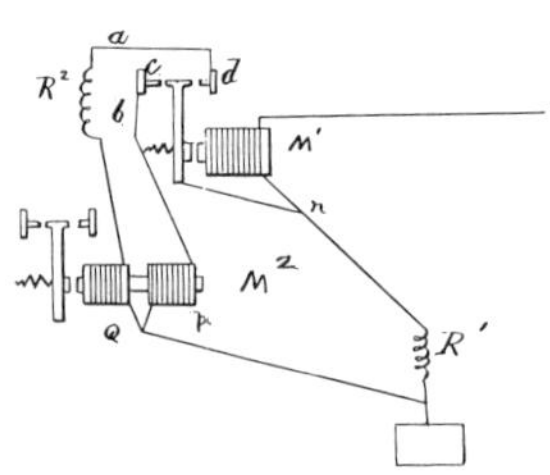

Inventor Thos. A. Edison L. W. Serrell atty.
Witnesses, Chas H. Smith Geo. T. Pinckney[b]

M^1 is adjusted to respond to the strong currents, when the lever is away from the magnet, the main circuit is comp~~el~~leted, thus first it passes through the coil of M^1 then to the earth through a resistance of several hundred ohms, in R^1, another route to the earth is through shunt wire n. to lever of M thence through point c. to the coil p. on the relay M^2 in this position the relay M^2 responds to the weak currents from the distant station. While the tension on the retractile spring of M^1 being great and the cores some distance from its armature, does not so respond, but when the strong battery is put to line the lever of M^1 is attracted and the strong current instead of passing through the coil p. ~~o~~in M^2 now passes through the resistance

R^2 and coil Q. the effect of its passage through the coil R^2 is to weaken the total current nearly as much as it was increased by addition of battery power.

It is obvious that these devices can be arranged within a Wheatstone bridge or by the use of extra differential coils to effect quadruplex transmission in the various manner described in previous applications and caveats.

My claims will probably be

First. The combination of a polarized relay operated by strong currents with a double coiled differential unpolarized relay, operated by a weak current for the purpose set forth.

Second. The combination of two transmitting sounders 5 and 4. arranged in the battery circuits as and for the purposes set forth.

Third. The relays A. B. and C. figure 2. and the circuit connections and sounders arranged substantially as set forth and for the purposes specified.

Fourth. The arrangement shown in fig. 3. of the magnets M^1 and differential magnetic relay M^2. and circuit connections, for the purposes set forth.

Signed by me this 18th day of January A.D. 1875.

Thos A Edison

Witnesses. Geo. T. Pinckney Chas H Smith

DS, DNA, RG-241, Edison Caveat 59. Written by Pinckney; petition and oath omitted. [a]Place taken from petition; date taken from text, form altered. [b]All figures are on a separate page, signed and witnessed at the bottom.

1. See headnote, p. 347.

2. This is Caveat 59. It formalizes aspects of earlier work, probably dating from September 1874 and possibly as early as the fall of 1873. See, for example, Doc. 485; Cat. 297:81(3) verso, 84(3), 85(3), 86(1), 144(2), Lab. (*TAEM* 5:760, 779, 784, 786, 960).

3. That is, diplex telegraphy. Edison designed a variety of diplexes. See, for example, Docs. 485 and 507; Edison Caveat 58; Cat. 297:128(1) verso, Lab. (*TAEM* 5:909); NS-Undated-005, Lab. (*TAEM* 8:381).

4. Combining a sequence of interactive relays that respond to signals of different strength was the most common approach to diplex telegraphy in the two decades before Edison's success with the quite different method in his basic quadruplex design. None of the earlier designs was put to practical use. Prescott 1877, 825–38.

5. Edison referred here to the sounder from which the operator took the message, and not to the particular kind of device known as a "receiving magnet." Page 1866.

–534–

Caveat: Multiple Telegraphy[1]

New York, January 18th 1875[2a]

To all whom it may concern.

Be it known that I, Thomas A. Edison of Newark in the county of Essex and State of New Jersey have invented an improvement in Quadruplex Telegraphs of which the following is a specification.

The invention consists, first in the method of neutralizing the permanent magnetism in the ordinary relay used for working by increased and decreased current; second, in making the permanent magnetism or polar directive force of the polarized relay or magnet less than the weakest signalling current; third, in the arrangement of the two relays in a bridge wire, a part of which bridge wire is formed of a resistance shunted with a condenser: fourth, the method of working the local sounder of the polarized relay; and fifth, a progressive rheostat for balancing.

Fig. 1. shows my method.

Fig. 1.

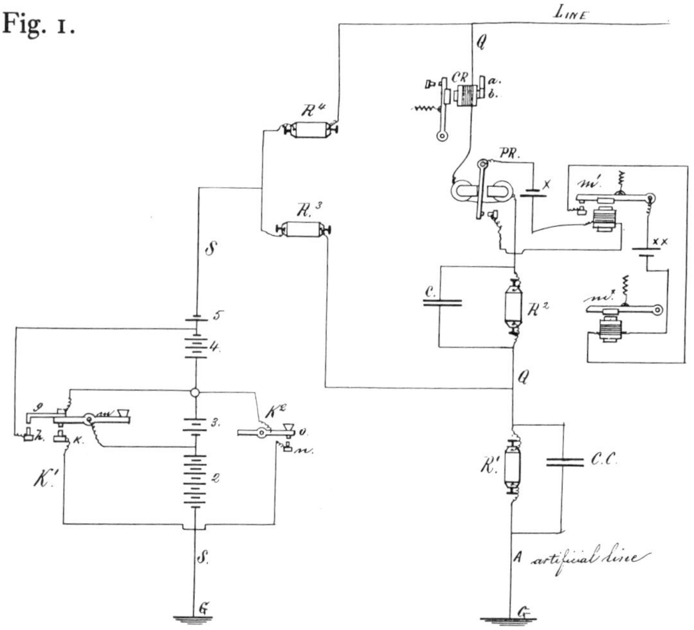

R^3. R^4 bridge wire, Q line and artificial line with wire s. form part of a Wheatstone bridge or balance. When R^3 and R^4 are equal and the resistance of the artificial line is made equal to that of the main line by adjusting the rheostat R^1 any current from the batteries 2, 3, 4 & 5, will pass to the two lines without any of it passing through the bridge wire Q containing the two receiving instruments, but any current from the distant station will come over the line part of it passing to earth through R^4 and the other part passing through the bridge wire, receiving instrument etc. to earth through R^3, R^1 and S.[3]

C.R. is a common relay operating the receiving sounder by contact with a back point in the manner already described in one of my applications.[4] The improvement which I have made on this relay consists in placing a small permanently magnetized bar of steel <u>a.</u> in contact with the yoke b. of the relay. Owing to the necessarily high resistance which I am compelled to put in the bridge wire by means of the rheostat R^2, to compensate for the static current, the current from the distant station which passes through the bridge is greatly weakened.

It is known that most all ordinary electro magnets acquire after a time a permanent magnetism in the soft iron cores, the effect of which is to turn the ordinary relay into a polarized one. This permanent magnetism ~~o~~is of little account when the relay is used upon lines in which the currents are always sent through it in one direction, but when it is worked in a circuit in which the current is reversed, the case becomes different and much trouble is caused by it especially when the current is very weak; for instance, in my present quadruplex, positive and negative currents are used to transmit one message, which is received upon a polarized relay, while an increase or decrease of the strength of these reversed currents are used to transmit another message which is received on a common relay.[5] Now if the permanent magnetism of the common relay be equal to 10 units and the signalling current equal to 20 units, then a negative current say will cause the magnet to attract its armature with a force of 30, but on the other hand when a positive current is sent 10 units of current is lost in neutralizing 10 units of permanent magnetism, hence the magnet will attract its armature with a force of 10 only. It is this constant change in the common relay, that makes signalling irregular especially when the permanent magnetism is strong and the signalling current is weak. Were it unnecessary to use the resistance R^2 the main current would be so strong that the permanent magnetism would be a very small factor and its effect would not be noticeable. The method of demagnetizing or rather balancing the permanent magnetism in the cores of the common relay is shown in fig. 2.

Fig. 2.

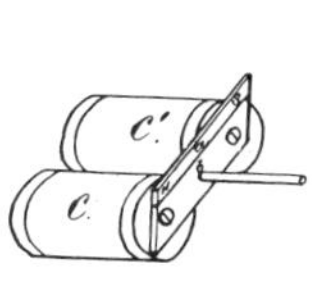

C & C[1] are the coils of the relay; b. is the back yoke, and a. is a bar of steel permanently magnetized and laid upon the yoke in such a manner that the polarity given to the two cores of the magnet is the opposite of the permanent magnetism in such cores. In practice I arrange this steel bar in such a manner that it can be made to approach or recede from the yoke b. so as to increase or diminish the amount of communicated magnetism. I do not wish to confine myself to any particular method of neutralizing the slight permanent magnetism of common relay magnets, by means of permanently magnetized bars, as the bars may be attached to the front of the two cores or a piece of magnetized steel inserted in the ends of the cores or the armatures may be made of steel and slightly magnetized or what is better made of iron to which is attached a piece of magnetized steel, or it may be magnetized by induced magnetism from a permanent magnet in its vicinity.

My second improvement consists in the employment of a polarized relay which has a permanent magnetism or polar magnetic directive force of less strength than the weakest signalling current.

Previous to this invention no account has been taken of the amount of permanent magnetism in polarized relays and they are made with all degrees of strength, as heretofore they have been used under conditions where the amount of strength of the permanent magnetism is a matter of indifference; but in the Quadruplex telegraph they are used under peculiar conditions: part of the time signals are made with two hundred cells and the next instant with only fifty cells;[6] this constant change in the strength of the current should not, theoretically, make any difference in the signals as the movement of the polarized tongue is dependent entirely upon the direction of the current and not upon its strength, but in practice it does make a difference and although the signals all come they are irregular. I have found that if the tongue be permanently magnetized with a magnetism less than that produced in the cores of the magnet by the weakest current used, the writing is perfectly even. This may be illustrated by supposing that the permanent magnetism in the tongue has a strength of 30 units and the signalling current a strength of 50 units. The tongue will be only attracted with a force of 30 units notwithstanding that the magnetism in the electro magnet is 50 units. Were the tongue of iron, it would not move to either side no matter what the amount of the magnetism in the electromagnet, but if one unit of permanent magnetism be put in it, it would move toward one of the poles with a force of one unit.

It will be seen by this that if I am signalling with 50 units of current that this will work the relay to its maximum and if this current should be increased to 200 units by the increase of the battery to send a signal to the common relay the additional current will produce no extra effect upon the polarized tongue as its maximum strength was obtained with 50 units; but on the other hand if the permanent magnetism was 100 units and the weakest signalling current 50 units, the 50 units would not attract the tongue with its maximum strength and the addition of 200 units would work it more strongly, hence the lever would be attracted strongly & weakly causing the signals to sound uneven.[7]

My third invention consists in arranging the common relay and polarized relay in a bridge wire or balance with an adjustable rheostat R^2, shunted with a condenser, the object of which is to compensate for the effect of the static current upon the received signals.[8]

My fourth invention consists in arranging a repeating magnet with the polarized relay.[9] $P.R^1$ is the polarized relay; x is a local battery; M^1 is a sounder or other electro-magnet provided with two contact points. This magnet is operated by the movement of the tongue of the polarized relay. The lever of M^1 opens and closes a second local circuit in which is included the receiving sounder M^2 and local battery x.x.

The object of using a repeating sounder, is to obtain more even signals than could be obtained by using the first magnet M^1 for receiving.

In very wet weather the current from the distant station is so weakened by defective insulation, that the tongue of the polarized relay has not sufficient power at times to close the local circuit without having a considerable resistance in circuit at the points owing to oxide etc. The effect of this slight resistance owing to want of pressure to make a perfect contact is to cause the magnet M^1 to attract its armature with a weak and strong current; hence if the operator was to receive from this magnet direct, he would be bothered by the lever making irregular sounds, but if very little play be given the lever of M^1. it will open and close the second circuit containing M^2 perfectly and the signals will be made with even sounds.

The fifth part of my invention consists in the method of increasing and decreasing the resistance of R^1. This method is shown in figure 3.

Fig. 3.

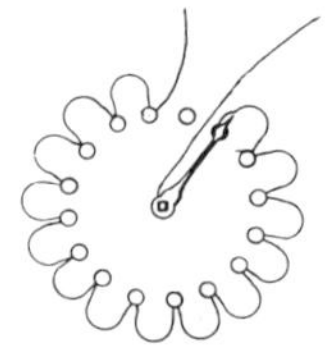

Inventor Thomas A. Edison. Lemuel W. Serrell atty.
Witnesses, Chas H Smith Geo. T. Pinckney[b]

Heretofore, rheostats used in duplex and quadruplex telegraphs, have been those in which the changes in resistance have been made by means of metallic plugs inserted between metallic squares or circles to which coils of various resistances are attached; adjusting the rheostat in this manner requires some experience and considerable time.

Instead of making the coils of various resistances, I make a great number of coils of equal resistance so that their total resistance will always be greater than ever required upon ordinary telegraph circuits worked duplex or quadruplex. I form a circle of metallic squares or circles and connect between each one a coil say of 100 ohms, the last circle or square being connected to the bridge and an arm carrying a platina wheel is connected to the earth. When not in use the arm and platina wheel rests upon a square not connected to the resistance coils, but when it is desired to balance the quadruplex apparatus as regards the outgoing current, the arm is turned to the right gradually until a perfect balance is obtained

I do not wish to confine myself to any particular method of progressively increasing the resistance, by means of a number of coils having about the same resistance as it may be done in a number of ways, but consider my invention to consist in the arrangement of a number of coils of similar resistance in such a manner that they may be brought progressively in circuit. The first coil might be 1000 while the remainder are 100 units as the resistance required on ordinary circuits is seldom less than 1000 ohms.[10]

Another portion of my invention consists in the method of reversing and increasing and decreasing the strength of the transmitting currents.[11] 5. is a battery of 20 cells; 4. is a battery 60 cells; this battery has copper to line. 3. is a battery of 40 cells, and 2. is a battery of 120 cells, this has zinc to line. When both batteries are allowed to become active by destroying the shunts at the key, one half of the battery 3.2. neutralizes the battery 5.4. and the other half sends a zinc current of

60 cells to line. If now this large battery be short circuited by depressing the key K^2, the battery 5.4. sends a copper current to line of 60 cells; thus it will be seen that the current is alternately reversed by depressing the key K^2. If key K^2 is closed and the key K^1 is open, 4. and 2. will be short circuited and only 20 cells copper will go to the line; if K^2 is now opened, battery 3. of 40 cells will overpower battery 5. of 20 cells and send 20 of zinc to line: g. is a contact spring insulated from M, and k and h. are back points.

My claims will probably be.

First. Neutralizing the permanent magnetism of a common relay or electromagnet by means of a permanent magnet for the purpose set forth.

Second. The combination of a polarized relay worked by reversals having a permanent magnetism less than the weakest signalling current with a common relay worked by increase and decrease in the strength of such reversals.

Third. The combination of a polarized relay and common relay or unpolarized relay in the same circuit with a resistance shunted with a condenser for the purpose set forth.

Fourth. Repeating the signals from the polarized relay P.R. to a receiving sounder by means of an intermediate magnet operated substantially as set forth and for the purposes specified.

Fifth. The combination of a rheostat composed of a number of coils having nearly the same resistance and operated substantially as described to form a balancing circuit or effect a balance of the outgoing currents in a duplex or quadruplex telegraph.

Sixth. The method of increasing or decreasing the signal current and reversing the same as described.

Signed by me this 18th day of January A.D. 1875.

Thos A Edison

Witnesses Geo. T. Pinckney Chas H Smith

DS, DNA, RG-241, Edison Caveat 60. Written by Pinckney; petition and oath omitted. [a]Place taken from petition; date taken from text, form altered. [b]All figures are on a separate page, signed and witnessed at the bottom.

1. See headnote, p. 347.

2. This is Caveat 60. Internal references in the group of caveats executed on this date indicate that this one was drawn up before others with lower numbers. Aspects of the design date from considerably earlier than the date of execution. See Doc. 515.

3. This should read "R^2, R^3 and S."

4. Doc. 472.

5. Edison's "present quadruplex" refers to the basic principle rather than to any particular design.

6. In the earliest trials the ratio of the stronger and weaker signal levels was not so high. This difference, or "margin," was found to be important in practice. Pope 1875, 337–39.

7. Edison had earlier designed a relay that relied on the principle of magnetic saturation (Doc. 40).

8. Edison used this configuration in circuit designs over four months earlier but did not need to employ it in practice until working on the New York–Chicago line (see illustration, p. 298). An earlier, related alternative using an electromagnet instead of a condenser is in Cat. 297:125(1) (*TAEM* 5:896).

9. This was apparently used in the circuit operated on the New York–Chicago route; see Doc. 515 n. 2.

10. See Doc. 494 nn. 2, 8.

11. Edison gave considerable attention to alternative transmission arrangements of this sort, which would not require specially constructed double-contact armature levers. See, for example, Cat. 298:48(2) verso; Cat. 297:111(1), 115(2); both Lab. (*TAEM* 5:98, 865, 872); and the accompanying illustration.

Edison tried many combinations of batteries and ordinary Morse keys as quadruplex transmitters.

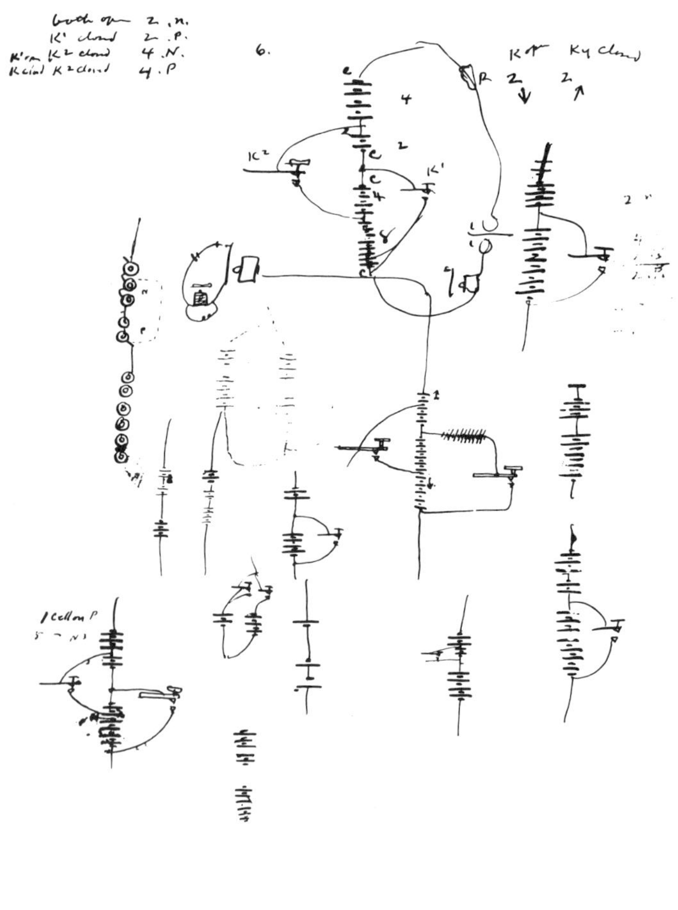

-535-

And George Prescott from William Orton

[New York,] JANUARY 19, 1875[1]

GENTLEMEN:

Referring to the negotiations and arrangements heretofore made between you and the Western Union Telegraph Company for the sale and transfer to that company of all your patents relating to the duplex and quadruplex telegraphy, subject to definite ascertainment of compensation to be paid, and especially to the two offers in writing made by you on or about the 30th day of December last, as follows:[2]

"1st. We will take twenty-five thousand down and twenty-five thousand in six months for all patents, and a royalty on quadruplex[a] of $166 per year for each circuit created."

"2d. We will take twenty-five thousand down for all patents and a royalty of $233 per year for each circuit created."[3]

I hereby notify you, on behalf of the Western Union Telegraph Company, that the proposition for compensation above quoted, and by you marked "2," is hereby accepted as made, and the company is ready to close the business at your earliest convenience, and to make all the payments called for upon receiving from you proper assignments and transfers of the said patents.[4] Very respectfully,

WILLIAM ORTON, *President.*[5]

PL (transcript), NjWOE, Quad. 72.16, p. 17 (*TAEM* 9:267). There are several versions of this document in the Quadruplex Case and another in LBO 14:120–21; except where noted, differences are insignificant. [a]Some versions have "duplex" here.

1. Marshall Lefferts handed this letter to Edison the next day.

> Immediately after reading the same, said Edison said first "No," and then further stated to deponent that he would not sell or convey any of his patents to the Western Union Telegraph Company, for any price whatever; that he denied that there was any binding contract between him and the Western Union Telegraph Company, because, as he said, the two propositions quoted in the letter of Mr. Orton, although in writing, were not signed by me, and declared, that it was his intention not to convey to the said Company any interest whatever in any of his Patents. [Lefferts's affidavit, Quad. TLC.7, p. 20 (*TAEM* 10:995)]

2. Here Orton quotes Doc. 521. Although he treats that document as a formal, joint offer from Edison and Prescott, its status was hotly disputed in the litigation over ownership of the quadruplex.

3. By February 1875 Western Union had seven quadruplex circuits working from the New York office alone. *J. Teleg.* 8 (1875): 51; see also Doc. 521 n. 4.

4. It is not clear whether Edison received any payment on the basis of this document. He had not by mid-August (see Doc. 603), and at the end of 1875 he released Western Union from any obligation for "further

payment" (Doc. 694), but that might have referred only to the $5,000 he had received in December (see Doc. 515 n. 6). However, Orton also arranged a $5,000 advance for Prescott to match Edison's December payment, and he received Prescott's written agreement to these terms (Prescott's receipt, 16 Jan. 1875; Prescott to Orton, 20 Jan. 1875; both Quad. 70.4, pp. 16, 19 [*TAEM* 9:333, 334]); see also Orton to Lowrey, 20 Jan. 1875, LBO 14:125–30).

5. Around this time Orton was also seeking to purchase any possible patent rights that Henry Nicholson might have had to the quadruplex. Orton to Anson Stager, 23 Jan. and 3 Feb. 1875, LBO 14:145–48, 171.

–536–

To the Commissioner of Patents

NEWARK, N.J., Jan. 23d, 1875.

SIR:

There are on file in your office certain applications for patents in my name, as sole inventor, accompanied with a request to issue the patents, when allowed, to myself and George B. Prescott.[1]

I made this arrangement with Mr. Prescott under an erroneous impression.

On the 1st October, 1870, I made a copartnership contract with Mr. George Harrington, and on the 4th April, 1871, in furtherance thereof, I made to him an assignment, coupled with an irrevocable power of attorney, reciting the main provisions of the contract, both of which are still in full force and effect, whereby the control of these inventions was placed in the hands of Mr. Harrington.[2]

Under and by virtue of this power of attorney and assignment therein, Mr. Harrington claims title in these inventions, and objects, and has objected, to any deviation from the said assignment and power of attorney. I therefore withdraw my request for the issue of patent for duplex and quadruplex transmission to Edison and Prescott, and unite with George Harrington in his petition, this day filed in your office,[3] requesting the issue of the letters patent to Geo. Harrington and myself, in the proportions set forth in the power of attorney and assignment of April 4th, 1871, and the contract therein recited.[4] Respectfully yours,

THOS. A. EDISON.

PL (transcript), NjWOE, Quad. TLC.3, p. 202 (*TAEM* 10:922). There are several essentially identical versions of this document in the Quadruplex Case.

1. Docs. 467–73 and Case 112 (see Doc. 449). Under the terms of his agreement with Edison (Doc. 466), Prescott believed he also had a claim to Cases 111 and 113, neither of which had yet been filed. Edison

had executed Case 111 on 18 January 1875 without any request for joint assignment to himself and Prescott; he filed it on 26 January. He executed Case 113 in February and filed it on 23 March 1875. On 20 and 21 January, Lemuel Serrell had sent to the Patent Office various amendments, clarifications, and arguments for Edison in connection with most of these applications. These included formal abandonment of Case F (Doc. 310) and presumably also Case E (Doc. 309). Serrell responded to examiner Zenas Wilber's objections and strongly urged "an immediate allowance." Pat. App. 168,385, 178,221–178,223, 180,858, 207,723–207,724; see especially Serrell's letter of 20 January in Pat. App. 178,221.

2. Docs. 109 and 155. Their applicability to these inventions was strongly contested; see n. 5 below.

3. For Harrington's petition, see Quad. 72.6, p. 40 (*TAEM* 9:113). Other practically identical versions also survive.

4. Following his receipt of this document, Commissioner of Patents John Thacher told Wilber to report to him on whether or not Edison's assignment to Harrington covered the applications in question. This turned on whether or not the phrase "fast telegraphy" had become standard, professional usage for all systems able to speed the flow of business in a busy telegraph office, or whether it really referred only to automatic telegraphy. Wilber reported on 29 January that Harrington indeed could claim Edison's inventions in duplex and quadruplex telegraphy (the report is in Edison's Pat. App. 178,221; printed versions also exist—e.g., Quad. 72.11, pp. 55–57 [*TAEM* 9:192–93]). The Commissioner did not find Wilber's view compelling, and Western Union and Prescott opposed Harrington's petition, so Thacher allowed both sides to prepare and present their arguments as fully as they could. See Doc. 554; on Thacher, a former patent examiner and future patent lawyer, see *NCAB* (rev. ed.) 3:530.

–537–

To William Orton[1]

WASHINGTON, D.C., Jan'y 23, 1875.[2]

Dear Sir:

When I opened negotiations[a] with you in regard to my inventions of duplex and quadruplex, I hoped, in case we could agree upon the terms, which, however, has not been done,[3] that I should be enabled to give a valid title thereto.

I am now advised that the claims of Mr. George Harrington, under a prior contract and irrevocable power of attorney, of which you were aware,[4] and which are still in full force and effect, will prevent a valid transfer of such title, either to you or other parties.

Under these circumstances, all further negotiations between us must necessarily cease, and any moneys expended or paid in anticipation of final action I am ready at any moment to return to you.[5] Respectfully,

(Signed), THOS. A. EDISON.

PL (transcript), NjWOE, Quad. TLC.4, p. 214 (*TAEM* 10:943). There are several versions of this document in the Quadruplex Case; except where noted, differences are insignificant. [a]"correspondence" in one version.

1. Though done in Edison's name, this letter was almost certainly prepared by Harrington, who resided in Washington, D.C. Under his power of attorney from Edison of 4 April 1871 (Doc. 155), Harrington prepared several documents relating to the sale of Edison's automatic and quadruplex patents to Jay Gould.

2. This was not received until 9 February and was postmarked in New York City at 6 A.M. that day (see n. 5). A letter from Edison to George Prescott, also dated 23 January 1875, offered to repay Prescott for any expenses related to the duplex patent applications and included a copy of this document. Quad. 72.16, p. 21 (*TAEM* 9:269).

3. Orton and Western Union repeatedly insisted that Docs. 521 and 535 legally constituted precisely that agreement.

4. Docs. 109 and 155. In his testimony in the Quadruplex Case, Orton claimed that before January 1875 he was unaware of these documents and of Harrington's claim under them (Quad. 70.7, pp. 196, 258–64 [*TAEM* 10:104, 136–39]). In specific reply to this letter, through Western Union's lawyers, Orton claimed Edison had insisted that his relations with Harrington concerned only technology totally distinct from that in which Western Union was interested. See n. 5.

5. Orton responded on 9 February 1875 by having Western Union's lawyers—the firm of Porter, Lowrey, Soren, and Stone—write to Edison. (Grosvenor Lowrey had been informed of Orton's view of the case and shown several documents, such as Docs. 445, 466, 517, 519, 521, and 535, on 19 and 20 January; see Orton to Lowrey, 20 Jan. 1875, LBO 14:125–30). Their reply to Edison noted the postmark, incorporated a full transcription of Doc. 535 (and thus also of Doc. 521), and presented the following points (Quad. 71.2, p. 144 [*TAEM* 10:300]). They asserted that Edison, together with Prescott, could indeed give a valid title to the quadruplex, and that Western Union would defend Edison against claims by Harrington if Edison would act appropriately. They claimed there existed a binding sales agreement between Western Union and Edison and Prescott. They said that further negotiations were not the issue, but rather that Edison was required to take action to obtain and deliver good patents on the invention for the company. The letter also threatened Edison personally with punishment for contempt if he or Jay Gould violated in any way the injunction they had obtained in the New Jersey Court of Chancery (see Doc. 538 n. 1). Western Union also had Lowrey act to oppose Harrington's and Edison's position in proceedings in the Patent Office. See Docs. 536 n. 4, 554 n. 9, and 577 n. 6.

–538–

Affidavit of Gerritt Smith

New York, January 27, 1875[a]

Gerritt Smith, being duly sworn, deposes and says:[1] that he resides in Astoria, Long Island City, and is well acquainted with the parties above named;[2] that deponent had a conversation with the defendant, Edison, at the City of New York, on the 19th day of January, 1875, in which he informed deponent that he, said Edison, was not going to the office of the Western Union Telegraph Company, the complainant herein; that its President, Mr. Orton, had not treated him right; that he was obliged to get help from some source, and, seeing no prospect of it from the Western Union Telegraph Company, he was forced to look for assistance elsewhere, and that he had found just the man to help him out; that he should take the quadruplex to the opposition; that the complainant could not use it, and if they did they would have trouble about it; and that he was going for the Western Union Telegraph Company red hot, and was interested with business men, men that slept with their boots on; that he had stopped making the quadruplex apparatus at the shop,[3] and the force were all engaged on automatic instruments; and that he had accepted the position of Electrician to the Atlantic and Pacific Telegraph Company;[4] that the Western Union Telegraph Company would give him $100,000 for the quadruplex, but that he would not accept it now; that he had done with them.[5] The said Edison also informed deponent that he, said Edison, had an arrangement in his shop, by a bell wire, by which he was notified when any one from the Western Union Telegraph Company called to see him, so as to avoid such person; and that he had caused himself to be denied to deponent and to said Prescott while at his shop in Newark.[6] And further deponent saith not.

GERRITT SMITH.

Sworn to and subscribed before me, in the said City of New York, this 27th day of January, 1875.[b] Witness my hand and official seal, (L.S.) WILLIAM H. CLARKSON, *Commissioner for New Jersey, in New York.*

PD (transcript), NjWOE, Quad. TLC.7, p. 22 (*TAEM* 10:996). [a]Date taken from text, form altered. [b]"Sworn . . . 1875" enclosed in a brace.

1. This statement and a similar deposition made by Marshall Lefferts were supporting documents in Western Union's petition in the New Jersey Court of Chancery filed the following day (App. 3, Western Union Bill of Complaint). The company sought and obtained an injunction forbidding Edison (and George Prescott) from selling the quadruplex telegraph inventions to any party except Western Union. The petition, injunction, and Lefferts's affidavit are in Quad. TLC.7, pp. 1, 19, 24 (*TAEM* 10:985, 994, 997); Edison's answer (filed 20 May 1875) is Quad. 72.14 (*TAEM* 9:204).

2. A formal heading for this affidavit names the complainant (Western Union) and defendants (Edison and Prescott) and indicates it is part of a bill in chancery. Though the court proceeding was directed only at

Edison, Western Union had to insist upon the validity of his partnership arrangement with Prescott and thus treat both as defendants.

3. Western Union had ordered twenty sets of quadruplex equipment from Edison and Murray. Docs. 517 n. 1 and 526 n. 8.

4. See Docs. 522 and 585.

5. Edison had not yet received Orton's agreement to any specific terms. The day after he met with Smith, Edison suggested to Marshall Lefferts that patents for the quadruplex might be bought by Western Union for half a million dollars. Lefferts's affidavit, Quad. TLC.7, p. 20 (*TAEM* 10:995).

6. Prescott had been trying to get in touch with Edison for some time. Prescott to TAE, 13 Jan. 1875, Cat. 1144:63, Scraps. (*TAEM* 27:597); "The Telegraphic Contest," *Telegr.* 11 (1875): 22.

–539–

Memorandum: Automatic Lines of Atlantic & Pacific Telegraph Co.

c. January 27, 1875[1]

Bedford[2] OK= Waterbury OK when gets local—ordered J. G. S.[3] New Brittain no relay ordered waterbury send him one he having two. West[a] Meriden all right= ~~Mis~~ Meriden no Auto relay ordered thro J. G. S Hartford all right= Warehous point[4] got one not in ordered him put it in & keep it there, through opr at Windsor Locks who promises to see it done Windsor Locks got his in long ago= "Strange" Springfield got auto relay in nothing else good to say about the place. Spencer[5] cant find out anything from here Operator cant understand a word I say evidently learne[d] some other alphabet=

Worcester all hunky= put full set of instruments in there never had a local or sounder Opr in Business 24 years. had two auto relays no use for 2nd I brought it with me. Lake Crossing[6] Just heard of the town tonight 430 PM Couldnt raise it—since. Brighton no auto relay 4 miles from Boston going send him one brought from worcester Southern route=[b]

~~Ind~~[----] In dark as to offices between NY & Br[idgeport?] good writing at Br any speed Br fitted up auto to receive all OK= New Haven fixed all right get NY good there= Hartford, all right for automatic any time= Notes=

Pass on to sg=[7]

Woman named Gouge[8] there as manager has little pen one side a stinking dye house keeept by Jew her husband a out of job 4th rate actor=loafs on her she has telegrapheic school in offic students filling small space to repletion pay her $2[-] 50c each per week Business men instructed correspondents not to send business to them via A.&.P act of dislike to go into institution manageress and Jew landlord have frequent heavy word combats much to amusement hilarious students = pl Contrast Mr Pearson[9] night signal Operator very gentlemanly young fellow well known among Business men and popular (Outside ~~the tstimony~~ testimony) He has in view and can consummate bargain for good office slight advance on present one number old customers will come back if he is appointed—

Side note = Universal testimony Sam Bowles[10] governs the press matters telegraphice of large percentage New England papers everybody says = To make your fortune secure Bowles, not difficult matter if work satisfactorily done. S[p?]ringfield[c] Corruption wiped out first & important step to that end

All WU offices in N England towns well stocked with Operators and doing app~~l~~arently handsome business. Our offices most part poor in every respect that would give impression of a business concern si~~n~~[--]gns wanted etc ~~make~~[b]

Quite prominent complaint of managers = "Utter impossibility get satisfactory replies or in fact any at all to office messages sent to NY concerning their business notable instance Largest Longest standing customer Hartford had complaint present month inquired for explanation modestly promised it = went away came back no answer Hufty Came again No ans Swore off eternally

Better operators at one or two way stations on no 5 essential to doing Automatic there = Cant adjust for fartherest office = ~~wanted Spencer~~[d] Offices near NYork habitually ground to work with NY offices near Boston habitually do the same thing. Offices in the middle swear loudly to NY & to Boston only result repetition of this offence

I get Boston splendid at Worcester just as fast as possible for him to turn the mark at very highest speed actual timing near 1400 min deepest blue black = showing good condition of wire bet[ween] ~~&~~here & W at W I get NY tip top on No 6 all tailed to devl on No .5. caused perhaps only by Warehouse point & one other man =

All instruments of this ~~cir~~Co bet[ween] Boston & NY either old Amn pattern or big box relays all have served over their time 9 years everyone should at once be recalled and auto relays substituted.

I find a relay here on wire no .6. that measures—four inch Spools, and two inches thick through— such is the inst at end of wire against which auto relays have to work majority of relays in this office very trifle superior to it all great lumbering things = 2 old style auto relays on auto tables. Ought be ~~subs~~ replaced with best pattern—[b]

Switches.

Only place that has anything at all equal to service and even moderately safe is Hartford & Middletown Each got new pin switch boards. I like them better than anything else because wherever one is found there you will invariably find office wires strung in nice style ~~wi~~ [-]& wires not in use safely cut out & everything about instruments in corresponding good condition =

Should be one sent to NHaven at once: he only has one of his own make which is tumbling to pieces and & not safe for 2 consecutive hours his office connections & instruments generally partake of the same general decay = Most dangerous point on the line & only saved from being a pest by energetic movements of manager who is a good worker and knows all the loose points in his outfit though not a morse operator has made nearly everything in his office by his own hands including a pigeon hole desk = a quartette table press stand and number of other things including a large ~~and~~ variety of Begging appeals for instruments & a board the latter he says would rescuscite his ofs & put everything etc in good shape

Phila[e] has old auto switch. have Cockey[11] <u>gobble</u> it

~~Switchboard at Worcester is a new one to me cant never explain it except~~

Boston Switch here now heres s[o]m[e]t[hin]g u love. = its marked 1867. its a black board 18 inches square with 2 full double rows of Binding posts and dem a ting else.

Br [~~Read Read~~?][c] 25 msgs pr da both papers 1–2 wires = Coleman 65 d asst his sister 25 .d. dissatisfied

Magnificent show window ofs plenty of Room good pelan rent 400 per year. Cole has telegh school advised him teach perf[o]r[at]ion got no perftr

Good slice A&P stock in this town no morning papers = Complains biz very much delayed in NY

Wy[e] OK got everything expct a local to run sounder with I sent to J G. S. dont k[no]w wh[e]t[h]er its been sent hvg to wk with Relay he dont hear Meridan say ~~aga~~—[12] I think he is doing so past 2 days n[ot] heard anything to contrary good man takes to it. had widen 18 inches to get table & chair in [aut?][c] room nuf for 2 to sit in & copy he has no assistant.

<u>Middlefield</u> bet[ween] meridan & Middletown great

t[rou]bl[e] is they dnt kw wt to do wi em wn ty get em   sm of em [-]dt no hw to pt in a relay   sm hv n[o]t[hin]g but a box relay & never saw a sounder & local   worcester a place like it had only a box relay   u want a detailed report

Meridan[f] wire should have battery at Middletown so as not to Change the adjustment so greatly at Waterbury and Meriden when through to NYork NY fails miserably to get Meridan Morse, =

Put Hulbert[13] on press till they, get used to it East =

You[g] want get ½ doz auto Relays some sounders and some Two and some four wire switches of that pattern in making ready for me to bring with me back in a bag

AD, NjWOE, Scraps, Cat. 30,096:41 (*TAEM* 27:576). Nine numbered, two unnumbered, and two partial pages of undated notes were pasted together in the scrapbook. [a]Interlined above. [b]Followed by centered horizontal line. [c]Unclear. [d]"~~wanted Spencer~~" interlined above. [e]Start of unnumbered page. [f]Start of torn page with "3" at upper right. [g]Start of unnumbered torn page.

1. This tour of Atlantic and Pacific offices on lines from New York through Connecticut to Massachusetts probably occurred after George Harrington's letter of 22 January 1875 to Thomas Eckert regarding the takeover of automatic telegraphy by Atlantic and Pacific Telegraph Co., in which Harrington noted that Edison would indicate his immediate assistants from the personnel of Automatic Telegraph Co. (1:41, Box 17B, *Harrington v. A&P*). Edison probably left for this tour after writing Doc. 537 but before the orders for relays mentioned in these notes appeared in Edison and Murray accounts on 1 February (Cat. 1223:12–13, Accts. [*TAEM* 22:137]). The takeover of Automatic Telegraph lines and equipment occured about 1 February (Bill of Complaint, 1:16, Box 17B, *Harrington v. A&P*).

Towns Edison visited on his tour of the Atlantic and Pacific line.

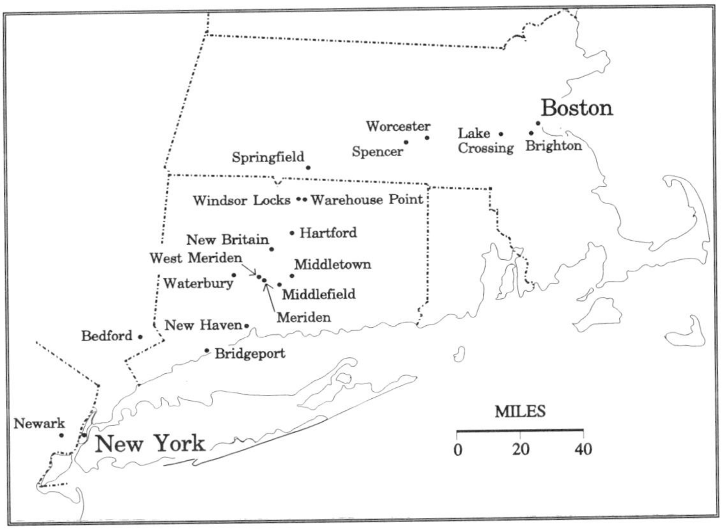

2. Presumably Bedford, N.Y.

3. Possibly a reference to Jay Gould's secretary, G. P. Morosini. The Atlantic and Pacific supply agent, Edward Cockey, appears not to have been directly involved with such orders until the end of February (see 75-018, DF [*TAEM* 13:652–705 passim]). A memorandum of 8 March notes that a new order of equipment for eastern lines to Omaha was provided for "in accordance with arrangement between Mr. Gould, Mr. Edison, and Genl. Eckert," suggesting that the earlier equipping of lines north of New York also took place under such an arrangement.

4. Warehouse Point is a town in north central Connecticut.

5. Spencer is a town in Massachusetts.

6. A railroad stop in what is now Wellesley, Mass.

7. Probably Springfield, which would be the next major city on the line.

8. Mrs. L. E. Gouge, manager of the Franklin Telegraph office, lived at 12 Court St. *Springfield City Directory and Business Advertiser* 1875, 213.

9. W. H. Pearson, listed as a U.S. Signal Service employee, lived at 79 Vernon St. Ibid., p. 308.

10. Samuel Bowles was editor of the *Springfield Republican,* which under his leadership became an influential provincial paper noted for fighting financial and political corruption. *DAB,* s.v. "Bowles, Samuel (1826–1878)."

11. See n. 3 above.

12. May be "ya."

13. Unidentified.

–540–

Caveat: High-Speed Telegraphy

[Newark,] February 4, 1875[a]

To all whom it may concern.

Be it known that I, Thomas A. Edison of Newark in the county of Essex and State of New Jersey have invented an Improvement in Accumulators for Receiving, Holding and Transmitting Telegraphic signals, of which the following is a specification.

The object of this invention is to transmit telegrams at high speeds over long lines of telegraph.

The invention consists in the use of accumulators of electricity, composed of a number of condenser plates[1] or secondary batteries arranged and combined together with a revolving contact arm moved regularly by clock work or other source of power and transmitting a series of waves of electricity necessary to form the characters and compose the message, into this aggregation of accumulators in such a manner that when such accumulators are discharged into a telegraph line progressively the dots and dashes of the messages will pass out of the combined accumulators in the same order as they were sent in.[2]

The invention further consists in employing a second accumulator at the receiving station into which the successive waves from the first accumulator are received.

The invention further consists of the method shown of storing up a message in the accumulator, also the method employed for discharging into the line and receiving from the line by the accumulators.

Fig. 1.

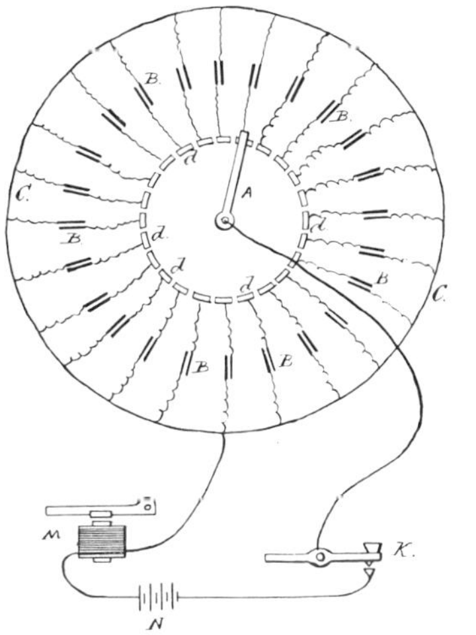

Inventor per. Lemuel Serrell atty.
Witnesses. Geo. D. Walker. Geo. T. Pinckney

In the drawing:

Fig. 1, shows the revolving arm and circle of contact plates each of which is connected to an inductive leaf of a condenser and facing another set of leaves all of which are secured to a common wire. These leaves are separated by an insulating material; they form in fact a large number of condensers or accumulators of electricity arranged in such a manner that any one may be charged or not at the pleasure of the operator.

A, is the revolving contact shaft, d, d, d, the contact points insulated from each other and connected to one leaf of the condensers, B. B. B. C, is a common wire to which is connected all the remaining plates of the condenser. The arm is arranged with a clockwork or train of gearing and can be made to start from the zero point and sweep around the full circle.

To this arm is connected one pole of the battery N, while the common wire is connected to the other pole. Within this circuit is placed the sounder M, for the purpose of allowing the transmitting operator to hear his own writing and also for the purpose of ascertaining that the charge from the battery passes into the various condensers as the arm progresses around the circle when the key is operated. This circle is composed of several hundred points connected to as many plates or condensers so that before the arm has passed completely around the circle an ordinary message can be sent upon the Morse key K. Suppose that the arm had started to revolve and the key k, was open none of the condensers would become charged; if the key is now closed to make a dot, one or even two or three of the condensers will receive a maximum charge of electricity, which they will hold for several hours according to the insulation of the diaelectric[b] which separates the plates of the condenser. If now the key is opened after making this dot the following condensers will not be charged. If the key is closed to make a dash, five or six condensers will become fully charged. It will thus be seen that some of the condensers are charged and others are not, which is controlled by the operator who works the key. After the message has been sent into the accumulator, it is disconnected from the sounder M, battery N, and key K, and by means of suitable switches is connected to the main line ready for transmission over the wire, fifteen to twenty times faster than the message was sent into it, the clockwork which rotates the arm A, being so arranged that any degree of speed of rotation may be given the arm.

At the distant station is another accumulator precisely like the one already described which is placed in connection with the line ready to be charged by the pulsations sent into the line by the above described accumulator. Of course the plates of the receiving accumulator are free from charge.

By means of the ordinary Morse relay and sounder, the operator at the transmitting station, signals to the operator at the receiving station that he has a message ready. The receiving operator says "go ahead": at the same instant both start their clock works going at high speeds and the message is transferred almost instantly from one accumulator into the other through the medium of the line. After being received it is disconnected from the line and placed in connection with a receiving sounder; the speed of the clockwork is reduced and set going, the currents discharge through the sounder in the proper order, allowing the receiving operator to read them by sound.

It will thus be seen that high speeds are obtained without employing any recording machinery or magnetism or devices whatever upon the wire and without the use of perforated paper or other such devices, and allows the use of Morse instruments and operators. It is obvious that innumerable modifications can be made in the application of the accumulation system of telegraphing and I do not wish to confine myself to any particular arrangement of the wires or batteries or any particular arrangement for progressive accumulation of the signals.

I shall probably claim.

First. The transmission of telegraphic signals by the system of accumulation substantially as set forth.

Second. The combination of a battery and key with a number of accumulators either chemical or inductive and a device for the distribution of the waves of electricity in these accumulators for telegraphic purposes, as set forth.

Third. A range or series of condensers or accumulators and a moving conductor to distribute pulsations of electricity to such accumulators in the same progression as they are sent, to be stored for after transmission or development.

Signed by me this 4th day of February A.D. 1875.[3]

Thos A Edison

Witnesses Geo. T. Pinckney Geo. D. Walker.

DS, DNA, RG-241, Edison Caveat 61. [a]Date taken from text, form altered. [b]First "e" is interlined above.

1. Condensers were very expensive; see Doc. 420. Later in the year Edison attempted to manufacture condensers himself (Doc. 698).
2. For the genesis of this idea see Doc. 360.
3. No patent application has been found.

–541–

From Stephen Field

San Francisco, Feby 10—1875[a]

My dear Sir:

Have you a patent on a punching machine, to be used in connection with Automatic Telegraphy, which will punch the tape in Roman letters?[1]

If so what are the chances for a persons getting the use of the Patent right?

We have an Enthusiast here who has made what he calls a discovery, (and for all that I know to the contrary it may be a useful one) in ~~a~~Automatics, it is that paper on being soaked with kerosene will allow pencil writing to strike through, and make a conducting film through the paper, of any character which may be desired.

Is there anything new in it?— Yours truly

S. D. Field[2]

P.S. Why dont you "see" those "Telegrapher" fellows a little and shut up their howling[3]—I know how it is myself—

ALS, NjWOE, DF (*TAEM* 13:399). Letterhead of The Electrical Construction and Maintenance Co. [a]"San Francisco" and "187" preprinted.

1. U.S. Pat. 151,209.

2. Stephen Dudley Field (1846–1913), electrical inventor and manufacturer, was the nephew of Cyrus Field, American promoter of the first Atlantic cable. Stephen became a telegraph operator in California in 1863 and, following the Civil War, was associated with the unsuccessful attempt to establish a Russian-American telegraph line. He was an inspector with the San Francisco Fire Alarm Telegraph Co. until 1872, when he organized the California Electrical Works to manufacture telegraph and electrical equipment, some of which he had invented. Field eventually obtained more than one hundred electrical patents, including a system for working telegraphs with dynamos in place of batteries (used extensively by Western Union), a high-speed stock ticker, and a submarine quadruplex telegraph system. During the 1880s Field joined Edison in the invention and manufacture of electric motors and railways. *DAB*, s.v. "Field, Stephen Dudley."

3. See Doc. 492.

-542- Newark, 11 February 1875[1a]

Invitation

Yourself and Company are respectfully invited to attend a

Masquerade Party

To be given by

Mrs. Thos. A. Edison,

Thursday Ev'g, Feb. 11th, 1875,

At her residence, 65 Bank Street, at 8 o'clock, sharp.

COMPLIMENTS OF

Mr. & Mrs. Thos. A. Edison.

CHAS. E. SMITH'S PRINT.

PD, NjWOE, Scraps., Cat. 1178 (*TAEM* 27:146). [a]Date of party.

1. This was Edison's twenty-eighth birthday.

–543–

From Ezra Gilliland

Cincinnati, Feby 22 1875.[a]

Friend Edison

I send you today a Cincinnati Enquirer Containing an article[1] on Dr. Nicholson[2] & his Quadruplex— I have made the most of his apparatus up to the time the WU Tel Co took hold of him[3] I am perfictly familiar with the history of his invention as I showed him how to work his two line alphabet over one wire & made him the polarized relays which led to this invention he now claims as his—[4] If there is any information you desire in regard to it[b] I would be pleased to furnish it—

We are working ten 10[c] men & doing a very fair business principally however outside of Telegraph & Electrical apparatus[5]

I left the WU Cos Employ about 2½ years ago since which time have devoted entire time to our business— My Brother[6] is my partner— My father through the Orton's, is connected it the WU Tel Co.[7] Hoping to hear from you I remain as ever Yours Very Truly

Ed. T. Gilliland[8]

ALS, NjWOE, DF (*TAEM* 13:192). On letterhead of E. T. Gilliland & Co. [a]"Cincinnati," and "1875." preprinted. [b]"in regard to it" interlined above. [c]Circled.

1. The article, "Old Flip-Flap," is in Cat. 1143, Scraps. (*TAEM* 27:245). It was also introduced into evidence in the Patent Office interference between Edison and Nicholson. During his testimony in the Nicholson interference, Edison incorrectly recalled that someone else had sent this article and that he had then shown it to Gilliland. Testimony and Exhibits on Behalf of T. A. Edison, p. 20, *Nicholson v. Edison*.

2. Dr. Henry C. Nicholson (b. 1834) was a Kenton County, Ky., physician whose interest in electricity led him to make experiments on telegraphy. As early as 1847 he had experimented with a printing telegraph using funds provided by American telegraph entrepreneur Henry O'Rielly. He conceived a multiple telegraph around 1867 and began experiments along this line in 1870. In 1871 and again in 1873 he communicated his ideas to Western Union president William Orton and company electrician George Prescott. Nicholson also demonstrated a system at the Cincinnati Exposition in 1873. After reading an account of Edison and Prescott's work on quadruplex telegraphy in the 15 July 1874 issue of the *Journal of the Telegraph*, Nicholson applied for a patent. Nicholson filed his application on 14 October 1874; Edison had filed his on 1 September (Doc. 472). Action on the applications was sus-

pended in view of a probable interference on 31 December 1874 (Doc. 526 n. 4). Nicholson believed that Edison and Prescott had adopted ideas communicated in his letters to the company, and this became the basis of his claims in the patent interference. The proceedings were delayed by other Patent Office and court actions; it officially began in 1878, the case was not resolved until 1885, and Edison's patent on the core of the quadruplex was not finally granted until 1892. Nicholson's testimony, *Nicholson v. Edison*; Commissioner of Patents to TAE, 16 May 1892, Pat. App. 480,567.

3. To protect its rights to quadruplex telegraphy, Western Union entered into negotiations with Nicholson in January 1875 for control of his invention, concluding an agreement on 12 February 1875. Orton to Anson Stager, 23 Jan. and 3 and 9 Feb. 1875, LBO 14:145–49, 171–77, 189–92; Nicholson to Western Union, 16 Dec. 1889, C. L. Buckingham Correspondence File, Box 12, WU Coll.

4. Nicholson initially sought to increase transmission speed by developing a modified telegraph code that could be sent by two keys working separate wires and received by two sounders or registers. One of Edison's notebooks contains Edison's 1875 drawings, based on Gilliland's explanations, of Nicholson's original system and Gilliland's suggested modification for working it over a single line with polarized relays (Cat.

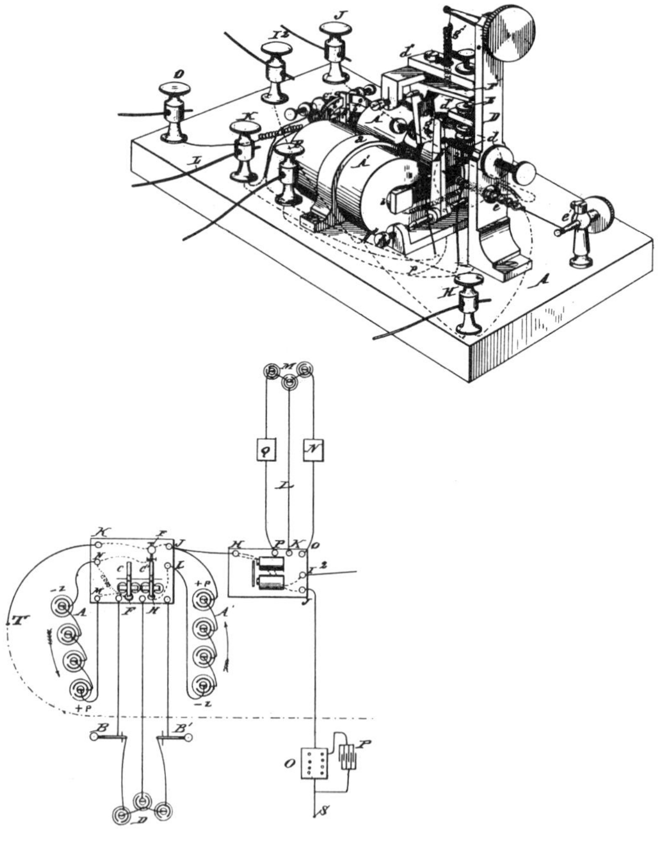

The heart of Henry Nicholson's 1874 quadruplex telegraph patent application (U.S. Pat. 332,549) was a complex, double-acting relay (shown in detail and at center right of the circuit diagram). In the circuit diagram, **Q** *and* **N** *are sounders,* **B** *and* **B′** *are transmitting keys,* **O** *is a rheostat, and* **P** *is a condenser.*

1176:65–66, Lab. [*TAEM* 6:64–65]). According to notes accompanying these drawings, Nicholson asked Gilliland in September 1870 to construct a two-pen register (also drawn by Edison), and it was at that time that Gilliland, who knew about multiple telegraphy designs from his earlier association with Edison, suggested working the system over a single wire (*TAEB* 1:48 n. 2). Nicholson soon focused his attention on sending two Morse messages over a single wire. See also Nicholson's testimony in *Nicholson v. Edison*.

5. According to the letterhead, E. T. Gilliland & Co. manufactured and sold electrical and telegraph machinery at the corner of Fifth and Elm Streets in Cincinnati. They were also agents for Geo. H. Bliss & Co., a Chicago electrical manufacturer.

6. Ezra had two brothers, James and William, but only James is known to have been a manufacturer of electrical equipment. He and Ezra later formed the Gilliland Electric Manufacturing Co. in Indianapolis, Ind. James Gilliland's testimony, *Edison v. Lane;* "In Memory of Mrs. Caroline G. Gilliland," BC.

7. Ezra's father, Robert Gilliland, later became associated with Edison's electric-pen manufacturing operation. His connection with Orton is not known.

8. Ezra T. Gilliland (1848–1903), an electrical manufacturer and inventor, had worked as a telegraph operator with Edison at Adrian, Mich., and Cincinnati, Ohio (*TAEB* 1:16). Later in the year he and Edison formed Gilliland & Co. (see Doc. 622).

–544–

Caveat: Automatic and Cable Telegraphy

New York, February 24, 1875[a]

To all whom it may concern. Be it known that I, Thomas A. Edison, of Newark, in the County of Essex, and State of New Jersey, have invented an Improvement in Telegraphic Transmitting and Receiving apparatus for Cables, of which the following is a specification.

The object of this invention is to transmit telegrams at high speeds over long lengths of submarine cables.

The invention consists in the method of transmitting the necessary waves and compensating currents to form telegraphic characters, the arrangement of the circuits, chemical instruments, compensating devices etc. for recording the same. It further consists in the employment of a spectroscope translator whereby the particles of metal set free or deposited upon the strip of paper by electro decomposition brought about by the waves of electricity from the distant station, can be detected with precision when the quantity is several times smaller than that which would be sufficient to give a colorable mark.

It further consists of several devices attached to the spectroscope to facilitate the reading of the characters.

Fig. 1.

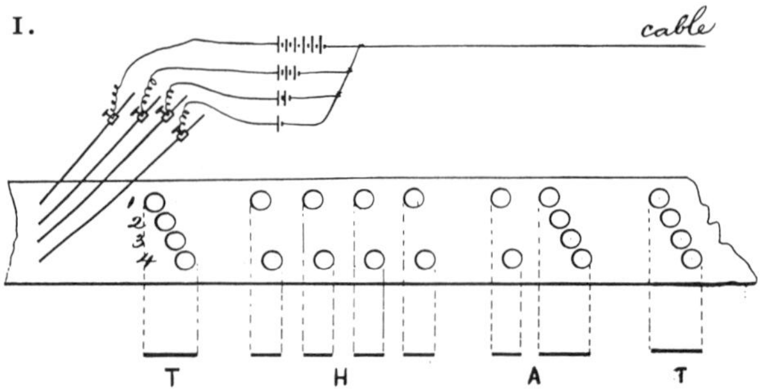

Fig. 1. shows one form of perforated slip for transmitting the waves of electricity necessary to produce telegraphic characters at the distant end. The word "that" is shown.

To form a dash I use fore holes 1. 2. 3 & 4. The hole 1. is in line with a contact roller to which is connected the largest signalling battery; hole 2. is in line with a contact roller which is connected to the next weaker signalling battery, while the holes 3 & 4. are in line with contact rollers having signalling batteries still weaker.

These holes are so arranged that before the contact in line with hole 1. breaks contact, the contact roller makes contact through hole 2. and so on.

It will be seen that the first part of a dash is made with a strong current, the next portion with a weaker current and so on and would be recorded upon a line having no electrostatic capacity as in fig. 2.

Fig. 2.

But the record would be different upon a line having great electrostatic capacity, for then the increased battery used to transmit the first portion of the dash would then be used to charge the line statically and the record would be more like that in Fig. 3. the first strong current being nearly all used in charging the line.

Fig. 3.

Fig. 4.

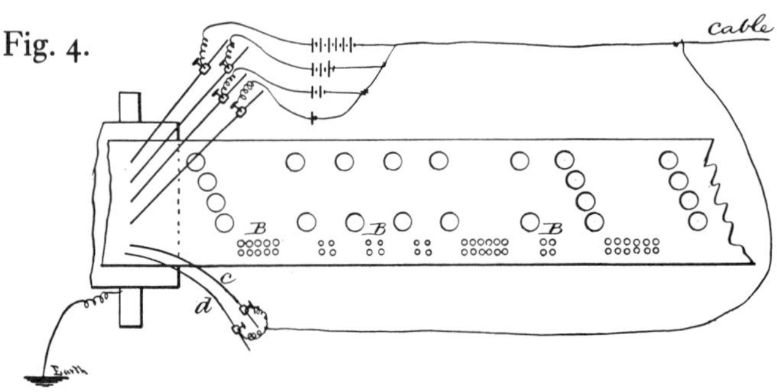

Fig. 4. shows another perforated slip wherein an earth wire is connected during the ti~~n~~me that no signal is being transmitted, the earth connection serving to discharge the cable at the transmitting end.

Between the letters is a double row of small holes B. B. punched in the paper by the space key, or by the letter keys.

In line with these holes are two contact rollers connected to the cable. These contact rollers are set one ahead of the other so that before one of them breaks contact through a hole the other will make contact, thus keeping the cable to earth during that period when no signal is being transmitted.

Instead of using the extra roller pens to put the cable to earth, a permanent shunt may be used in which is an electromagnet, as in Fig. 5.

Fig. 5.

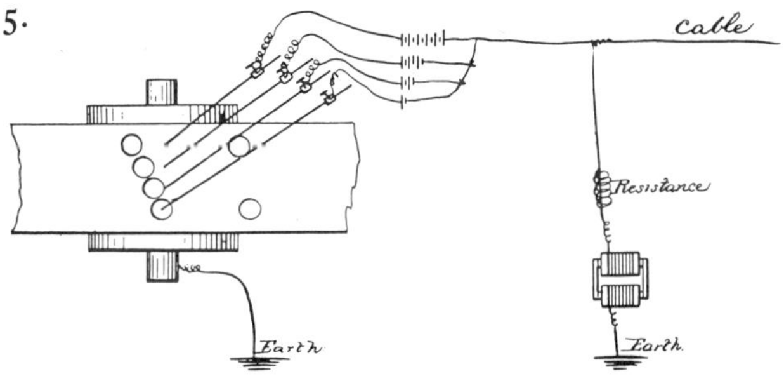

Fig. 6. shows another method by which the first part[b] of a signal may be made the strongest.

Fig. 6.

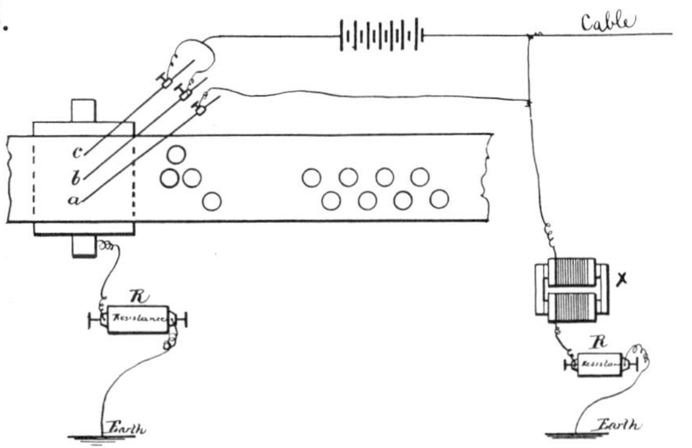

The p~~i~~ens b. and c. are connected together and to the battery and cable; when they pass through the first and second row of holes the battery is made to pass to cable and down the earth wire containing an electromagnet X. This diverted current passing through such electromagnet sets up a secondary current which is added to the current going on the cable this current is but momentary and does not last the whole of a dash on opening the battery or rather when the pens c and b. are not in contact with the drum the magnet would send a reverse current into the cable.

This reverse current is valuable upon ordinary land wires, it serving to neutralize the small static current but on submarine cables it produces secondary evils and mutilation of the signals, hence I use a third row of holes whereby with an extra point the electromagnet and resistance coil is shortcircuited at the moment the signalling battery is disconnected, thus preventing the reflex or contrary current from passing into the cable. I introduce a resistance into the earth wire connected to the drum so as to increase the resistance of that route and allow more of the inductive current go on the cable.

Fig. 7.

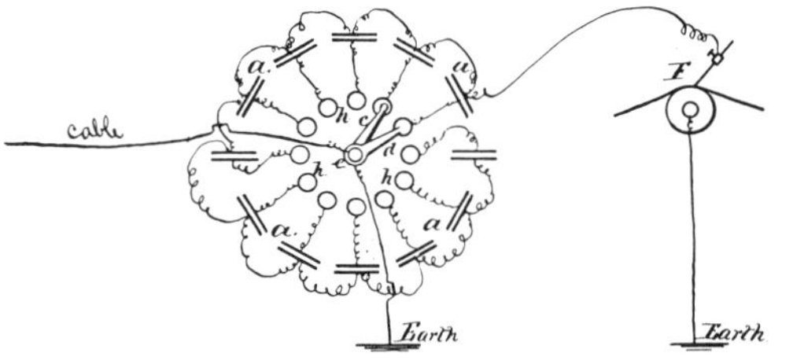

Fig. 7. shows a method for receiving a despatch from the distant station. a. a. a. are a series of condensers, each one connected to a contact pin h. h. h. in the path of the contact arms c. d. What is called the earth arms, of all the condensers, are connected together, thence to the recording point on the chemical instrument F.

The spring c. is secured to an insulating substance on the shaft e. which is revolved by clockwork, electric engine, or other source of power; this spring c. is in connection with the cable; when the shaft e. is rotated and waves of electricity are being sent of the cable the current rushes into these condensers and must necessarily pass through the chemical instrument recording its passage; were a single condenser used as is ordinarily the case it would soon be charged up to the same potential as the cable and marks would cease on the chemical recorder, and if the primary current was made to cease at the distant end of the cable the condenser would send a reverse current back into the cable and a mutilation of the next signal would take place but by the method shown in figure 7. fresh condensers are connected with the cable with great rapidity, hence a constant strong inductive charge is constantly passing through the chemical receiving instrument if the battery is to cable at the distant end; d. is a second contact spring following c. closely. This spring is connected to the earth and discharges the condenser previously in contact with cable thus putting them in a condition to receive a full charge when they are again connected to the cable and at the same time the reflex current is entirely prevented from passing into the cable and mutilating the signals.

I do not wish to confine myself to any particular method of arranging the various condensers to produce this result as it can obviously be arranged in many ways.

After the strip has been received no marks or colorations will be noticed on account of the very weak currents which I allow to pass through the receiving instrument to obtain powerful compensation. The decomposing point used may be any metal which is detected with facility in the spectrum with a spectroscope. for instance, a lithium, thallium or tellurium pen or decomposing point may be used, the slightest current passing through the point and ~~maintained~~ moistened paper serving to decompose the liquid and form oxygen, Hydrogen, or chlorine, which again attaches the point and an infinitesimal amount of oxide chloride, or finely divided metal is set free on the paper.

Fig. 8. shows the method of spectroscope translation.

Fig. 8.

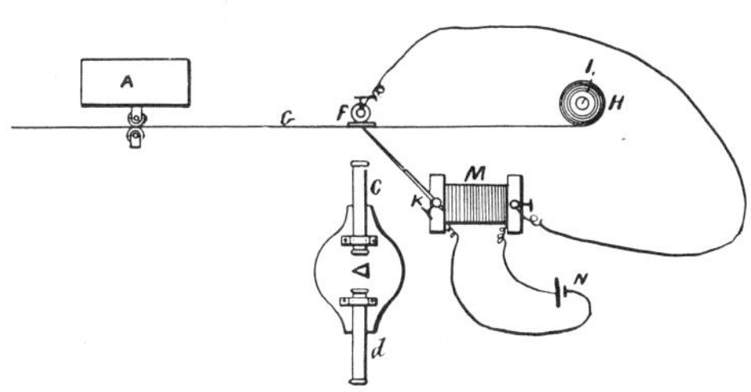

A. is a clockwork for drawing the paper along slowly in front of the slot in the spectroscop~~e~~ic tube C. F is a standard against which the paper passes. G. is the paper with the received message upon it. H. is a roll of the same. I. is a standard holding the roll. d. is the observing tube of the spectroscope. M. is an induction coil having one pole K. running to a point directly in front of the spectroscope slit. The standard F. is connected to the other end of the induction coil. N. is the induction coil battery. The operation is as follows.

Suppose the paper be moving slowly in front of the slit a stream of electric sparks is passing from the point K. through the wet paper to the standard F. right in line with the decompositions made on the wire; this continuous flow of sparks give a beautiful spectrum free from all lines except those due to the metals in the paper such as sodium. If we suppose a lithium pen be used the moment a dot or dash passes along, the spark volatilizes the metal and the very bright lithium line is observed by the operator, if it is a dot the line will last but an instant, but if a dash it will last a longer time and the telegram may be read with great facility. In practice I prefer to mask by a screen or other device all the lines except those due to the metal of the decomposing point.

It is obvious that the spectrum may be thrown on a white screen in a darkened box and be read from that; a small, intense battery may replace the induction coil to volatilize the metal.[1]

It will require several spectroscopic operators to translate the moistened strips to keep up with the rate at which it can be received over the wire.

My claims will probably be:—

1st Transmitting strong currents of electricity to form the first portion of the dot or dash and successively decreasing the strength for the purposes set forth.

2nd The arrangement of the perforations shown in Fig. 1. substantially and for the purposes set forth.

3rd A series of condensers arranged at the receiving end of a telegraphic circuit operated substantially as described and for the purposes set forth.

4th A spectroscopic apparatus to render visible the effect of electro decomposition for telegraphic purposes.

5th A spectroscop~~e~~ic electric spark apparatus, and device for ~~wa~~moving the paper, operated substantially as described for telegraphic purposes.

6th The use in telegraphy of a decomposing point whose oxide chloride or other salt or finely divided metal can be easily detected by the spectroscope.

Signed by me this 24th day of February A.D. 1875.[2]

Thomas A Edison

Witnesses Geo. T. Pinckney Chas H. Smith

DS, DNA, RG-241, Edison Caveat 62. Written by Pinckney; petition and oath omitted. [a]Place from oath; date taken from text, form altered. [b]Interlined above.

1. Spectroscopes were commonly used in chemical analyses. Knight (1876–77, 3:2259) shows one utilizing an induction coil to volatize metals in order to obtain their spectra.

2. Edison obtained a spectroscope in early January 1875, along with a copy of Henry Roscoe's *Spectrum Analysis*, but there are no extant notebook entries discussing Edison's experiments with a spectroscope (PN-75-01-05, Lab. [*TAEM* 20:31]; 75-002, DF [*TAEM* 13:224]). In late March 1875 Edison demonstrated a system like the one described in this caveat to a Mr. Dhyrr, secretary of the German telegraph system, who was also connected with the Direct United States Cable Co. Edison proposed to use his system on the lines of this company, which was associated with the Atlantic and Pacific Telegraph Co. The *Operator* article that reported on Dhyrr's visit to Edison's laboratory noted that Edison exhibited "his submarine cable system . . . based upon a phenomenon connected with spectroscopy" (1 Apr. 1875, 7), but the phenomenon is not identified. Edison never patented this system.

–545–

And Charles Batchelor Patent Application: District Telegraphy

New York, February 24, 1875[1a]

1 To all whom it may concern. Be it known that we, Thomas A. Edison and Charles Batchelor, both of Newark in the State of New Jersey, have invented an Improvement in Electric Telegraphs, and the following is hereby declared to be a full, clear and exact description of the same.

2 This telegraph apparatus is made for local or private lines, and for alarms and messenger calls, and is known as the District Alarm or Domestic Telegraph.

3 An instrument adapted to such uses has to be simple, inexpensive, durable, not easily deranged or put out of working condition and withal reliable in the signal sent.

4 We make use of a segmental weighted lever with conducting and non-conducting spaces upon its surface, so that by moving the segment in contact with a circuit closing spring or roller, the signal will be given, and the same is received at the central station by taps on a bell or sounder, or upon chemical paper; an instrument of this general character is shown in Edison's patent No. 154,788.[2]

5 We use in connection with this district telegraph, chemical paper upon which the mark is made by electrical decomposition as heretofore well known, and we employ in that connection automatic mechanism that sets the paper in motion, and peculiar devices for opening and closing the circuits in connection with the transmitting segments.

6 As an additional convenience a simple finger-key instrument is sometimes employed to spell out the communication letter by letter, especially upon local and private lines.

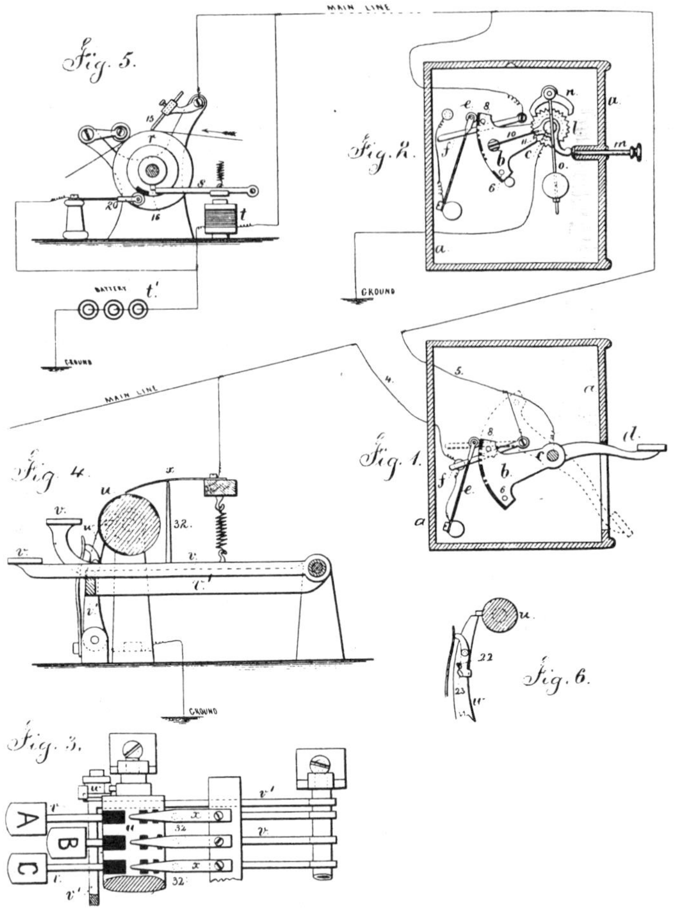

Inventors Thomas A. Edison. Charles Batchelor
per Lemuel Serrell Atty

Witnesses Chas H. Smith Geo. T Pinckney

7 In the drawing

Fig. 1. represents one of the transmitting instruments adapted to giving a signal in one direction only as the key is ~~depressed~~ restored,[b] instead of giving it in reverse as the key is liberated as in aforesaid patent.

Fig. 2. represents the signaling instrument with a regulator to determine the speed of movement.

8 Figs. 3. and 4. are a partial plan and side view of the finger-key transmitter, and Fig. 5. represents the instruments at the receiving station. These all are shown with circuit wire connections to illustrate the manner in which the pulsations reach the central station without interfering with other instruments.

9 The small box or case a. is adapted to being hung or fastened up against a wall, door-post, or window-frame, and within it is the segmental weight b. that is made to swing upon the axis or shaft c. and is raised by the lever or key d. and drops by its own weight.

10 At the curved edge of this segment there are pieces of non-conducting material such as hard rubber inserted so as to open and close the circuit through the spring e. when the segment is moved and these conducting and non-conducting surfaces are such that the pulsations will be sent in the pre-arranged order and length to denote numbers or other signs at the receiving station, and in some instances two or more of these transmitting segmental weights are employed side by side in the same instrument to send different pre-arranged signals.

11 A lever switch f. is introduced between the line connections 4. and 5. so that the circuit will remain closed between 4. and 5. and no pulsation will be made as the segment b. is raised, but just as the segment reaches its upward movement a pin 6. lifts the switch f. and the electric circuit is broken between 4. and 5. so that the pulsations will be sent on the line as the segment b. descends and opens and closes the circuit through b. and e. and the signal will thus pass to the central station.

12 A pin 8. restores the switch f. to place as the segment b. completes its downward movement.

13 The same thing is effected by the instrument fig. 2. but the main line is tapped by a branch circuit passing to the earth; this circuit in a normal condition is broken by the switch f. and that switch is moved by the pin 6. upon the extreme upward movement of the segment, so that the circuit will be complete through the spring e. segment b. and base of the instrument to the earth, and the pulsations will be made by the alternate conducting and non-conducting surfaces as the segment b. falls.

14 At the extreme downward movement the pin 8. moves the switch and opens the branch circuit to earth.

15 In order to prevent a sudden movement of the segment b. we apply on the same shaft c. the ratchet-wheel l. that is only connected to the segment b. through the spring 10. and a yoke 11. upon this ratchet-wheel is formed with a lever end to be pressed upon by the finger or by a slide rod m, and there is a pallet escapement n. and a pendulum-vibrator o. which is operated by the ratchet-wheel as the segment b. is forced up

and as it falls, and prevents too rapid movement of the segment, but the spring 10. lessens the irregular movement of the segment b. which would result if the wheel l. was firmly attached to the segment b.

16 At the central receiving station the instrument is provided with a drum r. over which the strip of chemically prepared paper passes, and upon this paper the stylus 15. rests, so that a mark will be made by the electricity as it passes, and this drum is held from revolving by the armature lever s. and stop 16.[3]

17 When the circuit is closed at the sending station, the current first passes through the magnet t. attracting the armature and lever and liberating the stop 16. so that the drum r. is instantly set in motion by clock-work and a weight or spring, so that the paper will be moved along beneath the stylus 15, and be marked by the passage of the electricity. The spring 20. rests upon a non-conducting surface so that when the drum r. has made one revolution and stops with the pin 16. against the armature-lever s. the circuit through the stylus is broken, thus compelling the first pulsation to act in the magnet t. The battery is shown at t′.

18 The transmitting key-instrument shown in figs. 3. and 4. is made with a cylinder u. the surface of which is provided with conducting and non-conducting substances in lines around the periphery as has heretofore been used, and this is driven by clockwork when liberated by depressing one of the keys v. The key acts upon the swinging frame v′. to move that down and by an inclined finger 22. press back the spring-stop w. so that the cylinder u. will be liberated and revolved by the clockwork and then the stop springs back to place upon the upward movement of the frame v′. and its finger 22. passes behind a spring projection 23.[c] on w. see fig. 6.[d] so that the stop w. remains in place to prevent more than one revolution of the cylinder u.

19 One wire of the line passes to the journals of the cylinder u. and the other to the range of transmitting springs x. and these are held up from the surface of the cylinder by the non-conducting studs 32. but when a key is depressed its spring x. rests upon the surface of the cylinder, and as the latter revolves, the pulsations of electricity are made to indicate the letter of the key.[4]

We claim as our Invention.

First. The combination, with the segmental circuit closing weight, of the switch f. for the purposes set forth.

Second. The combination with the circuit closing segmental weight, of the ratchet-wheel l. escapement n. pendulum o. and spring 10. for the purposes set forth.

Third. The receiving instrument composed of a drum and stylus for chemical paper, and electro-magnet and armature-lever stop, and a circuit closer insulated from the drum when the parts are at rest, for the purposes set forth.

Fourth. The combination, with the circuit closing cylinder u. and keys v. of the springs x. held up by the studs 32. and the stop w. actuated by the frame v′. substantially as set forth.

Signed by us this 24th day of February, A.D. 1875.

Thos A Edison | Chas. Batchelor

Witnesses Geo. T. Pinckney Chas H. Smith

DS and PD, MdSuFR, RG-24, Pat. App. 169,972. Written by Pinckney; petition and oath omitted. [a]Place taken from oath; date taken from text, form altered. [b]Interlined above, probably by Edison. [c]Interlined above. [d]"see" written in right margin and "fig. 6." interlined above.

1. Although Edison and Batchelor did not execute this patent application until February 1875, they appear to have completed the design it embodies by December 1874. At that time they made a patent model to accompany the Canadian patent (Cat. 1214:52; Cat. 1184:113; both Accts. [*TAEM* 21:589, 839]). The few dated drawings pertaining to this work date from November and early December. On a sketch of the domestic recorder dated 2 November 1874, Batchelor wrote, "Invented by T A Edison September 15 Designed by C Batchelor" (Cat. 30,094, Lab. [*TAEM* 5:331]). On a sketch of another transmitter he wrote, "Edison's Signal box for Fire & Domestic telegraphy Designed Nov 22nd 1874" (Cat. 30,095, Lab. [*TAEM* 6:417]), but on an identical drawing in his own notebook Batchelor noted that he had designed it (Cat. 1307:14 [*TAEM* 90:615]). See also Cat. 30,095, Lab. (*TAEM* 6:392, 420, 481).

Edison's patent model for U.S. Patent 169,972.

2. Edison had executed the patent application for this signal-box design on 2 April 1874 and the patent had been granted on 8 September.

It focused on the problem of interference between signal boxes on the same line, a serious problem in district telegraphy.

3. Doc. 615.

4. See Doc. 563.

–546–

From Jay Gould[1]

[New York,] Feby. 27, 1875.

T. A. EDISON,

You will proceed to experiment and perfect a set of one wire Roman Recording Printing instruments for operation between NY & Philadelphia & bill the same to the A.&P. Telegraph Co. & I will advance you the amount of such bill until the amount of bill is paid by that Company.[2]

JAY GOULD.

PL, *Harrington v. A&P*, Box 17B, Complainants' Exhibits 3:161.

1. Josiah Reiff later testified that Edison wrote the body of this letter; Gould only signed it. Reiff's testimony, 3:93, Box 17B, *Harrington v. A&P*.

2. Edison worked on his roman-letter automatic between January and May 1875. He first showed it to Gould on the night of 30 December 1874 (Thomas Eckert's testimony, 2:230–31, Box 17B, *Harrington v. A&P*) and demonstrated it on the Atlantic and Pacific's Philadelphia–New York line in late January (see Chapter 7 introduction, p. 374). Edison later claimed that Gould paid about $700 toward these experiments (see Doc. 648). This would be the amount ($687.47) indicated in bills of 22 and 29 March (74-011, DF [*TAEM* 13:471]; these bills are also presented as exhibits in 1:289–90, Box 17B, *Harrington v. A&P*). Gould appears to have been reimbursed for these payments (1:368, Box 17B, *Harrington v. A&P*).

–547–

R. G. Dun & Co. Credit Report

March 4/75

Edison & Murray[a] Manfrs Telegraph Inst'm'ts

#3700[1] It is impossible for us to make out anything difinite as to the means of this firm. They keep no firm bank a/c & do not come in the way of cr here at all. They seem to have met all engagements & to be dg a gd bus. E" was some few mos since [--][b] appointed "Electrician" to one of the main tel Co's, the Altantic[2] we think, & is highly regd as a theoretical & practical electrician & it is supposed that his mfrng bus is a paying one & that the firm contracts will be met.

D (abstract), MH-BA, RGD, N.J. 23:206. [a]Followed by "(Prevs see T A Edison 333/2[38])". [b]Canceled.

1. See Doc. 352 n. 1.

2. Edison's formal appointment as electrician of the Atlantic and Pacific Telegraph Co. did not come until June. See Docs. 522 and 585.

–548–

From Henry Bentley

Philadelphia, 3d Mo 5th 1875[a]

Dr Sir,

Am I right in understanding you to say that the Edison double type wheel printer (Stocks)[1] such as you altered for the G&S folks are the same price as the old ones $50? Of course I know you charge $25 for altering the old ones over but when you make a lot new I presume the cost is no more than to make them old style or not enough more to make it worth speaking about. I am contemplating a change in 125 or 150 printers in our G&S, Produce, Cotton & similar dep'ts[2] and I want greater speed &c but before starting out I wish to know just where I am.[3] A immediate reply will much oblige yrs truly

Henry Bentley[4] Prest

ALS, NjWOE, DF (*TAEM* 13:642). Letterhead of Philadelphia Local Telegraph Co. [a]"Philadelphia," and "187" preprinted.

1. Edison's universal stock printer.

2. The letterhead of the Philadelphia Local Telegraph Co. indicates that it furnished quotations on gold, stocks, government securities, cotton, naval stores, breadstuffs, petroleum, and produce, as well as general commercial and financial news from the principal U.S. and European cities, and shipping and marine news from the principal seaports in America, Europe, India, China, Japan, Australia, Cuba, and other places.

3. An account book entry of 12 May 1875 shows eight stock printers altered for Bentley at a cost of $25 each. Cat. 1220:130, Accts. (*TAEM* 22:63).

4. Henry Bentley (b. 1834) began his career as a newspaperman. He became connected with telegraphy in 1855 when he helped establish the New York City and Suburban Printing Telegraph Co., which used modified House printing telegraph instruments to provide service within New York City. Bentley later moved to Philadelphia, where he became connected with the *Philadelphia Inquirer* and acted as a Civil War correspondent for several other papers. Immediately after the war he formed the Philadelphia Local Telegraph Co. Bentley was also credited with a few telegraph inventions, although none of them was patented. Reid 1879, 597–601.

–549–

Article in the Journal of the Society of Telegraph Engineers

Newark, N.Y., U.S.A. March 10th, 1875[1a]

ON THE IMPERFECT CONTACTS WHICH OCCUR IN SIGNALLING WITH RIGID CONTACT-POINTS

By T. A. Edison.

In a paper by Mr. Latimer Clark describing some experiments on the retardation of signals in submarine cables, printed in the Report of the Submarine Telegraph Committee,[2] a strip of Bain chemical paper is shown upon which is recorded the return static charge from a long circuit formed of buried wires.

On this strip the record, instead of commencing strong and gradually weakening until invisible, shows a distinct break in its continuity in the first portion of it, as in fig. 1, and this break was observed in all the experiments of this character which he made.

This same phenomenon was observed by myself in some experiments with an artificial cable formed of resistances and condensers and puzzled me for some time, until the cause was traced to the apparatus employed.

The key which I used for charging and discharging the cable was an ordinary one, with front and back contact-point; the former was connected to the battery; the latter with the recording-point of the Bain instrument, while the lever was connected to the cable.

When the key lever was depressed the cable was charged by contact with the battery; if the key was then allowed to rise the cable was put in connection with the Bain instrument through the back point, and the return charge recorded.

With an iron stylus resting upon ordinary prussiate of potash paper the record was the same as that of Mr. Clark, shown in fig. 1; but when mechanical tailing or elongation of the mark was prevented by using a lead point, and paper moistened in a solution of pyrogallic acid and nitrate of ammonia, the record was that shown in fig. 2.

Fig. 1. 

Fig. 2. 

Fig. 3. 

Fig. 4. 

If more play was given the lever of the key the break was longer, with indications of a second break (fig. 3); by increasing the play the second break is made distinct, as in fig. 4. If the Bain instrument is shunted to decrease the volume of the current, and nearly dry paper used, several distinct breaks are observable; but when the lever of the key is firmly grasped and brought slowly in contact with the back point these breaks did not occur, showing that they are due to the rebound of the lever from the rigid contact-points when brought forcibly against them. A flat spring, adjusted in such a manner as that it would follow the lever up for some distance when rebounding, entirely prevented these breaks.

It follows from this that a Morse dash sent with an ordinary key is not one unbroken wave, but a series of waves, and that the minute breaks which occur, although not perceptible on ordinary relays, cause to be set up in the bobbins of the electro-magnet several inductive currents, which tend to prevent the instantaneous magnetization of its cores. The tongue of a polarised relay in closing a local circuit rebounds in the same manner and gives these breaks. It might at first thought be supposed that the pressure of the hand upon a key-lever would prevent its rebound, but on the contrary the greater the force and pressure used the greater will be the rebound, upon the same principle that it is almost if not impossible to bring a hammer down upon an anvil without several rebounds.

PD, *J. Soc. Teleg. Eng.* 4 (1875–76): 117–19. [a]Place and date not that of publication; see n. 1.

1. This paper was read at the Society's meeting of this date.

2. Clark's paper is appendix 2 of the *Report of the Joint Committee appointed by the Lords of the Committee of Privy Council for Trade and the Atlantic Telegraph Company to inquire into the construction of Submarine Telegraph Cables; together with the Minutes of Evidence and Appendix* (London: H.M.S.O., 1861), pp. 302–6.

-550-

Notebook Entry: Automatic Telegraphy

[Newark,] March 10th [1875][1]

Experiments on Pyrogallic Solution and Silver Pen. Chlorides combined with Pyrogallic show no reaction with Silver— Sulphate of Potash & Nitrates show the reaction. Nitrate of Amonia, better than Nit Silver. It is the hydrated base which does the business. Pyrogallic, Nitrate Soda & a small quantity ~~Nostic~~ Caustic Potash just sufficient to turn the solution a deep yellow— when well shaken gives the most delicate results with 3 cells on short circuit— the mark is as black Stygian gloom on 10,000 Ohms a slight mark of considerable delicacy is obtained nearly as good ferr~~e~~id. The paper has a light brown color. to a solution of Pyrogallic Acid & Sulphate of Potash a very small quantity of Caustic Potash is added, there will be a whitish precipitate or milki~~n~~shness & the solution does not become colored. Caustic Potash is to be add~~e~~ed, but not enough to re-dissolve this milkiness further the liquid becomes colored. Paper dipped in this milky solution is white, & although the marks are not so deep a black as some other of the Pyrogallic Solutions the have equal sensibility & after the paper has become dry there are as black as any.[2] This appears to be a 1st class Solution. Patent[3]

X, NjWOE, Lab., Cat. 1172:82 (*TAEM* 3:110). Written by Charles Edison.

1. This entry is preceded and followed by several pages of chemical solutions for automatic telegraphy from March 1875. They begin on page 68 and are headed "Chemical Experiments for a Sensitive & permanent solution March 8th 75." In the following pages of notes (to p. 89, 16 March 1875), the even-numbered pages are headed "Edison" and the odd-numbered "Batchelor," although all of the entries were written by Charles Edison. James Adams and Alexander Fox also assisted in the experiments.

2. Many of the chemicals Edison used in his automatic telegraph experiments—including most of those mentioned in this entry—were known for their role in early photographic processes.

3. Edison never patented this solution.

–551–

From P. H. Shaughness

New York March 11/75

Friend Edison

There is a fine opportunity now for a new Quotation Teleg— I have been feeling the Pulse of the Brokers since the new deal— They are indignant at increased rates & would sign an agreement for a year at $15.00 per month with new concern—[1] cant you shake a good Inst out of your Box & stir up the animiles" I am still out of business & in hard luck generally— lost a child last week— dont forget me if you can see a chance to to use me Yrs

P H Shaughness[2]

ALS, NjWOE, DF, 75-010 (*TAEM* 13:400).

1. On 11 February 1875 Gold and Stock Telegraph Co. gained control of Manhattan Quotation Telegraph Co., with the latter continuing to operate as a separate firm. For some time before this merger the two companies had engaged in a price war, Gold and Stock, for example, reducing its monthly rental price for quotation instruments from $26 to $10 per month in December 1873. After the merger, however, prices returned to near their previous level; five days after the merger the monthly rental rate for quotation instruments was raised to $20, and in July it increased to $25. Gold and Stock Telegraph Co. Executive and Financial Committee Minutebook, pp. 40–41, 73–74; Gold and Stock Board of Directors Minutebook, p. 157; Manhattan Quotation Telegraph Co. Executive Committee Minutebook, pp. 148, 166; all NjWU.

2. P. H. Shaughness had been superintendent of the Manhattan Quotation Telegraph Co. from the time it began operations in October 1873. Manhattan Quotation Telegraph Co. Executive Committee Minutebook, p. 3, NjWU.

–552–

George Preece to Charles Batchelor

London, March 11, 1875[a]

Dear Sir,

I have much pleasure in acknowledging receipt of your letter enclosing £2 for the subscription of self & Mr. Edison.

Kindly tell Mr. Edison that I have long been expecting a Communication from him about the Electromotograph.[1]

We shall be very glad at all times to receive Communications from our Telegraph friends in America. Yrs faithfully

Geo. E. Preece

ALS, NjWOE, Scraps., Cat. 30,096 (*TAEM* 27:600). Letterhead of Society of Telegraph Engineers. [a]"London," preprinted.

1. No such article appeared in the *Journal of the Society of Telegraph Engineers.*

–553–

D. G. Farwell to Edison and Murray

New York. Mch 12th 1875[a]

Mess Edison & Murray

Please send to this office, at as early a time, as next Monday—not later—an Estimate for furnishing us, the following

200 Signal Boxes—[1]

(Similar to the last one)[2] with combinations of numbers running from one to three hundred—to be japaned a dark ~~perle~~ pearly[b] color—lettered on front by a stencil (which we will furnish)—With plain screws and washers in place of the brass binding posts— Also estimate same without Japaning &c. The above to be Supplyed[c] upon a time stipulated—say 50 within one week from the date of order and 50 each successive week untill the order is completed[3]

Payment will be made upon receipt of each lot.

Your attention is called to the fact that the vibrating pendilum detatches from the verge, after a little use—[4] Cannot a more secure means of fastening be devised?

Please also inform us how soon you can complete the Six fire Alarm boxes with combinations furnished some months since,[5] together with[b] the necessary apparatus within for "Talking circuit"[6] also state price of said six boxes delivered ~~all~~ready for use[d]— Yours Truly—

D. G. Farwell, Secty[7]

ALS, NjWOE, DF (*TAEM* 13:515). Letterhead of the Domestic Telegraph Co. [a]"New York." and "187" preprinted. [b]Interlined above. [c]"l" interlined above. [d]"for use" interlined above.

1. On 1 March 1875 Domestic Telegraph Co. President Jarvis Edson had written to Edison, "At the rate we are getting subscribers we shall

want District boxes as soon as they will be ready." In that letter he also encouraged Edison to complete work on his fire alarm telegraph. DF (*TAEM* 13:514).

2. An order book entry of 25 January 1875 lists 5 district boxes for the Domestic Telegraph Co. Cat. 1223:6, Accts. (*TAEM* 22:134).

3. Account records indicate that Edison and Murray supplied 124 domestic boxes between 27 April and 24 May 1875. All of the boxes were numbered between 3 and 344, with the exception of three boxes numbered 411, 432, and 531. Cat. 1214:377; Cat. 1220:136, 138; both Accts. (*TAEM* 21:692; 22:66–67).

4. This probably refers to clockwork in receiving units.

5. An Edison and Murray order book entry of 25 January 1875 shows "6 boxes for Fire alarm circuit New York" for the Domestic Telegraph Co (Cat. 1223:5, Accts. [*TAEM* 22:133]). Account book entries indicate work on fire alarm boxes for the Domestic Telegraph Co. on 18 January 1875 (160 hours of work on fifteen fire alarm boxes) and on 1 February (381 hours of work through 29 January on fire alarm boxes for New York City) (Cat. 1184:141; Cat. 1220:30; both Accts. [*TAEM* 21:850; 22:13]).

6. The "talking circuit" was a Morse line. See Doc. 654.

7. Farwell also served as treasurer of the Domestic Telegraph Co. He may be Darius G. Farwell, who sold drugs at 697 Broadway. Wilson 1875, 396.

–554–

Patent Application Amendment: Multiple Telegraphy

Newark N.J.[a] March 18, 1875

Specification describing a new and useful improvement in Duplex Telegraph Apparatus.[1] invented by Thomas A Edison of the city of Newark in the county of Essex and state of New Jersey.[2]

The invention has for its object the simultaneous transmission of two different dispatches or signals over the same line wire from opposite directions, or in the same direction, and the invention consists in the transmission of positive and negative currents over the line to effect the reception of one message, and the increase and decrease of the strengths of these currents, either positive or negative to effect the reception of the other message.[3]

The accompanying drawings[4] represent a plan view of my improved apparatus in this case, part of the apparatus being placed at one end of the line and part at the other end.

A and B are electromagnets of equal strength, and are placed at an equal distance from and on the opposite sides of the armature lever L, pivoted between them.

Both magnets are arranged with seperate batteries. The battery MB being connected with the magnet A and with its positive pole to the line, and the battery MB′ is connected

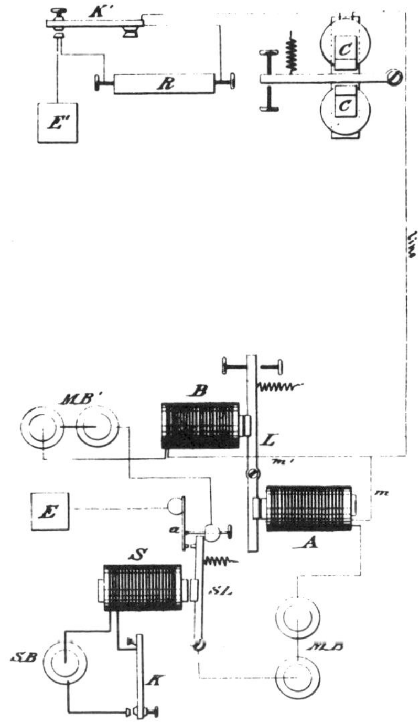

Inventor:[b] T. A. Edison Per[b] Munn & Co.[5] Attorneys.[b] Witnesses:[b] Chas. Nida.[6] [P?] Sedgwick.[7c]

with the magnet B and has its negative pole to the line The batteries MB and MB′ are alternately placed into the circuit by the sounder lever SL and the double spring contact a which is in contact with the earth.

The key lever SL. is operated as usual by the local battery SB Key K and magnet S.

The magnets A and B are connected to the main line wire and polarized relay C by the wires m and m′

The polarized relay C being either at the same station or at the distant station, is operated by the positive and negative currents sent over the line.

K′ is a morse key and R a resistance coil or rheostat, connected to the key K and earth E

The object of the key K′ and rheostat R is to increase and decrease the strength of the current upon the line, so as to effect the lever of the relay AB. The increase and decrease in the strength of the line current does not effect the polarized relay C as this is dependant for working upon the direction of the current, or polarity, independant of its strength, so that signals may be transmitted by SL batteries MB and MB′ or by any other [-]battery reversing device, and these currents acting upon the relay .C cause its tongue to be thrown to the right or left, according to the polarity of the current transmitted. At the same time another set of signals may be sent over the same wire, by the depression of the key K′ which causes a decrease in the resistance of the line and a consequent increase in the strength of the current on the line, and this increased strength of current produces sufficient magnetism in either A or B to overcome the attraction of its spring and it is drawn towards the magnets

On closing the key K the magnet attracts the lever SL. throwing the battery MB′ and magnet B out of circuit and the magnet B and battery MB with its positive pole is placed in circuit and towards the line. In a like manner when the key lever is not attracted by its magnet, the magnet A and battery MB is thrown out of circuit, and the magnet B and battery MB′ with its negative pole towards the line is thrown in circuit, the armature lever L remaining constantly attracted (if K′ be closed) by either A or B. as the instantaneous transfer of polarity permits no seperation of the same.

If both positive and negative currents were passed through one magnet only, a charge and discharge would be produced with a change of polarity in the iron cores and the armature would be drawn away from the cores of the magnet for an instant by its retractile spring, at the moment when the core were changing their polarity in this arrangment the wires upon the magnets connections and arrangement of batteries are such that the polarity of both magnets is never changed

The polarized relay is self adjustable and follows the positive and negative currents whether the tension of the same is suddenly increased or decreased

The polarized relay .C. can be placed at a number of stations on the line and each will be able to receive the signals from the station transmitting positive and negative currents

The relay A. B. may also be placed at a number of stations if A or B be dispensed with and other devices applied to prevent the mutilation of the signals by change in the polarity of its iron core[8]

I claim as new and desire to secure by letters patent

1st. The armature lever L pivoted between the magnets A and B and operated by an increase and decrease in the tension of the line current in combination with the Key K′ and rheostat R for the purpose set forth.

2nd. The sounder lever SL double spring contact .a. batteries MB and MB′ arranged substantially as set forth and for the purpose specified.

3rd The combination with the main line circuit of a receiving instrument operated by changes in the polarity of the current independently of tension with another receiving instrument operated by changes of tension independent of polarity, means of changing the polarity of the current and means of changing the tension of the current substantially as and for the purpose specified.[d]

Thomas .A. Edison[9]

ADS and PD, MdSuFR, RG-241, Pat. App. 162,633. [a]Place taken from letter accompanying amended application. [b]Preprinted. [c]Drawing and accompanying signatures taken from printed patent. [d]Followed by horizontal line across page.

1. This is an amended version of Doc. 315; it became U.S. Patent 162,633. See Doc. 577.

2. No assignment of the rights to this patent accompanied either the original or this amended application, but Edison filed an assignment to himself (a one-third interest) and George Harrington (two-thirds), as specified in Doc. 155, on 24 March 1875.

3. The claim here of both diplex and duplex transmission, the phrasing of this paragraph, and the third claim at the end of the application represent an attempt to broaden this case to cover the general possibility of any diplex of the sort that was basic to Edison's quadruplex. See Docs. 348, 472, and 569 n. 5; but see also n. 8 below.

4. The only drawing is the one filed with the original application (Doc. 315). Patent Office regulations allowed the amendment of specifications, claims, and models, but not drawings.

5. Munn & Co. was the patent agency that acted for Edison in preparing and filing the original application.

6. Unidentified.

7. Unidentified.

8. Thus, to turn the system into a diplex would actually require something like Edison's bug trap (i.e., "other devices . . .") to keep one message from ruining the other. See Doc. 348 n. 16.

9. The right and title to Edison's pending patent applications for multiple telegraphy, particularly Case 99 (Doc. 472), were contested and under review at the Patent Office (Doc. 536 n. 4). The claims of George Prescott (and Western Union), Harrington (and Jay Gould and Atlantic and Pacific), and Edison alone had been argued before Patent Commissioner Thacher, but his decision had not been announced. This amendment was filed on 20 March; the application was not among those considered by the commissioner. Thacher's decision, also dated 20 March, directed that any contested applications that were approved would be assigned to Edison and Prescott when issued (Quad. TLC.6, p. 35 [*TAEM* 10:980]). The decision was published in the 16 March issue of the *Patent Office Gazette* (7:423–25), and became publicly known about 29 March, whereupon Harrington's lawyers immediately petitioned the Secretary of the Interior to block or reverse it, the Patent Office then

being part of that department (Quad. 72.4, p. 30 [*TAEM* 9:60]; see also "Special Despatch to the Daily Graphic," 29 Mar. [1875], Cat. 1143, Scraps. [*TAEM* 27:247]). The Secretary, Columbus Delano, who had acted in accord with Gould's urgings in the past, promptly agreed to hear arguments on the issues and ordered the relevant application files sent to him (Delano to Commissioner of Patents, 30 Mar. 1875, Quad. 72.4, p. 31 [*TAEM* 9:60]; McFeely 1981, 373). Case H was not included. Prescott later contended that this constituted an attempt to get around Thacher's decision, claim the quadruplex for Harrington et al., and deprive him of his rights in Case 99 (Doc. 569; Quad. 72.3–5 passim [*TAEM* 9:38–91]). On Delano, an Ohio Republican politician, see *DAB*, s.v. "Delano, Columbus."

–555–

Memorandum: Automatic Telegraphy[1]

[Newark, Winter 1875?[2]]

The cost of perforating 40 words upon the Large Perforator is one cent=

Thus An average of 10 words per minute per day of 10 hours, (or 25 per minute when operated) gives 6000 words per day on each perforator The pay for which is 1.50 per day which gives 40 words for one cent=

The average message is 20 words The cost of the paper perforated and the Blank for delivery, ink etc envelope is 4 mills Salary of opr writing out message ~~4~~3 mills[a] The rent of office 4 mills ~~del~~repair of Line 1 mill battery power ½ mill wear & tear machinery ½ mill Making total cost of sending and receiving an ordinary message placed in the envelope ready for delivery ~~is~~13 mills or Little over 1 cent and a ½ the delivery 2c which makes 3 cents 5 mills for each message. If the Person sending the Message should perforate it & the addressed receive the dot & dash s~~t~~lip then the Cost would be reduced From 13 mills to 6 mills for sending & receiving & 2 cents for delivery. for repeating this message between one point and another it would cost 6 mills or Little Less than one cent consequently a message for San Francisco[b] if repeated Chicago Ohmaha Denver & Salt Lake the Cost would be 6 cents & 5 mills delivered= In repeating a message from one station to another for a distant Station the incidental expenses such as rent, repair Line, battery power, wear & tear of Machinery papers etc should be counted, if this is done then a Message from N York to San Francisco would cost 7 cents 5 mills delivered This estimate is based on an increased ~~of~~ business 3 to one as now done by WU[c] It costs the WU Company about 1 dollar and 25 cents alone for Labor, material, batty, keep line in order etc[d] ~~Say n~~ for transmitting on message between N York & San Francisco[e]

It will probably be necessary when the Washn office is in full blast to have a sufficient number of perforators & operatives to Transmit the ordinary commercial business and about 150 000 words press= to Transmit 150,000 words press & get it off in good season it will require about 15 perforators for transmitting The Commercial business 3 perforators This will require 30 females 1~~5~~4 in the daytime & 16 at Night— Office Manager Chief Opr Man & Lady Chief Opr The Cost of which will be

30 females at 1.50 per day	$45 00
Office Manager	4 00
Lady Chief Opr	4 00
Man Chief Opr	4 00
Clerk	2 00
3 Copyist Ladies	4.50
~~Total for~~	
Line Man	3.00
~~To~~Battery Man—	3 00
Total for Labor	69.50
Rent per day	7.00
Stationery—paper	
Batty power	
Ink pens gas etc	10.00
Total	86.50

With this force doing 600 messages per day at 25c each making $150 and transmitting to the press 150 000 words at 5 words for one cent making $300, making $450. per day ~~from this is to be deducted for N York~~ for N York alone if Boston Balto Cincinnati Chicago Philadelphia & St Louis take a copy then the total will be $~~3~~2,150.$^{100}/_{00}$. T~~h~~o be deducted from this is the Cost of writing out in each city which will amount to $55. which deducted from 2,150 leaves 1,765 net profit= per day[f]

AD, NjWOE, DF (*TAEM* 13:434). Two unrelated pages in Edison's hand and on similar paper were appended to this document in the file. [a]"Salary . . . mills" interlined above. [b]"for San Francisco" interlined above. [c]"3 . . . WU" interlined above. [d]Underlined twice. [e]Followed by centered horizontal double line. [f]An unrelated drawing is on the back of this page.

1. It is not known whether Edison sent a version of this to anyone.

2. This cost analysis of automatic telegraphy was probably prepared during the takeover of the Automatic Telegraph Co. by the Atlantic and Pacific Telegraph Co.

THREE-KEY PERFORATOR Doc. 556

Edison originally designed small, inexpensive perforators at the behest of Daniel Craig, who envisioned them as businessmen's "composing machines" for preparing perforated messages that would be delivered to the Automatic Telegraph Company and transmitted at a reduced rate.[1] Edison worked on such designs from at least the winter of 1871 and he patented a six-key perforator in 1872 (Doc. 255).[2] In the following production model, the only extant example of a three-key perforator design, one key was used for dots, another for dashes, and the third for spaces.

It is unclear whether this design was actually employed. According to an 1879 article on Edison's automatic telegraph system, it was not used before Atlantic and Pacific abandoned Edison's automatic system sometime in 1876.[3] Other evi-

A production model of Edison's three-key perforator.

dence, however, suggests that operators used three-key perforators on automatic lines for some time. A February 1875 article in the *Telegrapher* described Atlantic and Pacific operators "receiving by sound [probably from Morse branch lines] and punching the strips for repetition [on main lines] without the necessity of their being first copied."[4] This could only have been done with small perforators. The large perforators were too cumbersome, and the operators, knowing Morse code, could easily operate a three-key perforator. Although the *Telegrapher* does not describe the perforators used by the operators, account records indicate that ten three-key perforators were made for Automatic Telegraph as early as October 1873.[5] The design of three-key perforators used by both Automatic Telegraph and Atlantic and Pacific is unknown, but the design embodied in this production model may well reflect modifications made between March and May 1875.[6]

1. Craig mentioned the possibility of a "merchant's" perforator in Doc. 147, but did not explain how such a machine would be used. He later described the purpose of such a machine in a brochure written in 1879 for the American Rapid Telegraph Co., which was then using another system of automatic telegraphy to compete with Western Union (*Craig's Manual of the Telegraph, Illustrating the Electro-Mechanical System of the American "Rapid" Telegraph Co. of New York* [New York: John Polhemus, printer, 1879]).

2. For Edison's earlier work on three-key perforators see Docs. 149 and 153 n. 3.

3. The article was illustrated with a drawing that appears to be the same design as an instrument shown in one of Edison's photograph albums (Cat. 558), which is a slightly modified version of the production model (the machine in the album has a pressure roller to guide the paper as it leaves the machine and a tray for holding the message to be transcribed). The instrument in the article was described as:

> A small perforating machine . . . having three keys, [which] could be purchased from the company at a small cost. With this instrument any person might, after a few hours' practice, punch the dots and dashes of his message in a strip of paper. When thus perforated he could send the paper to the telegraph office for transmission. . . . For thus transmitting, the company proposed to charge not as is now the custom, so much per word, but so much per yard, the sender being at liberty to crowd a volume, if he could, within that space. [Fox 1879, 844–45]

4. "The Telegraphic Situation," *Telegr.* 11 (1875): 34.

5. Cat. 1214:34, Accts. (*TAEM* 21:581).

6. Cat. 1214:275, Accts. (*TAEM* 21:678).

–556–

Production Model: Automatic Telegraphy[1]

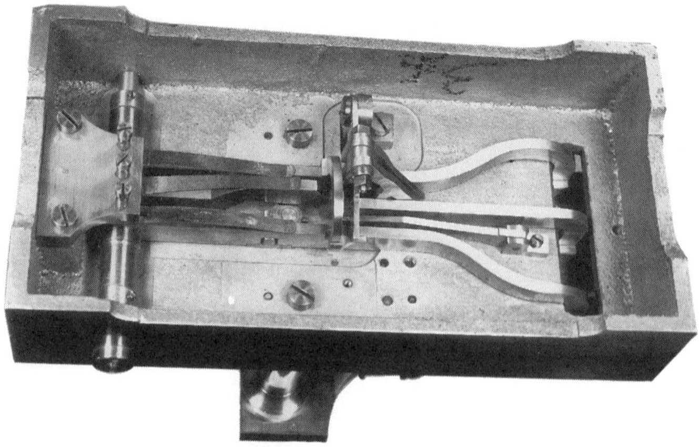

M (21 cm × 12 cm × 17 cm), NjWOE, Cat. 1523.

1. See headnote above.

-557-

Caveat: Automatic Telegraphy

New York, March 22, 1875[a]

To all whom it may concern. Be it known that I, Thomas A. Edison, of Newark, in the County of Essex, and State of New Jersey, have invented an Improvement in Automatic Telegraphs of which the following is a specification.

The object of this invention is to obtain a more perfect compensation for the effect of the static currents upon telegraphic signals.

The invention consists of devices, some of which are applicable to short lines of small electrostatic capacity and others to long lines having great electrostatic capacity. In Fig. 1. is shown a perforated slip having perforations arranged representing the word "The" E. is the drum which connected to clock work or other source of power gives movement to the perforated paper.

Fig. 1.

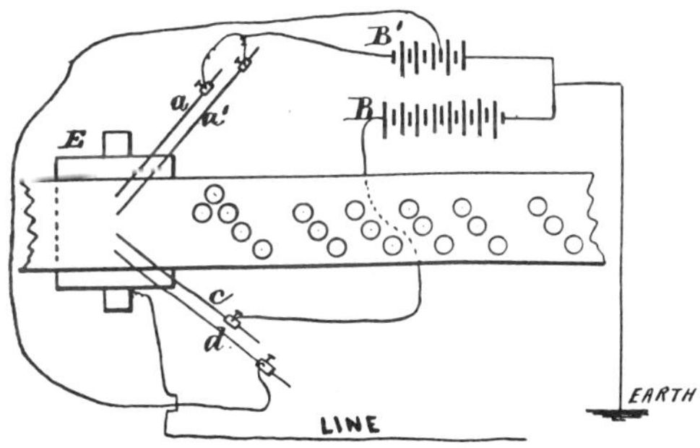

This drum is connected to the line wire; resting upon the perforated paper are four contact rollers.

The contact rollers a. a′. are connected together and are in line with holes exactly similar to my present chemical telegraph already patented.[1] These two contact rollers are connected to the battery B′. The other end of which battery is connected to earth.

Were the contact rollers C. and D. disconnected the word "The" would be transmitted over the line by a. and a′. in a precisely similar manner to that already described in previous patents.[2] The object of the point C. which is connected to a battery B. of nearly twice the strength of B′. and having an opposite pole to that of B′. connected to the line, is to transmit a short powerful contrary current immediately after the signalling current for the purpose of discharging the line by neutralizing the static electricity.

The object of the point D. which is insulated from C. is to transmit immediately after the compensating current from B.

a weak current of the same polarity as the signalling current which serves to prevent the current from B. from charging the line. The current put to line by contact of D. through the drum being just sufficient to prevent this re-charging but insufficient to record upon the chemical paper at the distant station.

On long lines of telegraph of great electrostatic capacity, a dash transmitted over the line encounters a static current and the first portion of it is nearly annihilated.

Fig. 2.

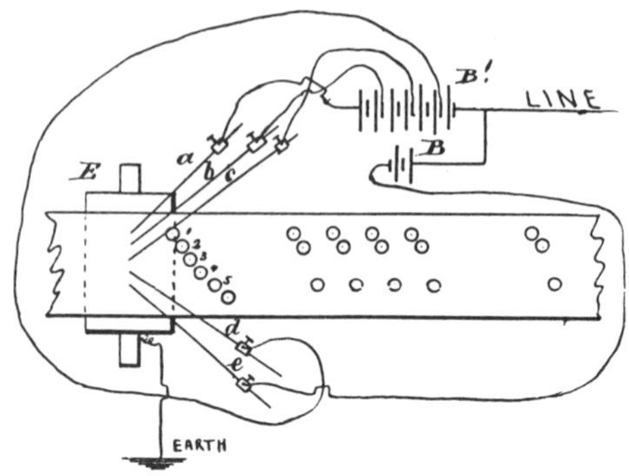

In Fig. 2. there is shown a device by which very strong currents are transmitted over the wire to form the first part of the dash, which gradually decreases in strength to the end. A. B. and C. and D. are contact rollers connected to the battery B′. E is a contact roller connected to a small battery of opposite polarity and is used to transmit a short wave to neutralize the static charge of the line.

The contact roller A. is connected to the full battery. B. to ¾ of the battery, C to ½ and D. to ¼ of the battery. The holes 1. 2. 3. 4. form the dash. The contact roller a. enters the hole 1. and puts the full battery to line. At the moment a. is about to break contact with the drum E. B. enters the hole 2. and the battery is suddenly decreased one fourth. As B. is leaving the hole 2. C. enters the hole 3. without interrupting continuity and decreases the battery to one half. At the moment that C. leaves the hole 3. D. enters the hole 4. and reduces the battery to ¼. At the moment that D. leaves the hole 4. the contact roller E. enters 5. and transmits from the battery B. a current of opposite polarity which serves to clear the line. Were this arrangement used upon a line having no electrostatic capacity the dash as recorded would resemble a wedge, the large end appearing first. Were this arrangement not used on a line having great electrostatic capacity the dash would appear as a double wedge, but by the use of this device upon a line having great electrostatic capacity the beginning and end of the dash are the same.

Were the full battery connected to the line when the contact roller C.^{3} was leaving the hole 4. the line would be charged by the full power of the battery B′. and the battery B. which is put to line by E. would have to be a very powerful one to neutralize the full charge which the battery B′. would give it, and it would re-charge the line so that the following signal would be entirely wiped out by being neutralized by this recharge

But in this device, the line is charged with but ¼ of the total cups or cells in B′ at the moment when C.^{4} leaves the hole 4. hence the reverse current from B. may be a very slight one. In transmitting dots the contact rollers A. B. and E. are used and but two strengths of current are transmitted to full battery B′. to transmit the first part, and ¾ of it to transmit the last part.

In Fig. 3. is shown another plan by which compensation is obtained by receiving the discharge of the line in a very large condenser C.

Fig. 3

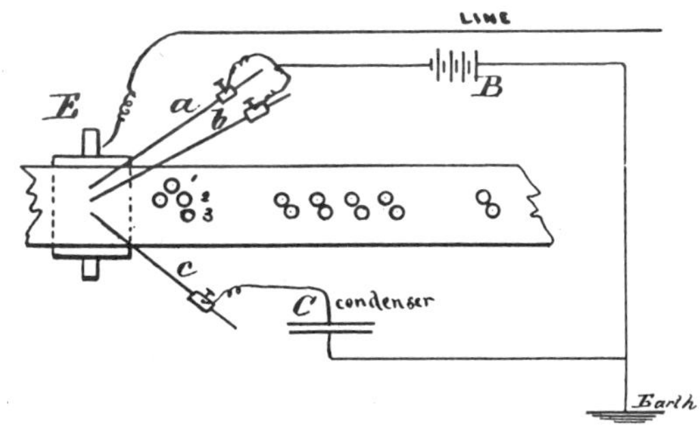

The roller pens a. b. holes 1. 2. and battery B. are arranged in the usual manner to transmit the character.

Immediately after and below the signalling holes are extra holes in line with a contact roller connected to a large condenser which may in some cases be shunted with a resistance coil or still another row of holes used to discharge it after the transmission of each signal the line is connected to the condenser C. and the return charge rushes in and fills the condenser in the same manner as it would pass to earth were no condenser used.

By the use of a powerful condenser it exerts a peculiar effect upon the return charge insomuch that it appears to discharge or free the line quicker than by the use of a simple earth connection probably offering an attraction. It also serves to in a great measure to prevent permanent currents from adjacent wires or from earth plates from passing over the line and mutilating the signals and also reduces the inductive currents or "cross fire" from other wires.

If a condenser be used at the receiving station also the effect of aurorial currents is also obviated.

In figure 4. is shown another device for the transmission of signals through a condenser and which produces characters in a manner similar to that in Fig. 2. The first part of a character being stronger than the last part.

Fig. 4

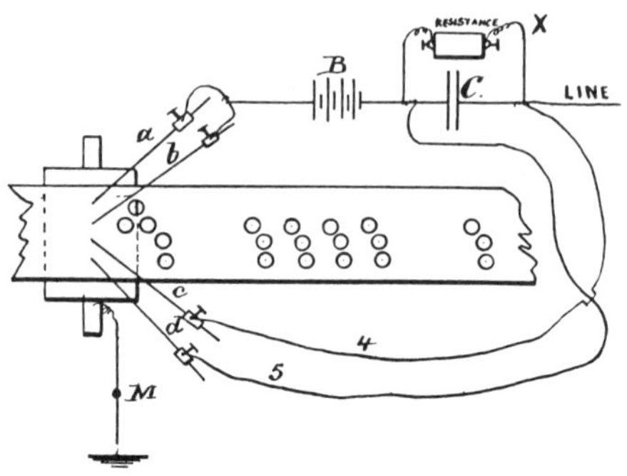

a. and b. are in connection with the battery B. and in line with a row of perforations made in the usual manner. C. is a large condenser in the line wire circuit. At the moment when the points a. b. connect the battery B. with the earth a strong charge rushes into the condenser and of course the line.

The first portions of each character being stronger than the last portions as in figure 5.

Fig. 5 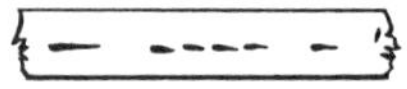

The strength being regulated by the shunt X around the condenser C. If the line is very long the first portion of, say a dash must be very strong and suddenly decrease; in this case the shunt X. is entirely disconnected. If the line be of small electrostatic capacity the shunt is added and its resistance so reduced that the increased strength at the beginning of a signal is barely perceptable.

After a signal has been transmitted, it is obvious that the line must be put to earth to discharge it of its static current and the condenser must also be discharged ready for the next signal, otherwise but weak currents would go on the line.

I effect this by means of two extra rows of perforations in line with two extra contact rollers c. d. which by means of the wires 4. and 5. short circuit the condenser C. immediately after a signal has been sent by a. and b. and at the same time the line is put to earth by the wire 4. point c. and drum, an electro magnet might be inserted in the earth wires at M. to transmit a reverse current into the line after the condenser was short circuited.

Fig. 6.

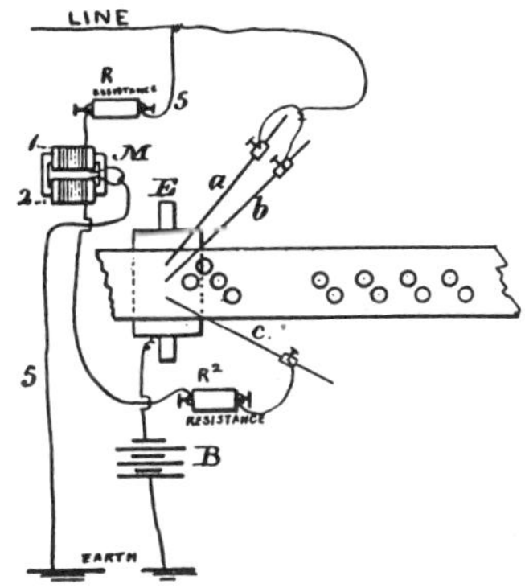

In Fig. 6. is shown a method of compensation by means of an electromagnet. Contact rollers a. and b. are connected to the line wire and are in line with a row of holes made in the usual manner. E. is the drum connected to the battery B. Wire 5. is a shunt connecting the line with the earth.

In this shunt is connected a resistance coil R. one helix 1. of an electromagnet M. which when the battery is connected to the line, sends or adds an inductive current to the first part of the dash or dot, and when the battery is disconnected send a weak reverse current into the line to neutralize the effect of the static current.

Immediately after and below the regular line of perforations is an extra row in line with a contact roller c. which is connected through a resistance coil R^2. to the free end of the helix 2. of the magnet M.

When the rollers a. b. leaves the regular perforations the roller c. comes in contact with the drum and battery B. through the extra row, this places the whole of the battery B. through the helix 2. and this sends a powerful induction current into the helix 1. and on the line of a polarity opposite to the signalling current. The resistance R^2 can be varied to increase or decrease the amount of battery current passing through helix 2. and a consequent increase or decrease of the strength of the inductive currents.

Fig. 7.

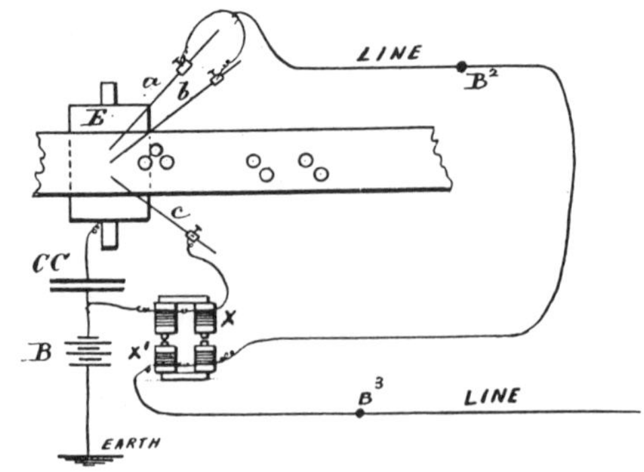

In Fig. 7. is shown a method similar to that shown in Fig. 4. A single extra point c. and extra row of perforations being used. a. and b. are in line with a row of perforations made in the usual manner.

E. is the drum connected to the battery B. through a powerful condenser C.C. At the moment that a. or b. place the line in contact with the condenser C.C. a powerful current rushes into the lines which quickly weakens as the condenser becomes charged; as a. or b. leaves the perforations the point c. comes in contact with the drum and short circuits the condenser within this short circuit is connected a high resistance electro magnet X. whose cores face another electromagnet X′ of smaller resistance in the line circuit. The condensers C.C. in the act of discharging its current in the short circuit powerfully energizes the magnet X. and this in its turn throws into the coils of X′ a powerful induction current which passes on the line in a contrary direction to the signalling current and neutralizes the effect of the static current; a shunt with or without an electromagnet might be inserted at B^2. or B^3.

Fig. 8.

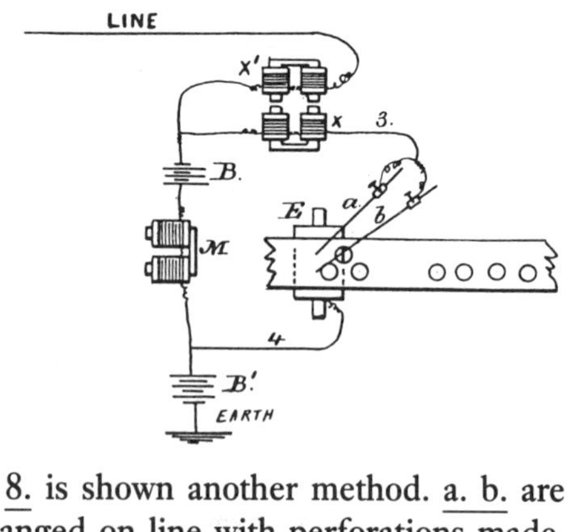

In Fig. 8. is shown another method. a. b. are the contact rollers arranged on line with perforations made in the ordinary manner. E. is the drum B. and B′. are two batteries with an equal number of cells and of the same electro magnetic force having their like ~~Having~~ poles connected together and producing no current.

M. is an electromagnet.

Wires 3. and 4. connected to the roller pens and drum are for the purposes of "shortcircuiting" the battery B. so as to allow the battery B′. to become active and transmit the signal over the line. When the battery B. is short circuited the whole power of B. passes through the electromagnet M. and ~~a~~ it becomes powerfully magnetized. Now at the moment that the point b. is disconnected from the drum E. the shunt is broken, both batteries become neutral, but the electromagnet sends a powerful inverse current into the wires which neutralizes the static current, within this shunt is placed a second electromagnet X. having large wires and whose cores ~~frace~~ face another electromagnet X′ connected in the main line circuit.

At the moment when the short circuit is completed the magnet X sends a powerful inductive into the coils of X′. and upon the line which strengthens the first part of the signal and when the shunt is broken the [-]reverse current is sent on the line in the same direction as the inverse current from M. one assisting the other.

If it is desirable that the inverse currents should be weakened an extra row of perforations can be employed and the extra magnets X X′ be arranged as in Fig. 9.

Fig. 9.

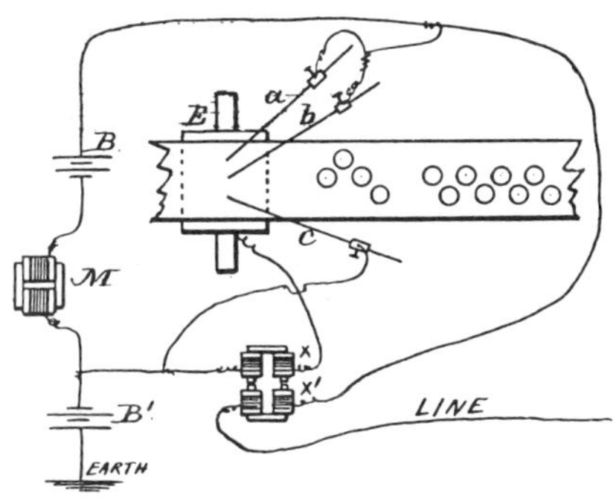

The extra row of perforations and point c. serving to short circuit the magnet X. and preventing an inverse current from being sent from X′. only that from M. being used.

I~~f~~t now remains to describe the various devices used at the receiving station to meet the requirements of the various lines on which the system is to be used.

In Fig. 10. is shown a device for the manipulation of the received signal so as to prevent as far as possible the effect of the static currents.

Fig. 10.

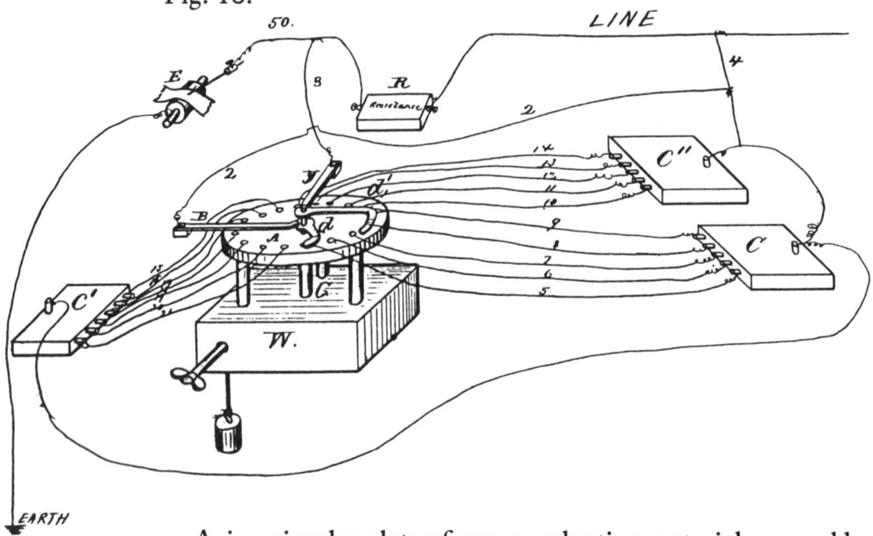

A. is a circular plate of non-conducting material secured by pillars to the top of the clock work W. which serves to give motion to the shaft G. which extends upwards through the nonconducting plate A. upon the end of this shaft is secured two contact arms insulated from each other, the arm d'. is in contact with the wire 3. and resistance coil R. the arm d. is in contact with the wire 2. which is connected with the wire 4. to the other side of the resistance coil R.

The spring B. serves to connect the wire 2. to d. while the spring y. serves to connect the wire 3. to d'. although this spring could be dispensed with and the wire 3. connected to the shaft by connecting it to the clockwork W. In the path of these two arms d'. and d. are a number of metallic pins each of which is connected to a condenser by means of the wires 5. to 21. The condensers being placed in the boxes C. C'. C''.

As the shaft is rotated the arm d'. alternately connects fresh condensers to the resistance coil R. in other words the resistance coil R. is being constantly shunted by condensers so that no one condenser is allowed to remain in contact with the resistance long enough to acquire a full charge.

The moment a condenser has become partly charged it is disconnected by the arm d'. breaking contact with the pen to which it is connected and coming in contact with another pen to which is connected a fresh condenser, the arm d. follows immediately after d'. and shortcircuits the condensers having been already charged and thereby discharges them ready to be again connected to the resistance coil R.

In a previous patent I have shown the resistance coil R shunted permanently with a condenser,[5] the effect of which is to prevent in a great measure the weakening of the first part of a received signal but when the line is very long the condenser sends back into the line a reverse current as strong as the signalling current, and the effect of this inverse current is to mutilate the following signal. In the device shown in Fig. 10. the condensers are disconnected at a moment before they are fully charged hence they cannot send a reverse current into the line to mutilate signals.

E. is the receiving instrument provided with chemical paper etc. in the usual manner.

Instead of condensers a number of electro magnets might be used connected in the same manner with the exception that the resistance R. could be dispensed with and the wire 3. connected to the ground wire E, or secondary batteries having plates of metal of the same character immersed in a fluid could be used and ~~the~~ be made to replace the condensers, the resistance R. being dispensed with; Even in some cases the condensers could be used without the resistance R. the connection between 3. and 4. where they connect with R be entirely broken.

Fig. 11.

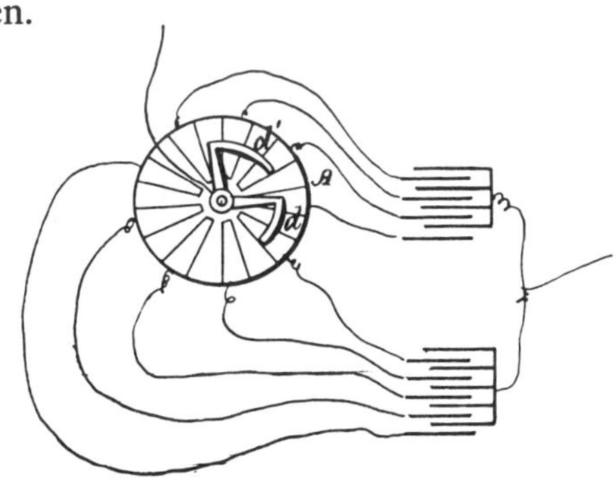

Figure 11. shows another method of arranging the contact points on the disk of insulating material. I do not wish to confine myself to any particular arrangement of contact points to throw fresh condensers in the line as two wheels divided into teeth separated by insulating material and from contact rollers connected to two condensers could be used one condenser being placed around R. while the other was being discharged.

Fig. 12.

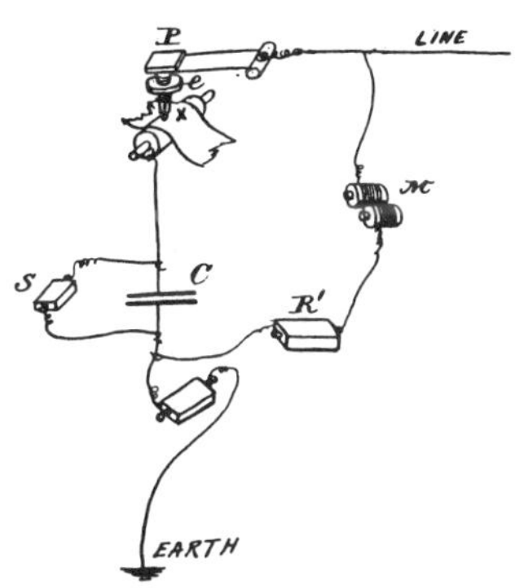

In Figure 12. is shown a device by which the back or reverse charge of the condenser is utilized and a very powerful compensation is obtained both by the condenser C. and the electromagnet M

In a previous patent I have described the peculiar action of metallic tellurium when used as a recording point,[6] unlike most metals or metals and chemical solutions it marks when it is connected to the zinc pole of the battery, hydrogen being evolved upon the tellurium point, forming hydrotelluric acid which is immediately decomposed by contact with the leaving finely divided tellurium.

This action takes place on paper saturated with almost any chemical solution.

It is also known that an iron point resting upon paper moistened with a solution containing a soluable chloride and ferridcyanide of an alkali metal gives a blue mark the oxygen or chlorine due to the decomposition of the chloride of sodium or water combining with the iron point to form a chloride or oxide and this in its turn combines with the ferridcyanide to form turnbulls blue or prussian blue.

I take advantage of these two dissimilar actions to obviate the mutilation and increase the speed of the signals.

The secondary pen P. Fig. 12. is formed of a taper split screw having a hole in the middle. e. is a jam nut which acting on the taper screw serves to tighten anything placed in the hole. Within this hole is placed a small pencil of metallic tellurium x which rests upon the paper; cast within this pencil of metallic tellurium is an iron wire which also rests upon the paper.

It is obvious that no matter what the polarity of the current may be, a mark will be made upon the paper either by the tellurium or iron.

C. is a condenser which may or may not be shunted by the resistance S.

M. is an electromagnet. R′ a resistance together forming a shunt around the receiving instrument. When a dash comes over the line the first portion is greatly weakened by the peculiar neutralizing effect of the static current and would so record itself were it not for the augumentative power of the condenser C. and electromagnet M. derived from their inductive capacity.

The consequence of this is to strengthen the first part of the dash to an amount equal to that lost by encountering the static current. If zinc be to the line the tellurium point will record before the current is taken from the line at the distant station the condenser C. would have nearly been charged and when the current is finally disconnected or made to cease the condenser sends a reverse current back into the line and the iron point records the discharge current thus it is obvious that dots and dashes of but half the length is necessary in this case to record dots and dashes of full length and strength and consequently the line does not become fully charged owing to the short period of time which the battery is allowed to remain in contact with it.

In Fig. 13. is shown a method to obviate the necessity of placing the iron wire within the tellurium pencil

Fig. 13

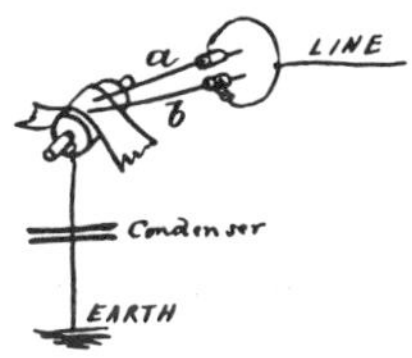

a. is a tellurium tipped point b. is an iron point. the former is placed immediately in front of the latter and in the same path, so that when one ceases to mark, the other will commence the natural elongation of the mark serving to bridge over the space between the two pens.

Fig. 14.

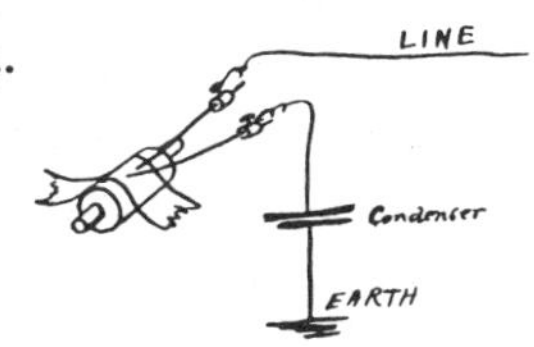

In Fig. ~~15~~14. is shown these two pens arranged as regard the line connections in a different manner from that in Fig. 12. that in Fig. 12. being more applicable by short lines where but little resistance should be offered to the paper the double surface of the points reducing the resistance while the drum forms the other pole and has a large surface.

Fig. 15.

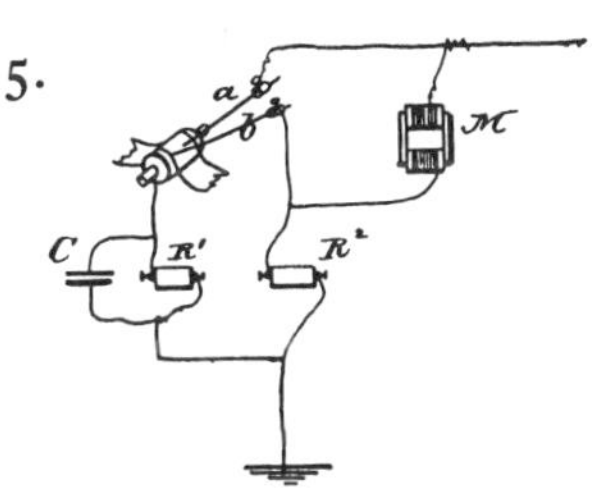

In Fig. ~~16~~15 is shown double recording points of the same metal, the main current passing part through the point a. across the moistened paper to the point b. thence through the resistance R^2. to earth and part passing through the compensating electromagnet M. to the earth while still another part passes through the drum to the resistance $R^{1\prime}$. and condenser C; by connecting in this manner I obtain a very powerful compensation both from the condenser C. and electro magnet M.

Fig. 1~~7~~6.

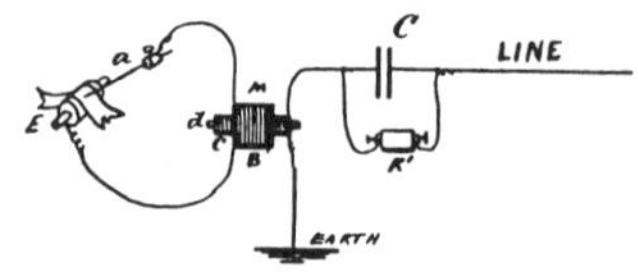

In Fig. 1~~7~~6. is shown the receiving instrument arranged with a peculiar inductive magnet or induction coil M. d. is the iron core of this magnet C. The primary coil of a great number of convolutions of very fine wire and offering a resistance equal to if not greater than that of the line wi~~s~~re. B. is the secondary coil composed of moderately large wire its two free ends being connected to the chemical recording instrument a. E.

C. is a condenser shunted around the resistance R^1. the action of which has already been explained.

By employing a primary coil having a great number of convolutions the effect of the distant current is to powerfully magnetize the iron core d. and allow of a strong quantative induction current to be thrown in the secondary coil to effect the recording of the signals.

If this arrangement was to be inserted at the point marked 50. in Fig. 10. the efficacy of the arrangement would be greatly increased. In Figs. ~~19~~17 and ~~20~~18. the condenser C. is arranged in different ways but the effect is almost the same.

Fig. 17

Fig. 18.

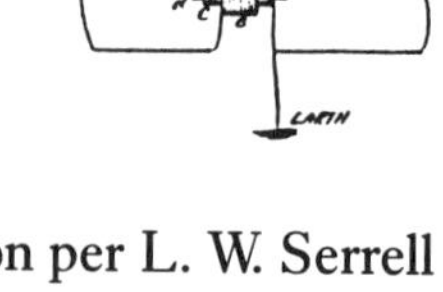

Inventor T. A. Edison per L. W. Serrell atty

Witnesses, Chas. H. Smith Harold Serrell.[b]

I will mention here that I have encountered considerable difficulty in casting pencils of metallic tellurium but have succeeded by forming the moulds of plaster of paris (It is likely a great number of substances can be used other than metals.) in casting these pencils by drawing the melted tellurium into the moulds by suction.

I shall probably claim

First. Transmitting dots and dashes by means of currents of unequal strength, the first part being of greater strength than the last, by employing several rows of perforations arranged substantially as described in Fig. 2.

Second. Transmitting a reverse current of a greater strength than the signalling current immediately after that has ceased and following the same with a current of the same polarity as the first signalling current but of less strength immediately after the transmission of the strong counter current in the manner substantially as shown in Fig. 1.

Third. The condenser C. shunted or not shunted, placed in the discharging earth wire operated by an extra row of perforations, substantially as set forth in Fig. 3.

Fourth. The condenser C. shunt X. points c. d. wires 4. 5. all arranged and operated substantially as set forth in Fig. 4.

Fifth. The electromagnet M. point c. resistance R. and R^2 arranged as in Fig. 6. substantially as described and for the purpose set forth.

Sixth. The condenser C.C. electromagnets X and X′ point c. shown in Fig. 7. and arranged substantially as described and for the purpose set forth.

Seventh. The two batteries B. and B′. magnets M. X and X′ point c. shown in Fig. 9. arranged as described and for the purpose set forth.

Eighth. The rheatome G.A.W. Fig. 10. condensers C. C′. C″ electromagnets or secondary batteries, resistance R. chemical recorder E all arranged connected and operated substantially as described and for the purpose set forth.

Ninth. The double recording point e. of iron and tellurium, condenser C. shunt S. electromagnet M. resistance R′. as shown in Figures 12. 13. and 14. and substantially as described for the purpose set forth.

Tenth. The double recording points a. b. Fig. ~~16~~15. magnet M. condenser C. resistance R^1. and R^2 arranged substantially as described and for the purpose set forth.

Eleventh. Casting metallic tellurium in non-metalic moulds by compression or suction with air.

Twelfth. The induction coil M. condenser C. shunt R. shown in Figures 16. 17.[c] 18. ~~19. and 20.~~ substantially as described for the purpose set forth.

Signed by me this 22nd day of March A.D. 1875.

Thomas A Edison

Witnesses Geo. T. Pinckney Chas H. Smith

DS, DNA, RG-241, Edison Caveat 63. Written by Pinckney; petition and oath omitted. [a]Place from oath; date taken from text, form altered. [b]All figures are on two separate pages, both of which are signed and witnessed at the bottom. [c]"16. 17." written in left margin.

1. U.S. Pat. 121,601.

2. U.S. Patents 121,601 and 132,456 for perforator designs show such an arrangement of holes. Although Edison had no transmitter or receiver patents at the time showing this arrangement, he had filed such a patent application in January 1875. It issued as U.S. Patent 168,242 on 22 September 1875.

3. Should be "D."

4. Should be "D."

5. U.S. Pat. 147,311.

6. U.S. Pat. 166,859.

–8– April–June 1875

In the spring of 1875 Edison consolidated his new business and technical alliances and established a laboratory independent of his manufacturing facilities. He continued as electrician for the Atlantic and Pacific Telegraph Company, a position the company formalized in June. In this capacity he focused on installation of his automatic system, sought designs for repeaters for the automatic, and worked further on his roman-letter system. Jay Gould formally acquired Edison's automatic telegraph patents for Atlantic and Pacific, laying the groundwork for future litigation. Edison remained in the public light as newspapers and technical journals took note of the conflict between Atlantic and Pacific and Western Union over control of his quadruplex patents.

Edison's principal concern in the spring of 1875 was the installation of his automatic system on the lines of the Atlantic and Pacific Telegraph Company. Having begun in the winter on the New York–Boston line, he sent automatic equipment to Atlantic and Pacific offices across the country, stocking major offices and smaller way stations with new apparatus. The company formally appointed Edison electrician in June, although he had been acting in that capacity since January. As electrician, Edison was responsible for the smooth working of the system. He spent considerable time trying to develop a repeater that could handle the rapid automatic signals. Edison hoped the electromotograph would solve the problem, but despite his best efforts, Atlantic and Pacific never installed any repeater on its automatic lines.[1] At the same time, Edison was trying to produce a commercially useful roman-letter system. He proposed several roman-letter schemes in experimental notes and in an extensive caveat, and by July he claimed to have developed a functional one-wire system. The company, however, never introduced any roman-letter apparatus, continuing instead to use the dot-and-dash system.[2]

In mid-April, acting as a trustee for Atlantic and Pacific, Jay Gould consummated the agreement of 30 December 1874 by purchasing the Automatic Telegraph Company's interest in Edison's automatic telegraph patents. In return, the Automatic investors were to receive Atlantic and Pacific stock worth a total of $795,000. Four days after the patent reassignment Gould settled with George Harrington, paying him the full value of the shares agreed upon but giving him cash and bonds rather than Atlantic and Pacific stock. Harrington, who had been seriously ill for some time, promptly retired to Europe, where he spent the rest of his life. Gould offered the other Automatic principals the same terms as those he'd offered Harrington, but only one accepted.[3] The remaining investors wanted the Atlantic and Pacific stock specified in the agreement; they never received anything. Moreover, Gould never assigned the patents to Atlantic and Pacific.[4]

At the same time, the legal maelstrom surrounding the ownership of the quadruplex intensified. Edison continued to meet with Atlantic and Pacific lawyers and to help the company with its case. He filed two lengthy legal documents. One was an answer to Western Union's New Jersey injunction; the other was an affidavit for the review by the Secretary of the Interior of the Patent Commissioner's decision about the assignment of Edison's quadruplex patent applications.[5] Although Patent Office action on Edison's disputed patent applications was officially suspended, in late April a telegram sent in Lemuel Serrell's name (by Josiah Reiff) resulted in the abrupt allowance of an old multiple-telegraphy application Edison had amended in March.[6] The issued patent was assigned to Edison and Harrington and immediately became the focus of intense interest. Atlantic and Pacific hoped to enjoin Western Union from using the quadruplex if the new patent's specification could be shown to cover it. However, Western Union contested both the patent's legitimacy and its scope, and Atlantic and Pacific neither employed the quadruplex nor sued Western Union at this time.[7]

Important developments occurred at the Ward Street shop, where the firm of Edison and Murray was busy manufacturing automatic instruments for Atlantic and Pacific and district signal boxes for the Domestic Telegraph Company. Edison, with substantial assets and the anticipation of a steady income, decided in mid-May to separate himself from the manufacturing business by dissolving his partnership with Murray.[8] Taking several key men with him, he established an independent laboratory in the same building. At the end of May, after drawing up a list of proposed inventions that ranged from a "sexduplex" (six-message) telegraph system to a chimneyless kerosene lamp, the group commenced experimentation. One of their first projects was a manuscript-duplicating process for business offices. The only available means of producing multiple copies of a document—the papyrograph—was a little-used, messy chemical process. Within a month the laboratory team conceived its first significant invention, the electric pen. This hand-held writing device pricked paper with a rapidly vibrating needle; the stencil thus created could produce hundreds of copies. Edison also continued to search for new forces, hoping to find something he might apply to telegraphy. Edison left developmental work on the quadruplex in the hands of Western Union's George Prescott and Gerritt Smith, although he claimed that several of their "developments" were his work of the previous fall.

Edison's public life was busy as well. W. S. Williams, publisher of the *Operator,* still wanted Edison to write for the paper, although nothing attributable to Edison appeared. The New York and Newark press—including the unceasingly abusive *Telegrapher*—continued to follow the quadruplex contest. In addition, Edison found himself publicly accused of stealing credit for the success of his automatic telegraph from Spencer M. Clark, who had been superintendent of the American Telegraph Works several years earlier. In a *New York Sun* article about the history of the Quadruplex Case, Clark claimed that he had perfected the automatic apparatus. Edison, questioned by a *Newark Evening Courier* reporter, denied that Clark had assisted him in the machines' development.[9]

As is so generally true of these years, little is known of the Edisons' private life during this period. They were in their new residence on South Orange Street, and money was no longer the problem it had been. One important development, however, is revealed through later events—at some point in the spring, Mary found herself expecting their second child.

1. In a standard relay or repeater, the signaling current had to create a magnetic field strong enough to move an armature, which in turn had to be agile enough to respond quickly and accurately. As sensitive as they were, such devices could not react fast enough to repeat automatic mes-

sages, which could easily generate more than 7500 signals per minute (500 words of five letters, each letter being three characters). In the electromotograph repeater, the electrochemical reaction created by the signal current simply allowed the arm of the instrument to slip, thereby closing a contact. When the current ceased, friction between the arm and the moving surface increased again, opening the contact. Because the armature motion did not depend directly on power from the line, Edison hoped it would be fast enough for automatic signals.

2. See Doc. 596 nn. 3–5; and TAE to Jay Gould, 2 Feb. 1877, 2:185–77, Box 17B, *Harrington v. A&P.*

3. In January, Gould had offered Harrington cash; he actually gave Harrington 10 percent cash and the rest in Northern Central Railroad bonds paying 7 percent interest. Bill of Complaint, 1:31–32; Harrington's testimony, 1:125; both Box 17B, *Harrington v. A&P.*

4. The lack of payment to the Automatic investors was the basis for *Harrington v. A&P.* Because Atlantic and Pacific never received the patents from Gould, but instead used them under another agreement, the company could claim to have received nothing from the investors, thus weakening the investors' claims in their suit. Quad. 71.2, pp. 69, 73–74 (*TAEM* 10:261, 263–64).

5. Quad. 72.14 (*TAEM* 9:204); App. 3, Edison's Affidavit; see also Doc. 577 n. 6.

6. Doc. 554.

7. The next year Atlantic and Pacific initiated two suits in which they used this patent to attempt to stop Western Union's use of the quadruplex. However, they abandoned both, pursuing instead a third suit that did not include this patent. Quad. TLC.3, pp. 176, 196; Quad. TLC.4, pp. 211, 226 (*TAEM* 10:910, 919, 926, 934).

8. The partnership did not formally end until July. See Docs. 593 and 594.

9. "Quadruplex Telegraphy," *New York Sun,* 16 Apr. 1875; "A Quadruplex Telegraph," *Newark Evening Courier,* 16 Apr. 1875; both Cat. 1144, Scraps. (*TAEM* 27:278). The *Sun* article identifies Clark as having been "the originator and First Superintendent of the Bureau of Engraving and Printing" of the U.S. Treasury Department before Harrington hired him for the American Telegraph Works. See also Doc. 272.

–558–

From David Brooks

Philadelphia, April 2d— 1875[a]

Dear Sir:

How do you come on with the induction experiments?[1]

You remember that I ~~e~~ventured an opinion that placed in poles, the usual hight from the ground the insulators would not show much.

My advices from Paris are that Stearns is doing splendid in the main circuits, with the double transmission using the printers.[2] My son[3] has arrived home and I expect to go there soon as I can get off—

What can you do for me with the P&A[4] in the way of insulators.[5] And what do you think of my cross-arm insulator ~~n~~fig.

no 3. other side.[6] Call at Tillotson and see a sample—[7] Yours truly

David Brooks[8]

AXS, NjWOE, DF (*TAEM* 13:401). Letterhead of The Brooks Patent Insulator Works. [a]"Philadelphia" and "187" preprinted.

1. Unidentified.

2. Stearns attempted to duplex a line employing Hughes printers between Paris and Lille, but his system failed to work dependably and was not adopted. Butrica 1986, 113.

3. Probably David Brooks, Jr.

4. These are the initials of the Pacific and Atlantic Telegraph Co., but Brooks was probably referring to the Atlantic and Pacific Telegraph Co., for which Edison was acting as electrician.

5. Edison apparently wrote Brooks the following day to inquire about the insulators. Brooks responded by defending them against claims that spider webs were a problem, noting that webs were easily cleaned off and that this should count for little in comparison with the improved insulation offered. He also noted that railroad companies were successfully using his insulators, which he claimed would not break with ordinary usage. Brooks to TAE, 5 Apr. 1875, DF (*TAEM* 13:402).

6. The following circular was printed on the back of Brooks's stationery.

David Brooks's patent insulator.

THE

Brooks Patent Insulator Works,

No. 22 SOUTH TWENTY-FIRST STREET,

PHILADELPHIA.

FIGURE 1.

This cut shows the Insulator as it is fastened to the pole, with the improved method of securing the wire.

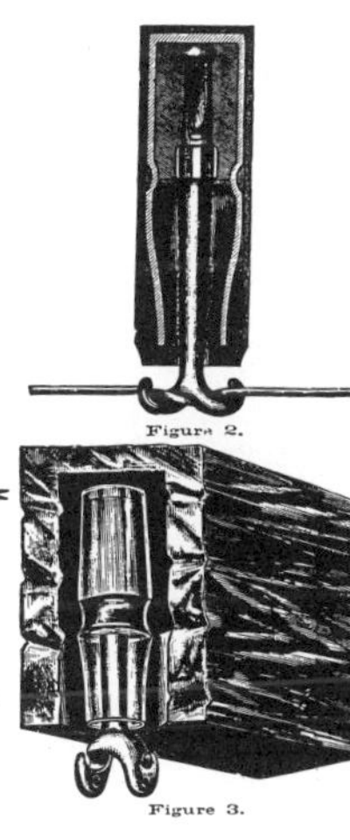

FIGURE 2.

This cut shows another form of the Hook entirely dispensing with the tie wire.

FIGURE 3.

This cut represents the Insulator in a cross arm without the iron shield. It also presents a view of the Hook, in which the wire is placed without bending or injury. This Hook can be used with or without a tie wire; the latter not being necessary to keep the wire securely in the Hook. It can, however, be used for every sixth or tenth Insulator as may be necessary to keep the wire from slackening or rending through the Hook.

REDUCTION OF PRICES.

THE INSULATOR, fastened to cross arm, will be sold at the price, or cost, of an ordinary glass Insulator and its supporting wooden pin.

They are warranted *to exceed in insulating properties* more than one hundred times any other Insulator offered in the market.

Their strength and durability are also superior to any other Insulator, and are guaranteed.

PHILADELPHIA, *March 15, 1875.*

7. L. G. Tillotson & Co. was a major manufacturer of telegraph and electrical equipment in New York City.

8. David Brooks (1820–1891), a prominent telegraph engineer and inventor, helped construct some of the earliest telegraph lines in the United States beginning in 1845. He had several inventions to his credit, including a well-known insulator. Brooks was also an early advocate of underground telegraph lines in the United States and developed a patented system that became the basis of the Brooks Underground Telegraph Co., founded in 1879 with help from Western Union. Reid 1886, 749–51; obituary in *J. Frank. Inst.* 132 (1891): 75–77.

–559–

From Patrick Delany

Buffalo Apl 4/75

Friend Edison—

The last style of Automatic transmitter & recvr is, infinitely superior to anything yet produced—[1] It works admirably, there is but one slight defect—the rubber wont do for transmitting channel—being unyielding it cant be adjusted so as to keep the paper flat on drum— I think you will find that in that channel a brush or some other retarding friction will be necessary— The new boxes are fine, plugs the proper thing—but unless handled carefully in shipment magnets get snapped off by being pitched about—[2] Please hurry one here to replace the one returned for repair— I think it would be a good idea to put a small single button switch on each table to switch on & off the Auto batty put it between the screw cup on table & the Auto inst— this would in great measure prevent bu[-]rning up of so many insts by careless people putting arms down—pulling wires out bad plan as wires get broken—

I think the balance & pully wheels on crank should be keyed—screws wont hold— Give us some stiff reels for transmitting strips—& some sort of an iron standard attached to Perforator or bench to place them in for Winding— It would be a good idea to make some scrapers for the receiving drum something the size of the channel—with smooth sharp surface that wont scratch the nickel plating off— Sandpaper & emory paper dont do it well & scratches the plate— make the standards for copying spools on table—higher—some "takes" too large for present sizes— try & make clock machines so that speed be regulated by screw that now puts on the brake[3] have taken weight off the flywheel vibrator & yet they dont run fast enough especially when writing elongated—have no weights to add— They mk a d——l of a racket—cant you obviate that too— I notice that filings allowed remain in connecting[a] holes of boxes get up a circuit when plugs out burns the insulation & varies the resistance— I had give mine thorough cleaning— Have been noticing these trifles so thot would let you have them Yours truly

Delany[4]

ALS, NjWOE, DF (*TAEM* 13:483). [a]Interlined above.

1. See p. 246 n. 2.

2. This is probably Edison's magnet box, which contained electromagnets arranged to bring any number of them into a shunt circuit to counter static discharge and prevent tailing. See Docs. 324 and 339.

3. For clockwork in automatic telegraphy see Docs. 418 and 481.

4. Delany was superintendent of the Automatic Telegraph Co. line before its transfer to Atlantic and Pacific. Delany's testimony, 3:2, Box 17B, *Harrington v. A&P;* see also Doc. 721 and *TAEB* 1:619 n. 9.

–560–

From Edward Johnson

New York Apl 5/75

My Dear Edison

Will you please make me Four Wheels of this size with your Metal rim[1] = & notify me when done ~~I~~So I can make requisition on you for them. If you have them "on hand" you know I'll have less talking to do to get my requisition approved. Cant you manage some way to send me one of your new transmitters = Ill manage to have one of mine condemned if neccessary— I should have one of them at least to study—so that I may be able to advise Others, as to possible location of trouble &c—!

Please hurry up the ~~P~~3 Perforators for which I have made requisition If I had had them Friday & Saturday Could have done an Immense auto biz = the 3 old ones I am going to return are as follows 1 Roman 1 Old one now used to Patch with & no more! = I want more Machinery = So as to be able to do a good days work, when the good day offers itself—

Blains wire did 937 messages 4228 w[ords] Press on Friday Hows that? = Had the Perforators been here it would have run over a thousand telegraph me if you are Coming Over tomorrow, If not I am Coming to Newark— want to borrow your Stereoptican for a private Exhibition at my Boarding House Wednesday evening.[2] Dont know how to work it— want Instructions. Yours

E. H. J.[3]

Somethings up. I was sick all night & today had lay down 2 Hours can now barely hold my head up

ALS, NjWOE, DF (*TAEM* 13:486).

1. Perhaps a flywheel for the automatic telegraph transmitter/receiver.

2. A stereopticon used a bright flame to project two coincident images onto a wall or screen for a stereoscopic effect (Knight 1876–77, s.v. "Stereopticon"). Edison had purchased 200 stereopticon views on 9 June 1874 (bill of Carman Peasell, 9 June 1874, 74-009, DF [*TAEM* 13:159]).

3. See *TAEB* 1:505 n. 13.

–561–

George Harrington to Jay Gould

New York, April 16, 1875.

Sir:

I hand you herewith a specific assignment of each and every patent and application for patents, covering all of T. A. Edison's inventions for automatic telegraphy, and whereby the full and complete title invests.[1]

The consideration to be paid therefor is thirty-one thousand eight hundred shares of the stock of the Atlantic & Pacific Telegraph Company.

I will thank you to withhold the within assignment until the Atlantic & Pacific Telegraph Company shall deliver to you the said shares of their stock, when the assignment will be delivered to them.[2]

These shares you please hold subject to delivery to the following named parties:

John McManus[3]	Reading, Pa.	43 shares.[a]
Seyfert, McManus & Co.,	Phila.	4,698 d[itt]o.
Wm. M. Seyfert	Phila.	320 do.
Wm. J. Palmer	Colorado	540 do.
John Elliot[4]	Riggs & Co., N.Y.	200 do.
H. C. Dallett, Jr.[5]	Phila.	60 do.
E. Corning	Albany	80 do.
James Dallett, Trustee	Phila.	120 do.
Alex. Morton	N.Y., 80 Broadway	40 do.
J. J. Marsh	Haverhall, Mass.	60 do.
Sam'l B. Parsons	Flushing	500 do.
J. C. Reiff	New York	7,057 do.
A.&P. Telegraph Co.		1,400 do.
T. A. Edison		3,000 shares.
J. C. Reiff, Sec'y.		1,428 do.
Geo. Harrington		12,254 do.
		31,800 do.

The receipts of said parties shall be your full acquittance. Very respectfully,

GEO. HARRINGTON.

Of the above sums there is the amount of $40,000 (about) currency, or about 1,600 shares (a little less) to be deducted from the account of J. C. Reiff, and redistributed to J. C. Reiff, Geo. Harrington, S. B. Parsons, Wm. J. Palmer, Edison and McManus. This redistribution, as it shall be agreed to, will be handed to you in the form of a paper signed by Reiff, Parsons and Palmer, and should be *approved* by Edison.[6]

With such paper please deduct and add to respectively as that paper will show.

PD (transcript), NjWOE, Quad. 71.2, pp. 35–36 (*TAEM* 10:244). Another transcription of this letter appears in *Harrington v. A&P.* [a]All entries in list aligned with dot leaders.

1. The assignment, dated 9 April 1875, appears in Quad. 71.2, pp. 31–34 (*TAEM* 10:242–44), and also in *Harrington v. A&P.*

2. Four days later Harrington accepted cash and bonds in lieu of his allotted shares and, shortly afterward, so did Samuel Parsons. Gould never delivered any stock, and in May 1876 the Automatic interests brought suit against him and the Atlantic and Pacific Telegraph Co. They sought to enjoin the defendants from using any of thirty-nine Edison patents involving the automatic telegraphy system, to have them account for profits made from the use of those inventions, and to have the deeds assigning Edison's patents and transferring the Automatic Telegraph Co. to Gould declared inoperative (Bill of Complaint, 1:31–32, Box 17B, *Harrington v. A&P*). At some point (probably before filing the suit), Edison drew up a list of investors in the automatic, noting how much they had invested (a total of $526,723) and how much each would receive on the basis of a $350,000 settlement (75-012, DF [*TAEM* 13:505]).

Gould and Atlantic and Pacific filed a demurrer in 1876 and answers in 1878 and the complainants took some testimony in 1879–80. During this time Josiah Reiff unsuccessfully attempted to negotiate a settlement. After 1880 the case lay dormant until 1895, when Reiff revived it against Atlantic and Pacific and Gould's estate, Harrington and Gould both having died in 1892. The case was finally tried in 1905 and decided in favor of the complainants in 1906. It was referred to a master for accounting to affix damages, and in 1910 Reiff and Edison were awarded one dollar. Both sides appealed, and in 1911 the decision in favor of Reiff and Edison was nullified. They appealed to the U.S. Supreme Court, but their appeal was dismissed. *Harrington v. A&P.*

3. On 10 May 1871 (Doc. 159), John McManus, William Seyfert, William Palmer, Samuel Parsons, and Josiah Reiff acquired one half of Harrington's interest in his 1 October 1870 contract with Edison (Doc. 109). J. J. Marsh, Erastus Corning, Jr., and Alex Morten later became parties to the agreement (Doc. 159 n. 7).

4. John Elliot was a banker with Riggs & Co. at 56 Wall St. in New York City. H. Wilson 1876, 382, 1118.

5. Henry Dallett, Jr., and James Dallett were both connected with

Dallett & Sons of Philadelphia, which dealt in sugar. Gospill 1875, 386.

6. According to Harrington, this paper was never delivered. Harrington's testimony, 1:131, Box 17B, *Harrington v. A&P.*

–562–

Agreement with George Harrington and Josiah Reiff

New York, *April* 16, '75.

I, Thomas A. Edison, owner of one third of my several inventions for automatic telegraphy, sold with my consent and approval to Mr. Jay Gould, do hereby make an allowance to Geo. Harrington and J. C. Reiff, from my ⅓ share of the proceeds obtained for said patents, for their time, trouble and services in connection with said inventions, and authorize such further deductions from my share as, with the ⅔ controlled by Mr. Harrington, shall be required to reimburse the several parties by whom money may have been advanced for automatic purposes, upon the basis of four in A.&P. stock to one of cash; that is to say, in the several amounts as herein set forth.[1]

THOS. A. EDISON.

PD (transcript), NjWOE, Quad. 71.2, p. 36 (*TAEM* 10:245). Another transcription is in *Harrington v. A&P.*

1. See Doc. 561.

–563–

Notebook Entry: Chemical Printer

[Newark,] April 17th 1875

Description of Edison's Dot & Dash Private line printer.[1a]

Fig. 1

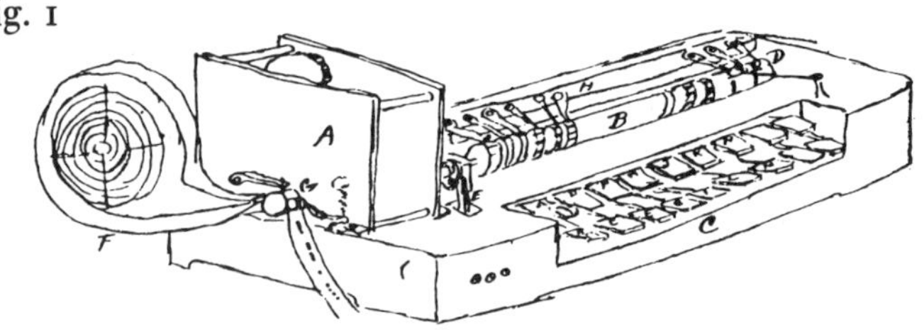

The object of the invention was to supply a cheap & reliable instrument for private lines. Those having type wheels being exceedingly dear & complicated It works without switches is always ready to receive & is always in open circuit It records on chemically prepared paper The Instrument consists of a clockwork A a transmitting roller B composed of brass washers with the 'morse character' cut on the periphery The contact springs H a Keyboard C a reversing arrangement[2] D a release E which is worked by a magnet under the base, a paper reel F enclosed in an iron box in order to keep the paper moist, a receiving drum G on which rests a stylus of Tellurium.[3]

Fig 2 shews the principle of the connections

[Fig 2]

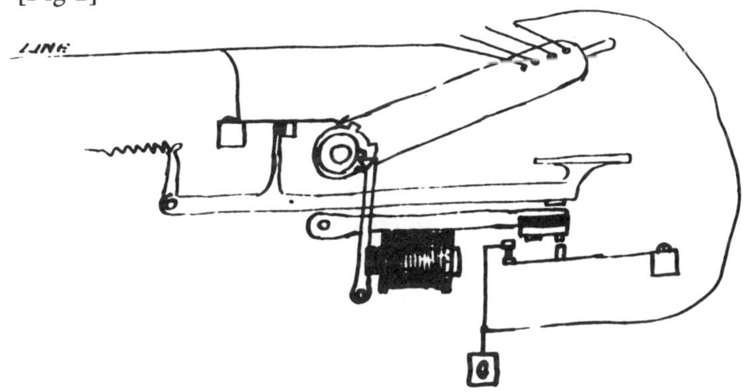

X, NjWOE, Batchelor, Cat. 1307:26 (*TAEM* 90:626). Written by Charles Batchelor. [a]Followed by a centered, double horizontal line.

1. Edison had designed this private-line printer the previous year as part of his domestic telegraph system, probably beginning work on it in the spring or summer; an account entry shows 185 hours of experimental work on "Private line printer to July 22nd 1874" (Cat. 1184:63, Accts. [*TAEM* 21:814]). Although the design was not patented until 24 February 1875 (Doc. 545), it was completed by December 1874 (Cat. 1214:51–52, Accts. [*TAEM* 21:589–90]).

2. For the reversing circuit see Batchelor's drawing of 11 April 1875 (Cat. 1307:23, Batchelor [*TAEM* 90:624]).

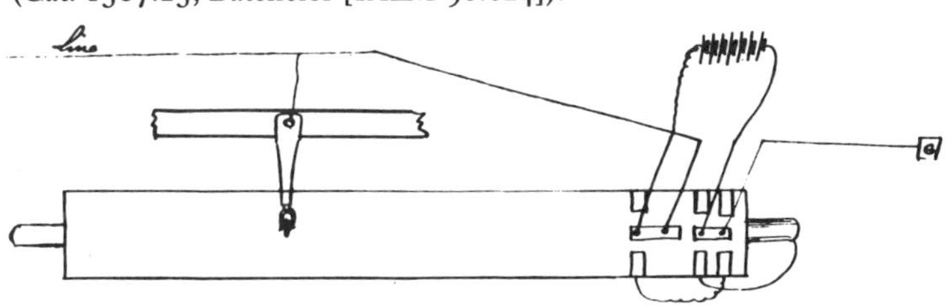

Charles Batchelor's drawing of the "Reversing arrangement for Morse character Private Line printer."

3. See Doc. 615.

-564-

Agreement with Gold and Stock Telegraph Co.

[New York,] April 20, 1875[a]

Whereas I am indebted to the Gold and Stock Telegraph Company in the sum of Forty two hundred & twenty two 26/100[b] standing upon the books of the said company to my debit as of date of[1c]

And whereas the said Company pending a settlement of the questions hereinafter mentioned has advanced to me the further sum of three thousand dollars.

And whereas questions have arisen ~~between~~ between the said Company and myself touching the amount which in equity and in fairness under the provisions of the agreement dated twenty sixth May 1871 made and entered into between myself of the first part and the said Company of the second part[2] I am entitled ~~to~~ to receive as extra[d] compensation for the several inventions of and improvements in Electromagnetic Printing Telegraphic Machines apparatus and appliances applicable to the business of the said Company invented discovered or made by me prior to the seventeenth day of April 1875 in addition to the regular salary of two thousand dollars per annum stipulated in said agreement and which has been regularly paid.

And whereas the said questions have been finally adjusted and settled upon the basis hereinafter mentioned. Now I hereby acknowledge to have received of the said Gold and Stock Telegraph Company the sum of Eight thousand Two hundred & twenty two 26/ [e] in full settlement satisfaction and discharge of all claims and demands on my part of every description against the said Company to the date hereof the said sum having been paid as follows—Forty two hundred & twenty two 26/ [f] thereof by the release by the said Company from the said sum standing to my debit on the books of said Company as of the date of Sept. 1st. 1873—[g] three thousand dollars thereof by the said amount so advanced to me as aforesaid and the remaining one thousand dollars thereof by so much cash paid to me by the said Company at the time of my executing these presents.

And in consideration of the premises and of the sums so paid as aforesaid I covenant and agree to and with the said Company its successors and assigns upon request to make sign execute and deliver to the said Company its successors and assigns all such assignments conveyances and instruments as may be necessary or proper or as the said Company its successors or assigns or their counsel may reasonably devise advise or require to vest in said Company the title to all such inventions modifications improvements and discoveries as may have been made by me prior to the date hereof and are or may be embraced within the said contract of twenty sixth May 1871 or to obtain letters patent for any such inventions, modifications improvements or discoveries or any reissues of any such letters patent and to do and perform every act deed matter and thing requisite and necessary to give the said Company the full benefit of all such inventions modifications improvements and discoveries.

Nothing herein contained is to be construed as terminating my relations to the said Company under said contract which is to continue in force for the period therein stipulated.

Witness my hand and seal this twentieth day of April 1875.

Thos. A. Edison[h]

In presence of M. Lefferts.

DS, NNC, Edison Coll. [a]Date taken from text, form altered. [b]"Forty . . . 26/100" written by Lefferts. [c]Space left for date. [d]Interlined above. [e]"Eight . . . 26/ " written by Lefferts. [f]"Forty . . . 26/ " written by Lefferts. [g]"Sept. 1st. 1873—" written by Lefferts. [h]Followed by wax seal.

1. An intermittent record of Edison's balance with Gold and Stock appears in Cat. 1185:13, Accts. (*TAEM* 22:568).

2. Doc. 164.

–565– [Newark,] April 20, 1875[a]

Summary Account

Assets Thomas A Edison		
Shop 10 & 12 Ward. machinery only	valued	50,000.00
160 Shares Pt Huron & Ft Gratiot St RR	"	22 000.00
2200 Shares Domestic Telegraph	"	10,000.00
20. 1000. Union Pac Sinking F. bonds[1]	"	18,600.00
Due from Gold & Stock	"	1 200.00
" Automatic Co patents	"	78,000.00
Real Estate at Port Huron[2]		1 000.00
Insurance Policies 20 000[3]		300.00
E H Johnson Note due Apl 7 75[4]		200.00
Liabilities Thos A Edison		
Small bills		500.00

AD, NjWOE, Accts., Cat. 1185:84–85 (*TAEM* 22:594). [a]Date taken from text, form altered.

1. Edison sold two of these bonds through Jay Gould in September 1875. Gould to Edison, 10 Sept. 1875, DF (*TAEM* 13:201).

2. Edison apparently owned two lots in Ft. Gratiot. For records of his real estate dealings there see Register of Deeds, Libers 45:455, 48:324, 49:230, and 50:501, MiPHStCo.

3. There are no extant insurance policies from before 1876.

4. In a small pocket notebook containing accounts for the first half of 1875 Edison wrote that this note was taken on 7 January 1875. He also loaned Johnson another $25 dollars on 8 January and $5 on 27 January.

It is unknown when or if Johnson paid off these loans. This same book also shows \$82.50 in other cash payments to Johnson. PN-75-01-05, Accts. (*TAEM* 20:29–32); see also Cat. 1185:200, Accts. (*TAEM* 22:646).

–566–

Caveat: Automatic Telegraphy

[Newark,] April 27, 1875[a]

To all whom it may concern.

Be it known that I, Thomas A. Edison, of Newark, in the County of Essex, and State of New Jersey, have invented an Improvement in Automatic Telegraphs, of which the following is a specification.

The object of this invention is to transmit by means of perforated paper or metallic types, electric waves, or currents to form roman letters at the distant station, upon chemically prepared paper.

The invention consists:—

First. in the manner of grouping the various perforations of the perforated slip, and the combination therewith of contact rollers, batteries to effect the transmission of the proper waves for forming the letter.

Second. In the arrangement at the receiving station of the recording points, compensating devices, and electro-voltaic corrector, for correcting any mutilation of the letter, which in one of my methods takes place, and which is corrected by other means which will be presently described.

In figure 1. is shewn a method for transmitting the waves necessary to form the letter at the distant station. In a previous application I have shown a perforated slip, where the whole roman letter was formed by groups of perforations.[1]

Fig. 1.

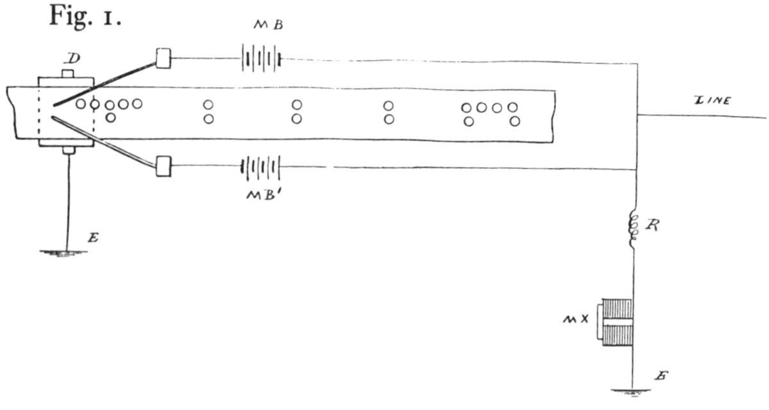

In the perforated slip shown in Fig. 1. but half of a letter is placed in one group while the other half is placed in another group, thus using but two rows of perforations to form the letters; Two contact rollers are used running on the paper in the path of the two rows.

The slip shown in Fig. 1. has the letters T. and H. The five holes in the first row and the single hole in the second row forms the top part of the letter T. The second group of two holes, one in the first row and the other in the second row serves to form the bottom part of the letter T.

The second and third pair of holes form the top of the letter H, while the last group of holes form the bottom of the letter H.

A. and B. are the contact rollers M.B. is a main line battery with the zinc pole to the line M.B′. is another main battery with the copper pole to the line.

R. is a resistance coil and M.X. a magnet combined together to form a shunt at the transmitting station for the purpose of compensating for the static currents of the line. D. is the drum which serves to carry forward the perforated strip, the rotation being obtained preferably by a clockwork movement regulated by a ball pendulum as will be described hereafter.

The pen A. is set one hole in advance of the pen B. hence the waves due to the contact through the first row of holes will send a negative current over the line, and the contact made through the second row of holes will send a positive current over the line.

While one of the points is in connection with the drum D. through a perforation, the other pen will be upon the paper forming a space between the holes in the second line.

In Fig. 2. is shown the receiving apparatus. A. and B. are the receiving pens. A. is connected to the line while B. is connected to the earth.

Fig. 2.

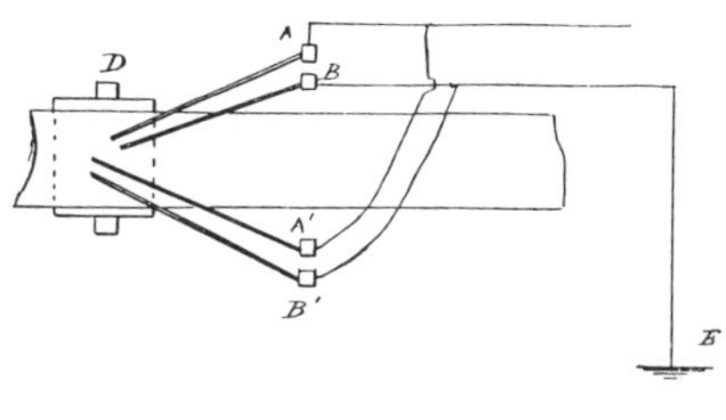

D. is the drum and C.P. the chemical paper.[2] This drum is rotated preferably by clockwork regulated by a ball pendulum as will be hereafter described.

A′. and B′. are also recording pens, A′. being connected to

the line and B′. to the earth. A′. and B′. are set one letter ahead of A. and B. I will now describe the method of transmitting and recording.

When the points A. and B. at the transmitting station pass alternately through the perforation forming the first part of the letter T. the points A. and B. record the same upon the chemically prepared paper and form the top of a perfect T. at the same time A′. and B′. duplicate this T. or in other words record false marks unnecessary to effect the formation of the letter.

After the points A. and B. at the transmitting station have left the first group of perforations forming the letter T. they come to the second group forming the bottom part of the letter T. at this moment that part of the letter T. recorded at the receiving station by the pens A. and B. has by the forward movement of the paper been brought directly over the pens A′. and B′. at this instant the currents from the second group of holes forming the bottom part of the letter T is sent over the wire and recorded immediately under the first part, thus forming a perfect letter T., but while A′. and B′. were recording the bottom part of this letter T. the points A. and B. were recording it also, forming what I call a false or unnecessary signal.

These false signals are shown in Fig. 3. between the perfect letters.

Fig. 3.

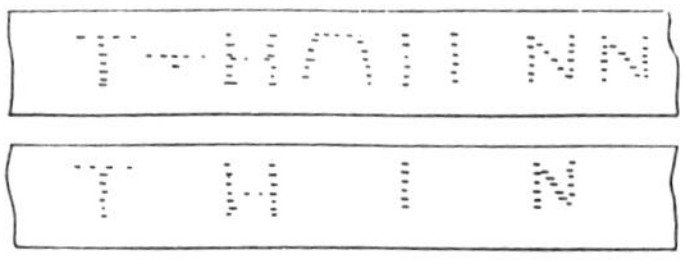

It remains now to describe the manner by which I am enabled to obliterate these false signals in an expeditious manner and render the message clear and legible.[3]

Fig. 4.

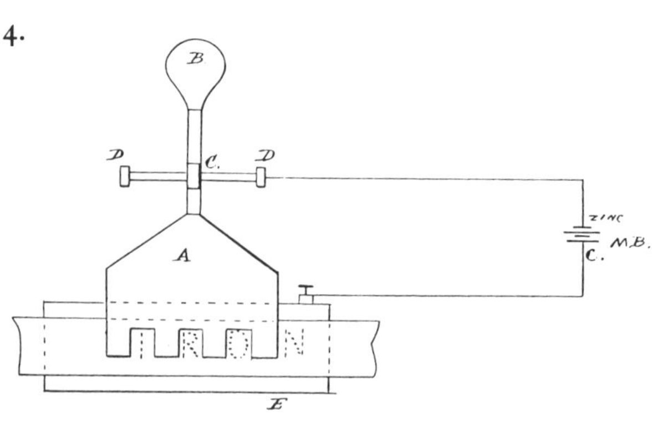

A. is a fork having 15 or 20 flat platina faced teeth operated by a handle B. pivoted upon the shaft C. which is connected to the bearings D. D. F. is a flat plate connected to the copper pole of the battery M.B. the zinc pole being connected to the bearings D. D. shafts C. to A.

Laid upon this flat plate is the wet chemical strip containing the mutilated message. The spaces between the letters owing to the nearly synchronous movement of the transmitting and receiving instruments and the equal space with which one letter is separated from another on the perforated slip the squares X.X.X.[4] when pressed down upon the strip by the handle B. come in contact with the paper having the mutilated signals between the properly formed letters.

While these teeth are resting on the paper the passage of the current through it obliterates these false signals by a chemical change brought about by electrolysis.

In practice the fork A. may have sufficient number of teeth to obliterate the false marks between the letters forming several words

The strip is then passed along automatically or by the hand of the operator and a fresh portion subjected to electrolysis.

Fig. 5.

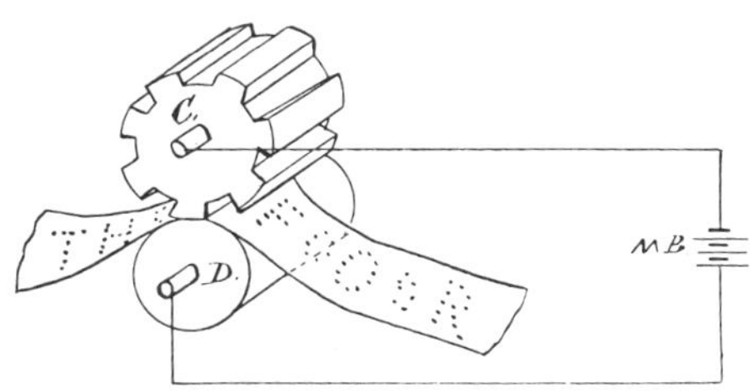

In figure 5. is shown another method by which the same object is accomplished automatically and more expeditiously. C. is a large wheel having a number of flatened teeth of a width nearly equal to the space between the letter and resting upon a drum D. which are geared together in such a manner that if either be rotated by hand, clockwork or otherwise the chemical strip is passed between them, one of the teeth being set (before the message is passed through) exactly in a space between the letters D and C. are then connected to a battery and then made or allowed to slowly rotate. Each tooth as it passes along fits between the letters and obliterates the false signal.

Fig. 6.

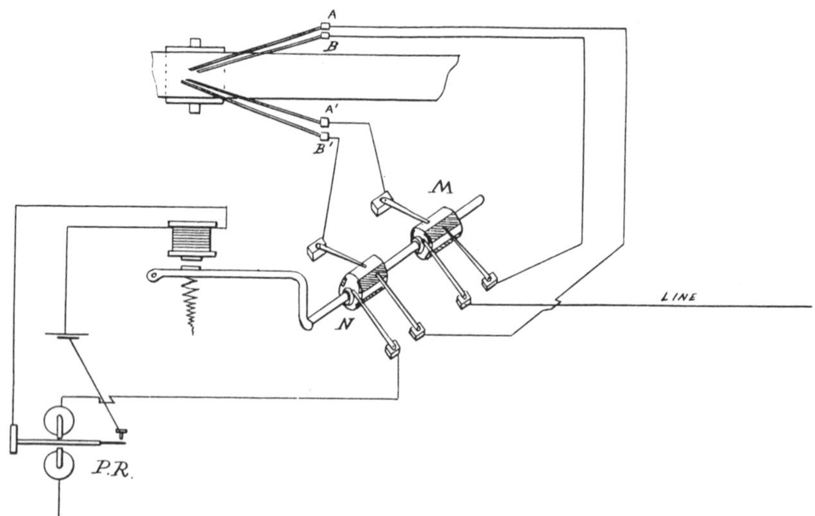

In Fig. 6. is shewn a method by which the synchronism of the two machines at the transmitting and receiving stations is connected at every few words, thus insuring correct spacing and allowing of a rapid obliteration of the false marks without the necessity of adjusting the obliterating devices every 15. or 20. words which is necessary when the machines do not run in any degree synchronously. It also allows me to prevent the formation of false marks altogether and dispensing with the correcting apparatus. I effect this object by means of a large extra hole in the perforated strip at the transmitting station, which hole is in line with a contact roller connected to a very powerful battery with its positive pole to the line.

The current transmitted from this extra hole which is placed at every 15. or 20. words at the pleasure of the perforating operator is to actuate a polarized relay at the receiving station which serves to release two circuit directing wheels N and M. which are arranged in such a manner with the recording points A. and B. and A′. and B′. that while A. and B. are recording the top part of a letter A′ and B′. are disconnected from the line and earth and prevented from recording a false signal, and while A′. and B′. are connected to form the other part of the letter the pens A. and B. are disconnected and also prevented from forming false signals.

Fig. 7.

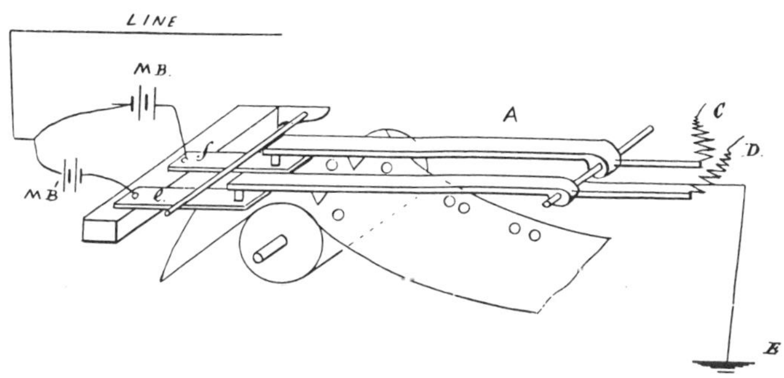

In figure 7. is shown a method for transmitting positive and negative waves in a manner similar to that shown in figure 1. except that the strip in this case is embossed or indented as it were, and in line with two levers A. and B. held down upon the paper by the springs C. and D. and both connected to the earth. Springs f. and e. are connected to the line, f. through the battery MB. having copper to line and e. through the battery MB. having zinc to line.

Immediately an indentation passes under the tip part of the lever it falls therein and allows the platina tipped point to come in contact with f. or e. as the case may be and transmit a positive or negative current over the wire. These indentations could be dispensed with and the ordinary perforated slip used to give motion to these levers, a slight groove being placed in the drum to allow the levers to fall a greater distance or the drum might be made stationary and two deep holes drilled in it immediately underneath the points near the end of the levers, the paper being drawn forward by another drum, or the embossed paper might give motion to these levers by raising them up but of course the springs e. and f. would in this case be placed above instead of below as is shown in figure 7.

Fig. 9.[5]

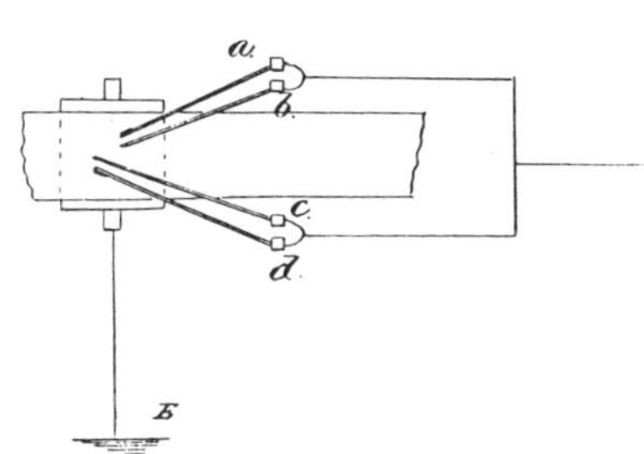

Figure 8. shows the clockwork and swinging pendulum with regulating ball E and stationary ball F. which can be

made to approach or recede from the fulcrum of the pendulum by means of a cord and adjusting nut h.

Fig. 9.

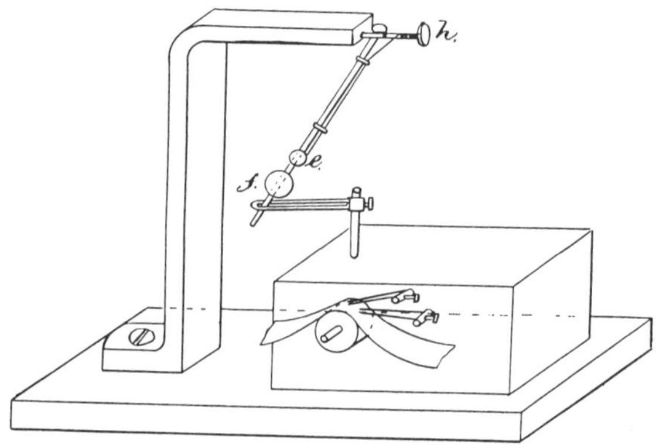

Inventor Thos A. Edison per L. W. Serrell atty.
Witnesses Chas H. Smith Harold Serrell[b]

Fig. 9. shows another arrangement of the receiving pens in which the four are connected and to the line the drum being connected with the earth. a. and c. are tipped with metallic tellurium which record by the passage of a zinc current over the line and b. and d. are iron points or any other metal which records by the passage of a copper current over the wire.

I will mention here that if the roman letter is recorded by means of a tellurium point that the flat teeth shown in figure 4. might be covered with an absorbent substance and moistened with an alkaline hyposulphite, preferably hypo-sulphite of sodium or a weak solution of nitric acid and many other chemical substances which coming in contact with the false marks will produce a chemical change and render them invisible, or if an iron recording point were used the absorbent substance might be moistened with caustic potash and many other chemical substances.

I shall probably claim.

First. groups of perforations either in the form of roman letters or in groups from which roman letters can be formed in combination with contact points and batteries and a telegraphic circuit arranged substantially as set forth and for the purpose specified.

Second. So arranging the groups of perforations in a strip of of paper that the waves necessary to form half of a letter can be transmitted at one time and the waves necessary to form the other half of the letter can be transmitted at another time substantially as and for the purposes specified.

Third. Transmitting from a group of perforations positive and negative currents to form one half of a roman letter and positive and negative currents from another group of perforations to form the other half of the letter.

Fourth. The batteries M.B. and M′.B′. contact rollers A. and B. arranged substantially as and for the purpose set forth.

Fifth. In a telegraph recording roman letters the transmission of voltaic[c] currents to record the same and of inductive currents to effect compensation of the static discharge.

Sixth. The use of two or more roller pens or contact rollers when insulated from each other and connected to batteries in a telegraph recording roman letters.

Seventh. Transmitting waves of current necessary to form roman letters from contact levers and springs operated by a motion derived from perforated or embossed paper.

Eighth. The combination in a chemical roman letter recording telegraph of four recording pens, two connected to the earth and two connected to the line arranged substantially as shown in Fig. 2.

Ninth. In a chemical recording telegraph the use of four recording points, two of which record by the action of hydrogen due to electrolysis and two which record by oxygen, also due to electrolysis when the four are connected to the line wire as shown in Fig. 8. for the purpose specified.

Tenth. In a chemical telegraph the obliteration of false signals by electrolysis.

Eleventh. The fork A. plate F. battery MB. as shown in Fig. 4. arranged substantially as described and for the purposes set forth.

Twelfth. The toothed wheel C. drum D. and galvanic battery M.B. as shown in Fig. 5. for the purpose specified.

Thirteenth. The combination of a clockwork regulated by a horizontal pendulum and adjusted as set forth for giving motion to perforated and chemical paper with batteries and a telegraphic circuit for the purpose specified.

Fourteenth. The use in a roman letter chemical telegraph of a circuit changing device controlled and regulated by an increased wave of current transmitted by an extra hole in the perforation at the transmitting station.

Fifteenth. The circuit changer N.M. and its contact point (fig. 6.) polarized relay PR recording points A. and B. and A′.

and B′ arranged substantially as set forth and for the purpose specified.

Signed by me this 27th day of April A.D. 1875.

Thos A Edison

Witnesses Geo. T. Pinckney Chas H Smith

DS, DNA, RG-241, Edison Caveat 64. Written by Pinckney; petition and oath omitted. [a]Date taken from text, form altered. [b]All figures are on two separate pages, both of which are signed and witnessed at the bottom. [c]"l" interlined above.

1. Doc. 374.
2. Label missing in drawing.
3. See Cat. 1143, Scraps. (*TAEM* 27:284) for experimental strips of this design.
4. Label missing in drawing.
5. Should be "8."

–567–

Notebook Entry: Automatic Telegraphy

[Newark,] April 2~~1~~7~~st~~th 1875

Platina Solution

Water 1 Litre
Pyrogallic Acid 3 dwt[1]
Nitrate Strontia 17 dwt
Salt 6 oz.[a]
Platina pen on H[b]

Very good solution & permanent use it on my printer[2] & it is O.K.

Johnson tried it last night to Phil & Washn & got writing as fast as he could turn

C B

X, NjWOE, Batchelor, Cat. 1307:27 (*TAEM* 90:627). Written by Batchelor. [a]"Water . . . 6 oz." enclosed by a brace. [b]"Water . . . H" enclosed in a box.

1. Pennyweight (1.555 grams).
2. On 7 February 1875 Batchelor wrote to Jay Gould offering him a "stock quotation printing telegraph"; on 10 February, Gould agreed to purchase it (Gould to Batchelor, 10 Feb. 1875, Unbound Documents, Batchelor [*TAEM* 92:248]). Account entries for 16 and 29 February 1875 show payments from Gould totaling $600 for work on Batchelor's printer (PN-75-01-05, Accts. [*TAEM* 20:33]). Batchelor executed a caveat on 9 August 1875 (Cat. 1238:1, Batchelor [*TAEM* 93:16]) and a patent application on 31 August 1875, which issued as U.S. Pat. 173,754 on 22 February 1876. He continued to work on his system through the fall of 1875 (Cat. 1172:290, Lab. [*TAEM* 3:171]; Cat. 1307:17, 24, and Cat. 1317:11, both Batchelor [*TAEM* 90:618, 625, 662]). An account

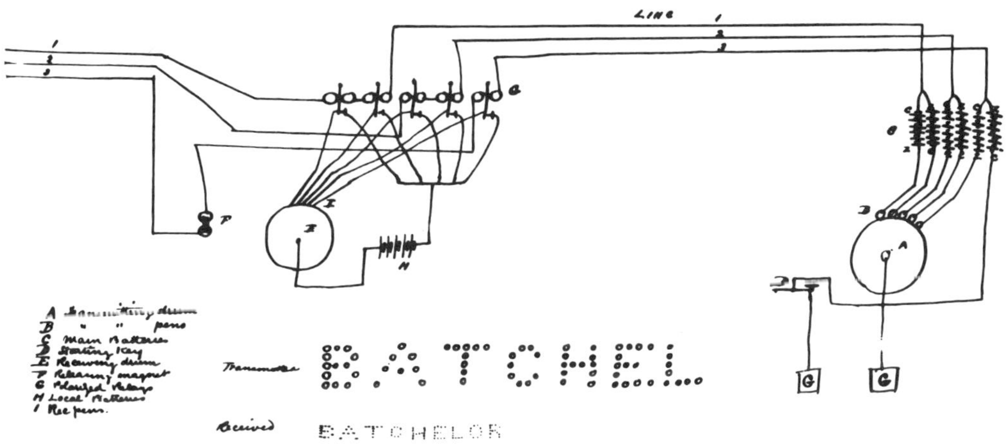

Charles Batchelor's chemical stock-quotation printer.

entry shows work on the printer between 31 July and 24 September 1875 (Cat. 1214:70, Accts. [*TAEM* 21:599]).

A 14 March 1875 drawing of the printer, on which Batchelor indicated that the invention dated from 8 February, shows a design similar to those of Edison's roman-letter automatic telegraphs (Cat. 1307:13, Batchelor [*TAEM* 90:614]). The transmitting drum is on the right and the receiving drum on the left. Positive and negative battery currents sent over the line wires by alternate transmitting rollers activated the polarized relays (located over the receiving drum in the drawing). These in turn opened and closed the local circuit connected to the recording pens.

–568–

Notebook Entry: Autographic Printing

[Newark,] April 30th 1875

Copying Experiment

Aniline strikes blue with chromic acid   tried to copy with it but with no success. Lacquered a platina plate & wrote on it with a strong sol. of caus pot. which dissolved shellac & washed off   Could then take a great many copies with battery & sensitive paper but they are wrong side about.[1]

X, NjWOE, Batchelor, Cat. 1307:27 (*TAEM* 90:627). Written by Charles Batchelor.

1. This represents Edison's first experimental work in document copying and multiple duplication. By the 1870s many businesses wanted to circulate instructions, announcements, and other information to large numbers of employees or customers. Having these documents printed was slow and expensive; letter-copying books made only one or two copies. Firms such as banks, railroads, and insurance companies needed a fast, inexpensive, and convenient means of making large numbers of copies of written documents on their own premises. When Edison turned his attention to this problem there was only one other potential

competitor on the market—Eugenio de Zuccato's papyrograph. This system required the use of a steel-nibbed pen and a solution of caustic soda in place of ink to write on a thin sheet of paper coated with an ink-impermeable lacquer. The caustic soda softened the lacquer along the lines of writing so that when the written sheet was pressed between water-wetted blotting paper, the lacquer was removed, leaving a lacquer-free stencil of the handwriting. Although this method proved complicated and messy, it was used in offices in Britain and the United States during the 1870s and 1880s. Proudfoot 1972, 40–42; Yates 1989, 50.

–569–

From Albert Chandler

New York [c. May 1,] 187[5][1a]

Edison:—

When were the seven inventions made, which are[b] covered by the seven specifications in your application for patent in which Prescott claims joint interest?—[2]

Before or after Harrington's contract with you of April 4, 1871?—[3]

Prescotts counsel are said to claim that your case "H." is for the same invention as that described in your application 99.—[4]

Is this true?—[5]

Is "99" the one referred to in yr. contract with Prescott?—

Will not "case H." enable us to enjoin W.U. from use of quadruplex?—[6]

Cannot you make new application entirely different from "99." yet covering its essential features, but based upon amended and consolidated specifications?—[7]

A. B. C.[8]

ALS, NjWOE, Scraps., Cat. 30,096:47 (*TAEM* 27:583). Letterhead of Atlantic and Pacific and Franklin Telegraph Companies. [a]"New York" and "187" preprinted. [b]"which are" written in left margin.

1. This letter apparently postdates the 27 April publication of Edison's U.S. Pat. 162,633 (Case H [Doc. 554; see Doc. 577 n. 10]), since George Prescott insisted he learned of the Case H application only after its allowance and this letter refers to his reaction. It is probably related to the preparation of documents filed with the Secretary of the Interior at the end of April and early in May in response to the 20 March decision of the Patent Commissioner (see n. 3). Brief of George B. Prescott, Quad. 72.4, pp. 18, 26; Statement and Brief of Z. F. Wilber, Quad. 72.5, p. 38 n. 16; and Robert Russell, Argument for Mr. Edison, Quad. 72.7 (*TAEM* 9:54, 58, 83, 114); see also n. 8.

2. The specifications are Docs. 467–73. For Prescott's claim to them see Doc. 466.

3. Doc. 155. The point in question was the date Edison conceived the designs at issue, not when the applications were prepared or filed. Doc. 155 was a legal *assignment of title* to patent rights only if the inven-

tion was legally in being at the time Doc. 155 was signed. Otherwise it was only an obligation to make such an assignment at the appropriate later time. That in turn affected whether Harrington could claim to have a *legal* title or just a title in *equity* to patents for those inventions. The Commissioner of Patents said that he based his 20 March 1875 decision—that any issued patents would be assigned to Edison and Prescott instead of Edison and Harrington—largely on his interpretation of this point. Commissioner Thacher's Decision, Quad. TLC.6, p. 13 (*TAEM* 10:980); for a thorough review of the legal issues see Argument on Behalf of Edison, Quad. 72.8 (*TAEM* 9:122).

4. Case H (U.S. Pat. 162,633) is Docs. 315 and 554; Case 99 (U.S. Pat. 480,567) is Doc. 472.

5. The amended version of Case H (Doc. 554) claimed that the apparatus shown therein could function not only as a duplex telegraph but also as a diplex (with straightforward rearrangements and additions). Also, its working involved independent operation of signals made by increase and decrease of current (or tension) on the one hand and by reversals of current (or polarity) on the other. Prescott therefore charged that the granting of that patent and its assignment to Edison and Harrington constituted an attempt to appropriate the heart of Case 99 and thereby defraud him (as well as Western Union) of his rights (Prescott to Commissioner of Patents, Quad. 72.3, pp. 2–5; Brief of George B. Prescott, Quad. 72.4, pp. 1–18, 28–30; Statement and Brief of Z. F. Wilber, Quad. 72.5, pp. 10–13 nn. 29–34, pp. 27–33 nn. 1–5, 10 [*TAEM* 9:39–41, 45–54, 59–60, 69–70, 77–80]). Years later a Patent Office Board of Examiners in Chief officially decided that Case H, even as amended and granted, did not really describe a practical way of sending two messages in the same direction, which was the crucial aspect of Case 99, and thus the two cases were not similar enough to interfere. H. C. Townsend to TAE, 8 Mar. 1879, Pat. App. 162,633.

6. The Atlantic and Pacific Telegraph Co. did not seek such an injunction against Western Union until the following year, when Case H was part of its suits in federal courts in Washington, D.C., and New York. Even then it did not use this approach in its suit in the New York state courts, which was the only suit pursued to trial. For the 11 April 1876 New York state suit see Exhibit 13 in Quad. TLC.2, p. 74; for the 11 May 1876 Washington federal suit see Quad. TLC.3; and for the 23 October 1876 New York federal suit see Quad. TLC.4 (*TAEM* 10:857, 909, 926).

7. This approach was discussed in detail in a confidential review of the case drawn up around this time, possibly by Robert Russell, one of the attorneys acting for Edison. Argument on Behalf of Edison, Quad. 72.8, pp. 28–34 (*TAEM* 9:136–39); see also Argument for Mr. Edison, Quad. 72.7 (*TAEM* 9:114).

8. About this time pneumonia disabled President Eckert, and Chandler temporarily took over the management of Atlantic and Pacific. Eckert's illness began at the end of April, though he managed to attend some company meetings as late as 3 May. Reid 1886, 596; Eckert's testimony, 2:249, Box 17B, *Harrington v. A&P*.

–570–

Notebook Entry: Miscellaneous

[Newark,] Sunday May 2 1875

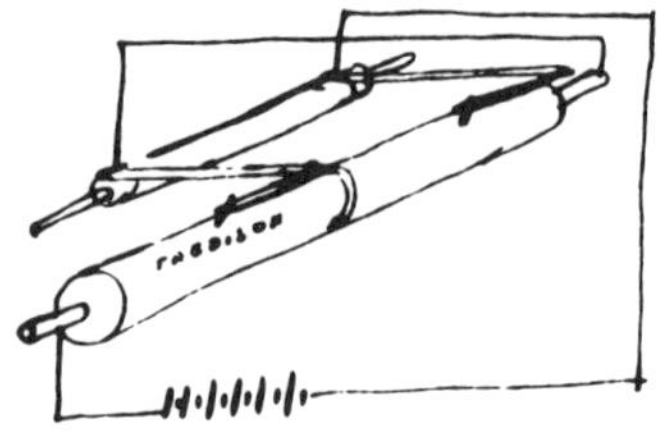

Made machine with one pen like drawing. A long drum split in halves & insulated from each other 2 Pens sliding together on an insulated rod 1 on each drum. Wrote with insulating ink on 1 drum & put sensitive paper on the other & moved the pens backwards & forward quickly over them. The result was pretty good. Tried it & got some copies off Pyrogal sol. also Prussian blue paper to be whitened by the current but the latter was not sensitive.

Took 109 Samples of static charge at 10 words per minute over the cable by shifting the condensers[1]

Tried (Alcohol & No 5) Sulphuric Acid & Carb Soda for a new force no go

X, NjWOE, Batchelor, Cat. 1307:45 (*TAEM* 90:630). Written by Charles Batchelor.

1. The samples and an accompanying notebook entry can be found in Cat. 1181:202–13, Lab. (*TAEM 3:131–36).*

–571–

Notebook Entry: Automatic Telegraphy

[Newark,] May 3rd 1875[a]

Solution

Found a good solution

Pyrogallic Acid
Nitric Acid in excess
Nitrate of Ammonia in excess[b]
Yellow mark comes out reddish & marks on Oxygen.[c]

One Wire Roman letter

Worked all night on this testing our new syncronous instruments.[1] Ferrid paper, was able to read it almost first trial. Had considerable difficulty throughout the evening in not getting the dots on line 4 but getting the duplicate found the difficulity in the perforations being wider & pens not being properly adjusted ~~from~~. Not tried it yet for perfect syncronism

Bad receiving pen, it must be altered.

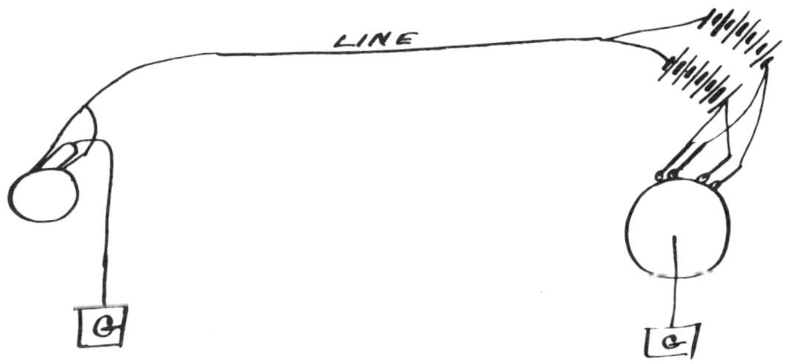

2 Batteries 24 cells each small Bichrom. No 4 Sol.
Worked all night ⟨12hrs.⟩[d]
Edison[a] Batchelor Adams Assistants[e]

X, NjWOE, Lab., Cat. 1172:219 (*TAEM* 3:138). Written by Batchelor. [a]Underlined twice. [b]Followed by centered horizontal line. [c]Followed by broken horizontal line across the page. [d]Marginalia written by Batchelor. [e]Names written by Batchelor and enclosed by a drawn rectangle; "Batchelor" and "Adams" enclosed by a brace and followed by "Assistants".

1. See Doc. 566.

-572-

Notebook Entry: Electromotograph

[Newark,] May 7th [1875][1]

Batchelor fixed the Electromotograph with single lever by midnight & then we went to work on repeating, ~~&~~ successfully repeated (300) three hundred words per Minute Electromotograph working with platina pen on No 4 Solution[2] samples not saved worked till 8 AM of the 8th of May ⟨12⟩[a]

Batchelor, Edison, Adams, & Gilliland[b]

X, NjWOE, Lab., Cat. 1172:223 (*TAEM* 3:140). Written by Batchelor. [a]Marginalia written by Batchelor. [b]Names written by Batchelor.

1. Dated by reference to surrounding notebook entries.
2. Unknown.

-573-

Notebook Entry: Electromotograph

[Newark,] May 10th 1875

Repeating by Electromotograph.

Worked all night. Had a great deal of trouble to get it going up till 1 a m could not tell what was trouble. it seemed ~~to gradually slip back when current was on & go forward too far~~

~~when space came~~ as if the normal friction lessened after working a short time after supper however it worked better with new paper & we kept getting it good till we reached (800) eight hundred words per minute perfect[a]

Phenomenon. Salt & Oil Sweet Almonds Jump! only on opening. a friction jump.[a]
Nit. Mercury—Salt. Fearful friction on O, with Pyrogallic strips on H making a splendid double current paper.[a]
NA Cl, HO, & Cajeput Oil[1] EMGraph nix[a]
Na Chl & Anise oil Friction on O. No mark[a]
Na Chl. & Rosemary oil Nix[a]
Na Chl & Lavender Oil Nix[a]
Na Chl & Lemon Oil Nix[a]
Na Chl. & Mirbane Oil[2] & K Bichrom. increase friction &c [-][b]
Coffee[c] eggs Sugar & milk Phenomenon decreased friction on oxygen.[a]
Pyro added to the above no good[a]
Na Cl. & Caraway Oil Nix[a]
Na Cl. & Cloves Oil Nix
Na Cl & Organum oil[3] Nix
Na Cl & Peppermint Oil not very good
Na Cl & N Bichron. gives fearful thump on first closing only.[a]
Na Cl. Cajeput Oil & Proto Hg. good apparently better than Pyro.[a]
Na Cl & Anise Oil nix
Na Cl Cajeput Oil Proto Hg & Pyro nix

Edison, Johnson, Adams Batchelor.[d] 12.

X, NjWOE, Lab., Cat. 1172:225 (*TAEM* 3:141). Written by Batchelor. [a]Followed by broken horizontal line. [b]Followed by "over" to indicate page turn. [c]Preceded by "May 10 continued. Electromotograph." [d]Names written by Batchelor.

1. Cajeput oil is a medicinal oil derived from an East Indian tree of the myrtle family.
2. Nitrobenzene.
3. Origanum oil is derived from plants in the mint and vervain families, particularly oregano.

–574–

Charles Batchelor Diary Entry

[Newark, May] 16 [1875][a]

Agreement verbal between Edison & Murray[1] Murray takes top room & Half next floor & ¼ cellar.[2]

AD, NjWOE, Batchelor, Cat. 1307:47 (*TAEM* 90:632). [a]"May" and "1875" are ditto marks in original.

1. See Docs. 593 and 594 for the formal dissolution of the firm.

2. Inventories of stock and machinery in the firm were made on 21 May 1875. Cat. 1123:204–16, Accts. (*TAEM* 22:207–13).

–575–

From Henry Spang

Reading, Pa., May 17th 1875.[a]

Friend Edison.

As promised, I herewith send you a small map of the P.&R.R.R.[1] and its branches, along which, the P.R.&P Teleg. Co.[2] have lines &[b] offices. This Co. has 28 teleg. offices in Philada. and 235 teleg. offices along P.&R.R.R. and its branches. We connect at Philada. with the A.&P. Teleg. Co. and give them a large number of messages daily, taken in at Philada, Pottsville, Reading and other points in our territory. We have recently erected a line from Reading to Slatington and will extend it to Catasauqua, Allentown Bethlehem, Easton & Phillipsburg N.J. in a month or two. Yours truly

H. W. Spang Asst. Supt.[3]

P.S. Do not forget to have the polarized relay made as soon as convenient and forward with bill to me via Del. Lack. & Western Express. I will remit amt.[c] as soon as recd. here. Coils should be of 10, 25, 50, 75, 100 & 200 ohms resistance and as I may want to try your new method of working polarized relays by the inductive current of a magnet[d] please have 3 connecting screws placed on the base of relay and a wire connected to the wires which join the two coils thus.

Did I understand you to say that Bastet used Bichromate Potash & Nitrate of Lime in his new battery solution,[4] if so, have you any idea what proportions of each he uses in the solution? Should you come across any solution for carbon battery, which is as strong as the ordinary[c] Bich. Potash & sulp. acid solution, but which does not crystallize on carbon & is more constant, please advise me. When you get time please read the notice of ethat [cell?][e] of chromate of Lime on page 111 of Sprague's work.[5] H W S

ALS, NjWOE, DF (*TAEM* 13:423). Letterhead of Philadelphia, Reading & Pottsville Telegraph Co. [a]"Reading, Pa.," and "187" preprinted. [b]"lines &" interlined above. [c]Interlined above. [d]"by . . . magnet" interlined above. [e]Obscured by ink.

1. Philadelphia and Reading Railroad.

2. Philadelphia, Reading & Pottsville Telegraph Co.

3. Henry Spang was assistant superintendent of the Philadelphia, Reading & Pottsville Telegraph Co. He had several patents on railway telegraphs and designed a system (described in his 1902 *Treatise on Perfect Railway Signaling*) to prevent such telegraphs' impairment by thunderstorms.

4. Louis Bastet sold his battery at 607 Broadway in New York City. H. Wilson 1876, 72; *Telegr.* 11 (1875): 228.

5. This is the Fitzgerald and Molloy cell, in which the cheaper chromate of calcium was substituted for bichromate of potash. Edison obtained a copy of Sprague 1875 on 11 May. Bill of D. Van Nostrand, 15 Nov. 1875, 75-002, DF (*TAEM* 13:240).

–576–

Henry Brainard to Edison and Murray

Phoenix, N.Y.,[a] May 20th 1875

Gents—

Some time ago—abt. 6 weeks I think—I sent to you for a couple of your Inductoriums—one for myself & one which I sold to go into a school. My own works first rate and is satisfactory—But today the one I sold came back to me with the complaint that it wont work

I have examined it and find that the Battery is all right and the vibrating circuit breaker works good, still there is only a barely perceptable induced current—which indicates I think a fault in the fine wire coil— I would like to have it made good in some way—and perhaps the cheapest will be for you to send me by mail just the bare coil and I will return the one I have— I can put it in place in the instrument all right if you leave sufficnt wire projecting

The express charges are so high that they are worth more than the instrumt. While I think athe coils can be exchanged by mail for a comparative trifle— I have one or two of your instrmts spoken for and I want them all right Yours &c

Henry A. Brainard[1]

ALS, NjWOE, DF (*TAEM* 13:544). Letterhead of Henry Brainard. [a]"Phoenix, N.Y.," preprinted.

1. Brainard's letterhead indicates that he was an attorney-at-law and a civil engineer.

-577-

Benjamin Butler to Josiah Reiff

Washington D.C. May. 27. 1875

My Dear Mr Reiff.

I read with care the very able and excellent brief of Mr Russell.[1] It covers the whole case from his stand point, that is to say, the power of the Commissioner and the rights of Edison.[2] I read Mr Latrobes remarks with care and agree with him that we had better make no cha~~r~~nge.[3] The first military [----] Maxim is never to charge fronts under fire.[4] I have been to day to see the Commissioner and have stated to the assistant Attorney General for the Interior Department, Mr Smith[5] that I did not think I would answer the very disgraceful and blackguard brief of Mr Conklin,[6] but I called his attention to one point You know the Commissioner decided that there was no evidence before the Department that Edison had ever thought of double transmitted or duplex telegraphy, until after the deed to Harrington in 1870.[7] Yet Conkling quotes in his brief a deed of Edison~~s~~ to Walsh[8] in 1869, more than a year before, where Edison~~s~~ deeds to Walsh a double transmit~~tible~~er" by name.[9] I have great confidence that we shall win our case at least I shall advise to hold where we are.[10] Yours very truly

Benj. Butler[11]

A contemporary caricature of Benjamin Butler.

ALS, NjWOE, DF (*TAEM* 13:707).

1. Robert Russell, evidently a lawyer, was identified in the New York City directory as the president of a firm at 29 Broadway. He was later associated with the American Automatic Telegraph Co. (H. Wilson 1876, 1151; Doc. 676 n. 2). His official brief for Edison, presented to Interior Secretary Columbus Delano on appeal from the decision of Patent Commissioner Thacher, is Quad. 72.7 (*TAEM* 9:114). It covers the topics mentioned below, but Butler probably refers here to the confidential argument for Edison, apparently also by Russell (Quad. 72.8 [*TAEM* 9:114, 122]; see also n. 3).

2. Apart from the question of whether Edison's agreements with Harrington (Docs. 109 and 155) covered his quadruplex work, the arguments for George Prescott (and Western Union) insisted that a decision of the Commissioner of Patents could not properly be reviewed, let alone overruled, by the Secretary of the Interior, to whom Harrington had appealed. Russell argued to the contrary—that the Commissioner had exceeded his powers in his decision of 20 March. Quad. 72.7, 72.18, and TLC.6 (*TAEM* 9:114, 280; 10:962); see also n. 4.

3. John Latrobe, prominent Baltimore lawyer and inventor, represented Harrington in the arguments before Commissioner Thacher and

Secretary Delano, in partnership with Leonard Myers, a Philadelphia lawyer and former chair of the House of Representatives' committee on patents (*DAB*, s.v. "Latrobe, John Hazlehurst Boneval"; "A Quadruplex Telegraph," *Newark Evening Courier*, 16 Apr. 1875; "Quadruplex Telegraphy," *New York Daily Graphic*, [17 Apr. 1875?]; the last two are also in Cat. 1144, Scraps. [*TAEM* 27:278–79]). A letter of Latrobe's to Reiff of 29 May 1875 (DF [*TAEM* 13:708]) indicates that the unsigned confidential argument for Edison (Quad. 72.8 [*TAEM* 9:122]) was written by Russell; that Latrobe had reviewed it; and that Latrobe opposed immediately following the strategy proposed therein (see n. 4 below), preferring to wait for favorable or at least definite action by Secretary Delano.

4. Russell suggested that all of Edison's applications be withdrawn in favor of new ones, differently combined and described, and including some additional matter; and that they still cover the same inventions, but not be accompanied by an assignment of a half interest to Prescott. This change was not attempted until the next year, when a new Interior Secretary, Zachariah Chandler, refused to block the decision of Commissioner Thacher. See Doc. 734.

5. Unidentified.

6. After hearing extensive arguments Secretary Delano, in a decision dated 14 April 1875 (Quad. 72.18, p. 13 [*TAEM* 9:286], reserved judgment on whether he could or should block or reverse the Patent Commissioner's decision regarding the assignment of Edison's disputed cases (Docs. 467–73). He allowed the parties to file written arguments and replies, and stated he would not decide the case until approved patents came before him for his signature.

Roscoe Conkling, U.S. Senator from New York and a major figure in the Republican party, was a lawyer, although he seldom practiced (*DAB*, s.v. "Conkling, Roscoe"). The brief in question (Quad. 72.18 [*TAEM* 9:280]) was also the work of Grosvenor Lowrey and J. Hubley Ashton. For Lowrey see Doc. 695; for Ashton see his obituary, *New York Times*, 16 Mar. 1907, 9; see also Quad. 72.6 and TLC.6 (*TAEM* 9:92; 10:962).

7. Doc. 109. See also Doc. 569 n. 3.

8. E. Baker Welch; see *TAEB* 1:71 n. 1.

9. Doc. 61; this was not recorded at the Patent Office until 30 January 1875.

10. Even after receiving written arguments (see n. 6), Secretary Delano kept the disputed patent application files in his office. Whether intentional (as claimed later in Western Union's *Journal of the Telegraph* 8 [1876]: 100) or not, this effectively prevented the Patent Office from either approving or rejecting the applications. However, on 24 April 1875 Reiff sent to the Patent Office a telegram in Lemuel Serrell's name that said, "Please advise me condition of Case H of T. A. Edison, filed April, 1873, and amended by himself. If ready, please issue immediately, and charge my account for final fee" (Serrell's testimony, Quad. 71.1, pp. 94–100 [*TAEM* 10:52–55]). Zenas Wilber allowed Case H the same day, and William Chandler—lawyer, prominent Republican politician from New Hampshire, and regular lobbyist for Jay Gould's Union Pacific Railroad—immediately paid the final fee. Thacher and Delano both signed the patent, and it was published on 27 April 1875 (U.S. Pat. 162,633), assigned to Edison and Harrington in accord with Doc. 155. Prescott's response was to bring charges of conspiracy, fraud, and cor-

ruption. Docs. 554, 569, and 584; Prescott's charges and Statement and Brief of Z. F. Wilber, Quad. 72.3–5 (*TAEM* 9:38–91); Eckert to TAE, 16 Apr. 1875, Cat. 1144, Scraps. (*TAEM* 27:279); *DAB*, s.v. "Chandler, William Eaton"; Klein 1986, 174.

11. Benjamin Butler, a controversial Northern general in the Civil War and a prominent member of the U.S. House of Representatives from Massachusetts, served as a lawyer for Edison, employed by Jay Gould (*DAB*, s.v. "Butler, Benjamin Franklin"; Thomas Eckert to TAE, 16 Apr. 1875, Cat. 1144, Scraps. [*TAEM* 27:279]). His argument for Edison before the Interior Secretary is Quad. 72.11 (*TAEM* 9:164).

–578–

Charles Batchelor Diary Entry

[Newark, May] 27 [1875][1]

Allen[2] came at night & offered Edison 3 pages of matter if he would take Scientific part of Operator I made Domestic Bills to June 1st $1650.31[3]

AD, NjWOE, Batchelor, Cat. 1307:47 (*TAEM* 90:632).

1. All of the entries on this page date from May 1875.

2. Tom Allen, a Western Union operator, was co-founder and editor of the *Operator,* which began publication in March 1874. W. S. Williams bought out his interest in February 1875, but he returned as co-editor in April. Reid 1879, 669; *Operator* mastheads of 15 Feb. and 15 Apr. 1875.

3. For Domestic Telegraph Co. accounts with Edison and Murray see Cats. 1220 and 1223, Accts. (*TAEM* 22).

–579–

Memorandum: List of Experimental Topics

[Newark,] May 31 1875[1]

Wanted

1 A Method of making '<u>Malleable iron</u>' out of cast iron

2 Making cast iron as hard as steel & to have some of the same properties[a]

3 A method of making '<u>sawdust</u>' soluble to form a cheap substitute for Ebony Hard rubber or Celluloid.[2a]

4 A cheap intense green equal to Aniline green without Iodine or Arsenic.[3]

5 An Electromagnet which does not require wire

6 The formation of organic substances for the decomposition of water under certain influences.[4]

7 A Kerosene or other oil lamp which burns without chimney & gives a bright light[5]

8 A new Engraving process

9 A Galvanic Battery of equal constancy to Daniels but with an electromotive force equal or nearly equal to Grove.[6]

10 A substance which will not pass or pass but little P. current & all of a N current
11 A New force for Telegraphic communication[7]
12 Make soluble peroxidyzed paper with less acid than the Celluloid Co.[8]
13 A cheap process for the extraction of loaw grade ores, decomposed earth like either Carb Ag. or H_2S Ag. ores.
14 Platina solution cheap & as delicate as Iodide of Potassium without drawbacks[9]
15 A Detector for Gold & Silver at a distance.[10]
16 A Polarized Electromotograph.[11]
17 A Sexduplex Telegraph.[12]
18 A copying press that will take 100 copies & system[13]
19 A cheap process of printing new.
20

D, NjWOE, Lab., NS-75-001 (*TAEM* 7:319). Written by Charles Batchelor. [a]Followed by broken horizontal line.

1. A longer, partly congruent list of experimental topics is found in the index to an account book, with corresponding ledger entries for many of the items. These entries were assembled retrospectively, at least in part. Cat. 1214:1–3, Lab. (*TAEM* 21:564–65).

2. Experiments on celluloid solvents can be found in Cats. 1172 and 994, Lab. (*TAEM* 3:75, 192), and Cat. 1317, Batchelor (*TAEM* 90:661); see also Docs. 395, 583, 586, and 645. An account entry for experiments on 15 July can be found in Cat. 1214:112, Accts. (*TAEM* 21:620).

3. For an experiment related to this see Cat. 1172:259, Lab. (*TAEM* 3:156).

4. Cf. "Chemical Decomposition by Electricity and Magnetism," in Cat. 1214:88, Accts. (*TAEM* 21:608).

5. Edison, Batchelor, and Adams conducted some related experiments on 3 June (Cat. 1172:233, Lab. [*TAEM* 3:145]); a corresponding account entry is in Cat. 1214:82, Accts. (*TAEM* 21:605).

6. "Experiments to find a strong constant battery" are found in Cat. 1172:241–53, Lab. (*TAEM* 3:149–53). The experimental account is probably the one headed "Experiments to find a Cheap Battery" in Cat. 1214:60, Accts. (*TAEM* 21:594).

7. Cat. 1214:80, Accts. (*TAEM* 21:604) contains accounting entries for 3 May and 2 (Doc. 581), 8, and 9 June which correspond with notebook entries related to these experiments.

8. Probably related to the entry for "Celluloid" in Cat. 1214:84, Accts. (*TAEM* 21:606).

9. An account entry for these experiments is found in Cat. 1214:76, Accts. (*TAEM* 21:602).

10. Cf. Doc. 650.

11. An account entry labeled "Electromotograph" contains entries dating back to Edison's earliest experiments in April 1874 and continues to 3 September 1875. Cat. 1214:72, Accts. (*TAEM* 21:600).

12. See Docs. 583 and 611. An account entry for experiments of 7 June is found in Cat. 1214:86, Accts. (*TAEM* 21:607).

13. See documents related to Edison's electric pen and autographic press. There is an account entry for "Copying Experiments" in Cat. 1214:78–79, Accts. (*TAEM* 21:603).

–580–

Notebook Entry: Autographic Printing

Newark N.J., ~~June 1~~ Monday May 31st[a] 1875

The samples shown on the opposite page[1] are the result of the first experiment tried in the new Labora~~es~~tory, and for the want of a better name it will be called the Autographic Copying Press—

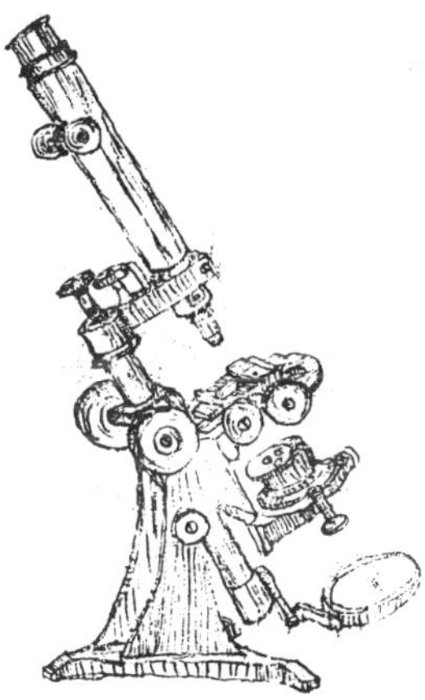

Tissue paper written upon by an ink composed of Aniline Vi~~e~~olet and proto chloride of Iron Saturated Solution. ~~t~~The paper which receives the copy is saturated with the Ferrid Cyanide of Potassium— The Ink may be the reagent for the solution in which the paper is satur~~atated~~ted[b] and produce coloration by a chemical reaction for instance, the ink may be a solution of perchloride of iron or any persalt of a metal and the paper on which the copy is to be taken may be moistened in a solution of a chemical which is a reagent for a persalt and thus give a coloration ⟨12⟩[c]

T. A. Edison Chas. Batchelor
Jas. Adams E. T. Gilliland[d]

X, NjWOE, Lab., Cat. 1172:229 (*TAEM* 3:143). [a]"Monday May 31st" interlined above. [b]Written by Gilliland to this point; the remainder written by Edison. [c]Marginalia written by Batchelor. [d]Followed by representation of a seal.

1. Included below in the document.

–581–

Notebook Entry: New Force

Newark NJ June 2d [1875][1]

9.05 Pm Experimenting endeavoring to discover a new force—[2] a circuit of No 12[3] Iron wire 25 feet in length. an electromagnet with charged battery[a] is brought in contact with this wire at one end and the following contrivances were used at the opposite end to detect any manifestations

a slip of paper suspended between two points made by breaking the iron wire circuits & turning up the ends having the ends about one half inch apart & the paper is suspended between—

a paper saturated with chloride of[b] maganese nothing that can be detected with the naked eye, using the spectroscrope gives no results— a jet of gas is introduced between the points nothing— a candle is used—nothing a jet of gas is substituted for the Electro magnet= candle used for veiwing— a paper saturated with salt is burned in the focus of the spectroscope nothing— Salt paper[b] is burned at the sending end nothing— Pyrogalic acid & many others are tried with same results paper is again introduced in receiving end Electro magnet at sending end— a magnifying glass is used at the receiving end—nothing

Magnet introduced at receiving end everything else remaining the same— paper used as an armature—nothing—

Edison Batchelor Gilliland Adams Johnson 12 hrs all night[4c]

X, NjWOE, Lab., Cat. 1172:231 (*TAEM* 3:144). Written by Gilliland. [a]"charged battery" written in margin with "charged" interlined above. [b]Interlined above. [c]"Edison . . . night" written by Batchelor.

1. Dated from surrounding entries in notebook. See also Doc. 579 n. 7.

2. The first entry for a new force related to telegraphy is Doc. 570; see also Cat. 1172:217, Lab. (*TAEM* 3:137).

3. Approximately 2 millimeters in diameter.

4. There are subsequent entries for the nights of 8 and 9 June on p. 235 of this notebook (*TAEM* 3:146). Edison, Batchelor, and Adams again sought a new force without success.

–582–

From Jarvis Edson

New York, June 4th 1875.[a]

Dr Sir,

I would remind you of your promise to go ahead and submit a working sketch of the repeater and alarm gong affair yet required for the Fire Alarm to enable us to make a proper exhibit—[1] We cannot possibly have it ready one day too soon—

Washington is going to advertise for bids & for parties to come on & put up something in working order to make an exhibit—

"Calls" are increasing steadily every day— We have only about ~~89~~ 96[b] boxes working. ~~h~~Have just had 10 duplicate ones altered today so they will make ~~about~~ over[c] a hundred— What we want is 300 working & then we will be making money— I think we can have them in by middle of July if Murray does not delay— Also I am in favor of starting in other districts at once, one in Fulton St—& the other in the Enemys camp in Broad St where it pays best—[2] Perhaps we could get in with the A&P. in their new office there[3] & stow the boys upstairs—

Please hurry up the Repeater— Yours &c

J. B. Edson, V.P.

ALS, NjWOE, DF (*TAEM* 13:519). Letterhead of the Domestic Telegraph Co. [a]"New York," and "187" preprinted. [b]"96" interlined above. [c]Interlined above.

1. See Doc. 654 for the design of Edison's system.
2. Probably refers to the American District Telegraph Co., which had its headquarters at 62 Broadway.
3. Around this time Atlantic and Pacific closed its office at 11 Broad St. and opened new ones at 8 and 87 Broad. H. Wilson 1876, 48; idem 1877, 49.

–583–

Notebook Entry: Celluloid and Multiple Telegraphy

[Newark,] June 7th 1875

Worked all night.

New Solvents for Celluloid:—[1]

Alcohol & Naphtaline, cold, very good.
Turpentine & Naphtaline heat good.
Alcohol & Gum Senegal cold, not very good[a]

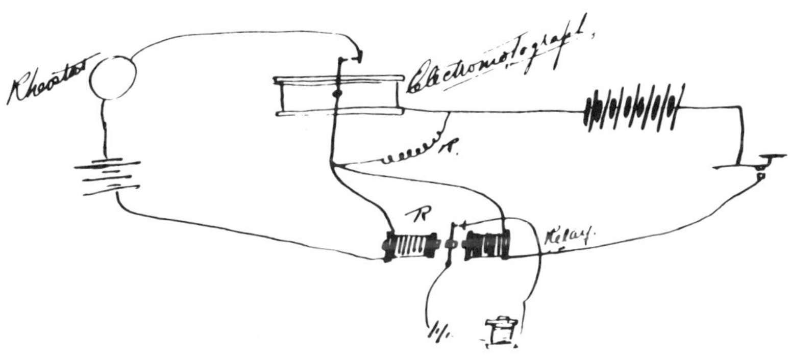

Tried a preliminary experiment on Sexduplex to see whether the Emg. would close quick enough as in Diagram. Owing to defective Instruments it did not work well The principle of sending 3 messages from each end is to send on to polarized Relay reverse currents with 50 cells Relay R[b] with 100 cells & Emg. with 150 cells & relying on the Emg to close quick enough to neutralize the relay [&c?][c] back spool to a certain extent

In working this experiment we found that when we shunted the electromotograph although there was a heavy mark by the stylus the movement of the stylus was killed.

Batchelor, Edison, Adams,[d] ⟨31 for all⟩[e]

X, NjWOE, Lab., Cat. 1172:234 (*TAEM* 3:146). Written by Batchelor. [a]Followed by broken, centered, double horizontal line. [b]Interlined above. [c]Illegible. [d]Names written by Batchelor and followed by broken centered horizontal line. [e]Marginalia written by Batchelor.

1. Developing better solvents was a key factor in the search for improved forms of celluloid, the first commercially successful plastic. A solution of nitrocellulose and camphor, celluloid was developed in 1869 by John Hyatt, an Albany, N.Y., printer. With his brother Isaiah, Hyatt established the Celluloid Manufacturing Co. in 1871, and the following year they moved its operations to Newark, N.J. Edison and Murray began manufacturing comb-cutters and other items for them in 1874. Friedel 1983, chap. 1.

2. This is the first dated reference to Edison's attempts to simultaneously send three messages in each direction over a telegraph wire. In Doc. 611 he uses the term "sextruplex"; by 1877 he was referring to it as the "sextuplex" (NS-77-004, Lab. [*TAEM* 7:457]).

–584–

To Charles Buell

Newark N.J. June 10th 1875

Dear Sir

Electrical examiner Wilbur of the patent office I have never had any business with except through my attorneys & I have never heard of any irregularity or misconduct in his practice.[1] I believe him to be strictly honest, he has had charges preferred against him by a party named Sawyer,[2] a drunken newspaper correspondent lately which have been investigated & proven to be nothing more than personal spite. Resptly Yours,

T A Edison per C. B.

L, NjWOE, DF (*TAEM* 13:381). Written by Charles Batchelor.

1. On 2 and 7 June, George Prescott wrote to the Commissioner of Patents claiming that examiner Zenas Wilber had been guilty of flagrant malfeasance (or, at best, gross incompetence). Wilber had allowed Edi-

son to amend Case H (Docs. 315 and 554, U.S. Pat. 162,633) to claim the basis of the quadruplex, and then he had issued it assigned to Edison and Harrington. The following December, after both the Commissioner of Patents and the Secretary of the Interior had been replaced, and after an additional charge had arisen over the isssue of Case 111 (U.S. Pat. 168,385; see Doc. 450), Prescott preferred charges, which led to official proceedings in the Patent Office. Quad. 72.3–5 (*TAEM* 9:38–91).

2. See *Arguments in the Case of Charges Preferred by William E. Sawyer, of Washington, Against Z. F. Wilbur, a Principal Examiner in the Patent Office* (Washington, D.C.: R. Beresford, printer, 1875), Library, NjWOE.

–585–

From Albert Chandler

NEW YORK, June 10, 1875.

MY DEAR SIR:

You are hereby appointed Electrician of the Atlantic and Pacific Telegraph Company.[1]

You will in that capacity have charge of the transmission of business entrusted to us: meaning the arrangement of circuits and the batteries and apparatus necessary to secure the most prompt and efficient transmission possible: such arrangements to be carried out through the General Superintends and with their co operation and approval.

All questions relating to battery, new instruments and machinery will be referred to you.

Mr. E. H. Johnson has been appointed as your assistant and his Experience in connection with both the Automatic and Morse systems will enable him to render very valuable service both to you and the Company[2]

By order of the President

(signed) ALBERT B. CHANDLER
Secretary & Asst. Genl Mgr.

D, NjBaFAR, *Harrington v. A&P*, Box 17B, Defendants' Exhibits, 1:388.

1. See Docs. 522 and 539. Edison had apparently been acting in this capacity since January.

2. On 1 July 1875 a notice of Edison's and Johnson's appointments appeared in the *Operator* (p. 7).

–586–

Notebook Entry: Miscellaneous

[Newark,] June 10, 1875.

Worked trying to dissolve sawdust soaked some in Boiling Caustic Potash & then tried to dissolve it also soaked some in Sulphuric & Nitric Acids & some in HCl. No result as yet

Batchelor & Adams[a]

Petrole~~o~~um ether caught fire while Jim had it in his hand no harm.[1b]

Edison & Johnson made statistics of A&P delay of messages[2]

Copying press does not work satisfactory

Resolved to work daytimes & stay at home nights.

New solvent for Cellulose Nitro Benzole.[c] Good.

Edison Batchelor & Adams.[a]

X, NjWOE, Lab., Cat. 1172:235 (*TAEM* 3:146). Written by Charles Batchelor. [a]Names written by Batchelor. [b]Followed by centered horizontal line. [c]Underlined twice.

1. Cf. App. 1.B7, D262, E11.

2. One of Edison's scrapbooks contains several pages of tables by Edison, with comments by Johnson, showing delays of messages at various Atlantic and Pacific offices (Cat. 30,096, Scraps. [*TAEM* 27:537–72]). The scrapbook also contains correspondence relating to the collection of that data and diagrams of Atlantic and Pacific circuits.

–587–

Notebook Entry: Autographic Printing

[Newark,] June 11th, 1875

Copying Experiments:—

No 1 was ~~Ink:~~— Solution:—

2~~2~~1 fluid oz water
2 dwt Ferridcyanide of Potassium

Ink Sat sol FeCl.

a little Violet aniline
a little Sugar[a]
Pomatum[1] grease on paper

took 21 copies.[b]

[1] 21 off

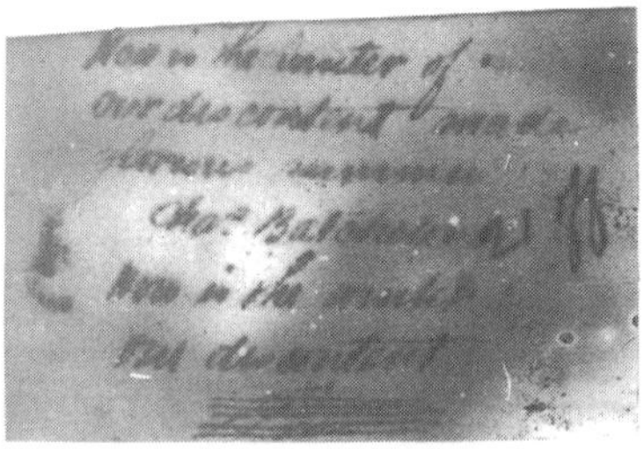

No 2.[b] differed only in the way of laying the roller paper on.

2　22 off 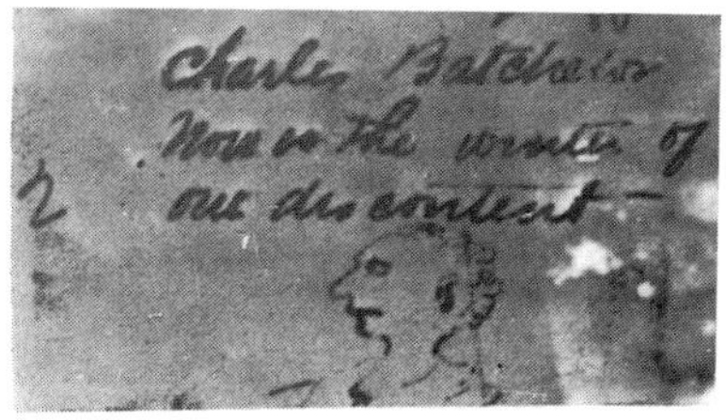

Charles Batchelor
Now is the winter of
our discontent—
2

3, 4, & 5 all took about 22 copies & differed slightly in the ink.[2]

5 & 6. Sat Sol Fe_2Cl & Sat Sol Aniline. proportion 2 Fe_2—1 Aniline　write with this & let dry or blot it & then coat over with Balsam fir　This will take 50 copies easily　55 good copies being taken with No 4 & 45 with No 6.

5th　55 copies 

Charles
Batchelor

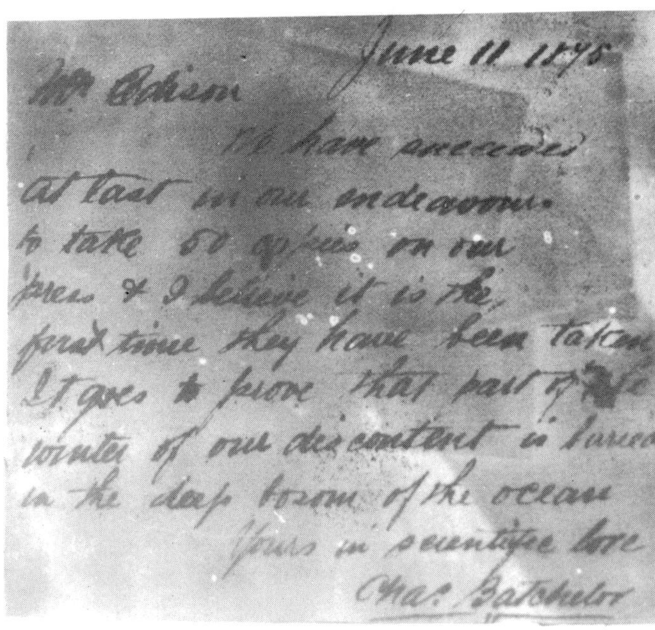

June 11 1875
Mr Edison
We have succeeded
at last in our endeavours
to take 60 copies on our
press & I believe it is the
first time they have been taken
It goes to prove that part of the
winter of our discontent is buried
in the deep bosom of the ocean
Yours in scientific love
Chas Batchelor

No. 6　45 copies
Bachelor & Adams[c]　6 hrs. each.[d]

X, NjWOE, Lab., Cat. 1172:237 (*TAEM* 3:147). Written by Batchelor. [a]"Ink . . . Sugar" enclosed by a brace. [b]Punctuation doubled. [c]Names written by Batchelor. [d]"6 hrs. each" underlined twice.

1. A perfumed ointment made from apples; also known as pomade.

2. Samples for trials 3 and 4 were not pasted in the notebook and have not been found.

–588–

Notebook Entry: Autographic Printing

[Newark,] June 30th 1875

We struck the idea of making a stencil of the paper by pricking with a pen & then rubbing over with an ink.

This best ink we found t be glycerine & aniline V[-]iolet. Cut a piece of steel 2 × 3 like a file but with points straight up & tried to perforate the paper by writing on this with a stylus it works but it takes such pressure to do it at all, that it is not much good. resolved to make a machine to go by clock work or engine to prick as we write, in the mean time we took numerous copies with the letter pricked by hand[1] ⟨12 hr each⟩[a]

Thos A Edison Chas. Batchelor
James Adams Saml Edison

X, NjWOE, Lab., Cat. 1172:239 (*TAEM* 3:148). Written by Charles Batchelor. [a]Marginalia written by Batchelor.

1. The following copies were pasted on the preceding page of the notebook.

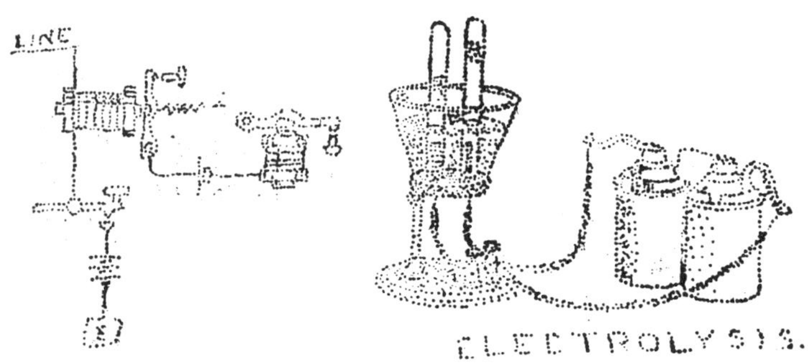

Samples from Edison's first stencil copying experiments, dated 30 June 1875.

–589–

Draft to the Editor of the Scientific American

Newark N.J. June, 1~~7~~875

<u>To the editor of the Scientific American</u>

I have just disccovered a very querious phenomenon and one which I am unable to offer any explanation as to its cause. It is this:—If a number of plates of tin be placed side by side with an equal number of plates of Vulcanite in such a manner that the whole will form a large smooth plate and upon this is laid a strip of paper moistened with a solution of pyrogallic acid and chloride of sodium, Those portions of the paper over the tin will upon passing ~~a blunt sm~~ the end of a smooth platina from one end of the strip to the other, feel rough and offer considerable friction to the passage of the wire while the paper over the Vulcanite will feel smooth & offer much less friction. This phenomenon cannot be attributed to electric action as in my "motorgraph" ~~o~~for none is possible under the conditions of the experiment.[1]

ADf, NjWOE, DF (*TAEM* 13:193).

1. This letter did not appear in the *Scientific American.*

-590-

Memorandum [to Jay Gould?]

[Newark, June 1875?[1]]

Some Ideas.

1st A neat little phamplet book should be printed upon cheap paper, and in large quantities, in which is a peculiar ~~m~~telegraph map (small) of the U.S. divided in parts by red lines, within which the tariff is so much, and so much as you radiate from a given square. A table of special tariffs, list of the principal branch offices in the main cities Some brief hints on, "How to write telegrams" cyphers etc, on giving immediate answers, and some interesting matter relating to telegraphy. The book to be universally distributed by messenger boys in business centres.

Make special contracts with large users of the telegraph.

Print a Tariff map on the back of the delivery message blanks, & special tariff list in very fine type. (pearl).[2]

By advertisments, phamplets, talk, and other devices, the A.&.P may be made to appear to the public as of greater magnitude than it really i~~t~~s, = hence more business.

Establish between distant points where the necessity is urgent, a <u>Letter Telegram service</u>. operated at night only with rates so low that it will attract urgent letters to the Telegraph— for instance ~~Er~~Receive these special telegrams during the day and night to be transmitted at night for delivery next morning, say 50 words or less from Chicago 50 cents & 1 cent for each additional word, no message less than 50 cents These telegrams will allow us to keep our perforating Operators busy all day: again ~~o~~the day business does not commence until 10 am hence our messager boys are idle, ~~d~~the delivery of these special telegrams received during the night could take place when the whole force was idle.[3]

1500—50 word telegrams can be received and delivered between Chicago & NY in one night on one wire as they now are, at a total cost of $145, including delivery gas, stationary, etc. Receipts $750.

The rate between San Francisco and NY for 50 words $1.00.

Under a given condition the cost of a 10 word message is 12 cents a 20 words message 13½ a 40 word message 16 cents, by increasing the number of words we give cheap telegraphy to the public while the increased cost to us[a] as you will see is trifling and will be returned a thousand fold by increase in the number of messages, or what is perhaps nearer the truth, diverting correspondence from the mail to the telegraph =

The WU Co have no eextra facilities to accommodate an increased number of words; for instance between NY and Chicago they work 4 wires and the business is greatly delayed were the number of words increased to twenty, they would require 2 more wires again the increase in the words would increase their present business ~~hence~~ and they would probably require ~~4~~6 or ~~6~~8 more wires. With us the case is different with one wire worked Automatic we have a capacity of 15 or more wires as worked by the WU The policy of the WU has been as far as I can see to put the tariffs at a point where it will bring enough business to fully employ their wires, and to raise it when it increase beyond that point. It is manifest that weith the Automatic no such policy is necessary with our Company

Transmit isf not inconvenient the NYork money markets free of charge to all points of any importance and place on a bulletin in the office. This will bring that class of Country people who dabble in Wall street to our offices and their aggregate business will be large.

Keep "semiweekly" inserted in dalies twice each week,[b] advertisements in all the trade papers & dailies.

Have no such offices as is now known as main offices at enormous rentals. Have a Central office situated in a side street in the top of some large building that cannot be rented for other purposes, receive no messages from the public at this office, and collect your business from low rental branch offices. Were the WU to do this they would save 500 000 annually without the loss of a message

Have solicitors of same address going among the business men and drum up business

When you do build a line anticipate the future and put up the best one that can be built. A single wire at 150 per mile which is always in order and upon which great speeds can be obtained is much cheaper than ~~ordinary routes with~~ three ordinary lines at 35 each in constant disorder, with low rate of speed, reliability in telegraphy is an important factor

In putting up a short[c] wire for instance between Oil City and Pittsburgh, do not use small wire etc with the idea that it will answer to carry all the business between those two points, but use the largest and best wire with the idea that in continguences which arise this short wire may form part of a circuit between NY and Chicago ~~e~~and full speeds and good conditions will be obtained

Do everything as far as possible by "peice work"—

Cut down high rentals, and put the money in advertising, druming etc.

Give great attention to the delivery department which as at present managed in a weak spot in telegraphy.

And finally aged men and Telegraphy are incompatible.

ADDENDUM[d]

[June 1875?]

JCR[4] as 3rd vice president will do more work in a day[e] than any 5 men I know.[f]

Do not put up wires on routes that we already have wires, ~~put~~ but use every mile to bring the several large cities with which we do not connect into our system. The old wires should be put in good order ~~as Great E~~[5]

AD, NjWOE, DF (*TAEM* 13:441). [a]"to us" interlined above. [b]"inserted . . . week" written at bottom of page as footnote. [c]Interlined above. [d]Addendum is an AD, in pencil on the back of the last page. [e]"in a day" interlined above. [f]Followed by centered horizontal line.

1. Edison probably wrote this after beginning work as electrician of Atlantic and Pacific (see Doc. 585), but before he proposed leasing Atlantic and Pacific lines for a night-message service such as the one mentioned here (see Docs. 596 and 598).

2. A type size of about 5 points (0.07″).

3. See Docs. 596 and 598 for Edison's attempt to put this night-letter scheme into operation.

4. Josiah Reiff.

5. Probably the Great Eastern Telegraph Co.

–9– July–September 1875

While experiments in the new laboratory took much of Edison's time during the summer, he was still occupied with old concerns and acquaintances. William Orton sought to bring Edison back into the Western Union fold by encouraging him to work on acoustic telegraphy. At the same time that the Atlantic and Pacific Telegraph Company used Edison's automatic telegraph in its battle for business with Western Union, Jay Gould sought negotiations to end the costly telegraph war. Edison and Murray formally dissolved their manufacturing firm, leaving Edison more time to devote to inventing. The most important product of that work was the electric-pen copying system, which Edison believed would be widely used in business offices.

In July, President William Orton of Western Union met with Edison to discuss an entirely new telegraphic venture. Several inventors, chief among them Elisha Gray, had been exploring the possibilities of acoustic telegraphy as a potential form of multiple telegraphy. Acoustic telegraphy used the vibration of tuned reeds or metallic tongues to create rapid, regular electrical pulsations. In theory, several differently tuned transmitters could send simultaneously to sympathetically tuned receivers over the same line. In practice, receiving the signals proved extremely difficult. Gray was company electrician at the Western Electric Manufacturing Company in Chicago, one of Western Union's principal factories. He had demonstrated his acoustic apparatus to Western Union in May 1874, and the company had considered supporting his work as early as November 1874.[1] Although Edison read an article on some earlier telephonic[2] research and made some preliminary sketches, he did not conduct any acoustic telegraph experiments until the fall.

Edison's time was taken up instead by the continuing contest between Atlantic and Pacific and Western Union. In early July he prepared a caveat for a quadruplex automatic, the sole purpose of which seems to have been to strengthen George Harrington's—and thus Jay Gould's and Atlantic and Pacific's—claim to the quadruplex. Edison also continued to experiment with the electromotograph as a repeater for the automatic, and even used it in experiments on a "sextruplex" telegraph (capable of carrying six messages at a time). The automatic system had not proven superior to Western Union's Morse technology, and rate cuts had failed to gain Atlantic and Pacific an increased market share.[3] With Eckert still ill, Gould began negotiating a merger with Western Union in early August. Within a month the negotiations had failed.[4] Edison, concerned about the Automatic Telegraph Company investors (particularly Josiah Reiff), offered Gould his own suggestions for a settlement. At the same time, Edison told Prescott that these negotiations might allow them to settle the ownership of the quadruplex.

From his position as Atlantic and Pacific electrician, Edison proposed several ventures. At the end of July, he asked to be allowed to lease Atlantic and Pacific wires from New York to Chicago between 2 and 6 A.M. for a night-letter business operated from his laboratory. Although Edison drafted an agreement with Atlantic and Pacific, he apparently did not put the scheme into operation. He also drew a circuit for a "duplex to prevent anyone tapping Jay Goulds wire."[5] In September he proposed duplexing the transatlantic cable of the Direct United States Cable Company, which was allied with Atlantic and Pacific.

These months also saw the development of Edison's electric pen and copying press from a laboratory experiment to a commercial venture. The laboratory staff made the first experimental pen in mid-July, and by early September production was under way. John Ott, one of Edison's experimental mechanics, manufactured the pen, while Gilliland & Company made the presses. By the end of September, Edison had prepared a caveat to protect the invention, and an agent was selling the pen in New York. In addition to the other work in the laboratory, Batchelor continued development work on the domestic telegraph,[6] and he and Edison worked on a hotel annunciator.[7]

Edison and Joseph Murray, having verbally agreed in May

to dissolve their manufacturing partnership, legally ended it in mid-July. Edison, who had provided funds to keep the shop working earlier in the year, was to receive all money still due from Atlantic and Pacific and Domestic Telegraph for manufacturing their instruments; Murray retained the business, including the machinery, most of the workers, and the use of about one and a half floors of the building. In late July, Ezra Gilliland was given room in the building, and Edison provided funds and machinery for him to establish Gilliland & Company as a manufacturing firm. The company was soon making electric pens and presses.

Edison's reputation as an inventor continued to grow during these months. Robert Lines, who was preparing the official report on telegraphs exhibited at the 1873 Vienna International Exhibition, asked Edison for a description of his quadruplex and electromotograph for inclusion in the report, even though Edison had not exhibited any instruments in Vienna.[8] Edison did plan to exhibit at the United States Centennial Exhibition in Philadelphia the following year, and he and Batchelor drew up plans for a display space that was larger than the one granted to Western Union.[9] They also proposed to exhibit the new electric pen at the Newark Industrial Institute and at the New York State Fair. In telling his brother Thomas of these plans, Batchelor noted that Edison "stands today the foremost inventor & electrician in this country by far." The R. G. Dun agent in Newark agreed that Edison was "a genius in this line."[10]

1. Gray 1977, 33; Western Union Executive Committee Minute Book B, p. 433, NjWU.

2. The word "telephone" was often applied to acoustic telegraphy.

3. Reid 1879, 585.

4. Atlantic and Pacific Telegraph Co. Stockholders' Minutebook 1875–88, 1–12. Competition continued for two more years until Western Union bought a controlling interest in Atlantic and Pacific (Reid 1879, 587; Klein 1986, 199–205).

5. Cat. 1307:16, Batchelor (*TAEM* 90:617). This design may be based on an earlier secret signaling system Edison developed in 1867–68. Docs. 47 and 49; *TAEB* 1, App. 1.D220.

6. Drawing of 24 July 1874, Cat. 30,100, Lab. (*TAEM* Supp. III); Cat. 1172:285 and Cat. 1168:24, both Lab. (*TAEM* 3:168, 327); Cat. 1307:48–49, Batchelor (*TAEM* 90:633–34).

7. NS-75-002, Lab. (*TAEM* 7:332–35). See also Partrick and Carter to Batchelor, 14 July 1875; and Jesse Bunnell to Batchelor, 15 July 1875; both DF (*TAEM* 5:521–22).

8. The final report also included descriptions of Edison's dot-and-dash and roman-letter automatic systems.

9. In the winter of 1876 Edison agreed to join his exhibit space with that of Western Union. See Docs. 730 and 732.

10. Docs. 619 and 616.

–591–

From Robert Lines

Washington, July 1 1875[a]

Dear Sir:

I have received your two letters[1] for which please accept my thanks. Since hearing from you I have seen Prof. Thurston of Stevens' Institute, Hoboken,[2] who is editing the reports, and he says that if the description of the Quadruplex is sent him within a couple of months he will insert it in the space I have left for it. Please address him directly & oblige.[b] I will also send him the description which you have kindly furnished me of the motograph—although it is a little out of the line to which I had intended to confine myself—i.e. instruments exhibited at Vienna, or those not so exhibited which are actually in use in telegraphy.[3] Very truly yours

Robt. B Lines.[4]

ALS, NjWOE, DF (*TAEM* 13:405). Letterhead of United States Senate Chamber. [a]"Washington," and "187" preprinted. [b]"Please . . . oblige." interlined.

1. Not found.

2. Robert Henry Thurston (1839–1903) was a pioneer mechanical engineering educator. After serving as first assistant engineer in the U.S. Navy during the Civil War, he headed the Department of Natural and Experimental Philosophy of the U.S. Naval Academy. Appointed professor of mechanical engineering at Stevens Institute of Technology in 1871, Thurston remained there until 1885, when he moved to Cornell University, where he served as head of the engineering school until his death. *DAB,* s.v. "Thurston, Robert Henry."

3. As a member of the U.S. Artisan Commission to the Vienna International Exhibition of 1873, Lines authored the *Report on Telegraphs and on the Telegraphic Administration* (Washington, D.C.: GPO, 1876). Included in the report were accounts of Edison's automatic telegraph system (pp. 15–22), electromotograph (pp. 34–36), roman-letter automatic (p. 37), and quadruplex (p. 81). The account of the quadruplex was published as an addendum, with a note that it should have been published on p. 34 following the account of Stearns's duplex and early attempts at quadruplex arrangements, but that it was "received too late for insertion."

4. Little is known of Robert Lines other than his connection with the U.S. Post Office Department in 1874, when he witnessed the Automatic Telegraph Co.'s transmission of the President's message and the Spanish protocol (see Doc. 393).

–592–

New York, July 6, 1875[a]

Caveat: Automatic and Multiple Telegraphy

To all whom it may concern.

Be it known that I, Thomas A. Edison, of Newark in the County of Essex, and State of New Jersey, have invented an Improvement in chemical Duplex & Quadrupl~~x~~ex Telegraphs of which the following is a specification.[1]

The object of this invention is to transmit two or more messages over a single wire at the same time.

The invention consists in the arrangement of a chemical receiving instrument or instruments and transmitting instrument wherein perforated paper is used with battery resistance coils and other devices to ensure duplex or quadruplex transmission.

In Fig. 1. is shown a method of duplex transmission with perforated and chemical paper.

Fig. 1.

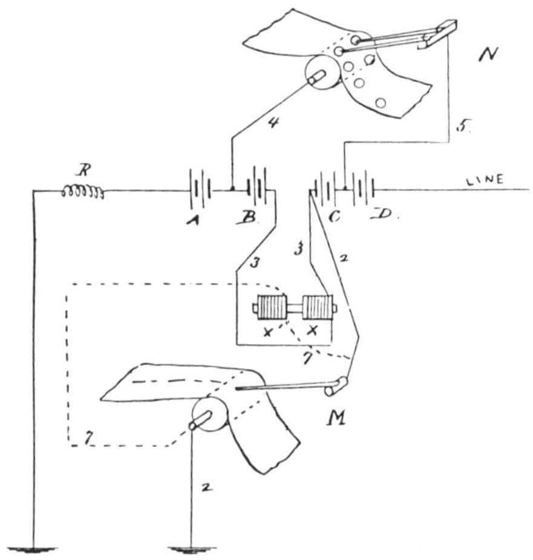

N. is the transmitting instrument which consists of a drum, strip of perforated paper and contact rollers.

A. B. C. and D. are four batteries of equal power. A and B. are placed in the artificial line made up by the resistance coil R. which has a resistance equal to that of the line or nearly so.

The like poles of the batteries A. and B. are connected together and neutralize each other so that no current is produced C. and D. are two batteries, both in the main line with their like poles connected together and they also neutralize each other and produce no current.

When the short circuit formed by the wires 4. and 5. and the contact of the roller through a hole in the slip is broken no current is produced either in the artificial line by A. and B. or in the main line by C. and D. The batteries C. and B. are connected together through a helix X of the inductive magnet Q. by the wire 3.3. between the batteries C. and B. is connected a wire 2. to the chemical receiving instrument M. and thence to the ground. This connection is shown more plainly in figure 2.

Fig. 2.

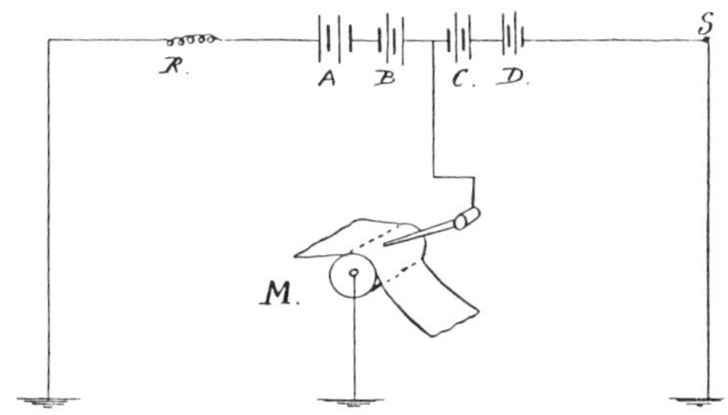

As the 4 batteries do not generate any current the chemical receiving instrument is free to record signals from the distant station S. while these signals are being recorded. If the batteries B. and C. are short circuited (see Fig. 1.) by the passage of a contact roller through a perforation in the paper, the balance of currents is destroyed, the battery D. sends a current through the line to the distant chemical recording instrument and the battery A. sends a current of equal strength through the artificial line R. as regards the receiving instrument M. the current from D. tends to pass through it in a direction opposite to that from A. as is shown in Fig. 3. hence if the two batteries are equal no effect will be produced upon M. which will continue to record the signals from the distant station. The current from D. will also pass over the main line and record upon an instrument similarly connected at the distant station.

Fig. 3.

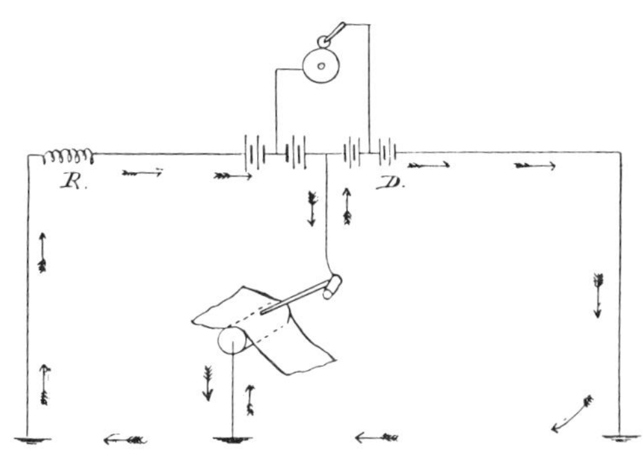

It will be seen by referring to Fig. 1. that in the wire connecting the batteries C. and B. is included a heliex of an electro magnet whiceh receives a powerful current when the batteries A. and C. are short circuited, on the same i~~n~~ron core with this helix X. is another helix X′. with its two free ends shunting the receiving instrument M. This device is used for the purpose of sending a short current of one polarity through the chemical recorder M. when A. and D. become active and another current of an opposite polarity when they become inactive for the purpose of neutralizing the effect of the static charge of the line, when a current is sent through it or abstracted from it.

The powerful current which X receives at the moment when it forms part of the short circuit serves to magnetize its iron core and this act of magnetization transmits into the helix X′. a powerful inductive current of one polarity and when the short circuit is removed the iron core is demagnetized and a current of the opposite polarity is sent into X′.

In Fig. 4. is shown a method of sending the two messages over the same wire in the same direction at the same time one by reversals of the current and the other by a rise and fall of tension.

Fig. 4.

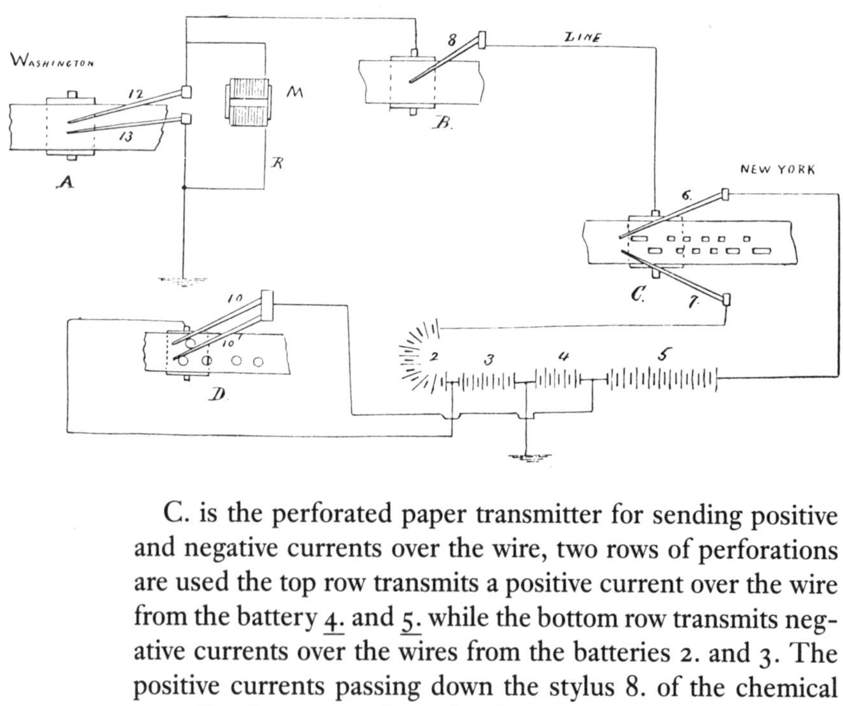

C. is the perforated paper transmitter for sending positive and negative currents over the wire, two rows of perforations are used the top row transmits a positive current over the wire from the battery 4. and 5. while the bottom row transmits negative currents over the wires from the batteries 2. and 3. The positive currents passing down the stylus 8. of the chemical recording instrument B. at the distant station record the characters punched in the top row but when a negative current is sent no record is made at B.

This second row of holes also serves to preserve the continuity of the circuit and allow the transmitter D. to increase and decrease the current.

It will be noticed that the ground wire is connected in between the two sets of batteries 2. 3. and 4. 5.

The battery 4. has 100 cells battery 5 125 cells, so arranged that when not short circuited by D. they oppose each other, 100 cells in 4. neutralizing 100 cells in 5. leaving only 25 cells to furnish current. The same arrangement is carried out with the batteries 2 and 3. so that when the transmitting pens 10. 10′. of D. are not in contact with the drum the short circuit is broken and partial neutralization of batteries go on leaving but 25 cells connected to 6. and 25 to 7.

The current from either of these batteries is insufficient to make a mark on either recording point of A. owing to the low resistance of the shunt R and magnet M. but the positive current from 5. of 25 cells will be sufficient to record on B.

If now the rollers 10. 10′. of D. come in contact with the drum the batteries 3. and 4. are "short circuited" and the neutralization of currents stopped hence both 2. and 5. will act with full power (ie) 125 cells and this will not produce any other effect at B. than to increase the depth of the colors of the marks, but the increased current will be powerful enough to allow a portion to pass through 12. across the paper to 13. and thence to earth recording on 12. if the current be positive and on 13. if the current be negative. If the two points are side by side the characters received by rise and fall in tension will be as in Fig. 5.

Fig. 5.

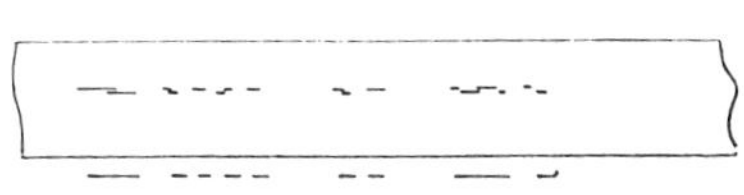

If the pens are set in line with each other but very close together the record will be as in Fig. 6.

Fig. 6.

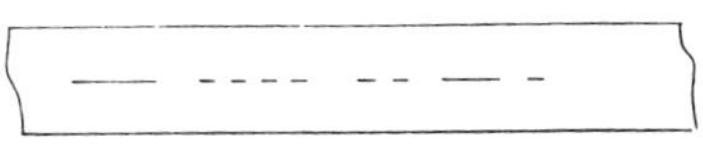

The double pens 12. and 13. may be dispensed with if a peculiar chemical paper is used such as a decoction of Brazil Wood, and a nitrate or chloride of an alkali, for then it does not make any difference whether oxygen or hydrogen is evolved a mark is obtained, certain combinations with pyrogallic acid, tannic and gallic acid will give the same result.

In Fig. 7. is shown a method of transmitting four messages over a single wire at the same time, two in one direction and two in another.

Fig. 7.

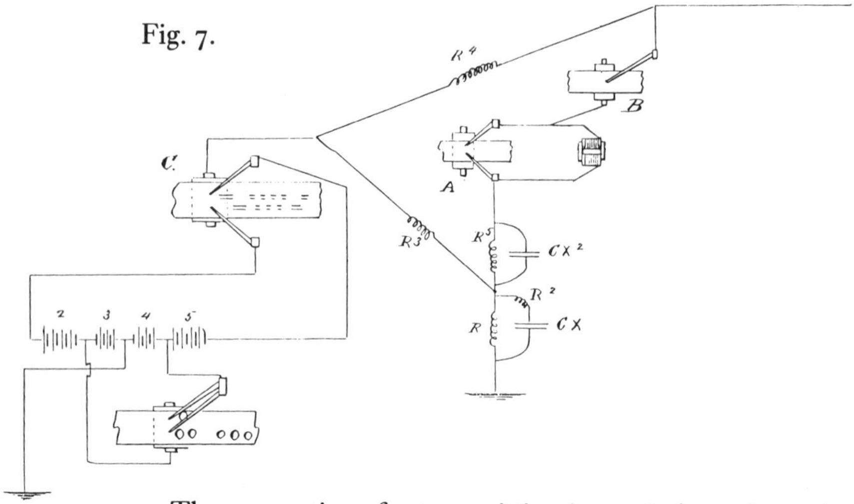

The connections for transmitting is precisely as shown in Fig. 4. but a set of receiving instruments are added for receiving from the distant station at the same time that messages are sent to the distant station.

These receiving instruments A and B. are placed in the bridge wire of a Wheatstone balance, and adjusted in the usual way as in duplex so that the effect of the outgoing currents from 2. 3. 4. 5. are balanced and produce no effect on A. and B. Morse instruments are also inserted in the bridge wire for signalling and keys are arranged with the batteries 2. 3. 4. 5. at each station so that they may be reversed or increased and decreased also for signalling.

For a simple duplex to be worked with my present system of automatic telegraphy I prefer to use the arrangements shown in Figs. 8. 9

Fig. 8.

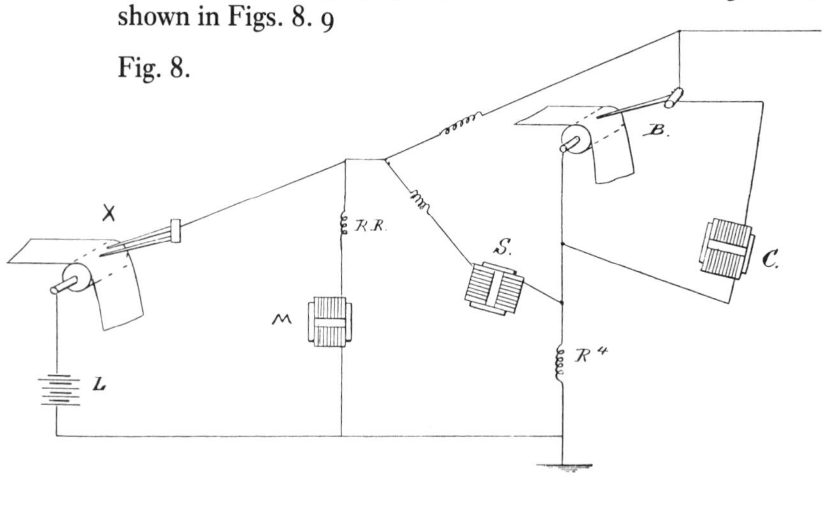

X is the ordinary perforated paper transmitter, L the transmitting battery M. the shunt magnet RR a resistance to regulate the shunt.

B. is the chemical receiving instrument placed in the bridge wire of the Wheatstone balance, S. is an electro magnet placed in one of the balance branches and is used for the purpose of setting up an induction charge to balance the static charge of the line.

R^4 is the resistance used to create an artificial line. C. is the shunt magnet used for neutralizing or compensating for the static charge due to the passage of the current from the distant battery over the line.

Fig. 9. shows a condenser X in the bridge wire to obtain a more powerful compensation for the distant current.

Fig. 9.

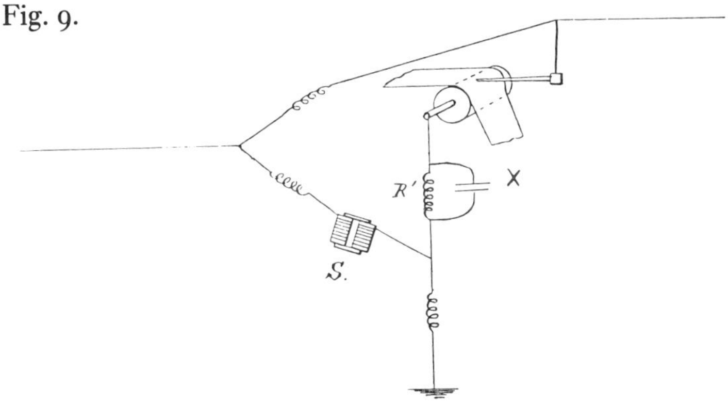

Inventor Thos. A. Edison. per L. W. Serrell atty
Witnesses Chas H. Smith Geo. T. Pinckney[b]

I will mention that any of my various methods of producing a compensation for the static charge both at the receiving or sending ends may be applied to the transmitting and receiving instrument in the chemical duplex and quadruplex above described.

I shall probably claim

First. The use of balanced batteries in a chemical duplex or quadruplex when ~~of~~ the balance is destroyed or partially destroyed by contacts controlled by perforated paper.

Second. Transmitting positive and negative currents over a wire by a double row of perforations and two batteries to effect the transmission of one message, and increasing and decreasing the strength of these reversed currents by contact controlled by perforated paper for the transmission of another message.

Third. The combination in a line with an apparatus for reversing the direction of the current therein and the increase and decrease of the same of two chemical recording instruments one arranged to record reversals independent of tension and the other to record by rise and fall of tension.

Fourth. In a chemical receiving instrument placing two recording stylus side by side or in line with each other and close together to record signals by a rise and fall of tension when the currents are being constantly reversed.

Fifth. In a chemical duplex or quadruplex balancing the effect of the static charge on the receiving instrument or instruments due to the outgoing current by magnetic induction.

Sixth. The combination of two chemical recording instruments at each end of a line wire, in the bridge wire of a Wheatstone balance, with apparatus also at both ends for reversing the current and increasing and decreasing the strength of the same.

Seventh. The arrangement for chemical duplex transmission in opposite directions shown in Fig. 1.

Eighth. The arrangement for chemical duplex transmissions in the same direction substantially as shown in Fig. 4.

Ninth. The arrangement for chemical quadruplex transmission substantially as shown in Fig. 7.

Tenth. The arrangement for chemical duplex transmission substantially as shown in Fig. 8.

Signed by me this 6th day of July A.D 1875.

Thomas A Edison

Witnesses Geo. T. Pinckney Chas H. Smith

DS, DNA, RG-241, Edison Caveat 67. Written by Pinckney; petition and oath omitted. [a]Place taken from oath; date taken from text, form altered. [b]All figures are on two separate pages, both of which are signed and witnessed at the bottom.

1. There is no evidence that Edison worked on a quadruplex automatic between the summer of 1873 and this caveat, although he had executed a patent for a duplex automatic in March 1873 (U.S. Pat. 156,843). The timing of this document suggests strongly that its sole purpose was to reinforce George Harrington's claim to the quadruplex. The Atlantic and Pacific Telegraph Co. introduced this caveat to that end in the Quadruplex Case. Quad. 70.6, following p. 49, p. 80h (*TAEM* 9:352, 389).

–593–

Agreement with Joseph Murray

Newark N.J. July 13 1875

The firm of Edison and Murray Telegraph instrument makers of 10 and 12 Ward Street Newark N.J. is this day Dissolved by mutual consent. T A Edison will settle all accounts of the firm.[1]

Thos. A. Edison J. T Murray

Witness, Chas. Batchelor

DS, NjWOE, DF (*TAEM* 13:560). Written by Edison.

1. See Doc. 594.

–594–

Agreement with Joseph Murray

Newark N J July 15th. 1875

For and in consideration of the sum of one dollar the receipt of which is hereby acknowledged I do hereby sell transfer and convey to Thos A Edison of Newark N J all my right title and intrest, in all moneies due the firm of Edison and Murray from the Domestic Telegraph Co. and Automatic Telegraph Company of New~~ark~~ York[1]

And I do hereby authorize the said Co's to pay the same to said Edison.

J. T Murray

Witness, Chas. Batchelor

DS, NjWOE, DF (*TAEM* 13:561). Written by Edison.

1. These would have been accounts outstanding for instruments manufactured by Edison and Murray and possibly for some experimental work. Although there is no summary of the money owed by Automatic Telegraph, the Atlantic and Pacific Telegraph Co. Executive Committee minutes indicate that on 29 May 1875 the company agreed to pay Edison and Murray $23,068.17 due for the manufacture of automatic equipment, some of which was probably carried over from the Automatic Telegraph Co. (Atlantic and Pacific Telegraph Co. Executive Committee Minutes, 1873–78, p. 97; see pp. 65, 88, and 92 regarding earlier payments by Atlantic and Pacific of money due Edison and Murray from Automatic Telegraph). An account of the Domestic Telegraph Co. is found in 75-013, DF (*TAEM* 13:512).

–595–

Notebook Entry: Autographic Printing

[Newark,] July 20th 1875

Copying Experiments

Have had a pen for pricking made to run with clockwork but found it no good so so had one made to run by electric Engine & it was finished today.[1]

Fig 1

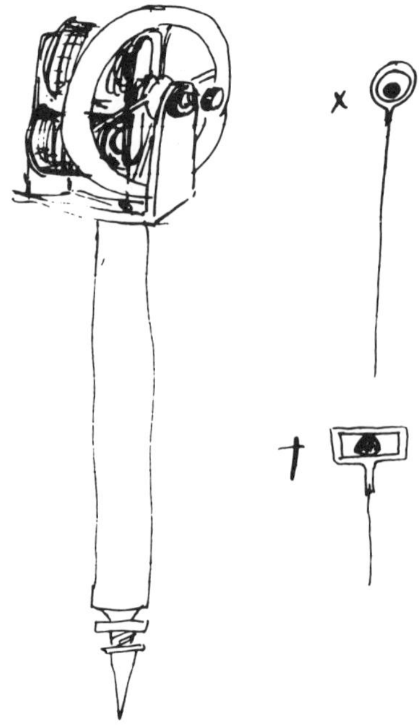

proceeded to night to try it, it worked well but ~~ran~~ pricke[d][a] too slow the cam that moved the needle was made in this way X[b] With wri[-]ting slow we got along well but ~~started~~ Wurth[2] making a new cam like this T which he finished & then we went to work to get the press right for working We found that if We cov[ered?][a] the roller of press with cloth saturated with ink & rolled it over the copy it worked very good, also fastened the ~~cop~~ letter round the cloth covered roller, worked well. also laying a flat piec[e of? -][a] saturated cloth [on the?][a] [ ][c] the letter over it then [tak?]ing[a] a copy [with?][a] the roller running over [it?][a] was good. Also in the common copying press it worked we[ll?][a] most any way. After this we foun[d it?][a] wanted very little pressure & then a[nother?][a] new idea dawned on us to take a s[mall?][a] printers roller with handle on it & cover [it with?][a] saturated cloth we also took a book & [---][a] the cover out large enough to take a shee[t?][a] of foolscap we then pasted our letter on the inside of cover,[d] shut up the book & rolled the rolle[r?][a] on the outside This was a perfect success & sample 1. was struck off this way with 100 others like it.[3]

We tried all the other anilines with Glycerine Olive oil, Molasses, some pretty fair as original,[e] but not so good Printers ink & oil no good.

We tried to write letter on parchment but the ink went right through it seems to combine.

Started Wurth on new [pen?][a] to be made direct acting thus[4] getting more po[-]wer like drawing, the barrel als[o?][a] to be shorter so that [the?][a] pen will be easier to ha[ndle?][a]

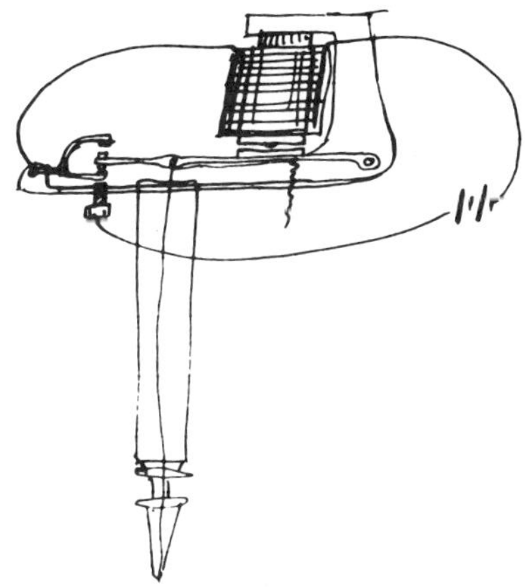

We propose to make ou[r?][a] press etc to consist of a cabinet with iron frame attached, to paste the letter to, the cabinet to have a cloth top.

A rubbing roller with handle
1 Bottle (pint) of ink.
1 Camel hair brush & porcelain boat the last three to go inside of cabinet.
1 Cell of Grenet battery
1 Pen & co[n]ne[ction?][a]
1 [ ][a]

Thos. [A Edis]on[a] Ch[as.][a] Batche[lor][a]
Jas. A[d]ams[a] Saml Edison ⟨12 hr. each.⟩[f]

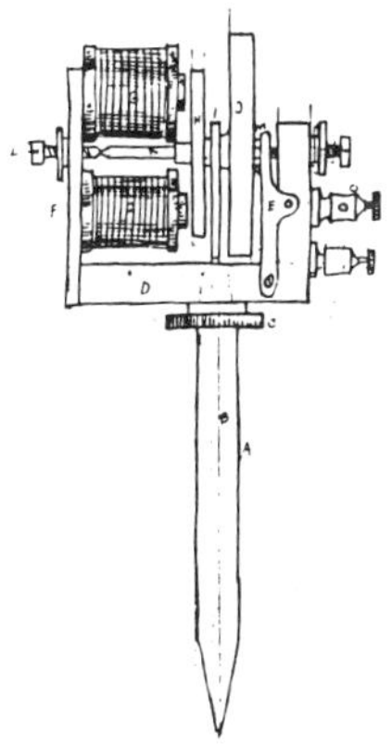

Edison's electric-pen design of 18 July 1875.

X, NjWOE, Lab., Cat. 1181:255 (*TAEM* 3:154). Written by Batchelor. [a]Document damaged. [b]"X" interlined above. [c]Document damaged; several words missing. [d]"of cover," interlined above. [e]"as original," interlined above. [f]Marginalia written by Batchelor; underlined twice.

1. In a separate notebook, Batchelor later made the accompanying drawing and labeled it "Edison's Pen July 18 1875 for Autographic Press." Cat. 1307:3, Lab. (*TAEM* 90:603).

2. Charles Nicholaus Wurth (1841–1921) was a Swiss machinist who came to the United States in 1869 as a toolmaker for the Singer Sewing Machine Co. in Elizabeth, N.J. He began working for Edison in November 1870 as a machinist. He left Edison's employ in September 1872 to return to Switzerland, at which time Edison wrote a letter of recommendation stating that Wurth had been in "charge of one of the floors our our manufactory for the past year, and has given satisfaction, we can recommend him as a steady and ingenious man" (letter of recommendation written by Edison on behalf of Edison and Unger, 21 Sept. 1872,

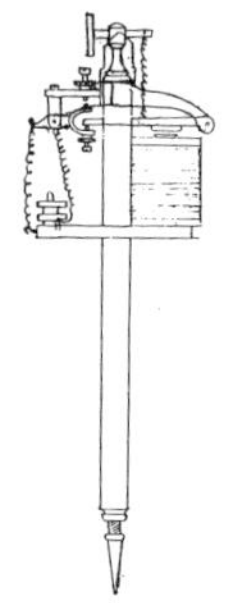

Another electric-pen design, from 6 August 1875.

Item 1-9, Box 1, EP&RI). Wurth came back to work for Edison in November 1873 and continued to work for him intermittently until 1910. "Wurth, Charles," Pioneers Bio.

3. Pasted on the facing page and labeled "Sampl 1" is an electric-pen copy of some of the opening lines of Shakespeare's *Richard III*:

> Now is the winter of our discontent made glorious summer by this son of york and all the clouds that lowered upon our house are in the deep bosom of the Ocean buried grim visaged war has smoothed his wrinkled front but now instead of mounting barbed [Cat. 1181:254, Lab. (*TAEM* 3:154)]

4. The accompanying drawing, labeled "Copying pen," went to John Ott on 6 August 1875. Cat. 30,100, Scraps. (*TAEM*) Supp. III).

–596–

Draft to Jay Gould

[Newark, July 26, 1875[1]]

Mr[-] Gould

I have sent the following letter to A B Chandler Secy A&P Tel[a]

If your board will act favorably upon this I propose to obtain 300 subscribers in NYork and Chicago who will pay me 3 per month (a number have expressed willingness to do so) for the privilege of sending 100 words ~~Nigh~~ Postal Telegrams for 50 cents.[2]

Telegrams for St Louis Milwaukee etc to be put in early mail at Chicago, and those from Chicago to Boston Phila Balto & Washn to be put in early mail here.

I propose to have[b] ~~an in expensive~~ an[b] office at NY Buffalo & Chicago—~~whe~~ and to work ~~it~~ the line[c] with my own machinery and appliances, and I think ~~it w~~ you wont have to wait until your grey headed to see what can be done with a telegraph ~~managed~~ even on a small scale

~~If the A.&.P do not wish to let me have this wire, an experiment of great moment to considerable value[d]to telegraphy will~~ The business which I would transmit would be created and entirely different from ordinary telegraphing & hurt neither the A&P or the W Union. I have apparatus by which I can work direct between NY & Chicago. You may think it strange that I have not placed a Repeater by which long circuits could be worked and the one wire Roman letter printing Telegraph upon the A.&.P wires.[3] The reason is that what they have got is not handled to make it of extraordinary use. The business is and always was badly delayed and the merchants have been bit too many times to patronize the line greatly & thus increase its business.[4][e] ~~I have worked night and day for many months~~ I have done all I could to reduce this delay but I have no authority to make changes necessary to bring it about.[5f] Without General Eckert makes some sweeping reforms ~~&~~[g] ~~changes~~, etc that boat will be a long time in reaching port.

Some people say ~~you~~ a person cant bore a Two inch hole with a little gimlet. I believe Mr. Orton is one who ma~~k~~de this assertion in reference to you use of A&P against the WU.[h] I have before me as I write a twisted wire & a little gimlet by which I can bore a beautiful round 2 inch hole, which goes to prove that it makes a great difference who has the handling of the gimlet. Yours

T A Edison

ENCLOSURE[i]

[Newark,] July 26, 1875

ABC[6]

I desire to lease ~~a~~one of the wires belonging to your Company between the Cities of New York and Chicago between the hours of 2 ~~A~~in the morning ~~till~~ and 6.30 morning, for which I am willing to pay ~~4000~~ a rental of four thousand dollars, yearly. ~~monthly~~ payable monthly in advance ~~I~~ to transmit free of ~~charg~~ cost to your Company ~~all busine~~ double the amount of business you now do ~~after~~ from 2 AM to 6 AM where but a single wire is OK ~~to Chicago~~ between NYork and Chicago. This circuit is to be connected with my offices by loops errected at my expense.[7] Yours

T A Edison

ADfS, NjWOE, DF (*TAEM* 13:426). [a]Followed by centered vertical line and space to indicate insertion of the letter to Chandler. [b]Interlined above. [c]"the line" interlined above. [d]"~~considerable value~~" interlined above. [e]"greatly . . . business" interlined above. [f]"necessary . . . about" interlined above. [g]"reforms ~~&~~" interlined above. [h]"in reference . . . WU" interlined above. [i]Enclosure is an ADfS.

1. This letter is part of a numbered, four-page document by Edison that begins with the enclosed, dated letter to Chandler. Edison wrote "Copy" at the top of the first page, indicating that he kept this draft as a copy.

2. An electric-pen copy of a draft agreement between Edison and subscribers specifying this amount can be found in 75-011, DF (*TAEM* 13:453–55).

3. It is not known what apparatus Edison proposed using to transmit automatic messages directly between New York and Chicago. His failure to install a repeater and to develop a practical roman-letter system was raised as an issue by Chandler and Eckert in their testimony in *Harrington v. A&P* (2:24–25, 32, 185, 231, 281; see also Edison's and Reiff's testimony, 3:36–37, 93; all Box 17B). Edison later complained that Eckert failed to provide the funds necessary to complete his experiments on the roman-letter system (TAE to Gould, 2 Feb. 1877, Complainants' Exhibit 4, 2:185–87, Box 17B, *Harrington v. A&P*).

4. Edison's complaint here echoes observations he had made during an inspection tour of offices in late January (Doc. 539).

5. Edison's actual authority over the operations of Atlantic and Pacific lines is unclear. He had been working in the capacity of company electrician since late January, though he was not officially appointed to that position until early June (Doc. 585). In February 1877 he would complain to Gould that Eckert had failed to give him the necessary authority as electrician and had hampered his efforts to improve the company's operations (TAE to Gould, 2 Feb. 1877, Complainants' Exhibit 4, 2:185–87, Box 17B, *Harrington v. A&P*).

6. Albert B. Chandler.

7. An electric-pen copy of a draft agreement relating to this proposal can be found in 75-011, DF (*TAEM* 13:456–57).

–597–

Notebook Entry: Electromagnetism

[Newark,] Monday 26 July 1875

Had Wurth make an experiment to find out whether we could make a column of water rise & fall by the expansion of a magnet in it when a curr[ent][a] passed round it[1]

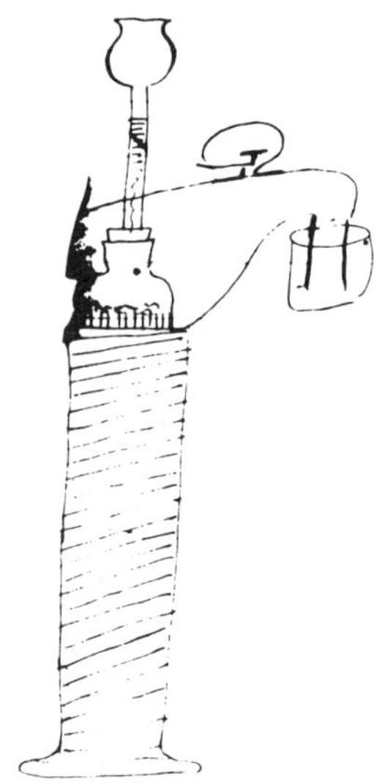

We took a glass measurer & put inside a number of iron wires & then wound 3 layers of coarse copper wire we then filled it up with water & pushed the cork with a glass tube fitted in it down in it so that the water was half way up the tube When we closed the key the water seemed to rise gradually in the tube from what cause we do not know as it did not recede when we opened the key Perhaps it may be that the iron is absorbing water [---]under the influence of electricity a'round it. Left for further experiment ⟨6 hrs⟩[b]

T A Edison TAE[c]

C. Batchelor[d]

Sa[muel][a] Edison

X, NjWOE, Lab., Cat. 1172:258 (*TAEM* 3:156). Written by Batchelor. [a]Document damaged. [b]Marginalia written by Batchelor. [c]Name written by Batchelor; initialed by Edison. [d]Document damaged; James Adams may have signed below.

1. In early 1841, James Joule had reported that iron wire lengthened when magnetized. Crowther 1962, 164–66.

–598–

Memorandum: Night Letter Telegrams

[Newark, c. July 26, 1875[1]]

Each copyist is to be paid 3 cents per 100 words each perforator 4 cents per 100 words; Buffalo perforators 5 cent 100 words ~~each~~ table man at NY to get ~~23~~ 2½[a] cents for Each 100 words;

Tableman at Chicago to get ~~3~~2½ cents for each 100 words;

Chief clerk accountenant & treasurer to get 1 cent on each 100 words sent on whole line, both ways.

Soliciter to get 25 cents per month from the $4 subscriptions.

Messenger boys to get 2 cents for each message

The General manager to get ~~43~~ 2[b] cents for every message passing over the line. Syracuse Buffalo Sandusky Detroit Repeater men to get 1 cent each for every message passing over line. ~~Ch~~Manager A.&.P NY & Ch ½ cent each on every message passes over line paid= Tableman to be manager=

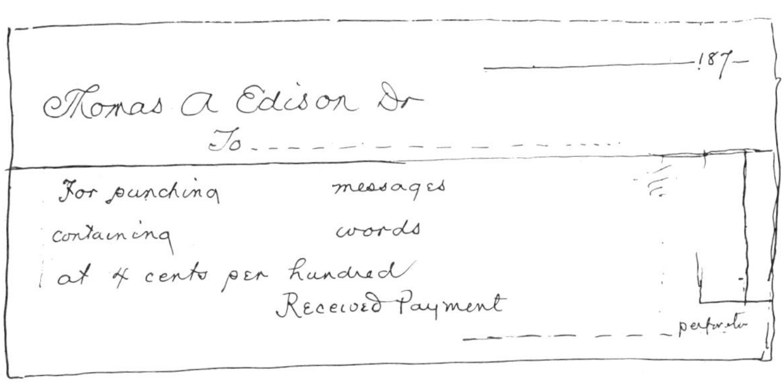

Money for Detroit. Sy Bu Sy remitted every saturday night=[2]

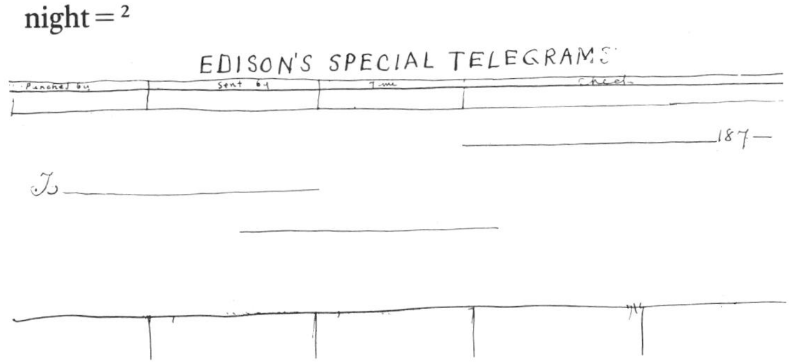

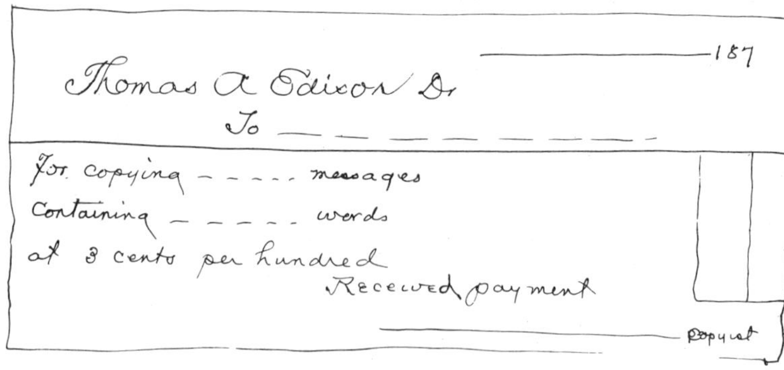

187

Thomas A Edison Dr

To

For copying ---- messages

Containing ---- words

at 3 cents per hundred

Received payment

Pay[?]

AD, NjWOE, DF (*TAEM* 13:449). [a]"2½" interlined below. [b]Interlined above.

1. This document was probably written about the same time as Edison's letters to Gould and Chandler asking for permission to establish a night-letter system on the Atlantic and Pacific lines (Doc. 596). Edison docketed the back, "NYork & Chicago Special Telegram Scheme."

2. Labels under "EDISON'S SPECIAL TELEGRAMS" are "Punched by," "Sent by," "Time," and "Check"; columns at bottom extend the full length of the original page.

REIS TELEPHONE DRAWINGS Doc. 599

Sometime in July 1875 William Orton called Edison to his office to discuss Elisha Gray's work in acoustic telegraphy and to ask if Edison would pursue the subject for Western Union. Edison agreed to do so, and shortly afterward Orton supplied him with a translation of a work on Philip Reis's 1861 telephone, a device to transmit sound by means of a contact point pressed against a vibrating membrane.[1] Edison, aware that Reis's apparatus was able to transmit only the pitch of a sound, sketched the following "rough ideas of how to carry out that which was necessary in my mind, to turn the Reiss transmitter into an articulating transmitter. They were notes for future use in experimentation."[2] Each of Edison's transmitters varies the resistance of a closed circuit by using a moving metallic contact point separated from a stationary one by a liquid. This change—using a closed circuit in which current strength varied rather than a make-and-break system—was the fundamental understanding crucial to successful voice transmission. In 1873 Edison had obtained a patent for a telegraph relay that operated on this principle,[3] and Gray's 1874 musical telegraph also worked this way. Edison did not build an ap-

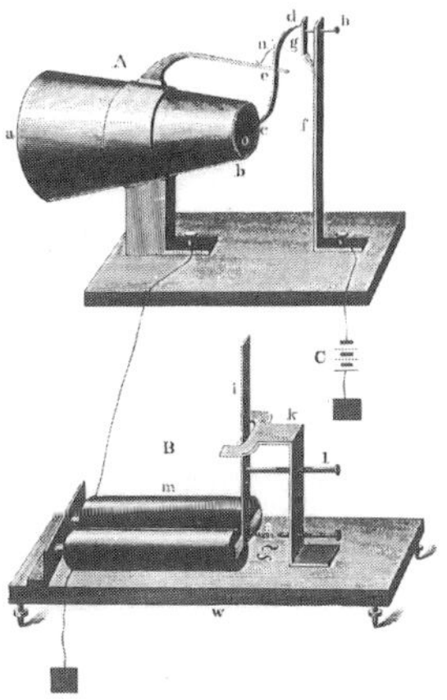

In Reis's transmitter (top) *the lever* **cd** *makes and breaks the circuit at* **g**; *the electromagnet* **m** *of the receiver* (bottom) *responds to the signal and causes the thin plate* **i** *to vibrate rapidly, reproducing the tone.*

paratus expressly to transmit articulate speech until the end of 1875, but when he did, it was a water telephone electrically like the device in the center of Doc. 599. Edison said that these particular instruments were never produced.[4]

Edison's first idea, at upper right of Doc. 599, involves a tuning fork with a resonant chamber (labeled "speak") attached to its upper arm by a needle in the center of a diaphragm. A needle attached to the fork's lower arm extends into a cup of mercury (labeled "mercury like Helmholz").[5] Sounds in the resonant chamber move the diaphragm, thereby setting the fork in motion, and the lower needle's vertical movement alters the resistance of the circuit.[6] In the central circuit, a metallic lever attached to the diaphragm at the end of the resonant chamber at left has its other end immersed in water above a second point of metal. As in the first design, the lever moves up and down in the liquid as the diaphragm vibrates, altering the resistance of the circuit. The tuning fork at right, placed in a resonant chamber, responds to the changing field of the electromagnet below it, which is in circuit with the lever. The diagram at lower left shows a chamber (labeled "parchment") with a diaphragm attached to a knife edge that is a short distance from a second knife edge. Over the gap between them is a funnel (labeled "adjust") containing water that drips slowly onto the knife edges. As the diaphragm vibrates, the movement of the attached knife edge elongates and compresses the water drops between the edges, varying the circuit resistance.

1. The article—"Reproducing Sounds on Extra Galvanic Way," by V. Legat (royal Prussian telegraph inspector in Cassel)—is reprinted in TI 1, following p. 509 (*TAEM* 11:635), and in Presscott 1879, 9–12. Edison remembered first learning of Reis's telephone in the early 1870s, though he almost certainly was referring to 1869 ("The American Institute: The Telephone," *J. Teleg.* 2 [1869]: 65; "The Telephone," *Telegr.* 5 [1869]: 309). He was aware of Gray's harmonic telegraph in 1874. Gray's work prompted Edison "to read upon the subject of acoustics, as I was interested in not having my Quadruplex replaced by another method." Edison's testimony, TI 1:4–6 (*TAEM* 11:23–24).

2. Edison's testimony, TI 1:9 (*TAEM* 11:26). Because Reis's telephone worked by opening and closing a circuit, it could transmit only the frequency of a sound, not any changes in volume or the tonal nuances carried by harmonic overtones. Edison recognized the shortcoming regarding volume (TI 1:10 [*TAEM* 11:26]).

3. U.S. Pat. 141,777.

4. TI 1:9 (*TAEM* 11:25); see Doc. 736.

5. Hermann von Helmholtz (1821–1894), perhaps the pre-eminent German physicist of his era, experimented and theorized extensively on

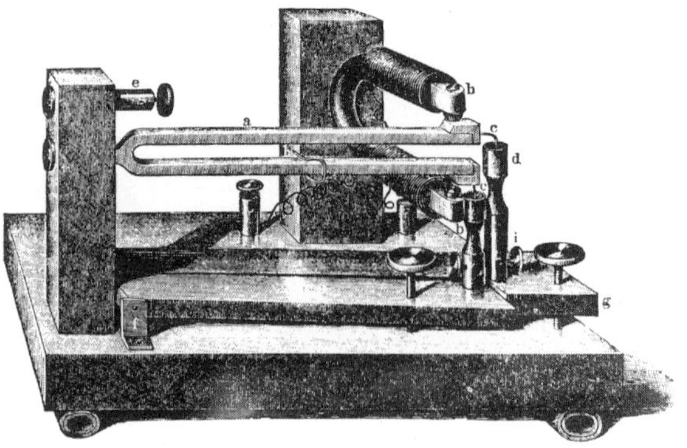

Hermann von Helmholtz used this tuning-fork transmitter to generate pure electrical frequencies for his acoustic experiments.

acoustics. The accompanying illustration appears on p. 178 of Edison's 1875 edition of Helmholtz's *Sensations of Tone* (presented by Mina Edison to the Massachusetts Institute of Technology; the Edison National Historic Site has a microfilm copy). Helmholtz arranged the needles **c c** to make and break the circuit at the surface of the mercury. In testimony describing these designs, Edison said that the first word in the label is "mercury." TI 1:8–9 (*TAEM* 11:25).

6. Edison seems not to have understood at this point that a given tuning fork responds strongly only to a particular frequency.

–599–

[Newark, c. July 1875]

Technical Drawing: Telephony[1]

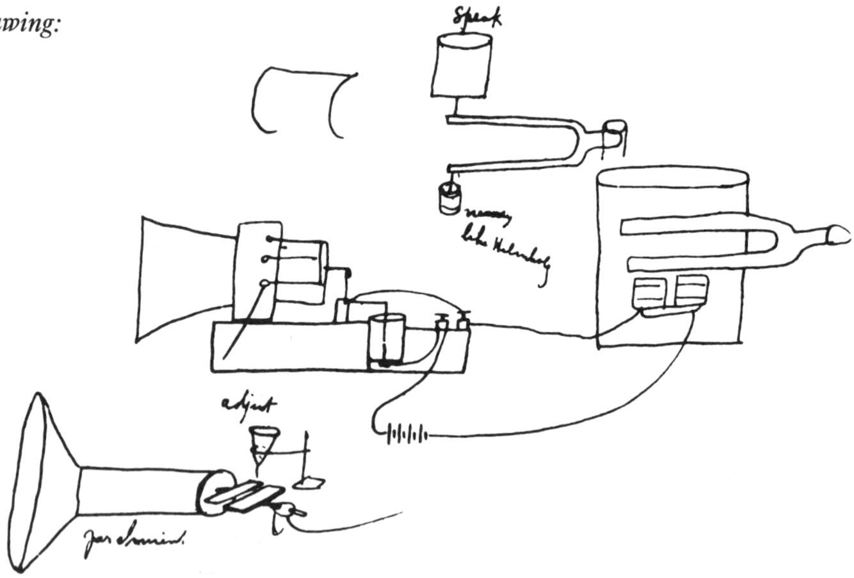

X (transcript), NjWOE, TI 2:509 (*TAEM* 11:634).

1. See headnote above.

–600–

Notebook Entry: Electromotograph[1]

[Newark,] Wednesday, Aug ~~3~~4th 1875

Tried electromoto-graph for repeating noticed a peculiarity about it that it would drop dots clean out.

Another thing we have noticed about it is that we sometimes work a long time before it will work & then it suddenly commences to work & it will work anyhow we fix it & we dont know ~~at present~~ how we have fixed it to make it work.[2] ⟨12 hrs⟩[a]

T A Edison T.A.E.[b] Jas Adams J.A.[c]
Chas. Batchelor Samuel Edison

X, NjWOE, Lab., Cat. 1172:259 (*TAEM* 3:156). Written by Batchelor. [a]Marginalia written by Batchelor. [b]Name written by Batchelor; initialed by Edison. [c]Name written by Batchelor; initialed by Adams.

1. This entry is continued in Doc. 604.

2. According to account records, the laboratory staff spent considerable time on the electromotograph in July. Cat. 1214:72, Accts. (*TAEM* 21:600).

–601–

List: Centennial Exhibition Exhibit Proposal

[Newark,] A[ug][a] 5th 1875

⟨Copy of what we propose to put in the Centennial[1] & the space we want[2] Aug. 9 1875⟩[b]

Quadruplex Telegraph (transmitting 4 messages over one wire at same time)
Duplex Telegraph (transmitting 2 messages over 1 wire at the same time)
Automatic Chemical Telegraph (high "speed" system)
Roman Letter Chemical Telegraph (Printing in Roman Letters)
Electromotograph (new discovery in Telegraphy)
Chemical Stock Reporting Telegraph
Type wheel printing Telegraph
Induction coils, Condensers, Electrical copying presses and various Telegraphic apparatus.
Fire Alarm Telegraph.
Domestic Telegraph Apparatus

I am the manufacturer and producer— These goods are entered for competition[3]

Th[os.][a] A E[dison][a]

ADS (letterpress copy), NjWOE, Lbk. 1:3 (*TAEM* 28:5). [a]Document damaged. [b]Marginalia written by Charles Batchelor.

1. In 1871 Congress authorized the mounting of an international exhibition of arts, products, and manufactured goods in Philadelphia under government auspices to celebrate the centennial of American independence. The exhibition opened on 10 May 1876 and ran through 16 December 1876. J. Wilson 1876–78, 3:xc, clxxxvi.

2. According to Batchelor, Edison applied for a 45′ × 30′ exhibition space (see Doc 619). But a letter from a Centennial official suggests that Edison applied for a 45′ × 35′ space (Henry Pettit to TAE, 29? September 1875, DF [*TAEM* 13:1005]).

3. Items placed on exhibit could also be entered into competition for medals. A small of group of competent individuals, half foreign and half from the United States, judged the exhibits. Instead of having several grades of awards, winning entries received uniform medals accompanied by a report and diploma that described the individual award. J. Wilson 1876–78, 3:cxi.

–602–

Charles Batchelor to Pearce and Jones[1]

Newark, N.J. Aug. 6th 1875

Gentlemen

We can make for you an artificial Atlantic Cable consisting of (25) twenty five condensers & suitable resistances for the sum of $(1800) eighteen hundred dollars.[2] We have one in our shop which cost us ($4000) four thousand dollars.[3] The condensers will all be divided up into 4 & more parts. Yours respecty

Charles Batchelor

ALS (letterpress copy), NjWOE, Lbk. 1:2 (*TAEM* 28:4).

1. According to their letterhead, Frederick Pearce and James Jones were telegraph-instrument manufacturers located at 61 Ann St. in New York City. They continued to be actively involved in telegraph manufacturing and invention through the 1880s. Pearce's and Jones's testimony, *Wiley v. Field.*

2. Two days earlier Pearce and Jones had written to Batchelor to inquire about placing an order for an artificial cable. DF (*TAEM* 13:410).

3. Edison had been experimenting with artificial cables since his return from England in the summer of 1873 and had made several. Because of the expense of purchasing the necessary condensers, he and Batchelor sought ways to produce their own. Later in 1875 they developed a method of making condensers with wax paper as the dielectric (Doc. 698). On the cost of condensers see Doc. 420.

–603–

From George Prescott

NEW YORK Aug 10 1875[a]

Dear Sir,

Referring to our interview of last Tuesday in which you stated that you thought Mr. Gould would reconvey to you your interest in the Quadruplex, so that there would be no obstacle in the way of the patents being granted to us, and of our obtaining the royalty from the Western Union in accordance with our agreement, I would like to know whether you have since had any conversation with Mr. Gould, and if he is willing to exchange his interest in your half of the Quadruplex for your interest in the Automatic as proposed by you.[1]

I have not yet received any replies from abroad to my proposition to sell the Quadruplex patents but expect to soon[2]
Yours respectfully Geo. B. Prescott

ALS, NjWOE, DF (*TAEM* 13:712). Letterhead of Western Union Telegraph Co., George Prescott, Electrician. [a]"NEW YORK" and "187" preprinted.

1. Edison was probably aware of Atlantic and Pacific Telegraph Co.'s interest in negotiating some kind of settlement of its differences with Western Union. On 2 August, the Atlantic and Pacific trustees appointed Gould, Thomas Eckert, and W. J. Syms to negotiate with a committee to be appointed by Western Union "for the purpose of determining whether, and if so, upon what basis an equitable arrangement can be made for the amalgamation or harmonious operation of the two companies" (extract from Atlantic and Pacific minutes, 2:252, Box 17B, *Harrington v. A&P*). On 11 August the Western Union Executive Committee received a letter from Gould requesting that it appoint such a committee (Western Union Executive Committee Minutebook B, p. 538, NjWU).

2. Nothing of consequence happened until 1877.

-604-

Notebook Entry: Electromotograph[1]

[Newark,] Aug. 10th [1875][2]

Electromotograph all night

Got it working on short circuit & repeating 600 words per minute with 10'000 ohms in transmitting circuit by adjusting lever of Electromotograph very close it required much less spring & wanted very slight ~~spring~~ also you could run any speed from 10 to 300 words without altering adjustment.

Called up Cogger of A&P[3] & had him transmit through a line copmposed of a wire to Philadelphia & back & then to Newark by old Aut. wire about (230) two hundred & thirty miles with 5 cables.[4a] We found now that although we got it perfect on the electromotograph paper we could get nothing intelligible on the repeated circuit. This was very strange as we had been able to get it on a 10'000 ohm circuit in the laboratory when no marks were discernible on the paper of the electromotograph.[b] Now the difference between the 2 lines the artifical & real was this; the artificial line had no sending shunt or receiving shunt & no static discharge whilst our outside line had a magnetic sending shunt & the static discharge from about 230 miles of wire & 5 cables The artificial was thus instantly opened after each closure leaving no route for anything such as static induction or whatever else may be the cause to pass Whilst the outside line was never opened owing to the sending shunt at New York being in. ~~It was between 2 causes Static induce or the magnet in the sending shunt.~~ We now put in a magnet[5] shunt[c] on our resistance transmitting circuit & we could not get anything although we got the marks on Electromot~~t~~ograph, very good. We now put in plain resistance shunt & we still could get nothing. We now came to the conclusion that there was a secondary action[6] from the Motograph something like the secondary action from a magnet & as long as the line was shunted or there was an escape on it which is the same thing it gave a route for this secondary action to work & mutiliate the signals. We shall now proceed to find out the nature of this secondary action

Thos A Edison — Chas. Batchelor.
Jas Adams — Samuel Edison

X, NjWOE, Lab., Cat. 1172:261 (*TAEM* 3:157). Written by Batchelor. [a]"with 5 cables" interlined above. [b]"when . . . electromotograph." interlined above. [c]Interlined above.

1. This entry is a continuation of Doc. 600 and is continued in Doc. 605.

2. Related entries in this notebook date from August 1875.

3. Unidentified, but probably an operator with the Atlantic and Pacific Telegraph Co.

4. That is, five underwater crossings.

5. Inductive.

6. That is, inductive discharge.

–605–

Notebook Entry: Electromotograph[1]

[Newark,] Aug 11, 12, 13, & 1~~3~~4th 1875

Electromotograph as a repeater.

Fig 1

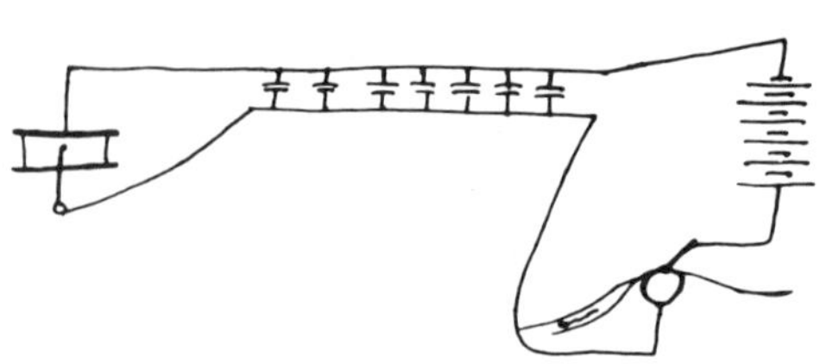

1. With no condenser can get about 600 or 800 words per minute even on 10,000 ohms with 25 cells carbon battery. 2, When shunt is added as in fig 2

with 200 ohms resistance we can get nothing although just as much current passes through the paper & the writing on Emg. paper itself is just as perfect no signals can be repeated at more than 25 or 50 words per minute. 3 Putting a 200 ohms magnet in the shunt besides the 200 plain gives no better result. 4 Putting in but 1 spool of the magnet does no better.

5 Adjusting the cores from & to the armature no better. 6[a] Putting magnet in main line at transmitting end without shunt no better

6 Taking out magnet, keeping in shunt & adding 10,000 ohms in line at receiving station no better

7 Adding magnet in main line with & without 10 000 ohms no better.

8 Putting in a 5000 German silver magnet[2] no result

9 Putting in 2 " " " " " "

10 Shunting Emg with magnet no result

11 " " " 2 " " "

12 Shunting Emg with plain resistance & adding magnet in branch containging Emg. a little better result.

13 Shunting Emg. with another Emg. in motion increases speed greatly but owing to the variation in the pressure which I held the Emg point it threw the other Emg out of adjustment constantly

14 By using a secondary battery of Platina & Platina immersed in Pyro sol.[3] we got very fair speed 15 but when we shunted Emg with a large condenser No 6 the speed was the highest we have ever got from 1000 to 1100 words per minute perfect

16 When condenser was taken off could get nothing

17 By taking shunt off at transmitting end could get 600 or 700, the resistance. The resistance of line was 1500 ohms. The shorter the circuit the greater the effect of a closed circuit in which the secondary current or other disturbing force from the Emg. We are not sure as to what kind of a force circulates from the Emg after the battery is disconnected. The condenser seems to completely neutralize this secondary effect if

there is any. The effect might perhaps be due to another cause to wit:—

We have always noticed that on Hydrogen solutions when there is no static charge or lateral induction[4] the first portion of a dot or dash is the strongest thus:—

We have also noticed that the power of the Emg. when worked by hand appears to be strongest on the first closing of the current & in working we have noticed that with certain adjustments when the circuit is first closed the lever moves with great power in a direction contrary to the revolution of the drum but immediately flies away from the contact point in the same manner as if the current had been as weakened that it had not enough power to hold it there though in this[b] case the current was of constant strength. This phenomenon prevents an equal adjustment & as we have previously noticed most solutions through a resistance will 'tail' even in laboratory the addition of a shunt will in most cases stop tailing & make sharp writing hence the addition of a shunt only aggravated in this ~~the~~ case the phenomenon making writing perhaps thus

~~18~~ Throwing the Emg more out of adjustment the effect from a condenser would be to counteract this thus the first part of the current coming over the line would pass through the Emg & condenser the portion which rushed into the Condenser to charge it would be subtracted from that portion which would pass through Emg were there no condenser, hence the first portion is weakened but the condenser almost instantly becomes charged ~~& the last portion of the~~ current passing through Emg hence instead of a dash being thus —— it is —— or perfectly even as regards the Emg phenomenon but in reality it is recorded thus —— first part being weak & as the first portion of the current acts strongest on Emg. the lever comes forward with a certain force just as it would tend to weaken & fly back as before mentioned the condenser is nearly charged & the current increases in strength counteracts & prevents it. the discharge of a current of opposite polarity through Emg. from condenser after battery is disconnected does not appear to effect it at all & may under these conditions be an advantage.

18. Tried a new lever on Emg thus:[5]—

Fig 1

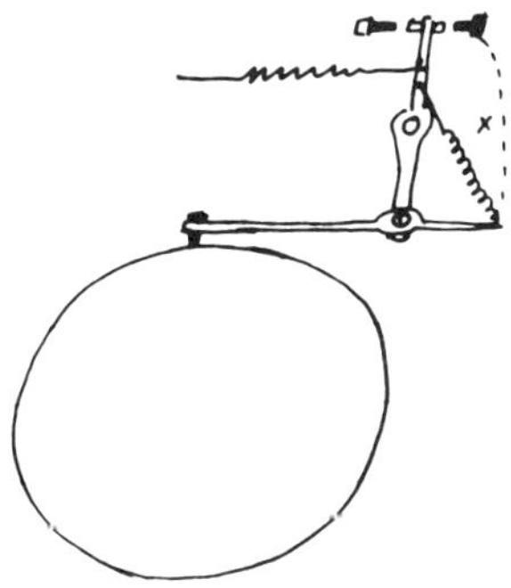

It could not be made to give a single movement with spring at X 19.[a] neiether could it by putting it as shewn isn dotted lines after almost giving it up as a failure found that by holding my finger on the platina friction points that at times when working a key I would get a very strong movement 20[a] after further experiment we found that if I held my finger on so as not to cant the lever from or towards me & thereby prevent friction in the fulcrum I could get the movement. We therefore concluded that slight frictions in the moveable lever of this Emg was of far greater importance than I had heretofore supposed. 21[a] I then had levers taken out & shoulder screws eased up & burnished so as to diminish the friction as much as possible although far too great after it was done 22[a] then timed it & on putting on spring as in fig 2

Fig 2

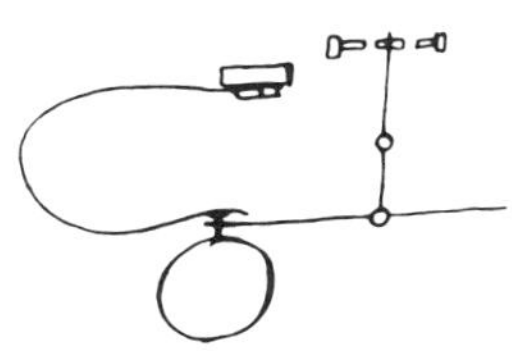

we succeeded in repeating without shunt about 600 words per minute this shows that at high speeds & with tailing etc there is great difference in the frictions although the lever responds so noticeably & conveys the idea of a great difference & that it takes but slight friction in the fulcrum of the moveable levers to neutralize completely this little difference 23[a] we also tried putting spring as in fig 3

Fig 3

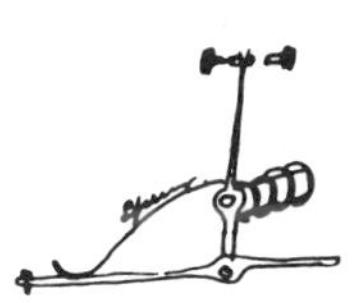

it did not work but it was before bad bearings were eased it will perhaps work with frictionless bearings

24. Best Emg solutin ever saw on H.

Lactate Zinc
Salt
Caustic K
Pyroxylitic spirit[6c]

note Caustic Potash is better than S~~t~~alt alone or together.

25 Silicate of Soda
Salt[d]

Note. Oxygen splendid ~~&~~ slips[b] most as good as reg. on H. On H it is very slight though plenty to work Emg. The trouble will be that the Sil Soda dries to crust & it becomes insensible in a few minutes. Peculiarity on H is that paper is quite slippery without current hence pen is not clutched by normal friction so fiercely when current is off as in ordinary sol.

26 Nitrate Lead
Caustic Potash[e]

A1 for Emg on H   Slight Slip on O.

27 Salt added to above but no better

28 Coffee
Salt
Sulphate of Quinine[f]

no good   Added Caustic Potash & decreases friction on both sides   Excess of Caustic K makes slip on O almost as good as reg on H but H is reduced.

29 Fustic[7]
Salt[g]

good as regular on H & cheaper

30. Borax
Salt[h]

middling good on H & slips very slightly on O. Caustic K added no better on O, good deal better on H   Bi Sulphate of K added makes slips on O better than any above

31 Glycerine water & Sugar:— Friction on H & O. O strongest   Added alittle K1 on Tin & Plat only. Ox friction increased very much, on both sides about equal   used plat flat plate   Salt added, slight slip on H but increased friction with O on all metals.

32 Boss solution   Coffee, milk, sugar, & eggs., Caustic K & salt, equally good on both sides   perhaps a little better

on H but by proper proportions can probably make them equal

T A Edison | Chas. Batchelor
James Adams | Samuel Edison

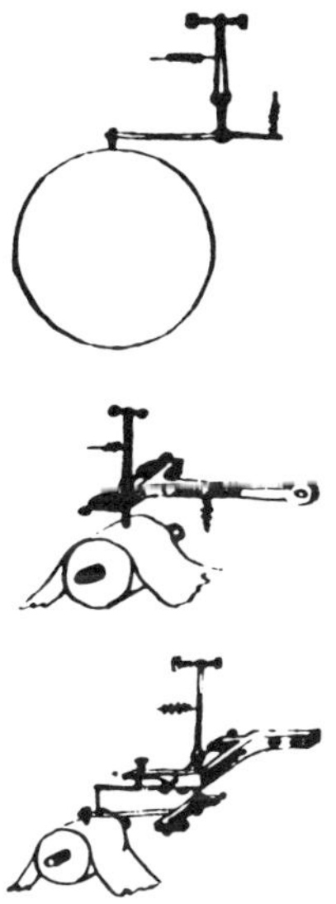

Batchelor's electromotograph stylus levers.

X, NjWOE, Lab., Cat. 1172:262 (*TAEM* 3:158). Written by Batchelor. Batchelor added section numbers 1–23 to body of text. [a]Written in margin. [b]Interlined above. [c]"Lactate . . . spirit" enclosed in brace. [d]"Silicate . . . Salt" enclosed in brace. [e]"Nitrate . . . Potash" enclosed in brace. [f]"Coffee . . . Quinine" enclosed in two-part brace. [g]"Fustic Salt" enclosed in brace. [h]"Borax Salt" enclosed in brace.

1. This entry is a continuation of Doc. 604 and is continued in Doc. 607.
2. A magnet wound with German silver wire, which was a high-resistance alloy of nickel, copper, and zinc.
3. Probably Edison's standard pyrogallic solution.
4. Unknown.
5. In another notebook, Batchelor drew and labeled these three "electromotograph motions." Cat. 1307:50, Batchelor (*TAEM*) 90:635).
6. Wood alcohol. Bloxum 1869, 134.
7. A yellow dye from the wood of a tropical American tree of the mulberry family or similar dyewoods.

–606–

Draft Memorandum [to Jay Gould?]

[Newark, c. August 15, 1875?[1]]

Plans for paying much from nothing[2]

1st Have A and .P. pay Automatic interests 5100 shares of Franklin stock now in A and P Treasury contract to deliver said 5100 shares to Western Union for 68½ per share The Franklin Co has about 3200 miles of wire and a million stock which is at rate of $311.52/100 stock per mile par or $218 at 68½ not including loops City wires & office fixtures etc W.U. being about 200 at 80.[3] This would be a good purchase for W.U. and they could buy enough more to make a valid lease cheaply= This divided equally gives you $95 586 31 cash

2nd Plan A and P. to pay Automatic interest 4400 shares Franklin and the Quadruplex patents your share is same as plan 1. I taking the Quadruplex at 40,000 in part payment & turning it over to W.U. on their old offer.[4] This will make them all safe from lawsuits in that direction. They can then readily pay 68½ for the 4400 shares Franklin

Plan 3rd Have A and P issue 2 million more stock to pay Automatic of which your share will be Contract to sell the whole to WU in a block at 18 This with amount held by Union Pacific (if they hold any?) will control.

4th Plan Have A.&.P issue coupon ~~bond~~ 7 per ct currency

bonds to amount of 600 000 to pay for automatic patents.

5th Plan A.&.P to pay 200 000 cash The Quadruplex 500 shares A&P stock now in treasury then issue 200 000 7 per cent bonds. This would realize 350,000. perhaps they havent the cash

6th plan If the .A.&.P cant pay for the Automatic system then sell to WU for 420,000. in their stock at the market return 60,000 advanced & expended on it & divide the proportion as in plan .1.

7th Reorganize the stock make 4 million prefd 6 million Common give present holders 3,500 000 prefd & 3 million Com Auto 500 000 pref 3. million Common

8th Increase Capital Stock of Franklin to 2,000,000. give Automatic people 500 000 of this for use of Automatic on Franklin & have A.&.P give Auto people 3000 shares & 50,000 cash[a] Total 8000 shares & 50 000 cash[a] then contract to deliver same to W Union This would give WU Control Franklin

This increase of the Franklin stock will give $622 stock at par per mile while the A&.P. even if we include ½ of the Franklin has 1116 stock for each mile.

9th Have A and P Lease auto system at ~~3~~540,000 a year for 10 years & ~~1~~520 000 a year for same from Franklin, WU to guarantee this should they lease

10th Plan Issue 3 million A.&.P Stock according to original programme to be paid 7 per cent on the stock at 24 by WU on a lease of whole company

11th Dont sell the system to the A.&.P or Franklin but transfer your mine & Reiffs & 1 or 2 other of his people[b] interest to the W.U. for ~~260~~300 000 stock at [-]market[c] & let Let[d] other ~~fellows~~ people[e] ~~whistle! try~~ go to the same place for theirs =

AD, NjWOE, DF, 75-012 (*TAEM* 13:507). [a]"& 50,000 cash" interlined above. [b]"& . . . people" interlined above. [c]"stock at [-]market" interlined above. [d]Written in left margin. [e]Interlined above.

1. This document was probably written during the August 1875 negotiations between Atlantic and Pacific and Western Union regarding possible consolidation or cooperation between the two companies. The Atlantic and Pacific trustees discussed formal terms at their 31 August meeting, at which time they also decided to purchase those rights to automatic telegraphy assigned to Jay Gould by George Harrington, George Little, and others. Atlantic and Pacific Telegraph Co. Stockholders' Minutebook (1875–88): 1–12, Box 5964, NjWU; extracts from Atlantic and Pacific minutes, 2:252–53, 305–8, Box 17B, *Harrington v. A&P.*

2. Edison wrote on the back of this document, "Plans for the Settlement of Automatic." Atlantic and Pacific had still not settled with most of the stockholders of the Automatic Telegraph Co.

3. The Franklin Telegraph Co. board of directors voted in November 1874 to lease the company's lines to Atlantic and Pacific. A lawsuit prevented formal consolidation until January 1876, although terms of the lease were agreed to in June 1875. The two companies appear to have operated jointly from the end of 1874. During 1875 Franklin had 3,262 miles of wire, which increased to 3,750 by April 1876, and a capital stock of one million dollars. See *Report of the Financial Affairs of the Franklin Telegraph Company, and its Relations with the Atlantic and Pacific Telegraph Company* (New York: Homer Lee & Co., printers, 1876), 13; and Thomas Eckert's report to the Stockholders of the Franklin Telegraph Co., 7 June 1876; both Box 5964, NjWU. See also Reid 1879, 594.

4. Doc. 535.

-607-

Notebook Entry: Electromotograph[1]

[Newark,] Aug 16 75[2]

Discovery of new phenomenon Hand EMG.[3]

metal of bat plates Aluminium Nickel Zinc Tin platina brass Silver—lead =

Pyro paper 2 cells new carbon battery ~~shunted with~~ Key & rheostat thus[4]

When all current passed through the pyro paper it slipped on all metals with Hydrogen by ~~sl~~ decreasing Resistance of the R[heostat] so that a small portion went through probably ¼ all the metals except ~~zinc~~ Tin showed an increase in friction on passing the current but tin still showed the slip, ~~on~~ but by still further shunting the current ~~Tin Cr~~ so that but little passed through Tin stopped slipping and increased its friction. Another noticeable thing was that when all metals except tin ~~w~~had an increased friction on the moment of closing the normal friction was increased say 10 degrees on opening the friction was increased to 30 deg but almost instantly subsiding, the instant the current being interrupted this friction was fearful strong.

This friction may be represented thus

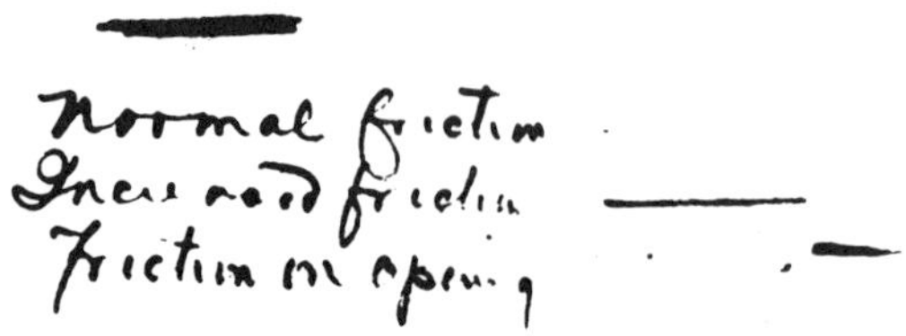

or[5]

I then tried paper wet in[a] a solution of Caustic potash but ~~with~~ on account of small current used it didnt show anything not being near as sensitive as pyro

I then put some salt NaCl in with the Caustic potash when I found that this strip far excelled in delicacy the Pyro—and gave an unexpected phenomenon.

No increase of friction could be noticed on closing but all metals after the current was very much weakened gave no results except zinc and this was very marked after all others even tin had ceased—and the slip was noticeable

This paper is the most valuable yet found & should be used on a zinc drum

Have found that Silver is even better than zinc on any low current though on an exceedingly fine current zinc may be better. Aluminum appears well on a fine current its all slip with strong momentary opening friction. Tried perforating paper ie[b] wood paper and it appeared better than Common Boston[6] perhaps thick wood paper will be better

I give Zinc the preference of tin with a low current on Wood paper thick.

Platina is exceedingly bad gives very little with full current on, tin middling

~~T~~Silver slips on both H & O but O very slightly—perhaps some others do but not noticeable =

We wet 2 rolls of ~~K~~Caustic K & NaCl & tried on EMG found that with sending shunt 1500 ohms & the condensers that we could by shunting EMG ~~o~~repeat slowly so it could be read though very ragged without shunt nothing. with this arrangement ~~w~~ words could be read but they were somewhat ragged they is something in it

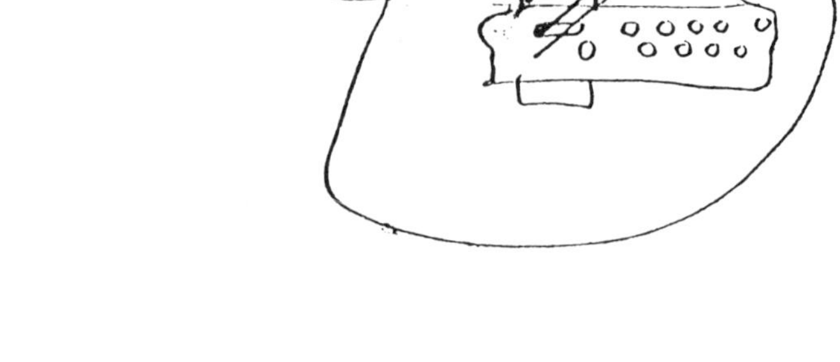

This weakens the first part of a dash & leaves end full strength as the first portion of battery current has two routes, the line & condenser & until condsr is charged full current dont go on line. the second roller pen short circuits the condenor through the discharging holes & prevents its charge going on the line.

With the potash & salt[c] paper highest speed straight (1500 ohms)[d] 25 cells batty[e] that was ever obtained about 1200 wds a minute =

With the new lever this evening we could get high speeds but the repeating contact was very inferior for some reason. we had no spring on platina point yet ~~Batchelor~~ I am under the impression that the old right angle vibrator would do good repeating without a spring ~~this~~ we intend ~~fixe~~ putting on a spring on new lever to test contact.

The lever used with the Potash salt paper was the old style =

after getting 1200 words per minute straight ie[b] 1500 ohms ~~w~~ I shunted EMG with 100 & also with 1000 ohms but could get nothing that was readable at 75 words minute This shows that some powerful secondary force action or otherwise ~~it~~ circulates within the closed circuit which throws whole thing out of adjustment

We have found that the think[7] wood paper used for perforating answers admirably for EMG it is more even & less friction than rag paper used for receiving = [8]

We now 350 am are going to make various EMG solutions for test on hand EMG to endeavor to obtain a paper free from this secondary phenomenon

Sulphat Copper increases friction on ~~O~~H. very powerful when not shunted = When not shunted on ~~H~~O it increases friction when shunted it "slips"[f] best on zinc

Bichromate K on ~~O~~.H. full slips slightly on ~~H~~O full friction great, vibrating shunted friction best on ~~tin~~ silver, shunted lower still friction good on Tilver.[9]

Sulphuret K nothing on any except slight friction thump[10] on silver on ~~O~~H.

Tartaric Acid nothing

Grape sugar friction great on ~~H.~~O = platina & silver best.

Borax friction incrsd on ~~.H.~~O platina & silver nothing on ~~.O~~H.

Cream Tartar nix on either

Carbonate Cu nix insoluble[g]

Hyposulphite Soda nothing on ~~H.~~O slips on ~~O~~H good both platina & silver latter perhaps better = even shunted ~~e~~does same

I have 2 cells, 75 ohms main Key and EMG (hand EMG) shunted with or without, 20 ohms

Sesquichlorade Fe nothing on either

Hyposulphite Soda & Sesqui Chl Fe mixed nothing on ~~H~~O Friction increased on ~~O~~.H. but nothing noticeable when shunted

Hypo ~~S~~Na—Sesqui Ch Fe & NaCl makes black mark on platina with platina might be built up = no good on EMG[11]

Tartaric A & Grape Sugar dont improve mark =[g]

Sul Zinc slight slip on ~~H~~O friction considerable on ~~O~~.H.[g]

Uric A slight (very) increase. friction on ~~H~~O.

Ordinary pyro acts one way & then another inside 2 mins from same roll— ~~must~~ phenomena must be in platina absorbing gas =

X, NjWOE, Lab., NS-75-001 (*TAEM* 7:322). [a]"paper wet in" interlined above. [b]Circled. [c]"& salt" interlined above. [d]"(1500 ohms)" written in right margin. [e]"25 cells batty" interlined above. [f]Underlined twice. [g]Followed by centered horizontal line.

1. These notes, later entered into a notebook by Charles Batchelor, continue Doc. 605. They are continued in Doc. 609. Batchelor's version (Cat. 1172:268–73, Lab. [*TAEM* 3:161–63]) is signed by Edison, Batchelor, James Adams, and Samuel Edison.

2. The following day Edison, Batchelor, and Adams conducted twelve more experiments using various electromotograph solutions. Cat. 1172:273–74, Lab. (*TAEM* 3:163–64).

3. This illustration of the hand electromotograph appeared in *Scientific American* (39 [1878]: 17). The instrument itself can be found at the Edison Institute, Dearborn, Mich. (Acc. 29.1980.20).

The hand electromotograph.

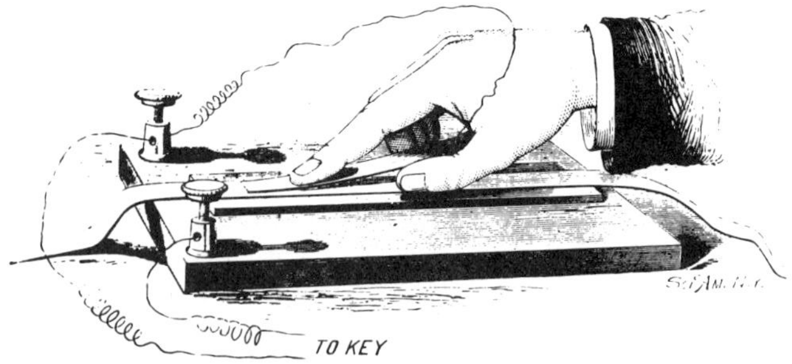

4. Figure labels are "R[esistance]" and "20 ohms." A clearer version of the drawing appears in Batchelor's notebook entry.

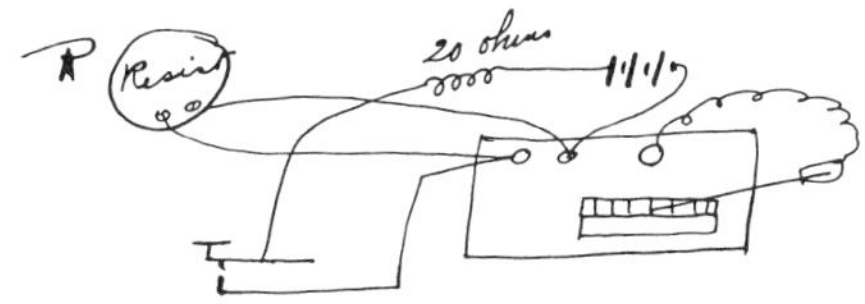

Charles Batchelor's schematic drawing of the hand electromotograph.

5. These figures are graphs in which the ordinate represents intensity of friction and the abscissa represents time. In the Batchelor notebook copy, it is clear that the thicker rectangle represents "normal"; the other rectangle, "increased"; and the triangle, "opening."

6. For Boston paper see Doc. 354 n. 8. This was apparently cotton rag paper (Batchelor wrote "common bibulous"). Following the Civil War wood paper began to replace cotton rag paper for most uses. McGaw 1987, 191–205.

7. Batchelor's copy has "thin"; probably meant "thick."

8. Batchelor's copy adds "but ruffles up after wet overnight with Nacl & KO with pressure receiving paper dont."

9. Batchelor's copy has "silver."

10. May be "thrown," but Batchelor wrote "thump."

11. This last statement may have been added after the experiments were completed. Batchelor's copy does not include it, but instead notes, "might be built up for a platina solution."

–608–

Notebook Entry: Autographic Printing

[Newark,] Aug 17 1875

Copying Experiments

Aug 10th–17th[a] Inc: Had press made like this[b]

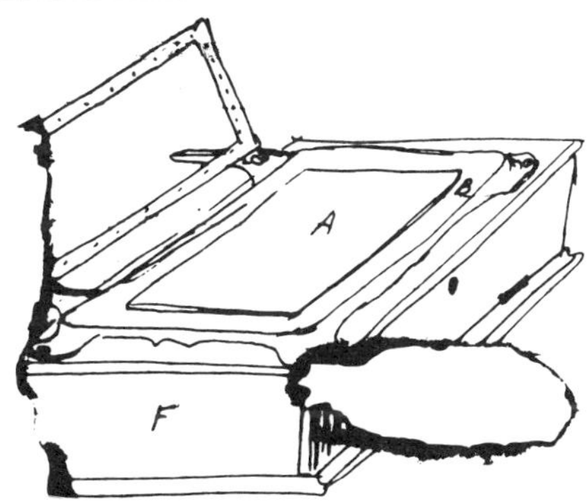

F is a cabinet B is an Iron frame cast C is a frame with needles in to put the copy down hinged at E & D is the support for C when you are putting in the letter.

The difficulty we found was that the pins would stick through the copies & pick them up so we wanted something [t?]hat[c] would hold the letter & not interfere with the copies. We altered the frame C ~~the~~ & put 4 clips in the corners like this

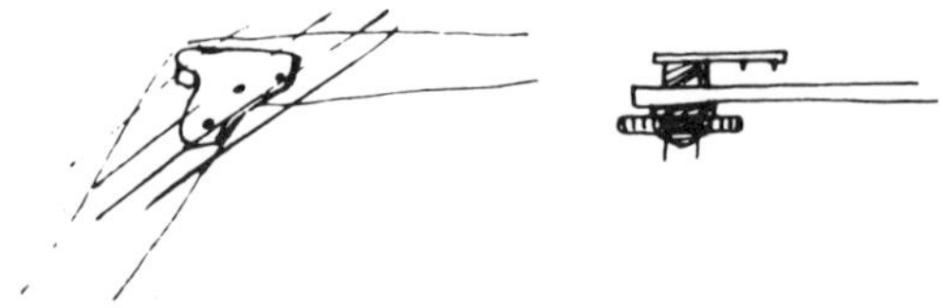

but we wasted no time on this as it was a palpable failure we then had it altered to just 3 little points in the corners but this would not do then Batchelor put springs on like this;[b]

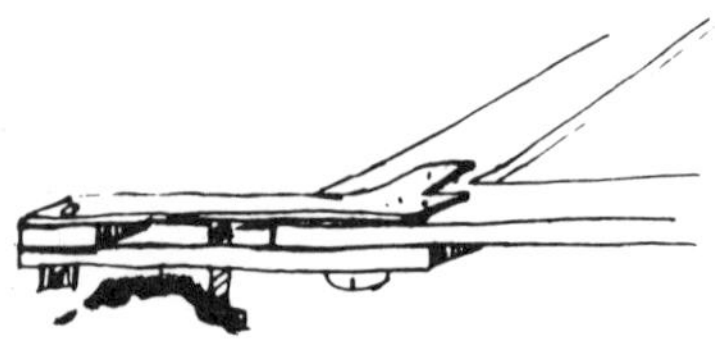

flat spring with points. it worked first rate when you put the paper in you raise the springs wi~~h~~th the adjusting screws [pu?]t[c] the paper under & press [ ][c] may be im[ ][c] the spring [ ]ple,[c] [ ][c] not do to [ ][c] [cov?]ered[c] with cloth [ ][c] it must be softer [ ]uld[c] be put on with a rag [ ][c] then rolled lightly with a [ ][c] [ma?]de[c] of rubber like a wringer [ ][c] [cov?]ered[c] with cloth

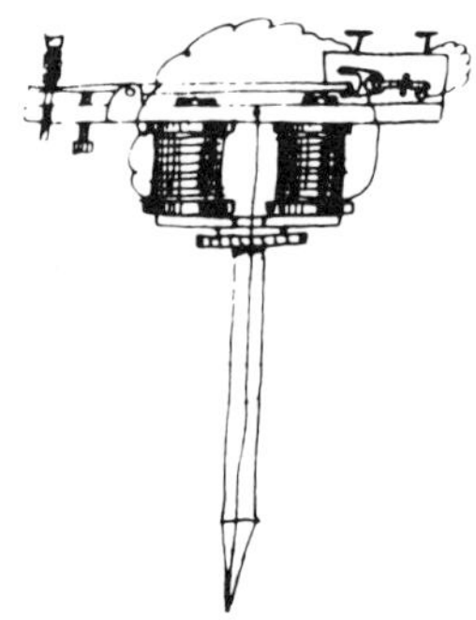

Ordered Gilliland & Co[1] to make a pen like this with coarse wire.[2d]

Altered Engine ~~to~~pen to double armature & ponts but no good.[d]

Tried taking copies from pa[d?][c] Wet pad with ink & lay lette[r?][c] on it now lay a sheet on & roll it with a plain dry [roller?][c] but not so handy as the other

We find we can take Photographs with this process We took a photograph & pricked i[t?][c] all over but not so deep as to go through bu[t?][c] to raise the surface we then smear with in[k?][c] & rub off again there will be enough ink in the holes to make a splendid copy by pressure[d]

Taken from Photograph   Sunday Aug 12th 1875[3b]

Tried soaking paper in resin dissolved in Alcohol b~~o~~ut no good for writing on[d]

We find the foolscap[4] we have here is not so good as the large letter because it is thinner & more liable to blot the copi[es?] a[s?] ~~it~~[5c]

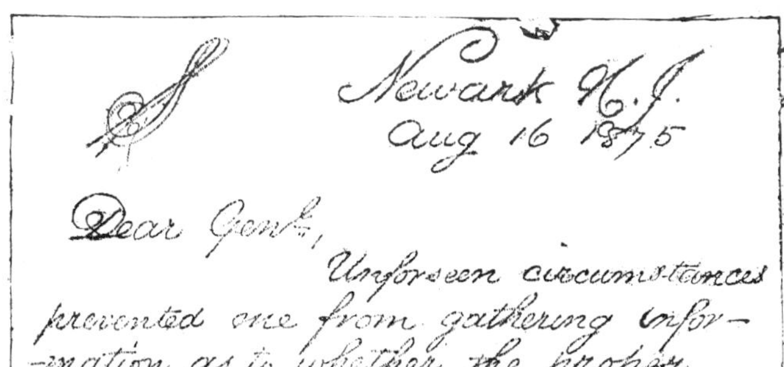

S

Newark N.J.
Aug 16 1875

Dear Gent.,

Unforseen circumstances prevented one from gathering information as to whether the proper inducements for the so-called cosmoghetaphicalighthornicosmopolitanischkum were really of the nature required by the parties interested in the propogation of a species of Bacterianalianchthintious under the efficient management of the members of the Society for providing the gospel to the guano islands. Thusting to hear from you on the subject soon We remain

very respectfully.
Simeon Shivers
Secty.

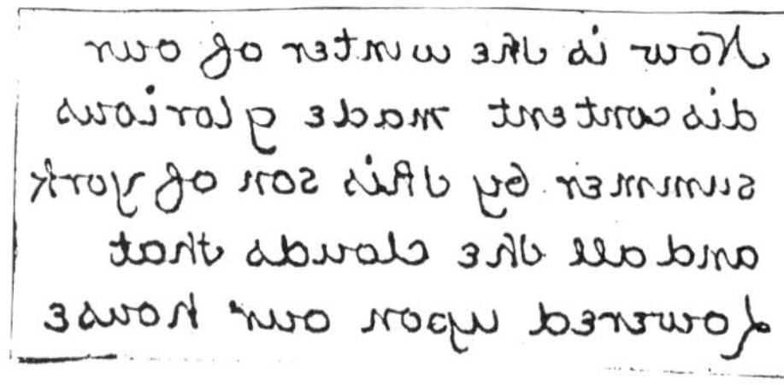

Thos. A. Edison	Chas Batchelor
James Adams	Samuel Edison

X, NjWOE, Lab., Cat. 1172:275 (*TAEM* 3:164; Supp. III). Written by Batchelor. [a]Underlined twice. [b]Drawing damaged. [c]Document damaged; letters and/or words missing. [d]Followed by centered horizontal line.

1. Ezra Gilliland formed Gilliland & Co. after leaving his brother and their partnership in Cincinnati (see Doc. 543; and R. G. Dun credit report for J. F. Gilliland & Co., Ohio 87:350, RGD). It appears that Edison, who later provided both machinery and money to Gilliland & Co., was a silent partner in the firm. Account records beginning on 1 April 1875 indicate that Gilliland & Co. was selling Edison and Murray products at 41 Dey St. in New York, the same address used later in the year by Edison's Electric Pen and Duplicating Co. (Cat. 1223:23, 171, 175, 200, 218, Accts. [*TAEM* 22:142, 192, 194, 206, 214]; see also Cat. 1220:104, 120, 123, 396, Accts. [*TAEM* 22:50, 58, 59, 91]). On 22 July 1875 Gilliland gave Edison a mortgage for machinery at 10 and 12 Ward St. (the Edison and Murray shop) as well as for more than $5,500 in notes due Edison (Miller [*TAEM* 28:983–86]).

2. In another notebook, Batchelor later made a similar, retrospective drawing (showing the electromagnets parallel with the shaft rather than perpendicular to it as in the standard design) and labeled it "Batchelor's Pen Aug 13 1875 Autographic Press." Cat. 1307:3, Lab. (*TAEM* 90:603); cf. Doc. 595, fig. 2.

3. Sunday was actually 15 August.

4. A paper size, typically 16″ × 13″.

5. The reversed writing is the back of a stencil.

–609–

Notebook Entry: Automatic Telegraphy and Electromotograph

[Newark,] Aug 18th. 1875

Extra Notes on Electromotograph.[1]

1 With the Caustic Potash & NaCl paper we can on plain line without shunt get 1200 a minute recorded on second circuit thus:—

2 With high resistance shunt: — ---- -- —- —-—

3 With high resistance[a] at slower speed: — ---- -- —- —-—

4 lower resistance in shunt — .... -- —. —.—

5 with less. - ••• • •• —• —•—

6 with 200 in shunt - •••• •• •• ••—

7 with less nothing dashes being ~~nothing~~ dots[b] & dots nothing. This shows that the Emg throws back a secondary charge, precisely like a condenser what appears to be so strange is that so small a surface as a platina point should send so powerful a discharge almost equal to the current which passes over 1500 ohms from 25 cells of battery. The problem thus resolves itself into some paper in which there shall be a chemical that will not polarize & give back charge & yet be sufficiently delicate to allow the lever to move at high speeds.

8 Last night tried the experiment of breaking the circuit very rapidly with a revolving breakwheel with half closure & half opening but it did not give very good results in fact not any. 9[c] then tried 9⁄10 closure & 1⁄10 opening no better.

10[c] revolved slow & fast no better[d]

11 Tried a shunt & connected a magnet of 200 ohms (Tillotson pat)[2] in thus:—

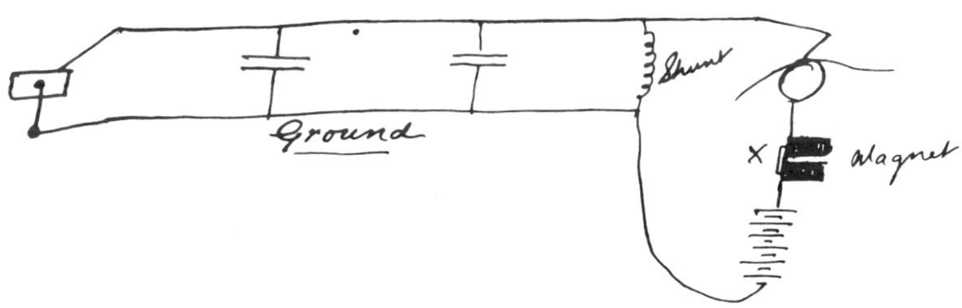

The object of inserting the magnet in the circuit between the battery & the instrument was to weaken the first part of every signal sent leaving the last part full strength & cut off sharp for instance on closing the circuit passing through the magnet meets the self induction current from it & is weakened but quickly gets full strentgth as induction current dies away, on opening the extra current from the magnet has no place to circulate it being broken entirely by the perforated paper cannot go on the line & prolong the mark the record should be thus:— —— ———— —— ——

As has been previously mentioned the Emg record where there is no static is thus with a straight current:—

—— ———— —— ——

12 I thought that the above would compensate & make it even, for some reason could do nothing probably because we had to use a shunt & the secondary charge from the Emg threw it out of adjustment as above described.

I find that it wont do to have too light a pressure on the lever of Emg. The delicacy is apparently increased by pressure

T. A. Edison Chas Batchelor

Jas Adams. Samuel Edison

X, NjWOE, Lab., Cat. 1172:278 (*TAEM* Supp. III; 3:165). Written by Batchelor. [a]"With high resistance" dittos in original. [b]Interlined above. [c]Written in left margin. [d]Sections 1–10 canceled with large "X".

1. See Doc. 607.
2. Unidentified.

-610-

Charles Batchelor to the Editor of the Telegraphic Journal

Newark, N. J. USA. Aug 20th 1875[1]

Dear sir

In No 59 of your Journal July 15, 1875 I find a system of Quadruplex Telegraphy described by Mr. Kempe which is a modification of that invented by Stark of Vienna.[2] Now I have tried every one of the so-called devices for the transmission of two messages in the same direction at the same time described by Düb Schellen, Blavier[3] & others including also Mr. Kemp's & will say that not one of these will allow of the transmission of a single word, & a very small amount of experimenting will make this evident

It seems to me to be a real shame that such theoretical devices are put in Standard telegraph works to be brought out against future inventors as an evidence of priority Respectfully Yours

Chas. Batchelor

ALS (letterpress copy), NjWOE, Lbk. 1:14 (*TAEM* 28:10).

1. This letter was published in the 1 November 1875 issue of the *Telegraphic Journal* (3:250).

2. H. R. Kempe published "On a System of Quadruplex Telegraphy" in the 15 July issue of the *Telegraphic Journal* (3:162–63). This followed by two weeks the publication of Francis Jones's paper on Edison's quadruplex system, which had been presented to the American Electrical Society in February. A series of other articles on systems of quadruplex telegraphy appeared in the *Telegraphic Journal* throughout the second half of the year.

3. See Doc. 293 n. 5.

-611-

Notebook Entry: Multiple Telegraphy

[Newark,] August 20 1875

SEXTRUPLEX

Edisons Sextruplex Telegraph[1]

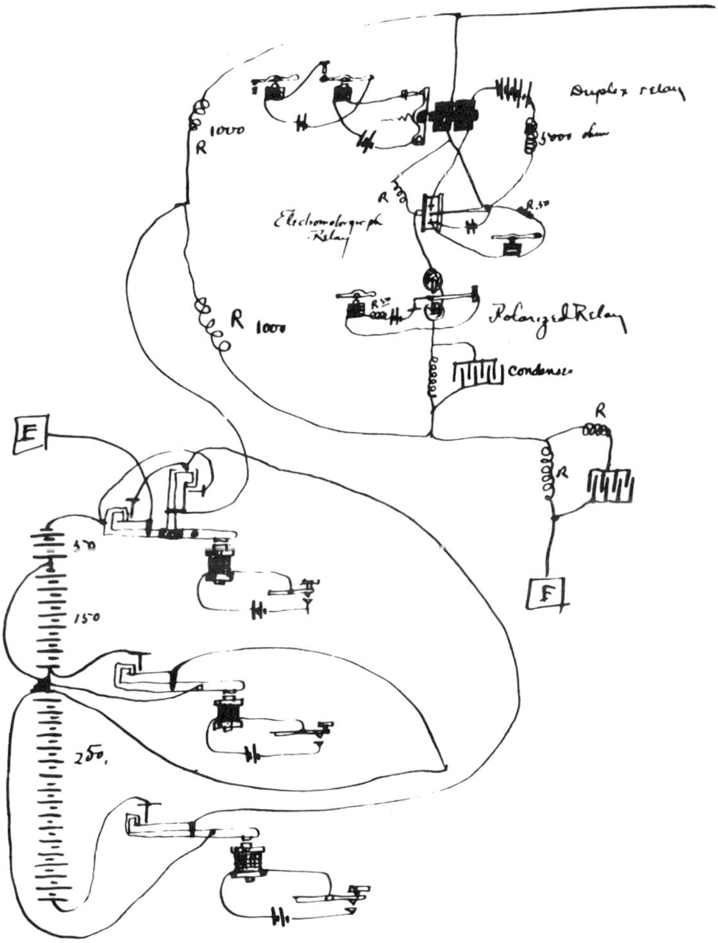

By this apparatus three messages may be transmitted in one direction and three in the opposite direction all at the same time over one wire. I send positive and negative currents from 50 cells to effect the movement of the lever of a polarized relay at the ~~dif~~ distant station and transmit one message and increase the strength of the current by the addition of 150 cells extra to effect the lever of a Duplex Common relay ie[a] a Common relay with two coils of wire wound on its Cores, to transmit a second message. I then add 250 cells more over the 150 to move the lever of my Electromotograph relay to transmit another message

The addition of the 250 cells is prevented from working the duplex relay by having the lever of the Emg close a local circuit in which is n battery whose current circulating in the extra coil on the duplex relay balances the current from the 250 cells.[2] The lever of the EMG moves very much quicker than ~~the Co~~ a magnet no jump is noticed=

Written & drawn August 20th 1875

In the above drawing I have shewn the Emg with a single lever for currents of one polarity. it should be thus:—

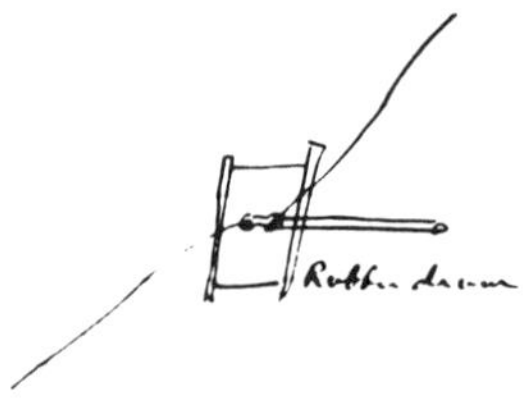

so that it responds equally to reversed currents Aug 20 1875=

X, NjWOE, Lab., Cat. 1172:134 (*TAEM* 3:117). [a]Circled.

1. See Doc. 583.

2. In Doc. 449 a polarized relay serves this function. The top of the electromotograph drum here is rotating out of the paper.

–612–

From Jesse Bunnell

[New York,] 8/23–75

To[a] Tom A. E.

It has occurred to me that a page or two or more of practical hash (as I would call it) from your pen would be very good thing to put in here— What I mean is twenty or thirty short items containing suggestions, practical information, practical science etc, each in a few words but pithy. I have no interest in the book beyond a disposition to help make it a good thing and this is, I think the best way I can do so—[1] Yours

J. H. Bunnell

ALS, NjWOE, DF (*TAEM* 13:197). Letterhead of L. G. Tillotson & Co. [a]Preprinted.

1. Bunnell was probably referring to the catalog of the telegraph manufacturing firm L. G. Tillotson and Co., which he had joined the previous spring (*Operator,* 1 Apr. 1875, 6). Later in the year the catalog appeared bound with a revised and enlarged edition of John E. Smith's *Manual of Telegraphy* (*Operator,* 15 Oct. 1875, 1).

–613–

Notebook Entry: Chemical Experiments and Electromotograph

[Newark,] Aug ~~23~~4[23][1] 1875

1 When certain chemicals are placed in a solution of Silicate of Soda water[2] there shoots up from the crystals beautiful forests or trees in some instances reaching to the top.

2 Chloride of Urinum gives very thick tru~~k~~nk like trees while nitrate of cobalt gives thin long [-]grass.

3 Sulphate Manganese shews growing from the crystal very fine thick white grass The Chl Urinum is like cactus & grows larger

We find that after tree is formed, if destroyed it dont grow again its the air bubbles carries the chemicals up & Silicate of Soda being thick goes slow & every part of the chemical that is carried up is precipitated forming hollow trunks or grasses

Electromotograph

We have found that when pyro paper is used on any of the metals platina included that by drawing the platina stylus connected to the zinc of the battery over the paper a great numbers of times that there is deposited an insulating substance on the metals underneath & in a very short time no mark can be obtained except the paper is very wet or the paper only damp but with great pressure Tin acts worse than platina

Thos. A. Edison — Batchelor
E. H. Johnson — Samuel Edison
Jas. Adams

X, NjWOE, Lab., Cat. 1172:284 (*TAEM* 3:168). Written by Batchelor.

1. This appears between two entries dated 23 August.

2. In the previous entry in this same notebook Batchelor noted that they had adopted ten parts water, two parts silicate of soda, and one part caustic potash as a standard solution for the electromotograph repeating experiments (Cat. 1172:283, Lab. [*TAEM* 3:167]). This solution allowed only 50 words per minute, and they continued to seek better solutions (see Cat. 1172:283–88, Lab. [*TAEM* 3:167–70]). This solution might have been suggested by a battery experiment the previous day in which silicate of soda and caustic potash were used as a "mucillaginous" solution in the porous cell of a battery to prevent polarization (Cat. 1172:280, Lab. [*TAEM* 3:166]).

–614–

Notebook Entry: District Telegraphy

[Newark,] Aug 23rd 1875

New stop movement for Domestic Recorder[1] Had the stop altered whereas before it was stopped by the lever being brought sideways to stop the vibrating pin as in fig 1

[Fig 1]

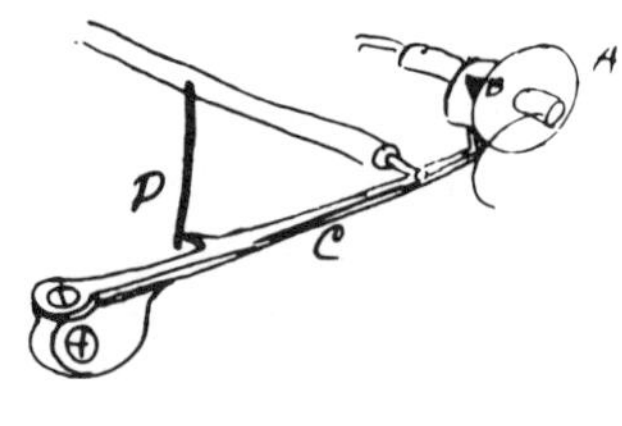

A is a shaft of recorder carrying a collar on which is a small cam of steel B which when it comes round to pin on end of lever C pushes it over so as to engage D the vibrating pin, however sharp the point is made on lever it will sometimes strike right on point & instead of going on one side of pin it strikes & bends it at such times when the circuit is opened it fails to release & consequently the recorder does not start soon enough in order to ob~~b~~viate this we have arranged it in this manner

Fig 2

When magnet is closed the screw D has thrown lever C out of v & consequently away from lever B which is kept away by a turning round & V rubbing on periphery By this arrangement that strain is done away with which occurs when the two strike on their extreme edges, in this case if they do it does not matter as the screw D ~~n~~knocks it out of the way entirely & does not depend on a spring

Thos. A. Edison	Chas. Batchelor
E. H. Johnson	Samuel Edison
	Jas. Adams

X, NjWOE, Lab., Cat. 1172:285 (*TAEM* 3:168). Written by Batchelor.

1. See Doc. 615.

DOMESTIC TELEGRAPH RECEIVER Doc. 615

Little is known about the development of Edison's domestic telegraph receivers. A few dated sketches show development work from 1874, and one undated sketch (probably from 1874) shows an instrument similar to one in an old photograph album.[1] The following domestic telegraph receiver embodies the design change described in a notebook entry of 23 August 1875 (Doc. 614).[2] The ornate decoration on the lid of

The commercial receiver of the Domestic Telegraph Co.

the paper holder indicates that this production model was probably used as the receiver for the private-line transmitter Edison and Charles Batchelor designed as part of their domestic telegraph system (Docs. 545 and 563). The receiver in the photograph album is a slightly different and much plainer version that was probably used in the district offices of the Domestic Telegraph Company.[3] A nearly identical receiver appears in Edison's fire alarm patent model.[4]

The domestic receiver recorded messages electrochemically as did those used in Edison's automatic telegraph system. In this instrument the current from the line first passed through an electromagnet that released a stop and allowed weight-driven clockwork to turn the receiving drum. This drew chemically treated paper under a stylus,[5] which marked the paper through electrolysis. In the private-line system the transmitter was a 26-key instrument that allowed an operator to transmit the Morse code for a letter by merely depressing the appropriate letter-key.[6]

1. Cat. 30,094 and Cat. 30,095, both Lab. (*TAEM* 5:319, 331; 6:405, 418, 420, 489); Cat. 1307:1, 3, 6, 15, Batchelor (*TAEM* 90:600, 602, 605, 616).

2. This may have been written on 24 August; see Doc. 613 n. 1.

3. Cat. 558.

4. U.S. Pat. 186,548; see illustration, p. 626.

5. According to Doc. 563, the stylus was tellurium.

6. See Docs. 545 and 563. In the district system a signal box automatically sent the appropriate coded signal for messenger, police, or fire.

–615–

[Newark, August 23, 1875?]

Production Model: District Telegraphy[1]

M (21 cm × 26 cm × 22 cm), NjWOE, Cat. 6895.

1. See headnote above.

–616–

R. G. Dun & Co. Credit Report

Newark Aug 25.75

Thos A Edison[a] Mfr Telegraphic Instruments

3700. J.[1] Formerly of E. & Murray, (4.206), but he recently sold his int[erest] in that conc[ern] to "M" taking Chattel mtges on stock &c am'ting to $12,589 for his int.[2] Is a reputable man, & a genius in this line. Has several valuable patents ~~fo~~rom wh he gets good royalty. His available assets are not known, but it is believed he will fulfill any engagement he makes. 4922.

D (abstract), MH-BA, RGD, N.J. 22:283. [a]Followed by "prev. (E. & Murray 4.206)"; "4.206" represents the volume and page numbers for related credit reports.

1. See Doc. 352 n. 1.

2. The Edison and Murray inventories of 21 May 1875 show a figure of $12,524.47: $11,618.24 total for the machinery and $906.23 for one-half interest in the stock. Cat. 1223:213, Accts. (*TAEM* 22:211).

–617–

Notebook Entry: Gas Pipe Threading

[Newark,] Aug 26 1875

Gaspipe threading

A few methods of putting threads on gas pipes & inside.

1st Heat the end & put on by pressure in the following manner

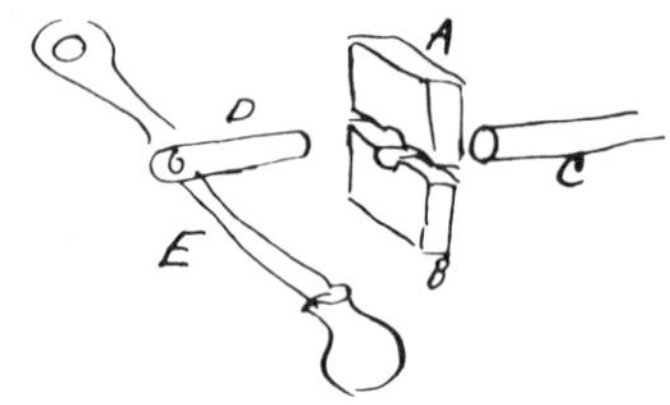

The two half dies A & B are worked by pressure push in pipe C under the dies A & B then pull ram D in the tube & put on thread by pressure on A & B

2 In drawing leave it bell mouthed & press it down

3 Find some way of eating it out with acid.

4 Make the end smaller than tube & drive plug in & force up into a die.

5 Use an iron disc running rapidly & make pipe feed the pitch of thread & also turn round

Thos A Edison — Chas Batchelor
E H Johnson — Jas. Adams.

X, NjWOE, Lab., Cat. 1172:290 (*TAEM* 3:171). Written by Batchelor.

–618–

To Jay Gould

[Newark, August 1875?[1]]

Mr. Gould

The automatic people are bothering me to death taking up my time from day to day. They want me to sign papers, etc. which I do not want to sign[2]

If it is not inconvenient I wish you would get me off the ragged edge, and settle amount due on "Auto" System so I can go ahead and carry out the programme marked out.

I hope you can fix "Auto" people up without the necessity of an increase in the Capital Stock of the A.&P. Co. as the extra amount would detract from the success of your plan, and put a damper on some of my calculations.

I think that McManus Dallett Parsons & others will take the same class of securities you gave Mr "H" as they are men of wealth.[3] I infer this from conversations heard But it would nearly ruin Reiff who I think will come out short any way you settle. He has borrowed money at ruinous rates for last 4 years to keep the "Auto" afloat and he deserves to have full amount.

I believe furthermore that he is thoroughly honest man and you can make good use of him in carrying out your plan.

The Little Craig National people, are not legitimately entitled to a dollar except $30,000 for National Line to Washington, which both me and the General[4] think we do not want. Little nor the National Company can not carry out any part of their contract, and the only object in Mr "H" bringing them in was to have peace and obtain a settlement. If you hold them to their Contract you need never pay a cent.[5] I am going to have Mr. Little's devices tested on the line for the General to show their utter worthlessness; I am arranging now with Reiff and the Domestic Co to endeavor to carry out the Stock Reporting Telgh[6] & Messenger Call plan without assistance from A.&P. this will increase A&P biz and worry the enemy.[7]

EDISON

Have written 2 column article on A&P Telegraphy with two cuts and I think it will be out in day or two.[8] have a 4 column article with illustrations nearly ready for Tribune, which ought to be put in morning after reduction of rates.[9]

PD, *Harrington v. A&P,* Box 17B, 21:344, Respondents' Exhibit 8.

1. This letter was likely written sometime after Edison's memorandum to Gould regarding the settlement of the automatic (Doc. 606). The papers the automatic interests wanted Edison to sign were probably those related to the organization of the American Automatic Telegraph Co., which they were attempting to form in August (see draft articles of association and bylaws, dated 31 Aug. 1875, DF [*TAEM* 13:489]). Edison did not agree to sign these papers until December (see Doc. 676).

2. See n. 1.

3. See Chapter 8 introduction, n. 3.

4. Thomas Eckert.

5. For the contract between the Automatic Telegraph Co. and George Little, Daniel Craig, and the National Telegraph Co., see Doc. 141.

6. This may be Charles Batchelor's chemical stock-reporting telegraph; Jay Gould was supporting its development. See Doc. 567 n. 2.

7. Connections with a company providing district telegraph messenger services provided a long-distance telegraph company with not only a messenger corps but also a monopoly over all intercity messages directed through the messenger service (Israel 1989, 177–79). Thus Atlantic and Pacific established connections with American District Telegraph of Philadelphia in December 1875 and early in 1876 it took over the Domestic Telegraph Co. (Atlantic and Pacific Telegraph Co. Stockholder and Trustees Minutebook [1875–88]: 73–74, Binder 796; Philadelphia Telegraph Co. [American District Telegraph]: 199, Box 5963; both NjWU). Domestic Telegraph already had close connections with Atlantic and Pacific through Thomas Eckert, who was president of both companies (see also Doc. 582).

8. Not found.

9. Not found. Edison probably refers to a proposed rate reduction. Atlantic and Pacific originally reduced its rates in February 1875. Rates returned to their original level during the summer, but were reduced again sometime before the end of May 1876. President's report to Atlantic and Pacific Telegraph Company Stockholders, 28 Apr. 1875 and 26 Apr. 1876, 2:183, 198, Box 17B, *Harrington v. A&P.*

–619–

Charles Batchelor to Thomas Batchelor

Newark NJ. Sep 1st 1875

Dear Tom.

Yours recd & glad to get a line from you[1] In regard to the press I think you could do something that would pay you well & as soon as we can get it established here I will write you & let you know what we propose to do. We shall try to form a company on it here, but in the meantime manufacture & sell what we can we shall exhibit it at our Newark Industrial Institute[2] & the New York state fair this fall. I am busy working up my new Stock quotation telegraph printer[3] I think we have secured a man who will furnish the necessary funds to do all we need.[4] We work all night experimenting & sleep till noon in the day. We have got 54 different things on the carpet & some we have been on for 4 or 5 years. Edison is an indefatigable worker & there is no kind of a failure however disastrous affects him. He stands today the foremost inventor & electrician in this country by far We have applied for 45 × 30 feet space in the Philadelphia Centennial Exhibition next year to exhibit some original inventions. We shall have a great many things in & if we get a medal it will be for something original.

I am sorry to hear of Wilfred Dearman[5] It seems to me that he must have very little grit! to be still working as a common mechanic. I left that business nearly 5 years ago with the intention of never entering it again except for my own pleasure & benefit. He ought to know by this time that manual labour commands the smallest smallest profit, & he that can fall back on it is the best man to branch out & see what he can do in other directions Tell Jim[6] I'm exceedingly glad to hear of his improvement & wish him every success. Our chuckey[7] is progressing well but the heat bothers her a little. Rosa[8] is well & sends her best love to all. Our kind love to Maria & children & believe me Your affectionate brother

Chas Batchelor

ALS (letterpress copy), NjWOE, Batchelor, Cat. 1238:16 (*TAEM* 93:30).

1. Not found.

2. This organization was founded in 1872 shortly after the first Newark Industrial Exhibit, in order to promote the industries of Newark through annual exhibitions held at a site acquired by the Institute in Newark. Ford 1874, 252–54.

3. Doc. 567 n. 2.

4. Jay Gould.

5. Unidentified.

6. Batchelor's brother, James Drew Batchelor.

7. Batchelor appears to have had a daughter named Emma who died, as another daughter, born in 1881, also was named Emma. Batchelor to James Batchelor, 27 Feb. 1877 (*TAEM* 93:92); passports; all Batchelor, NjWOE.

8. Batchelor had married Rosanna on 9 May 1871. Welch 1972, 2.

–620–

Notebook Entry: Autographic Printing

[Newark,] Sep 2nd 1875

Reduplication of Copies

Made new roller for press of washers of cloth punched out like Fig 1 & put on an arbre[1] in this manner Fig. 2.

Fig 1.

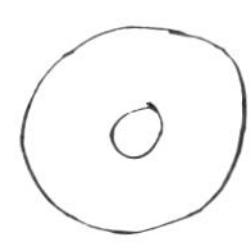

Fig 2.

The collars were made of Cast iron & pinned on The iron core left rough in middle & turned down in screw mach at each end the handle frame[a] was bent like Fig 3 & the handle made like Fig 4 & riveted into the frame & the handle wire made in screw machine like 5

Fig. 3

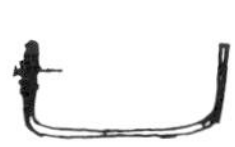

4

5

This was adopted as a model & work from. After all the cloth is on it must be shaved level.[2]

Made new engine with No 22 wire spools & it works on one cup but we shall have two for an excess

Our new vibrator pen made by Gilliland is very good but it has kincks which deter it from going out to any body but a man that understands a little [-]about electricity You could write much faster with it & it makes better copy but sometimes it will run for hours & then stop, as soon however you touch it or perhaps shake it runs off again.

Also improved the ink by adding molasses.

T A Edison Chas Batchelor.

Jas. Adams

X, NjWOE, Lab., Cat. 1182:40 (*TAEM* 3:70). Written by Batchelor. [a]Interlined above.

1. Arbor.

2. Cf. discussion of an earlier design of a perforator platen in Doc. 186, fig. 34.

–621–

Notebook Entry: Electromotograph

[Newark,] Sep 2nd 1875

Electromotograph.

Made a new style of Emg. We found that Acetate of Mercury & Caustic K were best for Emg. & we had an idea that if we could cover the drum with leather & keep it moist it would work just as well so we designed a machine like sample Fig 1

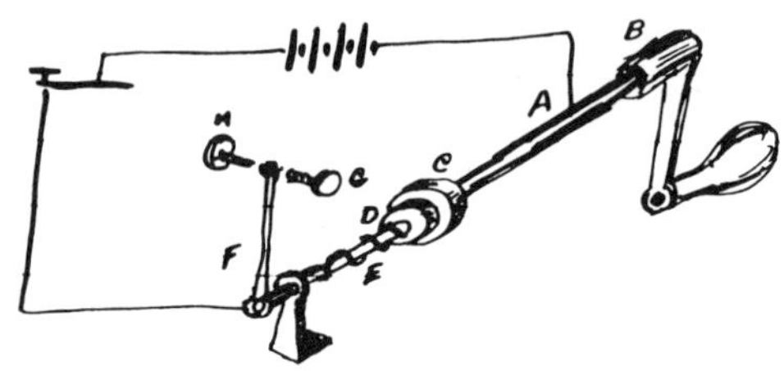

A is a spindle carried by handle B, on the end of which is a hard leather faced disk C, against which the plate D of Platina rubs. The handle being kept turned the friction of plate C against D keeps the point fF against one of the screws,[1] but when the current is put through & the friction discs kept moist with Acetate of Mercury & Caustic K. the lever will respond to every stroke of the Key. This worked very well the first time for a wonder.[a]

When freshly wet the mercury solutions work well on Emg repeating it as good as it comes but after laying by a while we find they do not work satisfactory

A small amount of either acetate of Mercury or of Caustic Potash does not work well

Thos. A. Edison

Chas Batchelor
James Adams

X, NjWOE, Lab., Cat. 1182:41 (*TAEM* 3:70). Written by Batchelor. [a]Followed by centered horizontal line.

1. E is a spring pressing D against C.

–622–

To Gilliland & Co.

Newark NJ Sept 3d 1875

Gents

You will please manufacture for me one hundred of the Autograph Presses one hundred Ink Rollers for same one hundred brushes one hundred Ink dishes one hundred writing boards at the following prices

Presses	three dollars fifty cents	($3.50)	each
Rollers	one dollar	1.00	"
Brushes	Thirty cents	30¢	"
Ink Dishes		10¢	
Writing boards		25¢	

Work to be Commenced at once, & rushed out Presses to be Jappaned nicely,[a] named & numbered, & ornamented,[b] Pine board used in presses otherwise same as sample Yours Truly

Thos. A Edison

L (letterpress copy), NjWOE, Lbk. 1:13 (*TAEM* 28:15). Written by Charles Batchelor. [a]"nicely," interlined above by Edison. [b]"& ornamented" interlined above by Edison.

–623–

To John Ott

[Newark,] Sep 6th 1875

John Ott,[1]

You will proceed to make (10) Ten pens same as model by hand to be delivered to the Laboratory by September 17th 1875 for which we will pay (4) Four dollars each.

You will then proceed to make the necessary tools for manufacturing them for which we will pay (20) Twenty dollars the said tools to belong to us.

You will proceed to manufacture from these tools (100) one hundred pens, equal to models made by hand & deliver to us as follows

(10) Ten on the 29th of September
(30) Thirty on the 6th of October
(60) Sixty on the 13th of October

We will pay for this hundred the sum of (1) one dollar each.[2] ⟨The first ten only taken⟩[a]

T. A. Edison per Batchelor

L (letterpress copy), NjWOE, Lbk. 1:18 (*TAEM* 28:20). [a]Marginalia written by Batchelor.

1. John Ott (1850–1931) was born in Jersey City, N.J., and attended the Cooper Union Institute, a technical school in New York City. He worked at the Pike Optical Co. before being employed by Edison and his partner William Unger in 1870. Ott, who was Unger's cousin, continued his association with Edison until his retirement in 1920. For most of his career he served as Edison's chief experimental-instrument maker. "Ott, John," Pioneers Bio.

2. This order for electric pens represents a common nineteenth-century manufacturing practice known as inside contracting. A shop foreman or other trusted machinist would receive a contract to manufacture a particular product. He would be responsible for making all special tools and hiring the necessary labor. The shop would provide the machinery, although in this case Ott might have used machinery and possibly labor in Murray's shop rather than in Edison's laboratory. On inside contracting see Englander 1987 and Buttrick 1952.

–624–

To Stephen Field

[Newark,] Sep 13 [187]5

Friend Field,

I enclose you a circular printed by my new autographic press, which will give you an idea of what it will accomplish.[1] The apparatus consists of an electrical pen, press, inking roller, Bottle ink, Cell Battery whole thing got up satisfactorily & neatly at the low price of $30. We have sold the general agency for the New England States, New York, Penna, Washn, Md, Ohio & Illinois, our agent in New York[2] is selling 8 per day. They are much liked by Com merchants, Lawyers, Ins Cos, etc & no difficulty at all is experienced in selling them. If you would care to take hold of the Pacific Coast or can recommend a reliable energetic person who would I should be pleased to affect an arrangement with you. We supply them to agents at $20 each will send you single sample at that price if you wish to go into it. I am exhibiting at nearly all the industrial fairs where it attracts great attention. Let me hear from you. This offer is open 30 days from date[3] Yours

Thos A Edison per Batchelor

P.S There is more money in this than telegraphy. It is to the Country house[4] what sewing machines are to the home circle.

L (letterpress copy), NjWOE, Lbk. 1:23 (*TAEM* 28:25).

1. The accompanying circular, labeled as one of the "first samples" taken with the pen, may be the one sent by Batchelor. 75-004, DF (*TAEM* 13:299).

2. P. Mullarkey.

3. Field's Electrical Construction and Manufacturing Co. took an agency for the West Coast (see Doc. 639).

4. Edison probably meant "Counting house."

An early electric-pen sample.

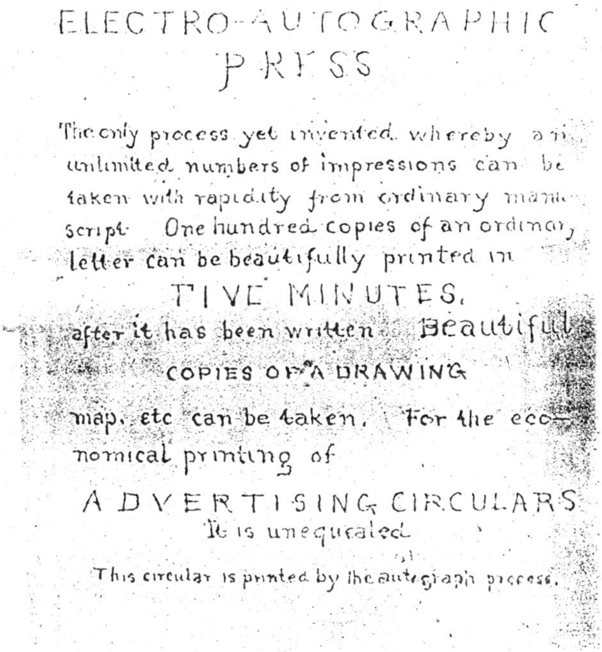

ELECTRO-AUTOGRAPHIC
PRESS

The only process yet invented whereby an unlimited numbers of impressions can be taken with rapidity from ordinary manuscript. One hundred copies of an ordinary letter can be beautifully printed in

FIVE MINUTES,

after it has been written. Beautiful

COPIES OF A DRAWING

map, etc can be taken. For the economical printing of

ADVERTISING CIRCULARS

It is unequaled

This circular is printed by the autograph process.

–625–

Charles Batchelor to Thomas Batchelor

[Newark,] Sep 13th, 1875.

Dear Tom

I write to give you a little better description of the press. We have sold agencies for 10 of the states & are hunting up reliable agents for the others.[1] Our press consists of an electrical writing pen, an inking roller, a press, a battery, ink & brush.

Our pen is a small electrical engine with a holder about the size of a good thick pencil down which runs a needle which is made to perforate the paper as you write like sample: there is no difficulty in writing with it. the engine is run by a cell of battery made in a nice shaped glass bottle & is an ornament on a man's desk.

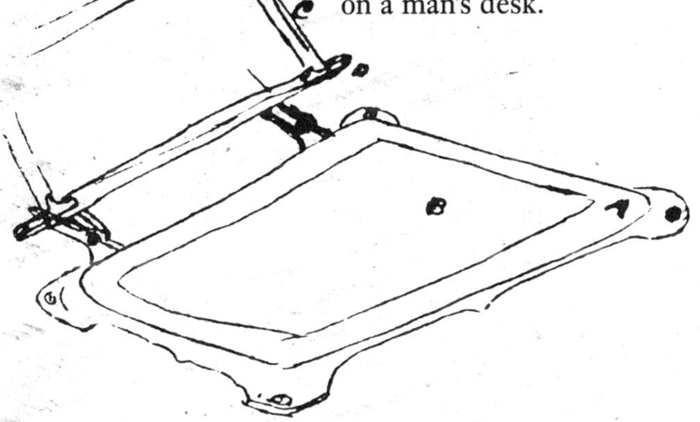

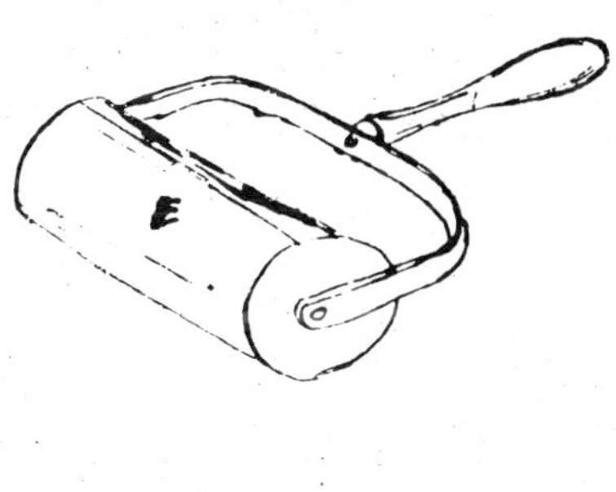

The press consists of an iron frame A with wood plate B let into it & hinged onto the frame is a light frame C with springs on the corners to hold the paper that is perforated. A sheet of paper is placed on B & the frame C let down onto it. The roller E made of cloth & filled with ink is now passed over it & the ink passing through the perforation leaves a fac-simile of[a] the copy written This can be repeated an indefinite number of times.

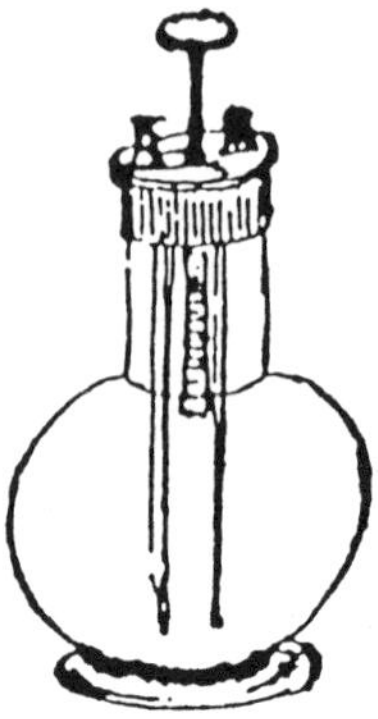

The battery I have tried to draw it stands about 6 inches high & now I will try to draw the pen on a separate sheet & the manner of holding it

The railroad companies here seem to have a great deal of reduplication & I think we shall supply them all Also, merchants, lawyers & Insurance Cos. want it bad. I've drawn the engine a great deal too large it will give you the idea that it is

cumbersome but the whole thing weighs less than 11 oz you know I'm not much as a draughtsman, & you can get an idea from it. We sell it here for $30 giving our agents 33⅓ per cent that is about the price of a copying press here.

We have not come to any arrangement yet about foreign ~~comp~~ countries We have thought that our best way would be to sell it right out as there is considerable trouble in working a foreign country, however we are in no hurry yet. Love to all.

Charley.

ALS (letterpress copy), NjWOE, Batchelor, Cat. 1238:18 (*TAEM* 93:32). [a]Repeated at end of one page and beginning of next.

1. See Doc. 639.

–626–

Notebook Entry: Autographic Printing

[Newark,] Sep 14th 1875

Pens.
square inside & ri[-]bbed & hubs put on where screws are

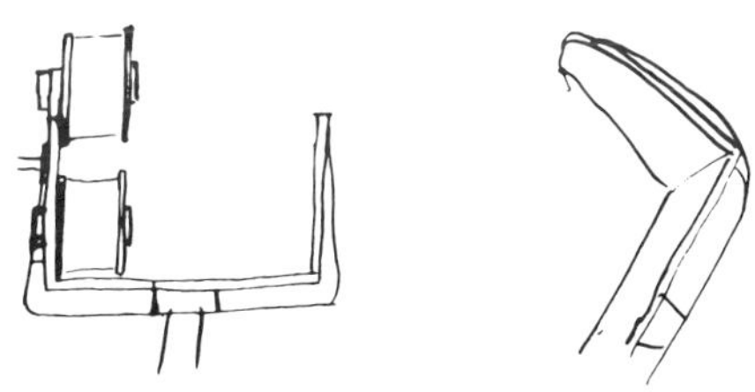

frame a thin square with more stock where the contact screw comes through & the spring scr[-]ew.[a]

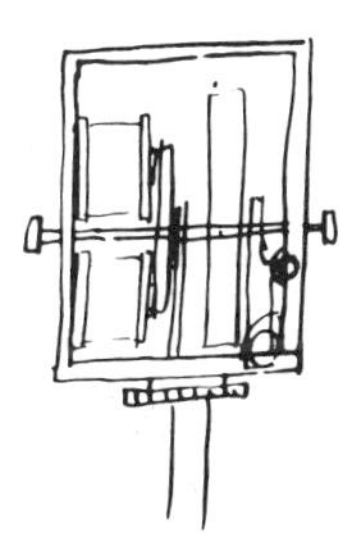

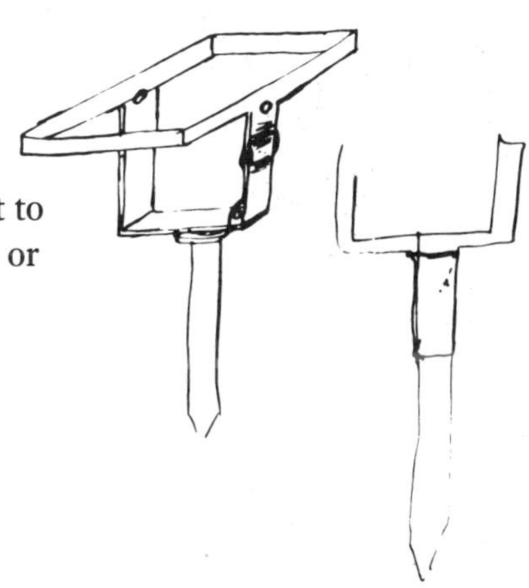

Frame made like this cast to prevent wheel being broken or bent when dropped

A Cast iron ribbed frame   B Wrought iron back

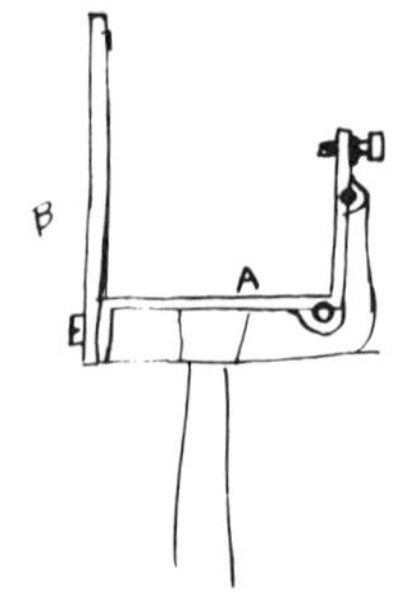

Cast iron frame with wrought iron magnet back screwed on

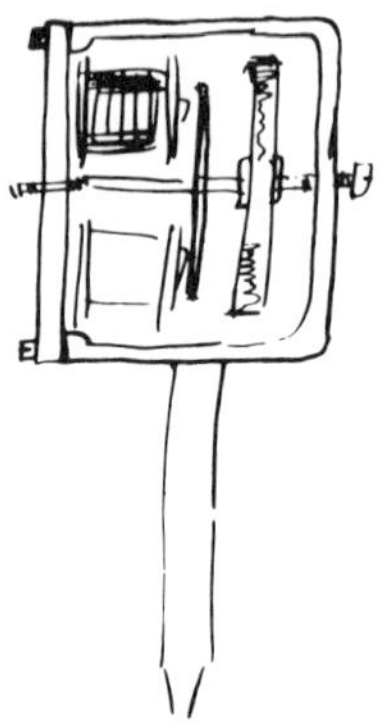

X, NjWOE, Lab., Cat. 30,100 (*TAEM* 6:608). Written by Charles Batchelor. [a]Followed by horizontal line.

-627-

Notebook Entry: Batteries

[Newark,] September 15th, 1875.

Tests of Grenet Batteries for Polarization[1]

1 1 Tablespoonful Bichromate K. Dry.
3 " So_3

Bradley Galvanometer[2]	No 4 coil	65½ degrees
	40 seconds	61½
	60 "	60
	75 "	58½
	120 "	57

By shaking constantly it stands at 90°.

2 By putting in 6 or 8 tablespoonfuls of SeO_3 it stood at

	75
10 secs.	68
20 "	65
90 "	57

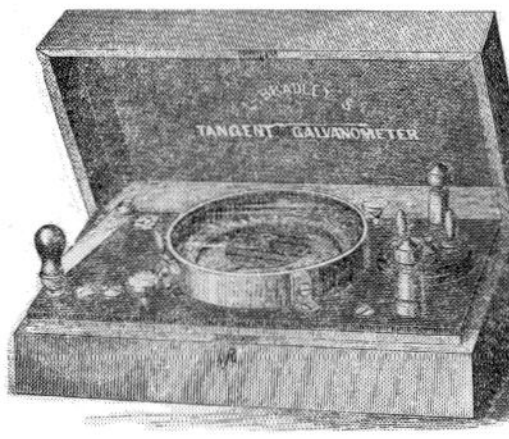

Leverett Bradley's galvanometer.

By shaking it stands at 90 all the time

3 Powdered Bichromate K, put in bottle & a little more H_2O than was necessary to wet it then 4 tablespoonfuls SO_3 poured in shaken ½ min then filled with H_2O.

		75
	5 seconds	60
	40 sec	56
	Shaken stands	68
	Shaken with another zinc	65
	½ min	55
3 Regular Electropoion[3]	Shaken	68
	10 secs	62 runs down
	18 secs	60
	30 secs	59

We here found Galv. wrong.

X, NjWOE, Lab., Cat. 1182:43 (*TAEM* 3:71). Written by Charles Batchelor.

1. The Grenet battery, which used a single fluid of potassium bichromate (bichromate of potash), was considered best suited for use when a powerful and constant but periodic current was needed. When used continously the battery rapidly polarized, greatly reducing its electromotive force (Prescott 1877, 72–73; see Doc. 625 for Charles Batchelor's drawing of a Grenet battery used with an electric pen). Polarization occurs as the ions in the electrolyte migrate to the electrodes and impede the passage of the current.

2. Leverett Bradley's galvanometers, like others, measured current strength in degrees of angular deflection.

3. Electropoion was a battery solution consisting of potassium bichromate and sulphuric acid. It was used in a modified Bunsen battery known commonly as a chromic acid battery. Such batteries were employed extensively on Gold and Stock Telegraph Co. lines in New York City. Reid 1879, 656; Prescott 1877, 71–72.

–628–

To Lawrence Oliphant[1]

New York Sept 16th 1875[a]

Dr Sir,

Anticipating that the present low rates will speedily tax your Cable[2] I hereby propose to practically Duplex its carrying capacity I do not want to enter into any discussion as to whether it can be done or not, but simply ask that you name a price which your company will agree to pay for a successful working apparatus. Payment to be made either ~~e~~wholly in cash or part in the Companys ~~stock~~ collateral.[3] Respy Yours.

Thomas A Edison

L, NjWOE, Lbk. 1, following p. 487 (*TAEM* 28:333). Written by Edward Johnson; letterhead of the Atlantic & Pacific and Franklin Telegraph Companies. [a]"New York" and "187" preprinted.

1. The internal address identifies Oliphant as the attorney of the Direct United States Cable Co. A newspaper clipping about the company, probably dating from the summer of 1875, identifies him as superintendent. Cat. 1144, Scraps. (*TAEM* 27:279).

2. The Direct United States Cable Co. was formed in 1873 by the naturalized British inventor William Siemens and others to provide direct competition with the Anglo American Telegraph Co., which had acquired a monopoly over Atlantic cable business the previous year. The Direct United States Cable line was originally intended to go directly to the United States at Rye Beach, N.Y., unlike other cable lines, which had landed in what is now Canada. However, concern over the strength of such a long cable led developers to change its landing point to Halifax, Nova Scotia. From there messages were transmitted over land lines to the United States. When the company began operation in 1875 it offered special rates to the press and contracted with Atlantic and Pacific for connection with its land lines. As a result of the competition, both cable companies kept general tariff rates low. Bright 1974, 129; Haigh 1968, 66; De Cogan 1986, 3–4; Thomas Eckert's report to the Stockholders of Atlantic and Pacific, 28 Apr. 1875, 2:179, Box 17B, *Harrington v. A&P.*

3. That same day Oliphant replied that he would refer Edison's proposal to the company's managers (DF [*TAEM* 13:410]). No action ensued until the following April, when the company's American superintendent, George Ward, communicated the managing director's request that Edison send his patent specifications and terms for further negotiation. Edison was dissatisfied by this response and refused further negotiations (DF [*TAEM* 13:1133–34]).

–629–

From P. Mullarkey

New York, 5 p.m. Sept 17th 75

My dear Edison,

Charlie Mixer last night received the distressing news of the death of his father and leaves tonight for his home in Illinois to return as soon as possible.[1] The poor fellow was, naturally, much distressed and desired me to notify you.

Now about the Press. I find that our greatest difficulty is to induce clerks in places where we call to give it a trial. They do not like to take the trouble to learn to use the pen. At that place of the organs the man objected to it on account of the noise but promised to let me know in a day or two if he could find use for the pen. I had an appointment yesterday with the Union Pac but he twice disappointed me. I was at the N.Y.C.R.R.[2] struck off a number of beautiful copies and explained the whole thing to a crowd there and received a promise that they would notify me of what they would do in a short time. This morning I went to the Merchants Exchange[3] and had a big crowd around me while I made the copies and distributed them but did not make a sale. It is pretty well advertised in that quarter. Am to call at U.P.R.R.[4] again tomorrow. Called at E. S. Joffray & Co[5] and Herring Safe Co[6] this P.M. With the former I have an appointment for tomorrow and with the latter when they have time. I have an appointment at Fairbanks[7] for Monday 10 A.M.

The thing is highly praised everywhere but it will be harder to sell than you anticipate. The box is infernally heavy and nearly drags my arms off to carry it.[8] I will see you about Tuesday or sooner if I can run across you in N.Y. In haste truly yours

P. Mullarkey[9]

ALS, NjWOE, DF (*TAEM* 3:258).

1. Charlie Mixer was a telegraph operator who had worked with Edison in Western Union's Louisville, Ky., office. Mixer to TAE, 11 Dec. 1889, 89-005, GF.

2. New York Central Railroad.

3. Probably the Merchant's Exchange National Bank, 257 Broadway. Wilson 1875, 890.

4. Union Pacific Railroad.

5. E. S. Jaffray & Co., a dry goods store located at 311 Broadway. Wilson 1875, 644.

6. Herring & Farrell, dealers in safes, 251 Broadway. Ibid., 581.

7. Probably Fairbanks & Co., a scale company at 311 Broadway. Ibid., 389.

8. The components of the electric-pen copying system (at the Edison Institute, Dearborn, Mich.) weigh as follows: pen, ¼ lb.; pen holder, 1¼ lbs.; battery cells (empty of liquid) and holder, 18¾ lbs.; roller, 1 lb.; press, 27 lbs.

9. P. Mullarkey worked as a night operator for Western Union at 145 Broadway. *Operator,* 15 Aug. 1875, 7.

–630–

From W. S. Williams

[New York,] Sep 21 1875

Edison

Please send me some "scientific" for next number. send early if at all

Williams

⟨Batch please mail Enclosed batch for operator⟩[1]

ALS, NjWOE, DF (*TAEM* 13:411).

1. The only items appearing in the next issue of the *Operator* (1 Oct. 1875) with which Edison might have been concerned are several extracts from scientific and technical journals. One contains an editor's note on

the similarity of the acoustic telegraph systems of Paul LaCour and Elisha Gray.

–631–

Charles Batchelor to Norman Miller

[Newark?] Sep 22nd [187]5

Dear Sir

Mr. Edison goes to Michigan today[1] & he instructs me to say, that when he returns next Wednesday he would like to see you, & will fix everything with you at that time[2] Yours Respy

Chas. Batchelor

ALS (letterpress copy), NjWOE, Lbk. 1:27 (*TAEM* 28:29).

1. Edison probably left the next day after executing a caveat (Doc. 633) in New York City. According to later testimony by Joseph Murray, Edison visited Michigan about every three months during 1873 and 1874. Quad. 71.1, p. 50.

2. See Doc. 639.

–632–

From P. Mullarkey

New York Sept 23d 75

My dear Edison

Not having had time to visit you I wish by this to report progress. I have been busy all the time. Mr. D'Infreville who called on me gave me his address as Room 25 109 Broad St and on going there I find no such number. I saw Mr. Clarke[1] and he said he would make arrangements with you. The Chief Engr of the U.P.R.R.[2] told me that the office at Omaha had written for a printer and on my showing him how this worked he said he would write recommending it and he had no doubt that they would order one. I suggested the general use of them on the road and he replied that that might follow. I know he was much interested and will try to introduce them where necessity may exist. I am promised an answer about the 30th.

This afternoon a crockery firm examined it and will give me an answer Monday. I have no doubt but I have made a sale to them, they being greatly pleased with it.

Bankers and Insurance people do not seem to want to take hold of it until it is established I have called at many places and find after showing the whole thing that the chief objection comes from clerks who do not want to have to use it— others offer such trifling objections as~~uch~~ the noise &— One man yesterday wasted three hours of my time and then pronounced it "quite a curiosity" I am so tired from todays work that I am hardly able to stand I am anxious to know if you can open your office here Monday so that I could have a base of operations.[3] As it is I have a world of trouble and some expense. I am trying a new class of men now and am in hopes of reporting better results. You may look for difficulty at first and plenty of it.[4] Hastily yours

P Mullarkey

ALS, NjWOE, DF (*TAEM* 13:261).

1. Possibly the same person listed in Doc. 639 as H. Clark, who had the electric-pen agency for Brooklyn.

2. Unidentified.

3. Edison opened a sales office for the electric-pen company at 41 Dey St., which Mullarkey was using by 1 October. See Doc. 635.

4. For Mullarkey's difficulties in selling the pen see Mullarkey to TAE, 24 Sept. 1875, DF (*TAEM* 3:264).

–633–

Caveat: Autographic Printing

[New York,] September 23, 1875[a]

To all whom it may concern. Be it known that I, Thomas A. Edison of Newark in the county of Essex, and State of New Jersey, have invented an Improvement in Electro magnetic Paper-perforators for Autographic Writing[b] of which the following is a specification.

The object of this invention is to print a number of copies of a drawing or manuscript. The invention consists in forming a stencil by the perforation of holes in paper or other like material and metallic sheets, through which holes ink is made to pass and deposited upon the material underneath.

The invention further consists in the methods of preparing the stencil, by a vibrating point, held within a stylus deriving an up and down motion from a small clockwork upon such stylus, or by a small magnetic engine driven by a voltaic battery, the current from which is led to the engine by flexible conducting cords.

The invention further consists in the various devices which serve to give a proper vibratory motion to the perforating needle or point.

The invention further consists in the combination of various devices with the perforating stylus to obtain a number of impressions or copies. It further consists of a peculiar ink and of various modifications to make the process applicable to various purposes.

Fig. 1.

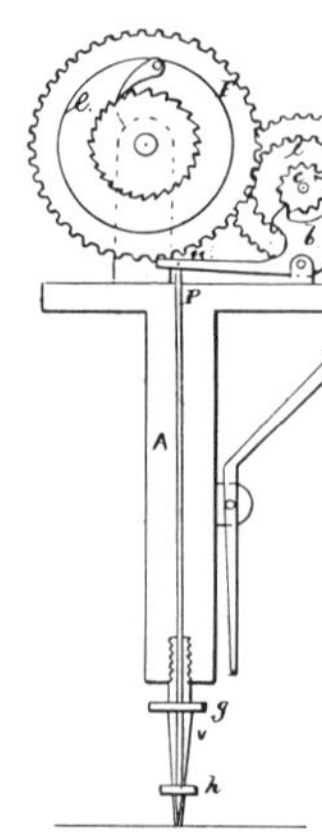

In figure 1, is shewn the vibrating writer, the point which punctures the paper being driven by clockwork. A, is the holder which is held in the hand like a pencil only perpendicular. P is the vibrating wire having on its extreme lower end a very fine point with just sufficient motion to pass through the material to be punctured. The extreme lower end of A is quite small and its end rests upon the material to be punctured. b, is a double toothed fork to which the vibrating wire p, is secured by a joint at n. The teeth on the fork are placed in the path of a ratchet wheel c, which is rotated by being secured to a shaft on which is the pinion e, which meshes in the large tooth wheel f, driven by a spring which is wound up from time to time. The rapid rotation of the ratchet wheel c, gives a vibratory up and down motion to the wire p, and as the rod A is passed rapidly along the material to be punctured ~~the~~ innumerable holes or punctures are made and which form characters etc. On the same shaft with the ratchet c, and pinion e, is another toothed wheel which engages in a worm x to which is secured a fan L which serves to regulate the motion of the train of gears and prevent it from running down too fast.

R, is a lever arranged conveniently on the rod A; this lever engages in a pin S, on the worm X and serves to stop and start the clockwork.

V is a hollow brass piece which screws into A: h, is a nut which serves to turn V, while g is a jam nut serving to secure V in any position required. The end of V resting on the paper serves as a stripper for the punctuating point on p, and as a convenience in drawing the whole apparatus along from point to point.

Fig. 2.

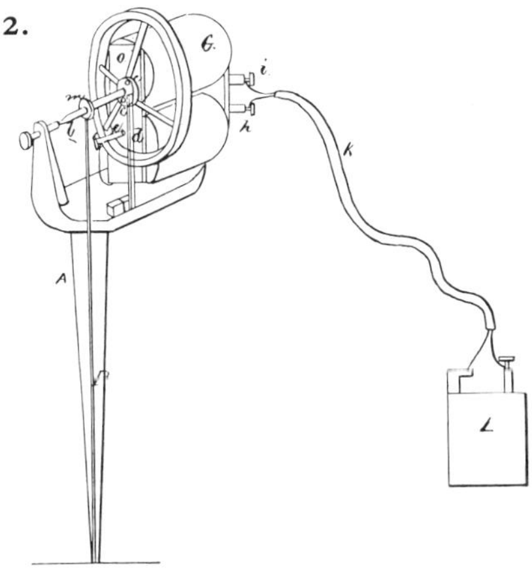

In figure 2, is shewn the same arrangement for puncturing the paper or other material except that the clockwork is replaced by a small electro magnetic engine, driven by a voltaic battery, connected electrically by a flexible conducting cord.

p, is the vibrating wire connected to the eccentric m, upon the engine shaft b, which at every revolution of the shaft forces the needle pointed wire p up and down, or if the needle is required to be vibrated faster a double cam may be used.

G. G. are the electro magnets of the engine, O its armature, F the fly wheel.[1] C. C, are two insulating points which serve to give motion to the circuit changing spring d, which acts against the current closing point e. K is the flexible conducting cord containing two conductors connecting the battery L to the magnets of the engine. In some cases where greater power is required such as punching very thick paper or thin metallic foil, I use two electro magnets (ie) four spools.

Fig. 3.

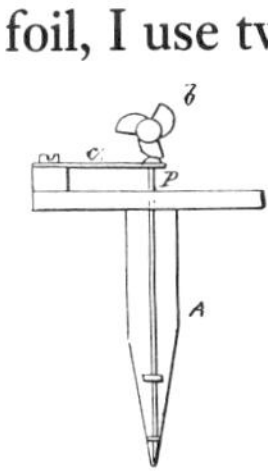

In fig. 3, is shewn the engine shaft b, provided with three or more cams acting on the needle p, direct and serving to give several vibrations to the rod p, at every revolution of the shaft b; c, is a thin spring which serves to bring p, back after being cammed downward by one of the teeth on b.

Fig. 4.

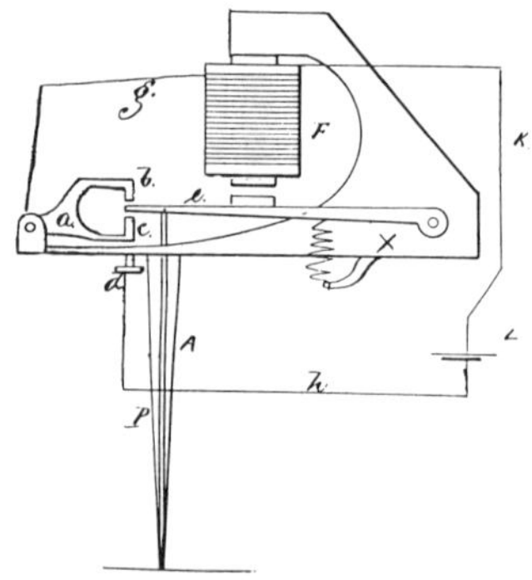

Fig. 4. shows another form where the needle p, is given a vibratory motion directly by the lever of an electro magnet. F. is the electro magnet, e, is its lever provided with an armature in the usual manner: near the extremity of the lever e, is secured the puncturing wire p, passing down through A, in the manner already described. a, is a double pronged lever, between the two prongs of which the end of c, is placed. d, is a contact point connected with the battery L by the wire h. The lever a, is connected to one end of the magnet F, by the wire g, while the other end is connected to the battery by the wire k. X is a spiral spring which draws e, from the magnet F, when the circuit is broken. When the lever e, comes in contact with the point c, on a, it places c, in contact with d, closing the circuit through the magnet, the lever e, is quickly attracted and just at the moment that it is about to touch the face of the magnet cores, the extreme point of e, comes in contact with b, of the arm a, breaking contact with c, and d, when the magnet instantly loses its magnetism and the spring X draws e, away from it until it replaces c, against d, when the same movement takes place again and so on with great rapidity.

Another method which I propose to employ for puncturing thin sheets of metal or other hard substances, consists in giving a vibratory motion to the puncturing needle from a separate clockwork or other source of power such as a treadle worked by foot, the motion being transmitted to the vibratory devices by universal joints and bevel gears, belts or other flexible joints, or the puncturing wire may be fitted in a cylinder with attached valve and worked by air, gas or steam led to it by a flexible tube, connected to a generator of such gases.

Fig. 5.

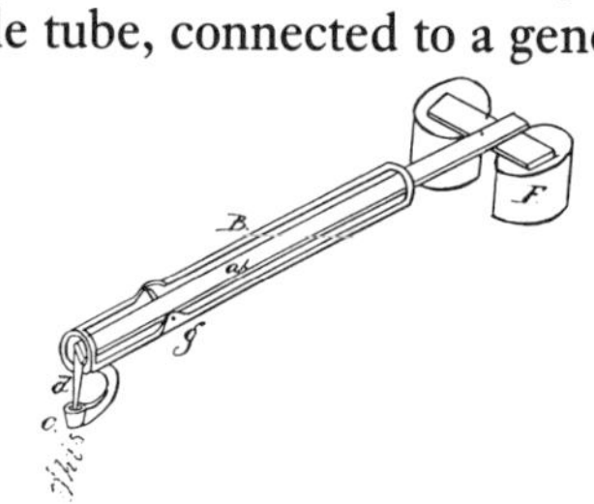

It is obvious that numerous modifications of the various devices shown may be made, for instance the plan shewn in fig. 4, may be so changed to that shewn in fig. 5, making the holding of the pen more convenient than perfectly perpendicular. B, is the tube in which is a lever a, pivotted within the tube and worked up and down like a walking beam by the electro magnet F, with its self acting devices or by the small electric engine shown in fig. 2, or by any other motor. On the extreme end of a, is secured by a joint the puncturing needle d, which enters the stripper c. Constructing it in this manner allows the pen to be held in the usual position that a common writing pen is held.

Fig. 6.

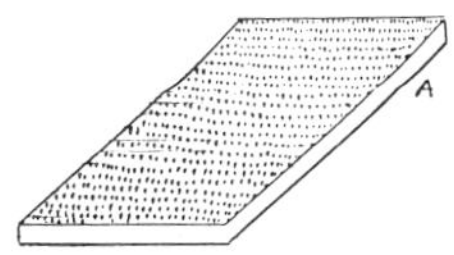

In figure 6, is shewn another method of puncturing the paper which for economy and simplicity excels some of the methods already described although not adapted for all kinds of work.

A, consists of a plate with innumerable sharp points over its entire surface and very close to each other and standing out in relief about ¼ of an inch. Upon this aggregation of points is placed a sheet of the material to be punctured, over this again is laid two or three sheets of soft paper and by passing a hardened steel or agate stylus over the paper bearing down with considerable pressure the points underneath are forced into and through the sheet of material in immediate contact with the plate at every point where the stylus presses. After the message has been written, the first sheet is separated from the plate and the writing will be found traced upon it by the innumerable small punctures on the points of the plate.

For drawing maps, diagrams etc, I prefer to use the puncturing device shown in fig. 7.

Fig. 7

C, is a small wheel provided with a number of puncturing points. D, is the holder with a flange extending nearly to the paper upon which the drawing is to be made, and serves as a stripper.

I will now describe the method of printing a number of copies from a sheet of paper which has been punctured.

Fig. 8

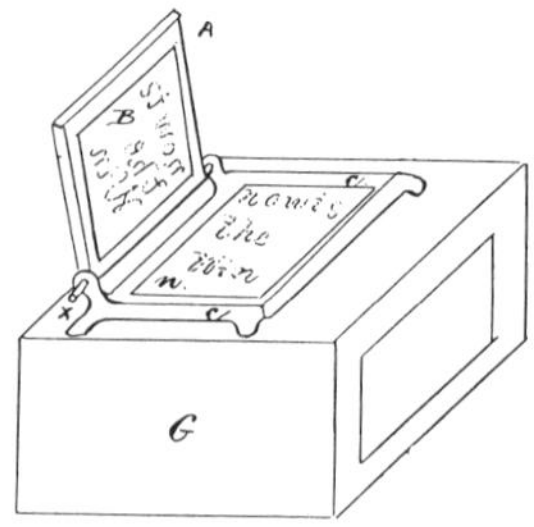

The press is shown in figure 8. G, is the cabinet upon which the press stands; c, is the iron frame of the press, upon which the sheet to be printed n, is laid. A, is a frame pivotted at xx′.[2] The punctured sheet B, is pasted or otherwise secured to this frame. E, is an inking roller: D the handle, F the roller covered preferably with felting or cloth or consists of a number of disks of cloth clamped together.[c] This roller is well saturated with an ink, and after the frame A, is brought down upon the bed of the press and the punctured sheet of paper is in contact with the paper n, to be printed upon, the roller is passed back and forward over the outside of B, until the ink passes through all of the innumerable holes or punctures, a portion of which is of course deposited on n, and an exact facsimile copy printed. The frame is then raised, a fresh sheet put in place of n, and the frame is again brought down and the roller E, passed over it, each time supplying what ink is taken in printing and also giving the necessary pressure.

Fig. 9.

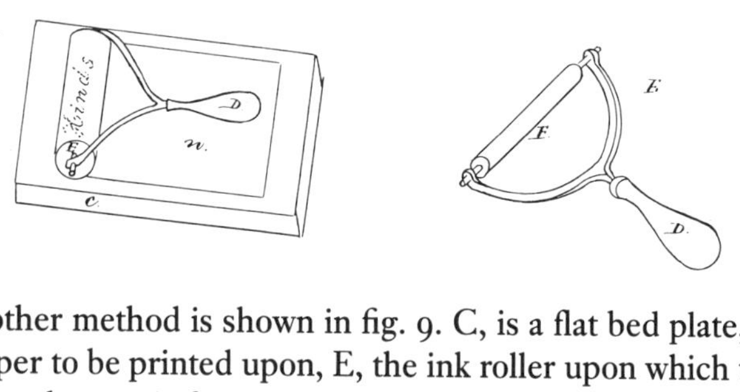

Another method is shown in fig. 9. C, is a flat bed plate, n, the paper to be printed upon, E, the ink roller upon which the punctured paper is fastened. Upon drawing the roll E, over n, the ink from the cloth on E, is forced through the holes of the paper and deposited upon, n. This is a very quick and economical method, but I prefer the method shown in fig. 8.

It is obvious that copies may be taken by hand or by direct pressure. It is also obvious that finely powdered coloring matter may be made to pass through the small punctures and be impressed on the paper underneath, or the paper underneath may be moistened with a certain chemical say, ferridcyanide of potassium or any other reagent and another chemical substance for which ferridcyanide of potassium is the reagent, for instance a protosalt of iron can be dissolved in glycerine, oil balsam liquid resin or water and passed through the punctures and coming in contact with the ferridcyanide of potassium in the paper underneath form prussian or rather turnbulls blue.[3] Ferrocyanide of potassium and a persalt of iron, sulphocyanide and a persalt of iron.

Logwood and a tin or aluminum salt, and in fact any chemical which with its reagent may be used.

Even gases may be used, for instance if the paper to be printed upon is saturated with acetate of lead and the punctured paper be laid upon it the holding of the two over a bottle or dish from which the fumes of sulphuretted hydrogen is coming off will produce a perfect copy of the punctures upon the acetate of lead paper underneath; the gas passing through each hole combines with the lead formish blackish sulphide of lead. Iodide of potassium and the fumes of nitrous acid, or any chemical which is colored by contact with a gas may be used in lieu of ink.

The ink I prefer to use consists of glycerine colored by any of the aniline colors, preferably equal mixtures of red and violet aniline: but the aniline colors may may be dissolved in other substances of great specific gravity and used: the glycerine gives the best results, as it does not pass through ordinary paper like water and oils, neither does it tend to loosen the fibres of the paper and destroy it before a large number of copies can be taken.[4]

If chemicals and their reagents are used to print with, the puncturing of the paper may be dispensed with and the letter, message etc can be written with a common pen upon very thin tissue paper, with an ink composed of a saturated solution of a protosalt of iron, colored slightly with any coloring matter, preferably aniline violet. This ink passes readily through the tissue paper, and if after the letter has been written the tissue paper is rubbed with grease, the ink will be prevented from spreading and if it is then laid upon a sheet of paper moistened with a weak solution of ferridcyanide of potassium, sufficient amount of the sulphate of iron solution composing the ink will have soaked through the tissue paper to form Turnbull's blue at every point where it touches the paper moistened with the ferridcyanide of potassium. A great number of copies may be taken before the sulphate of iron is exhausted, the ferridcyanide of potassium solution being so weak that there is not enough in the paper to combine at once with the iron but the aggregate amount of 50 or 100 sheets being enough to use up all the iron solution passing through the paper.

No particular form of press is required, the mere contact of the paper with the ferrocyanide paper forms the color without any considerable pressure.

If it is desirable to take a number of copies by the puncturing process the characters of which are like common type, the letters upon the end of the arms employed in ordinary type writing machines may be formed of groups of points.[5]

Fig. 10

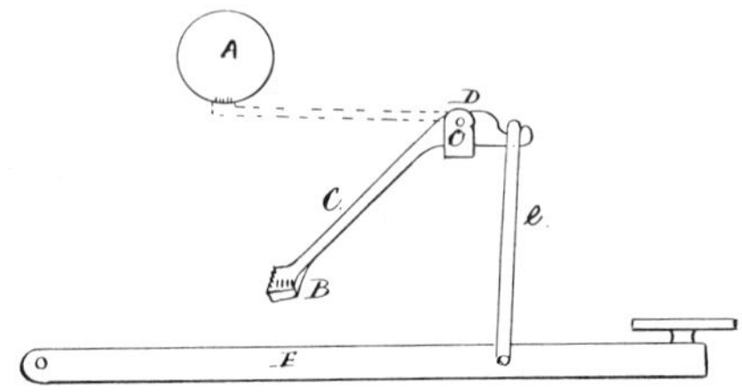

In figure 10, is shewn the arm and letter of points. C, is the printing arm or rather in this case the puncturing arm; upon the extreme end is the square B, with a number of points arranged to form the letter T. D, is the fulcrum of C; A, the drum covered with cloth and upon which is placed the paper or other material to be punctured: e, a connecting wire to the key F; 30 or more of the lever like C, being arranged in a circle all striking to a common center on A. It is obvious that the letters will ~~be~~ punch~~ed~~ the paper on A in place of printing with ink, when a type is used at B. After the paper on A has been punctured an unlimited number of copies can be printed by the process already described.

A much more simple and economical method of puncturing Roman letters is shown in fig. 11.

Fig. 11.

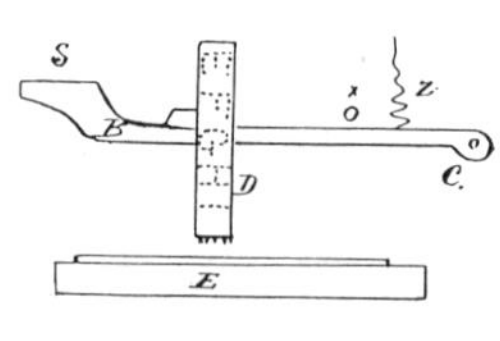

Inventor Thos. A. Edison per L. W. Serrell Atty

Witnesses Chas H. Smith Harold Serrell[d]

B, is a lever pivotted at C, and in its normal position is drawn up in contact with the limiting pin X by the spiral spring

Z. Upon this lever is a type wheel D, which can be revolved to any required position by hand. The letters on the periphery of this wheel are formed of needle points in the manner already described. Any letter can be brought immediately over the paper lying on the platen E, and by depressing B, with the knob S, be made to puncture the letter in the paper or other material. If it is desired to puncturate Roman characters on continuous strips of paper, the type wheel may be rotated by an escapement and lever operated by an electro magnet as in my printing telegraph for which I have numerous patents, or by any other source of power.

A number of needle points might be massed in a square like the punches in my perforators for perforating Roman characters for telegraph purposes and they could be connected and operated in the same manner.[6]

I shall probably claim.

1 Puncturing paper or other material by an electro-magnetic engine arranged and operated substantially and for the purposes specified.

2. Giving a vibrating motion to a puncturing point by the power of electro magnetism for the purposes set forth.

3. A puncturing instrument for preparing paper to duplicate copies from operated by a clockwork substantially as set forth.

4. A self acting electro magnet in combination with a puncturing point for the purposes specified.

5. The combination with an electrical puncturing apparatus of a Voltaic or other source of electricity when the same are connected by flexible conductors.

6. A vibrating point within a tapering tube whose limit of vibration in and out of the tube is just sufficient to puncture the paper, the end of which tube rests upon the material to be punctured in the act of puncturing for the purposes set forth.

7. An adjustable stripper tube for the purposes set forth.

8. The combination of a device upon the shaft of the electric engine which shall cause the puncturing point to vibrate several times at each revolution of the engine shaft.

9. A place whose surface is made up of a number of points and operated for the purpose set forth.

10. A toothed wheel with stripper extension for the purposes specified.

11. Forming Roman letters from groups of points substantially and for the purpose set forth.

12 Printing from a manuscript formed of numerous punctures in the manner set forth.

13. The use of glycerine colored with an aniline or other powerful color for printing from punctured manuscript as described.

14. Writing upon tissue or other thin paper with a chemical solution and printing from the underside of such manuscript by bringing paper in contact with it which is moistened with another chemical solution the reagent of the one in the tissue paper.

15. The combination of a punctured manuscript upon a roller covered with cloth and saturated with a colored ink for the purpose specified.

16. Multiplying copies from a punctured manuscript by gases and chemicals in the manner described.

17. The combination of the puncturing point with a cylinder and valve and operating it by air or other gases conveyed to it through flexible tubing for purposes set forth.

Signed by me this twenty third day of September 1875.

Thos A Edison

Witnesses Geo. D. Walker Chas H. Smith

DS, DNA, RG-241, Edison Caveat 68. Written by Walker; petition and oath omitted. [a]Place taken from oath; date taken from text, form altered. [b]"for Autographic Writing" interlined above. [c]"or consists . . . together" interlined above. [d]All figures are on two separate pages, both of which are signed and witnessed at the bottom.

1. The label is missing from the drawing.
2. The label "xx′" is missing.
3. An inorganic dye commonly used for making blueprints.
4. See Cat. 1172:259, Lab. (*TAEM* 3:156).
5. On 27 August 1875 Batchelor had John Kruesi "alter four letters out of Mr. Rieff's Shols & Glidden printer by taking out the letters & putting in spikes." Cat. 1182:40, Lab. (*TAEM* 3:70).
6. See Docs. 349 and 723; and U.S. Pat. 151,209.

–634–

From James Fuller

New York, Sept 28 1875[a]

Dr Sir

I send you herewith two samples of paper run through your instrument the current being furnished by my machine. One was done on a close circuit and the other on about 325 miles I can run it fast enough to show a want of uniformity of current but about 125 revolutions per minute shows no break— I do not know how fast you require to run for regular work, but it is quite probable that I may require some slight modifications, this being the first machine. in order to meet the requirement of this mode of operating I should like your opinion of the work done so far.[1] Yours truly

J. B. Fuller[2]

ALS, NjWOE, DF (*TAEM* 13:413). Letterhead of J. B. Fuller, Electrician. [a]"New York," and "187" preprinted.

1. In a 10 May 1875 letter to Edison and Murray, Fuller had inquired about ordering a table of telegraph apparatus connected up with a crank (DF [*TAEM* 13:647]). This was probably Edison's automatic table and instruments, which Fuller apparently hoped to operate by replacing the batteries with his magnetoelectric generator. Fuller may well have attempted to use his machine to operate automatic instruments on the lines of the Atlantic and Pacific Telegraph Co. The following year he did conduct experiments on the company's Morse lines. In 1877 Atlantic and Pacific used his generator to operate some of its Morse lines, but the company soon removed the machine, out of concern that it might damage the company's cable in the North (Hudson) River. Fuller also experimented with his generator on the lines of Western Union and the Gold and Stock Telegraph Co. (testimony of Georges D'Infreville and Stephen Field, *Field v. Lugo;* Fuller to the *Teleg. J. and Elec. Rev.* 5 [1877]: 24).

2. According to his letterhead James Fuller was an electrician and manufacturer of magnetoelectric machines, particularly for electroplating. Earlier in the year his letterhead had identified him as a manufacturer of metal lamps and fixtures, some of which would have been electroplated and probably would have prompted his interest in magnetoelectric machines (Fuller to Edison and Murray, 10 May 1875, DF [*TAEM* 13:647]). Fuller received his first patent on such a machine in October 1875 (U.S. Pat. 168,893). He later designed electric lighting systems sold by the Fuller Electrical Co. (*Teleg. J. and Elec. Rev.* 5 [1877]: 117–18).

–10– October–December 1875

Edison spent the autumn of 1875 in his laboratory. He was no longer working actively for the Atlantic and Pacific Telegraph Company, and he turned his full attention to a handful of the many technologies he had determined to investigate at the time he divorced his laboratory from the shop. He dabbled in several experimental directions, including facsimile telegraphy, mining, and celluloid, but three principal topics occupied his time. Through October and well into November, he developed his electric-pen copying system, refining the pen, copying press, batteries, and ink. In November he returned to the problem of acoustic telegraphy, and concomitantly to a working relationship with Western Union. Observations of his acoustic apparatus led him to announce the discovery of the "etheric force," which embroiled him in a public controversy at the end of the year. To show off his inventions, Edison prepared a grand exhibit for the 1876 Centennial Exhibition. And while all this was going on in Newark, he sent his father to locate land outside the city for a new laboratory site.

Edison's relationship with Atlantic and Pacific deteriorated. He felt they had reneged on their promise of support for his experiments, and in mid-October he complained about money long due him for work on his roman-letter system. On the other hand, President Eckert, active again after the pneumonia that had kept him resting all summer, complained about the way Josiah Reiff had interfered with Edison's work for Atlantic and Pacific. Edison, siding with Reiff, decided to assign his automatic telegraph patents—at that point still held by Jay Gould—to the American Automatic Telegraph Company, a new corporation formed by Reiff and the other disgruntled, unpaid investors of the old Automatic Telegraph Company.

Sometime in early November, Edison turned in earnest to acoustic telegraphy. It seems clear that he and Western Union president William Orton had reached an informal agreement on experimental funding. By the middle of the month Edison was experimenting, keeping accounts of his expenses, and being reimbursed by Western Union. The next month Edison and Western Union agreed to drop any reciprocal monetary claims that might stem from the quadruplex patent dispute, and at the same time Edison and Orton signed a contract committing Western Union to support Edison's acoustic experiments. One stipulation of the agreement bound Edison to appoint as his patent attorney Grosvenor Lowrey, general counsel for Western Union, who had been involved in the quadruplex fracas from the beginning.

Edison's first laboratory records of acoustic-telegraph experiments are dated 16 November 1875.[1] James Adams and Charles Batchelor were his principal assistants in acoustic work; in fact, they conducted many of the experiments because Edison was too deaf.[2] Machinists John Kruesi, John Ott, and James Eagan made most of the instruments. The team employed vibrating reeds and tuning forks in transmitters and designed a variety of receivers that used resonant tuning forks, iron diaphragms, and stretched animal membranes.[3] At the end of November, Edison again hired Robert Spice for instruction, this time in acoustical science. Spice stayed on into 1876 as a member of the laboratory team.[4]

In the course of his experiments Edison noticed sparks between parts of his apparatus that were not conducting currents. Such sparks had been observed before in other telegraph equipment subject to rapidly oscillating currents—acoustic telegraphy involved high-frequency vibrations on the order of several hundred cycles per second—but Edison investigated them as a potential new force, perhaps spurred on by his discovery of the electromotograph phenomenon. He devoted the last two weeks of November to experiments on what he soon called the "etheric force." Newspapers, now familiar with Edison and his work, publicized the new discovery, and Edison encouraged them, even allowing them to read and quote from his laboratory notebooks. Edison linked his discovery to the discredited "odic force" postulated by Karl Reichenbach earlier in the century, in which he had been interested the previous year and for which the *New York Times* ridiculed him. James Ashley's *Telegrapher* predictably castigated Edison's "new moonshine," and several papers reported the skepticism of electrical experts. Most of the press, however, including the *Scientific American*, reported the discovery favorably, echoing Edison's claims that the new force could revolutionize telegraphy. Edison demonstrated his new force privately and publicly, and found an ally in Dr. George Beard, a leading figure in American neurological medicine.[5]

Edison had hired his first agent for the electric pen at the end of September. Initial orders taxed the capacity of his machinists, but within a month he had sold about thirty pens.[6] By the end of December he had established agencies for New England, the Middle Atlantic states, much of the Midwest, and the Pacific Coast as far north as British Columbia. Norman Miller, who had long been associated with Edison through the Gold and Stock Telegraph Company, agreed in early October to oversee the operations of the electric-pen business. Meanwhile, the laboratory team continued to develop the various components of the copying system. Because this was a consumer product, the acid-filled batteries in particular had to be made as safe and long-lasting as possible. To acknowledge their contribution to the successful development of the invention, Edison gave Charles Batchelor and James Adams a percentage of any electric-pen profits.

Laboratory work included several smaller projects. Edison's work on the electric pen apparently stimulated his longstanding interest in facsimile telegraphy, and at the end of October he executed a long caveat that incorporated ideas and techniques from the pen, the electromotograph, and automatic and acoustic telegraphy. With the backing of a member of the New York Stock Exchange, Edison began to develop a printing-telegraph system with which brokers could send their orders privately from their offices to traders on the exchange floor, a scheme that never came to fruition. The Domestic Telegraph Company, already established in sixteen cities, began operations in Newark on 1 November.[7] Batchelor, who continued his work as company electrician, drew up detailed instructions for the Newark workers.

Edison also attended to the organization of the lab itself. Modeling himself after the British experimental scientist Michael Faraday, Edison began keeping laboratory records in notebooks labeled "Experimental Researches." Entries within the books were sequentially numbered, as were illustrations of the experimental work. Edison maintained this pattern until just after the new year. He also seems to have found the Newark site unsatisfactory, for in November he sent his father to look for a new laboratory location.[8] He considered Perth Amboy, a port south of Newark at the mouth of the Raritan River, but when he finally bought land at the end of December it was in Menlo Park, a rural community about twelve miles south of the city on a railroad line. He paid $5,200 for two tracts of land, one of which included a house.[9]

1. Much material on Edison's early work on acoustic telegraphy was lost when the laboratory moved to Menlo Park in 1876. TI 1:11 (*TAEM* 11:26).

2. Edison's and Batchelor's testimony, TI 1:38, 46, 248 (*TAEM* 11:40, 44, 101).

3. Although Edison himself testified that the transmission of articulate speech was not his goal at this point, patent examiner Zenas Wilber said that Edison mentioned the transmission of speech to him when he visited Edison's laboratory in late November. The possibility of such transmission had been a matter of public speculation for some time; nevertheless, Wilber said that this was the first time he had heard it suggested by an American inventor. TI 1:271 (*TAEM* 11:112); "A Wonderful Invention," 29 Nov. 1875, *Newark Morning Register,* Cat. 1144, Scraps. (*TAEM* 27:282); "Telegraphy. New Inventions in the Science of Electrical Transmissions," *New York Times,* 10 July 1874, 2.

4. In his testimony in the Telephone Interferences, Spice remembered working in the laboratory from 29 November to 20 January, but account records indicate he was employed until the end of February. TI 1:298–99 (*TAEM* 11:126); Cat. 1214:132, Accts. (*TAEM* 21:620); 75-020, DF (*TAEM* 13:722).

5. App. 1.G33. Edison started a caveat that described an etheric telegraph. NS-Undated-001, Lab. (*TAEM* 8:2–4).

6. TAE to P. Mullarkey and TAE to Maris and Smith, both 17 Oct. 1875, Lbk. 1:85 (*TAEM* 28:81); Cat. 1214:306–10, Accts. (*TAEM* 21:687–89).

7. Edison later recalled the difficulty of launching such service firms. App. 1.D265, G20.

8. "Edison's Newark Expense Book, 1875–1878," EP&RI. Edison's discontent sprang from a rental payment dispute with Thomas Slaight, a lock manufacturer and landlord for Murray and Co. at the old New Jersey Railroad Ave. shop. See App. 1.D154, G23; Ford 1874, 138–39; *TAEB* 1:148 (item 10), 378; Doc. 252; and Hopkins 1873, Fifth Ward map.

9. Edison took a $3,000 mortgage from the seller, George Goodyear (Deed, mortgage, bond, and agreement, TAE with Goodyear, 29 Dec. 1875, Kellow [*TAEM* 28:1180]). Robert Conot claims that Edison bought the land from the family of William Carman, later Edison's bookkeeper, but no information can be found to substantiate this (Conot 1979, 78).

–635–

From P. Mullarkey

New York Oct 1st 1875.

My dear Edison:

I called twice and sent a note addressed to you at the A.&P. office, being anxious to see you this afternoon.

I delivered an order to Mayer Bros[1] this morning and was told by them that there was much curiosity respecting the instrument among men in their business and that there would be a lively demand for it in that quarter should this one turn out well.

I made several calls yesterday and today and have been so successful in convincing them of its utility that I am certain of doing well in a very short time. I have discovered that the commencement must be gradual as each one appears to want to "wait until it comes more into general use." I have done so far all I could and have advertised it quite extensively— It is my purpose to advertise in Sundays Herald and start out six good men next Tuesday if you can supply me with the instruments for them.[2] I will not be able to do much outdoor work because I shall have my duties at the office which will be enough to absorb all my attention

Please send me on receipt of this one instrument complete with a new needle for my own and I will return the one I have. I also want an extra cup of battery for my instrument.

I have been running so much today that I can hardly sit up. After I get those canvassers started I am pretty sure we shall be able to do the business here in a manner pleasing to all concerned. Please send me some of that square paper and some blotting paper. Hastily yours

P. Mullarkey

ALS, NjWOE, DF (*TAEM* 13:266).

1. Mullarkey was probably referring to Mayer Brothers & Co., merchants at 142 Pearl St. Wilson 1875, 881.

2. No such advertisement has been found.

–636–

Agreement with Charles Edison and W. H. Blennerhassett

[Port Huron?,] October 1, 1875[1a]

<u>This agreement</u>[b] entered into this 1[c] day of October 1875[d] by and between Thomas A Edison of the city of Newark Essex Co State of New Jersey party of the first part and W H Blennerhassett[2] & Chas Edison of Port Huron Mich[e] parties of the second part witnesseth

In consideration of the sum of one dollar in hand to me paid the receipt of which is hereby acknowledged the said party of the first part does hereby grant an exclusive license to sell Edisons Autographic Press for which a patent has been applied for in the U S patent office[3] in the state of Michigan[c] for a period of one year from date provided the party of the second part shall work up the said state by the establishment of subagencies and dispose of such a number of presses as shall be satisfactory to the party of the first part

The said party of the first part agrees to allow no person other than the party of the second part or his agents to sell by personal solicitation in such state during the continuance of this agreement.

It is further agreed that the party of the second part shall allow his sub-agents a sum not less than six dollars on each press sold by them

This contract may be annulled by either on Thirty days notice in writing[4]

Witness our hand and seals this 1[c] day of October 1875[d]

Thos A. Edison[f] W H. Blennerhassett[f]

C. Edison[f]

Witness

ADDENDA

[Port Huron, October 1, 1875?][g]

The sale of one machine per week in the state of Michigan will be satisfactory

T. A. Edison

[Port Huron, October 1, 1875?][h]

Chas Edison to Receive $2. dollars on each machine and Blennerhasset $8.

T. A. Edison W H Blennerhassett

C. Edison

DS (electric pen copy), NjWOE, Miller (*TAEM* 28:990). Original written by Edison. [a]Date taken from text, form altered. [b]"This agreement" underlined twice. [c]Inserted by Edison. [d]"October 1875" inserted by Edison. [e]"W H . . . Mich" inserted by Edison. [f]Followed by representation of a seal. [g]Addendum is an ADS; "The . . . satisfactory" written by Edison; see n. 1. [h]Addendum is a DS; "Chas . . . $8." written by Edison; see n. 1.

1. This agreement, one of a number of standard electric-pen agency contracts that were produced with the pen (see Doc. 639 n. 8), was probably executed in Port Huron and postdated to 1 October. Edison had gone to Port Huron and was due to return to Newark on 29 September. An electric-pen sample in his handwriting and dated "Newark NJ Oct 1—75" indicates that he had probably returned by 1 October (Cat. 1144, Scraps. [*TAEM* 27:277]).

2. Unidentified.

3. Edison did not apply for a U.S. patent until 13 March 1876 (U.S. Pat. 180,857). He had filed a caveat on 27 September 1875 (Doc. 633) and applied for provisional protection in Britain on 29 October (Brit. Pat. 3,762 [1875]).

4. On 10 April 1876 Charles Batchelor wrote Charles Edison to inform him that Edison did not wish to renew the contract. Lbk. 2:74 (*TAEM* 28:409).

-637-

Agreement with Charles Batchelor and James Adams

NEWARK, N.J., Oct 2nd 1875[a]

It is understood that Thos A Edison Receives 6.13, Laboratory 1.15, Batchelor 2.45, Jas Adams .27.c on each $10 made on Autographic Press.

T. A. Edison

Chas. Batchelor

Jas. Adams

DS, NjWOE, DF (*TAEM* 13:268). Written by Edison; letterhead of T. A. Edison. [a]"NEWARK, N.J.," preprinted.

-638-

Draft Caveat: Autographic Printing

Newark Oct 3rd 1875[1]

Autographic Press. Caveat No 2[2a]

The object of this invention is to effect the multiplication of written characters.

The invention consists in various devices for perforating or puncturing paper whereby the multiplication of copies is brought about

It further consists in devices for applying the ink effecting the impression, and formulas for making ink

In a previous caveat[3] I have described various devices for puncturing the paper to [----][b] suit different wants and in[c] this caveat I shall describe various modifications many of which are undergoing experimental tests with a view of adopting the most practical & economical devices for use

In figure 1 is shewn a perforating pen the perforating needle being vibrated by a small electromagnetic engine placed upon the top of the hollow tube the action of which has been described in my Caveat No ______[4]

Fig. 1.

The modification consists in adding an extra tube or solid rod A so as to enable the writer to hold the pen in the usual position instead of attempting to learn to hold it perpendicular

The same object is attained in the device shewn in figure 2 the extra tube or rod being dispensed with and the engine inclined the puncturing needle remaining in a perpendicular position.

Fig. 2.

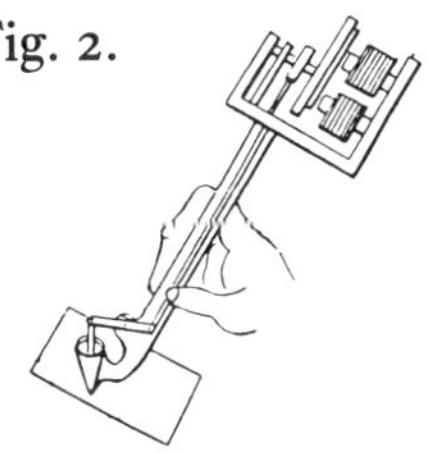

Fig. 3.

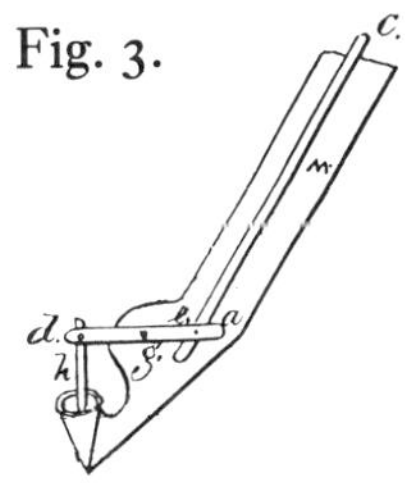

M is the hollow tube through which the reciprocating rod c plays this rod is pivoted [---][b] to the lever a at e The fulcrum of the lever a is at g the puncturing needle h is secured to a at d

It will be seend that although the whole apparatus may be inclined at an angle usual in writing with an ordinary pen[5]

In figure 4 is shewn a method for giving motion to the needle.

Fig. 4.

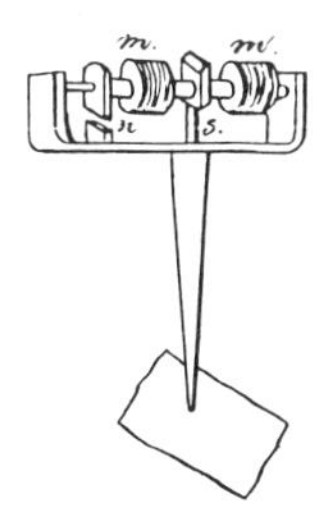

The helices M M. are hollow, through their centre passes the iron core a and runs in centers

Two iron tongues[6] n n′ are placed at each end of the spools; and serve to attract the core at the proper moment when it becomes magnetic by the passage of the current through the wire surrounding the helices, the cam and reciprocating rod[7] X is secured to the rotating core between the two helices by thus arranging the spools a very powerful effect is obtained and the device can be arranged in a very convenient manner.

In fig 5 is shewn the ordinary engine for giving motion to the puncturing needle: and the weight of the electromagnets which must necessarily be an adjunct of the pen is utilized to act as a fly wheel thus dispensing with the ordinary fly wheel and decreasing the weight of the pen. the magnet is secured to the revolving shaft and the current is led into the helices by contact springs and insulated collars upon the shaft in the usual manner. The armature is of course stationary

Fig. 5.

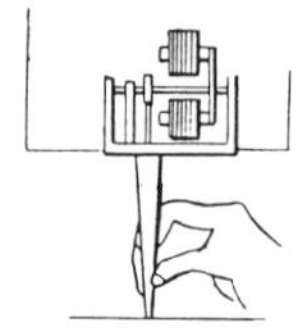

In figure 6 is shewn a set of vibratory apparatus—~~the~~

Fig. 6.

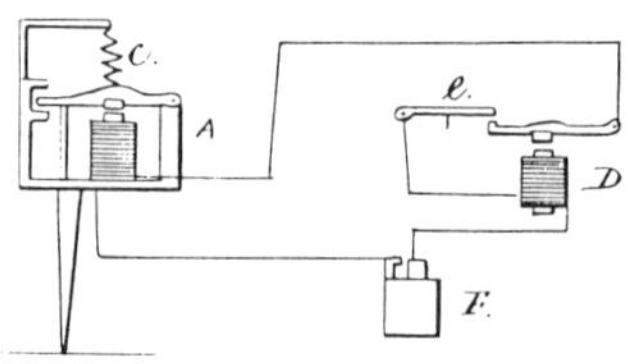

A is the pen electromagnet which serves to attract the lever b giving motion to the puncturing needle. C is a retractile spring serving to draw the lever ~~b~~ back after having been attracted by the magnet .A. D is a self vibrating secondary magnet arranged in the same circuit as A. F is the battery which furnishes power to both D & .A.

e is the contact breaking spring. When the lever of D comes in contact with e the circuit is closed and the lever of A is drawn to the face of A puncturing the paper at the same time the lever of D is attracted and just as it touches the core the contact at e is broken & the lever is drawn away by its retractile spring these vibrations go on with great rapidity=

Another modification consists in having a seperate electromagnetic engine for furnishing the breaks of the current which is in many respects more reliable than the self vibrator This is shewn in fig. 7.

Fig. 7.

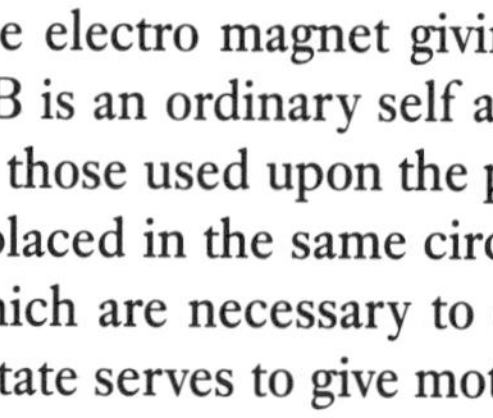

A is the electro magnet giving motion to the ~~n~~puncturing needle B is an ordinary self acting electro-magnetic engine similar to those used upon the pens themselves. The magnets of B are placed in the same circuit as A and the breaks of the circuit which are necessary to allow the armature of the engine to rotate serves to give motion to the lever of .A.

Fig. 8.

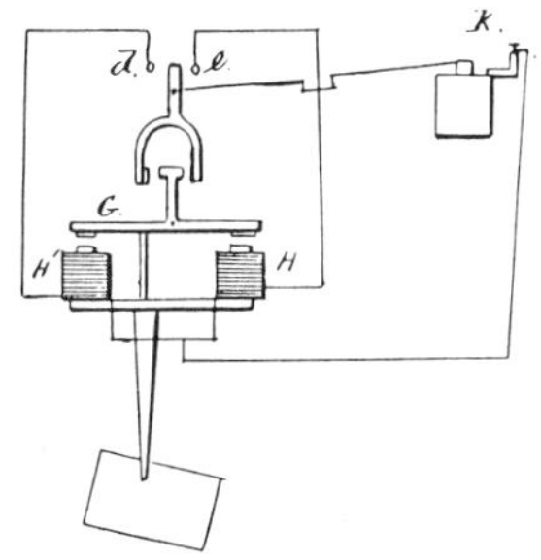

In fig 8 is shewn a self vibrating electromagnet or rather double electromagnet wherein the full play of the lever is obtained. H & H are the electromagnets.[8] GG [---][b] is the vibrating lever C is a contact lever for changing the currents from one electromagnet to the other. d and e are contact points. K is the battery. Suppose the lever G has just been attracted by the magnet H′ the lever C is thrown in contact with the point e breaking the current through H′ and throwing it into .H. ~~this again~~ the lever G is then attracted by H and just as it reaches its limit the lever C is thrown in contact with the point d disconnecting the current from H and placing it in ′H hence a very rapid motion is given the armature G to which the puncturing needle is secured

I will mention that all the electromagnets might be replaced by polarized magnets and the current reversed through them by suitable contact breakers either connected or disconnected from the magnets perhaps in practice polarized magnets would be found preferable in many instances as the lever would necessarily receive full play in being attracted by the positive or negative poles.

I will now describe several devices wherein magnetism is replaced by other sources of power

Fig. 9.

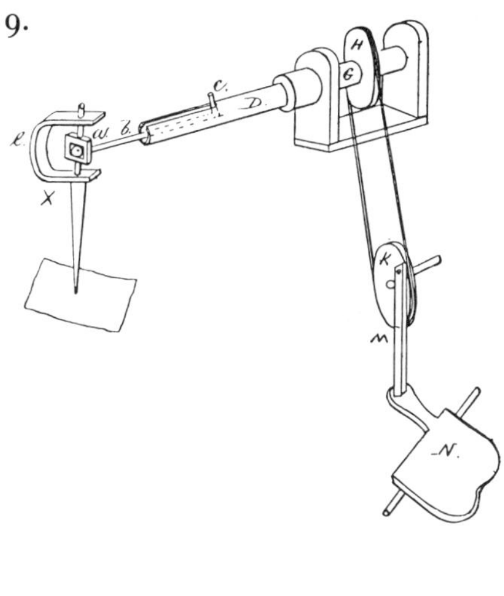

In figure 9 is shewn a device for puncturing the paper by mechical means. ~~e~~X is the holder to be grasped by the hand ~~e is a be~~ a is the cam and square for giving a reciprocating motion to the needlle ba[-]r b is a shaft upon which the cam is secured this shaft runs within a slotted tube .D. upon the extreme end of the shaft is a pin c passing through a long slot in the tube D allowing the shaft to be drawn in and out of the tube after the manner of telescope tubes, but is prevented from turning by the slot & pin .C. The tube D is connected to the [--][b] shaft G by a universal joint at f thus allowing the the pen X to be placed in every convenient position for writing while the tube d shaft b is rotated and the puncturing needle is reciprocated

The shaft G is rotated by foot motion similar to that employed upon sewing machines and similar apparatus, or a handle may be attached to the shaft G and rotated by the left hand of the person writing or rotated by any other source of power

In figure 10 is shewn a method of giving motion to the shaft connected to the reciprocating puncturing needle.

Fig. 10

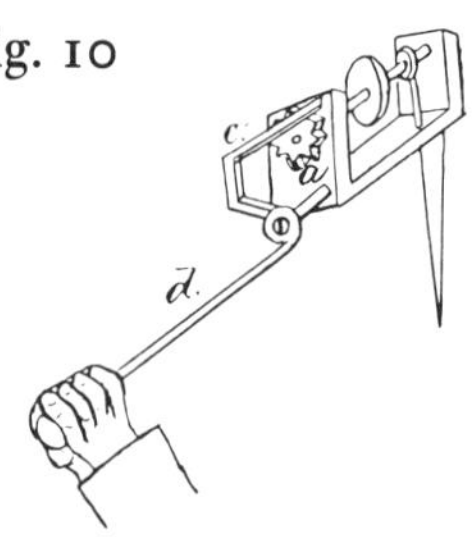

b is a fly wheel secured to the shaft a is a ratchet wheel, also connected to the shaft. e is a prong with a click engaging in the teeth of the ratchet wheel. this prong is secured to the [-] rod .d. The pen is to ~~p~~ be grasped in the right hand of the operater while with his left hand the rod d is turned at intervals giving a reciprocating motion to the prong c and & click which engaging in the ratchet wheel .a. gives motion to the shaft The fly wheel serves to keep the motion thus imparted even.

In fig 11 is shewn a device whereby the rotation of the needle shaft is effected by water power or other liquid falling from a high to a lower level

Fig. 11.

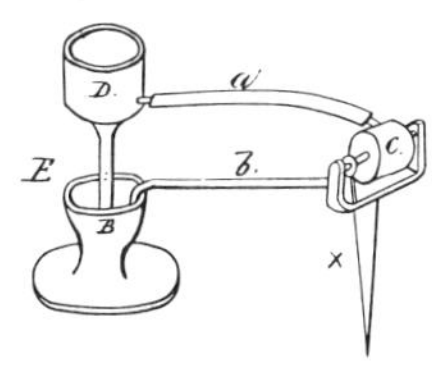

F is a stand [- -- ---- ---][b] D is a jar containing water or mercury: & provided with [---][b] flexible tube connected to the small turbine or water wheel within the tube C upon the pen .X. leading from C is another tube which delivers the water from a & D[d] into b and resevoir E The passage of the liquid through C serves to give a rapid motion to the leafed wheel inside. after the liquid in D is nearly exhausted, it is dipped out of E [~~back~~?] & replaced in D or the resevoirs may be so arranged that they may be exchanged

In fig 12 is shewn another mechanical device for vibrating the puncturing needle and at the same time giving it a freedom of motion

Fig. 12.

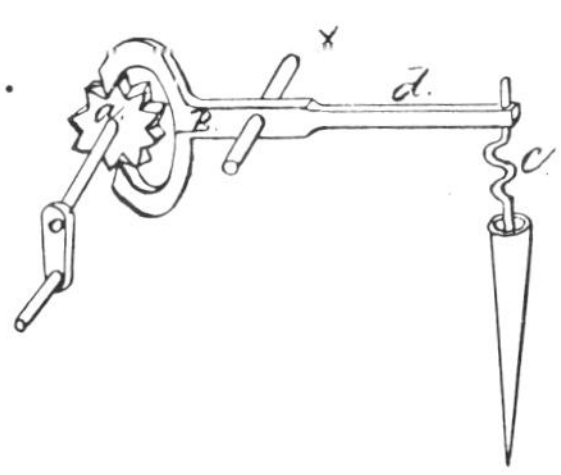

a is a star ratchet wheel rotated by hand or other source of power

B is an anchor pivoted at X upon the end of the lever d is secured a stiff spring which in its turn is secured to the needle inside the tube in such a manner that it will be vibrated without being withdrawn from the tube when it is moved about, the spring c stretching to accommodate itself to the movement of the needle tube.

In figure 13 is shewn a device whereby the movement of the needle tube in the act of writing is made to give motion to the needle=

Fig. 13

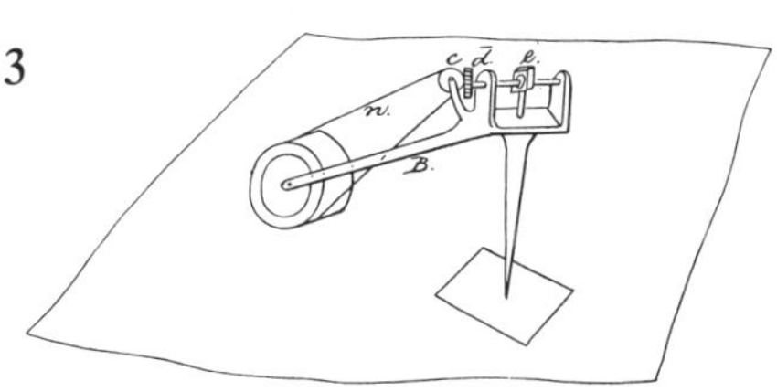

A is a heavy brass wheel laying on the table, and secured to the arm .B. which forms a part of the pen holder upon the side of this wheel is a thin pully C is a bevel gear wheel which serves to rotate the shaft carrying the cam E by meshing into another bevel gear wheel d connected to this shaft

The wheel C derives its motion from the belt n on the pulley by its side, the movement of the pen in the act of writing whether back or forward, turns the belt and this gives the shaft its motion

It is obvious that very fine writing cannot be done very well with this device.

In figure 14 is shewn a very practical method of driving the needle by treadle.

Fig. 14

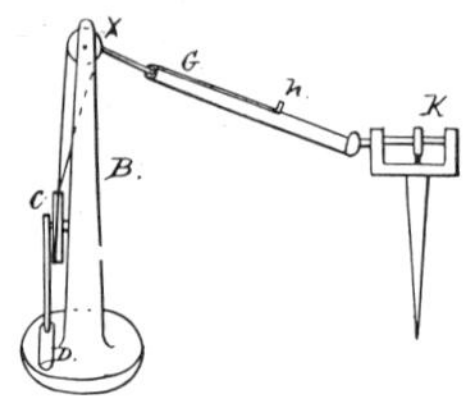

Another device consists connecting a very small steam engine similar to what is known as the dollar engine upon the top of the pen holder and connecting the puncturing pen to it in such a manner that a vibratory motion is obtained, the steam air or other gas being led to the cylinder by felexible tubeing [---][b] a[c] [---- friction?][b] Even the boiler ~~en~~Lamp and engine may be constructed so small that the whole can be placed upon the pen holder without much inconvenience.

I will here mention that when a simple electromagnet is used to move the needle in & out of the tube that the breaks of the current may be performed by a break wheel secured to the shaft of a constantly rotating clockwork or the break wheel may be rotated by hand like the transmitter of a printing telegraph.

I will also mention that the electrical engine may be strapped to the arm and connected to the writing pen by a flexible or universal joint shaft or ~~w~~spiral wire, which will lessen the inconvenence due to the weight of the engine on the pen itself=

For compressing air to be used in a cylinder for vibrating the needle I prefer to use the double bellows compresser ~~simila~~used by chemists for blow pipe analysis =

Having thus described the various devices for ~~punctur~~ obtaining motion for puncturing the paper I will now describe the methods for securing the puncturing needle in the needle bar.

[Fig.] 15.[e] [Fig.] 16. [Fig.] 17. [Fig.] 18 [Fig.] 19 [Fig.] 20.

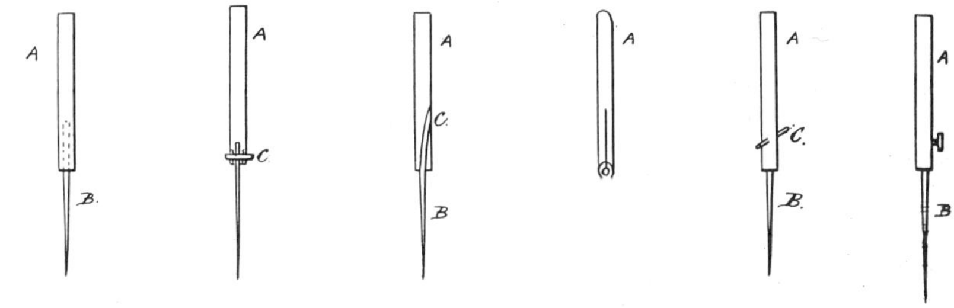

Fig 15[e] A is the needle bar B. the needle which is of the ordinary character a small hole is drilled in the bar and the needle inserted. The needle should be a shade larger than the hole. in fig 16 the bar A is slotted and a taper[c] screw cut upon its extreme end upon which is placed a nut C the needle being placed in the hole in the center of the bar it is clamped tightly by ~~th~~ tightning the nut.

In fig 17 the hole in the bar is ~~out~~ drilled out of true ~~&~~ the friction of the needle is sufficient to hold it in its place. in fig 18 a hole is drilled in the bar and the bar slotted making the two parts of the bar spring together thereby holding the needle= Fig 19 the needle is held in position by a pin passing through the bar A & eye of the needle Fig 20 the needle is secured by a small watch screw

[Fig.] 21 [Fig.] 22 [Fig.] 23 [Fig.] 24 [Fig.] 25 [Fig.] 26.

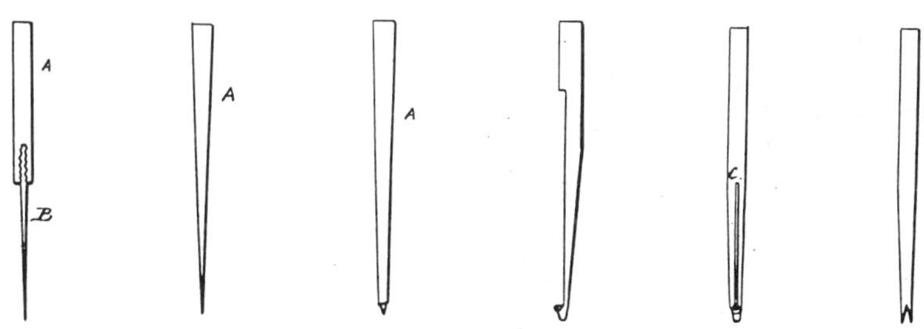

Fig 21 a thread is cut on the end of the needle Fig 22 the bar & needle is made in one piece Fig 23 the extreme end of the bar is made needle shape Fig 24. this bar is provided with a punch so as to cut out small disks of paper instead of mere puncturing, and is suitable for certain kinds of work when plenty of power is attainable to vibrate the punch Fig 25 same as 24 except that an extra needle C is used to keep the punch free of paper disks Fig 26 she[-]ws a double pointed needle suitable for certain kinds of work

The Roller which I prefer to used for placing the ink upon the punctured sheet is composed of a number of disks of cloth ~~feelt~~ felt or other soft material pressed closely together upon a shaft and held by a forked handle; the disks of cloth etc are turned almost true by a razor edged tool in a lathe =

The disks may be dispensed with and yarn wound upon a rod provided with flanges at each end. Even a wooden tube may be placed upon the shaft between its two heads upon which the yarn can be wound thus saving a quantity of yarn or the wooden roller may be felted a ¼ of an inch deep or covered with what is known as "rubber sponge" Compressed spong or soft leather may also be used Even the inking roller may be dispensed with and a brush provided with very short bristles used to force the ink through the holes of the punctured paper & give the requisite pressure Or a wooden roller may be used and a flat cloth soaked in ink may be placed upon the paper stencil and the ink forced from the cloth to the stencil by the pressure of the wooden roller or by a flat plate pressing on the cloth ~~all~~ at all points at once as in certain printing presses.

The battery which I prefer to used consists of a plate or plates of gas retort carbon and zinc in a solution of Bichromate of Potash & sulphuric acid, the zinc plate being arranged in such a manner that it may be lifted out out the solution when the battery is not in use =[9] The inks which I use have already been described in a previous caveat, relating to this subject.[10] I will here mention that the various chemical ingredients for the formation of ink may be placed in oils molasses, glycerine or other of great specific gravity and if soluable be used at once or if unsoluable ground in a paint mill.

Where it is desirable to trace drawings etc by puncturing from another drawing the puncturing paper may be soaked in a soil or substance which will give it a semitransp[arency][f] and placed over the picture or the picture etc may be posted upon the paper to be punctured and the puncturing point made to pass through ~~this~~ both this paper & the drawing itself = or the picture may be transferred to the paper to be punctured by soaking the picture if printed with printing ink moderately strong solution of a caustic alkali[g] and transferring by pressure =

I shall claim[h]

The various devices shewn for giving motion to the punching needle;

2nd The methods shewn for securing the needle to the needle bar

3rd The ~~m~~use of an inking roller composed of disks of soft material.

4th The inking brush for the purposes set forth =

T A Edison

ADS (letterpress copy), NjWOE, Lbk. 1:35 (*TAEM* 28:36); and DS, DNA, RG-241, Edison Caveat 69. Figures and figure labels are from Edison Caveat 69. A number of errors occur in the caveat figures. [a]"Autographic . . . 2" written by Charles Batchelor; followed by centered horizontal line. [b]Canceled. [c]Interlined above. [d]"& d" interlined above. [e]All figures are on two separate pages, both of which are signed and witnessed at the bottom; see n. 2. [f]From filed caveat. [g]First "al" traced over in pencil. [h]"I shall claim" traced over in pencil.

1. Edison wrote this date at the end of the document; Charles Batchelor wrote "Oct 2nd 1875." at the top of the document.

2. Edison filed this caveat on 29 November 1875. Edison's drawings in the letterbook draft are unclear, and the drawings from the filed caveat have been used instead. Draft versions of the drawings in Edison's hand are also found in Cat. 30,100 and NS-Undated-001, both Lab. (*TAEM* 6:12–14 and Supp. III).

Edison's draft drawings for the electric-pen caveat.

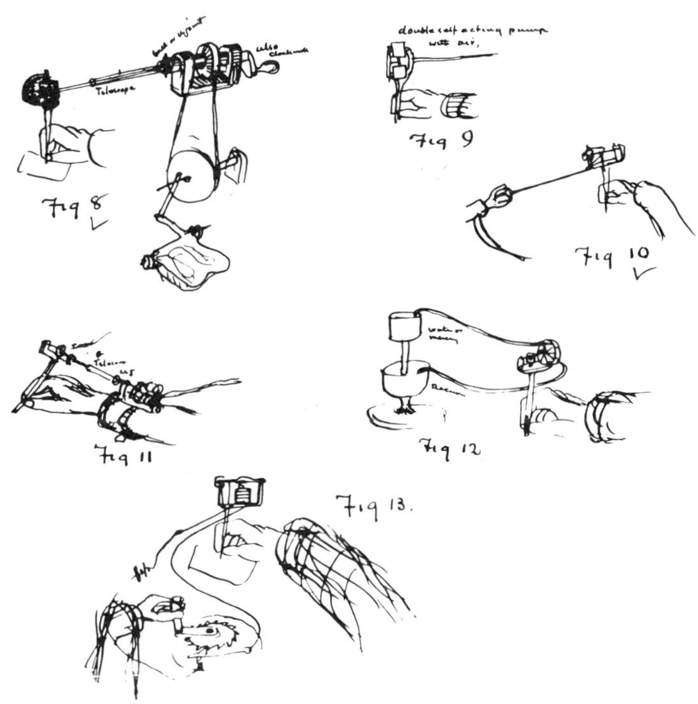

3. Doc. 633.

4. The filed caveat has "68" (Doc. 633) here.

5. In the filed caveat "the needle is perpendicular." follows here.

6. Both tongue n′ and its label are missing. Labels are also missing from figures 6, 8, 9, 10, and 13.

7. Mislabeled S. on figure.

8. Should be "H & H′."

9. This arrangement was illustrated and described by Edison in the filed caveat (Edison Caveat 69) as follows:

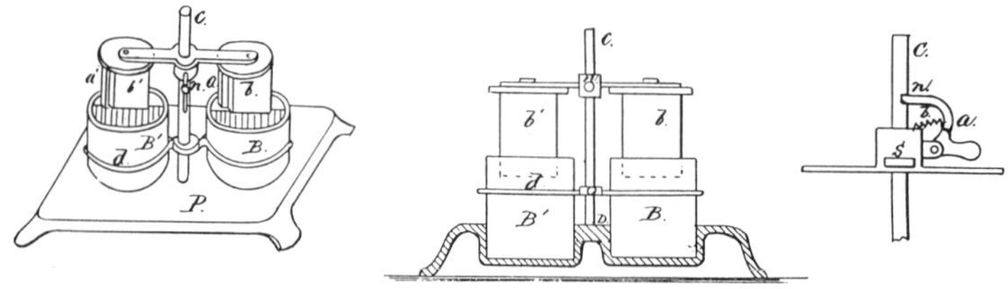

X. X. are hard rubber heads secured to the metallic arm s. secured to these heads are the carbon and zinc poles of the battery a′. b′. and a. b. immediately under are the two glass cells B′. B. each containing a porous cell having a half moon shape; n. is a set screw passing through s. into a long key seat or slot in the upright rod c. This screw by turning to the left unbinds the arm s. and allow it to be lifted up and then a turn or so of n. to the right secures it to the rod c. p. is the iron base, shown more clearly in fig 28.

The batteries or jars B′. B. are let down into the base for greater security.

d.d. is a double looped arm for retaining the cells in their proper position.

In Fig. 29. is shown a catch whereby the arm s. may be released from the bar c. and secured in another position more securely. a. is the catch. b. a spiral spring. n. a hole in the bar c. which is at a point where the elements are not in contact with the battery fluid.

10. Doc. 633.

–639–

Memorandum: Electric-Pen Business[1]

[Newark, c. October 4, 1875[2]]

Memorandum

If Mr. Miller will take general charge of the entire business I will give 50 cents nett on each machine sold in the U.S. After a time if it is found desirable to stock the Press, Mr Miller is to receive a portion of the cash & stock, as 30 is 1,000.[a]

Mr Miller is to arrange agencies, sign and complete contracts, keep all books accounts act as the treasurer, etc and generally give his sole time and attention to the business.[3]

In stocking the machine. Capital 200,000 divided in two thousand shares par value 100 each of money paid in Edison &[b] his partners receives 16,500 and 1,000 shares, full paid up. The remaining 1,000 shares to be sold for 20,000 cash 16 500 being paid to Edison etc. and the balance 3,500 to be used as fund to purchase 300 machines, office furniture etc and the making of a power press, patents =

WO[4] to receive 250 shares of Edison & partners stock for work in disposing of the 1,000 treasury stock.

The agents we now have are Electrical Construction & M Co[5] San Francisco, Nevada California, Oregon Washn Terr'y, B Columbia. P. Mallarkey[6] NYork City, Chas Mixer Maryland, B F Johnson Philadelphia.[7] W E Blennerhassett Michigan. W E Burton & Son[8] NYork State & New England States & H Clark[9] Brooklyn. ——— Wood,[10] Jersey City = [11]

Austria 3,000 all other Countries 2000 each = countries having no patent laws to be supplied by US company =

T. A. Edison

ADDENDUM[c]

[Newark, c. October 4, 1875]

With the present agents only they are bound to sell 48 presses each week and I estimate that when all the general agencies have been taken and each general agent has established subagencies that about 80 presses per day will be sold leaving $800 per day profit. The total cost of each complete apparatus is $10. with a margin

ADS (letterpress copy), NjWOE, Lbk. 1:70 (*TAEM* 28:67). [a]Followed by centered horizontal line. [b]Interlined above. [c]Addendum is an AD.

1. Edison probably wrote this for William Orton, Norman Miller, or both.

2. The document that precedes this in the letterbook is dated 4 October; one that follows it is dated 5 October.

3. Miller assumed the managership on 18 October but resigned on 13 January 1876. TAE to J. O. Clephane, 15 Oct. 1875, Lbk. 1:81 (*TAEM* 28:77); Charles Batchelor to Thomas Batchelor, 29 Mar. 1876, Cat. 1238:46 (*TAEM* 93:59).

4. "WO" is William Orton, who soon was promoting the electric pen. In a December letter to a prospective agent, he said:

> The Edison Autographic Pen is an apparatus that excites much interest wherever exhibited, and with a little practice can be, I think, made to serve a useful purpose in many cases. The apparatus has met with a very ready sale here. Indeed, it has been difficult to supply the demand. I incline to think you would do well to secure the agency for Pittsburgh and vicinity. [Orton to Thomas David, 22 Dec. 1875, LBO 15:384]

5. See Doc. 541 n. 2.

6. P. Mullarkey.

7. Benjamin F. Johnson was a telegraph operator. Gospill 1875, 779.

8. Unidentified.

9. Unidentified.

10. Unidentified.

11. Edison also entered into pen agency agreements with the following individuals: D. E. Drake, 14 September 1875, for Connecticut, Maine, New Hampshire, Rhode Island, Vermont, and New York, except for New York City and Brooklyn; Jacob C. Edison, 4 October 1875, for Minnesota; John Fox Gloyn, 16 November 1875, for the province of Ontario; William H. Barnard, 3 December 1875, for Delaware, Maryland, and Virginia; and James O. Clephane, 3 December 1875, for the District of Columbia. Miller (*TAEM* 28:988–98).

–640–

To E. C. Armstrong

[Newark, c. October 5, 1875[1]]

Armstrong[2]

Will obtain from Domestic Telegraph Co sole licence for Cincinnati to use the system during the life of the patents also 1 Recorder 1 back point relay and 1 box[3] for $2200. cash.

Will invent and make one pair of private line printers 60 revolutions per minute and assign whole patent for the US except the city of NYork for $3000[4] cash payable when the printers work satisfactory and the chief examiner (~~e~~Philosophical class) in Patent says he sees no infringement[5]

Cannot give you exclusive agency for press in Cincinnati as I have been in treaty with other parties[6] but can sell you any number at $24. cash

T. A Edison

ALS (letterpress copy), NjWOE, Lbk. 1:79 (*TAEM* 28:75).

1. The letter preceding this in the letterbook dates from 5 October 1875, the one following from 6 October.

2. Armstrong was superintendent of the City and Suburban Telegraph Assoc. Williams 1875, 85.

3. Transmitter.

4. Ink blurred; may be "$3600."

5. Nothing further is known of this offer.

6. Unidentified.

–641–

Draft Agreement with Edwin Coles and W. D. Clark

[Newark, c. October 5, 1875[1]]

This Memorandum of Agreement entered into this day of by and between Thomas A. Edison of Newark Essex County state of New Jersey party of the first part and Cole[2] of the City state & County of New York party of the second part, and W. D. Clark[3] of the City County & State of New York party of the third part Witnesseth

Whereas—the said party of the first part has invented a new printing Telegraph and transmitting Mechanism useful in transmitting telegrams from Broker's Offices to the stock Exchange[4] and is desirous of disposing of a portion of his interest therein, and to have it introduced, and

Whereas—the said part~~y~~ies of the second and third parts are desirous of acquiring a Certain interest therein—

Therefore be it Agreed that if the said party of the second part shall pay for the experimental Machines—a sum not exceeding one hundred and seventy five dollars, and will furnish all necessary funds to purchase one hundred complete sets of apparatus, or as many as shall be required, and to erect the necessary wires, & purchase the necessary appliances; He shall be entitled to a one third interest in and to the patent for such machinery, and one third of all nett profits arising from the use, rental or sale of such invention within the United States of America.

And if the said party of the third part shall give his whole time, talents and attention to the introduction and running of such Broker's Telegraph he shall be entitled to one third of all the nett profits arising from such sale, rental or use of said invention

Witness[a]

Df, NjWOE, DF (*TAEM* 13:418). Written in an unknown hand. [a]"Witness" written by Edison; followed by representations of seals.

1. Edison's draft of this agreement appears in a letterbook between a letter dated 4 October 1875 and a 5 October letter to Clark. In the letter to Clark, Edison states that he is ready to proceed with the stock-exchange project and encloses a draft letter that Coles sent in his own hand on 11 October (Lbk. 1:76–78 [*TAEM* 28:72–74]; DF [*TAEM* 13:413]). The letter from Coles stated:

> You can go ahead and manufacture one receiving Machine and one transmitting Machine for the transmission of Brokers orders, from their offices to Locked Boxes in the Stock Exchange.
>
> If the machine performs the Operation satisfactorily, I will pay your bills for the same, to an amount, not exceeding One Hundred and Seventy five dollars.

There is no evidence that this proposed scheme was ever carried out.

2. Edwin Coles was a stockbroker who had been secretary of the New York Stock Exchange Building Company since 1865 and a member of the exchange since 1870. *NCAB* 30:388–89.

3. Unidentified.

4. A drawing that may be related to this invention can be found in Cat. 30,100, Lab. (*TAEM* 6:617). It is reproduced at the top of the next page.

A drawing of a "Brokers Inst" from September 1875.

LABORATORY NOTEBOOK: EXPERIMENTAL RESEARCHES Docs. 642–647, 649–650, 655, 658, 662–663, 665–666, 668–669, 673, 679–680, 683, 686, 690, 701–702, 705, and 716

Edison began this notebook on 6 October 1875 as a formal record of experiments in his laboratory. He modeled it on Michael Faraday's famous *Experimental Researches in Electricity.*[1] Besides calling the notebook "Experimental Researches," Edison also copied the "Master Experimenter's" method of numbering entries.[2]

At first Edison wrote the entries and had Charles Batchelor and James Adams, his principal experimental assistants, witness them. He turned over the recording of the entries to Batchelor beginning on 19 October, although Edison and Adams often added their signatures.[3] That system of formal entries ended on 7 January 1876.[4]

The earliest entries in the notebook primarily concern the design of the electric pen and the batteries used to run it, although there are also a few entries on celluloid experiments and on mining ores. One entry relates to changes in the design of Edison's domestic telegraph receiver. On 16 November the entries begin to focus on acoustic and quadruplex telegraphy. Six days later these experiments led to observations of what Edison believed was a new force—the "etheric force." Most of the rest of the formal entries in this notebook concern experiments on the new force, with occasional entries on acoustic telegraphy.

1. Edison became acquainted with Faraday's work early in his career (see Doc. 7) and acquired a copy while in Boston (Dyer, Martin, and

Meadowcroft 1929, 101). At the Edison National Historic Site there is a copy of the three-volume *Experimental Researches in Electricity* which Edison signed and dated "Aug 21 1873." Marginal notes by Edison can be found scattered throughout the three volumes.

2. Edison applied this term to Faraday in Dyer, Martin, and Meadowcroft 1929, 101. The notebook entries are numbered consecutively from 1 to 61. After beginning the notebook, Edison decided also to number the drawings consecutively, and the numbers in Doc. 647 were correspondingly changed; some later drawings have both the consecutive numbers and an internal set of numbers. This is the case from the first etheric-force entry on 22 November (Doc. 665) until 26 November (Doc. 669), when the consecutive numbers were dropped. However, beginning in a 26 December entry, Batchelor again used the consecutive numbers.

3. Batchelor also kept his own formal record of some of the work in the laboratory in a separate notebook (Cat. 1317, Batchelor [*TAEM* 90:61]). He kept this record on an irregular basis through 13 September 1878, although he gave up numbering the entries after October 1876. Additional information regarding experiments recorded in the "Experimental Researches" can occasionally be found in the Batcheor notebook. The editors have provided such information through annotation or by selecting entries from Batchelor's notebook (Docs. 661 and 681).

4. The book was later used in 1878 at the Menlo Park laboratory by Edison's bookkeeper William Carman to copy some of the early experimental records on electric lighting.

-642-

Notebook Entry: Autographic Printing

[Newark,] Oct 6th 1875. All night

No 2[1] Electrical Pen

In testing a new model pen this evening, one with all the mechanism put in an iron frame (cast) we found that it did not approach in power or speed the rubber frame pens made by John Ott.[2] The spools of the iron frame pen were two sizes of wire longer than Otts magnets. Thinking perhaps that the resistance was too great or that owing to the greater number of convolutions the self induction delayed magnetization we unwound one layer from each spool, but this did not appear to increase the power or speed materially. we then tested R on Galvanometer of both pens & found them about the same. We used the cast iron frame as a yo~~i~~ke to the engine magnet and found by testing that there were free poles at the points where the cores are connected to the cast iron back. These free poles became nearly as strong when the armature was in contact with the regular poles, as the regular poles. We then in addition to the cast iron yoke inserted a small soft iron yoke about 3/64 thick. This made some difference in the speed still there were free poles. We then took the magnet out of the engine frame for testing. found that with the small soft iron back there was scarcely any free poles when the front armature was not in contact with the cores. This armature is 3⁄32 thick, but when in connection the free poles were very strong perhaps they would attract with a force almost equal to the regular ones. by making the armature thicker the free poles became stronger If the yoke was increased in thickness the free poles became weaker and it is very probable that if both the yoke cores and armature had the same mass of metal and all in good contact no magnetism or free poles could be detected but the moment the iron circuit, or perhaps we might call it the magnetic circuit[3] was weakened by decreasing the mass of the iron at any one point free poles would appear at that point. hence to obtain the maximum power the mass of iron in the yoke should be so great as to offer but little resistance to the passage of the magnetism so as not to cause free poles, thus compelling the poles to go to the point of greatest .R. those in front of the armature, and it may be found that a thin armature will increase this tendency by increasing the resistance We also discovered a Phenomenon. Thus with a thick armature and back or yoke considerably thinner a nail was attracted only on opening and closing the circuit[a] further tests should be given this and the armature & yoke proportions.

Thos. A Edison Chas. Batchelor
James Adams

X, NjWOE, Lab., Cat. 994:2 (*TAEM* 3:194). [a]"only . . . circuit" underlined twice.

1. See headnote above, n. 2.

2. On cast-iron frames, see Doc. 626; on Ott's manufacture of pens see Doc. 623.

3. In 1873 a young American physicist, Henry Rowland, had published an article in the British science journal *Philosophical Magazine* in which he derived for magnets an analogy to Ohm's law governing electric circuits ("On Magnetic Permeability, and the Maximum of Magnetism of Iron, Steel, and Nickel," 46:140–59; see Miller 1976). Edison knew *Philosophical Magazine*; whether he read Rowland's heavily mathematical paper is not known.

–643–

Notebook Entry: Autographic Printing[1]

[Newark,] October 7th 1875

No 3[2] Autographic Pen

Worked on pen all night= (model pen) found that cams gave vibrating pen too great a movement (ie) Three times more movement than was necessary to puncture the paper,

hence engine had to do more work, altered it so that it would give much less movement. There was a great improvement in the speed of the engine, but not enough to account for the unsatisfactory working of the pen. We tested the spring contact also in setting the armature in various positions.
Chas. Batchelor James Adams

X, NjWOE, Lab., Cat. 994:4 (*TAEM* 3:195). Written by Edison.

1. This entry is a continuation of Doc. 642 and is continued in Doc. 644.

2. See headnote, p. 600 n. 2.

–644–

Notebook Entry: Autographic Printing[1]

[Newark,] Oct 8th 1875

No 4[2] Autographic Pen

Went at model pen again. found that the trouble was principally in the back center screw. it was cupped too deep, and the magnet pulling the armature towards it made a great friction, not noticable when there was no magnetism. had a small center hole put in the screw, and the pen at once went perfectly satisfactory Also took out the magnet wound with sounder wire and rewound them with finer wire, perhaps three sizes smaller. Then engine went still faster, and the pen appears to be the most satisfactory yet produced Two inductorium cells[3] do not work it satisfactory. on all night=

Thos A Edison Chas. Batchelor
James Adams

X, NjWOE, Lab., Cat. 994:4 (*TAEM* 3:195).

1. This entry is a continuation of Doc. 643.

2. See headnote, p. 600 n. 2.

3. See Doc. 435.

–645–

Notebook Entry: Celluloid

[Newark,] October 8th 1875

No 5[1] Nitro-cellulose = Kid

The celluloid Cos[2] proxlin[3] containing a considerable quantity of chloride lime and perhaps some free SO_3 was found readily soluable in proxylic spirit, in the cold, even after so much proxlin or nitrocelluose had been added that it assumed the consistency of dough ~~T~~It would still continue to dissolve more

Phenomenon after it is thus dissolved it is easily and entirely soluable in water, a thin solution of which makes an excellent glue for fastening paper together. as the substance has an intense acid reaction I am inclined to think that the proxylic spirit is turned into acetic acid which is an excellent solvent for proxlin. I tried to vulcanize it by adding flower of sulphur and heating it in an iron watch glass dish—failure—It does not mix with gutta percha dissolved in Carbon Bisulphide

Thos A Edison

X, NjWOE, Lab., Cat. 994:4 (*TAEM* 3:195).

1. See headnote, p. 600 n. 2.

2. The Celluloid Manufacturing Co.

3. Pyroxylin, first developed by the French chemist Theophilé-Jules Pelouze, was a form of nitrocellulose used in the production of celluloid. See Doc. 667 n. 1; Friedel 1983, chap. 1.

-646-

Notebook Entry: Autographic Printing

[Newark,] October 14 1875

No 6.[1] Autographic Pen

Our original pen was faulty in these respects, hard rubber was not a suitable material for the frame and the whole pen was too expensive to manufacture, no two pen acted or were alike some would go right off and make a much larger number of revolutions in one minute than others, and work on a spent battery when others did not

We made a pen with an iron frame and it did not go well and for a time we thought the iron must have something to do with it but we have since found that it has no effect

We made a frame of brass and new parts and found that it did not run as well as some of the old pens. after considerable experimenting found following

1 The centres must be small.
2 The back centre must bottom thus 
3 The front centre must be thus 
4 Both centres and screws must be hardened steel
5 The wire on the magnets must be No instead of No
6 The cam must be of steel
7 The cam clip must be of steel
8 The contact vibrating spring must have a piece of platina wire thus

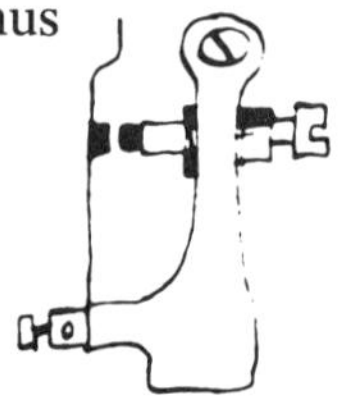

9 The armature must be half the thickness of the Keeper
10 The needle must have a very long taper, ie[a] not blunt
11 The cam must not move the clip but very little sideways.
12 The needle must be soldered to rod
13 centres must be very true to prevent armature from approaching magnets eccentrically
14 All wearing parts must have a looking glass burnish
15 The cam should vibrate the needle just enough to pass through the paper ie[a] the length[2] width[b] of this line .

Thos A Edison | Chas Batchelor
James Adams

X, NjWOE, Lab., Cat. 994:5 (*TAEM* 3:195). [a]Circled. [b]Interlined above.

1. See headnote, p. 600 n. 2.
2. Edison neglected to cancel this word.

–647–

Notebook Entry: Batteries for Autographic Printing

[Newark,] October 14 1875

No 7.[1] Batteries

We have experienced considerable trouble in obtaining a suitable battery to run our Autographic pen The French Grenet battery are easily tipped over and the rod which serves to raise and lower the zinc in the liquid becomes stuck in its holder by oxidation and by crystalization of salts. In order to obviate this objection we made a model of wood intending to make it of earthenware or glass like figure .~~1~~2.

Fig ~~1~~2[2]

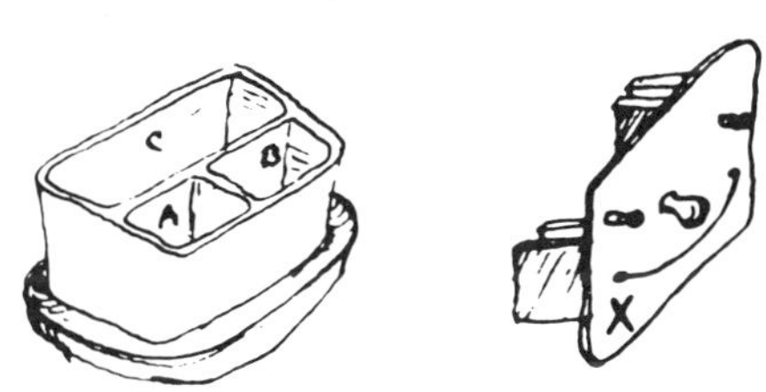

The appartments A and B for the battery and C for dripping when the cover X has been lifted up and turned around and the battery is not in use. We found that it was almost impossible to get the compartment jar made of earthenware, it contracts so much that the partition would seperate from the side, however a party in Trenton said he would try and make them for $1.25 each. We next went to the glass people found that the mould would cost $125.00/00 and the glass moulded with cost 25c per lb at lowest[3]

We then thought of making the cell shewn in fig .1. of cast iron placing glass cells in A B and C. also of forming the cell out of sheet hard rubber by mould and hot water or oils.

Also of taking two Grenet batteries and placing Them on a stand as in fig 2

Fig 2 3

which brought in another idea of putting two glasses upon a stand (iron) as in fig .3. and lifting both the zincs and carbons out using a porous cell. The glasses set in the iron base 1 inch The carbons are copper plated and soldered to the connection pieces. The zincs have a screw cast in them all fastened to thick rubber disks which are in turn fasted to an arm connected to the upright rod. The battery is shewn in fig 3

Fig 3 4

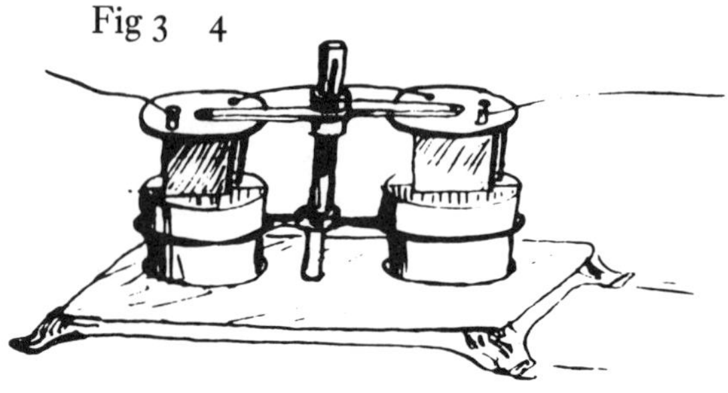

Fig 4 5

Fig 5 6

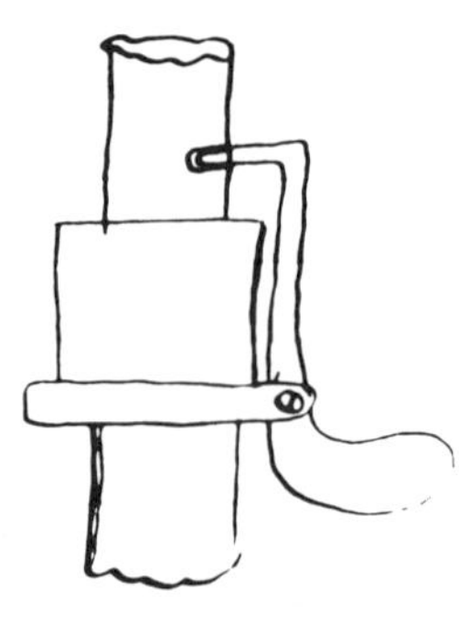

Fig 4 shews the porous cup Fig 5 a catch for holding the elements out of the liquid. The arm carrying the elements is prevented from Turning by a key seat in the upright in which a screw runs which can be used if necessary to hold the arm in any position in place of the catch shewn in Fig 5

We proposed to use electropoin of great strength in the porous pot.[4]

Thos A Edison   Chas. Batchelor
James Adams

X, NjWOE, Lab., Cat. 994:6 (*TAEM* 3:196).

1. See headnote, p. 600 n. 2.

2. Edison originally numbered the figures in this entry from 1 to 5, but Batchelor later changed them, numbering all the figures in the notebook consecutively.

3. See the correspondence between Edison and Hagerty Bros. & Co., Sept.–Nov. 1875, DF (*TAEM* 13:260, 263) and Lbk. 1:26, 32, 92, 121 (*TAEM* 28:28, 34, 87, 113).

4. Edison and his staff experimented on various aspects of the electric pen batteries during the next week. Cat. 994:6 18; Cat. 997:3 4; both Lab. (*TAEM* 3:196–202, 353–54).

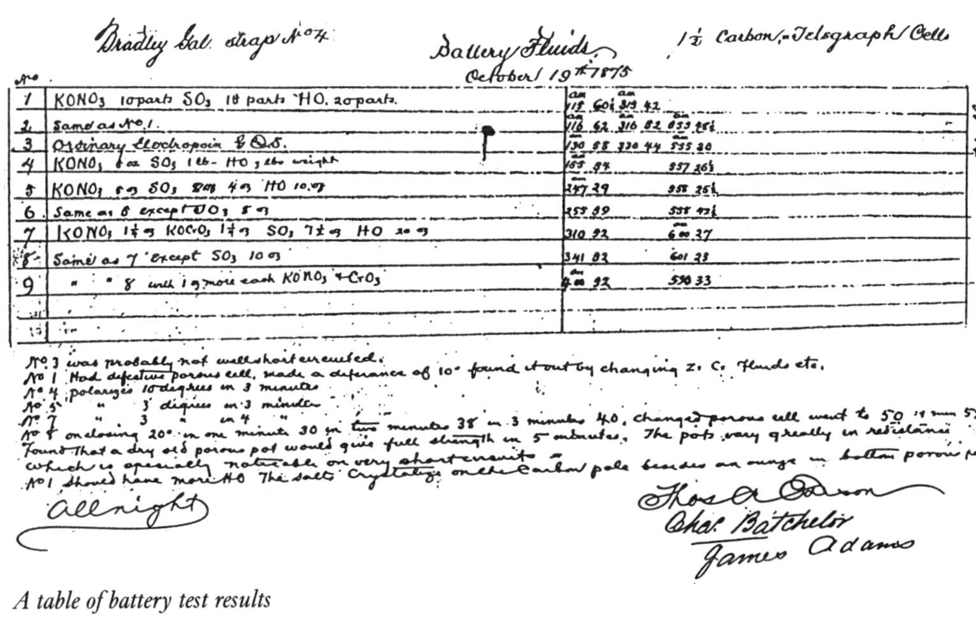
Bradley Gal. strap No 4   Battery Fluids October 19th 1875   1½ Carbon—Telegraph Cell

No			
1	KONO3 10 parts SO3 18 parts HO. 20 parts.	115 60½ 313 42	
2	Same as No. 1.	116 62 316 82	553 25½
3	Ordinary Electropoin & Q.S.	130 58 330 44	535 20
4	KONO3 6 oz SO3 1 lb– HO 3 lbs weight	165 94	557 36½
5	KONO3 5 oz SO3 ... 4 oz HO 10 oz	247 39	555 35½
6	Same as 5 except SO3 5 oz	255 89	535 47½
7	KONO3 1½ oz KOCrO3 1½ oz SO3 7½ oz HO 20 oz	310 92	600 37
8	Same as 7 except SO3 10 oz	341 83	601 23
9	" " 8 with 1 oz more each KONO3 + CrO3	400 92	570 33
10			
11			

No. 3 was probably not well short circuited.
No 1 Had defective porous cell. made a difference of 10° found it out by changing Z. C. fluids etc.
No 4 polarizes 10 degrees in 3 minutes
No 5 " 3 degrees in 3 minutes
No 7 " 3 " in 4 "
No 8 on closing 20° in one minute 30 in two minutes 38 in 3 minutes 40, changed porous cell went to 50 in 1 min 52
Found that a dry old porous pot would give full strength in 5 minutes, The pots vary greatly in resistance which is specially noticeable on very short circuits —
No 1 should have more HO The salts crystalize on the Carbon pole besides an ... in bottom porous pot
All night
Thos A Edison
Chas Batchelor
James Adams

A table of battery test results from 19 October 1875.

–648–

To Albert Chandler

Newark, N.J. Oct. 18th, 1875.

FRIEND CHANDLER

It is reported that I have received some $4,500 on account of experiments with the Roman Letter apparatus.[1]

Fearing that you may have the accounts mixed, I desire to say that I have received but about $700 for this purpose, and that from Mr. Gould.[2] The additional $1,450 necessary to complete the two machines was advanced from my abundant personal means.[3] I make this explanation to set the matter right and also because it instances one of the many reasons why I am so frequently hard up. Yours,

EDISON.

PL, NjBaFAR, *Harrington v. A&P,* Box 17B, Defendants' Exhibit 14, 1:317.

1. It is not known to what "report" Edison refers. The $4,500 may have included disbursements to Edison and Murray for manufacture of automatic telegraph instruments installed on the lines of Atlantic and Pacific (see Edison and Murray accounts with Atlantic and Pacific, 75-011, DF [*TAEM* 13:470, 474]; and Eckert's testimony, 2:241, Box 17B, *Harrington v. A&P*).

2. See Doc. 546 n. 2.

3. One unpaid bill for work on roman-letter apparatus, dated 22 May 1875, can be found in 75-011, DF (*TAEM* 13:476). It shows $351.60 for work on two one-wire roman-letter instruments. In 1877 Edison complained that his work on a roman-letter system had been hampered by company president Thomas Eckert's failure to provide the funds necessary to complete his experiments (TAE to Gould, 2 Feb. 1877, Complainants' Exhibit 4, 2:185–87, Box 17B, *Harrington v. A&P*).

–649–

Notebook Entry: Autographic Printing

[Newark,] Oct. 19th 1875

No 15[1] Printing Process

Use the Autographic pen to puncture the holes in the paper to form the drawing, writing etc. Wet the paper, or use paper previously made impervious to moisture by soaking in shellac or other material, or tin foil or other metallic foil. Lay the punctured sheet of tin foil or prepared paper upon a polished surface. Bring a frame work down onto it & pour in liquid 'plaster of Paris,' putting the frame work at such an angle that the weight of the extra Plaster of Paris will force it through the holes. After it is thoroughly set take it out & separate the paper from it either by pulling off or dissolving in H_2OSO_3 or burning Of course Shellac, sealing wax or any substance that can be poured upon the paper upon the paper in a liquid or semi-liquid state & hardens by time heat or other means will do.

After the plate is dry it is coated with plumbago or other conducting substance used by electrotypers to give a nonconducting substance a conducting surface, & then put in a plating bath & electrotyped in the usual manner, or it may be printed from direct.[2]

The punctured stencil done by the autographic pen may be used like a 'papier maché' matrix in stereotyping,[3] the stencil being placed in a mould & type metal poured into it, the pressure of the extra metal at the top of the mould serving to force it through every punctured hole to the smooth surface of metal underneath. When the metal is hard it is taken out the paper stripped from it & blocked, it is then ready for printing in the usual manner.

The punctured paper previously wet we tried & poured Plaster of Paris on it but found after hardenings that it did not go through the holes owing to them being closed up by the water swelling the paper.

We tried pouring lead on a stencil but it did not go through the holes. we then poured through a funnel shaped tube like Fig. 1 so that we could get a body of metal to press it in.

Fig. 1   9[4]

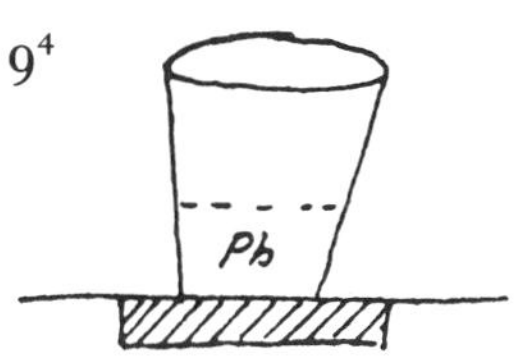

In this manner we could see the shape of the letters on the lead & there is no doubt that with care & lots of pressure we could get it perfect. We tried putting stencil paper on flat plate of Zn, & fastening down with Beeswax, then making a beeswax embankment around the writing & pouring in Hg but with little success as the holes seem to be too small for it to get through We also tried copperplating through the holes to get a raised surface on the metal to print from We tried writing with common ink & shaking plaster of paris on it to form a hard insoluble substance to print from it raises it up pretty high but it is probably no good.

Paraffine apparently goes through the holes & we took a good print from it.

T Thos A Edison Chas Batchelor
James Adams

X, NjWOE, Lab., Cat. 994:13 (*TAEM* 3:199). Written by Batchelor.

1. See headnote, p. 600 n. 2.

2. In electrotyping, a mold was immersed in a bath of copper sulfate; the passage of an electric current deposited copper on the mold. The resulting copper plate was used for printing. Knight 1876–77, s.vv. "Electroplating," "Electrotype."

3. Stereotyping was used in printing and employed a process similar to the one described by Edison except the mold was created by immersing type in the plaster of Paris. Knight 1876–77, s.v. "Stereotyping."

4. See Doc. 647 n. 2.

-650-

Notebook Entry: Mining

[Newark,] 12 P.Midnight Oct 19th 1875

No 17[1] Mining Ores.[2a]

To ascertain if a vein of smelting (ie) earthen ores is a pocket, connect an earth plate making good connection with the general earth to a telegraph wire of known resistance the end of the wire being at the mouth of the mine, which mouth must be soft earth all round, the wire is run into one end of a test battery thence to a galvanometer & Wheatstone bridge thence to a large plate (of given size previously tested together with other plate & wire in general earth) inserted in the body of ore in the vein By a little experience the fact that the vein is or is not a pocket can be determined also its size & richness. As the salts or oxides of the metals are better conductors than ordinary earth if the vein is a pocket the body of ore might be determined by charging the whole body statically or by secondary battery action.

Of course in these tests the condition of the county geological formation, hygroscopic condition, the nature of rock & ore must be taken into consideration as well as many electrical points. In a Quartz vein with walls of rock containing little or no moisture the ore being one mass of stained rock (ie) finely divided, the test can be made by charge, condenser, etc.

The H_2O oozing from the rocks of suspected localities might be tested with spectroscope to detect soluble chlorides of gold & silver salts the different parts of the rock showing different quantities from which the lay of the vein may be got by calculation.

We find by experiment that if porous earthenware is held on the tongue it soon clears it of all H_2O & the force of capillary attraction locks the tongue to it so that it is painful to separate it.

This principle may be useful in mining ores in the following manner:— Take earthen ore & roast it to turn the silver into a hygroscopic salt, it is then exposed to the air till it is moist, & afterwards thrown onto large trays or cylinders of porous earthenware when the silver salt will stick by capillary attraction & the sand & silica will not

Thos. A. Edison Chas Batchelor
James Adams

X, NjWOE, Lab., Cat. 994:15 (*TAEM 3:200*). Written by Batchelor. [a]"Mining Ores" underlined twice.

1. See headnote, p. 600 n. 2.

2. See Doc. 579. Entries of 21 and 22 October in the same notebook continue this subject.

–651–

From Pitt Edison

Port Huron, Mich, Oct 23 1875[a]

Der Bro

I would of rote you befor but have ben waiting from day [-]to day to hear from the new company thay have don nothing yet I told them to make up thair minds by next Wednsday or for ever after hold thair peace I told them[b] we would buy or we would build around to Gratiot thay could do as thay pleased[1] buisness is fair I put the two new cars on the Sarnia road tomorrow (Monday)[2] I have got the switch don and it works niseley I will send you statement of both roads on the first of every Month so you will see how wealthy you are getting[3] I think thay all want to sell but Wastell[4] so I think thay will get fighting among themselves before long shoud thay conclud to do any thing by Wednsday I will Telegraph you

(remeber me to wife & DOT[5])

WPE[c]

What States is thare left in the west for the Printer Partys her would like to opporate[6] Answer WPE

is any thing taken in Canada[7]

ALS, NjWOE, DF (*TAEM* 13:394). Letterhead of Port Huron and Gratiot Railway Co. [a]"Port Huron, Mich," and "187" preprinted. [b]Interlined above. [c]Followed by "over" to indicate page turn.

1. Pitt is probably referring here to a proposed merger between the City Railroad Co. and the Port Huron and Gratiot Railway (which by this time seems to have dropped "Street" from its name). City Railroad had been formed in 1873 after some of the original incorporators of the Port Huron Railway fell out with Pitt. The two companies sued each other in 1874, but these suits proved inconclusive by the spring of 1875.

Although they continued to compete for business, the town of Port Huron was too small to support two street railways, and both companies lost money. As a result they began negotiating a merger. Jenks n.d., 7–13.

2. Pitt was secretary of the Sarnia Street Railway Co., incorporated in 1874 by the Canadian Parliament. It connected the town of Sarnia with the Grand Trunk Railway depot in Point Edward, just across the St. Clair River from the depot at Fort Gratiot. That depot was connected to Port Huron by Pitt's Port Huron and Gratiot Railway. Bill of Incorporation and agreement between Sarnia Railway and Grand Trunk Railway, 20 Dec. 1874, DF (*TAEM* 13:26–32).

3. Edison apparently invested in the Sarnia Railway as well as the Port Huron line. Edison's own notes on the figures for both railways from May 1876 were attached to this letter in the file. Also in the file is a comparative statement of receipts and operating expenses of the Port Huron Railway for the month of November 1875. DF (*TAEM* 13:394, 396).

4. William Wastell (*TAEB* 1:306) led the revolt that resulted in the formation of the City Railroad Co. Jenks n.d., 7.

5. Edison's daughter Marion was nicknamed "Dot."

6. This refers to agencies for Edison's electric pen. At the time, Edison's nephew Charles and his cousin Jacob Edison had agencies for Michigan and Minnesota respectively. See Doc. 636 and Doc. 639 n. 11.

7. John Gloyn had the agency for the province of Ontario. Doc. 639 n. 11.

–652–

Draft Caveat: Facsimile Telegraphy

[Newark,] Oct 30 1875[1]

The object of this invention is to transmit and record over telegraphic wires autographic writing =

The invention consists 1st The method of transmitting from the original manuscript. 2nd The method of synchronously moving the various parts of both the transmitting and receiving machinery— 3rd in the method of manipulating the waves of current to obtain clear and distinct writing. The method of transmitting from punctured paper. The chemical solution for recording, ~~T~~and other devices, for performing the various operations.

In fig 1 is shewn the method of opening and closing the ~~e~~main circuit.

Fig. 1.

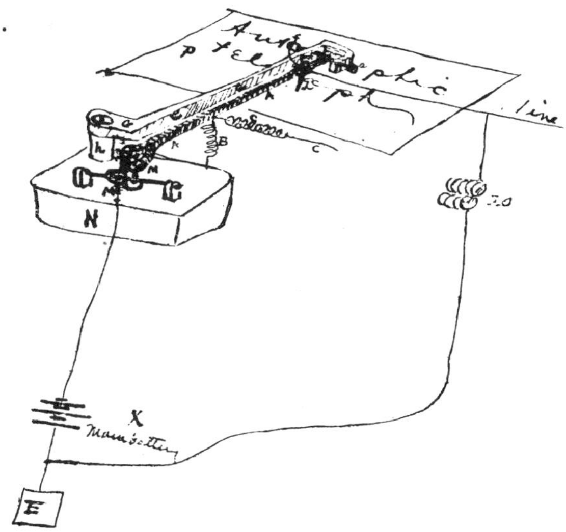

P is the original manuscript which is to be transmitted. It consists of high glazed pap[er][a] and the characters are w[ritten][a] with ordinary ink preferably copying ink This sheet is laid upon a plate which may be moved back and forward

A is a lever arranged in a similar manner to the lever of my Electromotograph already patented[2] it is pivoted at M and is free to swing either sideways or downward near its extremity is a finely [bu]rnished[a] pointed screw d resting upon the paper at its extreme end is the contact points which play between the contact screws e and f secured to the arm G

The spring .B. pulls the lever point down upon the paper while the spring c pulls it to the right

The operation is as follows

Suppose the platen upon which the paper is secured be moved always [from][a] right to left[b] or if it be a [cylinder rotates from][a] right to left. The tension of the spring .B. is such that there is insufficient friction between the point d and the paper to overcome the tension of the spring c hence the lever will remain in contact with the point e and open the main circuit but if an ink mark passes ~~un~~ in the path of d the ~~friction~~ surface is no longer smooth, but on the contrary is very rough & the friction is increased so greatly that the lever is moved away from e and put in contact with f thus connecting the main battery to the line for a period of time equal to the time which it takes the ink mark to pass under d. The faintest trace of ink is sufficient to move the lever A. with force and promptness.

It is obvious that the operation could be reversed the paper be soaked in a saturated solution [of][a] lead or other salt and [allowed to crystalize thereby making][a] the paper exceedingly rough & placing great friction upon the point d The ink used being of such a nature that it would either decompose the lead salt & lessen the friction or give a surface itself which would be smooth in this case ~~Anoth~~ The line should be connected to the point e and the platten or cylinder moved in the same direction as before or vice versa.

Another mode of increasing the friction consists in saturating the paper with a certain salt and writing with an ink which strikes a bulky precipitate with the same thus raising the marks up ~~and the raising of the lever wo~~ which striking against d would throw it to one side while ~~paper~~ passing—or the ~~marks~~ raised marks could be made to raise the lever and close a platina point.

In this connection I will mention that if ~~an~~a dispatch is written with an ink composed principally of a saturated solution of Bicarbonate of Potash, the passing it over a gas flame raises the marks 1/32 of an inch high a fact which I hope to take advantage of in this [-]connection =

The message might be written upon soft thick biblus paper with an agate stylus thus indenting it which would if used in the manner shewn in fig .1. would increase the friction when d dropped into an indentation, or the same could be of such a depth that they would raise and lower a lever sufficiently to open and close the main line circuit

As but very little movement of the lever A fig 1 can take place and as the spark would be very considerable if used upon a long line in the ordinary manner of connecting, I have inserted a shunt wire to earth in which is an electromagnet would[3] with a great number of convolutions and having a high resistance which effectually kills the spark.

When it is desired to repeat an autographic telegram from one circuit to the other ~~there~~ with the <u>friction</u> plan, the received telegram may be recorded upon paper saturated with a certain easily decomposed solution for instance, Chloride of Sodium, Potash or ammonium, the paper resting [---][c] upon a sheet of tin foil and so connected that the hydrogen will reduce it. The marks instead of being recorded upon the top (they could be if receiving by means of pyrogallic acid) will shew upon the surface of the tinfoil underneath which in this case forms one pole of the battery. It will be found that where the hydrogen has reduced the metal it is exceedingly rough and can be reemployed upon the second circuit in the same manner as the paper ~~was~~ is used in fig 1^{d}=

I do not wish to confine myself to ~~either~~ any particular metal to either reduce or oxidize or on any particular solutions for wetting the paper as there are many solutions which when acted upon at the surface of the metallic cylinder or plate upon which the paper rests would form a salt of the metal which tenaciously adheres to the metal causing great friction upon the passage of the point d Again if acetate of mercury or other soluable salt of mercury either with or without a conducting salt, be used to saturate the paper ~~free~~ mercury will be set free upon the tin foil or other metal and amalgamating with it gives indented writing after a few seconds when it is brushed away which of scarcely any depth [~~enormously~~?] serves to deflect the lever A when the point d passes over it=

Fig 2 shews a top view of fig 1

Fig. 2.

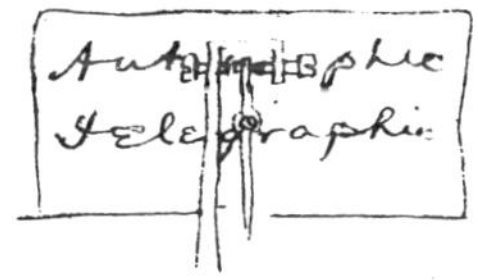

Fig. 3.

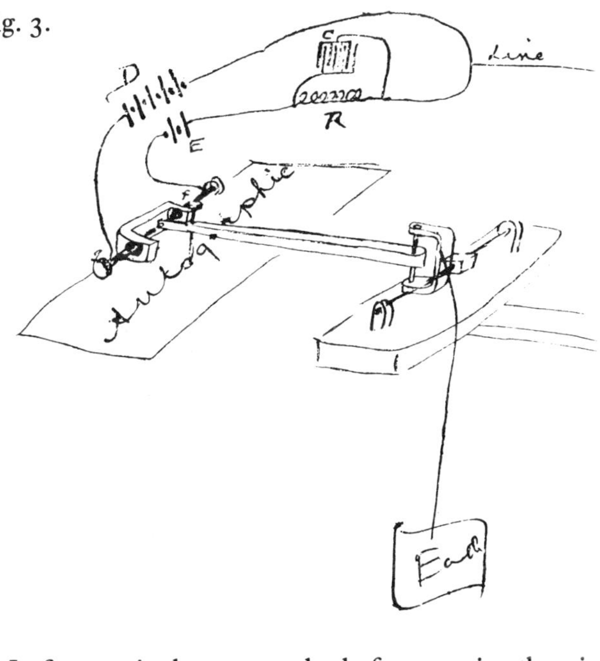

In figure 3 is shewn a method of connecting the wires so as to ensure correct and legiable recording at the distant station D is the recording battery connected to the point g which is connected & disconnected from the earth by the lever which in its turn is moved by the ink marks upon the manuscript underneath its point=

After connecting the battery D to the line for an instant the lever comes in contact with the point f, placing a smaller battery E ~~of~~ having opposite poles connected to the line in connection with it through a resistance R shunted with a large condenser

The object of the resistan[ce][a] coil & condenser is to cause the first portion of the wave to be very strong and gradually die away to almost nothing.

Fig. 4.

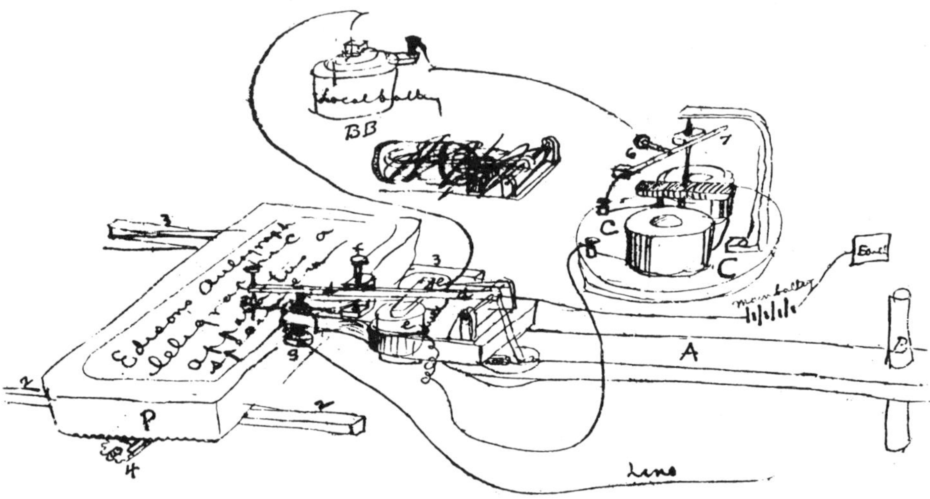

In figure 4 is shewn another method of transmitting the manuscript being in the form of stencil writing ie[d] the characters ~~of~~are formed of innumerable holes or punctures made with a vibrating needle which has already been described in my caveat relating to the electrical writing pen and autographic printing press for the reduplication of manuscript writing filed ____ 1875.[4]

A is a lever pivoted at B [-]swing back and forward a distance equal to the length of the plate .P. upon which the punctured manuscript is placed.

d is a lever pivoted like a sounder lever & playing between two points .f. and .g. d is a smooth straigh wire with a flat end and is a little smaller than the punctured holes in the manusc[rip]t.[a] The lever d is held against f by a suitable retractile spring= a rapid up~~on~~ & down motion is given this lever by [the][a] magnets e.e. which [are][a] placed in the same circuit with the Magnetic Engine .C. which is provided with a self acting circuit closer consisting of the cam upon the engine shaft the spring 7 & contact point .6. BB[5] is the local battery when the engine is in motion the contact spring .7. opens & closes the circuit with great rapidity and this causes the magnets ee. to vibrate its lever d also with great rapidity.

When no puncture is passing underneath the point of the wire d, upon the extreme point of d the thickness of the paper prevents the lever [---][c] from coming in contact with the point g. hence the main circuit remains open but the moment holes pass underneath d the thickness of the paper ceases to intervene and the point passed through the hole to the metal [or other][a] substance upon which the paper lays and the lever d comes in contact with .g. closing the main circuit & sending a wave to the distant station & recording a dot upon the chemically prepared paper

It is obvious that the engine may be replaced with a self vibrator like those used in induction coils or a clockwork & break wheel or apparatus turned by hand or other power or any device by which the current may be opened and closed with great rapidity Even these devices might be dispensed with and a self vibrating attachment placed upon the lever d so that it would vibrate of its self= [but I do not consider][a] the latter plan a good one=

The plate .P. is moved forward by the pinion 4 engaging in a rack upon each end of .P. the pinion shaft being moved either by an electromagnet actuated at the moment that A reaches its limit of movement to the right [or][a] left or by a movement derived from .A. for instance a ratchet wheel and lever with click

Fig. 5.

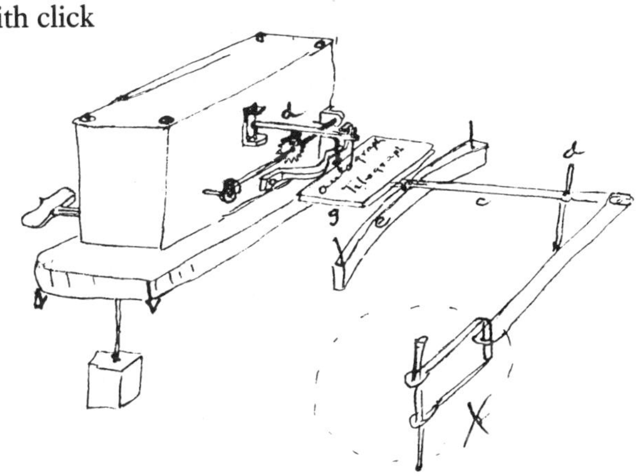

In figure 5. is shewn another method of vibrating the contact lever in this case a ratchet wheel a moved by bevel gear &

clockwork ~~op~~ rotates rapidly and moves the lever up and down by engaging in a single tooth upon the lever d g is the platen which is secured to the lever c which is pivoted [at][a] .d. is as given a back & forward motion by a second clockwork with crank x= e is a track upon which a wheel secured to c runs and [serves][a] to steady the plate .g. when it is rotated to right to left=

The feeding device is also [secured to the lever c.][a] [an]d[e] consists of a long [r]ack[e] engaging in pins at the extreme limit of the movement of .c. feeding the platen towards the fucrum d about the 1/100 of an inch at each movement This is not shewn in the drawing=

~~I will now describe the method which I adopt for obtaining a synchronous movement of the sending & receiving machines~~

I will now describe the method which I adopt for obtaining a synchronous movement of the transmitting & Receiving Machines.

Fig. 6

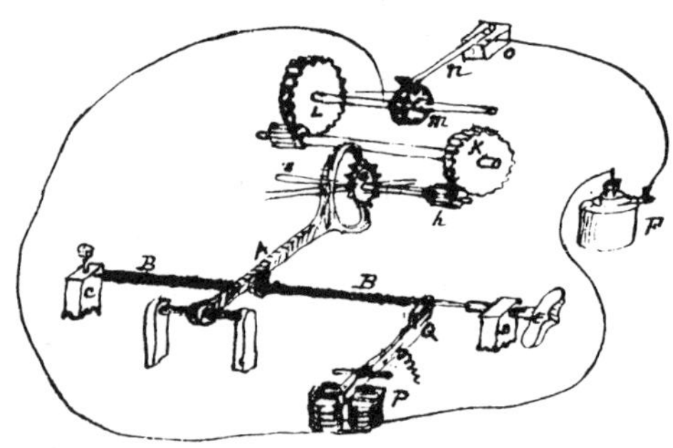

In fig .6. is shewn a clockwork which serves to actuate the various recording or transmitting devices.

m is a break wheel rotating in the path of a contact spring .n. placed in circuit with the battery F and magnet .P. which receives magnetism and attracts its lever at intervals upon the extreme end of the lever of the magnet P is a pad Q which hits against a pianoforte string B which may be made to give any required ~~number of~~ the required[f] vibrations by increasing [and][a] decreasing the tension upon it by the screw E in the upright D.

While the clockwork is rotating the string B is constantly vibrated by receiving impulses from the lever of P. This cord B is secured [to the escape][a] lever A. [g]iving[e] it a vibratory motion & releasing the escape wheel G & clockwork. As the cord will with a certain tension always make the same number of vibrations in a minute it follows that ~~th~~ two clockworks with cords tuned to give the same note will run exactly alike=

The wheel m might be placed upon either S or K ~~a~~ if a rapid vibration of the lever of P is required.

Fig. 7.

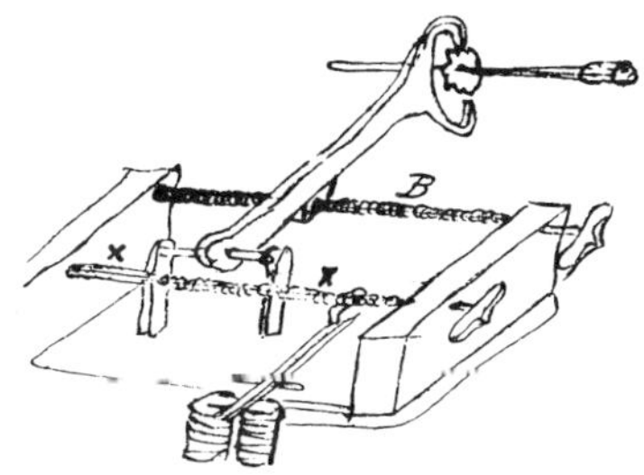

In figure 7 is shewn two cords the movement of X by the magnet lever setting the cord B in motion [by][a] sympathy =

Fig. 8.

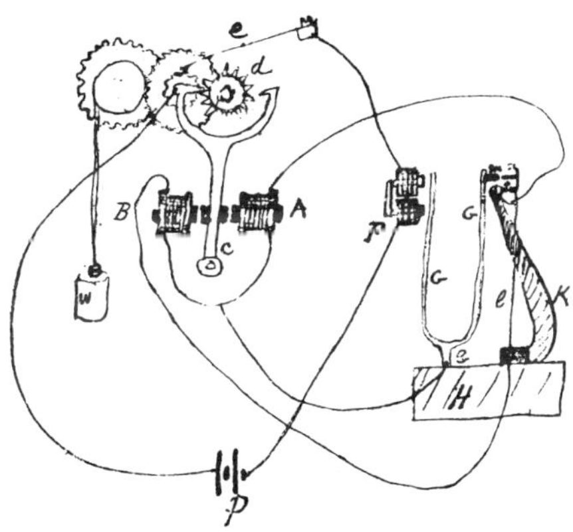

In figure 8 is shewn a tuning fork G kept vibrating by an electromagnet F which has its circuit opened & closed by the clockwork The contact springs m n so serve to throw the current from the battery .P. from the helix B to A & vice versa releasing the escapement at every vibration of the tuning fork = which has prongs of extreme length to ensure a slow vibration

Fig. 9.

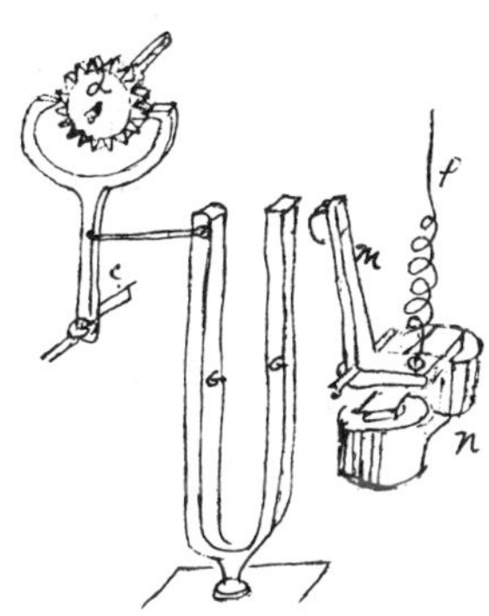

In fig. 9 is shewn the tuning fork connected directly to the escapement I will mention that one of[g] the prongs of the tuning fork might act as the armature of a magnet & being provided with a contact spring might open & close its own circuit dispensing with a break wheel upon the clockwork

[I][e] will here mention that it is not [n]ecessary[e] to vibrate the lever d Fig .4. if the spring is made to pull the ~~poin~~ lever d & point d down upon the paper the ~~poin~~ spring will cause the point d to enter the holes & close the circuit =

The manuscript may be held upon the cylinder or platen by perforating it with numerous fine hole leading into an air chamber inside the MSS being laid upon it the air is exhausted holding it by suction. This might be dispensed with & the plate covered with a non drying viscid liquid such as castor oil, glycerine the suction being sufficient to grasp & hold the paper down smoothly & firmly at all points =

My claims will probably be

1st opening and closing an Electric circuit by the movement of a lever when such lever is moved by ~~ink marks upon paper in the~~ a difference in friction between the written character & the material upon which it is written[h] sub[stantially] as described.

2nd opening & closing the circuit of an autographic telegraph by means of punctured paper ~~in the~~ sub as set forth.

3rd[i] giving a rapid vibratory[j] motion to a[i] lever by magnetism or other power for closing an electric circuit by means of writing formed of punctured holes sub as described =

4 The method of transmitting currents and compensating for the same shewn in fig 3

5th holding the MSS to the cylinder or platen by air suction or a viscid substance as [set forth][a]

[6th T]he[e] method of obtaining [s]ynchronism[e] by means of [t]uning[e] forks or vibrating cords as shewn in figures 6, 7, 8, & .9 for the purpose set forth =

7th The method of producing a second manuscript for retransmission by oxidation or reduction by hydrogen of the surface of a metal sub as described =

T A Edison

AD (letterpress copy), NjWOE, Lbk. 1:94 (*TAEM* 28:89). [a]From filed caveat; letterpress copy illegible. [b]"right to left" interlined above. [c]Canceled. [d]Circled. [e]From filed caveat; document damaged. [f]"the required" interlined above. [g]"one of" interlined above. [h]"a difference . . . written" interlined above. [i]Interlined above. [j]"a rapid vibratory" interlined above.

1. Edison executed this caveat on 11 November and filed it 29 November 1875 (Edison Caveat 70). What may be related drafts of caveats or patent applications for facsimile telegraphs can be found in NS-Undated-005, Lab. (*TAEM* 8:263–67, 462–66); see also Cat. 997:1 (*TAEM* 3:352). What may be experimental recorded messages can be found in Cat. 1144, Scraps. (*TAEM* 27:279).

2. U.S. Pat. 158,787.

3. Should be "wound."

4. Doc. 633 (filed 27 Sept.).

5. "M.B." in diagram.

–653–

Directions for the Domestic Telegraph

Newark NJ Oct 31st [187]5[1]

The Domestic Tel. Co. Market St Newark NJ[2]

Single Circuit.[3a]

Fig 1

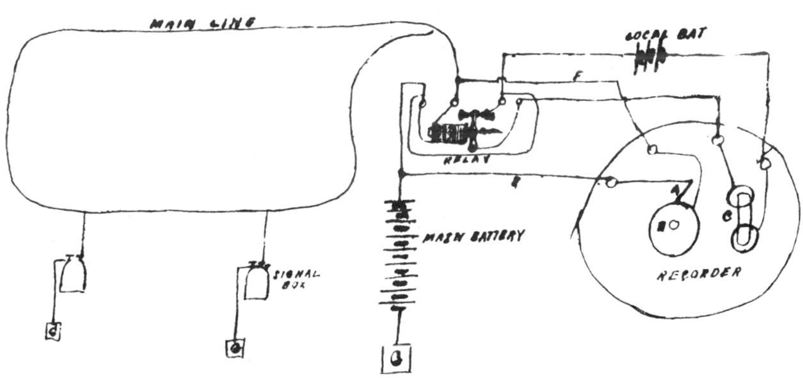

Fig. 2

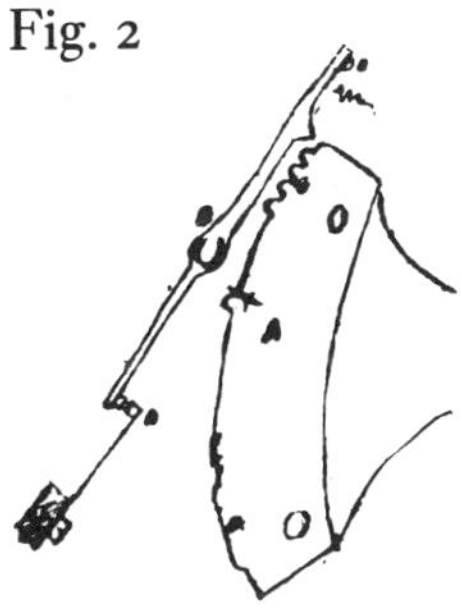

In Fig 1 is shown a single circuit complete which consists of a signal box, main line, main battery, relay, local circuit & battery, & recorder. Our principle is to send a current through a relay which connects through its backpoint a local circuit to start & stop the paper moving mechanism, & to leak off enough current to mark on chemically prepared paper by decomposition.

Our signal box works by gravity. It consists of a weighted segment with teeth cut on the periphery to correspond to the number of the call & an extra long tooth to start the mechanism. The signals are sent so fast that an ordinary magnet would not record them & this long tooth is necessary to insure the relay closing & allowing the recorder to make a revolution. The Signal box is also provided with a switch which connects the line only when the button is going up so that if a customer presses down, it matters not how long he keeps it down, as the signal will not be sent till he lets go of it. The main line should be a loop with both ends in the office & legs running off it to the boxes, these legs should be connected to the bindpost on the box that has a hard rubber insulation under it

The main battery should be a Leclanché as there is little or no action when the circuit is open. The local battery should be Callaud or Gravity battery as it is closed all the time.

The relay connects the local circuit on its backpoint. So that when it is at rest the recorder is stopped by the lever of magnet C Fig 1

The recorder [-]consists of a clockwork carrying a drum B for moving the paper with a magnet C for stopping & starting it & a platina pen A for decomposing. As will be seen in fig 1 the decomposing current is leaked off the main line by the wires e & F. Owing to the great resistance of the paper the greater part of the current goes through the relay & allows it to respond at first whilst there is amply sufficient to make a good mark goes through the pen.[4]

Where there are 3 or 4 circuits runnning into one office they can all be run out of the same main battery

With 4 main circuits & 2 Recorders I should fix them in the following manner I think:

Fig 3^{b}

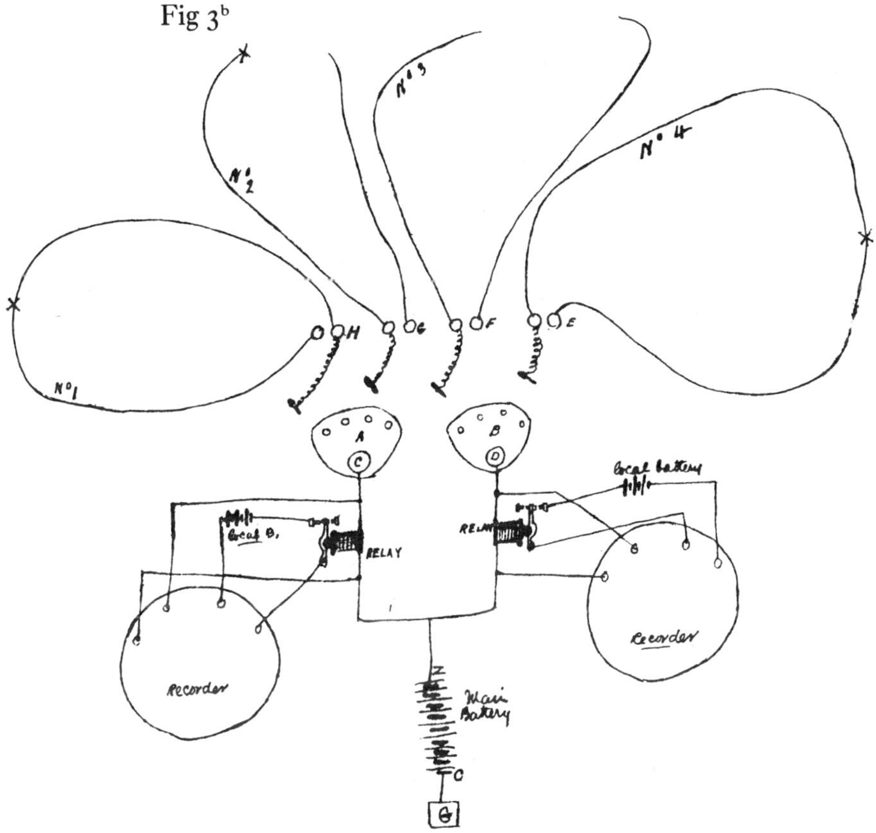

The circuits here terminate in the bindposts E. F. G. H. two to each circuit to one of which of each circuit a flexible cord with plug at the end is attached. A & B are two brass pieces with bindposts C & D. & each have four plug holes in, by putting all the plugs into either A or B all the circuits can be run on either recorder. It is necessary to be able to do this, as, in case of accident to one recorder, or changing paper, there is no loss of calls, & it is easily understood by the boy attending the Instruments

The four circuits ought all to be brought down within reach in the street at the point X & a single switch put there or the wires twisted together so that they can be easily disconnected, then if any trouble occurs on a circuit it can be disconnected at X & the good half connected to the instrument at the bindpost E F. G or H. In this way only half a circuit can be down at a time

Test the circuits once every hour.

Keep the paper in a crock.

Be sure to connect the leg of main wire to insulated bindpost on signal box if connected to the other one the line gets a partial ground through box & a great many such partial grounds might cause trouble

When the line is grounded the recorder will keep on running, if this occurs the man keeps pulling out the plugs till he finds the one that stops it when pulled out, he then sends a boy to disconnect the main line at X & connects the good half at the recorder & sends man to look at the other half.

In testing every hour at central station if there can be got no signal from the end of wire & ground you may conclude that the line is broken in which case connect the flexible cord for that circuit to both bindposts instead of one so that both sides are in & look for break

Caution people about laying anything on top of box or hanging anything on to it that might cause it to give a mutilated signal

Chas. Batchelor

ADS (letterpress copy), NjWOE, Lbk. 1:114 (*TAEM* 28:107). [a]Underlined twice. [b]Top of drawing damaged.

1. The Domestic Telegraph Co. began operating in Newark the following day. Advertisement of the Domestic Telegraph Co., *Newark Daily Advertiser,* 28 Oct. 1875.

2. An article on the Domestic Telegraph Co. in Newark indicates that the company's office was at "Fearey's, No. 180 Market street" ("Domestic Telegraphy in Newark," *Newark Daily Advertiser,* 28 Oct. 1875). The Newark city directory lists Jabez Feary as manager of a segar (cigar) store and the Atlantic and Pacific Telegraph Co. office at 180½ Market St. (Holbrook 1871, 256). Feary was later listed as Domestic Telegraph's superintendent in Newark (Holbrook 1878, 275).

3. A Charles Batchelor notebook entry of the same date shows this drawing with the following description: "My design for working four circuits with two recorders on the Domestic Telegraph Co's plan for Newark N.J." Batchelor's wife, Rosa, witnessed the entry. Cat. 1317:12, Batchelor (*TAEM* 90:663).

4. A version of this document, written by Batchelor with an electric pen and labeled "Directions for the Domestic Telegraph," ends here. Cat. 1144, Scraps. (*TAEM* 27:328–29).

–654–

Domestic Telegraph Co. Circular

New York, [October 1875?[1]]

The Fire Alarm System of the Domestic Telegraph Co. consists of a signalling and an alarm circuit: the former used exclusively for Fire signals and the latter for the transmission of such signals from Fire-Headquarters to Engine-houses, police stations and wherever else desired. The working principle of the system is the transmission over a wire of a signal sufficiently strong to release a paper moving mechanism and to take off enough current to decompose chemically prepared paper, and record the signal.

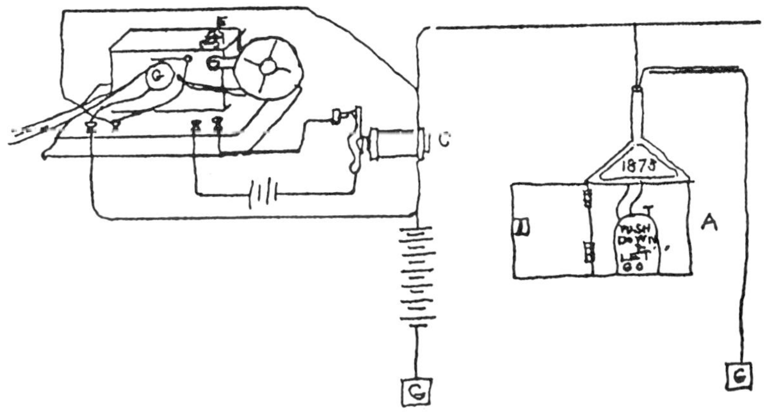

Fig 1 shows a circuit for Fire Signals. A is a street box with door open; a smaller box is seen inside with a push-button on it, and the words PUSH DOWN AND LET GO. This operation makes the signal which passes through the relay C. The latter completes through ists backpoint the local circuit J: this local circuit contains a magnet F which controls the paper moving mechanism and strikes a gong calling the[a] attention of an official in charge to the signal.

Chemically prepared paper is passed between the drum G and the pen and sufficient current taken from the main line to record in dots the number of the box.

In small places where there is only one engine house, this is all that is necessary, but where there are more, the second line must be used. This consists of a Repeater and Gong-striker shown in Fig. 2.

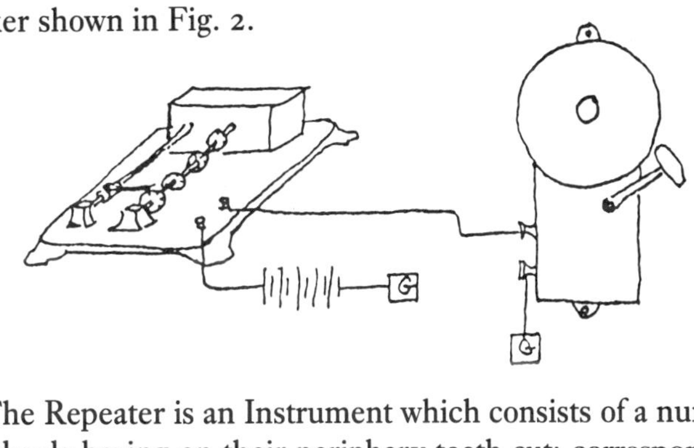

The Repeater is an Instrument which consists of a number of wheels having on their periphery teeth cut; corresponding to the numbers on the street boxes one for each box; a contact point which slides along a shaft can be placed on any one of these wheels and made to transmit through the line the signal the wheel represents.[2] After a signal has come in on the Recorder, the person in attendance sets the Repeater to the number recorded and puts the clockwork in motion, the wheel runs with a uniform speed and continues to repeat that number into the circuit till stopped; in the Engine-houses this is received on large striking gongs.

Keys & sounders or any other talking instruments can be used on this line. For very small places the Recording instruments and gong can be placed in the dwelling house of the official in charge.

The Superiority of the system consists in its Simplicity and Economy

Respy Yours
The Domestic Telegraph Co.
12 Vesey Street New York

Written with Edison's Electric Pen New York PO Box 3207[b]

D (electric-pen copy), DF, 75-013 (*TAEM* 13:534). Written by Charles Batchelor. [a]Interlined above. [b]"Written . . . 3207" written in left margin.

1. This document was probably created at the same time as the directions for the domestic telegraph system derived from Doc. 653.

2. Although Edison did not execute a patent for a fire alarm system until 9 May 1876 (U.S. Pat. 186,548), he had been working on such a system since the winter of 1874. The fact that the patent (see illustration of model [Acc. 29.1980.1607, MiDbEI(H)]) does not include this repeater, which had been developed by 24 July 1875 (Cat. 30,100, Lab. [*TAEM* Supp. III]), suggests that Edison continued to modify the system.

Model for Edison's fire alarm patent application (U.S. Pat. 186,548).

–655–

Notebook Entry: Celluloid

[Newark,] Nov 5th 1875

25[1] Celluloid

Took SO_3 24, NO_5 fuming 6, H_2O 4, & mixed & let them run down to 140° & put a little in a flat dish, then took brown tissue paper cut in strips & then with 2 glass rods kept turning it in the air to keep it cool after saturating it, threw it in large quantities of cold water violently shaking all the time to prevent heating In this way can peroxidize a large amount with a minimum quantity of acid after washing thoroughly neutralized it with NH_4 & washed & dried it. This was very combustible but insoluble in our alcohol. Probably it has got too much water in it. We found that by addition of Camphor it would dissolve Also on addition of Acetone, Wintergreen Oil, Oils of Myrbane, & Spruce & a few other of the essentials Gum Oleabanum dont seem to work. Thymol dont seem to work. Spruce & Myrbane seem to be the best the latter the better one of the two though not certain. We think the tissue paper ought to be pulped to get it to dissolve good. We called this No 1 sample & commenced to make some more

Sample No 2 was wet with acid sent in the form of spray from a reservoir it did not work well it seemed to be eaten up inside & like India rubber out

3 was oiled tissue paper worked in the same manner as No 1

4 was waste in the same manner.

5 We boiled some sawdust in KO & afterwards treated as No 1. 6 was wood paper in the same manner.

We now took No 1 (as none of these were soluble under the ordinary conditions) as being the best & proceeded to find a solvent for it.

We found that NH_4I & alcohol was a cold solvent

ZnChl & Alcohol good hot

Sulphocyanate of Potash & Alcohol partial hot.

NaI good hot.

Uric Acid & Alcohol partial but not much.

We bottled all these to try again[2]

Chas Batchelor James Adams

X, NjWOE, Lab., Cat. 994:20 (*TAEM* 3:201). Written by Batchelor.

1. See headnote, p. 600 n. 2. Another version of this note is in Batchelor's own notebook, witnessed by his wife, Rosa. Cat. 1317:13, Batchelor (*TAEM* 90:663).

2. The experiments are found two entries later in this notebook (Cat. 994:21, Lab. [*TAEM* 3:201]). Batchelor also made notes about them in his own notebook (Cat. 1317:14, Batchelor [*TAEM* 90:664]).

–656–

Charles Batchelor to Robert Spice

[Newark,] Nov. 8th [187]5

Dear Spice,

Only just recd yours in which you send sketch of tuning fork.[1] We do not do any outside work now I am in with Edison experimenting altogether & the manufacturing is done upstairs by Murray & another man I have some few men on our own experiments & I'll tell you what I'll do I'll furnish the steel cut it out, & turn & cut thread on shank for nothing if you will finish up two for us at the same time[2] If you think this is fair I will go ahead & send you four over & you can send us two back finished & hardened to tone Yours etc

Chas. Batchelor

ALS (letterpress copy), NjWOE, Lbk. 1:122 (*TAEM* 28:114).

1. Neither the letter nor the sketch has been found.

2. Spice apparently agreed; see Doc. 681.

–657–

To Henry Pettit[1]

[Newark,] Nov 1[3?][a] [187]5

Dear Sir,

As you cannot allot us more than 20 × 20′ we will be glad to use that & can fill it in the following manner.[2]

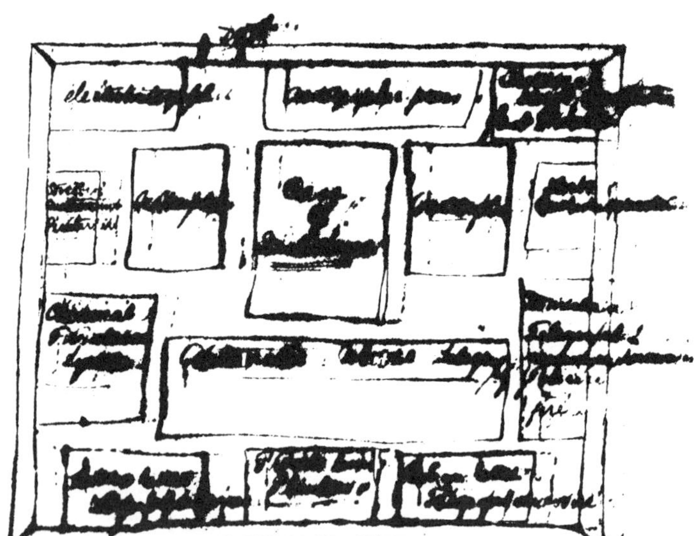

All these will be kept in operation. We could use more space if you can allow it as this is not enough to show these instruments properly & we have more to exhibit Yours Resply

Thos A Edison per Chas Batchelor

We expect to keep 8 or 9 men operating these instruments continually T. A. E.

D (letterpress copy), NjWOE, Lbk. 1:125 (*TAEM* 8:117). [a]Illegible.

1. Henry Pettit was chief of the Centennial Exhibition's Bureau of Installation and the person responsible for space allocation in the Main Exhibition Building (J. Wilson 1876–78, 3:cx). In September, after requesting a diagram of Edison's proposed exhibit, Pettit had informed Edison that he would not be able to have as large a space as he wanted. On the back of that notice Edison instructed Batchelor to make the drawing and inform Pettit they would take a space twenty feet square. Henry Pettit to TAE, 29? Sept. 1875, DF (*TAEM* 13:1005).

2. Cf. Doc. 601.

Edison's proposed Centennial Exhibition display.

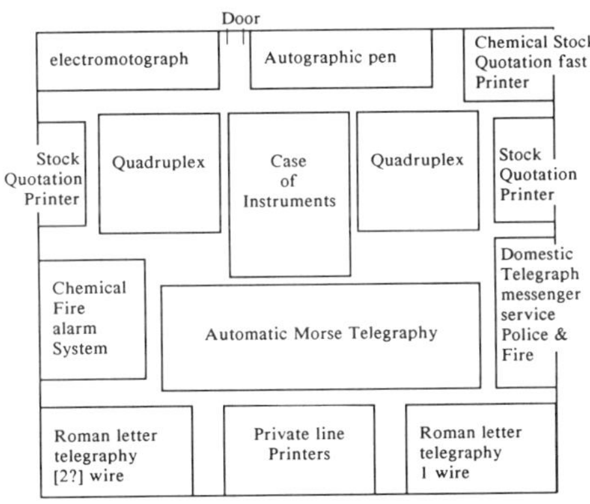

-658-

Notebook Entry: Acoustic Telegraphy

[Newark,] 5.35 AM Nov. 16 1875.

28[1] Acoustic Telegraphy.[a]

14

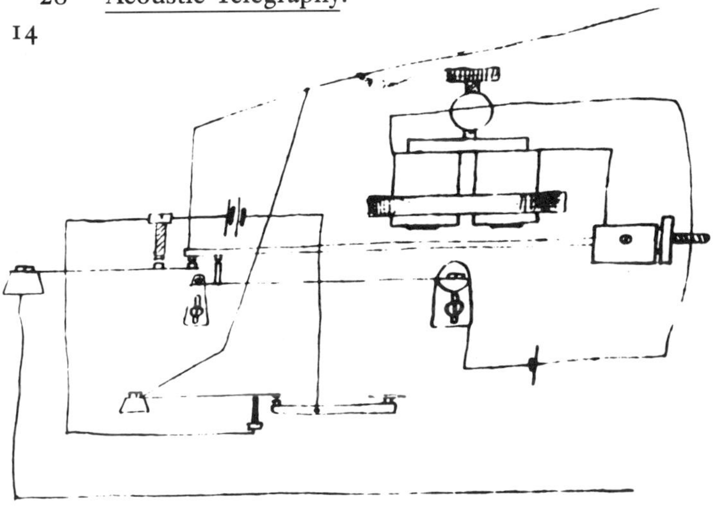

This device has a vibrating rod[2] always vibrated by magnet & local battery with great amplitude of swing. The extreme end works a duplex contact for throwing in & out the battery without breaking the continuity of the circuit & connected in circuit as shewn. The duplex key is used to throw the battery entirely out of circuit which would otherwise be constantly be thrown in & out by the duplex springs in the vibrating bar. not breaking the continuity of the circuit allows us to use a number of saparate batteries it also allows the wire to discharge to earth & allows of a condenser at the receiver to augment, quicken & sharpen the waves of current; it also allows us to transmit a greater number of messages, for by Gray's plan[3] 6 messages would keep the line mostly closed & of course the sounder to work badly.

We find that the arrangement we made last night gives a perfectly even break with great amplitude of spring & positive positive

15.

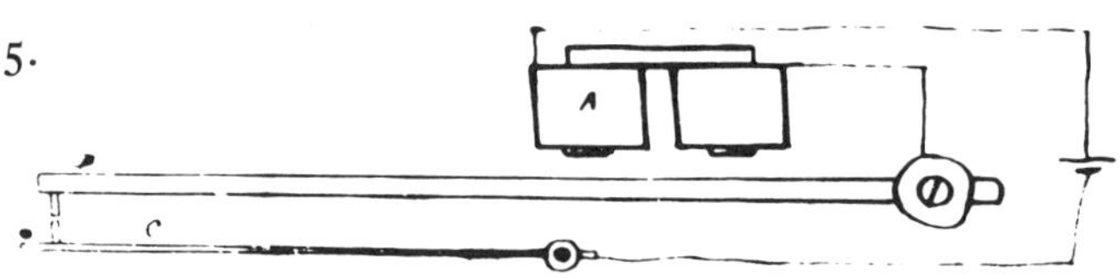

B 1/8 Stubbs steel wire,[4] C 1/32 stubbs steel wire A an ordinary 3 ohm magnet, 2 cells of carbon battery, 5/16 of an inch motion at the extreme end of bar B perfectly even on closing for a dot. The bar B vibrates in an arc about 1/16. On making a dash at ordinary morse speed the vibrations reach at least 5/32 of an inch. on making an L at the rate of 15 words per minute it reaches 5/16 of an inch. of course these extreme limits are not required. A similar apparatus to the above properly adjusted placed in circuit & operated by the breaks from the transmitting instrument answered promptly & vibrated about 1/32. & we could send at the rate of 7 words per minute but our contact breaker is inferior Batchelor is now making new spring device.

Chas Batchelor Jas Adams

X, NjWOE, Lab., Cat. 994:22 (*TAEM* 3:204). Written by Batchelor. [a]Underlined twice.

1. See headnote, p. 600 n. 2.

2. This device, called an automatic circuit-breaker or vibrator, was well known. The electromagnet at upper right and the horizontal lever at center are in circuit with a small battery (lower right). The magnet charges, attracting the lever; as the lever—which is pivoted at its right

end—responds, its (left) end moves off a contact point, breaking the circuit; the magnet discharges; and the lever moves away, once again making contact and energizing the magnet. This could produce "a continuous vibration of the armature of almost inconceivable rapidity." Prescott 1877, 103–4.

3. Edison had probably seen Elisha Gray's acoustic telegraph system at Western Union headquarters in New York in late August 1875, when Gray demonstrated it over the company's Boston line. Gray recalled that the system he demonstrated in New York was that embodied in his U.S. Patent 186,340, which issued 16 January 1877. Gray 1977, 58–60.

4. Unidentified.

–659–

Technical Note: Acoustic Telegraphy

[Newark,] Nov 16 1875[a]

Acoustic Telegraphy[1b]

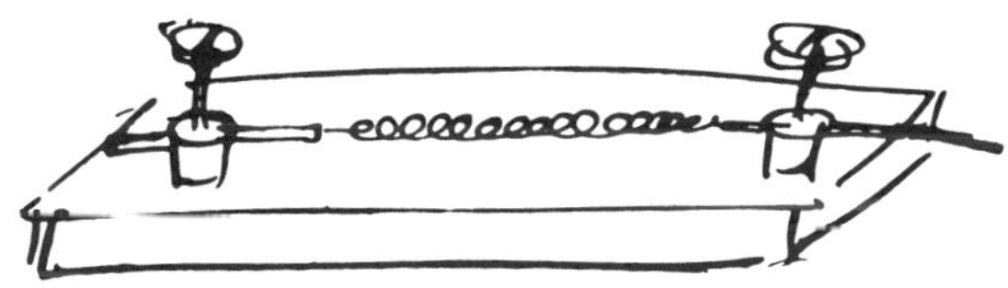

Acoustic Telegraphy

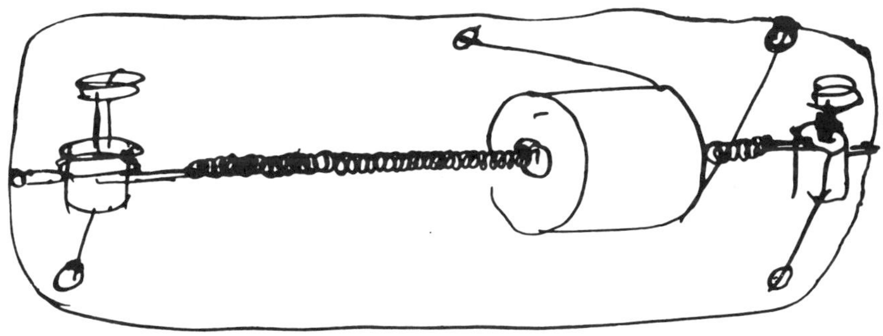

Iron wire delicate[2c]

Acoustic Telegraphy

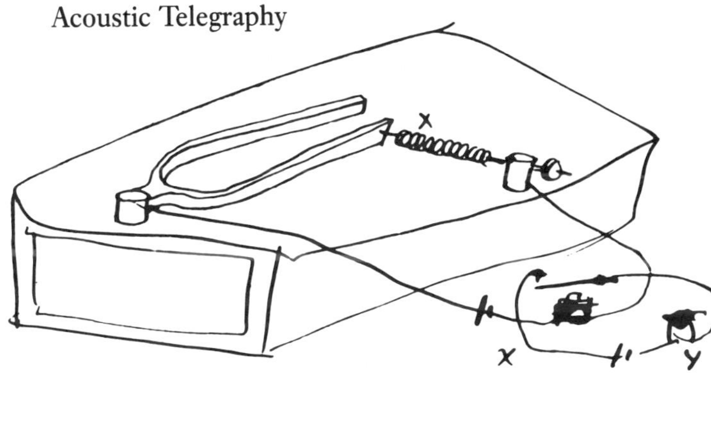

X german silver ~~which~~ in circuit with quantity battery & ½ ohm sounder when fork is[d] not vibrating each turn of the spring touches its neighbor and the sounder X is closed & Y is open. = when fork vibrates it sets the spiral in motion & separates each coil put in a large resistanse. this opens X & closes Y[3c]

~~Autograph~~ Acoustic Telegraphy[4e]

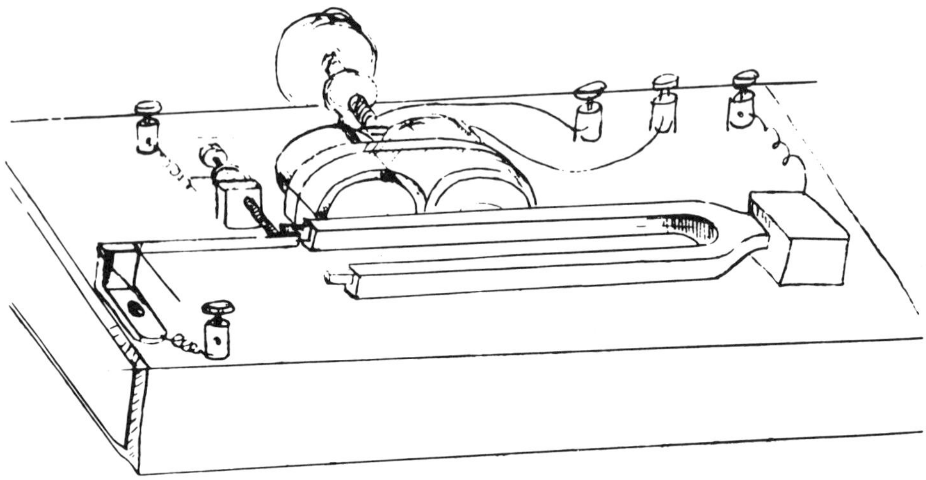

Acoustic Telegraphy[f]

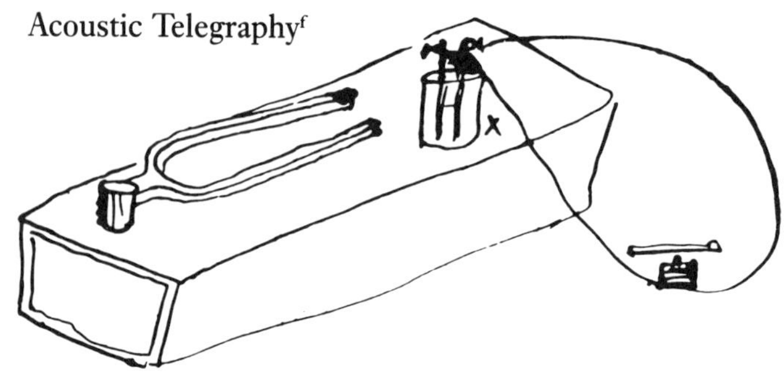

X is a quickly polarizing battery which depolarzes powerfully with the least jar. as the tuning for responds only to vibrations in union with its swinging time, it will be set in motion & this in its turn will set the Resonant box vibrating depolarzing or knocking loose the hydrogen bubbles on the plates of the batty X increasing the strength & closing the sounder. the S[ounder] should have very low resistance—[5g]

T. A. Edison — Jas Adams
Charles Batchelor — John Kruesi[6]

X, NjWOE, TI 2, Edison's Exhibits 9-9, 7-9, 9-9, 8-9, 6-9 (*TAEM* 11:208, 205, 207, 206, 204). Each page dated and labeled "Acoustic

Telegraphy" by Batchelor; text written by Edison. [a]Each drawing bears this date in some form. [b]Signed here by Edison, Batchelor, and Adams. [c]Signed here by Edison, Batchelor, Adams, and Kruesi. [d]Interlined above. [e]Signed by Batchelor and Adams. [f]Followed by centered horizontal line. [g]Signed here by Edison, Adams, and Kruesi.

1. On a similar drawing dated 15 November, Batchelor noted Edison's intent: "The jar of tuning fork would make German silver spring . . . shake apart & give greater resistance & work on the difference." Cat. 1307:54, Batchelor (*TAEM* 90:639); see n. 2.

2. One of the principal problems facing acoustic telegraphy was transferring rapid, generally weak line impulses to a local receiving circuit without damping the oscillations. Here Edison has the line current passing through the short, wide coil at right, through the center of which passes a long spiral of iron wire that is in the local circuit. As the current in the coil creates a magnetic field, the spiral compresses; the resulting short-circuits between adjacent turns of the spiral reduce the spiral's resistance, and the current increases in the local circuit. According to Edison and Batchelor, this apparatus received signals from several tuning forks and succeeded to some degree (TI 1:11–12, 226–27 [*TAEM* 11:26–27, 90]). Edison later included a variable resistance of this type in one of his telephone patents (U.S. Pat. 203,013).

3. This apparatus, mounted on a resonant box, is the final receiving component in a system that includes a transmitting tuning fork and a receiving fork mounted on a resonant box, similar to this one, with its opening facing the opening on this box. Excitation of the first receiving fork sympathetically vibrates this one. According to Batchelor, it worked "satisfactorily." TI 1:12–13, 228 (*TAEM* 11:27, 91).

4. In this acoustic transmitter, the electromagnet next to the upper prong of the tuning fork is in circuit with the tuning fork, the contact at the end of the upper prong, and a battery (not drawn). As the electromagnet charges, it attracts the prong, breaking the circuit, neutralizing the electromagnet, and allowing the prong to make contact again, thus reinforcing the natural period of the fork. Batchelor did not remember "just how these instruments were put in circuit." TI 1:227 (*TAEM* 11:90).

5. Here the "quickly polarizing battery" takes the place of the spring in the second and third designs above. The electrochemical action of this battery produces hydrogen bubbles on the surface of the cathode so quickly that the current rapidly becomes too feeble to activate the sounder. When a receiving tuning fork sympathetically excites this one, the bubbles are jarred loose and the current flows again until the vibrations stop. Batchelor and Adams experimented the next day with batteries of this sort to find a solution that would most rapidly shut off the current (Cat. 994:23, Lab. [*TAEM* 3:204]).

6. John Kruesi (1843–1899) was a highly skilled machinist who constructed much of Edison's experimental apparatus at Newark and later ran the machine shop at the Menlo Park laboratory. Born in Heiden, Appenzell, Switzerland, Kruesi worked in Zurich and Paris before coming to the United States in 1870. He worked for the Singer Sewing Machine Co. in Elizabeth, N.J., and joined Edison and Unger in Newark in 1871 or 1872. He later had charge of the manufacture of underground electric tubes for the first central station built by Edison in New

York City in 1882. In 1886 he became superintendent of the Edison Machine Works in Schenectady, N.Y., and he continued in that position after the creation of General Electric. *NCAB* 26:100; *DAB*,, s.v. "Kruesi, John"; "Johann Heinrich Krüsi," *Swiss American Review*, 27 Oct. 1976, 5, 9, in "Kruesi, John," Box 403, BC.

–660–

From Frank McLaughlin

NY November 19, 1875

Dear Sir:

The Technical Editor of "Scribners Monthly," Mr Charles Barnard[1] called on us to day and inspected the "Wonder of Electricity"[2]—he is highly delighted with the instrument and predicts an extended scope for its uses—

He wishes to greet you personally, and for that purpose will call on you next Tuesday afternoon (if stormy the next day) at No 12 Ward Street. Yours Very truly

Frank McLaughlin[3]

PS Dont forget the new pen.

ALS (electric-pen copy), NjWOE, Scraps., Cat. 593 (*TAEM* 27:670).

1. Charles Barnard (1838–1920) was editor of the "World's Work" department of *Scribner's Magazine* from 1875 to 1884. Among his writings were articles on scientific and technical subjects. *NCAB* 13:64–65; *DAB*, s.v. "Barnard, Charles."

2. The electric pen.

3. Robert Conot (1979, 490) identifies Frank McLaughlin (d. 1907) as a telegrapher acquaintance of Edison's. He was later involved with Edison's mining ventures in California, where he became a prominent figure in Republican party politics ("Major Frank McLaughlin Kills Daughter and Self," *San Francisco Chronicle*, 17 Nov. 1907, 1.

–661–

Notebook Entry: Batteries for Autographic Printing

[Newark,] Nov 21st 1875[a]

8[1] Autographic Press Battery.

We have experienced a great deal of trouble in the battery for running the pen for our autographic press. Up to Oct 14 1875 we have used the French small Grenet battery 2 cells to each pen. They are however very unsuitable for a man's desk & our experiments have at last terminated in one to our entire satisfaction.[2] It is easily cleaned & comes all apart, it cannot tipp over, we use porous cells & the carbon & zinc can be lifted up.

Fig 1

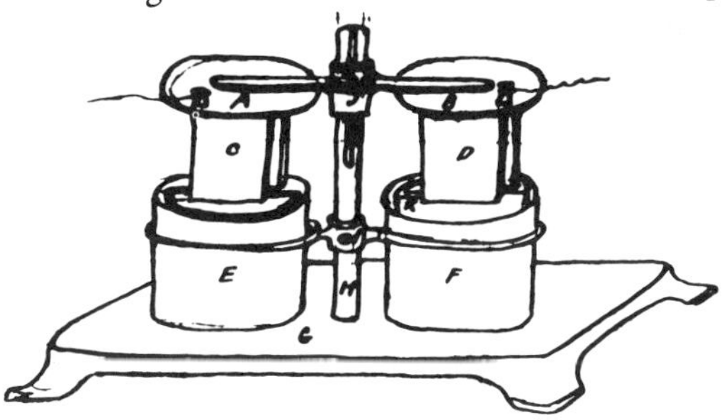

[Fig] 2

[Fig] 3

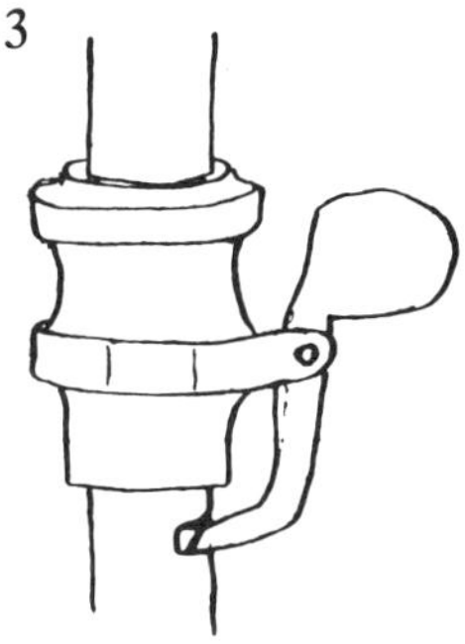

A & B are circular disks of thick rubber which hold the zincs & carbons of cells E & F which are glass & let into the iron base G about an inch. I is a spectacle secured to upright H & keeps the jars in proper position J is a sliding bar which holds A & B & it is kept from turning on the shaft by a screw which runs in a groove in upright. The porous cell is D shaped as shown in 2 on the back of J there is a click which drops in a slot across the upright when it is lifted up this is better shown in Fig 3. This we found to be the only battery that would do for the purpose & consequently adopted it.

Nov 21st written. adopted Oct 14 1875[3]

Chas Batchelor

X, NjWOE, Batchelor, Cat. 1317:15 (*TAEM* 90:664). Written by Batchelor. [a]Date taken from text.

1. Batchelor numbered the entries in this notebook.
2. See Doc. 647 n. 4.
3. See Doc. 647.

-662-

[Newark,] Nov 21st 1875

Notebook Entry: Acoustic and Multiple Telegraphy

(31)[1] Acoustic Telegraphy.

Tried the following experiment:—

18

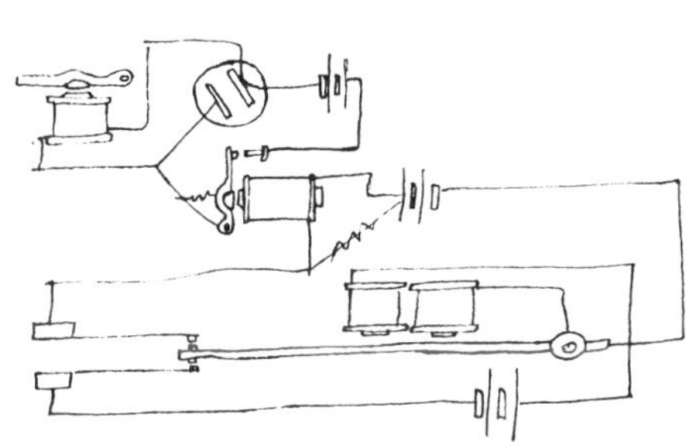

N is a secondary battery.[2] The slight closing of X of tuning fork against C[3] vibrates the relay badly but each dot from relay lever & local circuit charges the secondary battery N & closes sounder. When open the secondary N discharges in Sounder until next dot & keeps it closed.[a]

(32) Quadruplex Morse telegraphy

The following experiment whereby increased & decreased currents are made to work a common backpoint relay properly & reversed currents made to work also the ordinary front point relay.

19

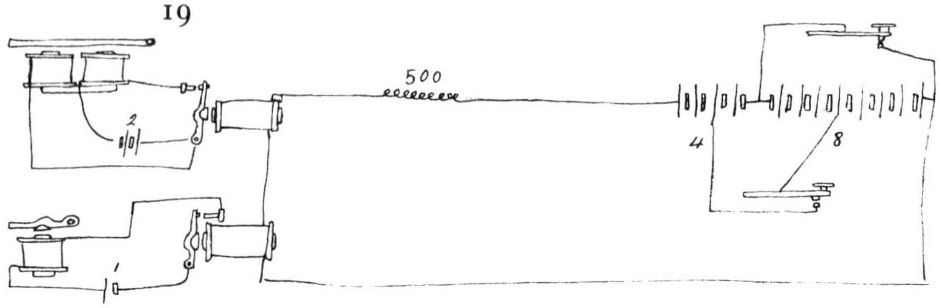

This works splendidly & would answer well for a Quadruplex.[a]

(33)

20

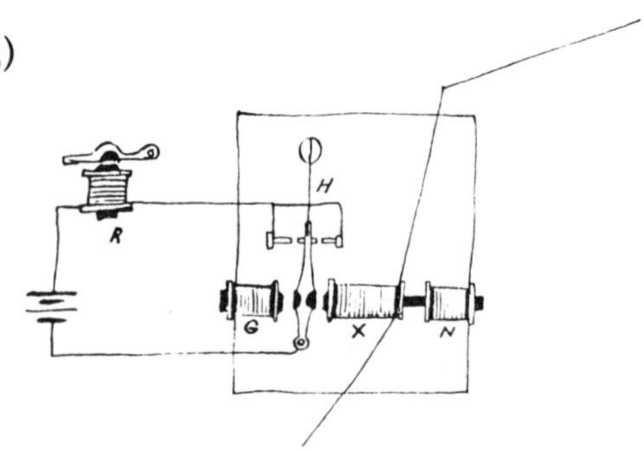

This works slow. The sounder works right but if you go fast sounder closes permanently. The induction from N closes G for an instant When X opens keeps R closed X close again keeping R closed[a]

34 Acoustic Telegraphy

21

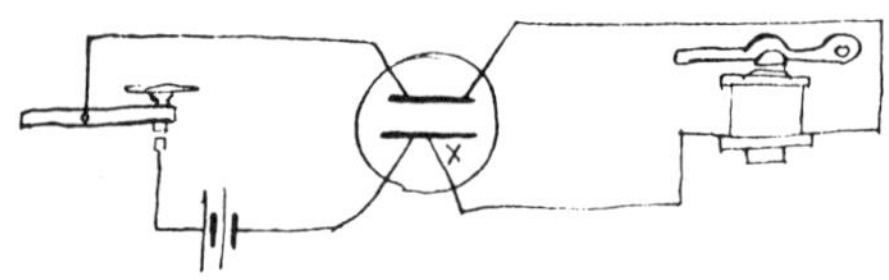

This makes a first class sounder dots far apart keeps sounder closed. The longer the dots the greater time sounder stays closed. Good for working by vibration.[b]

35 Quadruplex Telegraphy.[4]

22

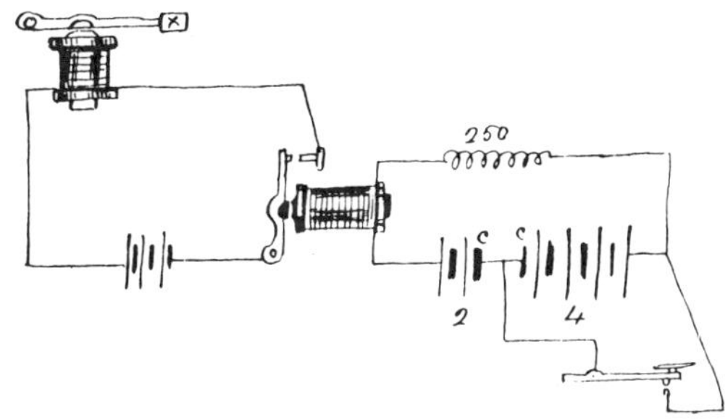

The weight at X dont work well but on closing relay the jump makes the down stroke on the sounder & on opening the other jump makes the back stroke. The writing sounds perfect & beautiful without the necessity of the sounder lever touching the backpoint This dispenses with the polarized relay in the quadruplex.[c]

(36) Morse Telegraphy

23

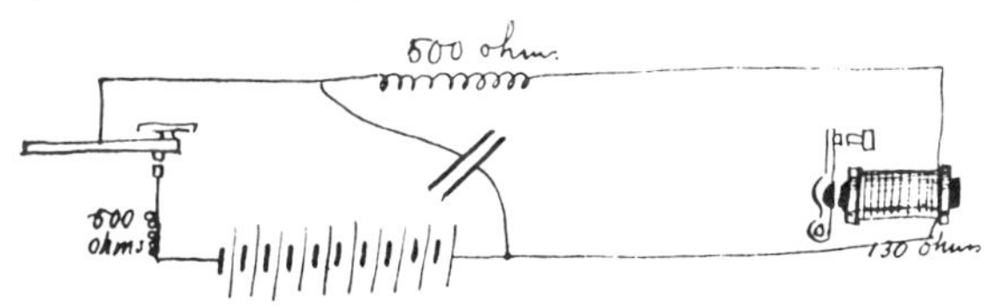

With or without resistance; putting a condenser in makes relay work sharp on an adjustment that without it would cause the relay to stick altogether (ie) about 2 secs after opening on ordinary local circuit with key open & no route for relay discharge. Apparently the condenser sends a reverse charge through relay & kills the residual magnetism.

Edison Chas Batchelor
James Adams

X, NjWOE, Lab., Cat. 994:24 (*TAEM* 3:205). Written by Batchelor. [a]Dated and signed here by Batchelor and Adams. [b]Dated and signed here by Batchelor. [c]Dated and signed here by Edison, Batchelor, and Adams.

1. See headnote, p. 600 n. 2.
2. **N** is the large circle at top.
3. **C** is the upper contact at the left end of the long bar below the magnet at bottom center.
4. See Doc. 663 n. 2.

[Newark,] Nov 22nd 1875

Notebook Entry: Multiple Telegraphy

37[1] Quadruplex Telegraphy.[2]

The experiment of Fig 1 works perfectly if magnet of A & spring A is so adjusted that it opens & closes in equal time; relay to open & close twice each forward & backward movement of lever of A & all down stroke on sounder works beautiful. The total breakage does not open C so bad as reversals. Shunting B with a condenser makes it worse.

Fig 1 24

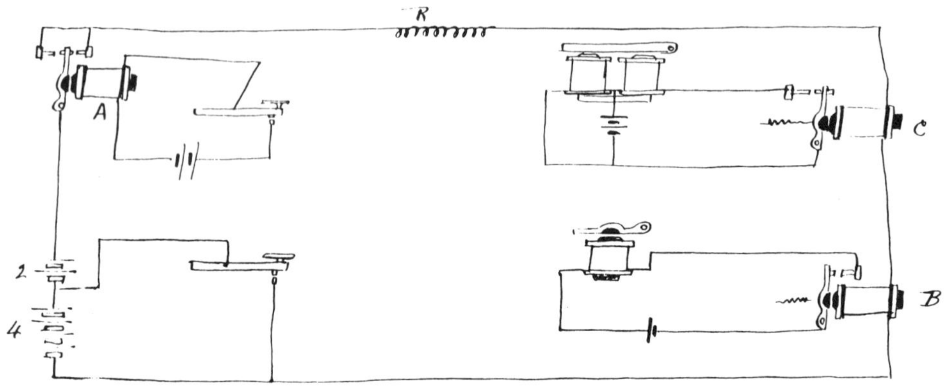

We tried a front point relay at C with total breaks disguised by shunting relay also by shunting sounder & use of secondary battery but breaks show too plain & we must use a polarized relay.

Fig 2 25

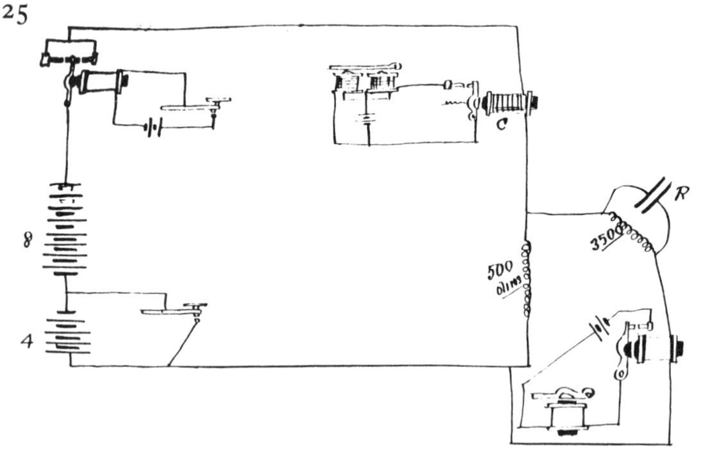

Fig 2 With 1000 ohms in condensers R not good but better than if no resistance or condenser used but when it is made 3500 the dead open relay works perfectly splendid & the backpoint relay works better than when reversed currents are used altogether this arrangement is a grand success.

Fig 3 26

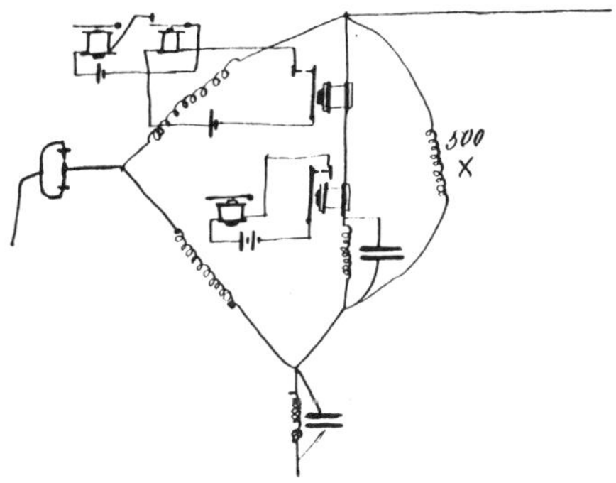

Fig 3 is another arrangement that works splendidly; it weakens back points but makes it work equally as well

Breaking circuit shunt X bursts up dead open relay as there is no route for the condenser.

Fig 4 shows one end of the Quadruplex on this new principle.

27

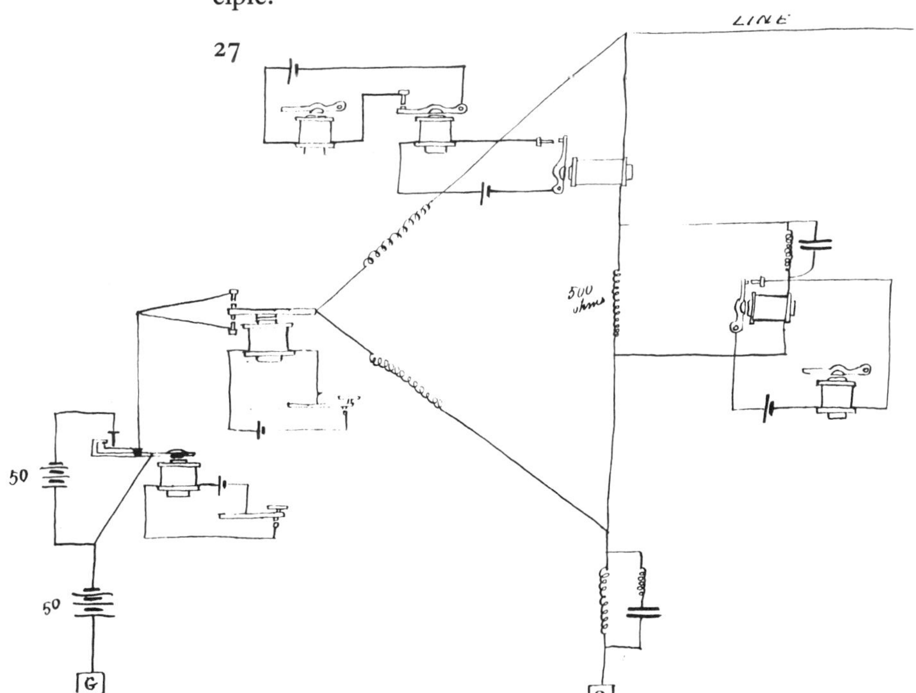

Chas Batchelor James Adams

X, NjWOE, Lab., Cat. 994:26 (*TAEM* 3:206). Written by Batchelor.

1. See headnote, p. 600 n. 2.

2. As in the previous entry (Doc. 662), Edison is working with diplex telegraph circuits of a general type that he had sketched in the fall of 1874 and developed in a caveat in January 1875 (Doc. 507; Edison Caveat 58). Rough notes on this work are in NS-75-002 (*TAEM* 7:336–39). Edison described these designs more extensively in a caveat that he

drafted the same day but did not finish until January 1876 (Lbk. 1:153 [*TAEM* 28:144]; Edison Caveat 72).

–664–

Caveat: Acoustic Telegraphy[1]

New York, November 22, 1875[a]

To all whom it may concern.

Be it known that I, Thomas A. Edison, of Newark, in the County of Essex, and State of New Jersey, have invented an improvement in Acoustic Telegraph Apparatus, of which the following is a specification.

The object of this invention is to transmit and receive several different messages over a single circuit at the same time without interference with each other.

The invention consists in transmitting waves of electricity into a circuit by means of "cut outs" operated by spring pendulums.

The primary motion of which is received by a jar or vibration from an electromagnet or other source of power.

It further consists in devices for throwing the battery in and out of the circuit without interrupting its continuity.

It further consists in arranging an number of spring pendulums set in motion by a mechanical device operated by the finger or an electromagnet which pendulums are of different lengths and make a different number of vibrations in a given time and arranging a main battery with the contact devices operated by the vibrations of such pendulums in such a manner that pendulum No 1. will at each vibration throw a main battery of say 50 cells in the circuit a given number of times in one second.

While pendulum No. 2. will also throw in and out of circuit another main battery a different number of times per second always preserving the continuity of the circuit.

It further consists in combining with the spring pendulum another spring pendulum of equal length or of such a length and thickness that it will make the same number of vibrations in a second; this second spring pendulum being used for the purpose of contacts were the pendulum used for contact purposes longer or shorter and made a greater or less number of vibrations in a second than the primary pendulum it would act as a damper and prevent it from vibrating if not altogether, at least to a great extent.

The invention further consists in placing at the receiving [end] in conjunction with ordinary Morse apparatus placed on resonant or sounding boxes, or boxes which will easily vibrate or communicate jars a number of spring pendulums, at different vibrating times, each of which has a pendulum at the transmitting station making an equal number of vibrations.

These pendulums unlike the vibrating strings used by Elisha Gray, of Chicago,[2] have no connection with the receiving magnet other than they are placed upon a sounding board with the electromagnets.

They are set in motion by jars or vibrations given such board by the lever of the magnet and not by the magnetic power of the magnet acting upon the spring as with Greay.

By using the vibrations transmitted through the resonant box, and by preserving the continuity of the circuit and transmitting the vibrations by various battery powers I am enabled to transmit a greater number of messages over a single circuit at once than is possible with the devices of Gray, who uses a single battery power and severs the continuity of the circuit.

The invention further consists in combining double isochronous contact springs on each side of the vibrating spring pendulum at the receiving station so that the local circuit is closed on both sides of the movement of the pendulum from a state of rest.

It further consists in the manner of connecting the transmitting and receiving instruments and battery at each station, whereby the battery at such station may have the same polarity as at other stations and be thrown in and out without breaking the circuit; also in placing each receiving instrument in a derived circuit from the main line by the use of resistance coils, and shunting the coil forming the derivation with a condenser the object of which is to produce a sharpness in the waves of current which on long lines would be apt to run together and produce a constant closing of the circuit without any breaks.

Fig. 1. Fig. 3.

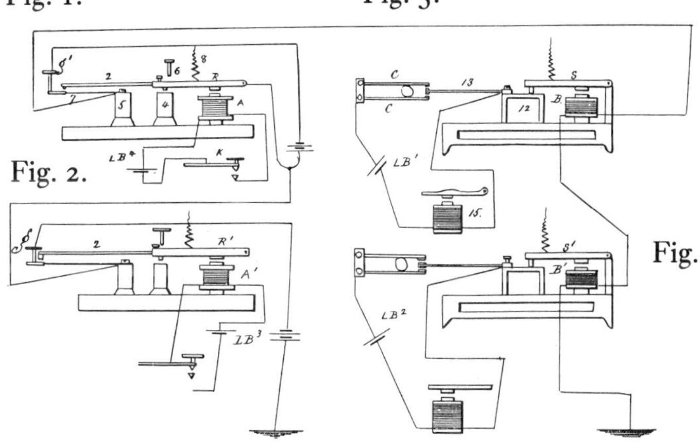

Fig. 2. Fig. 4.

In Figs. 1. and 2. are shown the transmitting instruments; both are alike, except that the spring pendulum 2. in Fig. 1. is longer and beats a lesser number of times than the pendulum[3] 2′ in Fig. 2.

The pendulum is secured to the lever of the sounder magnet A. which is operated by a local battery LB[4] and key K.

This lever R strikes upon the pillar 4. and has an upper limiting screw 6. also a retractile spring 8.

5. is a pillar to which is secured the isochronously vibrating contact spring 7. which rests against the contact limiting screw[4] 9′. when in a state of rest placing the battery in circuit.

If now the key K be closed the magnet A will attract the armature and lever R. and it striking upon the pillar 4. will set the pendulum 2. vibrating, thus striking 7. will also set that vibrating disconnecting it from the battery at 9′. and still preserving the continuity of the circuit through 2 and 7. which are in contact when 7. is separated from 9′. these waves going over the wire give a vibrat~~ing~~ory movement to the lever of B Fig 3. which has little or no play and the lever S. striking against the resonant bridge 12. sets the pendulum 13. vibrating and being adjusted make the same number of vibrations per second as 2. in Fig. 1. it will continue to vibrate as long as 2 is kept vibrating.

c.c. are two isochronously vibrating contact springs connected to the sounder 15 and local battery LB[1] and are so connected to the pendulum 13. that it touches neither when in a state of rest.

The apparatus shown in Fig. 4. is the same as in Fig. 3. except its pendulum vibrates a greater number of times per second and is adjusted to 2′ of Fig. 2.

Fig. 5.

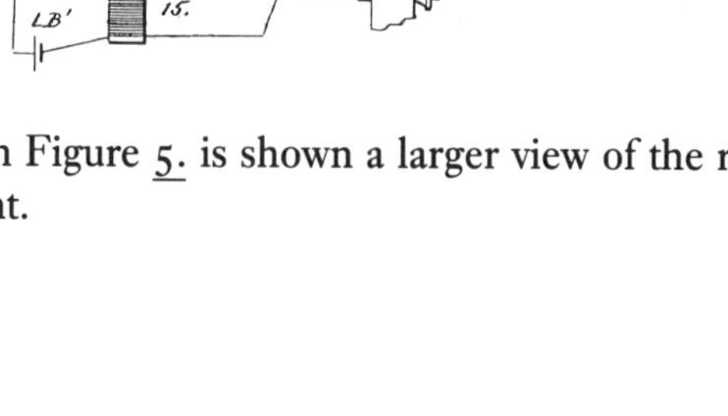

In Figure 5. is shown a larger view of the receiving instrument.

Fig. 6.

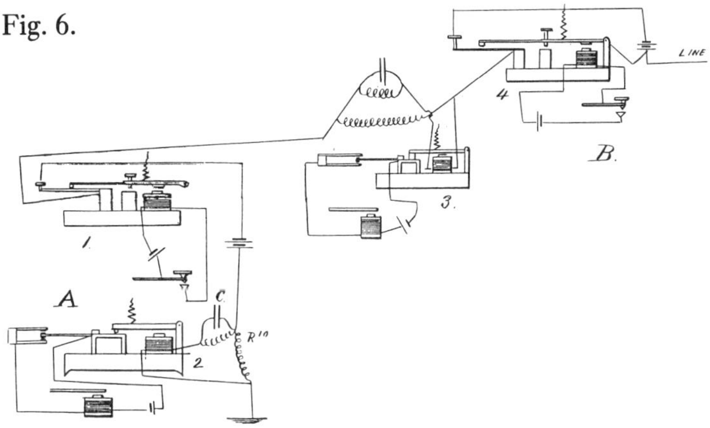

In Fig. 6. is shown the method of connecting the receiving and transmitting instruments at end and way stations.

At station A are two instruments 1 is for transmitting, serving to cut the battery in and [out] of circuit while preserving its continuity in the manner already described.

2. is the receiving instrument which serves to receive the vibrations from another station; the magnet of 2. is placed in a derived circuit from the line. a resistance 10. is made to form part of the main circuit, R. is a second resistance[5] forming part of the derived circuit in which the magnet of 2. is placed. This resistance is shunted by a condenser for the purpose already mentioned.

At Fig. B. where W is a way station[6] 3. and 4 are receiving and transmitting instruments connected precisely as at station A.

Any number of stations may be thus connected in the line, and several instruments like 3 and 4 may be placed at each station.

If it is desired to work a second circuit from any part of the line.

The sounder of 3. may be combined with devices like that on 4. and a second wire being connected to it the message may be transmitted upon that line, or, the sounder[7] O. of 3. may have an ordinary repeating points and open and close the circuit in the ordinary manner or the magnets of 3 and 4. may form part of a common automatic repeater, or the sounder of 3. may be a duplex or quadruplex transmitting sounder and that of 4. the receiving sounder of a duplex or quadruplex.

In Fig. 7. is shown devices by which the pulsations from 1 and 2. may be transmitted over the line wire without effecting the receiving instruments 3. and 4.

Fig. 7.

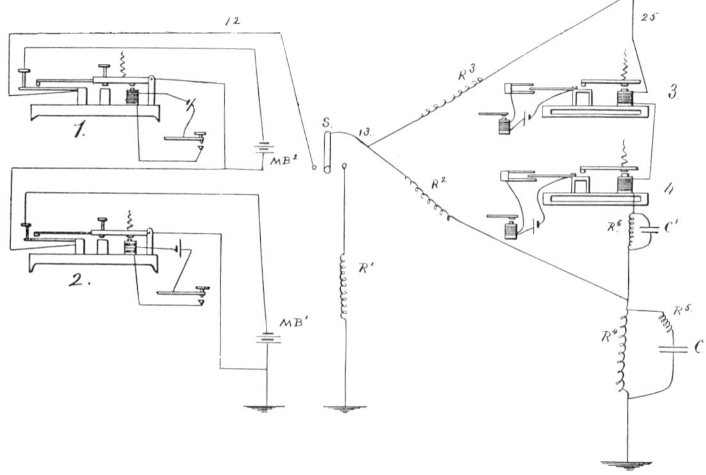

Inventor Thomas A. Edison per L. W. Serrell
Witnesses Chas H. Smith Harold Serrell[b]

Nos 1 and 2. are two transmitting instruments with spring pendulums making a different set of vibrations per second each of which cut in and out of circuit the main batteries MB^1. MB^2.

S is a switch for cutting out 1. and 2. and connecting the main wire with the earth through a resistance coil R^1. for the purpose of equating or balancing the circuit.

R^2. and R^3. are the stwo sides of a Wheatstone bridge; 25. is the bridge wire in which is placed the two receiving instruments 3 and 4. R^6 is a resistance forming part of the bridge wire and is shunted with a condenser to cause the vibrations to be more clear. R^4. is a resistance equal to the resistance of the line and distant apparatus.

R^5. is a resistance connected to the condenser C. which is for the purpose of balancing the static charge of the line the resistance being used to lengthen or shorten the charge from C. [-]When R^4. condensesr C. R^2. and R^3. are properly balanced, the circuit from MB^1. MB^2. pass over the line and move the distant instruments without passing through 3 and 4. of course the number of transmitting and receiving instruments can be increased with the number or sets of vibrators which can be practically sent over the wire.

The connections at the distant station are the same as here shown. If it is desired to retransmit into a second circuit at the distant station the sounder of the receiving instruments are made to actuate another set of vibrating pendulums like as in 1. and 2.

I do not wish to confine myself to any particular form of vibratoring pendulum as a tuning fork or string secured at both ends, or wind instrument may be used.

What I shall probably claim is

First. The combination of a number of vibrating pendulums each making a greater or less number of vibrations in a given time than the others with contact devices and separate batteries so that each shall by its vibration place in and withdraw from circuit its own battery without breaking the continuity of the same.

Second. Communicating a vibratory motion to a vibrating spring pendulum mechanically by the movement of the lever of electro magnets or otherwise both in the transmitting and receiving instruments.

Third. In combination with the spring pendulums of isochronously vibrating contact springs for the purposes set forth.

Fourth. The double springs c. c. for the purpose set forth.

Fifth. The combination of an electromagnet with a vibrating spring pendulum, tuning fork or other isochronously vibrating mechanism upon a resonant box or device and communicating a motion to the latter by jars of vibrating, each having no electrical connection with one another.

Sixth. The method shown for producing sharpness and the vibrations by placing the receiving instrument in a derived circuit and resistance and shunting the same with a condenser.

Seventh. The method described for repeating into side levers at way stations.

Eighth. The method shown for transmitting vibrations upon the line from station A. without passing through the receiving instrument at that station.

Signed by me this ~~20th~~2nd day of November A.D. 1875.

Thos. A. Edison

Witnesses. Geo. T. Pinckney Chas H. Smith

DS, DNA, RG-241, Edison Caveat 71. Written by Smith; petition and oath omitted. [a]Place taken from oath; date taken from text, form altered. [b]All figures are on two separate pages, both of which are signed and witnessed at the bottom.

1. There is an undated draft of this caveat in Lbk. 1:127 (*TAEM* 28:119).

2. Gray did not use any strings in his apparatus. Edison might have meant "springs." See Gray 1977, 43.

3. Labeled **2** in the figure.
4. Mislabeled **g′** in the drawing.
5. Not labeled.
6. Not labeled.
7. Not labeled.

–665–

Notebook Entry: Etheric Force[1]

[Newark,] Nov 22nd 1875

(38)[2] A New Force.

In experimenting with a vibrator magnet consisting of a bar of Stubb's steel fastened at one end & made to vibrate by means of a magnet we noticed a spark coming from the cores of the magnet;[3] this we have noticed often in relays, in Stock printers when there were a little iron filings between the armature & core & more often in our new electric pen & we have always come to the conclusion that it was caused by strong induction. But when we noticed it on this vibrator it seemed so strong that it struck us forcibly there might be something more than induction We now found that if we touched any metallic part of the vibrator or magnet we got the spark. The larger the body of iron touched to the vibrator the larger the spark.[4]

Fig 1 28

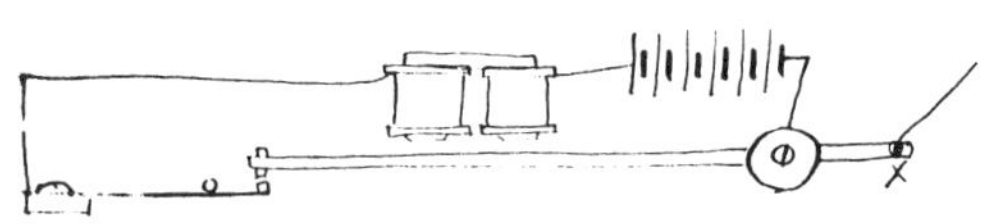

We now connected a wire to X the end of the vibrating rod & we found we could get a spark from it by touching a piece of iron to it[5] & one of the most curious phenomenons is if you turn the wire round on itself & let the point of the wire touch any other portion of itself you get a spark[6] By connecting X to the gas pipe we drew sparks from the gas pipes in any part of the room by drawing an iron wire over the brass jet of the cock. This is simply wonderful & a good proof that the cause of the spark is a true unknown force[7a]

T A Edison

Chas Batchelor
James Adams

X, NjWOE, Lab., Cat. 994:28 (*TAEM* 3:207). Written by Batchelor. [a]Underlined twice.

1. This entry is continued in Doc. 666.
2. See headnote, p. 600 n. 2.
3. In his own notebook, Batchelor described the sparks as appearing between the magnet's cores and the vibrating rod. Cat. 1317:16, Batchelor (*TAEM* 90:665).

4. Batchelor also noted that they then "fastened a screwdriver to the end of a glass rod 3 feet long & drew sparks from X with that." Ibid.

5. Batchelor stated that this wire was connected "to the stove 15 feet away." Ibid.

6. Batchelor also noted that "it would not move a Bradley Galv[anometer] It had not taste." Ibid.

7. Batchelor wrote, "Mr. Edison thinks this is a new force." Ibid.

–666–

Notebook Entry: Etheric Force[1]

[Newark,] Nov. 23 1875

(39)[2] New Force.

The following very curious result was obtained with it:— The vibrator shown in Fig. 1 (Art 38)[3] & battery were placed on insulated stands, & a wire connected to X (tried both copper & iron) carried over to the stove about 20 feet distant. when the end of the wire was rubbed on the stove it gave out splendid sparks. when permanently connected to the stove sparks could be drawn from the stove by a piece of wire held in the hand the point X of vibrator was now connected to the gas pipe & still the sparks could be drawn from the stove.

Fig 1 29

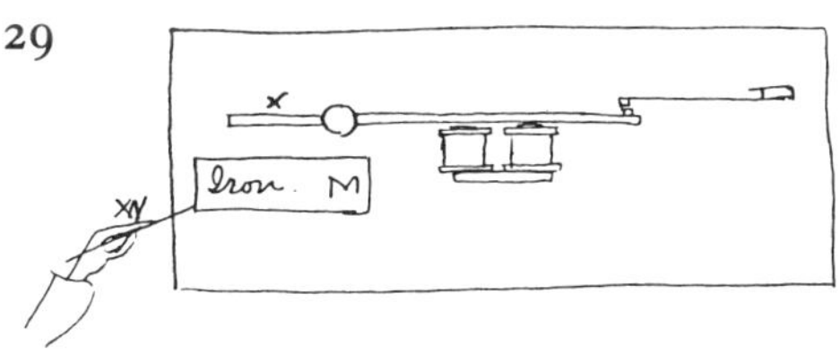

Whilst the rod was vibrating an iron piece M weighing about 8 pounds was placed near to X but not touching it in any way or connected to it except by the wood of the base & sparks could be drawn from M with a piece of wire as at W. This would leave it to be inferred that sparks can be drawn from iron bodies in a magnetic field; & that these sparks or force can be transmitted over metallic conductors & do not follow the laws of either voltaic or Static electricity.

Fig 2 30

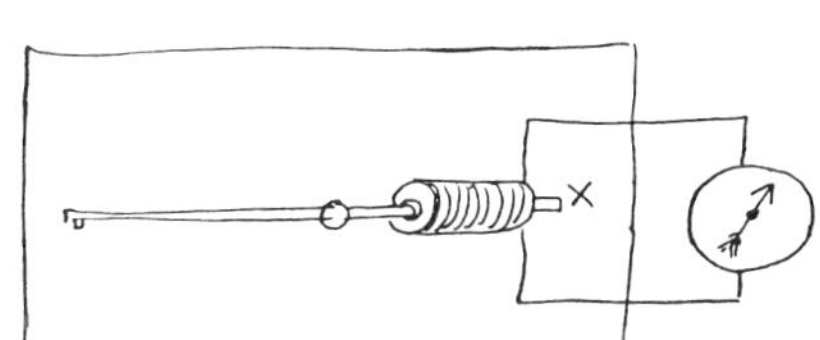

Put a coil of wire[4] over the end of rod X & passed the ends of spool through galvanometer[5] without affecting it in any way tried a 6 ohm spool & a 200 ohm.

We now tried all the metals touching each one in turn to the point X with

Antimony	we got a spark	
Lead	" " " "	
Iron	" " " "	very good
Carbon	" " " "	more actinic
Copper	we got a spark	
Manganese (perox)	we got a spark.	
Platinum	we got a spark.	
Mercury	we got a spark	
Tin	we got a spark	
Palladium	we got a spark	
Manganese	we got a spark	very poor
Potassium	we got a spark	green
Sodium	we got a spark	
Tellurium	we got a spark also a strong & disgusting smell of garlic	
Boron	we got no spark	
Aluminium	we got a spark.	
Zinc	we got a spark	
Cadmium	we got a spark	good scintillates
Bismuth	we got a spark	scintillates
Nickel	we got a spark	
Gold	we got a spark	
Rohdium	we got a spark	good
Cobalt	we got a spark	good
Arsenic	we got a spark	
Selenium	we got no spark	
Thallium	we got a spark magnificent green color very actinic little scintillation.	
Indium	we got a spark	
Silver	we got a spark magnificent green	

When standing on an insulated stand we can draw sparks from the following arrangement shewn in Fig 3

Fig 3   31

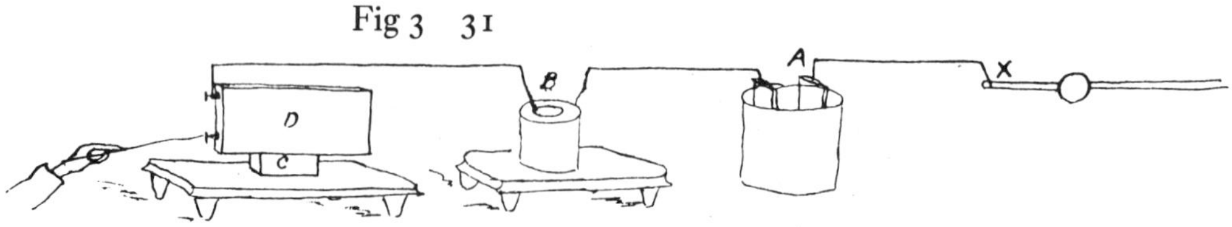

X is the end of vibrator which is insulated as well as battery A is a secondary battery B is a 200 ohm coil of copper wire, C is a block of iron, & D is a condenser all well insulated except A which is glass & stands right on table.

Fig 4 32

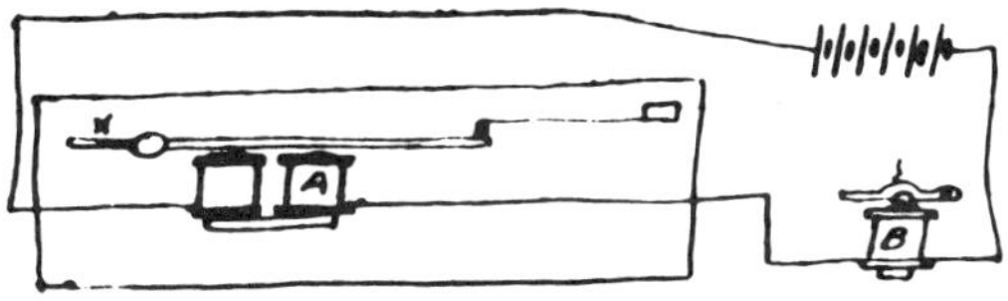

Another magnet B was now put in circuit with the vibrator magnet A & well insulated. A large piece of iron well insulated from everything was now placed 3 inches from A & another the same distance from B & sparks were drawn from both.
Chas Batchelor Jas Adams

X, NjWOE, Lab., Cat. 994:29 (*TAEM* 3:207). Written by Batchelor.

1. This entry is a continuation of Doc. 665.
2. See headnote, p. 600 n. 2.
3. Doc. 665.
4. This was a 6-ohm spool. Cat. 1317:16, Batchelor (*TAEM* 90:665).
5. This was a Bradley galvanometer. Cat. 1317:17, Batchelor (*TAEM* 90:665).

-667-

Technical Note: Condensers

[Newark,] Nov 23rd 1875

Idea just occurred to me that the parafine paper used in Condenser should be previously turned into tri-nitro cellulose ie[a] Explosive paper by nitric & sulphuric acid[1] whereby it becomes the most easily electrified substance [know]n[b] and would [p]robably[b] increase the Electrostatic capacity of the Condenser enormously
[Witness:] C Batchelor

AX, NjWOE, Lab., Vol. 9:18 (*TAEM* 3:805). [a]Circled. [b]Document damaged.

1. Treating paper with these acids to make nitrocellulose sheets began with Christian Schönbein's researches in 1846. The trinitrate form—called guncotton—was highly explosive, and no reliably safe methods for handling it were yet known. Less fully nitrated paper, called pyroxylin, when dissolved in ether and alcohol and dried, made sheets of collodion, a crucial photographic material that was less explosive but still highly inflammable. Pyroxylin paper was the basis for the production of the Celluloid Manufacturing Co. Friedel 1983, 4–6, 13–19.

–668–

Notebook Entry: Etheric Force[1]

[Newark,] Nov 24th 1875

(41)[2] New Force.

We tried our new force over a telegraph wire running from Newark to New Brunswick, then to New York & back to Newark. We connected X of vibrator shewn in prevous articles[3] to the end of line & obtained sparks from the other end of wire with a piece of iron wire held in the hand, showing that this force can be transmitted over long telegraph wires.[4] We then grounded the wire where the vibrator is connected to it & still obtained a spark, showing that this force may be transmitted over uninsulated iron wires buried in the earth for instance the sheathing of the Atlantic Cable.

When we take a spark from the end of the rod X it is much less than when we connect the gas pipe system to it.

We tried passing the force through water one pole being iron wire connected to vibrator X & the other to gas pipe but got nothing noticeable Tried solution of Alcoholic Sesquichloride of Iron also solution of Protochloride of Iron also a solution of Nitric Acid with no result. Tried gem of the organics, NIX with Nitric Acid NIX, Camphor floating in water, NIX. Tried piece of Ferrocyanide of Potassium in Protochloride of Iron, NIX. Tried connecting a large Iron plate weighing 80 lbs with X to try adhesion, Nix also tried connecting gas pipe for same purpose, NIX.

We found that two iron wires stick together quite strongly probably due to the effect of the spark as the same thing occurs with an electric spark. We shunted the vibrator with 3 ohm & could get no spark as it circulated in form of Inductive Electricity but could get spark when there was 20 ohms in shunt[5]

By increasing the battery from 8 to 12 cells we get a spark when the vibrating magnet is shunted with 3 ohms.

Fig 1 34

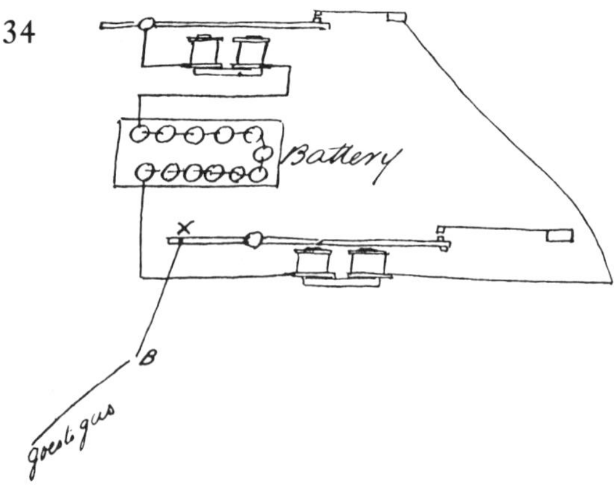

Cannot taste the least shock at at B yet between carbon points the spark is very vivid As will be seen from Fig 1 X has no connection with anything

With a glass rod four feet long well rubbed with a piece of silk over a hot stove, with a piece of battery carbon secured to one end we received vivid sparks into the carbon when the other end was held in the hand with the handkerchief, yet the galvanometer, chemical paper, the sense of shock in the tongue, or a Gold leaf electroscope which would diverge at two feet from an ½ inch spark plate glass machine were not affected in the least by it

A piece of coal held to the wire shewed faint sparks.

We had a box made thus:

2 35

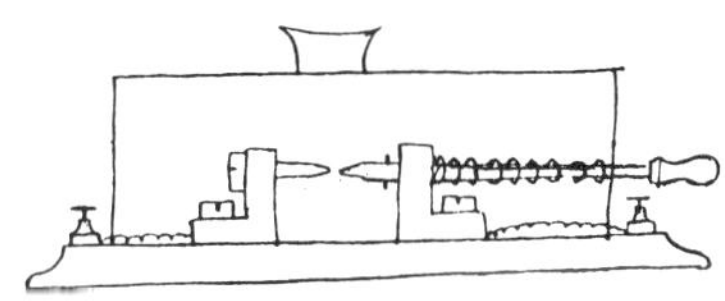

whereby 2 points could be brought together within a dark box provided with an eye piece The points were iron & we found the sparks were very irregular after testing some time two lead pencils were found more regular & very much more vivid. we then substituted the graphite points instead of iron

Fig 2 36

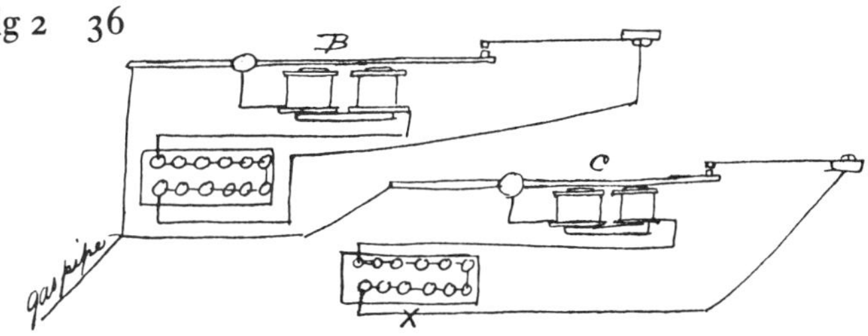

Tried Fig 2 but the sparks were not so vivid. We then reversed battery X & the sparks were more vivid but not so great as when B was used alone

4 37

We now found that when we had box in circuit & & a magnet operated by a key it would operate as well as a vibrator & with a piece of iron laid on cores at X Fig 3[6] & the wire attached & carried the length of the tab~~b~~le into the box we could get the spark every time we opened the key. We tried different metals at X with the following result:—

Iron at X very good
Gold at X good
Brass at X good
Antimony at X very good
Carbon at X very good
Electropoion fluid at X very good
Manganese at X good
Nickel at X good
Copper at X very good
A piece of Coal at X good
Cadmium at X very good in fact Iron & Cadmium surpass all others in this respect[7]

5 38

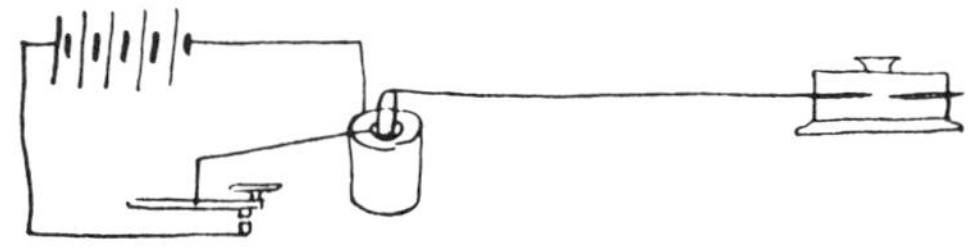

We got an iron wire spool & placed a piece of cadmium inside as in Fig 5 with very good result. Tried it also in a copper spool of 6 ohms resistance with good results.

6 39

We took a piece of No 16 copper wire bent in form of a horseshoe & placed it in two spools as in fig 6 & get the spark when the spools are connected so as to give no magnetism[8] we get spark just the same

With a magnet having a resistance of about twenty five ohms connected to make no magnetism & placed in a circuit of 10,000 ohms with 20 cells of 1½ Electropoion battery sparks were obtained when the ends of the wire running to the box containing the graphite points was held across the poles of the magnet by the finger

With 2 6 ohm spools with horse shoe copper wire put in 15 ohms resistance kills the spark.

With a 200 ohm magnet connected to give no magnetism, but which did give a little, we got spark on 10 000 ohms good.

It appears that when the Spools are connected to give no magnetism the electricity goes off into this new force With this same magnet I get it with only 12 cells battery & the magnet does not attract a fine iron wire suspended by a thread.

T A Edison Chas Batchelor

Jas Adams

X, NjWOE, Lab., Cat. 994:32 (*TAEM* 3:209). Written by Batchelor.

1. This entry is continued in Doc. 669.

2. See headnote, p. 600 n. 2.

3. Docs. 665 and 666.

4. In his own notebook Batchelor was more cautious, stating that "it might be that the force travels across the table instead of going out on the line." Cat. 1317:17, Batchelor (*TAEM* 90:665).

5. Batchelor noted that these were magnetic (i.e., inductive) shunts. Ibid.

6. Should be "4."

7. Batchelor noted that a bar of cadmium was preferable. Cat. 1317:18, Batchelor (*TAEM* 90:666).

8. That is, the spools are wound to create opposing magnetic fields.

-669-

Notebook Entry: Etheric Force[1]

[Newark,] Nov. 26th 1875.

(42)[2] New Force.

Started experiments in the following manner

1 40[3]

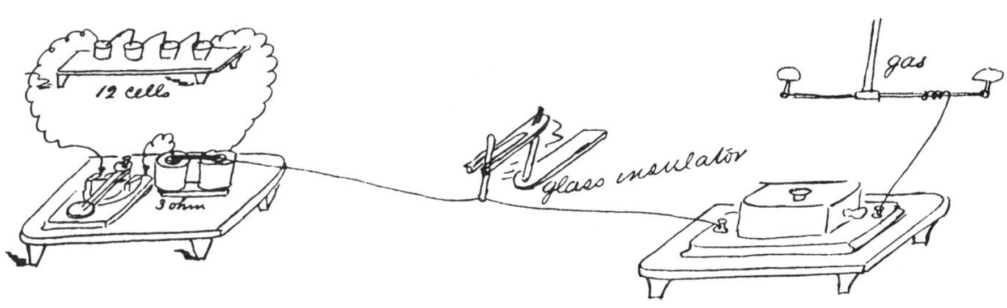

Of course we get the spark splendid with a piece of iron across the spools.

2 41

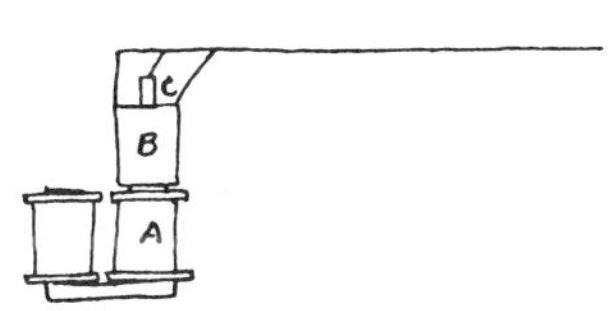

With magnet as in 2 weak spark. A is a 200 ohm magnet. B a 6 ohm spool with an iron core placed in & the two ends of 6 ohm spool together with the core connected to line.

3 42

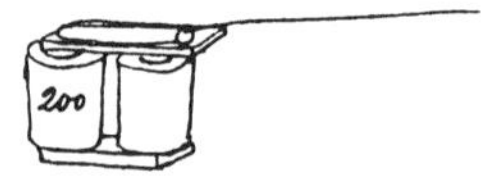

The 200 ohm magnet closed on top with an armature & then the iron laid on that very good.

4

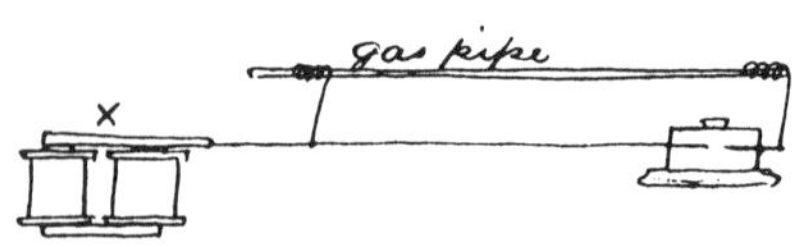

A piece of Cadmium at X & the line fastened to gas pipe on each side of the receiver & we get spark just same.

5[4]

With an inch of Hard rubber in line we get spark at intervals

We then put a piece of glass in place of rubber & we got spark at intervals

6

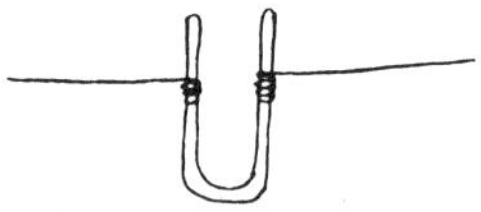

We put in a piece of glass that was bent as in 6 & we got spark at intervals

7

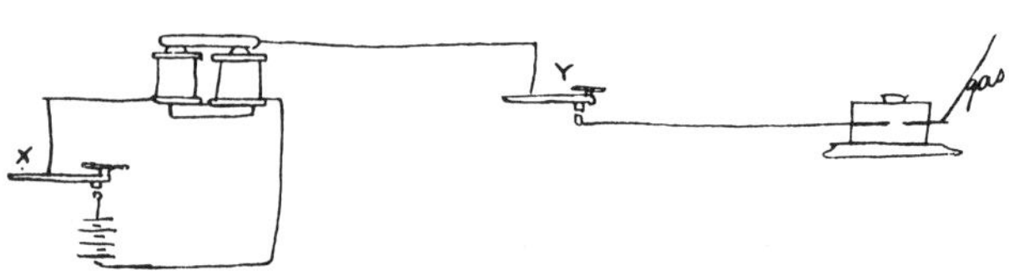

By closing X & working Y we could get no spark when fixed as in fig 7.

8[5]

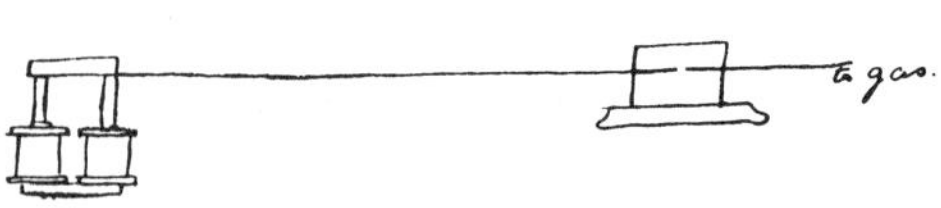

With the cores lengthened as in Fig 8 & the cadmium placed on top of them & the magnet actuated by 3 cells of battery we could get spark at times faintly.

We tried cores 6½ inches long in magnet shown in 8 & laid cadmium on top & got sparks just as brilliant as if on short cores

9

With mirror galvanometer made of copper wire of 3000 ohms resistance, director very weak & near top scale 2½ feet off could not get the slightest movement, but by connecting one end of our graphite pencil box to X we got brilliant sparks inside

With the cadmium bar stood up between the spools no result. With the cadmium bar placed across the spools in between cores no result.

We put an Epinus condenser[6] in circuit & get a splendid spark when both plates are in contact & weaker ones as they recede when both brass plates are six inches away from each other we get good sparks—

We placed the cadmium in local circuit & part of it with no result.

One of Kidder's Magneto Electric Machines[7] had both handles connected together & a wire taken from them to the pencils & got a good spark

When a six inch brass disc was placed on the poles of magnet & the cadmium laid on to that we get a good spark

When 30 inches of glass rod made part of line no spark wass visible

10

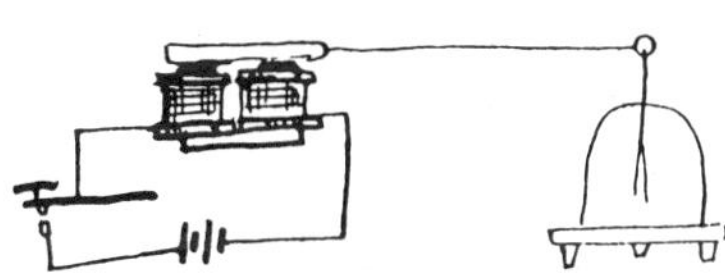

A gold leaf electroscope that diverges its leaves 3 feet away from a common old fashioned plate glass machine gave no indications of 'Etheric force'[8]

We tried to obtain Etherodes between which the spark will be more vivid than between graphite points. we fixed as in 11

11[9]

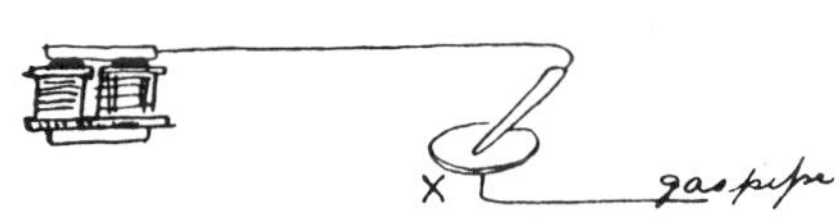

X is a plate covered with tin foil.

Sesquioxide lead no good Iodide lead more brilliant than on tin foil This would probably be good if hard pressed in to a pencil. We now discarded tin foil plate & used two lead pencil points thus[10]

Sulphuret Antimony no spark Spongy platinum spark no good. Nitrate Calcium spark small. Bone black pressed down with the finger in conjunction with the pencil point is better than the points alone. It seems probable that pencils of bone black pressed hard in a mould would be very delicate.

Benzoic acid no spark
Caustic Magnesia spark but no good
Lythargium no spark
Carbonate of Lead no spark
Sesquioxide of iron no spark
Arseniate of Copper improvement on Carbon points it makes longer spark

We now put on a self breaker thus:—

12[11]

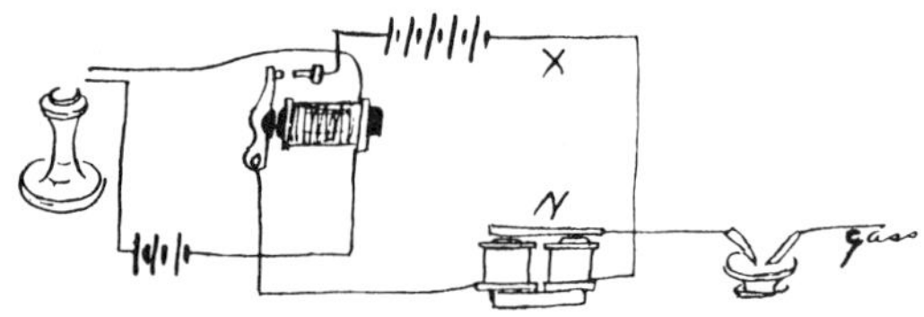

We found that if engine was connected in X circuit it either added resistance or the spark of the break sets up a path for the self induction of the magnet N & prevented the generation of the Etherism.

Cast iron filings between 2 carbon points give the best sparks yet.
Tar no spark
Ferrocyanide of iron no spark
Caustic Potash slight spark

We tried placing iron filings on a piece of paper over an electromagnet but the result was no better than if the magnet was not used

Iron reduced by Hydrogen gives no spark hardly

Tried the Etherism from several magnets connected up for quantity thus:—

13 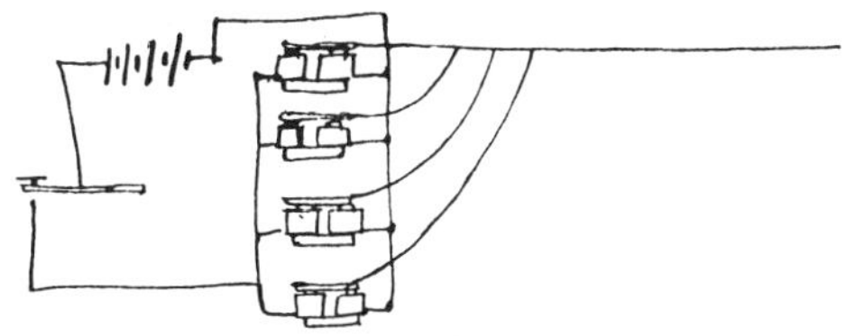

Spark not so good as when a single bar of cadmium is used. The magnets were not very even resistance. Took out one magnet which was fine when spark was not equal to single spool & cadmium. We tried middle power of our microscope & spark appeared about twice size. The glass lens or condenser at a certain focus makes the spark small but exceetingly sharp & beautiful with various colors. We arranged 3 6 ohm magnets thus

Fig 14

14 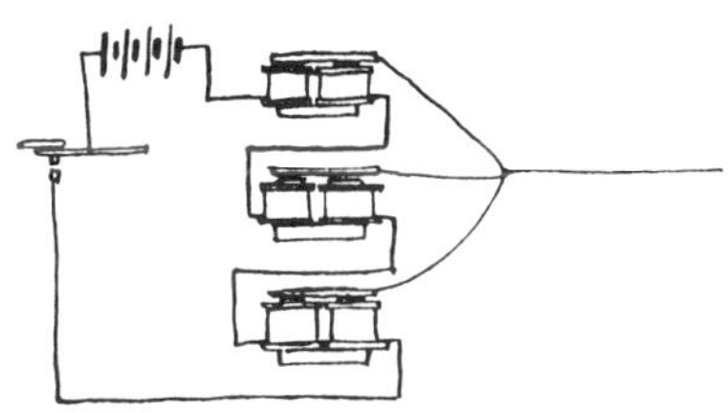

& the spark was better than with the 25 ohm magnet & cadmium bar that we had been using all along.

We now put in a single spool magnet of about 400 ohms & a cadmium bar on it & got a moderate spark, very irregular

By connecting the top & bottom of core to line also improved it immensely, made it as good as 25 ohm spool & magnet

Chas Batchelor

X, NjWOE, Lab., Cat. 994:35 (*TAEM* 3:210). Written by Batchelor.

1. This is a continuation of Doc. 668.

2. See headnote, p. 600 n. 2.

3. The figure label on the left table is "3 ohm."

4. Figure labels are "cadmium," "Hard rubber 1 inch," and "gas."

5. Figure label is "to gas."

6. Aepinus's ("air") condenser is a laboratory apparatus for demonstrating the accumulation of static charge. Atkinson 1910, 826.

7. Jerome Kidder, M.D., manufactured electromedical apparatus in New York City (H. Wilson 1876, 710). Edison corresponded with him during the summer about batteries for the electric pen. At that time Kidder asked if Edison might want to use his electrical machine. Kidder had three patents on arrangements of induction coils for electrotherapuetic apparatus that might have encompassed the design of the machine referred to here (U.S. Pats. 41,927, 58,105, and 93,625).

8. This is the earliest recorded use of Edison's preferred name for the "new force."

9. Figure label is "gas Pipe."

10. Figure label at left is "generator."

11. Figure label is "Gas."

–670–

Technical Note: Etheric Force

[Newark,] Nov 26 1875

Eatheroscope[1]

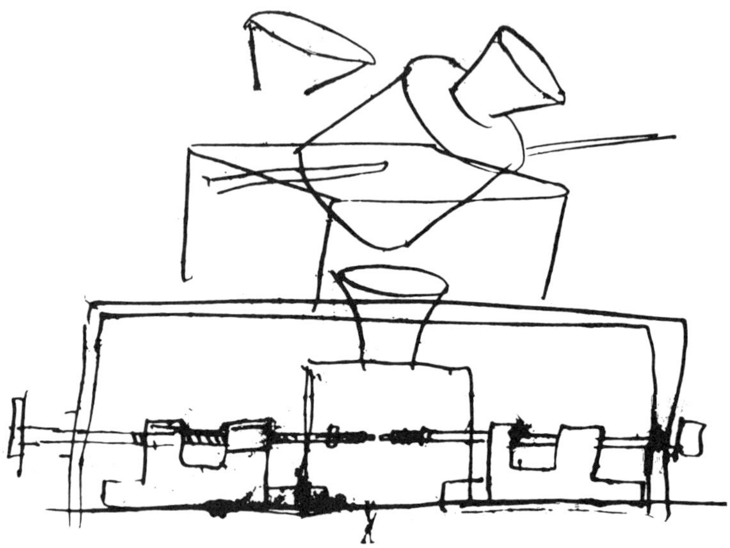

The "Etheroscope," for viewing etheric-force sparks.

T A Edison Chas Batchelor

X, NjWOE, Lab., Vol. 9:18 (*TAEM* 3:804).

1. This is a modification of the black box for viewing etheric-force sparks made two days earlier (Doc. 668). Edison later designed another box with an extension eyeshade. The Edison Institute in Dearborn, Mich., has an etheroscope, but it is in poor condition.

–671–

Receipt for Western Union Telegraph Co.

[Newark,] Nov 27 1875

Form of Receipt given Western Union Telegraph Co on the Acoustic Telegraph experiments:—

I herby acknowledge the receipt of two hundred $200 dollars paid to me by the Western Union Telegraph Company on account of expenses incurred in conducting experiments for said company for the purpose of ascertaining the practicability of Acoustic Telegraphy, and I hereby agree that in case I shall at any time hereafter secure letters patent of the United

States, for any of my inventions in Acoustic Telegraphy as explained on pages 141 142 143 and 144 of a book entitled "Wonders of Electricity" translated from the French of J Baile published by Scribner Armstrong & Co New York 1872, that I will assign such inventions so covered by letters patent to the Western Union Telegraph Co If in the event of failure to agree with said Company as to any further compensation than I have previously received on account of such inventions, which I may demand, then it is hereby agreed that the question as to the amount of such compensation if any shall be settled by competent arbitrators appointed in the usual manner[1]

Thos. A. Edison

Another recpt of which this is a copy was signed a week previous to this—[2] Edison

ADS, NjWOE, DF (*TAEM* 13:713).

1. This appears to represent a first attempt to formalize Edison's relationship with Western Union in regard to his experimental work on acoustic telegraphy. A contract was signed by both parties on 14 December (Doc. 695).

2. Not found. Edison had begun to keep an account of labor costs on acoustic-telegraph experiments on 19 November. Cat. 1214:132, Accts. (*TAEM* 21:620); DF-75-020 (*TAEM* 13:722).

-672-

Thomas Eckert to Uriah Painter[1]

New York[a] Novem 30th 1875.

My dear Painter:—[2]

About 4 P.M. on Saturday I began a letter to you which I was not able to finish, owing to a blinding headache and raging fever, the result of a cold taken at Philadelphia while visiting the centennial buildings and which I neglected to attend to.[3] I remember your admonition, but failed to heed it. Thinking I would be better after dinner and could then finish the letter, I folded the sheet and put it in my pocket, but before reaching the hotel I had a severe chill which decided my going directly to bed and sending for Dr [Crosby?][4b] who came promptly and began operations by prescribing 60 grains of quinine the first 24 hours and to continue at that rate until the fever was broken. I am happy to say the first 60 grains broke the fever since which I have been taking at the rate of 30 grains per day. I am without fever today and feel well with the exception of soreness of the chest. Am sitting up today and could go out if it were not for the severe change in the weather. Shall try to be prudent.

I was much gratified at the result of Union Pacific suit yesterday.[5] Am told Mr Gould looked and felt quite happy Have not seen him yet.

Now to the subject of your letter, received last night to which I am at a loss to know how to reply.

My dear Painter, I must tell you frankly that after what I have done and more that I would have been willing to do, ~~but~~ and could have done, but for the course of Mr. Reiff savors very much to me—but I will not finish this sentence. I have no inclination to be unjust, but the course he takes will drive me to desperation. I should ~~h~~ never have known Mr Reiff, but for you, and should never have trusted him but for my trust in you. Do you realize the fact my dear friend, that if I depended upon what these Automatic people (some of them, and Edison I believe to be one who would do right, but for improper influences) would or could do for or with[c] me, that I would be irretrievably ruined. I have become thoroughly discouraged and forced to look at the dark side of the picture whenever they feel that they[d] cannot drive me. I am ready for that issue when it is presented.

Mr Reiff has made all the trouble with Mr Edison. I had an interview with the latter and told him I was anxious to settle with him and asked to name his price which he said he would give. This was in the presence of Chandler. The next day he saw Reiff and the whole thing was knocked in the head since which I have not been able to get anything out out[d] of him.

You are mistaken about Mr Reiffs putting up for Mr Edison. The reverse is the fact.

I have never taken anyone more kindly to my heart than I have taken Mr Reiff. I felt a great sympathy for him, and for that matter do now, but his unfortunate temperament and condition not only annoy me, but they absolutely anger those who have and would again (but for his reckless and unjust criticisms) If he could be made to feel and act aright. You have been made to suffer improperly in connection with these matters, indirectly.

The inclosed note from Mr R—[6] came last week and annoyed me very much. What intrusion is there when a duty is involved and why make the matter so much a mystery?

I sent for him, he came and told me of some probable action by Prescott in reference to Quadruplex, what might be done re to counterbalance the move, all of which was in proper spirit and thankfully received. Following this he told me of Edisons very lukewarm feeling toward Gould & A.&P. and that W.U. were after him &c. I thought what a strange man you are, certainly no one could sustain a greater damage in such an event, than Mr Reiff himself and no one a greater individual interest to prevent it.

Now I beg of you, my dear Painter tell me what to do to close with these men, separately not jointly, and not on account, for that I have tried and it has caused reflection upon me, as well as to further delay a settlement.

I owe Mr Edison some bills which I am going to pay and at the earliest moment possible, but it is difficult to make some people understand that my construction account is necessarily heavy and that I cannot do exactly what I might wish under other circumstances. Do you reflect that within eleven months we have paid Mr. Edison and Mr. Reiff over 70,000$? This I think should have some weight.[7]

Money has been thrown away upon that roman letter machine to the neglect of a repeater which has from time to time been promised but never touched.[8] I know it would have been but for Mr Reiff.

AL (fragment), PHi, Painter. Letterhead of The Windsor [Hotel]. [a]"New York" preprinted. [b]Illegible. [c]"or with" interlined above. [d]Interlined above.

1. Although the signature page is missing, the content and handwriting of this letter identify Thomas Eckert as its author; cf. Eckert to TAE, 2 May 1876, DF (*TAEM* 13:1221).

2. Uriah H. Painter (1837–1900), a Civil War correspondent for the *Philadelphia Inquirer*, remained the paper's Washington correspondent for many years. A Washington insider connected to the Republican leadership, he also served for a time as Clerk of the House Committee on Post Offices and Post Roads, which held hearings on the subject of a postal telegraph. Painter first became acquainted with Edison in the early 1870s in connection with the Automatic Telegraph Co., possibly through his friend Thomas Scott (see Doc. 522). He was later associated with the Edison Speaking Phonograph Co., formed in 1878 to exploit the new invention. *New York Times*, 21 Oct. 1900, 11; *Philadelphia Public Ledger*, 22 Oct. 1900, 10; finding aid, Painter; Painter's testimony, *Nicholson v. Edison*.

3. The Centennial Exhibition was not yet open. Presumably Eckert was inspecting the building, possibly because Atlantic and Pacific Telegraph Co. had received the telegraph concession for the Exhibition. "Western Union and the Centennial," *Operator*, 1 June 1876, 7.

4. Possibly Henry Crosby, the only male physician of that last name listed in the New York City directory. Wilson 1875, 280.

5. It is not clear which of the many suits involving the Union Pacific Railroad Eckert referred to here. No decision regarding such a suit appears to have been handed down on this day; on 17 December, however, the *New York Times* (p. 3) reported a U.S. Supreme Court decision in favor of Union Pacific over the issue of state taxation.

6. Reiff's note has not been found.

7. No full accounting of this money has been found. However, exhibits and testimony in *Harrington v. A&P* do indicate that large sums were paid to Edison and Reiff, including a loan from Gould to Reiff. See, for example, Eckert to George Harrington, 14 Apr. 1876; and Reiff receipt to Eckert, 30 Mar. 1875; both 1:382–85, Box 17B, *Harrington v. A&P*.

8. In his 14 April 1876 letter to Harrington, Eckert claimed that Edison had failed to supply an adequate repeater for the automatic system. As a result, Atlantic and Pacific had been unable to use the system on a circuit longer than New York to Buffalo. For Edison's work for Atlantic and Pacific on a roman-letter automatic see Docs. 546, 566, 571, and 648.

–673–

Notebook Entry: Etheric Force

[Newark,] Nov 30th 1875

(44)[1] Etheric Force

We found the addition of battery to the Stubb's wire vibrator greatly increased the volume of spark. Several persons could obtain sparks from the gas pipes at once each spark being equal in volume & brilliancy to the spark drawn by a single person. If sparks are drawn off by a knife held in one hand, the placing of the other hand on another part of the gas pipe will materially reduce the spark

A very confusing & inexplainable experiment we tried thus:— At the extreme end of a line of gas pipe at the other end of room brilliant sparks were drawn from the brass tube of the jointed burner, three feet from this burner was another which when grasped by another person standing on the dry floor reduced the volume of spark drawn from the first burner to a point where they were imperceptible or nearly so, when the hand was taken off the spark acquired its full brilliancy. Water was then poured ŧ over the floor & a large piece of iron laid on it & connected to No 2 gas pipe by copper wire thus giving it a good route if it was electricity. The sparks on No 1 Burner were undiminished in brilliancy. a person now grasped No 2 burner & stood on the wet part of the floor & instead of diminishing as before it did not affect it at all.[2]

Several persons now grasped the pipes & Edison with a knife in one hand & a piece of wire in the other connected to the gas pipe obtained brilliant sparks between knife & end of wire

A wire from the pipe was led to one of the iron pillars, to the stove & other metallic objects all of which showed brilliant sparks

Edison now grasped the pipe & with the other hand holding a piece of metal he touched several other metallic substances obtaining sparks showing that the force passed through his body[3]

Touching the Cadmium at intervals with a red hot iron gave no result

The spools were taken out of a kidder magneto electric machine & the armature revolved at different speeds in front of the permanent magnet but without result.

Chas Batchelor

X, NjWOE, Lab., Cat. 994:40 (*TAEM* 3:213). Written by Batchelor.

1. See headnote, p. 600 n. 2.

2. In his own notebook Batchelor wrote, "This seems very paradoxical it would seem to be reduced by the current running to ground through the human body yet when you put better conditions for it doing so it apparently would not pass at all." Cat. 1317:20, Batchelor (*TAEM* 90:667).

3. In another experiment recorded in Bachelor's notebook, "three persons standing on blocks of paraffin & joining hands, one end of the wire held by the first man the third one could draw sparks from the stove with a knife in hand showing that it would pass directly through 3 human bodies although the spark was diminished a little" (ibid.). On 5 December similar experiments were performed with three and four persons. It was found with four persons that "the spark was weakened to a point where it was difficult to see especially if the pipe was rusty" (Cat. 994:49, Lab. [*TAEM* 3:217]).

ACOUSTIC TELEGRAPH RECEIVER Doc. 674

In November 1875 Edison built three acoustic telegraph receivers like the one illustrated in Doc. 674.[1] He made them of different diameters and put them in the same circuit with three transmitting reeds of various frequencies. The rapidly varying currents from the transmitters passed through the electromagnet, intermittently attracting the iron diaphragm. Edison intended that the smallest tube would resonate loudest to the highest frequency and the largest tube resonate to the lowest. Although this is the basic form of the transmitter used in 1876 by Alexander Graham Bell in his successful telephone, Edison in his experiments "never used [this form] for talking with, only for receiving."[2] In August 1877 Edison installed receivers of this type on a commercial telephone line.[3]

Like Edison's other acoustic experiments of this time, the ultimate purpose of these receivers was to allow multiplex telegraphy, not just to make transmitted frequencies audible. In these experiments, operators sent messages using regular

Morse keys, but instead of intermitting a steady current, they sent the undulating current provided by the transmitting reeds. These receivers, by selectively reinforcing particular frequencies, allowed several operators to use the same line. Receiving operators listened to the long and short sounds produced by the receiving tubes, just as they would listen to the intervals between the clicks of a regular sounder.[4] Edison soon tried to get each tube to click like a sounder or to activate a sounder.[5]

1. Edison entered this model and a second one as Exhibits A and A1 in a patent interference proceeding in late 1880. TI 2:513–14 (*TAEM* 11:641–42).

2. TI 1:114 (*TAEM* 11:78). To use this as a transmitter, the operator spoke into the tube.

3. TI 1:89 (*TAEM* 11:65).

4. TI 1:17–18 (*TAEM* 11:29–30).

5. See Doc. 708.

–674–

[Newark, November 1875?]

Experimental Model: Acoustic Telegraphy[1]

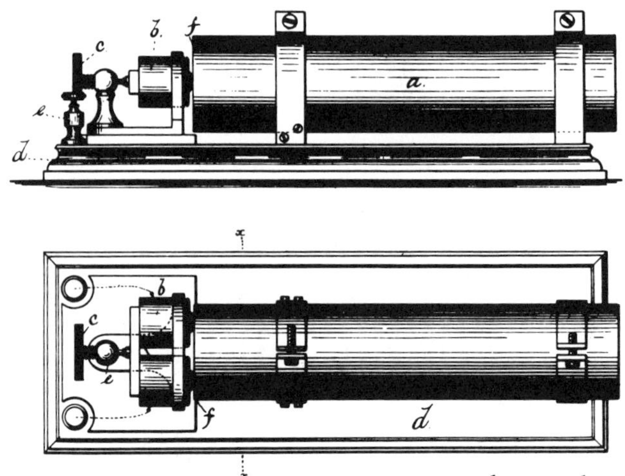

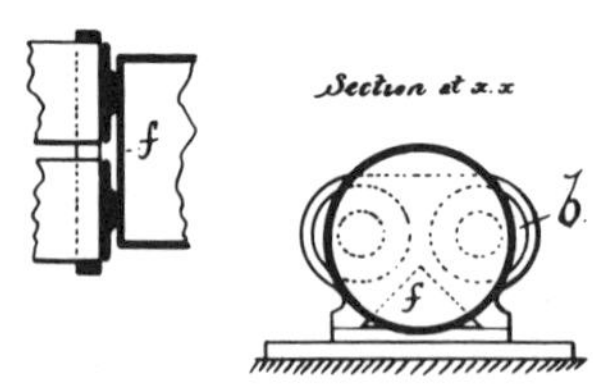

a. brass tube
b. electro magnet
c. adjusting screw
d. wooden base
e. binding posts
f. iron diaphragm

M (historic drawing) (32 cm × 11 cm × 10 cm), NjWOE, TI 2:513 (*TAEM* 11:641). These drawings were made from an instrument entered as an exhibit in the patent interference.

1. See headnote above.

POLARIZED ACOUSTIC TELEGRAPH RECEIVER Doc. 675

In this receiving instrument—"made within one month of the 15th of November, 1875, before or after"[1]—Edison replaced the diaphragm with a strip of metal passing across the mouth of the resonant tube. The transmitted current, modulated by a tuning fork or other acoustic transmitter, passed through the primary of an induction coil (shown in the accompanying drawing but not in the drawings in Doc. 675), thereby inducing currents of variable direction in the coil's secondary circuit. These currents in turn passed through the polarized relay **b**, which acted on tongue **e**, itself polarized by contact with one pole of the horseshoe magnet **c**. The vibrations of **e** produced sounds corresponding to the transmitted frequencies.

"It is easy to be seen," Edison testified, "that the proportioning of the faults of this instrument are such that it could not have worked satisfactorily and it did not do so." It was not loud enough, and the experimenters "had difficulty in getting a correspondence between the different plates used upon the receiving instrument, and the corresponding plates upon the transmitters."[2] However, Edison said, "this was not the fault of the instrument as an invention, but in the making of an instrument," and he pursued this design in a caveat of January 1876 and in an unsuccessful patent application of December 1877.[3]

1. Edison's testimony, TI 1:29 (*TAEM* 11:35).
2. Ibid.
3. Ibid.; Edison Caveat 74 (see accompanying drawings); Case 145, TI 2:25 (*TAEM* 11:198).

Two drawings, from Edison Caveat 74, of an acoustic telegraph receiver related to Doc. 675.

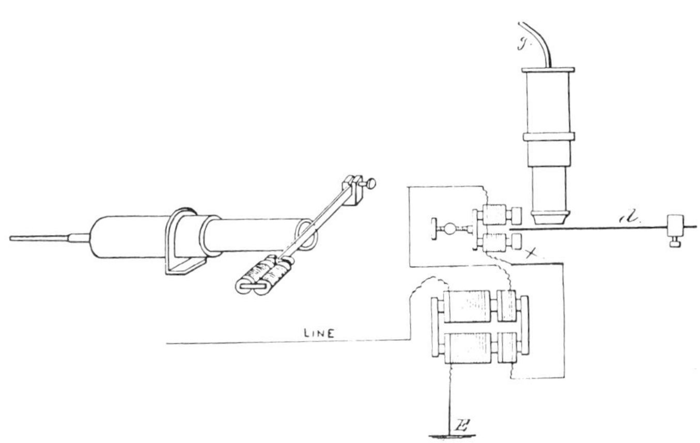

–675–

[Newark, November 1875?]

Experimental Model: Acoustic Telegraphy[1]

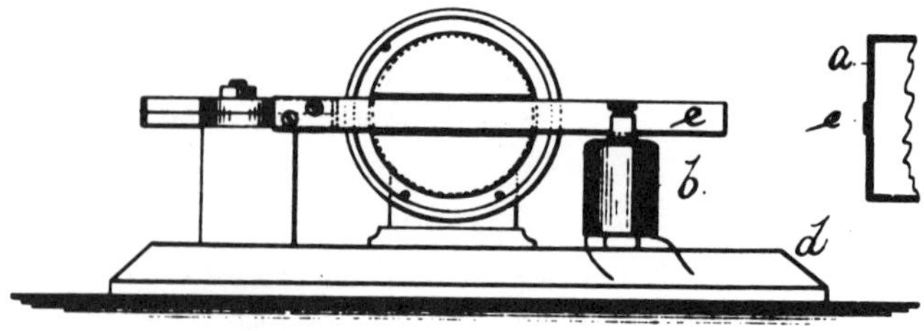

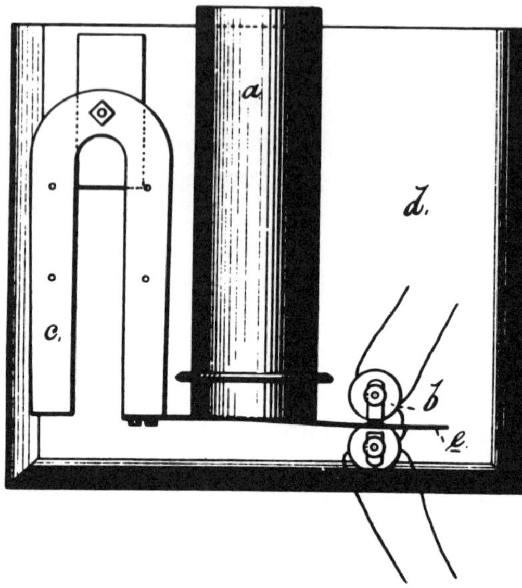

a. brass tube
b. electromagnet
c. horse shoe magnet
d. wood base
e. polarized tongue

M (historic drawing) (36 cm × 33 cm × 14 cm), NjWOE, TI 2:516 (*TAEM* 11:644). These drawings were made from an instrument entered as an exhibit in the patent interference.

1. See headnote above.

–676–

Draft to Josiah Reiff[1]

NEWARK, N.J., Decr 1st 1875[a]

Sir:

I will consent to approve the organization of the American Automatic Telegraph Co,[2] and ~~agree to~~ transfer to that Co all my right, title & interest in the Patents mentioned in the schedule furnished by Geo Harrington in his assignment of April 75[3] & accept therefore the 3000 shares of the stock proposed to be allotted; provided I am made a Trustee & the present Trustees will agree in advance to absolutely & formally release me from any ~~further~~ other[b] responsibility under the Harrington Edison Contract of Octo 1/70[4] & the assignment & Power of atty. based thereon dated Apl 4/1871,[5] which Contract Expired by limitation Octo 1/75.

I am compelled to make this special protective provision as I hear ~~int~~ intimated there may be an effort to lay claim to ~~b~~my ~~e~~services after Octo 1/75, without compensation. ~~or~~ [~~g---~~][c]

Df, NjWOE, DF (*TAEM* 13:498). Written by Reiff; letterhead of T. A. Edison. [a]"NEWARK, N.J.," and "187" preprinted. [b]Interlined above. [c]Heavily canceled.

1. Reiff's name appears as an internal address.

2. The investors in the old Automatic Telegraph Co. formed the American Automatic Telegraph Co. to contest the Atlantic and Pacific Telegraph Co.'s use of Edison's automatic patents without proper recompense. Although attempts to form American Automatic had started in August, Edison did not assign the patents until 23 December 1875 (draft articles of association and bylaws, dated 31 Aug. 1875, DF [*TAEM* 13:489]; Libers Pat. D-20:73–81). On 6 May 1876 the company received an assignment from George Harrington conveying his interest in the patents (TLC.2 [*TAEM* 10:898]); eleven days later Edison filed the Bill of Complaint in *Harrington v. A&P.*

3. The assignment, dated 9 April 1875, is in Exhibit 3, Bill of Complaint, 1:49–52, Box 17B, *Harrington v. A&P.*

4. Doc. 109.

5. Doc. 155.

-677-

From Charles Buell

New Haven, Conn Dec 2d 1875

Dear Sir

Can a fair sized spark be produced by the allotropic form of Electricity, with which you are Experimenting?[1]

And is the spark dense and of a bluish color? Very Truly Yours Etc

Charles. E. Buell Car Office N.Y. N.H. &. H. R.R[2]

⟨It has no relation to electricity it follows none of its Laws Yes. We have photographed it⟩[3]

ALS, NjWOE, DF (*TAEM* 13:311).

1. "Allotropic" is usually used to refer to a chemical element that appears in two forms. Here Buell seems to refer to the etheric force as a second form of electricity.

2. New York, New Haven & Hartford Railroad.

3. No photographs have been found. Edison's answering letter is dated 8 December 1875 (DF [*TAEM* 13:320]).

–678–

Article in the New York Herald

New York, December 2, 1875.

ETHERIC FORCE.

Mr. Edison's Discovery of a Supposed New Natural Force.[a]

Electricity Superseded by a More Docile Agent.[a]

Mr. Edison, whose name promises to become famous as the discoverer of a new natural force, was at his laboratory in Newark last evening, attended by his assistants and surrounded by a little company of interested persons, among whom were several expert electricians and a representative of the HERALD. Mr. Edison is a young man, of about the medium height, with full oval face, a large head and a manner that bespeaks the utmost devotion to his business. He was formerly a telegraph operator in the employ of the Western Union Company, and, during the war, was engaged in the Associated Press operating room, serving with excellent skill and fidelity. Possessed of natural inventive genius, he began to turn his attention to the betterment of the telegraphic instruments then in use, and succeeded in making a number of additions and improvements with such pecuniary benefit to himself that he was enabled to abandon the laborious occupation of an operator and devote his time to the study of electric science. His most important invention up to this time is that of the quadruplex system, by which four messages can be transmitted simultaneously through a single wire. It was while engaged in his occupation of experimenting that he made the discovery of what he is pleased to term, without much appropriateness, "etheric" force.

This force or principle is the direct offspring of electricity and magnetism. The operation of an ordinary telegraphic circuit affords the simplest means of deriving it and it is by this method that the present experiments are being conducted. When the circuit is open or broken the electricity stored in the magnetic coil by the operation of a continuous current is withdrawn artificially and it is in this manner of relieving the magnet that the invention consists. Hitherto the power stored in the magnet has been permitted to dissipate itself unnoticed in the air, in ignorance of the fact that it has assumed a new nature and possessed properties more valuable than those of electricity itself. Edison noticed that the accidental contact of a wire with the core of the magnet caused the production of a peculiar bright spark when a metallic substance was applied to it. Though this manifestation has been frequently observed, he was led to investigate it to its origin, and was amazed to find that the new manifestation failed to respond to the tests applied to discover the presence of electricity, either inductive or static. The galvanometer was unmoved by it; the dainty gold leaf of the electroscope exhibited no signs of deflection; the tongue could detect no sensation; the Leyden jar charged with it was possessed of no property that is contributed to it by electric contact. In a word, the manifestation was nonelectric. It was the display, recognized for the first time, of a principle until then buried in the depths of human ignorance.

Since the moment of this discovery Mr. Edison has worked night and day to ascertain more intimately the nature of the new "etheric" principle, and with results which are strongly confirmatory of his original theory. A voltaic battery, of several cells, is applied to an ordinary telegraph instrument. The trial of twenty-eight different metals having shown that cadmium is best adapted for the transmission of the new power,[1] a bar of that substance (analagous in its nature to zinc) is placed across or near to the magnetic coil, either directly or with an insulating substance intervening, and from this bar a copper wire conducts the "etheric" force to any desired point. Connected with the gas pipe it charges every gas fixture in the building so that the application of a metallic substance to any one of them will produce the etheric spark. A connection with the street main enabled Mr. Edison to draw the sparks from a gas fixture in his residence, some distance away.[2] Hence it seems that the new force is not affected by the contact of the earth, but performs its operations independent of the necessity of insulators. Herein lies its immense practical value. The cumbersome appliances of transmitting ordinary electricity, such as telegraph poles, insulating knobs, cable sheathings and so on may be left out of the problem of quick and cheap telegraphic transmission, and a great saving of time and labor accomplished. Ocean cables operated by "etheric" force need cost but a fraction of the present clumsy lines. Wires may be laid in the earth or in water. The existing methods and mechanisms may be completely revolutionized. The experiments witnessed by the HERALD representative were of an exceedingly interesting nature. Attaching the wire upon which the etheric current is conducted to a gas fixture in the large laboratory, it was found that the peculiar sparks could be obtained from every other gas fixture in the building simultaneously by the application of a file, knife blade or other metallic substance. A confusing and inexplicable experiment was as follows:— The current was transmitted through the gas pipes, and the experimenter proceeded to a distant burner and drew brilliant sparks by applying his knife blade. Three feet from this burner was another burner, at which a second person was stationed with instructions to grasp it firmly, which was no sooner done than the sparks at burner No. 1 became so feeble as to be almost imperceptible. When the grasp on burner No. 2 was released the sparks were evolved from No. 1 with renewed brilliancy. Water was then placed upon the floor and a piece of iron laid in the wet spot and connected with burner No. 2 by means of a copper wire, thus furnishing ample opportunity of escape had the current been an electric one. But the brilliancy of the sparks at No. 1 was found to be unaffected by this test. The person who first grasped burner No. 2 was then made to resume his grasp, standing on the wet spot, but the former effect of diminishing the force of the spark at No. 1 was now imperceptible. Standing upon the dry floor his grasp of burner No. 2 was fatal to the manifestations at No. 1, but when the floor was moistened the interruptive effect ceased. Still another experiment consisted in grasping the gaspipe into which the etheric current was being led, the person grasping it being made to hold a knife in his other hand. Upon touching a file to the knife blade sparks were freely drawn, showing that the human body is a good conductor of the new force. No peculiar sensation was observed by the person through whom the current passed.[3] When the writer left the laboratory Mr. Edison had just brought the microscope to bear upon the etheric sparks, magnifying them several thousand areas.

It has been suggested that this discovery throws new light upon the nature of the auroral phenomena. Telegraph operators are familiar with the difference between the effect of the electricity developed by a thunder storm and that which affects the wires previous to an auroral display. The former passes through the wire, emitting both spark and sound; it is instantaneous in action, seldom travelling far on the wire before discharging itself. But the electricity produced by the aurora passes along the wire in a continuous stream, with no sudden discharge, effecting the same result as that produced by the galvanic battery. A colored mark upon the paper is made by the positive current of the aurora as by the positive pole of the battery. When these effects have been observed the aurora follows, and so familiar have the operators become with the manifestations that they can predict an auroral display with absolute certainty. They regard the electricity which precedes its appearance as of precisely the same nature as that of the electro-galvanic battery, which is distinguished by its voluminous current without intensity of action, differing from ordinary atmospheric electricity or the kind produced by friction, which may be dissipated by means of a wire conductor leading to the ground. Now, it is an old theory that the cause of auroral displays is the accumulation of quantities of electricity in the earth (which may be regarded as a vast magnet) and by the liberation of this electricity, which is dissipated in the form of a magnificent display of light. But this is precisely the mechanism by which the etheric spark is produced. The core of the magnet corresponds to the magnetic pole of the earth; the cadmium and wire act in correspondence to the peculiar conditions of the atmosphere under which the earth's electricity is liberated; the galvanic battery and electric current are analogous to the vast currents of the electric fluid which encircle and pervade the earth; the auroral flame is kindred to the etheric spark. In short, the newly discovered manifestation, producing no effect but light, is believed to be a miniature of that grand and mysterious illumination which has excited the admiration and wonder of men for so many centuries.[4]

Mr. Edison is constructing a quantity of special apparatus for the purpose of experimenting with the new force. Gutta percha rods will be suspended so that the instruments may be more thoroughly insulated and all possibility of electric induction prevented; an "etherioscope" will be contrived by means of grathite points so that the etheric sparks may be more advantageously studied and a spectroscope analysis will be made of the spark in order to ascertain the affinity of the light with that of the auroral display. The discoverer will labor unremittingly to develop the etheric manifestations until he succeeds in putting them into more tangible shapes and evolves a force which shall be as docile and adaptable to the uses of man as those faithful drudges—steam, heat and electricity.

Mr. Edison's theory of "etheric" energy is as follows:—"Under certain conditions heat energy can be transmitted into electric energy, and that again, under certain conditions, into magnetic energy, this back again into electric energy, all forms of energy being interchangeable with each other. It follows that if electric energy under certain conditions is transformed into that of magnetism under other conditions it might be transformed into an entirely unknown force, subject to laws different from those of heat, light, electricity or magnetism. There is every reason to suppose that etheric energy is this new form. The only manifestation of its presence previously recorded with scientific accuracy is that of the German chemist Ruchenbach, who noticed that an electro-magnet, under certain conditions and placed in a dark room, became luminous when the hands of peculiarly constituted persons were made to approach the poles. This phenomenon, inexplicable to Ruchenbach, is easily to be accounted for on the etheric theory."[5]

PD, *New York Herald,* 2 Dec. 1875, 4. [a]Followed by centered horizontal rule.

1. See Docs. 666 and 668.

2. Edison was living at 89 South Orange Ave., about a mile from the Ward St. laboratory and shop. See Chapter 7 introduction, p. 375; gas bills, 75-002, DF (*TAEM* 13:229, 242).

3. These experiments had previously been conducted by Edison and Batchelor and recorded in Doc. 673. At points this passage reproduces the text of that notebook entry; articles in the *Newark Daily Advertiser* (29 Nov. 1875), *New York Tribune* (30 Nov. 1875, 2), and *New York Sun* (1 Dec. 1875, 4) also quote directly from notebook entries.

4. For a discussion of the aurora borealis in connection with telegraphy, see Prescott 1863, 309–32.

5. Edison referred here to Karl Reichenbach; for Reichenbach's claimed discovery of the odic force see Doc. 499 n. 1. The negative reception given to Edison's announcement by many in the scientific community may well have been related to such associations with Reichenbach's claims and with sensationalized press reports generally. Before this, for example, the 29 November 1875 *Newark Morning Register,* in reporting Edison's "Wonderful Invention" and "Startling Discovery," introduced the subject through references to Reichenbach and to William Crookes's investigations of spiritualist phenomena (for a clipping see Cat. 1144, Scraps. [*TAEM* 27:282]). On 3 December the *New York Times* ridiculed Edison's discovery, comparing it with that made by the "maligned and discredited" Reichenbach (p. 4). Clippings from at least a dozen newspapers in Newark, New York, and other cities are in Cat. 1144, Scraps. (e.g., *TAEM* 27:282, 290–91, 297–300).

–679–

Notebook Entry: Etheric Force

[Newark,] Dec 3rd 1875

(45)[1] Etheric Force.

Further experiments were tried to endeavour to prove that Etheric force was electricity of a certain degree of tension.

Smith from Chester's[2] came over with a small induction coil & made the statement that it was induction.

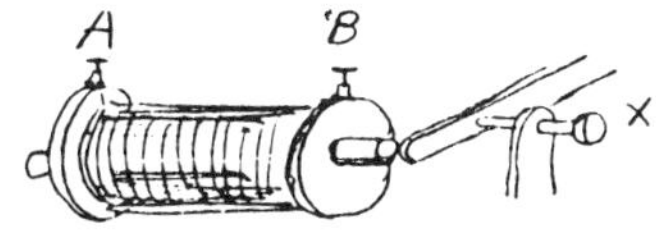

From X with 2 cells he could get no spark (because that was not enough battery to get a spark between copper & copper) but by holding the hand near A or the outside of the coil the spark immediately appeared upon X This was nothing more than induction as the spark could be felt & seen jumping to the hand from the outside layers of the coil thence through body to other hand & so to X thence to top coil of primary to lower coil of secondary connected to B. This could be illustrated thus:

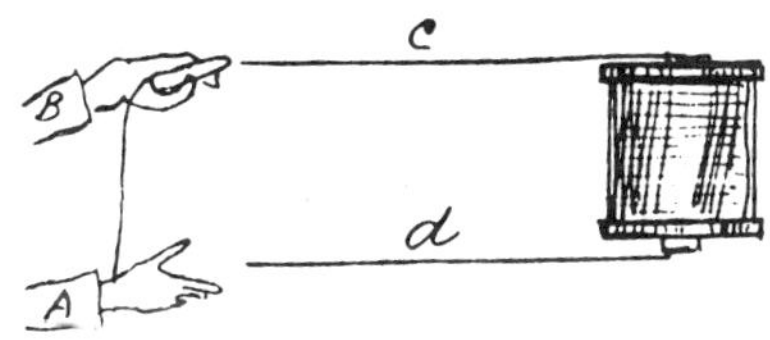

the distance between d & c was too great to jump[a] but when the space was filled by a partial conductor the spark was obtained. The spark between A & B terminals of the secondary was ¼ on an inch. Mr Smith did not stop to witness the production of Etheric force under conditions which preclude induction. Mr Edison however gave him a diagram shewing how to obtain it from a sounder magnet with wire laid over the poles of magnet. At 8 PM we took our induction coil which gives ¼ spark & could produce same result as Smith (ie) produce the spark from X by electrical induction, we then connected one end of secondary to the gas pipe leaving the other in air no spark could be obtained on pipes. We then connected the terminal in ai~~n~~r with a piece of metal on the table & moistened it still no spark on pipes. Used 8 cells on primary. Mr Edison then left both terminals in air of secondary & connected X to gas pipe & got good sparks.

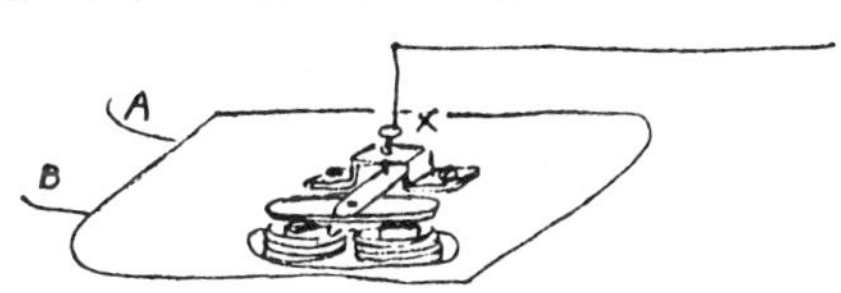

He then connected A & B terminals of the secondary coil together Still got spark but weakened as it should be part of the force going off in electricity though a small quantity of great tension

I then disconnected the gas pipe wire from X & attached it

to the secondary terminals which were fastened together could get no spark at all then from gas pipe. We now tried two 200 ohm spools thus:—

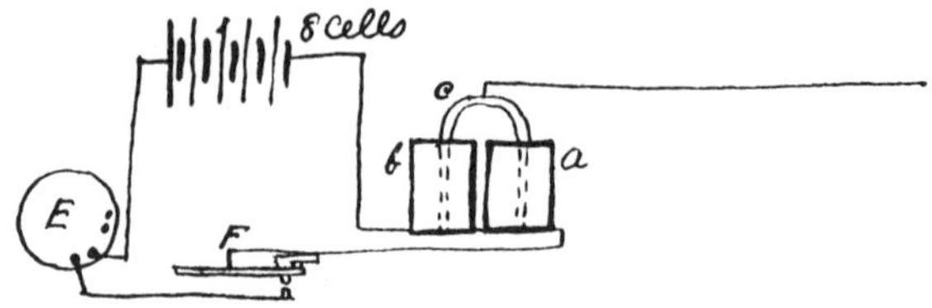

a & b are the two spools connected with their two inside ends together, c is a thick copper rod bent in the form of a horseshoe & placed in the centre of each helix b & a but not touching either C is connected to our graphite points & to the gas pipe. E is a Bradley rheostat F a key & everything thoroughly insulated. To prove the insulation we connected the prime conductor of our plate glass machine to the key F charging all the apparatus, upon disconnecting the prime conductor after ½ minute of time we got a spark ¼ in long from F notwithstanding the numerable points for the static electricity to dissipate from

We now received sparks on the graphite points with 1000 ohms resistance in the rheostat or from a strength of 1/100 of a Bunsen cell on an internal resistance equal to itself. We now reversed the connection in the spools so as to make no magnetism (we used no iron) so to speak & we could the spark though the electric circuit had 2500 ohms resistance in it or the 1/250 of a bunsen cell. We have noticed that when using an iron core & it is allowed to produce magnetism we get less etheric force & in proportion as we pull out the iron & insert the copper in its place we gain etheric force & if we connect the spools to give no magnetism we obtain the greatest etheric effect

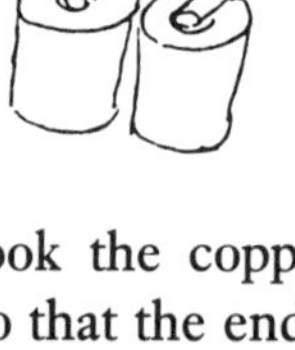

We took the copper horseshoe out & laid them on the spools so that the ends just covered the holes in the spools & we could get the spark when the electric circuit had 1000 ohms resistance.

We now put in the vibrator used in our first experiments & increased the battery from 8 to 20 cells & connected X (see article 38 Fig 1)[3] to gas pipe. Charley Edison hung to the gas pipe with feet above the floor & with a knife got a spark from the pipe he was hanging on. We now took the wire from vibrator in one hand & stood on a block of paraffine 18 inches square & 6 in thick holding a knife in the other hand we drew sparks from the stove pipe. We now tried the crucial test of passing the etheric current through the sciatic nerve of a frog just killed. Previous to trying we tested its sensibility by putting the current from a single bunsen cell; we put in resistance up to 500,000 ohms & the twitching was still perceptible. We tried the induced current from our induction coil having 1 cell on primary the spark jumping about 1/50 of an inch terminal of the secondary connected to the frog & it straightened out with violence. We arranged legs to pass etheric force through it; we placed legs on an inverted beaker & held the two ends of the wires on the ends of glass rods 8 inches long: on connecting one to the sciatic nerve & the other to the fleshy part of the leg no movement could be discerned although brilliant sparks could be obtained when the frog was in circuit on the graphite points. Dr Beard was present when this was tried. The frog was now tested for delicacy with an electric circuit & although it had lost a little of its delicacy it was still exceedingly sensitive.[4]

Chas Batchelor

X, NjWOE, Lab., Cat. 994:41 (*TAEM* 3:213). Written by Batchelor. [a]Repeated at end of one page and beginning of next.

1. See headnote, p. 600 n. 2.

2. John E. Smith was an inventor whose printing telegraphs were manufactured by Charles Chester, a noted telegraph inventor and manufacturer located in New York City. George Beard's published report states that "Prof. J. E. Smith" collaborated with him in many experiments on the new phenomenon over the next few days at Chester's establishment (see n. 4 below). Smith's instruments were used by the Manhattan Quotation Telegraph Co. in its attempt to compete for stock-quotation business with the Gold and Stock Telegraph Co. Manhattan Quotation Telegraph Co., Board of Directors Minutebook, NjWU; Reid 1879, 626–29.

3. Doc. 665.

4. George Beard wrote Edison on 1 December to inform him that he had contacted a man in Newark who would bring a frog over to Edison's the following day, and that Beard would come over to try a number of experiments (DF [*TAEM* 13:309]). On 4 and 5 December, Edison continued his etheric-force experiments using a frog (Cat. 994:47–50 [*TAEM* 3:216–18]). He published the results of these experiments in the *Scientific American* (Doc. 685). Beard also continued experimenting, with and without Smith, on his own and at Chester's. He prepared a long report on 6 December covering the trials done in Newark and New York and sent it to the *New York Daily Tribune*. Beard recounted the

experiments they did to show that something novel was involved, and he described the precautions he took to rule out trickery. He suggested that the force was a new and very different form of electricity and sharply attacked claims that identified the phenomenon with the "weak sparks" discussed by the German physicist Petrus Riess and recently reported in the English *Electrical News.* He also rejected emphatically and at some length any relationship to the supposed odic force of Karl Reichenbach. See "Mr. Edison's 'New Force,'" *New York Daily Tribune,* 9 Dec. 1875; and, for example, "The 'Etheric Force,'" *New York World,* 11 Dec. 1875; both Cat. 1144, Scraps. (*TAEM* 27:290, 298).

–680–

Notebook Entry: Etheric Force

[Newark,] Dec 3rd 1875.

47[1] Etheric Force

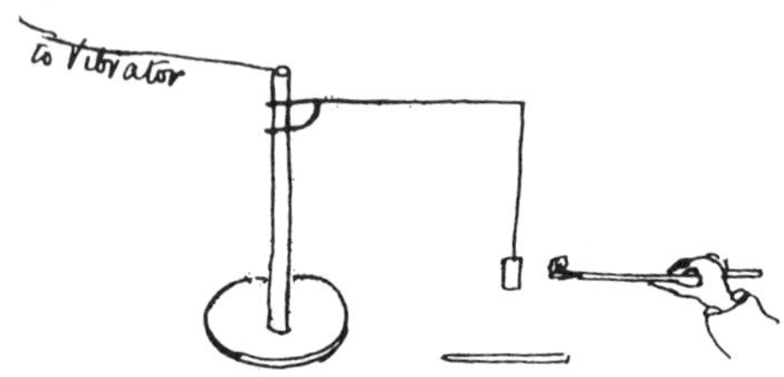

Tried several different kinds of chemical crystals to ascertain if attracted or repelled by ethericity but with no result

Tried experiments with induction sparks to obtain scintillations & found that they were nothing like what they were from the etheric yet the spark jumped ⅛ of an inch. We tried an experiment with the large coil giving ⅛ jump spark connected thus

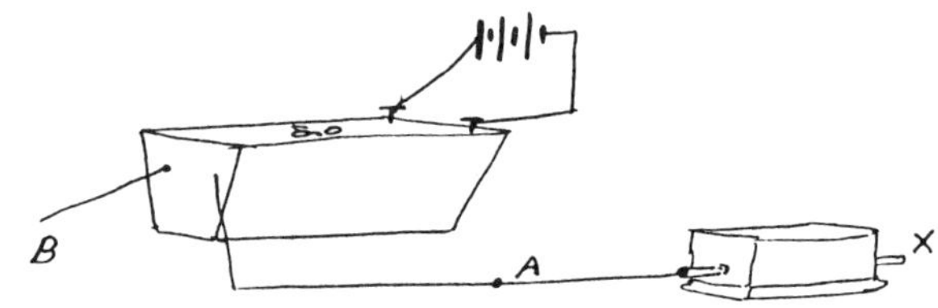

One end of the secondary coil was connected to the graphite points & everything well insulated, placed one hand at B & touched X got no spark but on tasting X we got a spark & strong shock on tongue; detached the hand from B but still kept tongue on X no spark; put the hand to X & got faint sparks; stood on insulated stool & got shocks; (put the wires on glass tubes 3 inches apart no spark touched the hand to the other wire & could draw sparks ⅛ of an inch long from it still no sparks could be drawn on the graphite points) connected a wire from A to gaspipe no spark; put the hand to X got a faint spark & shock on tongue, shewing it went through air. Tried this

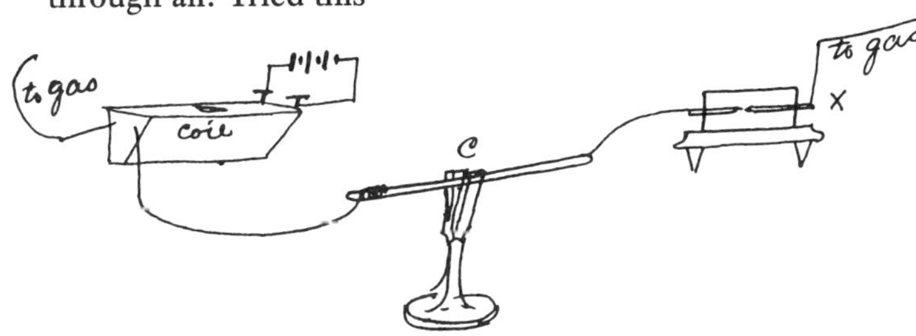

C is a glass rod 3 feet long ½ inch in diameter well warmed & wiped with silk; with induction current on full we got no spark, put the wires glass tube 3 inches apart, no spark touched the hand to the wire on left & could draw sparks ⅛ of an inch long from it still no spark stood on insulating stool with hand on wire to left felt strong shock, no spark; made the hand approach within 8 inches of the wire on left got spark tried it a number of times the approach of the hand rendering sparks visible. This proves quite conclusively that the Etheric force that cannot jump through space will travel through a length of glass many times greater than a high tension spark jumping ⅛ of an inch

We tried the following substances made into solution & paper wetted with them to see if we could get the force to mark on any. Caustic Magnesia & Silicate of Soda no mark Oxalate of Cerium[2] [no mark][a] Citrate of Potash Carbazotic Acid & Silicate of Soda Ferrocyanide of Potassium Sulphate of Manganese Bichloride of mercury & Sil of Soda Phosphate of Soda Lycopodium Muriate of Beeberin,[3] Acetate of Lime Sulpho carbolate of Soda, Sulphate of Morphia Sucromate of Ammonia, Sulphurate of Copper Tartrate of Ammonia, Iodine resublimed Bicarbonate of Potash, By Iodide of Mercury Chromate of Potash, Camphor. Carbonate of Copper, Arsenite of Manganese Carbarozate of Ammonia, Sodium Hyposulphite Chloride of Nickel, Bisulphate of Potash Thymol, Iodide of Sulphur, Antamonic Acid Sulphur, Protochloride of Iron, Iodide of Antimony Morphine, Acetate of Morphine, Sulphate of Ammonia Iodide of Silver, Brucine, Citrate of Quinine, Cyanide of Silver, Titanic acid, Tannate Quinine Asparagin, Sulphate of Caffeine, Hypophosphate Quinine, Lactophosphate of Bismuth Lactate of Bismuth, Urea, Bromide of Silver, Sulphate of nickel, Muriate of Quinine, Sulphuret of tin.

Chas Batchelor.

X, NjWOE, Lab., Cat. 994:45 (*TAEM* 3:215). Written by Batchelor.
[a]Ditto marks in original.

1. See headnote, p. 600 n. 2.
2. Cesium.
3. Probably berberine.

–681–

Notebook Entry: Acoustic Telegraphy

Newark NJ. Dec 3 1875

16[1] Acoustic Telegraph.[2]

Mr Edison proposed sometime ago that the study of acoustic telegraphy would be remunerative as it seemed to be very difficult to make practical the inventions of Gray & others so we have worked this up & intend to follow it for some result.[3] Our experiments so far have been confined to reeds of Stubbs steel but have had a pair of tuning forks made giving 256 vibrations per minute & resonant boxes for them.[4] These we placed on bases as in Fig 1 & the first principle we have tried to work was direct entirely stopping the transmitting fork which would not work well.

Fig 1

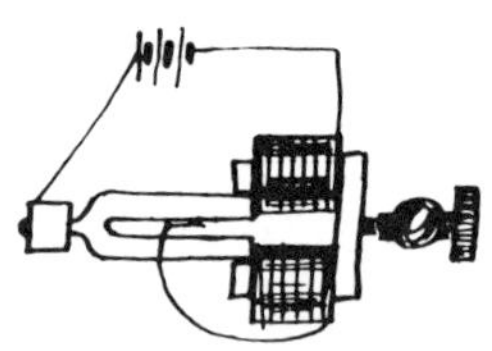

Fig 1 is the transmitter it differs only from the receiver in that the receiver has its contact outside the fork. Mr Edison hit a very good idea for working the sounder in place of closing the local circuit whilst the fork is vibrating which is only a series of very quick contacts he does it in the following manner

Fig 2

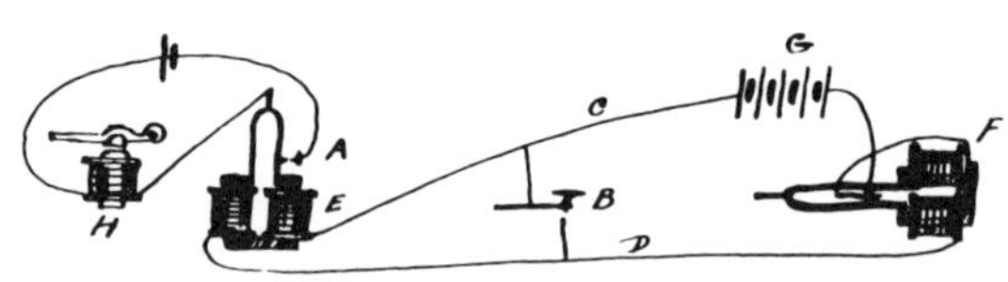

F is the transmitting magnet, G main battery, B Key, E receiving magnet, A repeating points, H ~~r~~sounder. In this case when Key B is open magnet E works & vibrates fork opening & closing sounder circuit so fast that the induction from spool will not let it close but when B is closed it shortcircuits main battery & the points A stop on closed circuit working sounder This works very well.

Chas Batchelor

X, NjWOE, Batchelor, Cat. 1317:21 (*TAEM* 90:667). Written by Batchelor.

1. Batchelor numbered the entries in this notebook.

2. There is another version of this entry in the "Experimental Researches" (no. 46, Cat. 994:44, Lab. [*TAEM* 3:215]). The version reproduced here provides some additional explanation.

3. See headnote, p. 524.

4. See Doc. 664.

–682–

From Osgood Wiley

Orange N.J. Dec 6th 1875

Dear Sir

I am desirous of making an electrician of myself and wish to know if there is any chance of my getting into your factory where I can get a superior advantage— I do not desire a paying position as I am not dependent on it— I merely want to be where I can learn more about the practical part of Telegraphing— I went into Mr. Prescotts office last May But Mr Prescott was always to busy & Mr. Smith away so much I had no chance to learn much— I have been employed in the main office of the W.U. Co and am considered a good operator but I do not desire to make my business merely an operator

I would be willing to pay for by being allowed to enter your place & be allowed to pick up what I can— My Father is a Scientific Publisher in N.Y. (John Wiley & Son 15 Astor Place) and I can precure any books whatever on the subject by replying you will greatly oblige me[1] Very Respy

Osgood S Wiley

ALS, NjWOE, DF (*TAEM* 13:306).

1. On 8 December, Edison asked Wiley to come and see him regarding this request; by the end of February 1876, Wiley was working for Edison in a paid position. Charles Batchelor to Wiley, 8 Dec. 1875, DF (*TAEM* 28:160); Cat. 1214:135–36, Accts. (*TAEM* 21:629–30); 75-020, DF (*TAEM* 13:725–26).

–683–

Notebook Entry: Acoustic Telegraphy

[Newark,] Dec 6th 1875.

Acoustic Telegraphy.

We found that the amplitude of the vibration of the large 6″ forks was practically insufficient to work the sounder. The fusion of the points at so close an adjustment being one of the objections We moved the spring contact on the self acting fork nearer fearing that the pressure dampened the fork & threw it out of unison, but found that placing the contact here where the swing was less was not a success although it appeared to give the fork a better tone. It made the receiving fork very weak owing to the lightness of the transmitted waves; attempting to lengthen them made the play of the spring in transmitter so close that it stuck & sent bad waves

We concluded to make thinner forks & whilst they are making Batchelor arranged thin reeds as temporary devices till the forks are completed.

Chas Batchelor

X, NjWOE, Lab., Cat. 994:51 (*TAEM* 3:318). Written by Batchelor.

–684–

To S. V. C. Smith[1]

[Newark,] Dec 8 [187]5

Dear Sir

Many thanks for the interest displayed in 'Etheric force' Our apparatus is not the kind to show it to advantage at present & we are making some more suitable We shall be happy to show it at your next meeting if convenient to you.[2] Respectfully yours

Thos. A. Edison per B.[3]

L (letterpress copy), NjWOE, Lbk. 1:167 (*TAEM* 28:158).

1. According to his letter to Edison (see n. 2), S. V. C. Smith was chairman of the Polytechnic Association of the American Institute of New York.

2. Smith wrote Edison on 4 December inviting him to exhibit the etheric force before the Polytechnic Association, which had its office in the Cooper Union in New York City. DF (*TAEM* 13:314); see also Doc. 692.

3. Charles Batchelor.

–685–

To the Editor of the Scientific American

Newark, N.J., December 8, 1875.[a]

To the Editor of the Scientific American:[1]

I have ascertained that the slight contractions in the legs of a galvanoscopic frog, upon the passage of the etheric current through it, were not caused by electricity, but by the transmission of mechanical longitudinal vibrations by the electrical vibrator over the conducting wire, as was proved by a tuning fork and by a resined pad. In fact the tuning fork, during vibration, would produce contractions when held one inch away from the frog. When the etheric current was passed through the wire and frog, without the transmission of mechanical vibrations, not the slightest effect was produced, although the frog exceeded in delicacy any of those previously tested.[2]

Further experiments in other directions have been made which resulted in rendering the phenomenon more inexplicable. An uninsulated wire proceding from the source of power (highly insulated, of course) was taken into the street and laid in the gutter around a whole block, and back into my laboratory by another door, and up to the floor above the one where the generator was. Excellent sparks were drawn from that end of the wire, although the ground the wire laid on was wet, it having rained all night.[3]

THOMAS A. EDISON.

PL, *Sci. Am.* 34 (1876): 2. [a]Place and date not those of publication.

1. This letter, published in the 1 January 1876 issue of the *Scientific American,* was preceded by an editorial article written before Edison's letter was received. Entitled "Etheric Force and Weak Electric Sparks," this article briefly described Edison's earlier etheric-force experiments and discussed the differences between them and Petrus Riess's earlier observations of sparks generated by static electricity. See also a *Scientific American* editorial article of 25 December 1875 (33:400) for another discussion of Edison's experiments and their relation to Riess's observations.

2. See Cat. 994:49–50 (Lab. [*TAEM* 3:217–18]) for the related experiments. Batchelor drew the accompanying diagram of this experiment, which was performed on 4 December (Cat. 994:48, Lab. [*TAEM* 3:216]).

An etheric-force experiment using a "galvanoscopic" frog.

3. Another experiment from 4 December involved transmitting between Charles Batchelor's and James Adams's houses. The staff made other attempts at transmitting the etheric force through lines laid in the streets on 6 and 9 December (Cat. 1317:23, Batchelor [*TAEM* 90:660]; Cat. 994:53, Lab. [*TAEM* 3:219]). On 9 December they also experimented over an Atlantic and Pacific Telegraph Co. line between the laboratory and New York City (Cat. 994:52, Lab. [*TAEM* 3:219]).

[Newark,] Dec—9th 1875

Notebook Entry: Acoustic Telegraphy

Acoustic Telegraphy.

Tried the four reed vibrators & had them so adjusted that by sending one set of vibrations its corresponding reed gave an a~~p~~mplitude of ⅛ of an inch while the other scarcely moved & by sending the other set of vibrations the other fork did the same, but when both came together it was mixed up badly The connections were thus

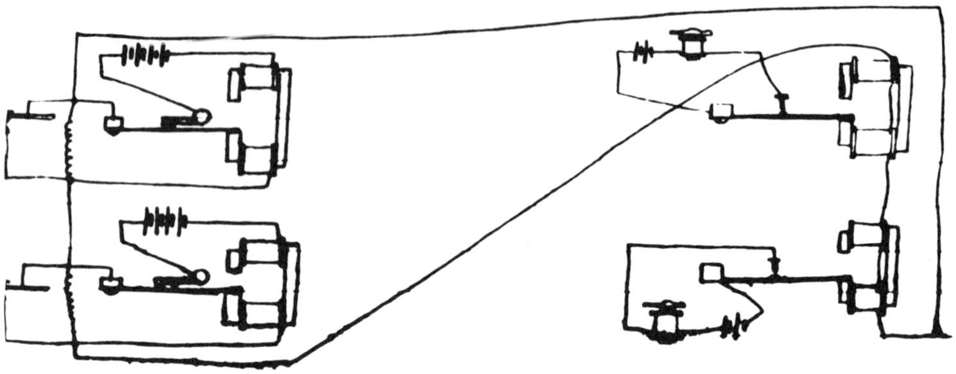

We also found that every time we altered the self make & break spring on the transmitting reed it altered the receiving reed, throwing it out so much sometimes that we were compelled to lengthen or shorten ¼ of an inch. At one time we received separate signals quite readable but there was no margin & the means of adjustment were entirely inadequate. Tried putting in a WU Relay & receiving vibrations on that. It is probable that Helmholtz's resonators would pick out the signals.[1] tried the following connections but they did not work well no 1

1

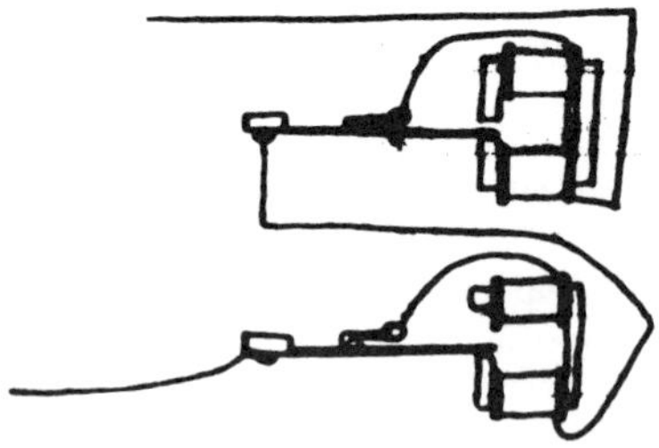

We also tried Fig 2 & it worked fair but not so good I think as the method first shown

[Fig 2]

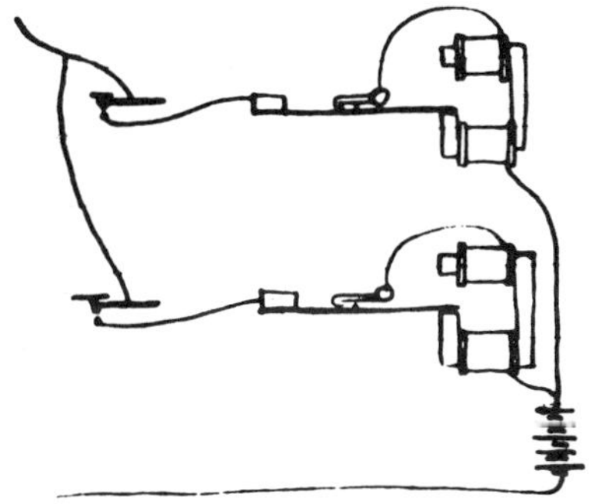

Chas Batchelor

X, NjWOE, Lab., Cat. 994:51 (*TAEM* 3:218). Written by Batchelor.

1. On Helmholtz resonators see Doc. 708 n. 5.

–687–

From Norman Miller

Bloomfield, N.J. Decr. 10/'75.

My dear Sir.

When I was in this P.M. I found O. & L.[1] going over the contract.[2]

I think that you had better bring in a Statement of expenditures and such vouchers as you have ready, also drawings, etc, and any thing that shows work done and progress made.

The papers are so full of "new force" that I want you to show that it has not taken up too much of your time. I shall try to be at 41 Dey[3] most of the time to see you before you go for money, but if I am not when you call send Geo.[4] down to Room 14, No. 61 Liberty St.[5] for me

The plan of putting on Agents for N.Y. each to have the full comm. is going to work well. If the 26 remaining Pens are not done till last of next week as the workman said this A.M. we shall run out again. Yours truly

Miller

ALS, NjWOE, DF (*TAEM* 13:212).

1. William Orton and Grosvenor Lowrey.
2. Doc. 695.
3. Headquarters of the electric-pen business in New York City.
4. Unidentified.
5. Unidentified.

–688–

From James Richardson

New York,[a] Decr. 10th [1875][1]

President Morton[2] will be happy to furnish any apparatus that may be needed for the proposed experiments. Have appointed Tuesday afternoon next—two o'clock—[b]as he wishes to witness the experiments, and will be out of town Saturday. He suggest that it would be well to have some one come over Monday and select the apparatus that will be needed, so that everything may be ready and in order.[3]

Jas Richardson[4]

ALS, NjWOE, DF (*TAEM* 13:321). Envelope addressed to "A. T. Edison". [a]Place taken from postmark. [b]"—two o'clock—" interlined above.

1. Dated by docket on envelope.
2. Henry Morton (1836–1902), a professor of chemistry, had been appointed as the first president of the Stevens Institute of Technology in 1870. He served in that capacity until his death. *BDAS*, s.v. "Morton, Henry."
3. See Doc. 693.
4. Unidentified.

–689–

Technical Note: Acoustic Telegraphy

[Newark,] Dec 10 187~~6~~5[1]

Table of Reed Lengths[2]

1	1¼	1¾	2¼
2	2½	3½	4½
4	5	7	9
8	10	14	18
16	20	28	36
32	40	56	72
64	80	112	144
128	160	224	288

T. A. Edison[3]

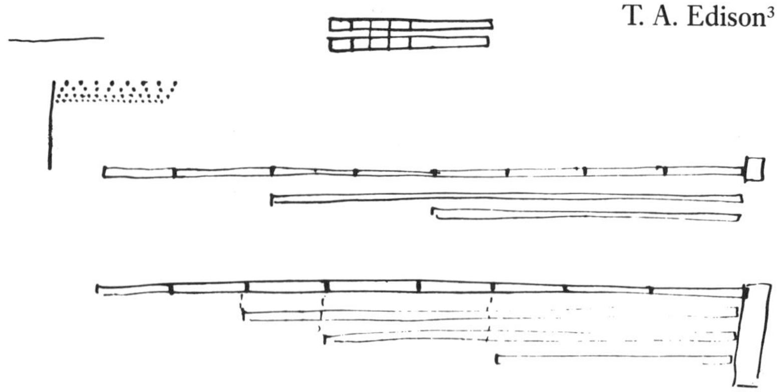

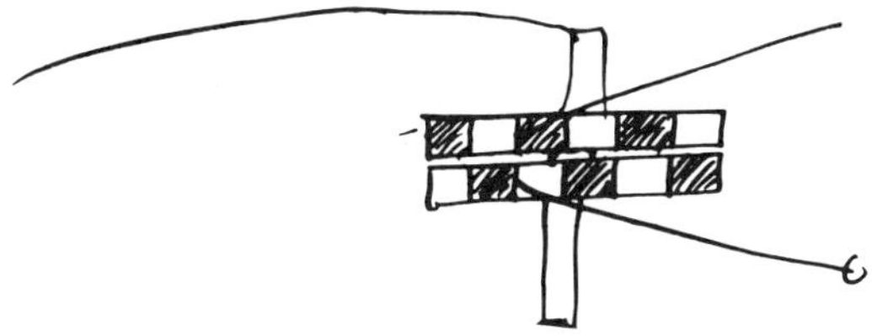

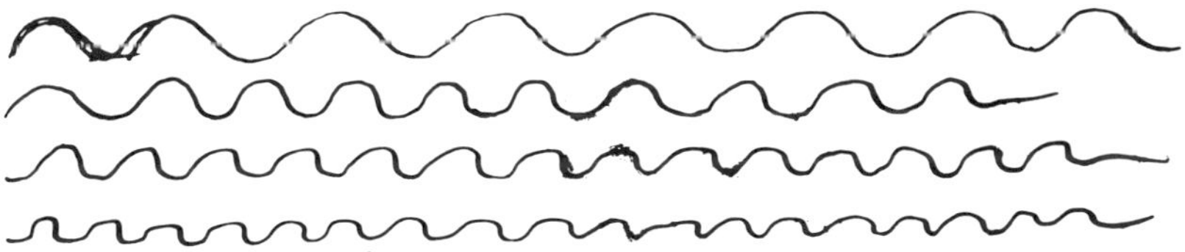

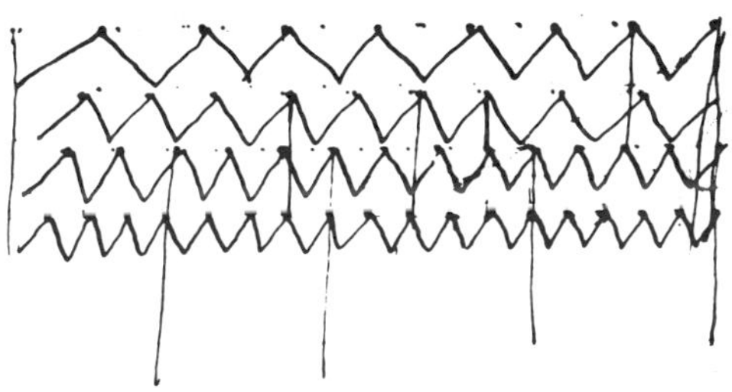

X, NjWOE, Lab., Vol. 9:20 (*TAEM* 3:807).

1. Edison apparently dated this document sometime after creating it.

2. Acoustic telegraph systems used vibrating reeds or tuning forks to transmit different frequencies of electrical impulses over the line. Receiving these frequencies separately was a key problem.

In this note Edison was concerned with establishing the relationships between the harmonic frequencies of different reed lengths in order to find those that would not interefere with one another. On the front of a sheet of paper Edison drew a table representing the relative overtone frequencies of different reed lengths. Beneath it he drew long reeds followed by shorter reeds that were simple fractions of the long reeds' lengths. He followed this with a sketch of what he called a harmonic engine—a double breakwheel using alternating contacts to transmit rapid vibrations in place of a reed or tuning fork (see Docs. 708, 709, and 712). On the reverse side of the paper he drew sound-wave diagrams that illustrated interfering frequencies. Some apparently unrelated calculations following these diagrams have been omitted.

3. In the original the numbers "500," "8," and "80" are scattered to the left of Edison's signature.

–690–

Notebook Entry: Etheric Force

[Newark,] Dec 11th 1875

55[1] Etheric force

Dr Edwards president of a scientific association in Newark[2] was present & stated that the phenomenon was far from a new force that there was not any more new forces. The sparks were simply due to inductive electricity of low tension Notwithstanding that the spark could be obtained through glass a greater distance with etheric than with ⅛ jump spark from an induction coil or from a plate glass machine. He could give no idea how to prove it.[3] The following experiments were undertaken to prove if possible Mr. Edwards assertions. Two condensers of about six microfarads capacity each were connected thus:—

Fig 1

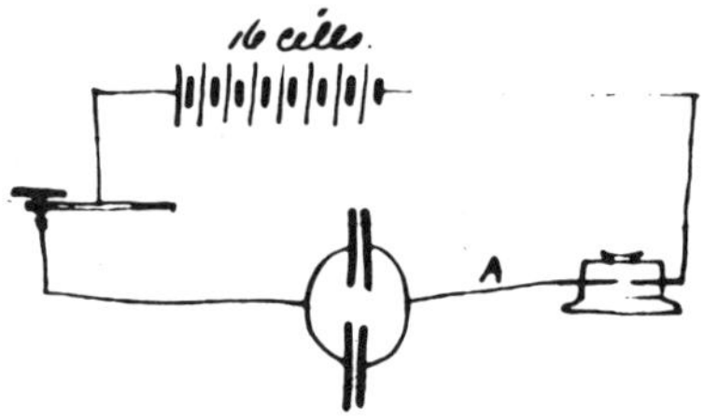

We obtained the spark as was to be expected

We now put in 10 000 ohms at A & still obtained the spark. We then opened the coil A so that the spark could go across the insulation <u>No spark</u> on trembling the key. We then connected it thus

2

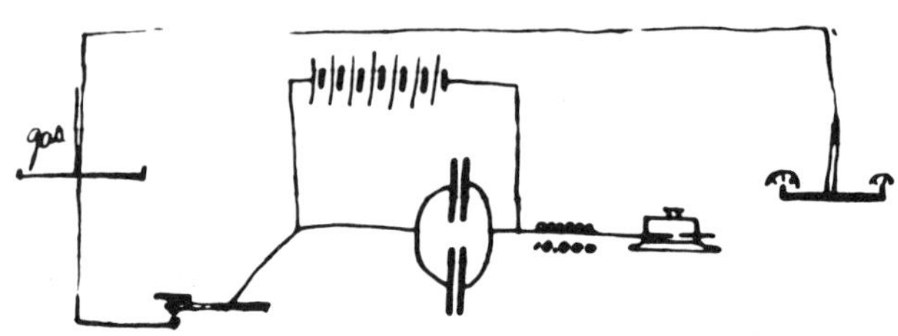

Got the spark through 10 000 ohms but when it was opened <u>no spark</u>. The opening & closing of the circuit with etheric makes no difference. Again a glass rod was used at A Fig 1 with wires ⅛ of an inch apart no spark while etheric gives it through 3 feet with only 1 sounder vibrating. Even induction coil current jumping ⅛ inch did not give it or plate glass machine

Another experiment was tried with low tension inductive electricity thus:—[4]

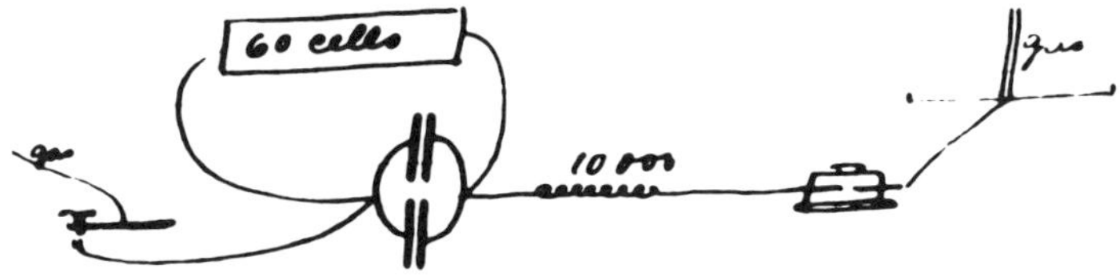

The battery of 60 cells gave powerful electric light with carbons ¼ inch apart. Working key gave faint red spark between pencil points as large amounts of battery electricity passed through besides the static.

The snap was much louder from the condensers than from a Leyden jar throwing ⅛ inch spark. This did not jump the air only gave spark on contact. When the resistance coil of 10 000 ohms was opened got no spark by the most careful adjustment

Connected it thus:—but got nothing.[a]

56 Condensers

We connected 8 condensers having an aggregate capacity of 22 microfarads to our 60 cell battery in the following manner:—

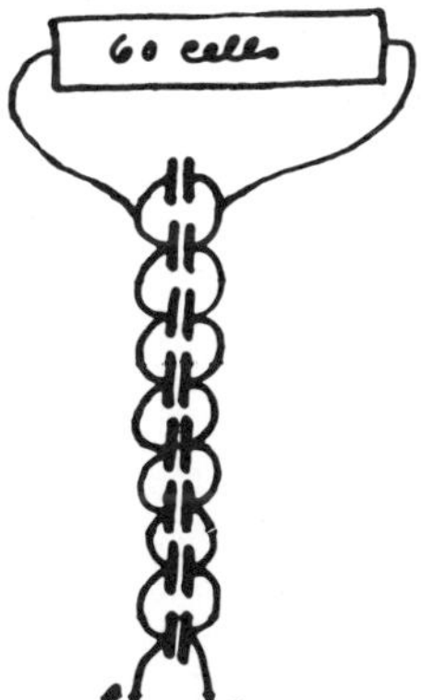

By putting a & b together the most brilliant spark that I ever saw was given off. The scintillations went out from the connect points at least 3 or 4 inches. Thousands radiating from the centre like the rays of the sun each scintillation ended in a fork like a hen's foot;[5] the snap was like the crack of a cap.

Yet with all this power no sparks were seen on the graphite points.[a]

57 Electric light

Thirty one cells of 1½ carbon battery was put up connected to an apparatus with carbon points it was very brilliant & sufficient for lecture purposes We found that when Sodium, Potasium, or Iodide of Cadmium was used the light could be drawn out much longer & was more brilliant

Chas Batchelor.

X, NjWOE, Lab., Cat. 994:54 (*TAEM* 3:220). Written by Batchelor. [a]Dated and signed here by Batchelor.

1. See headnote, p. 600 n. 2.

2. Dr. A. Mead Edwards was a Newark physician. On 23 December, Edison attended a meeting of the Newark Scientific Association presided over by Dr. Edwards at the Baptist Chapel; nothing more is known about the society. Edwards spoke and Edison (with an assistant) exhibited apparatus but did not speak, claiming his deafness prevented him from hearing Edwards's discussion and any questions. "Puzzling the Wise Men: The Electric What Is It They Have Found in Newark," *New York Sun,* 24 Dec. 1875, Cat. 1144, Scraps. (*TAEM* 27:300).

3. Edison, who was apparently not present during Edwards's visit, wrote him a letter asking him "to cite some authority on the subject and also indicate what kind of apparatus would probably generate static electricity of low tension that I may have one made" (TAE to Edwards, undated, Lbk. 1, following p. 487 [*TAEM* 28:334]). Static electricity, in contrast to voltaic or current electricity, was normally regarded as electricity of high tension, though small quantity. On 19 December, Edwards observed experiments at Edison's laboratory in company with George Barker, who apparently learned of Edison's work upon receiving a paper from George Beard describing some of the experiments Beard had performed with Edison (see Doc. 679 n. 4). Barker also informed the noted electrical inventor Moses Farmer of his visit to Edison (Edwards to Barker, 21 Dec. 1875; Beard to Barker, 11 Dec. 1875; Farmer to Barker, 21 Dec. 1875; all Barker Papers, PU). In his 21 December letter to Barker, Edwards suggested a means for confirming his ideas regarding the etheric force. In a talk two days later he indicated he had done various experiments with Edison, but it is not known if they included that test (see n. 2 above).

4. Both figure labels are "gas."

5. Batchelor illustrated this "hen's foot" spark in the account of this experiment entered in his own notebook. He also indicated that one experiment was done with **A** connected to the gas pipe and **B** held in the hand (Cat. 1317:23, Batchelor [*TAEM* 90:668]).

Edison drew the accompanying sketches of etheric-force sparks in a set of undated notes. NS-Undated-General, Lab. (*TAEM* 8:10).

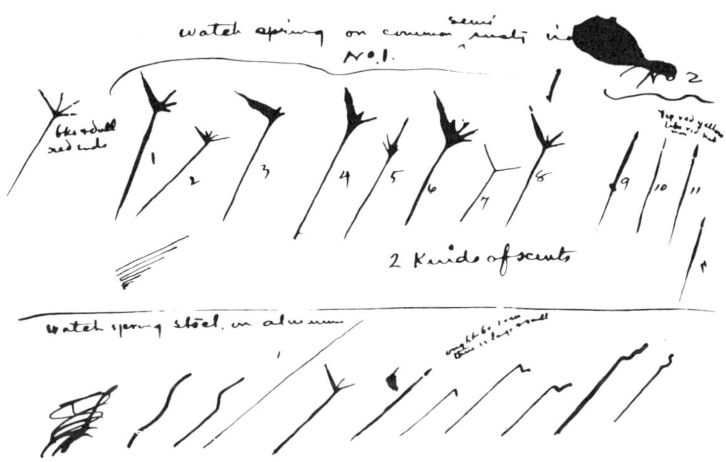

Edison's drawings of etheric-force scintillations.

–691–

Technical Note: Acoustic Telegraphy

[Newark,] December 12th 1875

Acoustic Telegh

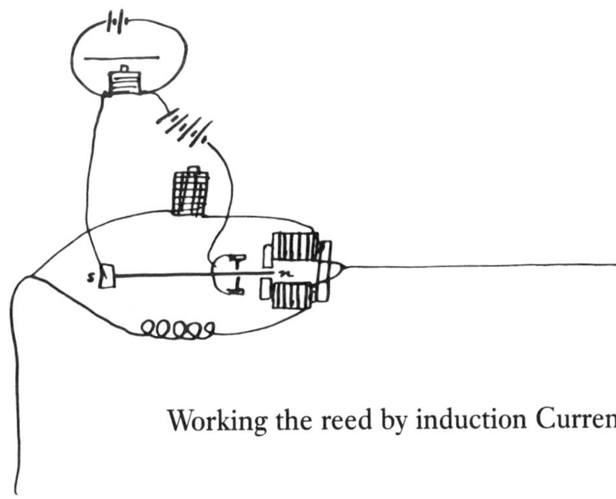

Working the reed by induction Currents =[1]

X, NjWOE, Lab., Vol. 9:21 (*TAEM* 3:809).

1. Edison has apparently shunted this polarized-relay receiver with a "magnetic" (high-induction) shunt above and a normal resistance shunt below. Assuming that to be the case, an incoming signal splits at the polarized relay. The upper, inductive resistance opposes the current, creating a net clockwise flow through the loop formed by the shunts. Thus energized, the relay attracts the long, permanently magnetized reed to one side (up or down in the drawing). When the signal ceases, the inductive resistance prolongs it, creating a net counterclockwise current in the loop. This reverses the magnetism of the polarized relay and moves the reed to the other pole. As the reed vibrates back and forth, it touches the contacts on either side, actuating the local circuit at the top.

–692–

From George Beard

New York, Dec 14, 1875[a]

My Dear Sir:

I have seen Dr Smith,[1] and he says that the notice has been given, and that [-]you had better give drawings on the board this week, and have a lecture on them;—and next week give the demonstrations with your improved apparatus.[2]

[-]I agree with him It will disappoint the audience to fail now that the announcement has been made

So please be on hand at the room in Cooper Institute Thursday evening promptly at 7.30 and I will meet you there, and will do all I can to make it go off well.

Come with drawings on paper, to be copied on the blackboard.

Bring also a helix or so in your pocket. The one you ~~firSet used~~ first used, or a sounder will answer.[3]

Dont fail, and I will not. Very Truly

Geo M. Beard

ALS, NjWOE, DF (*TAEM* 13:325). Letterhead of *Archives of Electrology and Neurology.* [a]"New York," and "187" preprinted.

1. S. V. C. Smith; see Doc. 684.
2. Only the first lecture was given.

Drawings reproduced in a newspaper account of George Beard's lecture on the etheric force.

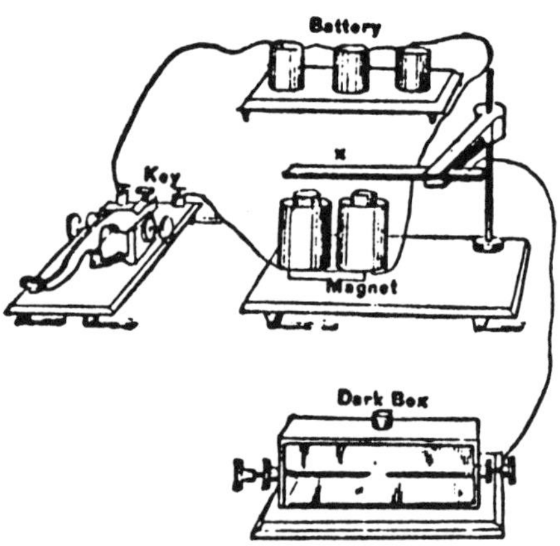

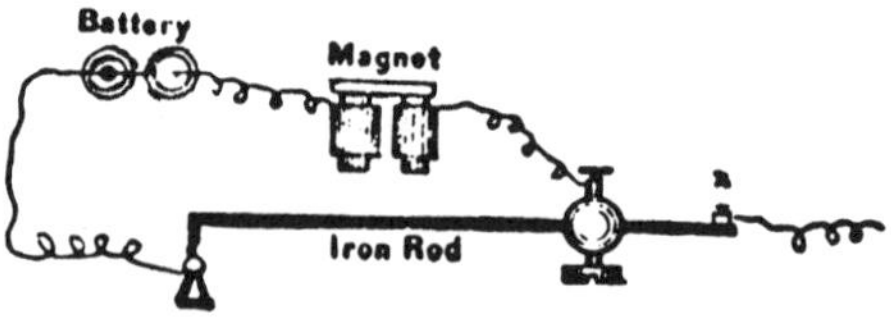

3. The day after the lecture the *New York Daily Tribune* published a detailed report (17 Dec. 1875, 5). The accompanying diagrams ap-

peared with the article and may represent those drawn by Edison on the blackboard. Beard summarized the experiments he and Edison had performed and their various theories to account for the effect. Edison conducted an experiment showing a spark in the dark box.

–693–

Notebook Entry: Etheric Force

[Newark,] Dec 14th 1875

20[1] Etheric force

Mr Edison, I, & Dr Beard went to the Steven's Institute of Tecnology to try some experiments. We found we could not use their large electromagnet as it took 15 seconds before it got its full charge. Their electroscope was out of order & we could not use it. We got the spartial vacuum[2] of 3 inches in length. Could not get it through a perfect vacuum

Chas Batchelor

X, NjWOE, Batchelor, Cat. 1317:24 (*TAEM* 90:669). Written by Batchelor.

1. Batchelor numbered the entries in this notebook.
2. Probably should read "spark through a partial vacuum."

–694–

Agreement with Western Union Telegraph Co.

[New York,] December 14, 1875[a]

Agreement for mutual Releases,[1b] this day entered into between the Western Union Telegraph Company of the one part and Thomas A. Edison of the other part.

—whereas[b] legal controversies have arisen and now exist between the parties in respect to certain inventions made by said Edison, which inventions were the subject of an agreement between said Edison and George B. Prescott, dated August nineteenth one thousand eight hundred and seventy four,[2] and subsequently of negotiations between the Company upon the one part and the said Edison and Prescott upon the other part whereby the said Company sought to purchase from Edison and Prescott the sole title to the same and all Letters Patent to be issued therefor, which negotiations the Company claims did result in a contract binding upon said Edison and Prescott for the conveyance of all their right, title and interest in and to said inventions and letters patent issued thereon, but all which claims said Edison denies and resists, and refuses to make any further conveyance to the said Company of his interest in said inventions or letters patent, or any part thereof, in consequence of which refusal an action has been begun by the Company in the State of New Jersey against the said Edison and Prescott to enforce specific performance of the contract which it asserts.[3]

Now, in consideration of the premises and the mutual promises and releases herein contained, it is agreed by each of the parties hereto, as follows:

First. The Western Union Telegraph Company hereby releases the said Edison from any claim which it may now or at any time hereafter have against him for pecuniary damages for the breach of contract by it alleged as above stated, over and above the amount of Two hundred and fifty dollars, and agrees not to prosecute the said Edison for such pecuniary damages for any amount exceeding Two hundred and fifty dollars; and not to prosecute him for any amount whatever unless such prosecution shall be, in the opinion of its Counsel, necessary or convenient in order to determine at law some of the rights of the parties in respect to the premises.

Second. The said Edison hereby releases the Western Union Telegraph Company from all claims which he now has or may hereafter have against it for any further payment on account of the inventions or letters patent above referred to, or any right or interest therein which the said Company may succeed in establishing or maintaining; it being understood that neither party intends in any way to waive or compromise his claim or defense in respect to the controversies above recited, except as herein expressly stated. And said Edison also hereby consents that said George B. Prescott may assign any interest which he may now or hereafter have in said inventions or letters patent thereon to the Western Union Telegraph Company: and does for himself hereby assign and set over to the Western Union Telegraph Company all his now remaining interest, if any, ~~of~~ in[c] the said inventions or letters patent granted or to be granted thereon; and does further stipulate upon the request and at the charge of the said Western Union Telegraph Company to execute and deliver to them complete assignments of all the remaining interest if any which may by the final judgment of any competent Court be decided to be in him at the date of this instrument; but it being also further understood that this contract does not and is not intended to affect the rights or interests of any other person or corporation to whom the said Edison has heretofore assigned interests in the said inventions or letters patent.

In Witness whereof the said Western Union Telegraph

Company has caused its common seal to be hereunto affixed and its corporate name to be hereunto subscribed by the hand of its president and the said Edison has hereunto set his hand and seal the fourteenth day of December in the year one thousand eight hundred and seventy five.

The Western Union Telegraph Company By Willm Orton President[d]

A. R. Brewer Sec'y.[e] Thos. A. Edison[f]

Witnesses Norman C Miller, George F. Fagan[4]

DS, NjWOE, Miller (*TAEM* 28:1000). An essentially identical version of this document is in Quad. 71.2, p. 20 (*TAEM* 10:237). [a]Date taken from text, form altered. [b]Underlined twice. [c]Interlined above. [d]"The Western . . . President" written by Orton. [e]Company seal in left margin. [f]Followed by seal.

1. Edison's draft of an agreement encompassing the main points here and in Doc. 695 is found in 75-002, DF (*TAEM* 13:715).

2. Doc. 466.

3. See App. 3, Western Union Bill of Complaint.

4. Unidentified.

–695–

Agreement with Western Union Telegraph Co.

[New York,] December 14, 1875[a]

This Agreement[1b] entered into this fourteenth day of December in the year one thousand eight hundred and seventy five by and between the Western Union Telegraph Company, party of the first part and Thomas A. Edison, of the City of Newark, County of Essex and State of New Jersey, party of the second part, witnesseth:[b]

The said Edison agrees to invent, construct and place in practical operation upon the lines of the said Company, a system or method for telegraphing by vibrations of the kind referred to and more particularly described under the head of "The Acoustic Telegraph" in a work called the "Wonders of Electricity," published by Charles Scribner & Co. in 1872, whereby not less than five or more distinct and seperate messages can be transmitted on a single wire in opposite directions at one and the same time, without interference with each other, said system to be hereafter called the Acoustic Telegraph.

It is further agreed by the said Edison that he shall cause to be made from time to time "Caveats" describing his ideas and experiments, to be filed in the Patent Office at Washington, copies of which are to be delivered to said Company as soon as practicable after each part of such system shall be invented or discovered, accompanied by assignments of the invention or discovery so described, to the Western Union Telegraph Company as sole assignee, with directions that the patent shall issue to said Western Union Telegraph Company when issued, and will also sign and deliver applications for letters patent in due form to be used by the attorneys hereinafter named at their discretion.

The said Edison shall from time to time construct all such models as may be convenient or necessary for the purposes of such applications, or such of them as in the opinion of William Orton, President of the Company, are necessary to protect such system of Acoustic Telegraphy, or any part thereof.

The said Edison shall also from time to time upon the request of the Western Union Telegraph Company make and sign such new and different applications as may become necessary or desirable for each invention discovery or improvement made by him in said system of Acoustic Telegraphy; and in like manner new assignments of the subject matter of each application with like request for issue of Letters Patent to said Company; it being the intent and purpose of the said Edison, for the considerations hereinafter named or provided, to secure to the Western Union Telegraph Company the sole title to and monopoly in the United States of all inventions or discoveries made or to be made by him which are necessary to perfect Acoustic Telegraphy. For the purpose of securing more fully the object herein stated, the said Edison shall also nominate and appoint Grosvenor P. Lowrey his attorney irrevocable,[2] with power of substitution and revocation, to prosecute all applications in the Patent Office covered by this agreement, and to represent said Edison in all matters respecting such applications; all papers necessary to the full performance of the respective stipulations herein, and all costs for legal service, to be at the charge of the Western Union Telegraph Company.[3]

The said Edison further agrees to use the utmost diligence in carrying forward his experiments for perfecting and introducing the said system of Acoustic Telegraphy upon the lines of the said Telegraph Company. The said Company agrees to furnish full facilities in all its offices and shops and over all its wires for testing such invention and to pay all the actual expenses of experimentation, not exceeding two hundred dollars per week, to be accounted for by vouchers,[4] until such invention has been completed to the satisfaction of William Orton, representing the Company, and in satisfactory operation upon its lines. But such weekly payments may be discontinued at any time at the option of the said Orton, if in his opinion, the results obtained at the time do not warrant such expenditures, or if he disapproves of the method of disbursing such monies. All necessary expenses for patent-office and Solicitors fees for obtaining letters patent in the United States for said Acoustic Telegraph are to be paid by the Company.

For and in consideration of the stipulations herein by Edison, and of the perfecting introducing and assigning such inventions, and of due performance of the provisions of this contract, the Company agrees, in addition to the payment of the charges and expenses above provided for, to pay to said Edison, upon the completion of a satisfactory system of Acoustic Telegraphy, and the issue of Letters Patent therefor to it as herein stipulated a sum not less than Six thousand dollars, and a further sum, the amount and the manner of payment of which is to be left to the fairness and judgment of the said Orton; provided however that in case of refusal or failure by said Orton to fix and determine such price within three months after demand by said Edison, the price shall be fixed by a Board of Arbitrators, consisting of three skilled persons, one to be chosen by each party hereto, and the third by such two, and whose award shall be final. The said Edison shall have the right in like manner to appeal to arbitration if in his opinion any price fixed by said Orton is grossly inadequate or unfair: but in all cases the arbitrators in forming their judgment of value shall consider only the commercial value of such invention to the Company, in view of its existing facilities, means of business and also the cost of applying and economy in using the same in comparison with any other method of telegraphy known at the time, as well as the expenditures already made by the Company in respect of Acoustic Telegraphy. Upon these and all such considerations as ought to control the Company acting as a purchaser, the Arbitrators shall decide what further sum, if any, shall be paid to the said Edison. In case the Company shall be dissatisfied with the award of any board of Arbitrators, it may discharge itself from obligation to pay the same, by offering to re-assign to Edison, his legal representatives or assigns, the subject matter for which the award was made, on receiving repayment of the sum (not including expenses) formerly paid thereon, and by making such re-assignment on receiving such repayment.

It is also agreed that while experimenting upon the subject of Acoustic Telegraphy, the said Edison shall not make, invent or operate any system or apparatus for telegraphy for any other Company opposed to or competing with the Western Union Company. Provided however that this period shall not continue for more than one year.

It is also agreed that if said Edison shall at anytime hereafter make or discover any process, device, instrument or method which is an improvement upon any of the processes, devices or apparatus used in Acoustic Telegraphy and covered or intended to be covered by this agreement; assignments therefor and of all letters patent thereon shall be immediately, or as soon as practicable, and as above provided, made to William Orton, personally, to be by him held in trust, subject to a satisfactory or legal adjustment of price in the same manner as hereinbefore provided. The said Edison shall in like manner through an attorney satisfactory to said Orton, file caveats and applications and prosecute the same in good faith, but at the expense of said Orton.

In Witness whereof[b] the said party of the first part has caused its common seal to be hereunto affixed and its corporate name to be hereunto subscribed by the hand of its President; and the said party of the second part has hereunto set his hand and seal the day and year first above written.

The Western Union Telegraph Compy, By Willm Orton President.[c]

A. R. Brewer Sec'y.[d] Thos. A. Edison[e]

Witnesses: Norman C Miller, George F Fagan

DS, NjWOE, Miller (*TAEM* 28:1004). Other versions of this with no significant differences are in TI 2:600 (*TAEM* 11:721) and NNC. [a]Date taken from text, form altered. [b]Underlined twice. [c]"The Western . . . President." written by Orton. [d]Company seal at left. [e]Followed by seal.

1. Edison's draft of an agreement encompassing the main points here and in Doc. 694 is in 75-020, DF (*TAEM* 13:715–16).

2. This marks the start of a lengthy association between Edison and New York lawyer Grosvenor Porter Lowrey (1831–1893). Lowrey's firm (Porter, Lowrey, Soren, and Stone) served as the law department for Western Union, and Lowrey was the company's general counsel. Raised in Massachusetts, Lowrey graduated from Lafayette College in Pennsylvania, served as private secretary to the territorial governor of Kansas during some of the violent unrest there in the 1850s, and attained the bar in New York. His connection with William Orton dated from their collaborative work on U.S. Treasury business during the Civil War. Lowrey had gotten involved in Edison's and Western Union's relations at least as early as January 1875 and had acted for Orton and the company in the dispute over rights to the quadruplex. Orton to Lowrey, 20 Jan.

1875, LBO 14:125–30; Doc. 537 n. 5 and Doc. 577 n. 6; Taylor 1978; "Grosvenor P. Lowrey Dead," *New York Times,* 22 Apr. 1893, 11.

3. Edison executed a power of attorney for Lowrey the same day. Miller (*TAEM* 28:1011).

4. Statements of Edison's expenses for acoustic experiments between 20 November 1875 and 8 January 1875 are in 75-020, DF (*TAEM* 13:720–21). The weekly amounts usually exceeded $200. Most of Edison's costs were for labor, and the weekly labor accounts show even larger sums than those listed in the statements (Cat. 1214:132, Accts. [*TAEM* 21:620]; 75-020, DF [*TAEM* 13:722]). There are also receipted bills upon which "Acoustic" has been written (75-002 and 76-002, both DF [*TAEM* 13:223–51, 799–902 passim]).

–696–

Draft Addendum for Western Union Telegraph Co.

[Newark, c. December 14, 1875]

Copy

I Thos A Edison party to the foregoing instrument of December 14, 1875[1] do hereby declare that the inventions, my then remaining interest in which if any I did by said instrument assign and set over to the Western Union Telegraph Co are all those inventions which are respectively described in applications relating to Duplex Telegraphs numbered 94 95 96 97 98 99 & 100,[2] which were severaly filed in the US patent office on the first day of September 1874 and also the inventions described in applications 112 & 113 filed in said Patent office Dec 28th 1874, and March 23, 1875 respectively,[3] and also the inventions described in application 111[4] filed in said patent office. And I further declare that the letters patent granted or to be granted or to be granted, referred to in said instrument of Dec 14 1875 my remaining interest in which if any I assigned by said instrument to the Western Union Telegraph Co for the letters patent of the U.S. are all the inventions described in the several applications above mentioned and especially the letters patent of the U.S. No 162 633, dated April 27, 1875[5]

In witness whereof I have hereto set my hand and seal, this day of in the year

AD, NjWOE, DF, 75-020 (*TAEM* 13:717).

1. Doc. 694.
2. See headnote, p. 254.
3. See Doc. 449.
4. See Doc. 450.
5. Doc. 554.

–697–

William Orton to George Phelps

New York Dec. 16, 1875[a]

Dear Sir,

Please confer with Mr Edison about the practicability of conducting some experiments at the shop. Also permit him to look over your collection of old instruments to ascertain what it will yield, & oblige. Very Respy

Willm Orton

ALS, NjWOE, DF (*TAEM* 13:719). Letterhead of Western Union Telegraph Co., Executive Office. [a]"New York" and "187" preprinted.

–698–

Notebook Entry: Condensers

Newark N.J Dec 20th 1875

21[1] Condensers[2]

Having occasion to make 24 condensers[3] we used for the first time an apparatus made last January for the same purpose but not used then it consisted of:[a]

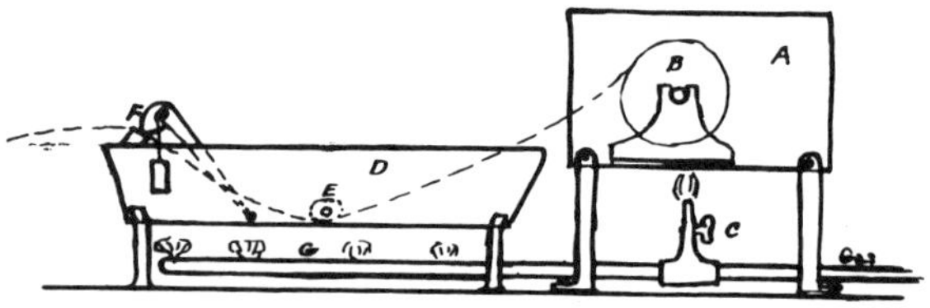

A is an oven kept about 180° Fahr., B is the roll of paper, D is the trough for paraffin. E is a roller under which the paper passes, F is the stripper for taking off the surplus paraffin G is the gas jets which keep the paraffin hot. The principle is to bake the paper till it gives up its moisture & then directly drag it through hot paraffin & draw it out along a board about 20 feet long holding it above board until it has cooled & then dropping it on it. The board is marked where to cut the paper into sheets. As it passes under the rubber F it strips it beautifully leaving a perfectly even thickness. It more than answers our expectations[4]

Chas Batchelor

X, NjWOE, Cat. 1317:24, Batchelor (*TAEM* 90:669). Written by Batchelor, who also wrote a large "7" over the entry with a pencil. [a]The drawing was numbered "14" with a pencil.

1. Batchelor numbered the entries in this notebook.
2. Batchelor wrote another version of this note, but that entry does not describe the machine as fully. Cat. 994:56, Lab. (*TAEM* 3:221).
3. This may refer to the order from Pearce and Jones (see Doc. 602).
4. Waxed paper was a common dielectric in condensers (Prescott

1877, 949). Edison later said that he was involved in its wider introduction (App. 1.D158, G22 [repeated in Dyer, Martin, and Meadowcroft 1929, 168]).

–699–

Technical Note: Acoustic Telegraphy

[Newark,] Dec 20th 1875[a]

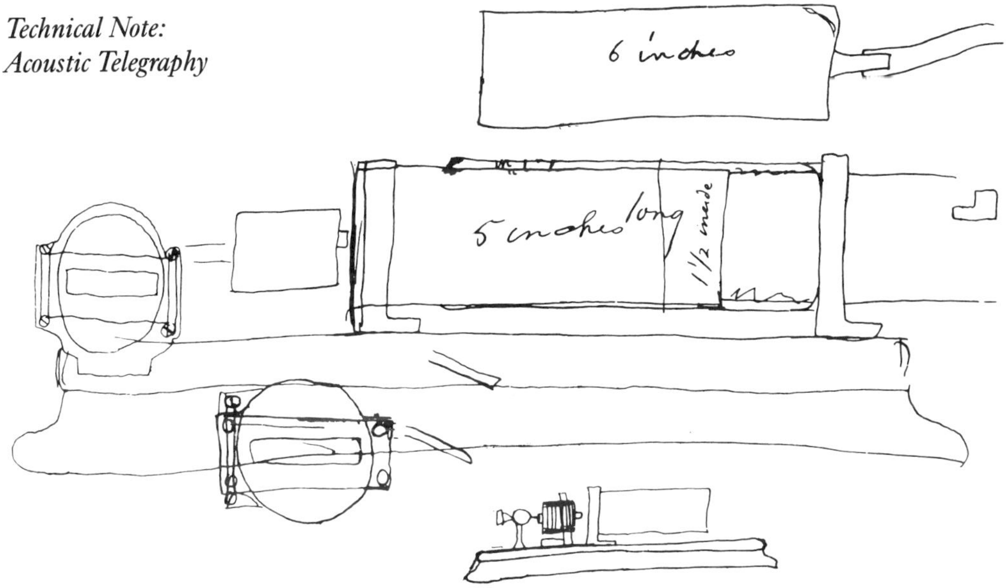

T A Edison

Robt Spice
J. Kruesi

Kruesi Please start on this right away with Jim.[1] If you want brass there is an order you can fill on my desk[2b]

K Tin will do if you cant get right brass Edison

X, NjWOE, Lab., Vol. 9:22 (*TAEM* 3:810). [a]Dated by Charles Batchelor. [b]"Kruesi . . . desk" written by Batchelor.

1. James Adams.

2. Frustrated by the generally weak response of circular iron diaphragms, Edison tried cutting away the top and bottom of the diaphragm, leaving the rectangular strip seen in the drawing at the far left. He first thought to try a brass diaphragm; he wound up using an iron plate coated with tin. The smaller rectangle in the center of the diaphragm is a piece of iron soldered to the diaphragm to act as an armature for the electromagnet. The tube at the upper right (labeled "6 inches," with a flexible rubber listening tube attached to its right end) fit inside the stationary cylinder and slid in and out as needed to tune the overall resonating chamber to the frequency of the incoming signal. Additionally, Edison intended "the width of the strip [to bear] some relation in its vibrating times, to that of the particular series of waves which it was desired to render audible." TI 1:15–16, 104 (*TAEM* 11:28–29, 73).

Edison remembered making "this instrument, and I think two others . . . within 24 hours after the sketch was made and dated." He claimed it "was capable of rendering audible any sounds whatever, transmitted telegraphically" (TI 1:16 [*TAEM* 11:29]). He included this design (with a piece of steel spring as a diaphragm) as figure 14 in Edison Caveat 75 (Doc. 708; see headnote, p. 709).

–700–

Daniel Craig to Josiah Reiff

New York, Dec. 24, 1875.[a]

Dear Sir:

Wishing to let Mr. Harrington see your views, I handed all the papers you gave to me, last winter, over to him, which, he said, he had handed to you. If you have got them, you can see what I am now justified in believing of your schemes promises of friendship and co-operation. If you have not got the papers, I will furnish you with copies, so that you can see what sort of a figure you may cut before an admiring world.

You know, of course, that you personally and solemnly assured me that my $50,000, cash, should have the same consideration as $50,000 cash actually expended by yourself or any other person in the development of the Automatic system.[1] I expect you and your associates will make this promise good, by entering into binding obligations as to the outcome of Automatic Telegraphy. My impression is that self-interest, to say nothing of honor, will prompt you to hasten to satisfy my demands upon this point.

Then, you know you solemnly promised that I should be one of three or four people to share in ~~th~~Mr. Edison's new things.[2] If you want me to respect you as an honorable gentleman, you will make your promises good in this respect, and without delay.

I think it right and proper that in all future arrangements or investments, looking to the control of Automatic Telegraphy, I should have as much influence and power as any other one person, & my judgment is that it will be beneficial to you and your associates, to assure me fully upon this point, and at once Respectfully yours

D. H. Craig

ADDENDUM[b]

New York, Dec. 24, 1875[c]

Confidential note to Mr. Edison.

It has suited the purposes of Harrington & his immediate

associates, to array you and others against me—but apart from my protestations against the villainey of that man & his associates in his deliberate purpose to rob me after I had put him in control of Automatic Telegraphy, and brought him and you together, I have never done anything but good to you, and have never had the least feeling of unkindness except what has oozed out of my pen when I have been recounting, really, Harrington's villainies.[3]

I have waited five years for something good to comc out of Automatic Telegraphy. In trusting to Harrington & his associates, I have trusted to a crowd of incompetent rogues, and I am seriously resolving upon a movement which, I am confident, can be made reasonably and immediately productive of profit to myself and to all who will follow my lead. Our mutual and good friend Lefferts knows something of what I can do to benefit you & others as well as myself, and he perhaps will tell you, if you ask him, what he thinks of my ability to take care of interests which Harrington & his associates have done their best to destroy.

I should be pleased to have a full, frank & friendly interview with you at your own convenience, & until that time, I am very certain that you will serve your own interest by holding all you have got in reserve. Faithfully yours,

D. H. Craig

ALS, NjWOE, DF (*TAEM* 13:499). [a]"Copy—to be returned to D. H. C. P.O. Box 988." written above. [b]Addendum is an ALS on same sheet. [c]Place and date taken from main document.

1. It is not clear whether Craig's claim is based on a private understanding with Reiff or on an earlier claim of payment due in stock of the Automatic Telegraph Co. In an 1874 bill of complaint filed against several parties involved in automatic telegraphy, Craig claimed that George Harrington had promised him half a million dollars in Automatic Telegraph Co. stock for his part in forming the company and in promoting automatic telegraphy. Quad. 70.8, p. 52 (*TAEM* 9:790).

2. Craig claimed that a one-third interest in Edison's 1 October 1870 agreement with Harrington had originally been reserved for him. *TAEB* 1:280 n. 6.

3. Cf. headnote, p. 307 n. 6.

–701–

Notebook Entry: Etheric Force

[Newark,] Dec 26 [1875][1]

59[2] Etheric Force.
An experiment tried tonight gives a curious result

Fig. 82

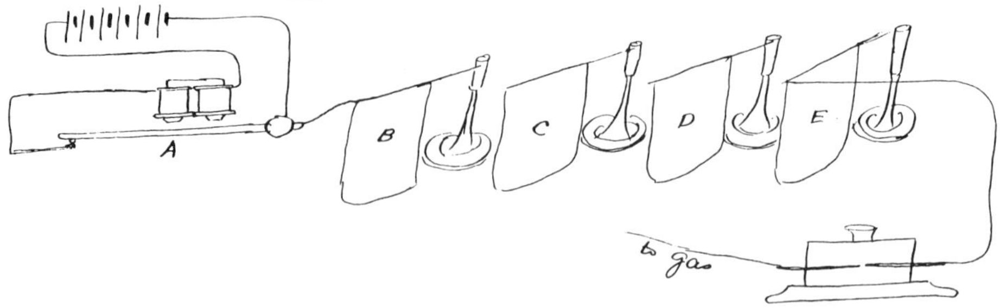

A is a vibrator B C D E are ~~vibra~~ sheets of tin foil hung on insulating stands the sheets about 12 × 8 B & C are 26 inches apart C & D 48 inches, & D & E 26 inches. B is connected to the vibrator & E is connected to point in box, the other point to ground. We received sparks at intervals although insulated by such space.

Chas Batchelor

X, NjWOE, Lab., Cat. 994:57 (*TAEM* 3:221). Written by Batchelor.

1. This is the last of the etheric-force notebook entries from November and December 1875.

2. See headnote, p. 600 n. 2.

–702–

Notebook Entry: Acoustic Telegraphy

[Newark,] Dec 26 1875

60[1] Acoustic Telegraphy.[a]

Worked 3 messages in the same direction on this principle

Fig 83

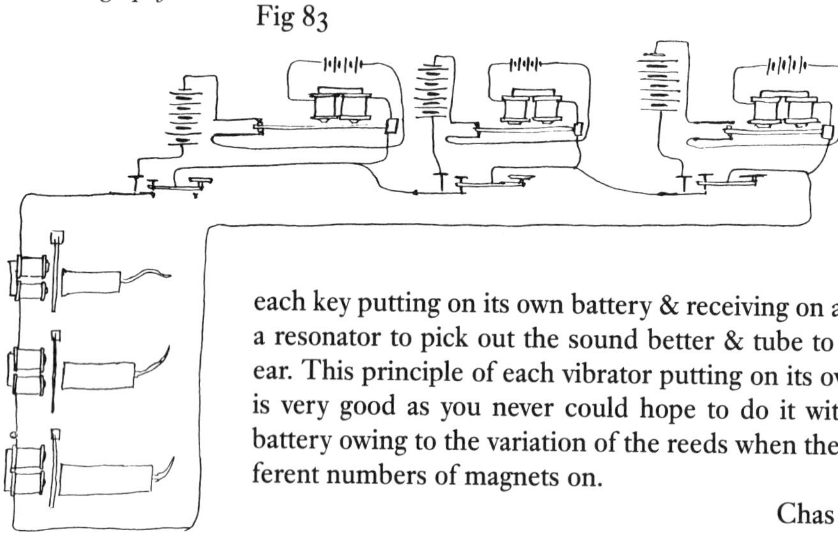

each key putting on its own battery & receiving on a reed with a resonator to pick out the sound better & tube to put in the ear. This principle of each vibrator putting on its own battery is very good as you never could hope to do it with a single battery owing to the variation of the reeds when there are different numbers of magnets on.

Chas Batchelor

X, NjWOE, Lab., Cat. 994:57 (*TAEM* 3:221). Written by Batchelor. [a]Underlined twice.

1. See headnote, p. 600 n. 2.

–703–

To William Orton

Newark[a] Dec 29, 1875

Wm Orton Esq

I struck the right principle for the transmission of vibrations last night, and received two different messages on my adjustable resonators without interferance. Tonight I am trying to transmit three. I have only one resonator and its hasnt range enough to pick the third message out, but will have some tubeing tomorrow for new ones.

will keep you advised hereafter by letter. If you had some place over there (spare room) I could bring app over & show you the two messages. =[1] perhaps 3 in couple more days. Yours

Edison.

ALS, NjWOE, DF (*TAEM* 13:719). [a]Bottom half of a word appears above on torn edge of paper.

1. See Doc. 704.

–11– January–March 1876

During the winter of 1876 Edison devoted his laboratory work principally to acoustic telegraphy. The electric-pen business grew, and by the end of March was taking all of Charles Batchelor's time. The etheric-force controversy drew in some new combatants and moved into the pages of the *Journal of the Franklin Institute,* one of America's more prestigious technical forums. The paper war over the quadruplex again involved Edison and soon precipitated major lawsuits. Starting in January, Samuel Edison oversaw the building of a new laboratory in Menlo Park, and at the end of March, Edison and his staff shifted their operations out of Newark. Edison's family moved to Menlo Park as well.

The complexities of acoustic telegraphy kept Edison and his staff occupied in the laboratory. The technology was theoretically straightforward, but the practical problems involved in separating and amplifying the individual signals proved refractory. Edison tried many approaches and described them in five extensive January caveats. One of the transmitters Edison experimented with during this time was a "water telephone" by which he later claimed to have transmitted speech. He also tested his acoustic apparatus extensively on Western Union lines from the company headquarters in New York.[1] At one point in mid-February, Edison notified Western Union president William Orton that he had the "animal tamed,"[2] but in fact none of the trials proved satisfactory. Charles Wurth, one of Edison's staff machinists, assumed a more important role in the acoustic experiments, making and altering many of the instruments. Ezra Gilliland and Samuel Edison also participated in laboratory trials.

The new Western Union headquarters building in New York City.

Western Union was sponsoring Edison's acoustic work out of concern that alternate forms of multiple telegraphy might outperform the quadruplex. Edison himself had experimented with his electromotograph in a system for sending six messages and in January refined a form of quadruplex he called a "doubler," which did not provide better performance so much as it presented the possibility of an alternative system not covered by his quadruplex patent applications. Those applications remained tied up in the Secretary of the Interior's office until March, when the secretary sent the papers back to the Patent Office to resume the normal process. At that point Edison, through his attorney Lemuel Serrell, requested the withdrawal of all his applications for those inventions.

Earlier in February, Edison had protested a patent application of Gerritt Smith's. Smith had worked with Edison on Western Union's early quadruplex tests and had himself refined the system on the company's lines. Edison maintained that Smith had stolen ideas represented in his patent application from Edison's work of 1874. George Prescott, who owned an interest in Smith's patent application, charged that patent examiner Zenas Wilber had improperly apprised Edison of the content of the application. Edison was having other problems with Western Union's staff, particularly Prescott, who was pressing him to file comprehensive patent applications for the quadruplex system. When Edison finally cemented arrangements with Western Union for the combination of their exhibit spaces at the upcoming Centennial Exhibition, he asked President William Orton to "protect me from any sinful games on the part of my friends your Electricians."[3]

The Western Union electricians were not the only ones with whom Edison had complaint. Outside the telegraph community he was embroiled in the continuing dispute over the nature of his etheric force. Elihu Thomson and Edwin Houston, science teachers at Philadelphia Central High School and members of the Franklin Institute, took exception to Edison's characterization of the phenomenon as a new force and proceeded with their own set of experiments. They published their results, which they felt proved the new force to be nothing but induction, in the January issue of the *Journal of the Franklin Institute.* The *Scientific American* reprinted the article and Edison responded, thereby giving rise to further experiments by Thomson and Houston and further published arguments. Many other letters and articles about the etheric force appeared in the *Scientific American* during the winter.

Apart from experimental work on acoustic telegraphy, the laboratory staff was primarily engaged in the manufacture and sale of the electric-pen copying system. In mid-January, after Norman Miller resigned as manager of the business, Charles Batchelor took on this role, moving into it full-time at the end of March and spending most of his time in New York City. When Batchelor took over, the company was selling two pens every five days in New York; by the end of March, sales averaged six per day.[4] The laboratory machinists also devoted part of their time in January and February to fabricating pens and presses for Gilliland & Company, which had charge of the manufacturing.[5] Edison, meanwhile, sought to bring new money and management to the enterprise. In February he sold an interest in the invention to Ezra Gilliland's father, Robert. After executing a patent application covering the system—his first application in more than a year—he arranged an offering of stock for the Edison Pen and Press Company. He and Batchelor also sought agents to sell the system overseas.

During the winter Samuel Edison supervised the erection of a laboratory building on the property his son had purchased at the end of December in rural Menlo Park, N.J. The new laboratory, which cost nearly $2,700, was finished by 25 March, and in the next few days the staff machinists moved their equipment and their work from Newark to Menlo Park. Edison moved his family—Mary, now twenty; Marion, just over three; and baby Thomas Alva, Jr., born 10 January 1876—into a house that had been standing on the land when he bought it.[6]

1. In some of these experiments Edison tried to work Morse and acoustic instruments on the same lines, according to an undated memorandum (77-015, DF [*TAEM* 14:518]) that Edison probably wrote in preparation for a patent interference with Charles Haskins (see Lemuel Serrell to TAE, 3 Dec. 1877, DF [*TAEM* 14:515].

2. Doc. 728.

3. Doc. 732.

4. Batchelor's testimony, TI 1:229 (*TAEM* 11:91); Charles Batchelor to Thomas Batchelor, 29 Mar. 1876, Cat. 1238:46 (*TAEM* 93:59).

5. Cat. 1214:146, Accts. (*TAEM* 21:634).

6. It is not known if Mary's sister Alice was living with the family in Newark at the time.

–704–

From William Orton

New York Jany 4 1876[a]

Can give you ample accommodations tomorrow with all necessary local and line wires. Bring in all your traps.[1]

William Orton. President

L (telegram), NjWOE, DF (*TAEM* 13:1253). Message form of the Western Union Telegraph Co. [a]"187" preprinted.

1. That is, experimental acoustic telegraph apparatus; see Doc. 703.

–705–

Notebook Entry: Acoustic Telegraphy

[Newark,] January 7th 1876.

(61)[1] Acoustic Telegraphy

We found a means of receiving on a Sounder off a reed by the following means. Fig 84 shews it.

84

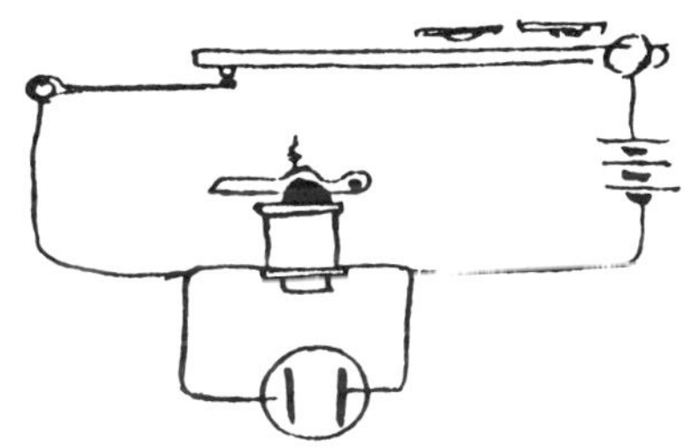

It is by putting a secondary battery or salt water shunt round the sounder. This secondary battery sends a current in the same direction on opening the electrical circuit. The vibrator strikes its contact point so rapidly that without shunt the induction from magnet would not allow the lever to come down but when the shunt is on its current operating in the same direction tends to hold it down. It seems strange that when you put 6 cells on to a local like this it will not respond but when you shunt ½ or ¾ of the current it works perfectly.

We have also found that it is necessary to have a contact on ~~transmitters~~ receivers[a] like 85

85

a is a vibrator, b is the contact which is solid & rests against the contact screw c When the right note comes the lever drives the contact lever away from screw.

Chas Batchelor

X, NjWOE, Lab., Cat. 994:58 (*TAEM* 3:222). Written by Batchelor. [a]Interlined above.

1. See headnote, p. 600 n. 2.

–706–

Technical Note: Acoustic Telegraphy

[Newark,] Jan 10th 1876[a]

Batch have reeds altered for tonight this way—(Transmitting)

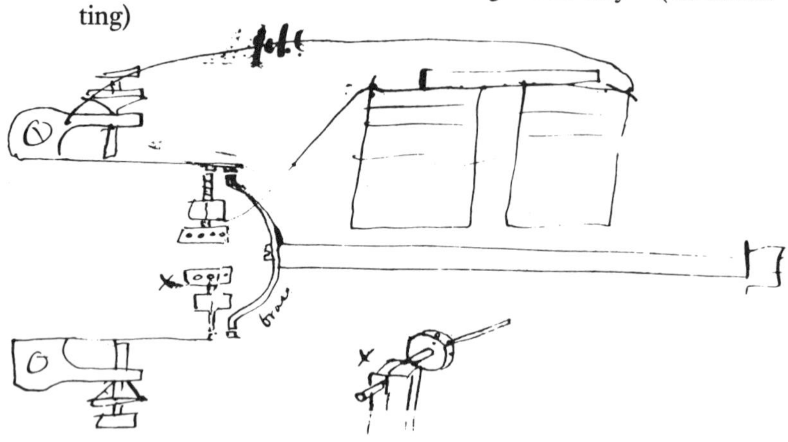

This will keep the spring off of the reeds except at extremities[1]

Kruesi attend to this[b]

Chas Batchelor J Kruesi

X, NjWOE, Lab., Vol. 9:11 (*TAEM* 3:800). [a]Dated by Batchelor. [b]"Kruesi . . . this" written by Batchelor; underlined twice.

1. On 15 January, Edison drew the accompanying diagram showing the wiring connections for a modification of this design. He called it a "New Style Vibrator" and directed Charles Wurth to make it. Vol. 10:11, Lab. (*TAEM* 3:826).

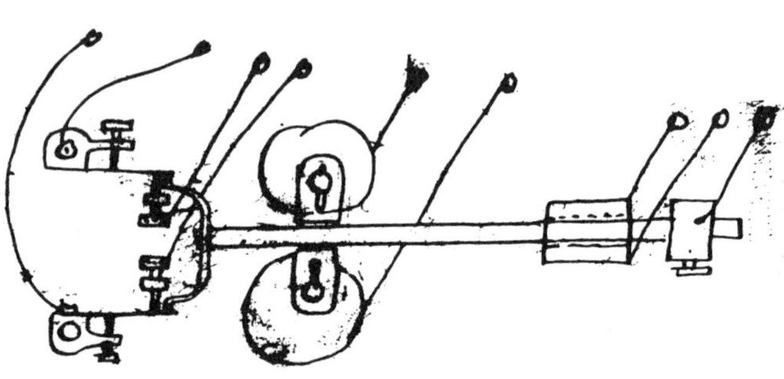

Edison's modification of the 10 January 1876 acoustic transmitter.

–707–

Technical Note: Acoustic Telegraphy

[Newark, c. January 11, 1876[1]]

Altered[a] Rec'g reeds so

but they ought to have been made so as not to produce a node in them in vibrating so

Started tonight with

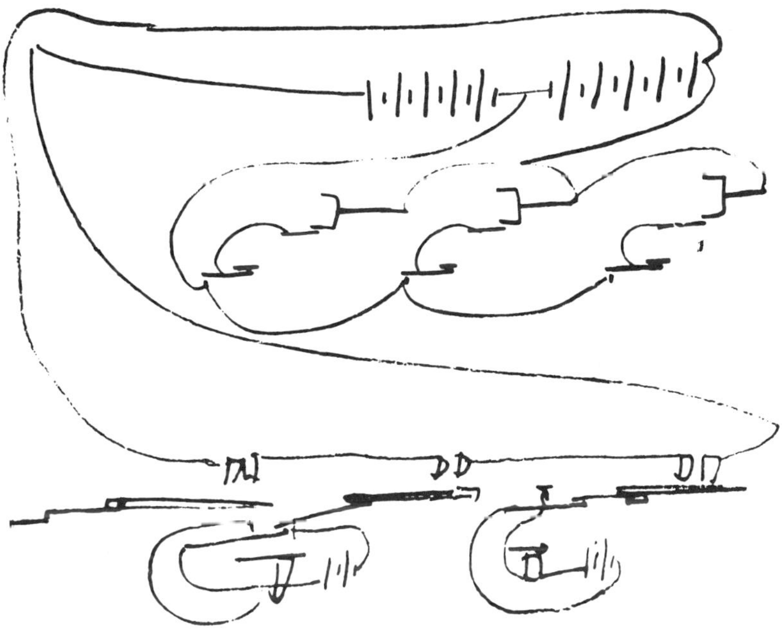

& receivers

Three messages OK on this principle[b]

T A Edison

X (fragment?), NjWOE, Lab., Vol. 9:7 (*TAEM* 3:820). [a]Top of page torn; some text appears to be missing. [b]Underlined twice.

1. Edison developed the transmitter circuit depicted here on 10 January. Since this experiment required alteration of the receiving reeds, it probably does not represent work from that day. Another drawing dated 11 January shows a similar circuit design. Vol. 10:5–6; Vol. 15:133; both Lab. (*TAEM* 3:817–19; 4:440).

ACOUSTIC-TELEGRAPH CAVEATS

Docs. 708–709 and 715

As a product of his experiments in acoustic telegraphy for Western Union, Edison executed five caveats in January 1876.[1] Three of these caveats, representing the character and range of Edison's acoustic experiments from November through January, are included in this volume.[2] This work was primarily developmental in character. Edison concentrated on exploring various practical alternatives for implementing acoustic telegraphy, assuming that the basic principle was already established.[3] In many of these designs Edison tried to ensure that transmitted signals would be brief, sharply distinct, and strong when received so that the receiving instrument would produce a sharp burst of sound somewhat like the click of a telegraph sounder. Alternatively he had the receivers actually operating sounders.

Acoustic telegraphy shared many problems with other forms of multiple telegraphy, and in these caveats Edison drew heavily on his earlier quadruplex work.[4] In February the Patent Office notified Edison of potential interferences involving one of these caveats (Caveat 74) and Doc. 664 (Caveat 71), which probably spurred Edison to execute a series of five patent applications in April and May 1876.[5] Those applications, all of which resulted in patents, included many of the designs described in the five January caveats.[6]

1. Edison executed Caveats 73–76 on 13 January and filed them the following day; he executed Caveat 77 on 26 January and filed it on the 29th. The former were mostly based on work done in 1875, while various designs and experiments from early and mid-January 1876 were incorporated in Caveat 77 (see, e.g., Doc. 650 nn. 5, 17–19; and Docs. 705, 706, and 715). Following the terms of Edison's agreement with Western Union (Doc. 695), these caveats, unlike the resultant patent applications and Edison's earlier acoustic caveat (Doc. 656), were prepared jointly by the offices of Lemuel Serrell (Edison's usual attorney) and Porter, Lowrey, Soren, and Stone (the attorneys who also served as the law department for Western Union).

2. Edison Caveats 75, 73, and 77; some drawings from Caveats 74 and 76 appear as illustrations. When Edison's drafts were prepared for filing they were assigned numbers at Serrell's office; these did not match the order in which Edison drew them up, as indicated by internal cross-references. He drafted Caveat 75 first, then 74, then 73; Caveat 76 may have followed or preceded 73. Copies of the drawings and transcriptions of the texts of Caveats 73–75 are in TI 2:554–77 (*TAEM* 11:680–99).

3. See headnote, p. 663.

4. For example, see Docs. 467–73, 512–13, and 531–34.

5. Commissioner of Patents to TAE, 5 Feb. 1876 (Edison Caveat 71) and 5 and 21 Feb. 1876 (Edison Caveat 74).

6. U.S. Pats. 182,996, 186,330, 198,087, 198,088, and 198,089.

–708–

Caveat: Acoustic Telegraphy[1]

New York, January 13, 1876[a]

To all whom it may concern.

Be it known that I, Thomas A. Edison of Newark in the State of New Jersey have invent~~ion~~ed an Improvement in Acoustic Telegraphs of which the following is a specification

The object of this invention is to transmit and receive several messages at the same time over a single wire without interference with each other.

The invention consists in various devices whereby the battery current may be thrown in and out of a telegraphic current by a self vibrating electro magnet in a manner to meet the various conditions found in practical telegraphy.

The invention further consists in various devices for stopping and starting or transmitting combination of waves to effect telegraphic transmission by dots and dashes. The invention further consists in a method of analyzing several sets of vibrations received upon one or more electro magnets or other sounder devices by the use of resonators each of which has its column of air of such a length that it will only respond to a certain series of waves.

The invention further consists in various methods for moving the column of air within the resonator.

The invention further consists in destroying the spark upon the contact points of the self vibrating reeds or forks by shunting such points or the magnet by a condenser.

The invention further consists in the adaptation of a Trevillyian rocker for the transmission of electric waves.[2]

In figure 1, is shewn a method for transmitting the different sets of vibrations or waves.

Fig. 1.[b]

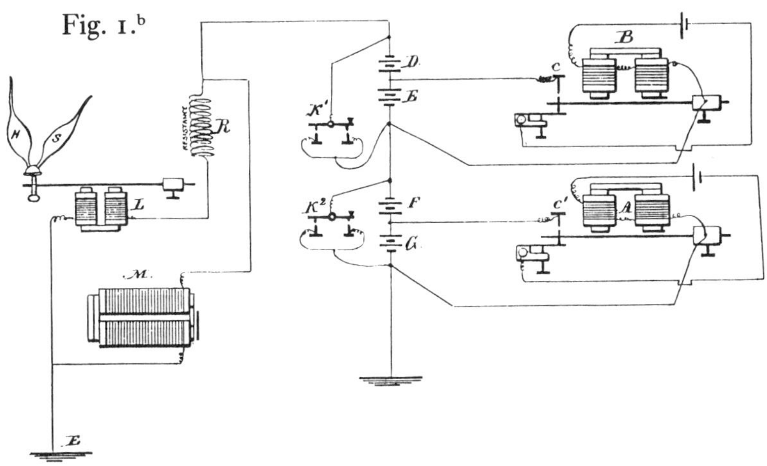

A. and B. are two electro magnets which serve to give motion to two reeds, tuning forks or bodies following the law of the pendulum.

Upon the extremity is secured a platina contact point opposite to which is placed a platina tipped contact spring which is so connected that the movement of the reed is made to open and close the current of the electro magnet automatically, thus giving a steady and powerful movement to the reed. C. and C' are two rigid contact points in contact with the batteries placed in the main circuit. D. and F. are two batteries which serve to transmit the waves of the wire. The current from D. is transmitted by the contact of the reed of B. against C. which short-circuits the balancing or neutralizing battery E. and allows the full current from D. to pass over the line at the rate of say 256 waves per second, while the reed of A coming in contact with the point C' short-circuits the battery G. 300 times per second and allows F to transmit that many pulsations or waves over the wire. K^1 and K^2 are transmitting keys which serve to short circuit both D. and E. and also F and G. preventing the transmission of waves altogether by the action of the reeds while the keys are closed on either the front or back point but allows the transmission by the reeds of a series of waves while the key lever is passing from the back to the front point or vice versa, and in contact with neither. The object of thus transmitting a short series of waves on closing and also on opening is to imitate the action which takes place in a ~~m~~Morse sounder and thus allow persons who are familiar with Morse sound reading to read from the resonators with equal facility. Upon a Morse sounder a dot is composed of two sounds following each other very closely, for instance in making the letter E, which is a single dot two sounds are necessary, first that produced by the lever of the magnet striking the lower limiting point, bridge or screw and the second when the lever raises and comes in contact to the upper limiting point or screw.

In the case of a dash the two sounds are necessary but do not occur so close together as with a dot; upon the downward movement of the lever a sound is produced and the lever is allowed to rest against this point for a certain length of time when it raises and coming in contact with the back point or screw produces the second sound.

In the case of the keys K^1 and K^2 figure 1, upon depressing say key K^1 waves of current pass over the wire and are heard in the resonator for an instant only when they cease, the key lever having touched the front point: this produces the first sound necessary to form a dot, as is the case with the Morse sounder; now when the key K^1 is allowed to rise, another series of waves pass over the line during the interval that the key lever is passing from one point to the other and this sound heard in the resonator gives the second sound necessary to complete a perfect dot. By thus transmitting the waves, messages can be transmitted in the usual manner and copied with remarkable facility from the resonator, or from the sound produced from the movement of reeds or tuning forks beating isochronously with the transmitting reed.

L. is the magnet at the receiving station. 4. is a thin strip[3] or bar of iron in contact with the cores. The magnet L is placed in one branch of the main circuit with an adjustable resistance coil R. The other branch has a large electro magnet placed within the second derivation, the object of which is to transmit inductive currents through the smaller magnet L. to cause it to act quicker. The resistance coil R. may be shunted with a condenser which increases the effect[4] and even the electro magnet L may be shunted with another condenser, which will also increase the effect.

H. is a Helmholz resonator[5] of a size which allows its column of air inside to be set in motion by the waves from B. S. is a second resonator having a different column of air which is only set in motion by the waves from A.

In figure 2. is shewn another method of throwing batteries in and out of circuit by devices operated by the vibrating reeds.

Fig. 2.

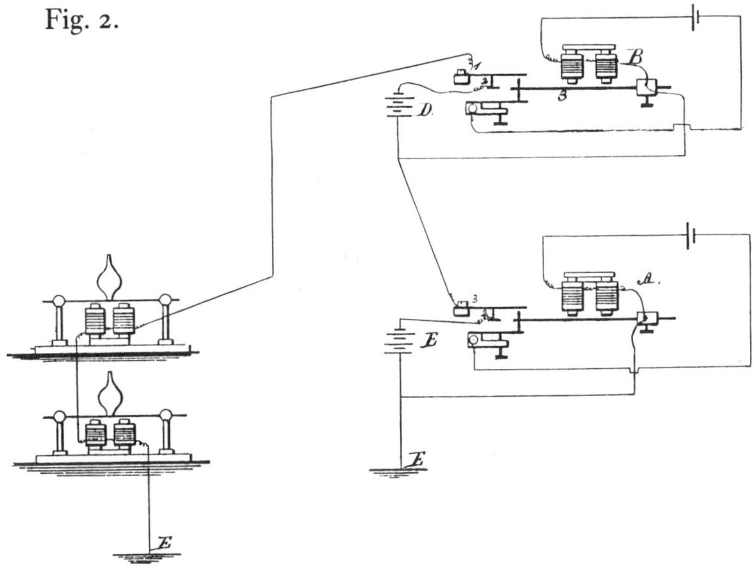

When the reeds are not in contact with the spring points 1, the battery is in the main circuit but when the reed is drawn forward by the action of the electro magnet it separates the spring 1. from the point 2. and disconnects the battery D.

from the line, at the same time preserving the continuity of the circuit. The same action takes place at A.

The transmitting keys may be made to short-circuit the batteries D. and E. in the same manner as in fig. 1, or the batteries may be thrown in and out of circuit in the same manner as in the ordinary duplex telegraph

In figure 3. is shewn another method for throwing currents in and out of circuit, and devices for transmitting the waves in a proper manner to produce dots and dashes.

Fig. 3.

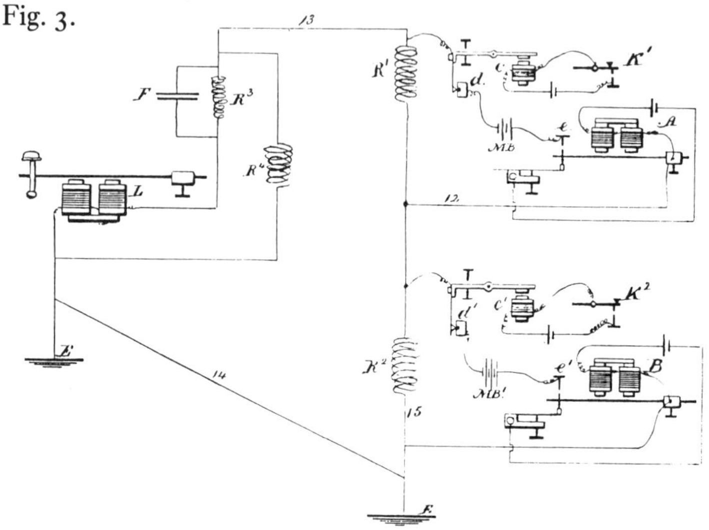

A. and B. are the self vibrating transmitting reeds; e. and e′. are the main line contact points which are connected to the batteries MB. and MB′. the other ends of which are connected to a piece of metal inserted in the insulating blocks d. and d′: passing over the surface of the block and insulated metal is the contact spring moved up and down by the electro magnets C. and C′. placed in local circuits each with a local battery and key.

The operation is as follows.[6] When the keys K^1, and K^2 are open, the lever of the magnets C. and C′ are in such positions that the point on the spring rests against the insulating substance of the blocks d. and d′. and although the reeds are making contact with the points e. and e′. no current is transmitted, but when either key is closed the magnets C. or C′ attracts its lever and an upward motion is given the contact spring which passing over the insulated metallic block in d. or d′ connects the battery to the line for an instant only as the spring, whether up or down rests upon the insulating substance and only connects the battery to the line in its passage from one point to the other.

R. is a resistance coil[7] having resistance perhaps four times greater than the internal resistance of the battery. When the reed of say A is not in contact with the point e. the continuity of the main circuit is preserved through the coil R^1 and when contact is made at e. and the spring of C. is in contact with the inserted metal of d. the current from battery MB circulates within a short circuit formed by R^1, wire 12, reed point e, battery MB, d, and spring, but a large portion passes into the longer circuit or main line formed by the line wire 13, wire 15, and earth[8] 14.

The same action takes place with the devices of B.

L. is the receiving instrument. R^4 is a shunt to preserve the continuity of the circuit. R^3. is a coil of large resistance shunted with a condenser the object of which is to compensate for the static charge of the line and render the waves sharper and thus give sharp action to the cores of the magnet L.

In figure 4 is shewn another device whereby the proper transmission of the signals can take place without increasing the power of the battery for each series of pulsations, and a device whereby the waves are manipulated similarly to those shown in figures .1. 2 & 3, to insure a proper transmission of the dots and dashes

Fig. 4

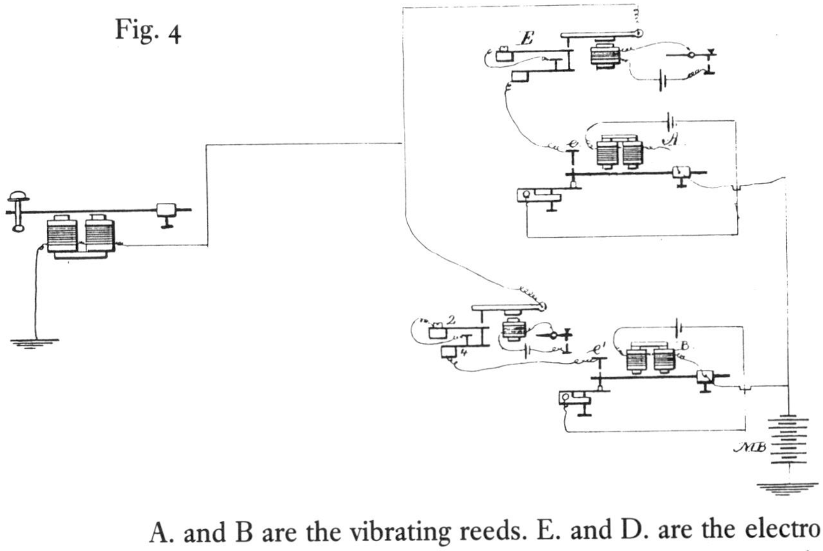

A. and B are the vibrating reeds. E. and D. are the electro magnets[9] for transmitting the series of waves in short intervals to produce dots and dashes at the distant station. The contact springs[10] 2. 3. and 4. shown in fig B. are so arranged that when the sounder lever touches 2, the waves of current pass into the line up to a point where the lever has reached the limit of its downward motion when the spring 4. is separated from the point[11] 3. by the rubber projection[12] from spring 2. when the waves cease; when the lever is allowed to rise the reverse action takes place.

Fig. 5.

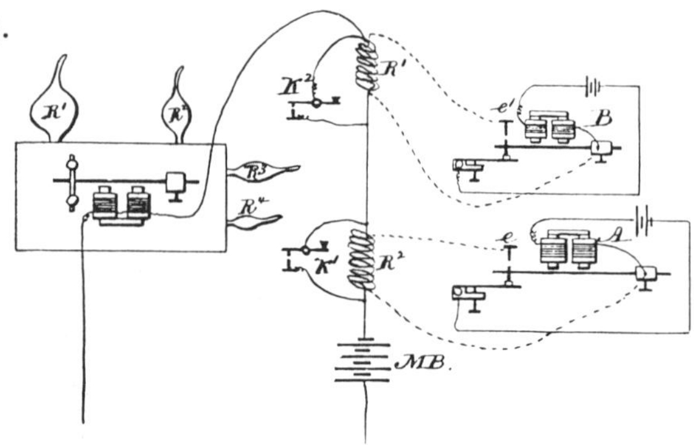

In figure 5. is shewn another device for throwing the battery current in and out of circuit or rather increasing and deceasing the amount of current in the circuit. A. and B. are the vibrators the reeds of which short circuit the resistance coils R^1 and R^2. when they come in contact with the points e. and e′. thereby increasing and decreasing the current from the main battery MB. K^1 and K^2 are keys for short circuiting the resistances so as to signal. The devices shewn in figs. 3. and 4. may be used instead of the keys K^1 and K^2 to effect more perfect signalling.

Fig. 6

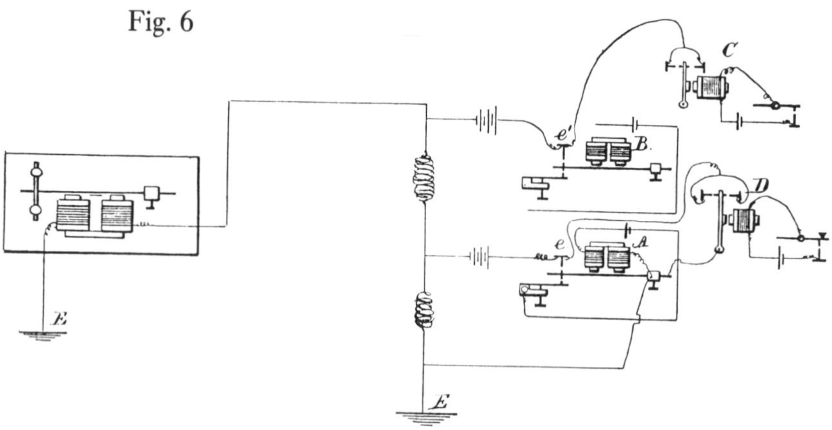

In fig. 6. is shewn another method similar to that shewn in fig. 3. except in the manner of sending the short series of pulsations. Extra magnets C. and D with levers playing in between double contact points are used to short circuit the make and break of the circuit between the reeds and their contact points e′ and e. when the levers of C. and D. are in contact with either their front or back contacts, but allows the reeds to transmit waves for that instant of time that it takes the levers to pass from their back to their front contact points.

Fig. 7.

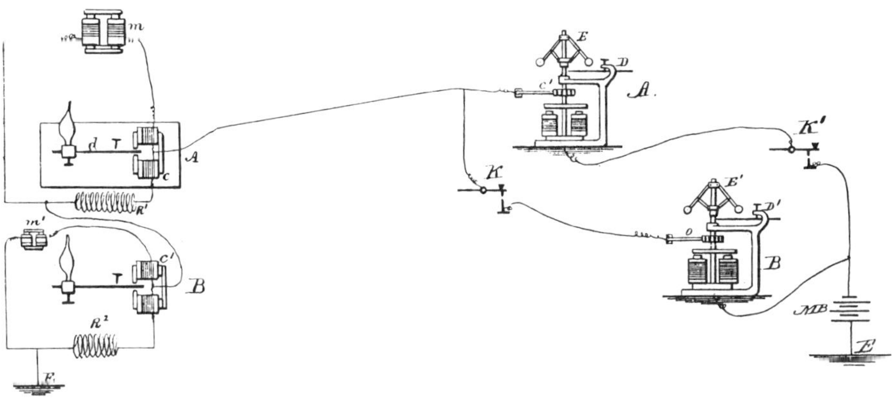

In fig. 7. is shown another method of transmitting different series of vibrations. A. and B. are two small electric engines provided with governors E. arranged with an electrical cut out operated by the rise of the governor balls. Both engines run at the same speed: upon the shaft of each is placed a contact wheel and springs the wheel having a certain number of metallic teeth with spaces filled with insulating substance, or the break wheel may be of any character which will serve to give a certain number of contacts at each revolution of the shaft. The break wheel of A has say 50 contacts, while the wheel on B has 65.

K. K′ are keys for controlling the vibrations or waves of current. M.B. is the main battery. I will mention that the keys might be replaced by the devices shewn in figs. 2. and 3. and a single engine could be used having upon its shaft several break wheels, each of which has a different number of contacts or will move a lever to give a different number of contacts.[13] I will also mention that the engine device could replace any of the reed vibrators shewn in the previously described methods.

The receiving apparatus shewn in fig. 7. is somewhat different from those previously described.

A. and B. are the receiving instruments, A. being adjusted to respond to the vibrations transmitted by the engine A. while

B. is adjusted to respond to the vibrations sent by the engine B.

C. is an electro-magnet (fig. A), the main line being connected to it on the wire which connects one spool with the other. M. is an electro magnet with its keeper permanently attached to its cores and is used for inductive purposes. R[1] is a resistance equal to the resistance of the electro magnet M: d. is a delicate vibrating reed impinging against a contact point which serves to operate a local circuit; this reed is made of steel hardened and magnetized permanently, its extreme end being placed between the two poles of the electro magnet C. If the current is passed over the main line permanently it divides where it enters C, one half passing through one spool to the resistance coil R[1] to the other instrument, while the other half passes through the other spool and magnet, to the next instrument and as each current acting in the spool tends to set up the same kind of magnetism in each iron core, the result is that no magnetism is created in the cores of C, but now if the current be interrupted the cores of the magnet M discharging within the circuit formed by R[1] and the spools of C. passes through the spools in a direction to make the iron cores magnetic. If now the circuit be closed or a current is sent over the wire, a contrary induced current is set up by the magnet M which magnetizes the cores of C. in an opposite direction hence the reed is made to vibrate back and forward like the tongue of a polarized relay and as the directive force of the magnetism in the reed may be less than the weakest current passing over the line it follows that it will always vibrate with a certain strength independent of the strength of the current upon the line. I will mention here that an induction coil might be used with its primary wire of greater resistance and having a greater number of convolutions than the secondary wire, the latter being connected to the magnet C. in the usual manner,[14] or the magnet C. might be placed in a derivation, so to speak, formed by a condenser of great electrostatic capacity.

Fig. 9.

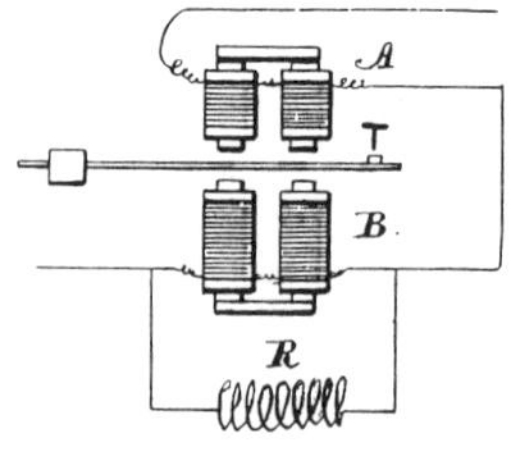

In figure 9. is shewn a modification. A. is a very short magnet placed upon one side of the reed and B. a long magnet placed upon the other side of the reed. The latter magnet is shunted with a resistance coil R. about equal to the resistances of the magnet itself.

At the moment the circuit is closed the cores of the short magnet A. become instantly magnetic, but the inductive charge from B. having a route or short circuit in which they can circulate, act to prevent the cores from instantly acquiring their magnetism, hence the reed is attracted by A. yet when the cores of both have acquired their maximum magnetism, the pull on the reed is equal on both sides. Upon opening the circuit the small magnet instantly loses magnetism while the secondary induced current from B. circulating through R. delays the demagnetization of the cores of B and this residuary magnetism draws the reed to the other side.

Fig. 10.

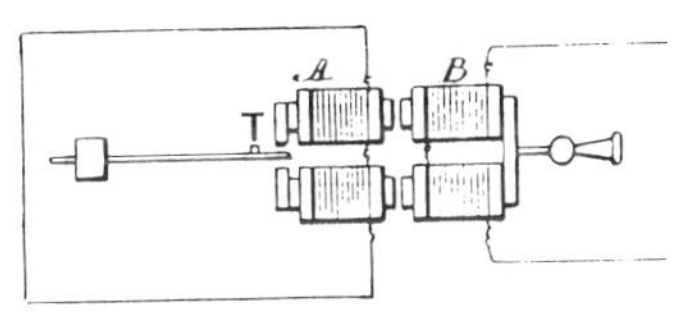

In figure 10. is shewn another modification. A is the magnet which gives motion to the polarized or magnetized reed by inductive currents, set up within its spools by the action of the iron cores of B. upon those of A. B is placed in the main circuit.

In figure 8. is shewn several iron cords adjusted to different rates of vibrations, but all set in motion by the electro magnet N. placed in the main circuit.[15]

Fig. 8.

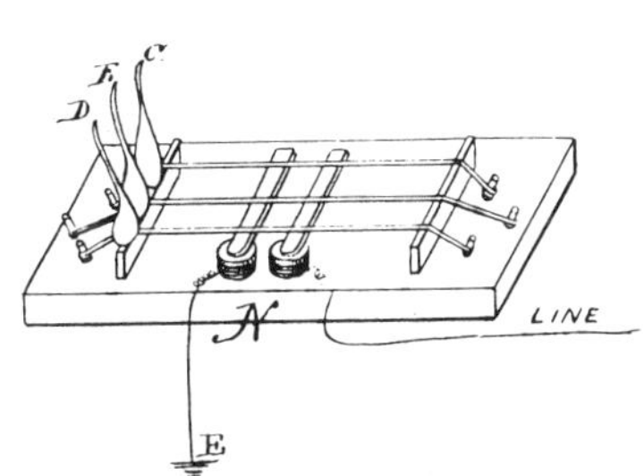

C. D. and E. are resonators to pick out the sound produced by the different vibrations.

In fig. 11 is shown a self acting vibrator for transmitting definite vibrations over the line wire, or controlling the current thereon.

Fig. 11.

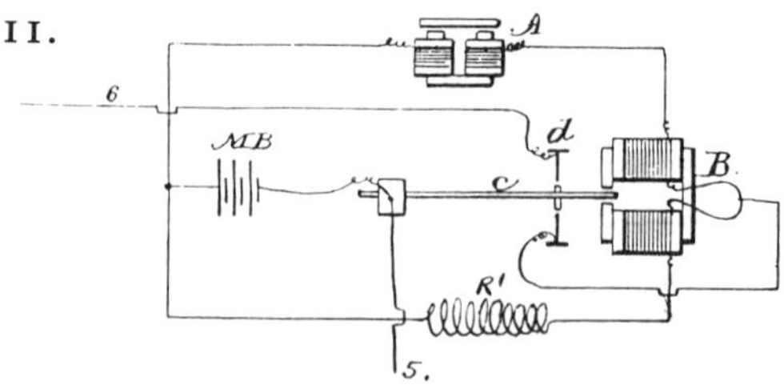

B. is the electromagnet which serves to give motion to the permanently magnetized reed C. playing between the two contact screws[16] d. and .e.

A. is an electro magnet which by its inductive currents gives motion to the reed through B. by changing the polarity of its cores, in opening and closing the circuit at the point e: the operation is as follows. When the reed C. touches e, the circuit is completed by passing through both spools of B. and the passage of this current sets up an inductive current in A. which in its turn acts upon B to give it a polarity necessary to attract the reed towards d. which it touches but just as it leaves e. the magnet A. discharges an inductive current of the opposite polarity through B. which tends to throw the reed to e. and so on. The contact of the reed with d. connects the battery wire 5. with the line wire 6.

Fig. 12.

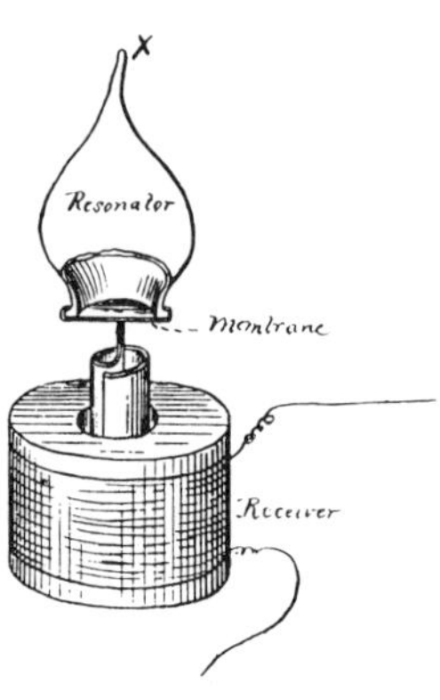

In figure 12. is shewn a Helmholz resonator with a membranous strip nearly covering the receiving end: attached to this membrane is a thin rod connected to a thin tube of iron passing within the hollow helice of wire connected to the line wire.[17] When certain vibrations corresponding to the vibrating rate of the air in the resonator pass over the wire, the movement of the membrane sets the air in motion and a large sound is heard when the ear is placed at X.

Fig. 13.

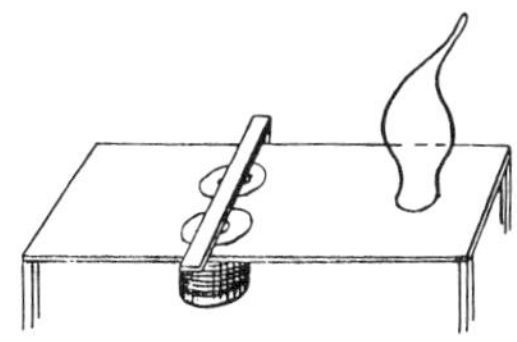

In figure 13. is shewn a modification, the electro magnet being provided with a flexible armature communicating vibrations to a resonant stand upon which is secured a resonator made of wood which taking up the vibrations communicate them to the column of air within and produce the requisite sound.

Fig. 15.

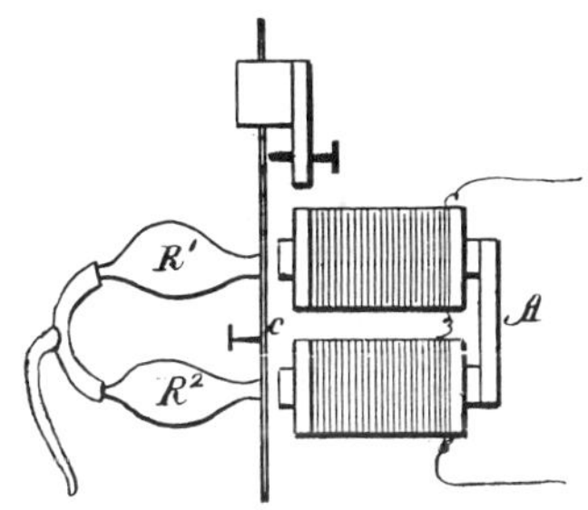

In fig. 15. is shewn a modification. A. is an electro magnet connected to the line wire, the cores of which act upon a delicate steel spring c. placed in front of the mouth of two Helmholz resonators, R^1 and R^2 both adjusted to sound to the same set of vibrations.

Fig. 16.

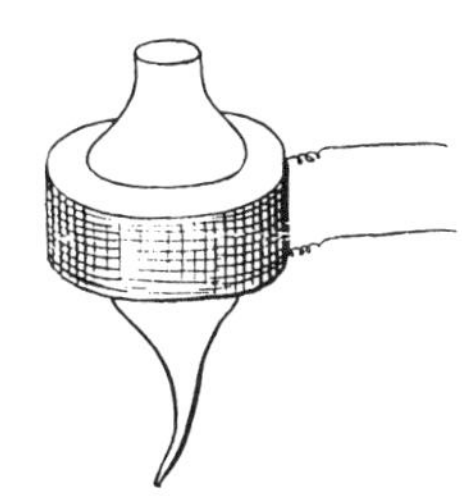

In fig. 16. is shown Helmholz resonators made of iron and having a coil of wire (connected to the line) wound around them. The magnetization and demagnetization of the iron of the resonators sets its column of air in motion and produces the requisite sound.

Fig. 17.

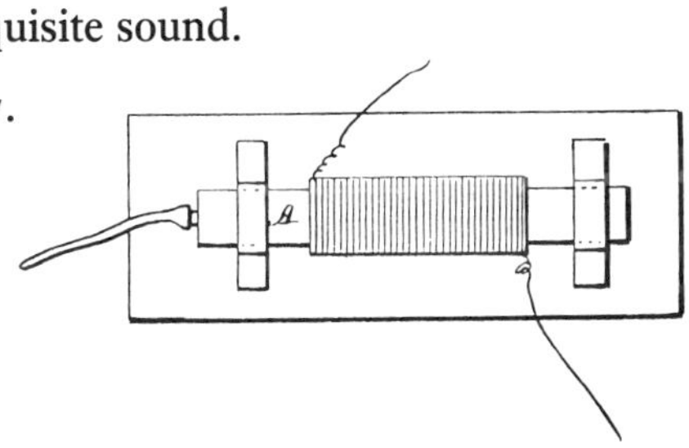

Fig. 17. shews a slight modification in so much as the column of air is in a long iron tube A. surrounded with the helice connected to the line wire.[18]

Fig. 14.

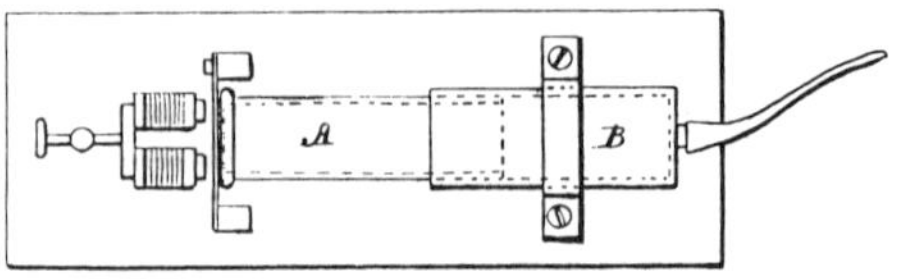

In fig. 14, is shewn another modification.[19] A. and B. are two tubes sliding over one another, after the manner of a telescope, so that the column of air can be so adjusted that it will respond to any given sets of vibrations: In front of the mouth of the same is a thin steel spring receiving a vibrating motion from an electro magnet placed in the main circuit.

My claims will probably be.

1st. The various devices shown in figures 1, 2, 3, 4, 5, 6, and 7, for increasing and decreasing or throwing in and out of ~~action~~ the main circuit the battery currents.

2nd The method described of transmitting dots and dashes by sending short waves composed of many vibrations, two short waves occuring close together to form a dot and two short waves not occuring so close together to form dashes.

3rd Translating from composite sounds a given sound by means of resonators having columns of air for telegraphic purposes.

4th. The use of polarized vibrating reeds tuning forks or bodies following the laws of the pendulum when the same are set in motion by the action of reverse currents transmitted from a battery or derived from induction substantially as shown in figures 7, 9 and 10.

5th The method of automatically vibrating the reed substantially as shewn in fig. 11.

6th The method shown in figures 12, 13 14, and 15, 17, and 18,[20] whereby the column of air in a resonator is set in motion for the purpose specified.

Signed by me this thirteenth day of January 1876.

Thos. A. Edison

Witnesses. H: M: Haigh[21] Wm. C. Ostrander.[22]

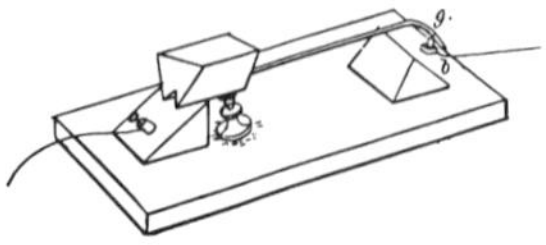

Edison's "Trevillyian rocker."

DS, DNA, RG-241, Edison Caveat 75. Petition and oath omitted. [a]Place taken from oath; date taken from text, form altered. [b]All figures are on four separate sheets.

1. See headnote above.

2. In Caveat 74 (fig. 2, shown at left) Edison defined the "Trevillyian

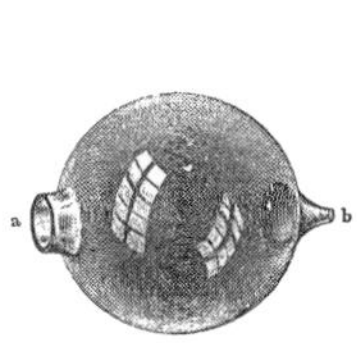

One of Hermann von Helmholtz's acoustic resonators.

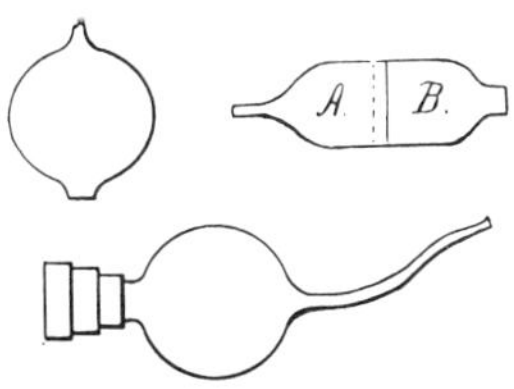

Drawings of resonators from Edison Caveat 74. (Top left): *"resonators with vanes for tuning."* (Top right): *"telescopic: for tuning."* (Bottom): *"telescopic tubes for tuning."*

rocker" as "a vibrating rocker toned to vibrate at a required rate, the temperature which it is necessary to keep the rocker up to, being supplied by a spirit lamp or gas jet; the lever g. vibrates against the contact b. serving to connect the battery to the main line in the manner shown in Fig. 1. or in any other manner, many of which are described in a previous Caveat upon this subject." The "previous caveat" is probably this document.

3. Not labeled.

4. Cf. figure 3.

5. These resonators were small, tuned, glass or metal vessels with two openings, a larger one directed toward a source of sound and a smaller one for the ear. They were designed to ease detection of faint sounds of a particular tone. Those featured by Helmholtz in his *Sensations of Tone* and his popular lectures were globes (Helmholtz 1875, 68; idem 1962, 49–50), but Edison applied the term to resonant chambers of any shape, as in the accompanying drawings from Caveat 74.

6. For clarification, cf. Doc. 715, figure 8 and its description.

7. "R." should be "R[1]."

8. In the diagram, the wire labeled **14** is an error; the vertical wire between **E** and **15** should be **14**.

9. **D** is not labeled.

10. **3** is not labeled.

11. Not labeled.

12. Not labeled.

13. This design for an acoustic engine appeared in Caveat 74, figure 6 (below left).

14. This configuration of an induction coil was included in Caveat 76, figure 5 (below right), and in another form in Caveat 74, figure 4, which appears as an illustration for Doc. 675.

(Left): *Edison's "acoustic engine" with two break-wheels. (Caveat 74, figure 6).* (Right): *In this sketch of Edison's induction coil (Caveat 76, fig. 5), the primary winding of induction coil* **m** *is in the main circuit; the secondary is connected in a local circuit with the magnet above the resonator.*

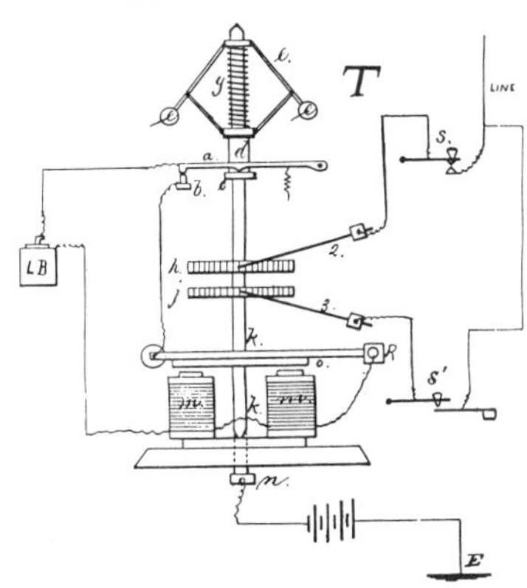

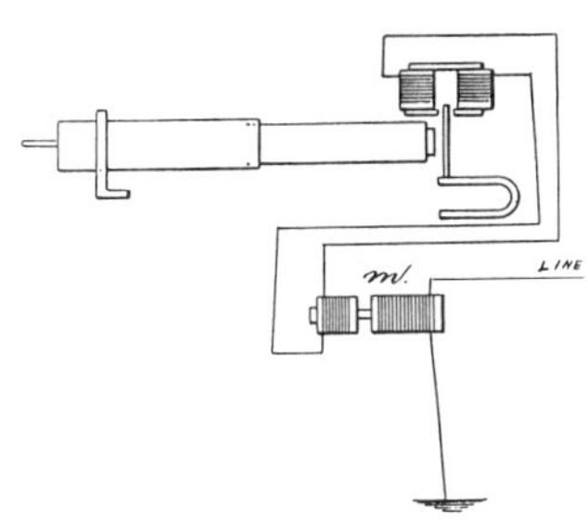

15. In May 1876 Edison received notice of a possible interference on this point, which also figured in Caveat 73 (Doc. 709, fig. 9). However, he did not claim this design in any known patent application. Commissioner of Patents to TAE, 18 May 1876, Edison Caveat 75.

16. **e** is not labeled.

17. Edison testified that he designed and tested a device like this during November 1875 (TI 1:29–30 [*TAEM* 11:35–36]). A later drawing of it was labeled "First Telephone on Record 1 month before Bell or Gray" (NS-76-003, Lab. [*TAEM* 7:403]).

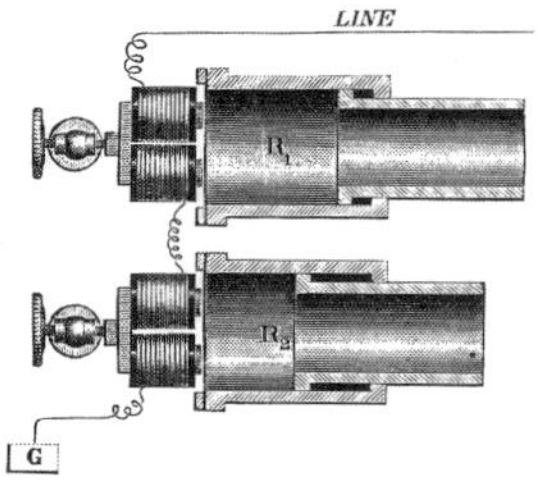

Edison's telescopically tuned receivers as drawn for George Prescott's 1879 book The Speaking Telephone (above), *and for his Caveat 76* (right).

18. Edison testified that he designed and tested a device like this during November or December 1875, in contrast to the version shown in figure 16, which was not made until March 1877. TI 1:29–30 (*TAEM* 11:35–36).

19. Edison testified that he designed and experimented with devices like this in November 1875, although in Prescott 1879 (pp. 219–20) he claimed to have designed them in September (TI 1:17–19 [*TAEM* 11:29–30]; Doc. 675). In an 1878 draft of that Prescott material (78-005, DF [*TAEM* 17:180]), Edison sketched a circuit including tuning forks and telescoping receiving tubes. He described it as having been designed in September, but all other evidence makes that unlikely. Edison designed many variations of the devices, such as Doc. 699 and those from Caveat 76.

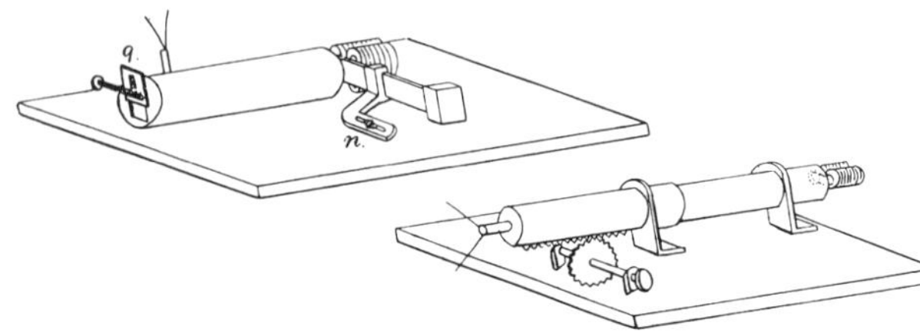

20. The number "18" should be "16."

21. Henry Haigh, a lawyer working for Porter, Lowrey, Soren, and Stone, was also the notary public before whom Edison signed the oath accompanying this caveat. Wilson 1874, 525.

22. This was probably the William Ostrander listed as a lawyer in Wilson 1877, 1088.

–709–

Caveat: Acoustic Telegraphy[1]

New York, January 13, 1876[a]

To all whom it may concern

Be it known that I Thomas A. Edison of Newark in the County of Essex and State of New Jersey have invented a certain Improvement in Multiplex Telegraphs and the following is declared to be a description of the same.

The object of this invention is to transmit several messages over a single wire at the same time without interfering with each other.

The invention consists in the various devices employed in transmitting and receiving the signals all based upon the transmission of several series of vibrations or electric waves each series having a different number per second.

Fig. 1.[b]

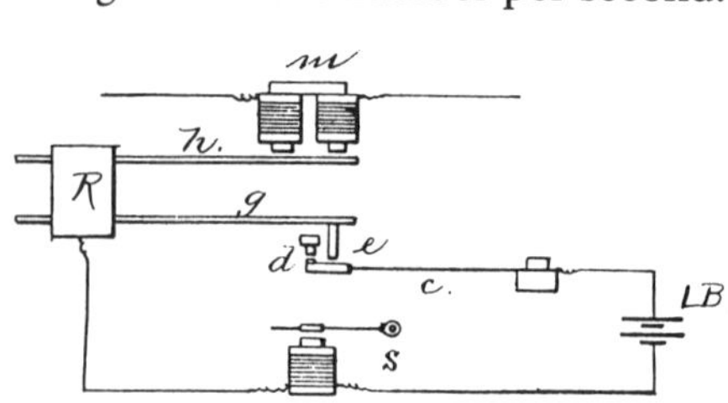

Fig. 1. shows a double reed receiving instrument. M. is an electromagnet placed either in the main telegraphic circuit and serving to give motion to the reed h. by direct current or arranged with its prongs or poles on each side of the reed and actuating the magnet by induction currents in the manner described in a previous caveat relating to this subject.[2]

g is another reed which is screwed to the rigid iron block R as with h. and is moved by sympathetic vibrations from h.. e. is a long piece of platina which serves to close the local circuit of the sounder s. upon the spring c. which is provided with a limiting screw d. this spring is of sufficient strength to dampen any tendency of the rod to an excessive amplitude of vibration.

The reason why the second reed is used and worked by sympathetic vibrations is because the transmission of several series of currents or waves through the magnet M. tends to cause the reed h. to vibrate in different planes and cause it to be a difficult problem to open and close the local circuit while g. not having the magnetic power of the magnet due to waves belonging to other system or reeds has its vibrations in the same plane hence closing and opening the local circuit is more complete.

It is to be understood that when the sounder is to be normally opened the vibrations are transmitted over the wire and they are made to cease when the sounder is to be closed: though the reverse action may be used with a different arrangement of the ~~so~~reed contact devices and sounder.

Fig. 2.

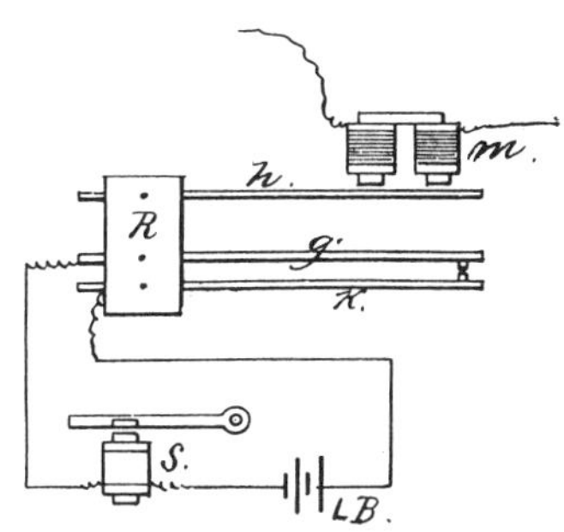

In Figure 2. is shown a slight modification of that shown in figure 1. which consists in dispensing with the spring contact points and substituting another reed k. tuned to respond to the same number of vibrations as g. and h.

When the vibrations to which these reeds answer or vibrate are not transmitted over the wire the two reeds g and k have their platina points in contact with each other closing the local circuit and sounder S′. but when the vibrations to which these reeds correspond are transmitted over the wire the two forks or reeds are set in motion receding and approaching each other like the prongs of a tuning fork which the two reeds are extended to be: this breaks the local circuit with great rapidity and the action of the self induced currents within the magnet keep it from attracting its lever. If desirable the reed k. may be secured to another standard as in figure 3. then both reeds unlike a tuning fork follow each other and if the points are separated a very small distance from each other the local circuit is opened but when the series of waves which set these reeds in motion is transmitted both ~~in~~reeds almost instantly reach an amplitude of vibrations necessary to bring the points in contact and close the local circuit permanently while the waves are sent over the wire.

Fig. 3.

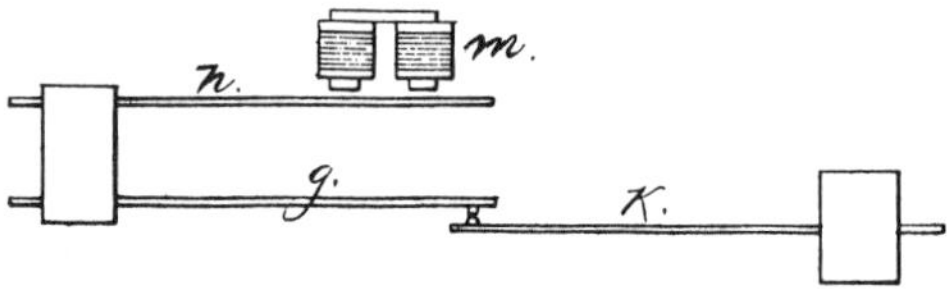

I will mention that the electro magnet M. fig. 3. might be placed directly in front of reed g., and h. dispensed with but I apprehend that the result will not be so satisfactory.

Fig. 4.

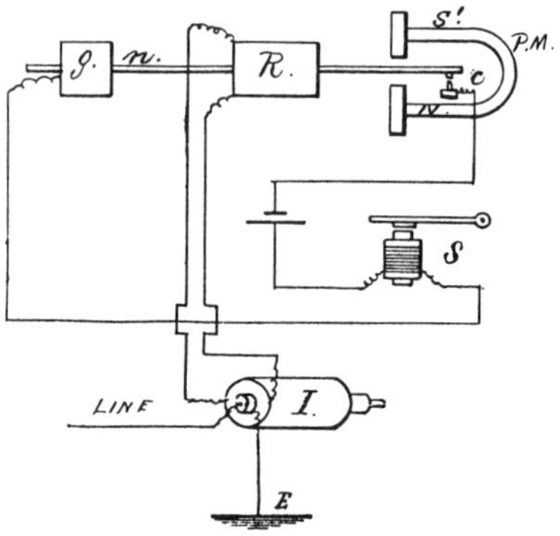

In figure 4. is shown a ~~r~~soft iron reed passing through a hollow bobbin of wire R. connected to the secondary wire of the induction coil I the primary of which is connected on the main circuit and is composed of finer wire and a greater number of convolutions than the wire upon the secondary coil.

On both sides of the reed n. at its extremity is placed the tongues of a permanently magnetized horseshoe steel magnet P.M. C. is a contact point which serves to open and close the sounder and local circuit.

When the proper series of waves are transmitted over the

wire each wave on closing sends say a positive current onto the secondary coil of I and through[c] the bobbin R. which giving a directive force to the soft iron reed it is attracted to say the left and when the main line wave causes[3] a negative current passes through R from the induction coil giving the reed a contrary directive force and deflecting it to the right of[4] the periodic time of the waves and the pendulum or reed are the same it will almost instantly be set in vigorous vibration to cease when the waves cease to pass over the wire.

Fig. 5.

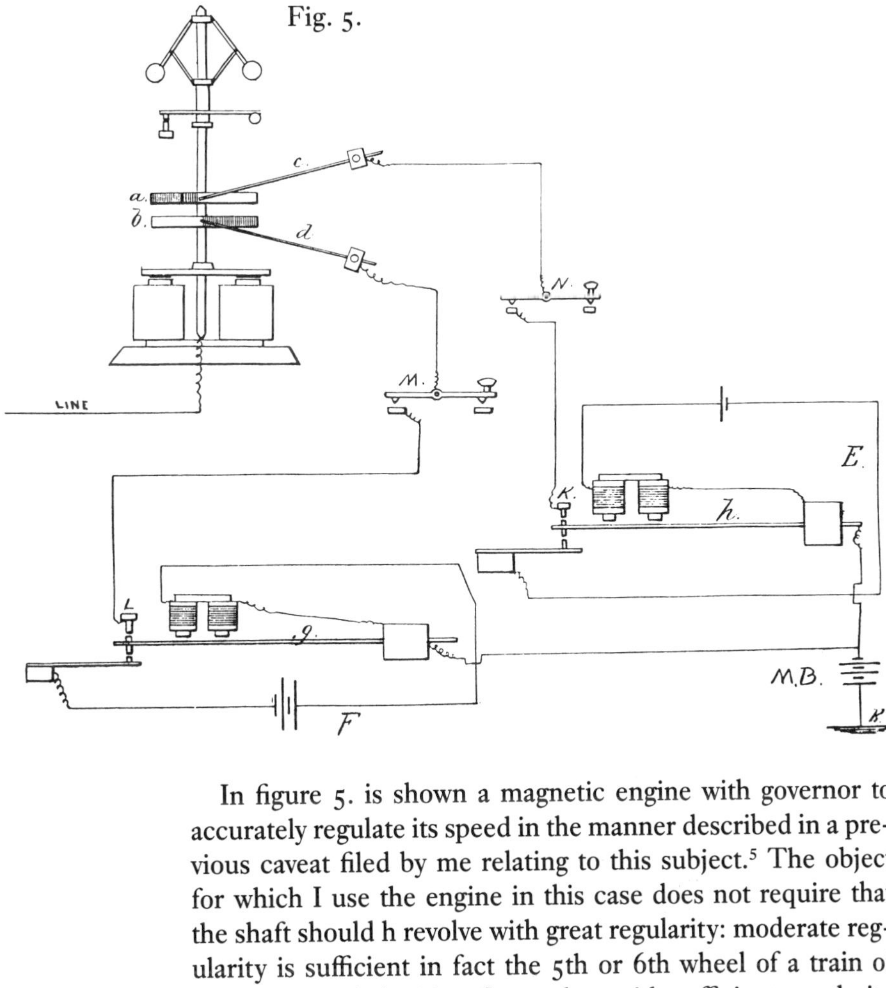

In figure 5. is shown a magnetic engine with governor to accurately regulate its speed in the manner described in a previous caveat filed by me relating to this subject.[5] The object for which I use the engine in this case does not require that the shaft should h revolve with great regularity: moderate regularity is sufficient in fact the 5th or 6th wheel of a train of gearing provided with a fly revolves with sufficient regularity to serve the purpose. a and b. are two wheels half of the circumference of each being inlaid with an insulating substance and the wheels are so arranged that the insulating substance of one wheel is immediately opposite the metallic or contact surface of the other[c] wheel.

These awheels are rapidly rotated by the electric engine. C.

and d. are two contact points connected to K. and L of the reeds through the keys N. and M. the reeds g. and h. being both connected to the main battery M.B. when both keys are closed and both reeds are vibrating two series of waves are sent out, but not at the same time: while the point c. is in contact with the metallic portion of the wheel a. the reed h. transmits into the line waves at the rate of 144 per second while no waves from g. can pass into the line as d. is resting upon the insulating surface of wheel b. but immediately afterwards the position of c and d change d passing on the metallic part of the wheel b. (a and b. are connected to line) and allowing the reed g. to transmit waves at the rate of 96 per second.

Now as the revolution of the wheels a and b. is extremely rapid the consequence is the transmission of rapid "beats" composed of different numbers of waves in the same time.

The effect on only transmitting one series of waves at a ~~distance~~time is to cause an easy adjustment of the receiving reeds at the distant station for in this case they respond to their proper vibrations and vibrate in the same plane whereas if both sets of vibrations were permitted to go together the plane of vibration is changed and the actuating of the local circuit becomes difficult. I will mention that the adjustment of the points L and K. may be such that the magnetic eff~~or~~ect on the magnets at the distant station will be the same the slow vibrating reed making a shorter contact or longer contact as the case may be.

I will also mention that several wheels may be placed upon the shaft of the electric engine each so divided that the metallic face of one will be opposite all the insulating faces upon all the other wheels, and several reeds used each of which transmits a different series of waves.

Fig. 6.

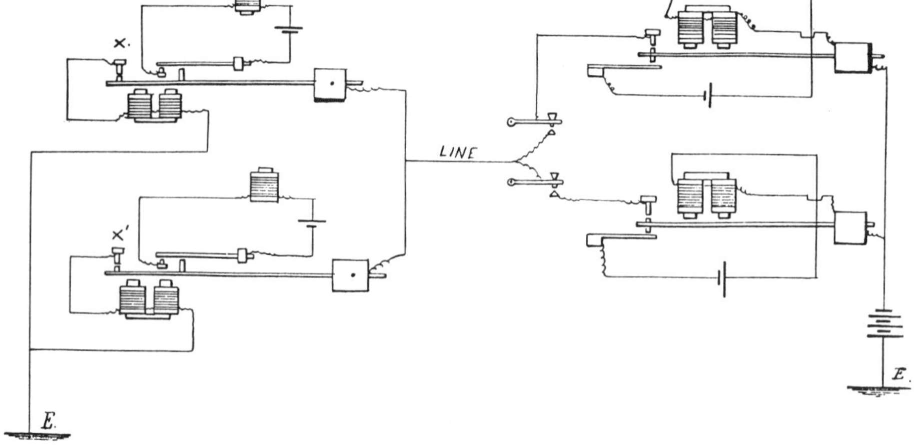

In figure 6. is shown an arrangement of the receiving reeds to prevent their changing their plane of vibration when several series of waves are being transmitted over the wire each reed has its magnet connected in a derived circuit in such a manner that the reed and contact points X and X break the derivation in which the magnet is placed. if the waves passing over the wire occur just at the moment that the reed touches say X the magnet is charged by it and attracts the reed away from X opening the derivation again and falling back on X and if another wave comes in the right time another movement is given but will not respond to other series of waves occuring at different times.

Fig. 7.

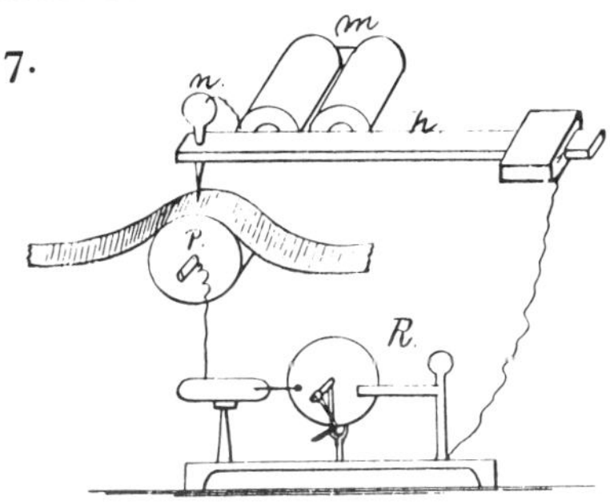

In figure 7. is shown a method of recording the waves whose periodic time is that of the vibrating time of the reed h. M. is an electro-magnet in the main line circuit and which serves to give motion to h. N. is a small bulb containing ink secured in the end of h. and ending in a small tube through which the ink cannot pass owing to the small diameter of its hole. P. is a metallic drum which serves to carry the paper forward: this drum is positively electrified while the ink is negatively electrified by a plate glass machine R. or other source of static electrification and this static electrification serves to draw the ink through the fine hole of the tube of n. in fine spray which is deposited in a finely divided line upon the paper upon the drum P. when the reed is at rest a faint straight line is recorded upon the paper but when the reed is set in motion a greater amount of ink is spurted upon the paper partly by the vibrating movement imparted to n. and by the effect of static attraction.

The record of Morse characters would be as in strip a.

A

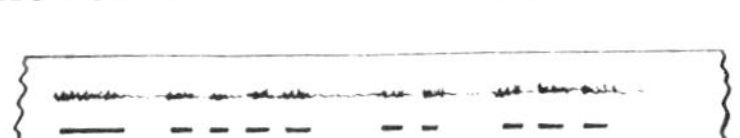

The lower line showing the dots and dashes as they ought to be and the upper line as they are recorded.

Fig. 8.

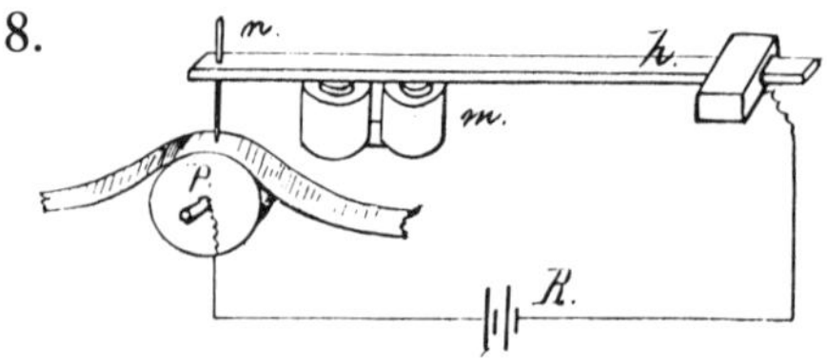

In figure 8. is shown a recording device whereby chemical decomposition is made to serve the purpose M. is the magnet placed in the main line, h. the reed vibrating up and down, n. is a metallic recording point, P. the drum carrying the chemical prepared paper B. the local battery to effect decomposition.

When the reed is at rest the point n is not in contact with the paper and makes no mark but when the reed vibrates the point touches and receeds from the paper with so great a rapidity that a slight line is formed the dots occuring so close together when the paper is moved at the rate which would record 40 words per minute to be practically continuous.

Fig. 9.

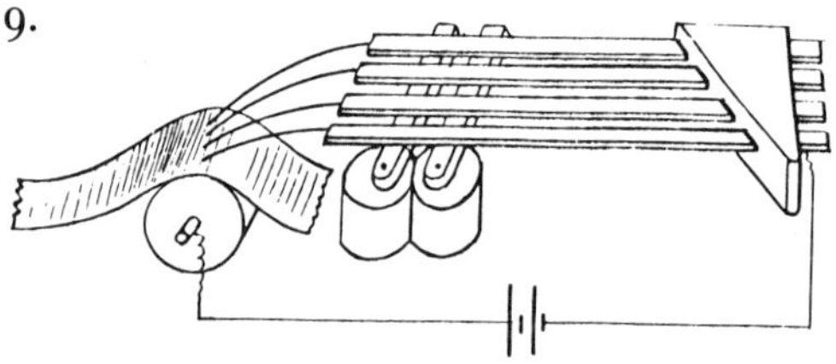

In figure 9. is shown the combination of several reeds each of which responds to a different set or series of waves all arranged to be actuated by a single magnet and provided with decomposing points grouped together over chemically prepared paper upon a revolving drum. The different series of waves being transmitted from the sending station by groups of perforations in paper.

Strip. B.

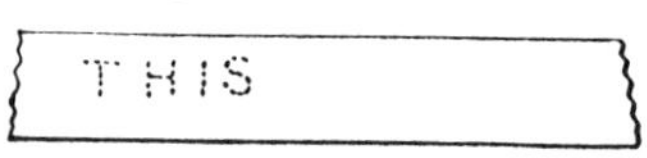

The record is[d] shown in strips B.

C.

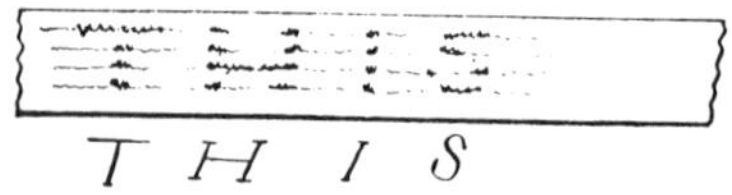

The record shown in strips C. is that when the ink bulb n. fig. 7. is used upon a series of reeds arranged to record roman letters.

Fig. 10.

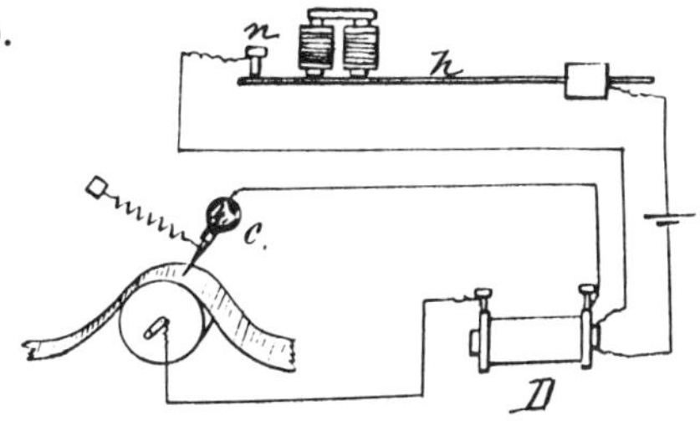

In fig. 10. I show another method of recording by ink the reed h. serving to interrupt the primary coil of an induction coil D̄ whose secondary wire connects one to an ink bulb and the other to the drum carrying the paper.

The extreme point of the ink bulb c. is not in connection with the paper the jump of the spark suffices to carry the ink from the bulb to the paper.

I shall probably claim.

1st The sympathetic reed as shown in Fig. 1. wherein one reed vibrates by sympathy and from the same base as the other substantially and for the purposes set forth.

2nd The methods of operating the local circuit and preventing the change in the plane of vibration of the reeds substantially as shown in figures 2. 3. and 6.

3rd The method of giving motion to the reed by passing the same through a helix as in fig. 4. substantially and for the purposes set forth.

4th The transmission of separate series of electric waves each series having a greater or lesser number in a given time following each other but not at the same time into and over a telegraphic circuit in substantially the manner described with reference to fig. 5.

5th The methods of recording Morse characters by ~~r~~vibrating reeds or bars as shown in figures 7. 8. and 10.

6th The method of recording Roman or other letters shown in fig. 9. by a series of vibrating reeds.

Signed by me this 13th day of January 1876[e]

Thos. A. Edison

Witnesses H. M. Haigh Wm. C. Ostrander

DS, DNA, RG-241, Edison Caveat 73. Petition and oath omitted. [a]Place taken from oath; date taken from text, form altered. [b]All figures are on two separate sheets at the end of the caveat. [c]"r" interlined above. [d]Interlined above. [e]"Signed . . . 1876" written in a different hand on two lines and enclosed in a brace.

1. See headnote, p. 709.

2. This requires that the reed be continuously magnetized. A full

description occurs in connection with figures 3 and 4 in Edison Caveat 74; the latter figure appears as an illustration accompanying Doc. 675. See also figure 5 from Edison Caveat 76, which accompanies Doc. 708 n. 14. Figure 4 of the present document shows a variation of this pattern.

3. Should be "ceases."

4. Should be "If."

5. See the description related to figure 7 in Doc. 708.

–710–

To Francis Nipher[1]

[Newark,] Jany 14 75[1876][2]

Dr Sir

Mr Beard has shewn me your letter of the 6th =[3] The following experiment will remove your objection that it is Extra Current =[4] I am fully acquainted with the phenomenon of the Extra Current as it was the greatest obstacle to the success of my Quadruplex Telegh now used by the W U Tel Co.

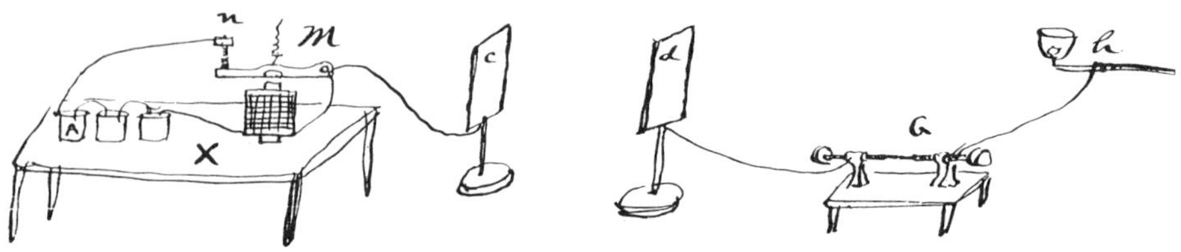

A is the battery, (6 cells) of Bunsen, M is a Common Telegraph Sounder with its back point n insulated and the wires so connected that it will make & break its own circuit like the interrupter of an induction coil the whole is placed on an ~~wood~~ insulated table X having hard rubber legs about 6 inch high c & d ~~is~~ are[a] ~~an~~ pieces of metal one foot square secured to glass rods, G are two screws insulated from each other and have lead pencil points on their ends pointed at the ends so they exactly meet = h is the gas pipe. by careful manipulation of the points you can obtain sparks when c & d are separated 1 to 3 feet = Yours

T A Edison

ALS, MoSHi, Nipher. [a]Interlined above.

1. Francis Nipher (1847–1926) had just begun his distinguished career as professor of physics at Washington University, St. Louis, Mo., in 1874. He was interested in electromagnetism and later conducted important work on magnetic measurement and electric discharge. *DAB*, s.v. "Nipher, Francis Eugene."

2. Edison misdated this letter; the etheric-force experiments did not take place until November 1875.

3. This letter has not been found. Nipher learned of the etheric-force experiments from Beard's lengthy piece in the *New York Daily Tribune*

of 9 December 1875 and responded in a letter to the editor dated 17 December ("No 'Etheric Force,'" *New York Daily Tribune,* Cat. 1144, Scraps. [*TAEM* 27:300]; on Beard's article see Doc. 679 n. 4).

4. "Extra current" referred to effects of self-induction. Nipher's 17 December letter (see n. 3) concluded that "the effects observed by Mr. Edison appear to be due to the passage of the 'extra current' through the insufficient insulation of his coils."

-711-

Edison's Electrical Pen and Duplicating Press Co. Account

41 Dey St New York Jan 14 1876

Thos A Edison[a] Edison's Electrical Pen & Duplicating Press Co.

Statement.

Assets		Liabilities	
Saleable stock	$ 56.35	Rent of office 2½ months at $400.	83.32
Office Furniture	59.75	Gilliland & Co	30.20
Cash in hand	30.54	Thos A Edison	391.89
Outstanding Bills	255.87		
	$402.51		$505.41

Chas Batchelor Genl Manager

ADS, NjWOE, DF, 76-007 (*TAEM* 13:935). Letterhead of Charles Batchelor. [a]Underlined twice.

Edison's draft of an advertising bill for the electric pen.

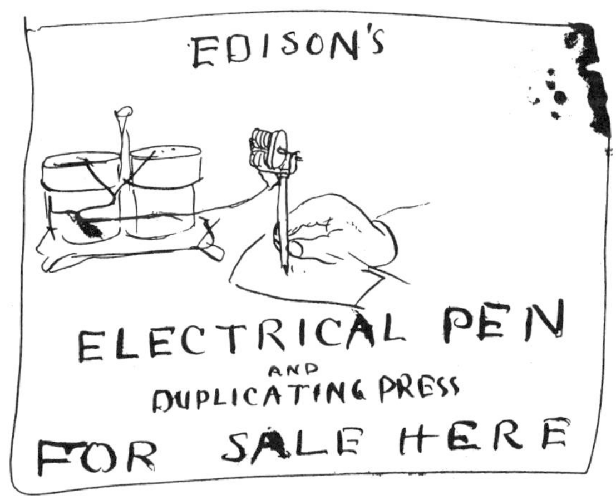

-712-

To William Orton

[Newark,] ⟨Jan 17/76⟩[a]

William Orton Esq

I send boy with Caveat for Lowrey.[1] It goes to Serrell for preparation.[2] I have the Engine working in place of tuning forks or pendulus bodies. The contacts are perfect but I am having trouble in keeping an even speed.[3] if I cannot succeed in making my engine run even I think Phelps has on that will.[4] Yours

Edison

ADS, NjWOE, DF (*TAEM* 13:1253). [a]Dated (with initials) by Orton.

1. Doc. 715.

2. See headnote, p. 709 n. 1.

3. This type of acoustic telegraph transmitter was included in Edison's Caveats 73–75; see esp. Doc. 708 n. 13. A few days earlier he had worked out the modification of the electrical contacts with the rotating shaft later incorporated in his U.S. Patent 198,087. See Vol. 9:11; and Vol. 15:133; both Lab. (*TAEM* 3:799; 4:440).

4. George Phelps's new main-line printing-telegraph instruments, then being tried on one of Western Union's major eastern lines, were driven by electric engines capable of maintaining a precise, high speed. Prescott 1877, 647–71, esp. 663–68.

–713–

Technical Note: Acoustic Telegraphy

[Newark,] Jany 25 1876

Tonight, replace the spring selfvibtg points on transmitter with rigid points and make front point more rigid.

Make both higher notes, and much nearer together as it is now we have[1]

of course they interfere. Should be thus

Jim[2] Gill[3] Batch dad spice,[4] Edison.

E Gill Batch, rec dad spic & jim send on three reeds

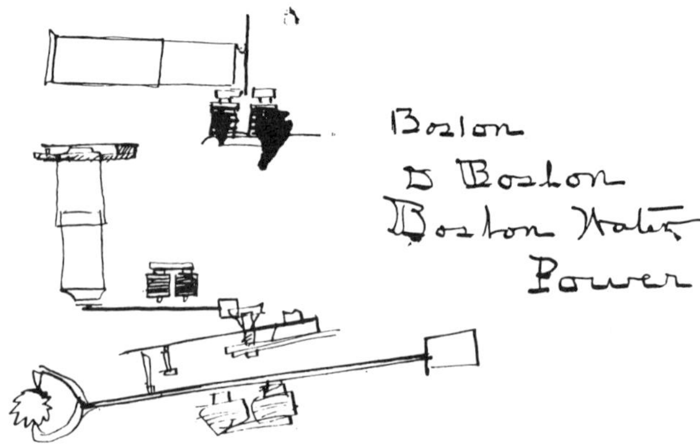

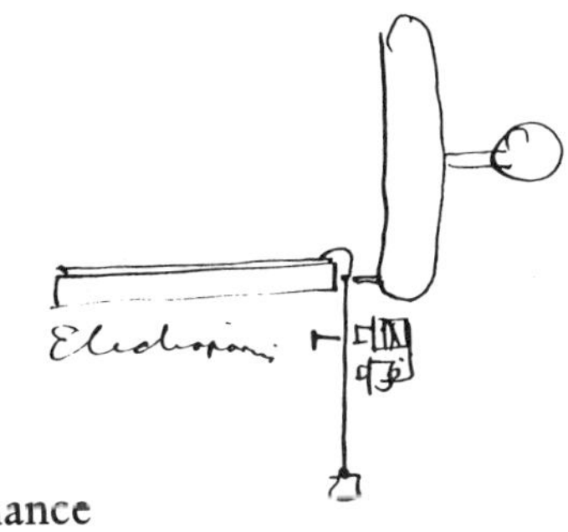

Powerful[5] Rcsonance

T A Edison

X, NjWOE, Lab., Vol. 10:15 (*TAEM* 3:830).

1. Figure labels are "High note" and "low note."
2. James Adams.
3. Ezra Gilliland.
4. Robert Spice.
5. Figure label at bottom is "Electropoion."

-714-

To William Orton

[Newark,] ⟨Jan. 26/76.⟩[a]

It will require eight days at least to finish the Acoustic Instruments.[1] In the meantime, I could go ahead with the "Doubler"[2] and shew you its merits if conveniences are obtainable.[3]

ALS, NjWOE, DF (*TAEM* 13:1254). Letterhead of Gold and Stock Telegraph Co. [a]Dated by Orton.

1. Edison planned a trial on a Boston line (TAE to Orton, 24 Jan. 1876, DF [*TAEM* 13:1253]). It is not known exactly which acoustic telegraph apparatus Edison was testing; but see Docs. 719 and 725.

2. This refers to an alternative type of diplex and quadruplex telegraph. Edison had worked on this design for more than a year, had recently taken it up again, and had executed a caveat for it on 12 January 1876. Docs. 507, 663, and 716 (esp. n. 1); Edison Caveats 58 and 72.

3. Presumably Orton agreed. An undated note from Edison to Orton (probably written in late January or early February 1876) accepts the use of particular rooms, requests keys, and requests apparatus necessary to test the doubler in an arrangement with four simulated way stations. Edison specified the equipment he needed in a separate list; such a list, unlabeled, is in one of Edison's collections of undated notes. TAE to Orton, [Jan. 1876], DF (*TAEM* 13:1255); NS-Undated-005, Lab. (*TAEM* 8:416).

–715–

Caveat: Acoustic Telegraphy[1]

New York, January 26, 1876[a]

To all whom it may concern

Be it known that I Thomas A. Edison of Newark in the County of Essex and State of New Jersey have invented an Improvement in Multiplex Telegraphs and the following is declared to be a correct description of the same.

The object of this invention is to transmit several messages over a single circuit at the same time without interference with each other, by means of tuning forks or bodies following the law of the pendulum.

The invention consists in various methods of transmitting the waves of electricity over the wire and of devices for opening and closing the local circuit of a telegraph Morse sounder.

Fig. 1.[b]

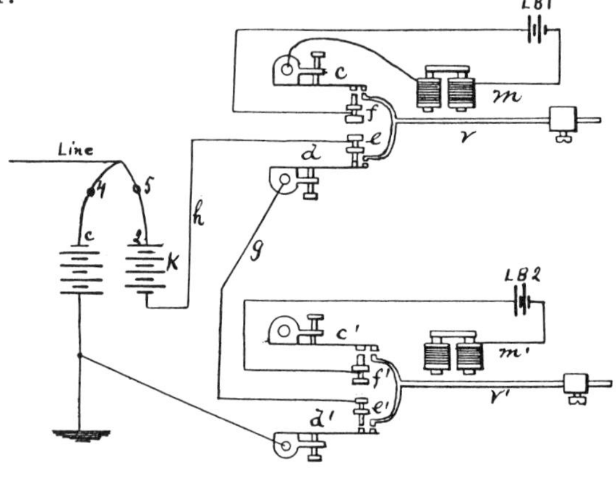

In figure 1. is shown a method of transmitting short waves of current by means of balanced batteries[2] L and K: L having its Carbon pole towards the line while K has its negative pole towards the line; r and r′ are the two vibrating reeds (there may be several) each of which vibrate a more or less number of times per second than the other.[3]

c. and f. are the contact points which serve to automatically open and close the local circuit containing the battery LB′ and magnet M.

The points d and e. of the same reed serve to open and close for a very short period of time; the ballancing battery K. thus allowing the battery L to transmit a wave of current over the line of a length equal to the length of time that the points d and e. are separated from each other by the movement of the reed r.

Included in the same circuit are the points d′. and e′. of the other reed which also open and close the battery K. allowing another series of waves to be transmitted over the wire, if both d and e. and d′ and e′. open at the same time it is just the same as if one opened, but as this does not occur very often two distinct series of waves each of which is of equal strength are transmitted over the wire; and reeds, forks or other pendulous bodies at the distant station are set in motion by the waves which occur periodic with its time of vibration.

The method of balancing the batteries so that no current shall pass over the line I believe to be peculiar to myself By tracing the flow of the currents from L and K. it will be noticed that their whole power circulates within the circuit in which they are placed and as the line and earth form a bridge wire across the circuit any tendency of a current to pass over the line from one battery in one direction is met by a current from the other battery in a contrary direction hence no current passes when the batteries are equal b~~y~~ut if one of them be interrupted for an instant the balance is destroyed and the other battery transmits a current as long as the other battery is so interrupted of the full strength of that battery.

In practice I shall insert a resistance of one or 200 ohms at 4 and 5. to diminish the spark upon d. e. and d′ e′.

In figure 2 is shown another method for transmitting short waves of current over the wire.[4]

Fig. 2.

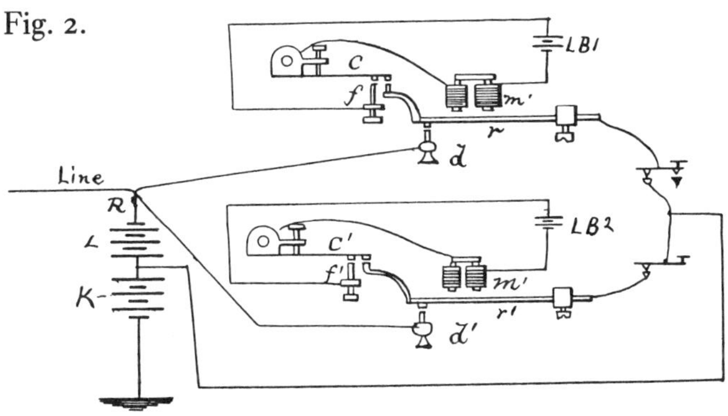

r and r′ are reeds c. f. and c′ f′. are the self breaking contact points which serve to keep the reeds vibrating. d and d′ are the short circuiting rigid contact points which with the reeds themselves serve to short circuit for an instant the balancing batter~~ies~~y L and thus allow the battery K. to transmit a wave of current over the wire R is a resistance of several hundred ohms which serves to lessen the spark upon the points d and d′. the object of the rigid points are to cause very short waves to be transmitted and thus prevent interference with the receiving reeds by the transmission of long and short waves which take place when each reed allows a long wave to be transmitted.

It will be seen that each reed short circuits the battery independent of the other reed.

Fig. 3.

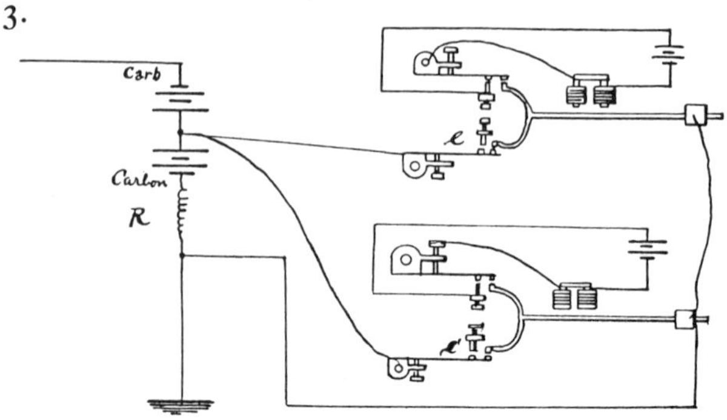

The same method is shown in fig. 3. except that the rigid point is replaced with spring points e. and e′.

Fig. 4.

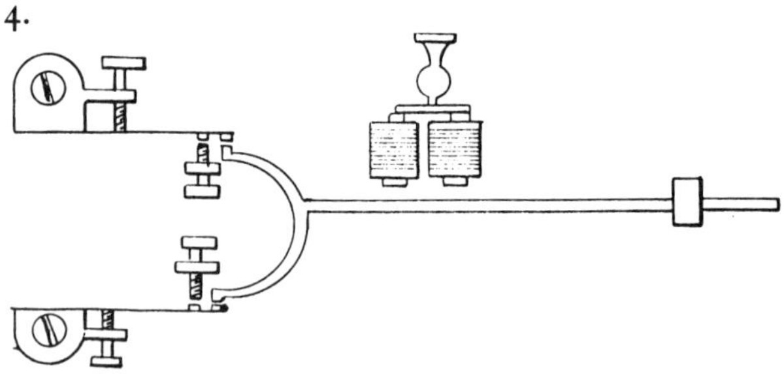

An enlarged drawing of the reed and contact points are shown in fig. 4.

Fig. 5.

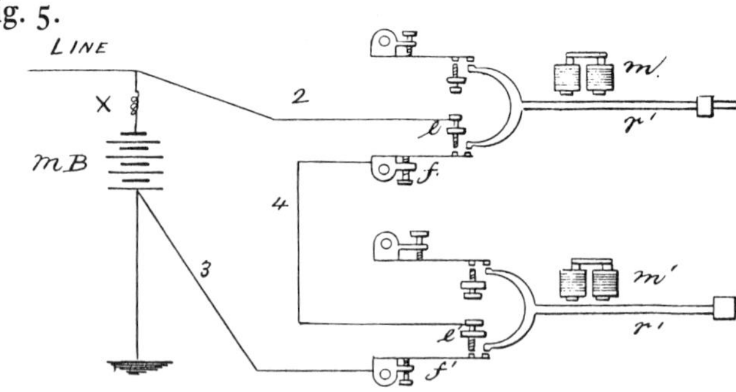

In figure 5. is shown a method of transmitting the waves by short circuiting the main battery. r and r′ are the reeds[5] provided with their self make and break contact springs; e and f.

and e′ and f′. are the short circuiting contact springs connected together with the battery by the wires 2. 3 and 4. so that when e. and f. and e′ and f′ are in contact the short circuit is complete but is broken where either the points are separated from one another by the movement of the reeds thus allowing the battery to transmit a short wave over the wire. A resistance may be inserted at X to lessen the spark upon the contact points.

Fig. 6.

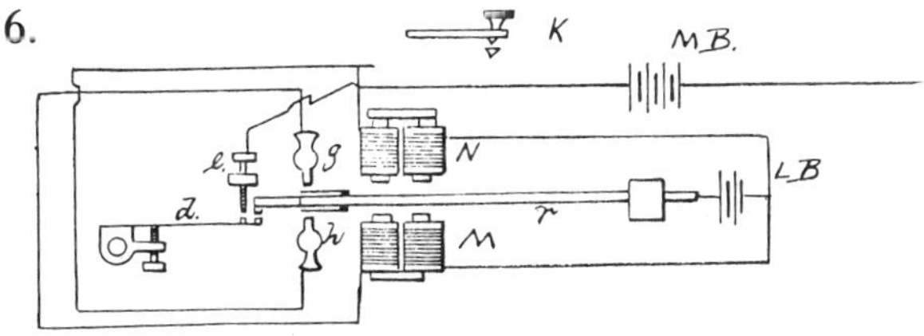

In figure ~~1~~6. is shown a transmitting reed which has some advantages over reeds provided with only one magnet r. is the ~~b~~vibrating reed placed between two electro magnets M. and N g and h. are two contact points opposite two spring points upon the reed r. and are so connected that when the reed goes to the right the current from the local battery LB is thrown through the magnet on the left which attracts the reed which coming in contact with the point h on the left throws the current through the magnet N on the right attracting the reed to that magnet and so on thus giving a very even vibration to the reed; d and e. are the line contact points by which the current can be made to flow upon the line[6] in any of the methods already described.

K is a key which serves to short circuit the contact points and main battery M.B. or it may short circuit the contact points only.

It is obvious that several of these reeds having different vibrating times may be combined together for transmitting several series of waves over the wire.

Fig. 7.

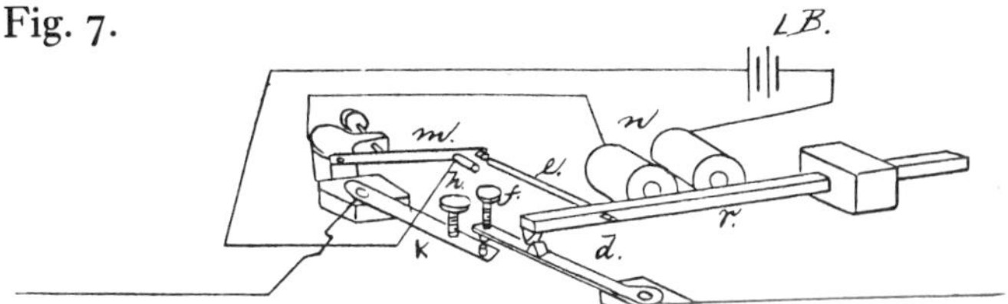

In figure 7 is shown another method of transmitting the vibrations or waves of current over the wire. The peculiarity of the device consists in arranging the contact points in such a manner that the length of the waves is independant of the amplitude of vibration of the reed r thus preventing one reed from interfering with another at the receiving station by the transmission of waves of different length. The reed r. in this case makes half the number of vibrations as there are waves transmitted.

d. is a contact spring or lever as the case may be provided with a V shaped tooth near its extremity and immediately underneath a similar shaped tooth upon the reed r. When the reed goes to the right the spring d. is forced downward in contact with the point K closing the line circuit or the circuit of a balanced battery etc. and when it goes to the left the spring d is again forced downward in contact with K. h and X are limiting screws,[7] it will thus be seen that the time which d. remains in contact with K. is independent of the amplitude or swing of the reed r to a great extent.

e. m. and g. are the self acting contact breakers[8] of the local ~~is~~battery L.B. & magnet N.

I do not wish to confine myself to any particular reed or method of keeping it in motion as this device can be combined with any of the forms already shown.

Fig. 8.

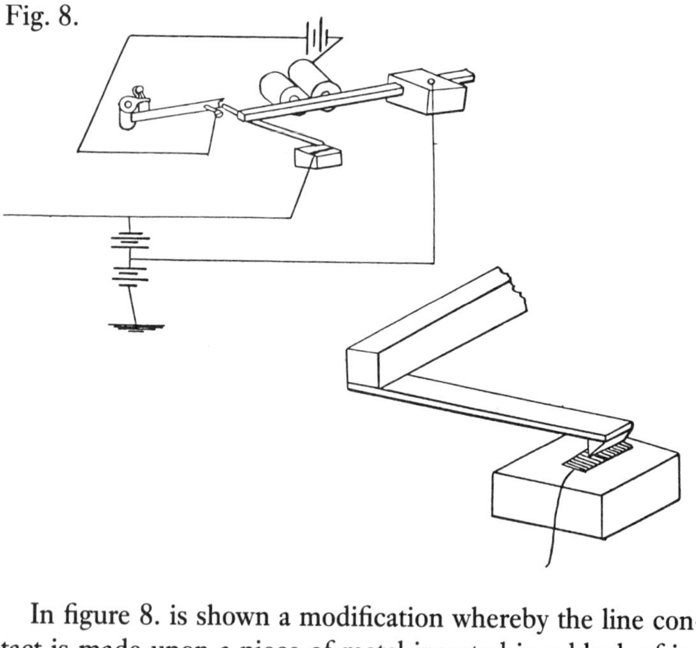

In figure 8. is shown a modification whereby the line contact is made upon a piece of metal inserted in a block of insulating substance which in this case serves to short circuit the battery g. both when the reed passes to the right and left.

Fig. 9.

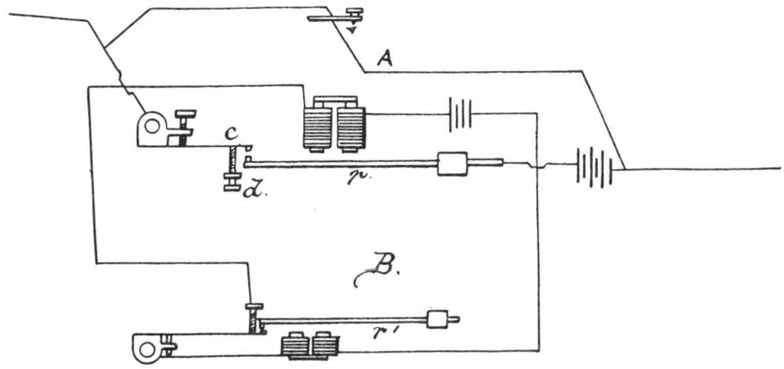

In figure 9. is shown[9] a repeating reed the object being to lessen the dampening effect of causing one reed to operate both the main circuit and its own contact springs. B is the primary reed breaking its own circuit at c and d included in this circuit is the reed A which is of course vibrated.

The reed r of A operates the main line contact points c. and d.

Fig. 9½.

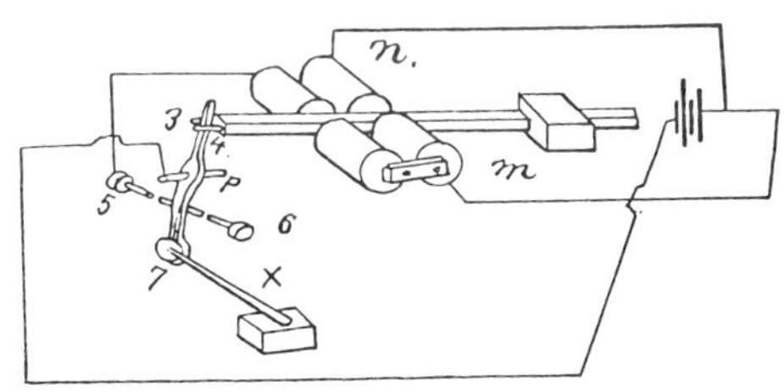

In figure 9½ is shown a peculiar form of contact points for operating the reed and its double magnets. r. is the reed[10] operated upon by both the magnets N and M.

P. is a thin light lever provided with double platina contact points playing in between contact points 5 and 6. thus causing the current from the local battery LB to be thrown through first M and then N. and so on.

3 and 4 are pins secured to the reed between which the lever p. is placed 7 is a wheel secured to the spring X and which serves to hold the lever p. against the point 5 or 6 as the case may be when the reed is not actuating p. which it does by being thrown over the centre of the axis of the wheel by the use of these contact points dampening the reed and the alteration of its time of vibration is greatly lessened and a better and longer contact is obtained. It is obvious that the main line contact point can be attached either to the lever p or to the reed itself.

Fig. 1~~1~~0.

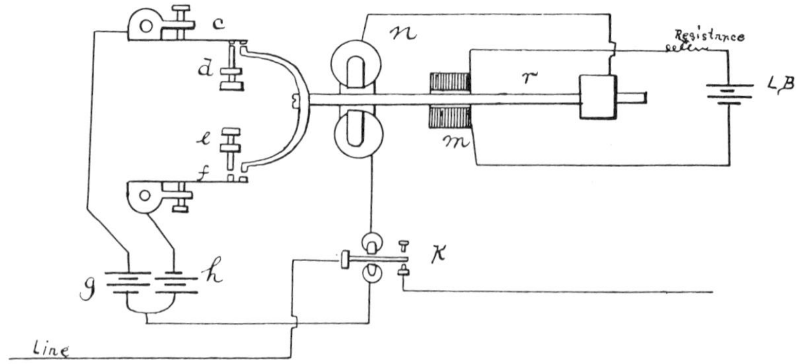

In fig. 10 is shown a method by which motion is given to the reed by reversed currents, and the line contact operated by a polarized relay placed in the same circuit as the magnet that operates the reed[11] r is the reed. M. is a spool of wire through the centre of which the reed r passes, this spool has a constant current from the battery LB passing through it and causes a constant magnetization of the reed. N. is the magnet through which reversed currents are transmitted by the contact points c. d. & e. f. g and h. are two batteries with opposite poles connected to the springs f. and c. the other ends being connected to a polarized relay K. and magnet N thence to reed.

When the reed comes in contact with say c. the battery g is thrown in circuit and this current tends to throw the reed towards f when it comes in contact with f the tendency is towards C. and 10[12] on reversed currents passing through K & N. the line contact is made by the lever and contact point of the polarized relay.

Fig. 11.

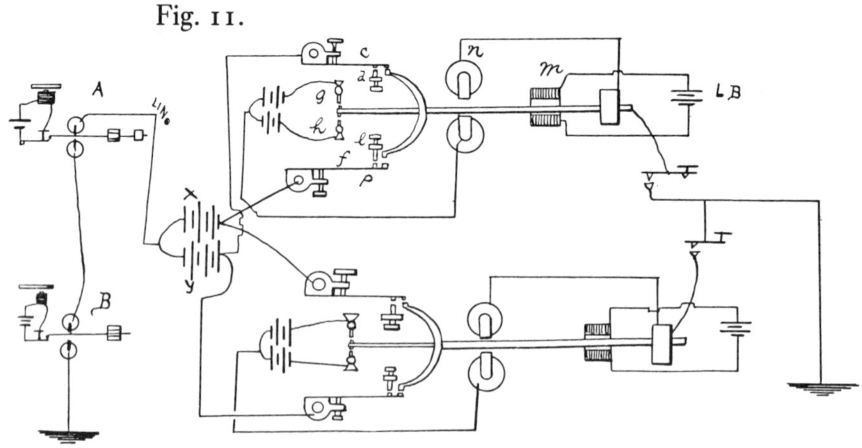

In figure 11 is shown a method of transmitting reversed currents over the main line circuit by polaried reeds though any reed operated by any of the methods described in this or former caveats may be used.

r. is the reed[13] operated in the same manner as that described in fig 10 except the position of the contacts are changed. c. and p are the main line contact points on both reeds.

X and Y are the two main line batteries with their opposite poles connected to the line: when the reed r comes in contact with c. one current is sent over the line and when it comes in contact with p an opposite current is sent over the line.

The other reed or as many more as are required act in the same manner. $K.K^1$. and $K.K^2$ are keys[14] by which the reeds may be prevented from making contact with the battery that is to say prevented from transmitting waves over the wire.

At the receiving station A. and B are two polarized reeds operated by the currents which come over the wire.

I shall now describe some of the receiving reeds: their methods of operation and devices for answering the repetition of signals into a local circuit containing a ~~m~~Morse sounder.

Fig. 12

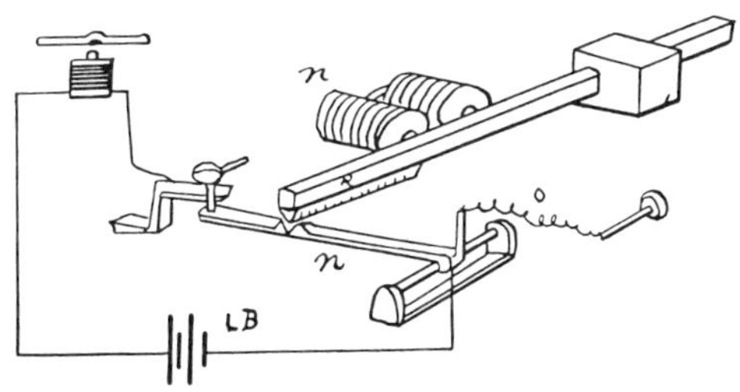

In figure 12 is shown a receiving reed R is the reed set in motion by the waves passing over the line[15] and through the electromagnet N upon the end of the reed R and underneath it is a V shaped extension one end of which lies between a V upon the local contact point N. p is another contact point[16] when the series of waves which set the reed in motion are not coming over the wire the V upon the reed is equidistant between the two teeth of the V upon the lever N. allowing the spring O to draw ~~in~~n in contact with p. and thereby closing the local circuit but when the waves come over the wire whose periodic time is the same as the reed it is set in motion and soon reaches a sufficient amplitude of vibration to throw the lever n away from p. on passing from right to left and from left to right thereby practically keeping the local circuit open and preventing the magnet of the sounder from attracting its lever: if now the key at the distant station be closed the waves are prevented from being transmitted the reed ceases to vibrate and the lever n comes permanently in contact with p. closing the local circuit and sounder.

The reason why the V shaped piece upon the reed is of some length is to allow of the reed being lengthened or shortened.

Fig. 13.

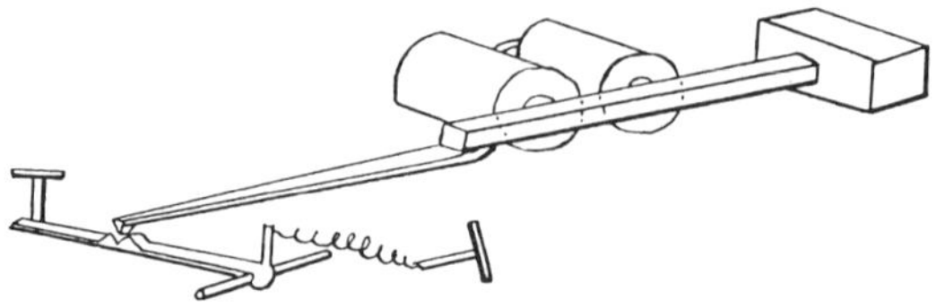

In figure 13 is shown a modification. X is an extension[17] of the reed proper made of very thin steel or aluminum so as to not materially lower the note of the reed and at the same time magnify the amplitude so as to give more movement to coperate the contact devices.

Fig. 14.

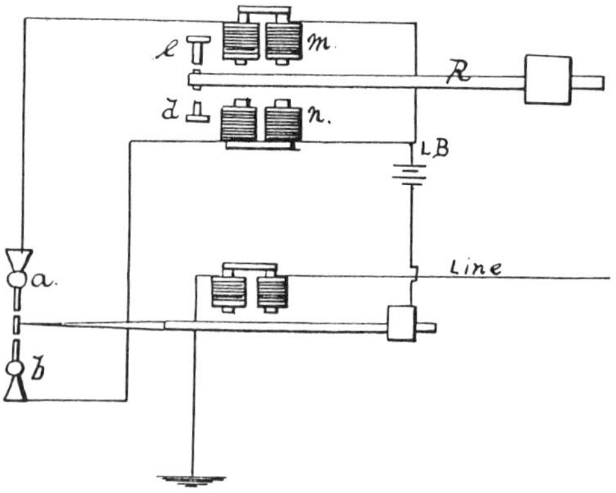

In figure 14 is shown a repeating reed and local sounding reed.

It has been found in practice that owing to interference with the reed by the peculiar effect of other series of waves not periodic with the reed that its movement is irregular and causes the sounder to work irregularly when its ~~is r~~ circuit is opened and closed by the reed. r is the main line reed with an aluminum extension[18] g upon the extreme end of which are two contact springs both centered between two contact points a and b.

R is a longer but heavier reed having the same vibrating time as the main line reed, and is placed between two powerful electromagnets M. and N.

LB is the local battery. when the reed r. is vibrated by the transmission of waves over the wire it practically speaking immediately reaches an amplitude of vibration sufficient to cause its contact springs to touch first a. then b. and so on the same number of times as it vibrates, these contacts first transmit currents through n. then m. of the local reed ~~circui~~causing it to vibrate in the same time but its vibration is very much more powerful and although the contact between g or the reed r and the points a and b. are irregular it makes scarcely any difference with the reed R The reed R knocking against the rigid points e and d give a powerful sound and if two series of waves are transmitted over the wire following each other closely they will cause R to give out the sound of a dot and two waves following each other not so closely will sound as a dash in the manner previously described in one of my caveats relating to this subject.[19]

But if desirable any of the contact points shown in this or previous caveats may be combined with R and cause a local circuit to be operated containing a ~~m~~Morse sounder.

Fig. 15.

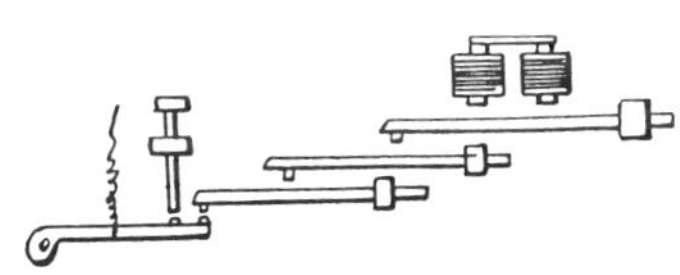

In figure 15 is shown three reeds all combined together and having the same vibrating time for the purpose of increasing the amplitude of vibration.

Fig. 16

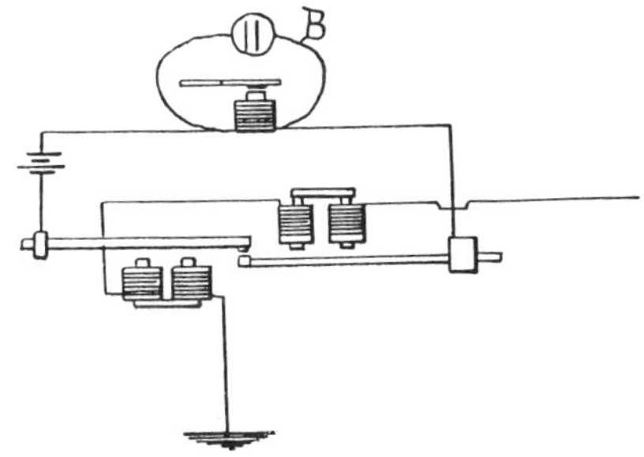

In fig. 16 is shown double reeds having the same vibrating time and operated by separate magnets the local circuit being closed at each vibrating, which although very slightly is sufficient to change the secondary battery B and cause the sounder to close, one of the objects of this is to correct the tendency of one reed being drawn towards its magnet upon the transmission of other sets of waves and throw the local contact points out of adjustment.[20]

Fig. 17.

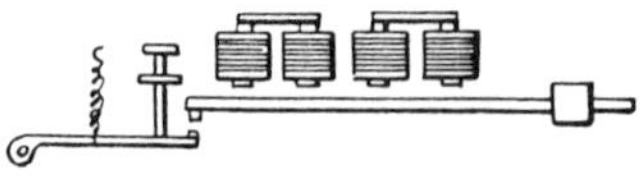

In figure 17 is shown a reed with two magnets both in the same circuit and acting together to give motion to the reed, thus causing it to be more sensitive to the short and feeble waves transmitted over the wire

Fig. 18

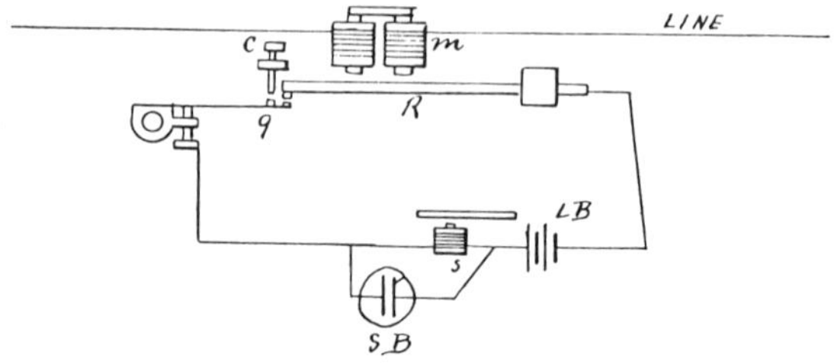

In figure 18 is shown a spring contact point arranged in the usual manner to open and close the local circuit containing the sounder s local battery LB.[21]

R is the reed which in a state of rest does not come in contact with the spring point g. but when vibrated makes infinitessimal contacts with g. closing the local circuit a great number of times per second by the use of the secondary battery S.B. consisting of two similar elements immersed in a conducting fluid, these infinitessimal contacts cause the battery to be charged and when R. is not in contact with g. while vibrating the secondary battery discharges through the sounder thus bridging over the period between the contacts.

Fig. 19

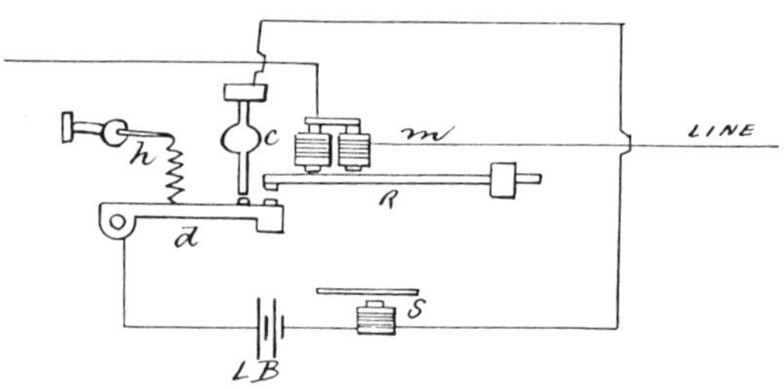

In figure 19. is shown a contact lever d which closes the local circuit upon the screw c when the reed R. is not vibrating but when it is vibrating the knock of R against d keeps it away from c. partly from its momentum and partly by the rapidity with which the reed R. hits it.

Fig. 20.

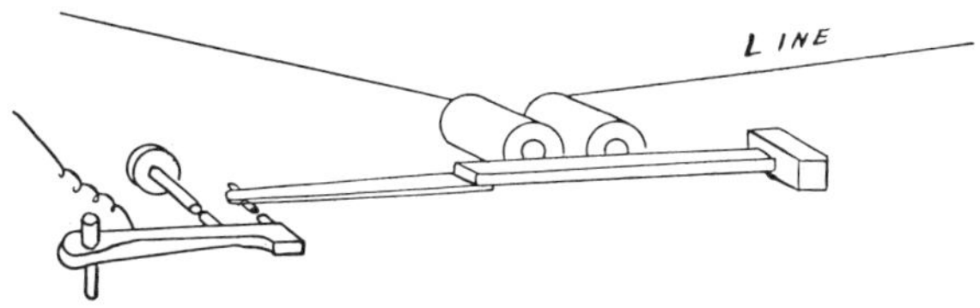

In fig. 20 is shown an extension of the reed R. by a thin aluminum plate[22] n.

Fig. 21

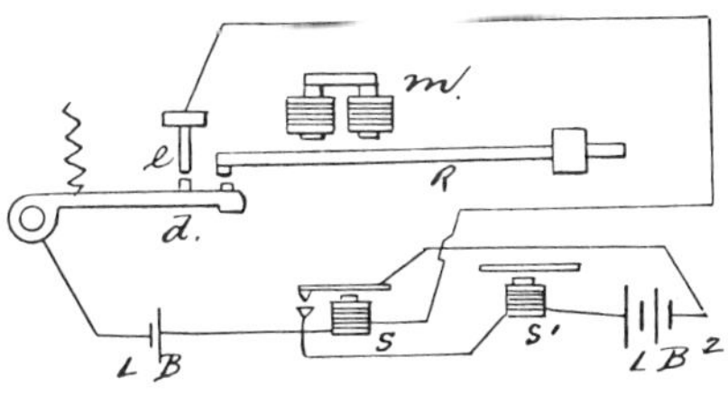

In fig. 21 is shown the same device as in fig 19 except that a repeating sounder S is interposed to intercept extra dots and cause a more even closing of the receiving sounder S′.

Fig. 22.

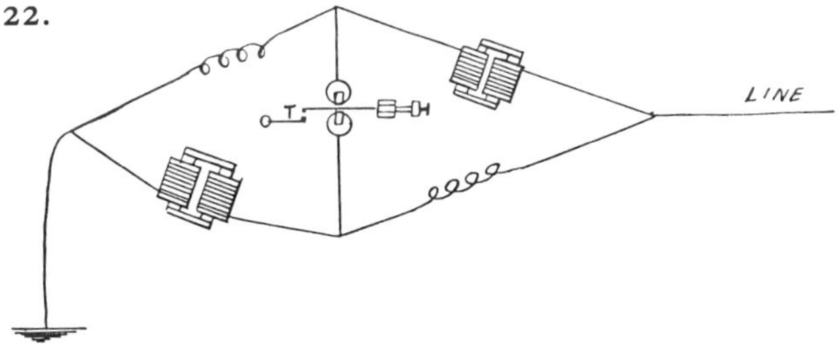

In fig. 22. is shown[23] a method of working a polarized reed by reversed currents derived from the discharge of powerful high resistance magnets. M. and m′ in the two sides of a wheatstone bridge the other two sides being formed of resistance coils sending no discharge currents.

The polarized reed R being placed in the bridge wire.

I shall probably claim

1st The method of transmitting short waves of current by interrupting the continuity of a balancing battery by a vibrating reed or reeds for the purpose set forth shown in fig. 1.

2nd The method shown in figs 2 & 3. of transmitting short waves of current by the short circuiting of an opposing battery in the manner and for the purpose set forth.

3. The method shown in fig. 5 of transmitting short waves of current ~~by~~ interrupted by a reed or reeds and[c] the short circuiting wire around a main battery in the main circuit for the purpose set forth.

4th The reed r. fig. 6. operated substantially as described and for the purposes set forth.

5. The methods shown in figs 7 & 8. of transmitting short waves of current independent or nearly so of the amplitude of vibration of the reed or reeds for the purpose set forth.

6th The self acting contact devices reed and magnets shown in fig 9½ for the purposes set forth.

7th The repeating reeds shown in fig. 9. & operated subsantially and for the purposes specified.

8th The polarized reed shown in fig. 10. operated substantially as described & for the purposes set forth.

9. Transmitting short waves of current by a movement of the tongue of a polarized relay operated by a polarized or unpolarized transmitting reed as shown in fig. 10 for the purpose set forth.

10th The transmission of positive and negative battery currents by contacts operated by reeds having different vibrating times for telegraphic purposes.

11th The method of transmitting and receiving reversed battery currents shown in figure 11 substantially & for the purposes set forth

12th The method shown in figures 12, 13, 15, 16, 18, 19, 20 & 21 of operating the local circuit and sounder at the receiving station.

13th In an accoustic telegraph the use of a local vibrating reed or body following the law of the pendulum when the same is set in motion by another reed placed in the main line circuit substantially as described and for the purposes set forth.

14th The method shown in fig. 17 for increasing the magnetic power over the reed by the employment of two magnets both in the main line circuit

15th The employment of the self induction of electro-magnets placed in a wheatstone balance for operating a polarized reed substantially as set forth.[d]

Signed by me this twenty sixth day of January A.D. 1876.

Thos. A. Edison

Witnesses: H: M: Haigh. E. W. Griffiths[24]

DS, DNA, RG-241, Edison Caveat 77. Petition and oath omitted. [a]Place taken from oath; date taken from text, form altered. [b]All figures are on six separate pages at the end of the caveat. [c]Interlined above. [d]Remaining text written in a different hand.

1. See headnote, p. 709.
2. L is not labeled.
3. For this form of transmitter see Doc. 706.
4. Concerning this and the circuit in figure 3, cf. Doc. 707.
5. In the drawing both are labeled **r′**.
6. Not shown.
7. "X" should be "f."
8. **g** is not labeled.
9. Text and drawing do not match.
10. Not labeled.
11. For this arrangement see Doc. 706 n. 1.
12. Should be "so."
13. Not labeled.
14. Not labeled.
15. Not shown.
16. Not labeled.
17. Not labeled.
18. Not labeled.
19. Doc. 708 (Edison Caveat 75).
20. Concerning such arrangements, cf. Doc. 709.
21. For this arrangement and that shown in figure 19 see Doc. 705.
22. Not labeled.
23. The figure lacks all labels.
24. Unidentified.

–716–

Notebook Entry: Multiple Telegraphy

[Newark, January 1876?][1]

Doubler[2]

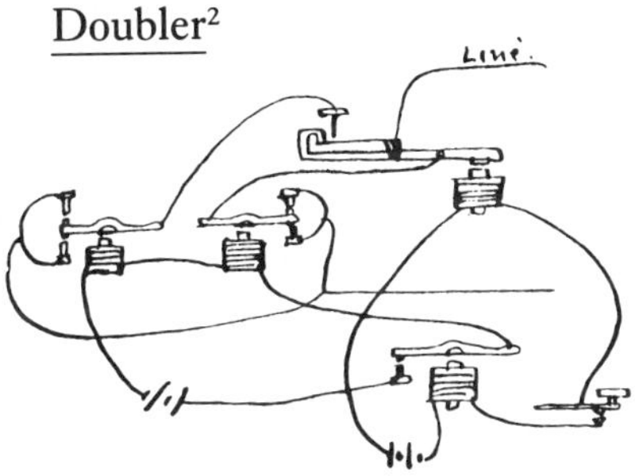

~~In~~ making both the back and front stroke on the down causes the operator to very easily get the "backstroke" I obviate this by transmitting a long opening of the circuit for closing and a short for opening[3] this causes the sounder at the receiving end to give a different sound on closing than on opening The following are various devices for converting two slight openings of a relay into regular dots and dashes on a sounder

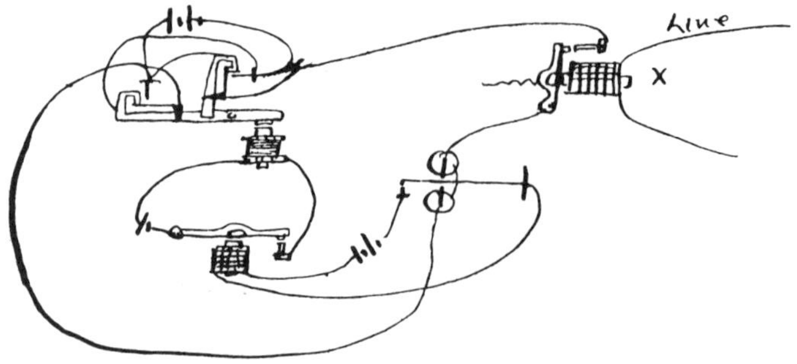

The relay X may work a back pointed sounder and the sounder work the polarized relay.

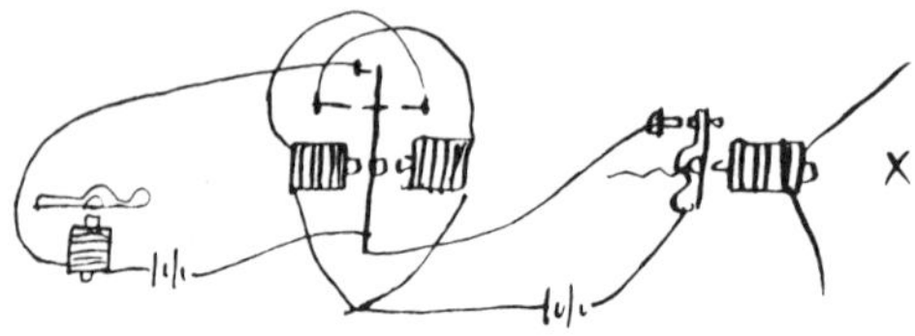

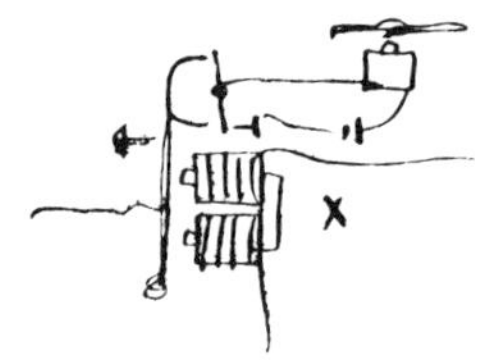

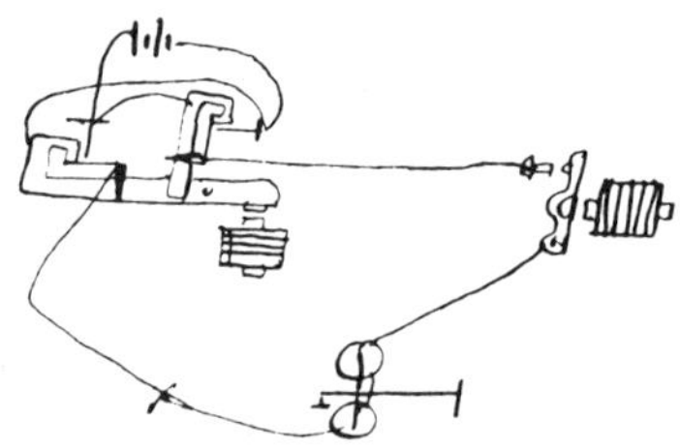

works very well

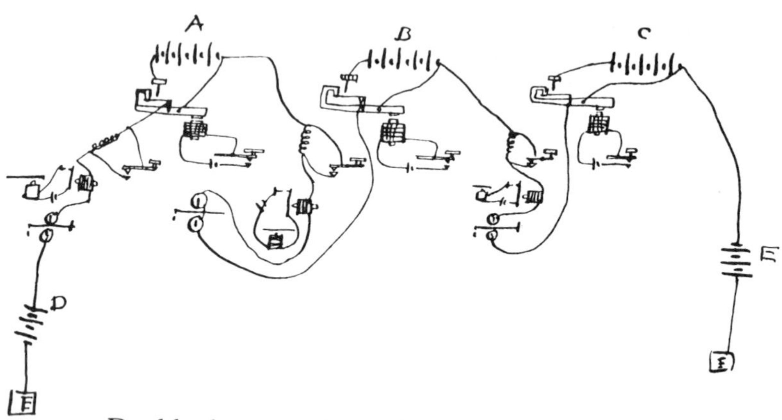

Doubler by reversals and increase and decrease[4] batteries A .B. C are at way stations[5] and have twice the strength as both D & E combined so that when D & E are in the line alone the current is say 100 in one direction but if either A B or C be inserted the current of D E is neutralized and 100 flows in the opposite direction

X, NjWOE, Lab., Cat. 994:59 (*TAEM* 3:222).

1. The preceding entry in this notebook is Doc. 705 (7 Jan. 1876), and Edison did not include these designs in his Caveat 72, executed on 12 January. However, he was actively concerned with this kind of system late in January (Doc. 714, esp. n. 3) and was at the same time engaged with similar problems in acoustic telegraphy.

2. This was Edison's name for a diplex/quadruplex design that differed significantly from those in his filed patent applications (e.g., Doc. 472). If successful, it would circumvent any such patents, and Edison had been interested in it for some time (see Doc. 714 n. 2).

This document shows variations on the basic doubler design. That design, shown in the accompanying December 1875 sketch by Batchelor (Cat. 1307:56, Batchelor [*TAEM* 90:641]), created one type of signal with pairs of very brief, total breaks in the circuit instead of polarity reversals as in Edison's standard quadruplex; the other type of signal was still made by alternating between strong and weak current. The equipment in the illustration, while including two keys and receiving relays at each end, is not actually arranged to work both ways simultaneously; standard duplexing techniques would have to be added for quadruplex operation. Another drawing of a doubler circuit—apparently a tracing of an Edison sketch—is in NS-Undated-005, Lab. (*TAEM* 8:381).

Charles Batchelor's drawing of the doubler circuit.

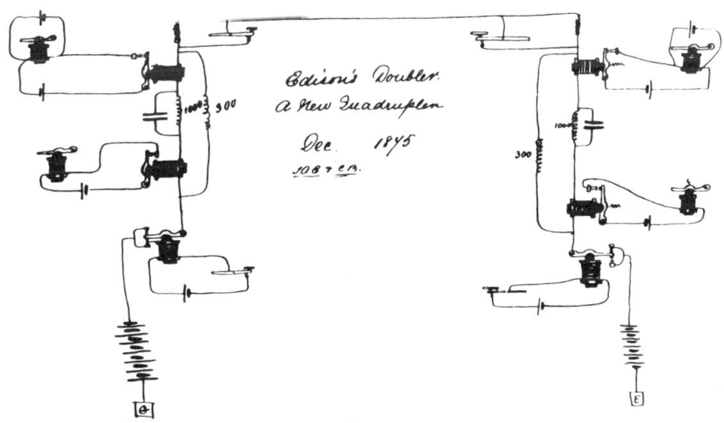

3. On a standard Morse sounder, each received signal consisted of two clicks—for a dot, two clicks close together; for a dash, two clicks farther apart—as the sending operator closed and opened the circuit and the sounder lever moved down and then up, each time hitting a stop and making a click. "Getting the backstroke" meant mistaking one click for the other and falling out of sync with the sending operator (see Edison's discussion in Prescott 1879, 221). The doubler used breaks instead of clicks; by making the closing and opening breaks different lengths, Edison hoped to avoid the backstroke problem.

4. Since this design uses polarity reversals for one type of signal rather than just brief circuit interruptions, it is more like Edison's basic quadruplex approach than the other doubler designs. However, any reversal must at least momentarily reduce the current to zero, and hence can produce the kind of action used in the other designs.

5. It is plausible that provision for way stations was an important feature of the doubler designs. Western Union had an interest in equipment that was not limited to terminal stations, and Edison's proposed demonstration of the system for Orton featured the use of four way stations on one line. See Doc. 714 n. 3.

-717-

Lemuel Serrell to Commissioner of Patents[1]

New York, Feb. 3 187~~5~~6.[a]

Sir

Mr Thos A Edison requests me to say that he believes Mr Gerritt Smith is seeking a patent upon a device for Quadruplex Telegraphs shown in Edisons English patent No 384, Feb. 2, 1875.[2] I send one of my copies of the drawings of that patent and direct attention to Fig. 20.[3] I have not received the printed specification, but presume it has reached the Patent Office.[4]

If desired the French patent showing the same will be sent for reference Respectfully yours

Lemuel W. Serrell

ALS, MdSuFR, RG-241, Pat. App. 185,589. [a]"New York," and "1875" preprinted.

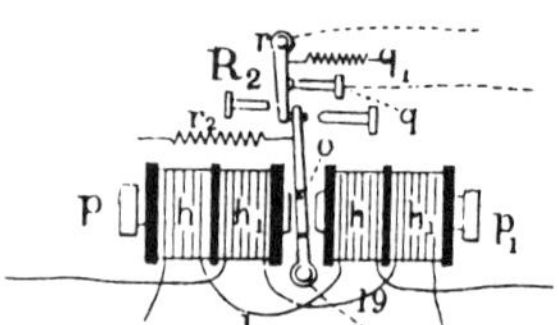

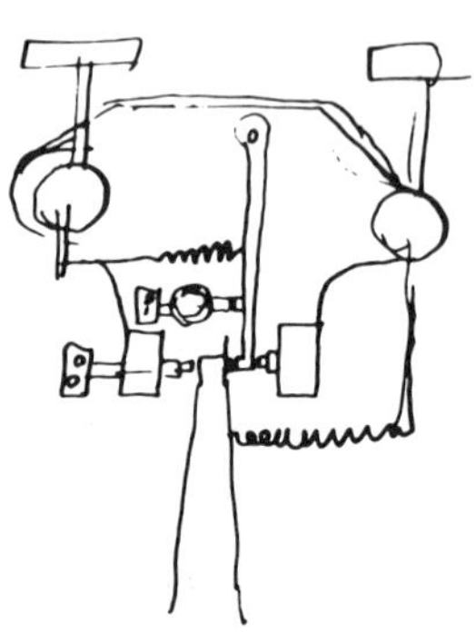

Gerritt Smith's patent drawing and Edison's undated sketch of a contested quadruplex device.

1. Rodolphus (or Robert?) Holland Duell succeeded John Thacher as Commissioner in October 1875, at the same time that a new Secretary of the Interior (Zachariah Chandler) replaced Columbus Delano, who was forced out in the course of corruption scandals. *NCAB* 12:285; *WWW-HV*, s.v. "Duell, Robert Holland"; McFeely 1981, 430–32; Nevins 1957, 769–81.

2. The dispute concerned Smith's application, executed on 3 December 1875, which later issued as U.S. Patent 185,589 after significant amendment, and Edison's designs stemming from Doc. 512. In the margins of Smith's application the patent office examiner noted several other possible conflicts between the application and prior caveats, patents, and pending applications by Edison and other inventors (see also H. C. Townsend and Zenas Wilber to Gerritt Smith, 14 Jan. 1876, Pat. App. 185,589). The first of the accompanying illustrations is from figure 1 in Smith's printed patent; the second is an Edison design (NS-Undated-001, Lab. [*TAEM* 8:46]). See also Edison's Brit. Pat. 384 (1875).

Smith assigned to George Prescott a half interest in this and several subsequent applications; he also employed Franklin Pope as his patent attorney instead of Serrell, whom he had used for earlier applications. Prescott charged in March that Edison's protest indicated malfeasance by Wilber; he claimed Wilber had improperly revealed the contents of Smith's application to Edison (Prescott to Commissioner of Patents, 20 Mar. 1876, Letters Received 1872–1882, RG-241, DNA). This was related to other charges by Prescott against Wilber regarding Edison, which were also connected with the Quadruplex Case (see Quad. 72.3–5 passim [*TAEM* 9:38–91]; cf. Doc. 727). However, Edison could easily have had other sources for this information; he had friends at Western Union, and possibly another in H. C. Townsend, the first assistant patent examiner (Townsend to TAE, 16 and 28 May 1878; TAE to Townsend, 26 Feb. 1878; all DF [*TAEM* 15:285, 715; 18:1000).

3. Serrell was asked for copies of all the British patent drawings and apparently furnished them. The examiners found that other parts of the patent also were related to Smith's claims. H. C. Townsend and Zenas Wilber to TAE, 16 Feb. 1876; Townsend and Wilber to Gerritt Smith, 8 Mar. 1876; both in Smith Pat. App. 185,589.

The disputed relay as it appears in Edison's British Patent 384 (1875).

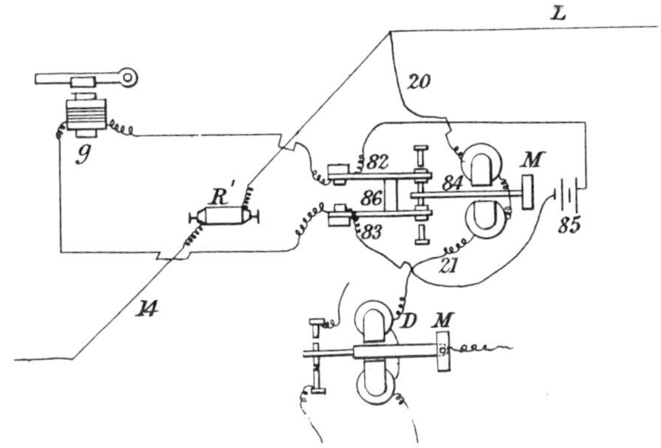

4. The printed specification was not yet available in the United States, and this led to claims by Smith, Prescott, and Pope that the examiners were not following proper procedures. Smith to Commissioner of Patents, 28 Mar. 1876, Pat. App. 185,589.

-718-

To the Editor of the Scientific American

Newark, N.J.[a] February 5, 1876

MR. EDISON'S NEW FORCE.

To the Editor of the Scientific American:

I notice in your issue of January 29, 1876, some experiments conducted by a gentleman signing himself "Electron," who attempts to prove that the phenomenon observed by me, and which I have called etheric force, is due to the "extra" current from an electromagnet.

There are several sources of error, I think, in his experiment, among which may be mentioned bad insulation of battery or leading wires and binding posts, and the close proximity of the galvanometer to the vibrating apparatus.

Owing to the extreme delicacy of a mirror galvanometer, the sources of error are extremely numerous, and it requires long practice and careful manipulation to eradicate them. If "Electron" will use large gutta percha wires, take the reading of his galvanometer in another room, suspend his battery by insulating cords, use a large hard rubber base for his vibrator, and keep a sharp look out for possible sources of error, he will obtain a brilliant etheric spark right through his galvanometer without a tremble of the spot of light.[1]

In reply to your correspondent J. P. H., who sees nothing inexplicable in obtaining a spark from an uninsulated wire laid for a long distance upon wet earth and connected to a highly insulated source of power, I will state that his telegraph experiment is not a similar one. If he had disconnected one pole of his battery from all connection with anything except air, and placed his battery upon an insulated stand, he would hardly have succeeded in working with a relay or any other electric instrument under the conditions he mentions. With this, there is no chance for circuit, and it is the same with the source of etheric force.[2]

THOMAS A. EDISON.

PD, *Sci. Am.* 34 (1876): 81. [a]Place not that of publication.

1. In a mirror galvanometer, a small mirror reflected a spot of light onto a scale. Any current caused the mirror to turn, moving the spot of light.

2. The letters to which Edison is replying were among several letters and articles that appeared in the *Scientific American* during January and February 1876 in response to that journal's many articles and editorials discussing Edison's etheric-force experiments. See also William Sawyer's letters (15 Jan. and 19 Feb. 1876); George Beard's article (22 Jan. 1876); Edwin Houston's article (29 Jan. 1876 *Supplement*) and Edison's answer (Doc. 726); P. H. Vander Weyde's article (5 Feb. 1876); letters by M. B., A. S. G., and C. H. A. (12 Feb. 1876); C. H. R.'s letter (26 Feb. 1876); and S. K. G.'s letter (4 Mar. 1876).

–719–

Technical Note: Acoustic Telegraphy

[Newark,] ⟨February 6th 1876⟩[a]

Wurth =

I find that I can make the new acoustic receiver work all right if I do not tighten the screws. It does not want to be so rigid make it thus

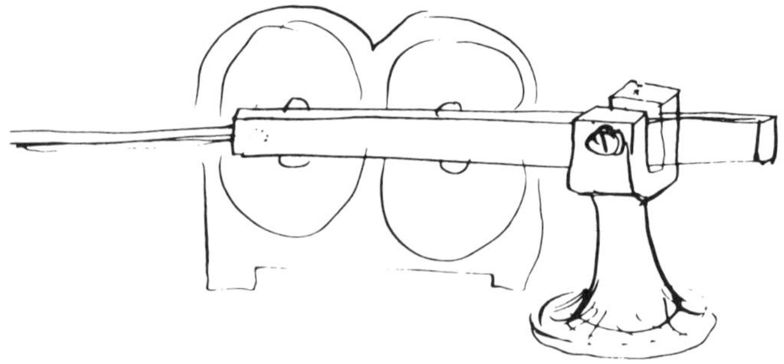

It wants to be a pillar that will give a little I think the same Size pillar as is used on the magnet for holding the adjusting screw = only fix the one down stairs with the pillar =[1]

AX, NjWOE, Lab., Vol. 10:16 (*TAEM* 3:831). [a]Dated (with signature) by Charles Wurth.

1. In another note for Charles Wurth on this date Edison called for an altered transmitter: "Also make me several boxwood peices—so as to make the transmitting reeds less rigid." Vol. 15:72, Lab. (*TAEM* 4:394).

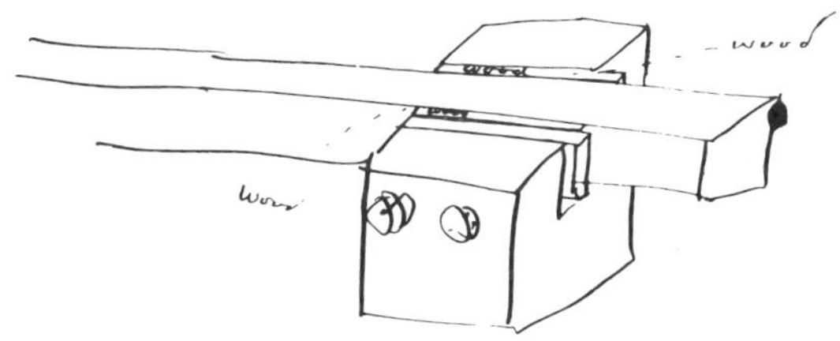

Edison's sketch of an acoustic transmitter modified with boxwood bushings.

-720-

Notebook Entry: Autographic Printing

[Newark,] February 7th 1876.

24[1a] Autographic Pen

We made a great improvement in the pen by giving the cam plenty of side shake and putting a guide above the cam A so that it could only move the needle up and down and not at all sideways.

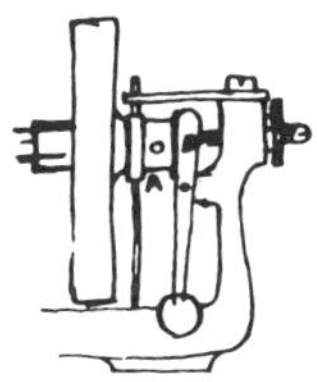

The spring B is also improved by being made stiffer at the point so that it hardly springs at all between the platina point and the end of spring

Chas Batchelor

X, NjWOE, Batchelor, Cat. 1317:26 (*TAEM* 90:670). Written by Batchelor. [a]"26" written across entry in pencil.

1. Batchelor numbered the entries in this notebook.

ELECTRIC-PEN COPYING SYSTEM Doc. 721

Edison and his laboratory staff had developed the electric-pen copying system during the summer and fall of 1875, and man-

The electric-pen system in use.

ufacturing and sale of the instrument had begun in September. They changed the basic design of the system little during the winter, but they made one significant modification to the pen. On 7 February (Doc. 720) they placed a guide above the cam to prevent the needle's moving sideways.[1] The following day several of the laboratory machinists began altering production models, including the one shown in Doc. 721, for Gilliland & Company, which was manufacturing the copying system at the Ward Street shop.[2]

1. The patent application Edison executed on 7 March 1876 also included this design (U.S. Pat. 180,857).

2. Cat. 1214:146, Accts. (*TAEM* 21:634). No 1876 order books or production records exist for the pen, batteries, or press aside from the time records of the laboratory machinists' work for Gilliland & Co.

–721–

[Newark, February 7, 1876?]

Production Model: Autographic Printing[1]

M (press: 38 cm × 31 cm × 8 cm; batteries: 35 cm × 19 cm × 33 cm; pen holder: 9 cm dia. × 9 cm; pen: 14 cm × 5 cm [flywheel dia.]), MiDbEI(H), Acc. 32.735.1, 00.1382.707, 00.1382.706.2, 00.1382.706.1.

1. See headnote above.

–722–

From Uriah Painter

W[ashington?] 2–8–76

Dr Edson

Yours rec'd. OK— I am going to have a show Satdy night & will then return your pictures—[1] Do you want any[a] of mine for your show!

I am sorry you are "not" mine in "A & P" as you say you are not— I hope you are not any mans man, but I hear the gloating of your bitterest enemy that he has almost got you in his toils![2] I see & feel his heel grinding those who he thinks have befriended as you against him— His hope is to strangle in its infancy the only rival he has ever[a] had, he expects to use you to do it or rather to aid him in so doing & then you have no one to buy your inventions ~~i~~or improvements but himself— It is possible he may succed— I have worked a long while to try & convert your work into money—but it is use~~e~~less to continue giving advice that is never taken except with some amendment which impairs or destroys its value— You & Reiff ought to have realized[b] long ago— It is not my fault it has not been done, & instead of drawing nearer to the desired end you are drifting farther & farther away— It is all very nice for you both to sacrafice the present, for the unknown future, to grope in the dark instead of making sure of what is within your reach, but its infernaly monotonous— Col Reiff is too good a fellow to be rolling barrelles up an unknown mountain to come down some day like Rip Van Winkle & find his creditors dead, his friends scattered, & the best part of his life sworse than slept away— But such is life—

I never saw my letter to American printed—[3] How fearfully Harper got sold on deadbeat Sawyers confidence game!—[4] Come down & I'll give you a show as is a show! I have a full set of Yosemite & across the continent— Yr

U H Painter

ALS, NjWOE, DF (*TAEM* 13:1118). [a]Interlined above. [b]Underlined twice.

1. Painter had apparently borrowed sixty-four stereopticon slides from Edison in October 1875. The list of slides, dated 22 October 1875, includes many religious scenes. 75-001, DF (*TAEM* 13:206–8); see Doc. 560 n. 2.

2. Painter is probably referring to Western Union president William Orton.

3. Painter is probably referring to the newspaper the *Philadelphia North American.*

4. This reference is unclear. No contemporaneous article concerning William E. Sawyer's telegraph has been found in either *Harper's Weekly* or *Harper's Monthly.*

–723–

Charles Batchelor to Thomas Batchelor

[Newark,] Feb. 8th [1876]

Dear Tom,

Your letter received. I have delayed answering because I heard Mr Gloyn's[1] friend[2] has accepted the proposition. I have been waiting to see if it was true or not and still do not know I have sold my interest in the press in the United States & Canada Still retaining my foreign interest & managing the whole thing until they can find someone else to intrust it to. have so many things to attend to that to continue to run that would take up more of my time than I could spare. I have appointed agents for China & Japan & sent them their first shipment.[3] I am now working a point to introduce them in South America. Your sample sent was written very good & you evidently handle the pen well but the printing is not such a success your ink wanted stirring well & a little more rolling with less ink at first will make better copy I send you some samples of copies taken by parties using the pen only a short time

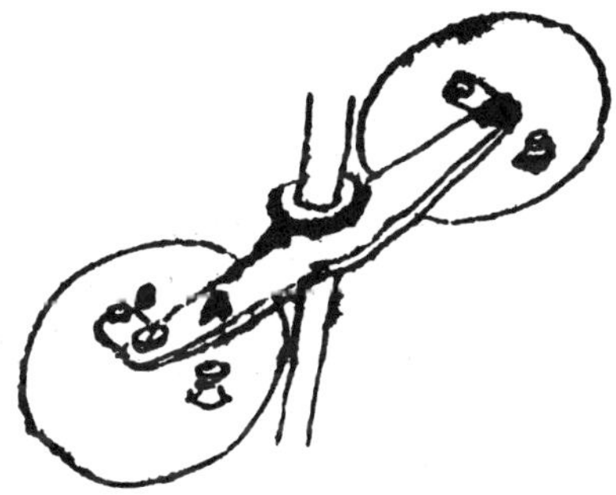

You are evidently using only one cell of your battery. perhaps the Japan[4] is not scraped off the under side of bar A where it connects with link B or it may be that the brass piece on top of the carbon does not make good connection You have extra carbons put them in & see if they work I send a few samples. We have selected a place to build a laboratory [-]at Menlo-Park on the Pennsłylvania RR 24 miles from New York and 15 from here the building is going up fast I shall buy a piece of land and build a house too I think during summer

The patents for Great Britain will be all secured by ~~the~~ us in shape of the 'Provisional Protection' & first cost All other payments must be paid by the parties holding the patents at the time they are due.[5]

Rosa had a little girl[6] Feb. 5 at 10 PM & as soon as we are all straight you can count on us for 'Cabinet pictures' in exchange In regard to your coming out of engagement in Spring I will say my prospects are good at present & if I should be able to throw anything in your way You may be sure Ill do it ~~Respectfu~~ Yours as ever

Charley

ALS (letterbook copy), NjWOE, Batchelor, Cat. 1238:42 (*TAEM* 93:55).

1. John Fox Gloyn had acquired the electric-pen agency for Ontario. Agreement of 11 Nov. 1875, Miller (*TAEM* 28:994).

2. George Walter, secretary and one of the principal promoters of the National Telegraph Co., signed an agreement with Edison on 28 January 1876 to sell the electric-pen patent in Great Britain. Three days later he informed Edison that he had negotiated the sale. Copies of the letter and agreement are in Lefferts; Walter's letter without the agreement is in DF (*TAEM* 13:965).

3. Batchelor appointed William H. Fogg & Co., a New York City tea

company, as pen agents for China and Japan. The agreement stipulated that sales should average at least 200 per year. H. Wilson 1876, 434; Batchelor to William H. Fogg & Co., 28 Jan. 1876, Lbk. 3:22 (*TAEM* 28:359).

4. Enamel.

5. Under British patent law, an applicant could obtain a patent by filing either a very general "provisional specification" or a more detailed "complete specification" of an invention. A provisional specification provided patent protection until the submission of the complete specification, which had to be done within six months after the granting of the patent. Davenport 1979, 30.

6. This child apparently died in infancy.

–724–

Patent Assignment to Robert Gilliland

[Newark,] February 8, 1876[a]

Whereas I, Thomas A. Edison of Newark in the State of New Jersey have invented certain apparatus for Autographic Printing for which I have filed Caveats in the U.S. Patent Office one filed Sep 27, 1875 and two others on Nov 29, 1875.[1]

And Whereas Robert Gilliland[2] of Hudson Michigan is desirous of acquiring a certain interest in and to the said inventions and any Letters Patent that may be granted for the same.

Now this Indenture Witnesseth that for and in consideration of the sum of ______ dollar to me in hand paid, the receipt whereof is hereby acknowledged, I have assigned and sold, and by these presents do transfer, convey and set over unto the said Robert Gilliland th[re]e[3] undivided tenths of the entire right, title and interest of every character, in and to the said invention and any Letters Patent that may be granted for the same: it being understood that the profits resulting from the said invention and Letters Patent, whether from joint or separate efforts are to be divided between myself and the said Gilliland or our legal representatives in proportion to the interest held in the Letters Patent and that neither party shall dispose of or encumber his interest in the Letters Patent without first offering it to the other party at the same price as can be obtained from third parties.

The said rights are to be held and enjoyed by the said Robert Gilliland for his own use and behoof and for the use and behoof of his legal representatives, to the full end and term for which said Letters Patent may be granted as fully and entirely as the same would have been held and enjoyed by me had this assignment and sale not been made.

In Testimony Whereof, I have hereunto set my hand and

affixed my seal this Eighth day of February in the year one thousand eight hundred and seventy-six.

Thos A. Edison.[b]

Sealed and delivered in the presence of Chas Batchelor[c]

D (transcript), MdSuFR, Libers Pat. D-20:236. [a]Date taken from text, form altered. [b]"Rec'd Feb 25, 1876 F.C.T." written in right margin. [c]Sentence written on three lines enclosed by brace.

1. Docs. 633 and 638; Edison Caveat 70.

2. Robert Gilliland was Ezra Gilliland's father.

3. "Three" appears in a copy of Edison patent assignments in Digest of Patent Applications, p. 32, WU Coll.

–725–

Technical Note: Acoustic Telegraphy

[Newark,] February 8th 1876

Wurth

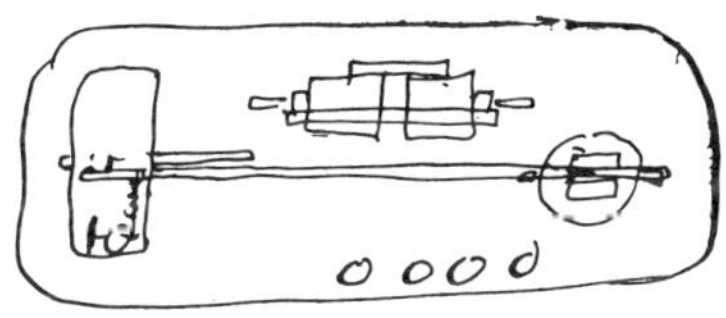

Fix the transmitters so I can put paper in all the same as you have done on the transmitting reed having the highest note.[1] It is a success

Also put the ~~th~~ two other Receiving reeds together, making the pillar adjustable in a long slot

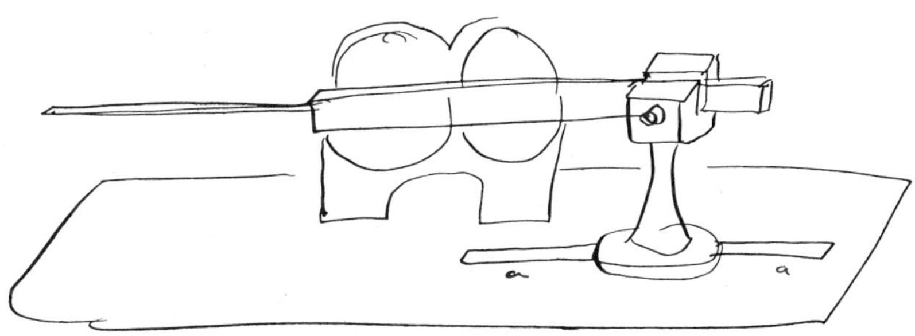

also make the magnets adjustable also the local points. I have marked in ink on the receiver downstairs about the length of the slots for magnet pillar & local points. =

Edison

You better put an extension of the hard rubber on so that the adjusting spring[a] pillar will be on it and all be moved together = Put the binding posts in front of the magnets on this side of the reed, and ~~solder~~ stiffen up the thin spring extension of the reed by adding another piece E

AXS, NjWOE, Lab., Vol. 10:17 (*TAEM* 3:832). [a]Interlined above.

1. See Doc. 719 n. 1; evidently Edison tried paper after using the boxwood called for there.

–726–

To the Editor of the Scientific American

Newark, N.J.[a] February 12, 1876

MR. EDISON'S NEW FORCE.

To the Editor of the Scientific American:

I notice in your SUPPLEMENT No. 5 an article upon the "phenomenon of induction," by Professor Houston of Philadelphia, copied from the *Journal of the Franklin Institute,*[1] in which he claims that etheric force is nothing but inductive electricity, and that he observed the same phenomenon in 1871. He attributes my failure, to obtain indications with the test instruments used, to the fact that the positive and negative currents from the vibrator followed each other with great rapidity, and thus prevented the instruments from responding. In reply, allow me to state the gentleman is entirely wrong in his conclusions, and that he cannot be familiar with the extra currents of low resistance magnets: otherwise he would have known that, upon connecting the battery, the extra current is provided with a circuit in which it may pass, consisting of the battery, connecting wires, and electromagnet. Under the conditions by which I obtain etheric force, no spark should theoretically be obtained, even if it were due to extra current upon closing the circuit; and in all my experiments none has ever been obtained. Neither is the brilliancy of the spark reduced by replacing the iron core of the electromagnet (used in one form of experiment) with a copper one, which should be the case were the spark due to extra current.

In regard to the Professor's claim of priority, I have on every occasion stated that the spark has been observed by electricians for many years, and attributed by them to inductive electricity; and all that I can lay claim to is that perhaps (if that is not too strong a word) I was the first to discover that it was not due to electricity.

In conclusion, I suggest that, as I have freely laid myself open to criticism by presuming to believe in the capacity of Nature to supply a new form of energy, which presumption rests upon experiment, it is but fair that my critics should also back up their assertions by experiment, and give me an equal chance as a critic.

T. A. EDISON.

PD, *Sci. Am.* 34 (1876): 101. [a]Place not that of publication.

1. Houston's article, which appeared on pages 77–78 of the 29 January 1876 supplement, was reprinted from the January 1876 issue of the *Journal of the Franklin Institute* (101:59–63). Houston and his partner, Elihu Thompson, were particularly critical of Edison's claims regarding the discovery of a new force and responded to Edison's letter by conducting further experiments and publishing their results in the April issue of the *Journal* (101:270–74). These were also reprinted in the *Scientific American Supplement* on 20 May (1:326). For the controversy between Edison and Houston and Thompson, see Carlson 1984, 106–31; and Bryant 1987.

–727–

To William Orton

[Newark, c. February 15, 1876[1]]

I have recd from the patent ofs notfication that another party[2] has just made application for an Acoustic patent which interferes with claims in my Caveats 71 and 74. I am allowed 3 months to complete my patent from date of this notification[a]

Prescott is going to get out a mandamus to compel me to complete my Quadruplex patents.[3] Also to compel me to make application for new Quadruplex patents shewn in Caveats;[4] He says if there I any hell I'll go there or words to that effect = abuses me to my face and then asks me if I wouldnt be so kind as to go ahead I[5] do all he wants; says he wants his money from the Co that the Income is 15 000 yearly; = and many other things =[6]

At the same time he wouldnt lend me a polarized relay out of his case; said go to Phelps =

AL, NjWU, WUTAE. [a]Followed by centered horizontal line.

1. Orton docketed this as "Feb. 1876." The Patent Office notification mentioned by Edison dates from 5 February; another notice, of 21 February, probably postdates this letter. Edison Caveats 71 and 74.

2. Probably Elisha Gray.

3. That is, to follow through on the applications in which Prescott claimed an interest, rather than to withdraw, abandon, or replace them. Edison did indeed soon try to withdraw them. See Doc. 734.

4. See headnote, p. 347.

5. Probably should be "&."

6. See Docs. 521 and 535.

–728–

To William Orton

[Newark,] ⟨Feb 19/76⟩[a]

I have at last got the animal tamed[1] a Seance will be held monday or tuesday[2] whereat the professor of "Acousticity" will exhibit the latest theft ~~in~~ from the German Telegraph books.[3]

AL, NjWOE, DF (*TAEM* 13:1255). [a]Dated by Orton.

1. The previous day Edison had sought from Orton a night pass for the operating room in the Western Union building to make a test on one of the regular lines (TAE to Orton, DF [*TAEM* 13:1255]). It is not known to what acoustic-telegraph design Edison refers.

2. That is, 21 or 22 February.

3. Edison was still being vilified in print by James Ashley, who described him in the *Telegrapher* as the "professor of duplicity and quadruplicity." In July 1874, in reponse to the initial glowing report of the invention of the quadruplex, Ashley had proclaimed that it was no invention—Edison had merely stolen from much earlier, published Austrian and German experiments in telegraphy that had produced nothing practical. See Doc. 581 n. 1; and headnote, p. 305.

–729–

Josiah Reiff to Thomas Eckert

N.Y., Feby. 25—76.

Confidential

GENL.

I hear you desire Mr. Edison to see Mr. Gould with view to a settlement of his personal interest in Automatic Telegraphy, based upon the understanding of Decr. 30—74.[1]

The individual interests of all parties (except Harrington & Parsons, assigned to Mr. Gould) have been transferred to the American Automatic Telegraph Co. of which I am Presdt.[2] This of course includes Mr. Edison. A settlement however can be readily accomplished through a third party whose action Mr. Edison and I will approve. Respectfully,

JOSIAH C. REIFF.

Approved, THOS. A. EDISON.

PL (transcript), NjBaFAR, *Harrington v. A&P,* Box 17B, Defendant's Exhibits 12, 1:308.

1. Doc. 522.

2. See Doc. 676.

–730–

From Henry Pettit

Philadelphia, Feb 26th 1876[a]

Dear Sir:

I have read letter from Wm Orton Esq Prest of Western Union Telegraph Co saying that you were willing to give up the space alloted to you, so that it might be transferred to them, provided they should show the instruments patented by you but of their manufacture.[1] Please advise me if this is the case. or if you have some instruments of your own that you wish to exhibit and are willing that your exhibit should be in connection with theirs— we can give you a permit for space in the area granted to the W.U Co—the latter Co to fix the space, or it is to be agreed upon between you.

Please advise me at once if you want to make this arrangement with them and return the Permit—[2] Yours—

Henry Pettit Chf of Bureau

ALS, NjWOE, DF (*TAEM* 13:1007). Letterhead of United States Centennial Commission. [a]"Philadelphia," and "187" preprinted.

1. In this letter Orton noted that Edison had been allotted forty-two more square feet than Western Union. Orton to Pettit, 21 Feb. 1876, LBO 16:265–69.

2. See Doc. 732.

–731–

Stock Offering: The Edison Pen and Press Co.

New York, March 7, 1876.

To whom it may concern,

The undersigned offers for sale to capitalists, through Mr Chas C. Yeaton, Manager of 27 Union Square,[1] his whole undivided interest for the United States of America, of an invention, with improvements already made or hereafter to be made known as "Edison's Electrical Pen and Duplicating Press," upon which invention Letters Patent have been applied for in said United States of America.[2] It being his purpose and understanding with Mr Yeaton that a joint stock company shall be formed to be known as the "Edison Pen & Press Co"[3] with a full paid capital of one hundred and fifty thousand dollars say ($150,000) of which $25,000 shall be the cash working capital, and the remainder to be paid the undersigned for said patent, when granted, executed and delivered as is more particularly set forth in the prospectus and subscription papers as follows viz:—

One hundred thousand dollars in cash and twenty five thousand dollars in stock in said proposed Company.

Pending the organization of the Company named, and until the execution and conveyance of the title of said patent is made by me to such Company, Mr Yeaton has my authority to represent me and my said patent (applied for) in all negotiations relative to the disposition of the invention as aforesaid and no one else has or shall have any right in the premises.[4]

Thos A. Edison[a]

Prospectus.[b]

Edison's Electrical Pen & Duplicating Press.[c]

Whereas Thomas Alva Edison, of Newark N.J. has invented a useful and practical devise which he calls the "Edison Electrical Pen and Duplicating Press" upon which he has filed in the United States Patent Office, at Washington D.C. two Caveats dated respectively, September 27th 1875, and November 29th 1875, and has recently filed his petition, according to Law, for a Patent upon said invention.

And Whereas, the object of said invention is to write out and re-duplicate, to an almost unlimited extent, Letters, Circulars, Price Lists, Market Quotations, Pamphlets, Catalogues, Lawyers' Briefs, Contracts, Manifests, Labels, Billheads, Envelopes, Maps, Architectural and Mechanical Drawings, Bills of Fare, Music, Insurance Policies, Cypher Books, Press Reports, Financial Exhibits, Tracings and Artistic Work.

And Whereas, as can be shown, this Invention is most useful and practical in cases above set forth; and promises to become as indispensable to such interests as the Sewing Machine is to the household. This invention is already in great demand is a monopoly and cannot fail to yeild large profits to an enterprising Company, managed by persons of good executive ability and commercial experience.

And Whereas the said Edison proposes to sell all his right, title and interest in his invention upon the issueing and procurement of his Letters Patent, so far as related to or embraces the United States of America, together with the improvements as made or hereafter to be made, for the sum of one hundred and twenty five thousand dollars as hereinafter set forth.

And Whereas it is now proposed to organize a joint Stock Company under the "Laws of the State of New York Limited" for the purpose of manufacturing and selling said Electrical Pens and Duplicating Presses, and transacting such other business as relates to the introduction or development of all which said Patent shall cover. The title of the Company to be

"The Edison Pen and Press Co" with a capital of $150,000, Divided into 150 shares of $1,000 each, Twenty five thousand of the above ($150,000) to be the Cash Working Capital of the Company, and the remainder, one hundred and twenty five thousand dollars to be paid to the Patentee for said Patent, as follows viz:—one hundred thousand dollars in Cash and twenty five thousand dollars in said proposed Company's stock.

It is therefore proposed to open subscriptions at once and invite a respectable class of capitalists to take stock in the purchase of said Patent as herein described and in the furnishing the Cash Working Capital as aforesaid; and when the full amount of stock shall have been subscribed, and upon the issueing of the Patent to the Patentee, and the same shall have been passed upon as valid, by a Committee of three, appointed by the shareholders, then the subscribers shall pay into the hands of the Manufacturers and Merchants Bank, No 561 Broadway, in the City of New York, in escrow their respective amounts therewith subscribed, and the same to be held by said Bank until the whole amount shall have been paid in and the title to said Patent assigned over through the hands of the President of said Bank, for the interest and benefit of its purchasers; then the said Bank through its President shall make over to the Treasurer of said proposed Company, the Cash Working Capital and the assignment of said Patent; also the amount of Cash and Stock to the Patentee or his order for the purchase of the said Patent as herein set forth.

Subscriptions.

Now therefore for and in consideration of one dollar in hand paid, the receipt whereof is hereby acknowledged, we the subscribers, severally agree, each for himself, to take the proportionate amount set opposite our respective names, and pay the cash therefor as aforesaid, upon the subscription of the total Amount as herein set forth.[a]

Names.	Adress.	Shares written	Amount.
Frank W. Allen[5]	145. E. 37	100	100,000.00
Thos C. Doremus Jr[6]	70 Union place	25	25,000.00
Chas C. Yeaton for L G. Yeaton[7]	41 West 16 St.	25	25,000.00

D (transcript), MdSuFR, Libers Pat. D-20:346. Another version of the prospectus, written with an electric pen, is in Lefferts. [a]"Recorded June

14, 1876. F. C. T." written in right margin; followed by line across page. [b]Underlined twice. [c]Followed by centered double horizontal line.

1. Charles Yeaton was a sales agent for the Fuller Magneto-Electric Telegraph machine (see Doc. 634) and for Doane & Wellington's Street Gas Lamps. Charles Yeaton business card, Cat. 1144, Scraps. (*TAEM* 27:335).

2. Edison executed the patent application—his first in more than a year—the same day as this agreement. It became U.S. Patent 180,857.

3. In subsequent correspondence Batchelor also referred to the company as "Edison's Electrical Pen & Duplicating Press Co." and the "Electrical Pen & Press Co."

4. Batchelor began using Yeaton's Union St. address for pen-related correspondence the next day, signing for Yeaton "per Batchelor" on the initial letter but using his own name on subsequent letters. Charles Yeaton to William Wheeler, 8 Mar. 1876, Lbk. 2:59 (*TAEM* 28:394).

5. Unidentified.

6. Thomas Doremus, Jr., was president of the Safeguard Fire Insurance Co. Wilson 1875, 27.

7. According to the New York City directory, Lily Yeaton owned the street lamp company that employed Charles Yeaton. The two lived at the same address; their relationship is not known (Wilson 1875, 1492). On 4 March 1876 Edison signed an agreement with Lily Yeaton for foreign sales of pen patents, to be negotiated by Charles Yeaton acting as agent. Miller (*TAEM* 28:1014).

–732–

To William Orton

[Newark, March 13, 1876?[1]]

Wm. O.

I will transfer my space to you[2] providing you will protect me from any sinful games on the part of my friends[a] your Electricians—[3]

Edison

ALS, NjWOE, DF (*TAEM* 13:1008). Written at bottom of Doc. 730. [a]Underlined twice.

1. This letter was probably written in response to Western Union's inquiry of 13 March, which indicated that Edison had not yet formally notified either Pettit or Orton that he agreed to the transfer. J. W. [Schmults?] to TAE, 13 Mar. 1876, LBO 16:282.

2. See Doc. 730 for the proposed transfer. According to Orton's letter to Henry Pettit of 9 March, Western Union, its subsidiary the Western Electric Manufacturing Co. of Chicago, and Edison would "combine their spaces, amounting in the aggregate to 30 feet by 56 feet." LBO 16:280.

3. That is, George Prescott and Gerritt Smith. Edison's relationship with them was strained by continuing controversy over the quadruplex and by his suspicions that Smith was claiming some of Edison's telegraph designs in his own patent applications. See Docs. 512, 513, and 727.

INVITATION Doc. 733

On the evening of 16 March 1876, shortly before the Edisons moved to Menlo Park,[1] Mary Edison's older sister Alice Stilwell, half-sister, Hattie Van Cleve, and Hattie's husband, Cornelius Van Cleve,[2] hosted a surprise farewell party for Mary at the Edison home. The invitation was produced with an electric pen. No other information about this event is known.

1. The family moved to Menlo Park sometime between 26 March and 28 March. See Doc. 737 n. 2.

2. See the Edison Family Genealogy prepared by John Deissler for the Charles Edison Fund, EBC.

-733-

Invitation[1]

[Newark,] March 16, 1876

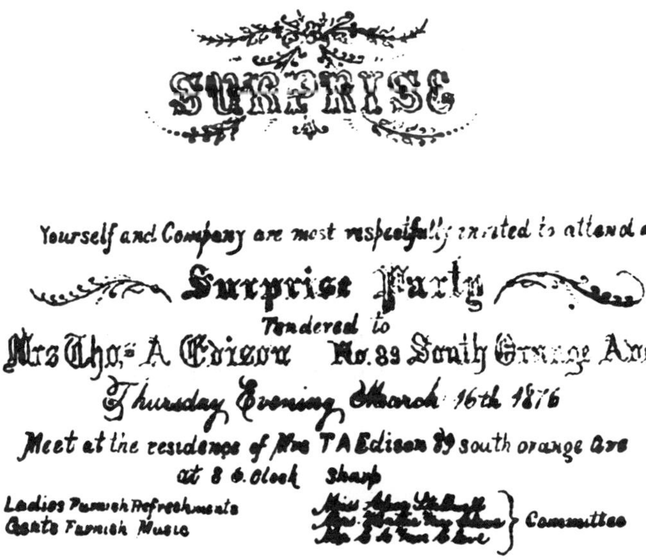

SURPRISE

Yourself and Company are most respectfully invited to attend a

Surprise Party

Tendered to

Mrs Thos A Edison No. 89 South Orange Ave,

Thursday Evening March 16th 1876

Meet at the residence of Mrs T A Edison 89 south orange Ave

at 8 o.Clock Sharp

Ladies Furnish Refreshments

Gents Furnish Music

[illegible] } Committee

D (electric-pen copy), Scraps., Cat. 593 (*TAEM* 27:653).

1. See headnote above.

-734-

To Commissioner of Patents[1]

NEWARK, New Jersey, Mch. 21, '76.[2]

Sir:

Referring to my applications for patents for my inventions in duplex and quadruplex telegraphy, described as follows:

Case 112,[3] filed in the patent office on or about December 28, '74—

& Case 113,[4] filed in the patent office on or about March 23, '75.

I beg leave to inform you that I have revoked the power of attorney given by me to L. W Serrell, to make such applications on my behalf. Enclosed is Mr. Serrell's acknowledgement of his receipt of my revocation.[5]

And I also hereby withdraw all of the said applications, it being my intention to file new and amended specifications.[6]

And I also hereby withdraw all applications made by me or in my behalf, for patents to be issued to myself and George Harrington jointly, whether allowed or not, and all applications made by me or on my behalf, for patents to be issued jointly to me and George B. Prescott—[7]

And I hereby revoke all powers of attorney or other authority to whomsoever granted, authorizing any applications to be made for patents for inventions in automatic or duplex and quadruplex telegraphy to be issued to any person or persons jointly with me:[8] and I hereby revoke all requests heretofore made by me, and all directions heretofore given by me to the Commissioner of Patents, respecting the issue of any patents for any of my said inventions for which patents have not been heretofore issued.[9] And I absolutely withdraw all pending applications for patents for any of my said inventions, whether such applications have been made by me alone, or by me in conjunction with any other person or persons claiming to be assignee or assignees of any part, share or interest of any of my said inventions in Automatic or duplex and quadruplex telegraphy.[10] Yours truly,

THOS. A. EDISON.

PL, NjWOE, Quad., TLC.3, p. 205 (*TAEM* 10:923).

1. R. H. Duell.

2. Although the transcription printed in the records of the Quadruplex Case bears this date, its contents indicate it was sent along with another paper that was not available until 28 March. See n. 5 below.

3. This covered one kind of quadruplex; it eventually led to Edison's U.S. Patent 207,724. See Doc. 449.

4. This covered repeaters for quadruplex telegraphy. See headnote, p. 347 n. 5.

5. Edison's letter to Serrell of 21 March and the acknowledgment, dated 28 March, are both transcribed in Quad., TLC.3, pp. 204–5 (*TAEM* 10:923–24). This letter also covered Cases 107 and 108, which issued in 1877 as U.S. Patents 195,752 and 195,751.

6. Edison here began to implement a strategy outlined in detail in a confidential memorandum by Robert Russell the previous spring (Quad. 72.8 [*TAEM* 9:122]; see Doc. 577, esp. nn. 2–4). This action was an

attempt to forestall the effects of an 8 March 1876 decision by Secretary of the Interior Zachariah Chandler, who had replaced Columbus Delano in October 1875. Chandler, a former senator from Michigan and a major figure in the Republican party, was a frequent ally of Western Union's lawyer Senator Roscoe Conkling. Chandler emphatically rejected the long-pending appeal of Edison and Harrington against the 20 March 1875 decision of Patent Commissioner John Thacher regarding the assignment of any future Edison quadruplex patents ("Edison and Harrington vs. Edison and Prescott. *Appeal to the Secretary of the Interior,*" *Patent Office Gazette* 9 [1876]: 403–4; *DAB*, s.v. "Chandler, Zachariah"). The applications had been held up by the inaction of the previous interior secretary; Chandler's action sent them back to the Patent Office for normal processing, whence they would issue, if at all, jointly assigned to Edison and George Prescott. Any further title claims would have to be settled in the courts.

With this document Edison meant to clear away most of the existing dispute, and to start over with patents issued to himself alone. That would have left all other claimants—including Harrington, Gould, and Atlantic and Pacific Telegraph Co. as well as Prescott and Western Union—with the task of suing for title in a court of equity. However, Edison did not file new and amended specifications, because his maneuver promptly generated protests from Prescott, and the applications soon were the subjects of two lawsuits: one by Atlantic and Pacific against Western Union, and another by Harrington and Edison against Atlantic and Pacific and Gould. Prescott to Commissioner of Patents, 22 Mar. 1876, Pat. App. 178,221; Quad. TLC.2, p. 74 (*TAEM* 10:857); Quad. 71.2, p. 44 (*TAEM* 10:249).

7. This repeated and superseded another letter of Edison's to the Commissioner, also dated 21 March 1876, in which Edison withdrew his Cases 94–100 (Docs. 467–73) and announced the revocation of powers of attorney regarding them. TAE to Commissioner of Patents, Quad. TLC.3, p. 204 (*TAEM* 10:923).

8. See TAE to Lemuel Serrell, 21 Mar. 1876, Quad. TLC.3, p. 203 (*TAEM* 10:922), which includes an acknowledgment by Serrell dated 21 March.

9. Among other things, this was an attempt to revoke Edison's 23 January 1875 letter to the Commissioner (Doc. 536).

10. Edison's inclusion of his automatic inventions makes it plausible that he designed this document with regard to disputes both with Jay Gould and the Atlantic and Pacific Telegraph Co. as well as with Western Union and George Prescott (Doc. 729).

-735-

Technical Note: Chemistry

[Newark, Winter 1876?[1]]

Make a bath of molten lead, immerse two carbon or platina electrodes connect on a powerful ~~p~~ quantity battery i.e[a] 40 cells 1½ battery all connected as 1 cell— See if cant decomp it[2] also try zinc, antimony= also try molten Ar[3] See if it dont take it out of Pb—

AX, NjWOE, Lab., NS-Undated-001 (*TAEM* 8:12). [a]Circled.

1. This is one of four similar pieces of paper which were apparently stuck together on a spindle (NS-Undated-001 [*TAEM* 8:11–12]). Two of the pages are a set of notes and drawings on acoustic-telegraph experiments related to Edison's January 1876 Caveats 73 (Doc. 709) and 76.

2. Soon after the invention of the electrochemical cell at the end of the eighteenth century, several investigators (notably William Nicholson, Anthony Carlisle, and Humphry Davy) used it to decompose previously inseparable compounds into constituent elements. *DSB*, s.vv. "Davy, Humphry," and "Nicholson, William."

3. By "Ar" Edison probably means silver (chemical symbol Ag), which he sometimes called by its Latin name, "argentum." Silver and lead commonly occur together in ores.

WATER-TELEPHONE TRANSMITTER Doc. 736

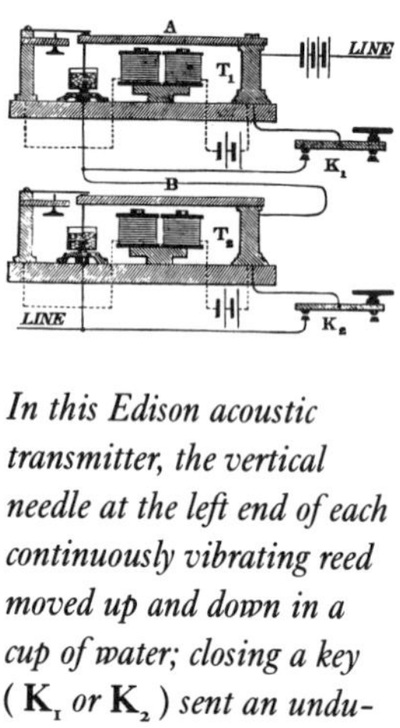

In this Edison acoustic transmitter, the vertical needle at the left end of each continuously vibrating reed moved up and down in a cup of water; closing a key ($\mathbf{K}_1$ *or* $\mathbf{K}_2$*) sent an undulating signal over the line.*

Edison had been thinking about using a movable electrode in liquid as a variable resistance since his first conversations about accoustic telegraphy with William Orton in mid-1875.[1] He used illustrations of a related transmitter "devised in September, 1875," in the account of his early telephone work that he wrote for George Prescott's history of the telephone.[2] In testimony Edison claimed that in February or March 1876 he "took up the subject of transmitting articulate speech, using transmitters which varied a volume of water."[3] The instrument shown in Doc. 736, unlike the September 1875 transmitter, depends on an external source of sound to move the electrode attached to the diaphragm.

In December 1877 Edison included several water-telephone designs in a patent application that became U.S. Patent 203,018.

1. See Doc. 599. In the fall of 1875 Edison built a sensitive instrument for measuring the change in resistance caused by "exceedingly slight motions of immersed electrodes." TI 1:20–21 (*TAEM* 11:31).

2. Prescott 1879, 219. This design (shown at left) is much like Edison's earlier relay patent (U.S. Pat. 141,777).

3. TI 1:31–32 (*TAEM* 11:36–37). Edison remembered making this instrument "in Dec., 1875, or any month between that time and November, 1876" (TI 1:101 [*TAEM* 11:71]). He also testified that he had made a water telephone to transmit speech in December 1875 (TI 1:21 [*TAEM* 11:31]). Lawyers for Alexander Graham Bell argued that the instrument shown in Doc. 736 had to date from the fall of 1876 (TI 4:195–98 [*TAEM* 11:801–3]).

–736– [Newark, Winter 1876?[1]]

Experimental Model: Telephony

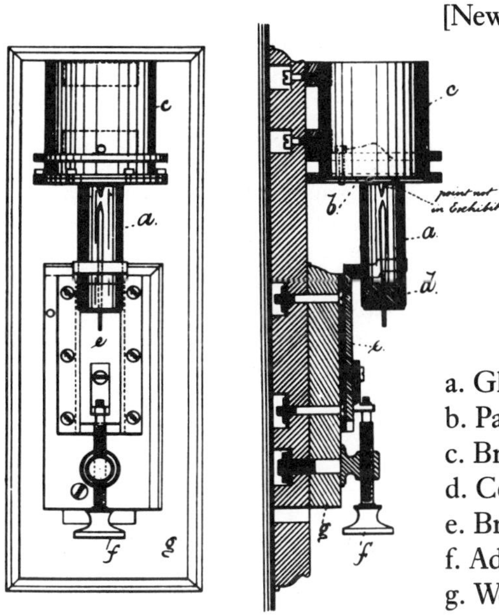

a. Glass for water
b. Parchment diaphragm
c. Brass tube
d. Cork
e. Brass slide
f. Adjusting screw
g. Wood base

M (historic drawing) (31 cm × 11 cm × 11 cm), NjWOE, TI 2:518 (*TAEM* 11:646). These drawings were made from an instrument entered as an exhibit in the patent interference.

1. See headnote above.

–737– NEWARK, N.J. March 25 1876[2a]

Account[1]

T A Edison	Dr
To one bilding	$2500.00[b]
TAE & Bachlor	37 00
bill of Pant[3]	21 75
Sundries	18.30
window & Door frames	9.00
Piasir[4] Lumber & woork cement & Lime	13 81
Jingine[5] foundation 1 days work	11.00
12 Peers[6]	24 00
Work and Materal on little bilding	27 00
to unloding cars 2 Teams	8 00
Labor in all	12 25
Team to factory 1½ Days	6 00
2 days help B Acker[7] & Nigger	5 50
B Acker ~~3~~2½ days work	3 75
	$2697 36

By lumber Hoover Harris &[8]	$1297 95	
" cash	869 00	
Loder & Iliff[9]	240 58	
	$2407 53	
Ballenc		$289 73
To[c] amount brought over		$2697.36
By " " "	$2407.53	
" cash	60 00	
To Siden Bords & Shingels		58 07
" Paid G. W. Ayers the Panter[10]		41 00
" 800 ft Hemlock Bords		16 00
" 250 f Studing of Soper[11] for little Bild		5 00
		$120 07

D, NjWOE, DF, 76-001 (*TAEM* 13:742). Written by Samuel Edison; letterhead of T. A. Edison. [a]"NEWARK, N.J" and "18" preprinted. [b]Ditto marks used as placeholders in front of subsequent numbers in this column. [c]Begins new page.

1. This account is for the building of the Menlo Park laboratory main building and one of the small outbuildings (possibly the carpenter or blacksmith shop). Besides this account record and some individual bills in 76-002, DF (*TAEM* 13:735), there is a Samuel Edison account book pertaining to some of the building expenses, especially labor accounts that indicate the construction took place between January and March 1876. ("Edison's Newark Expense Book, 1875–1878," EP&RI).

2. In a letter written the next day to William Wheeler, Chicago agent for the electric pen, Batchelor noted that the move to Menlo Park would occur on 27 March (Lbk. 2:72 [*TAEM* 28:407]). Two unsigned notes also suggest that the move took place about this time. One states that "Chas Worth was here with our machinery the 25th of March 1876," and the other, dated 28 March 1876, says, "Moved to Menlo Park" (76-001, DF [*TAEM* 13:744–45]).

3. Paint.

4. Possibly plaster.

5. Should probably read "Engine."

6. Piers.

7. Benjamin Acker, one of the day laborers, probably a carpenter. Receipt dated 8 Apr. 1876, 76-002, DF (*TAEM* 13:831); entry of 7 Feb. 1876, "Edison's Newark Expense Book, 1875–1878," EP&RI.

8. According to its bill and letterheads, Hoover, Harris & Co. of Phillipsburg, Pa., manufactured and dealt in timber, flooring, siding, shingles, lath, and other wood products, and also had an office in Philadelphia. The partners, who included G. W. Hoover, J. P. Harris, E. C. Humes, and W. V. Hughes, were also proprietors of the Union Steam Packing Box Manufacturing Co. of Philadelphia, superintended by T. R. Hamilton. Hoover, Harris & Co. to TAE, 18 Oct. 1876 and n.d. [1876], and bill of 28 Jan. 1876, 76-002, DF (*TAEM* 13:779, 794).

9. Alfred Loder and Danial Iliff were Newark carpenters and builders who built the door and window frames. Bills of 29 Feb. and 18 Mar. 1876, 76-002, DF (*TAEM* 13:816, 824).

10. Painter George Ayers. Receipt dated 1 Apr. 1876, 76-002, DF [*TAEM* 13:820]).

11. Soper was one of the day laborers, probably a carpenter. Entry of 7 Feb. 1876, "Edison's Newark Expense Book, 1875–1878," EP&RI.

Appendix 1

Edison's Autobiographical Notes

In 1908 and 1909 Edison wrote a series of autobiographical notes whose origin and nature are described in Appendix 1 of Volume One.[1] Four of the six documents discussed there contain sections related to events of the period of Volume Two; those sections are published here.[2] Portions of a seventh document written at the same time but found after the publication of Volume One are included here as well. The parts selected relate to the time periods of both Volumes One and Two. Edison sometimes referred in the same paragraph to periods covered by more than a single volume; such notes will be reprinted as appropriate. Each document has been designated by a letter and each paragraph sequentially numbered.

1. Pp. 627–28.
2. The autobiographical documents designated A and F do not refer to the period of this volume.

B. FIRST BATCH

The following is from a typescript that Edison revised; see Volume One, p. 646. Almost all of its eighty-one sections relate to later periods. The third of the four sections included here probably concerns the *Operator*; the last refers to Henry Thau and David Hermann; see Volume One, pp. 448 n. 3, 622 n. 14.

THE WEARY WALL STREET MAN

[6] One day while I was carrying on my shop in Newark a Wall Street broker came from the city and said he was tired of the "street" and wanted to go into something real. He said he had plenty of money. He wanted some kind of a job to keep his mind

off Wall Street. So we gave him a job as a "mucker" in chemical experiments. The second night he was there he could not stand the long hours and fell asleep on a sofa. One of the boys took a bottle of bromine and opened it under the sofa. It floated up and produced a violent effect on the mucus membrane. The broker was taken with such a fit of coughing he burst a blood vessel; and the man who let the bromine out got away and never came back. I suppose he thought there was going to be a death. But the broker man lived and left the next day, and I have never seen him since, either.

The Green Assistant's Fire.

[7] Some of my assistants in those days were very green in the business, as I did not care whether they had had any experience or not. I generally tried to turn them loose. One day I got a new man and told him to conduct a certain experiment. He got a quart of ether and started to boil it over a naked flame. Of course it caught fire. The flame was about four feet in diameter and 11 feet high. We had to call out the fire department, and they came down and put a stream through the window. That let all the fumes and chemicals out and overcame the firemen, and there was the devil to pay. Another time we experimented with a tub full of soapy water and put hydrogen into it to make large bubbles. One of the boys who was washing bottles in the place had read in some book that hydrogen was explosive, so he proceeded to blow the tub up. There was about four inches of soap in the bottom of the tub 14 inches high, and he filled it with soap bubbles up to the brim. Then he took a bamboo fishpole, put a piece of paper at the end and touched it off. It blew every window out of the place.

Starting a Technical Paper.

[78] I was interested in journalism and started the Telegraph Journal and got out about a dozen numbers when it was taken over by W. J. Johnson, who afterwards founded the Electrical World on it. I started Science and ran it for a year and a half. It cost me too much money to maintain, and I sold it to Gardiner Hubbard, the father-in-law of Graham Bell. He carried it along for some years.

Trying to Help People.

[80] I had two men working for me, one a German the other a Jew. They wanted me to put up a little money and start them in a shop in New York, to make repairs, etc. I put up $300 and was to get half the profits, and each of them a quarter. I never got anything for it. A few years afterwards I went to see them and asked what they were doing and said I would like to sell my interest. They said: "Sell out what?" "Why," I said "my interest in

this machinery." They said "You don't own this machinery. This is our machinery. You have no papers to show ~~no~~anything.[a] You had better get out!" The percentage of crooked people was smaller ~~what~~ when[b] I was young. It has been steadily rising and has got up to a very respectable amount now. I hope it will never reach par.

TD (transcript), NjWOE, Meadowcroft. [a]"no" cancelled and "any" interlined above, both in pencil. [b]"what" cancelled and "when" interlined above, both in pencil.

C. SECOND BATCH

The following is from a typescript that includes Edison's revisions; see Volume One, p. 647. Only two sections relate to the period covered by Volume Two; however, Edison misattributed them to his years at Menlo Park instead of Newark.

The Sailor in Search of a Job

[15] Down at Menlo Park a man camc in one day and wanted a job. He was a sailor. I hadn't any particular work to give him, but I had a number of small induction coils, and to give him something to do I told him to fix them up and sell them among his sailor friends. They were fixed up and he went over to New York and sold them all. He was an extraordinary fellow. His name was Adams. One day I asked him how long it was since he had been to sea, and he replied, two or three years. I asked him how he had made a living in the meantime before he came to Menlo Park. He said he made a pretty good living by going around to different clinics and getting ten dollars at each clinic, because of having the worst case of heart disease on record. I told him if that was the case he would have to be very careful around the laboratory. I had him there to help in experimenting. The heart disease didn't seem to bother him at all.

[17] Adams said that one time he was aboard a coffee ship in the harbor of Santos, Brazil. He fell down a hatchway and broke his arm. They took him up to the hospital—a Portuguese one—where he could not speak the language and they didn't understand English. They treated him for two weeks for yellow fever. He was certainly the most profane man we ever had around the laboratory. He stood high in his class.

TD (transcript), NjWOE, Meadowcroft.

D. BOOK NO. 2

This undated notebook contains a mix of narratives, questions, and notes in Edison's hand; see Volume One, p. 648. Only twelve notes pertain to the period covered by Volume Two.

[101] Jay Gould & quarrel with G & Stock had his indicator taken out. Thought it foolish— He explained great value of U ~~S~~Pacific to me 2 hours thought this strange— How his men couldnt beat him insanity in family— ~~had sour stomach~~—[a]

[102] Come over to shop 1873 plug hat Ulster reached to ground— looked like boy— ~~trench coat a~~ hadnt any sense humor— far off look & sour stomach 4 yrs labor lost Lawsuit lasting 30 yrs—Took line & everything. congratulate myself I had my life—[a]

[154] Reason left Newark—Sleight Law suit rent.

[157] $5. day to sheriff—

[158] Invention parafine paper

[224] Of patents acquired on tickers & teleghs by WU are not several lying dormant Can you give us a genl summary of the situation why should Co go on with 95% hand signalling when it has all these automatic systems at its command are the reasons financial or technical or due to Conservatism your discussion of this would be extremely interesting

[251] Is Dixons story true of forgetting name at tax ofs Newark—

[260] More subjects—

[261] Wall st broker as mucker. Bromine under lounge—

[262] Adams boiling ether on ~~na~~open flame—

[263] Explosion tub suds with Ox & H. broke glass—

[264] Gould visiting Lab[a]

[265] Browne & starting Domestic telegh— Auto Tel test station[a]

[371] Started the telgh journal # meadowcroft find 1st number[a]

[380] Sailor scrub Lab— fell in Hatch in Santos took to Portuguese Hospital & treated him 3 weeks for yellow fever

[382] Adams— heart disease, selling Induction Coils to Sailors— was ~~a pirate~~ a Slaver = fell in Hatchway in Santos treated him 3 weeks in Portugeese hospital for Yellow fever. had worst kind hart disease got 10 at Students Clinic NY finally died in London of Delirium Tremens—

AD (photographic transcript), NjWOE, Archives Office. [a]This paragraph canceled with a large "X".

E. NOTES

Taken from a notebook that has five pages in Edison's hand, these "Notes" are numbered consecutively from 1 to 33.

> [10] 10 = Wall St broker as mucker, bromine under lounge[a]
> [11] 11 Adams boiling ether naked flame.[a]
> [12] 12 Explosion blew windows out, tube soap Hydrogen[a]

AD, NjWOE, Lab., N-09-06-27. [a]Note canceled with a check mark.

G. MR. EDISON'S NOTES

The following is a transcription of relevant portions of a typescript titled "Mr. Edison's notes in Book No. 2," probably prepared by William Meadowcroft in 1908 or 1909. It had been separated from the Meadowcroft files and put in a general file of anecdotes in the Edison Biographical Collection; it was relocated after the publication of Volume One. Nineteen of the first twenty-three paragraphs and one of the remaining fifteen involve events from the years covered by Volumes One and Two.

[1] On black Friday, we had a very exciting time with the indicators. The Gould and Fiske crowd had cornered gold and had run the quotations up faster than the indicator would follow. The indicator was composed of several wheels on the circumference of each wheel were the numerals and one wheel had fractions. It worked the same as an ordinary counter, one wheel made 10 revolutions and at the tenth, it advanced the adjacent wheel 1, and this in its turn having gone ten revolutions, advanced the next wheel 1, and so on. On the morning of Black Friday, the indicator was quoting 150 premium, whereas, the bids by Goulds' agents in the gold room was 165 for 5 million, or any part. We had a paper weight on the transmitter and by 1 o'clock, reached the right quotation. The excitement was prodigious. New Street, as well as Broad was jammed with excited people. I sat up on the top of the W.U. Tel. booth to watch the surging crazy crowd. One man came to the booth, grabbed a pencil and attempted to write a message to Boston. The first stroke went clear off the blank, he was so excited that he had the operator write the message for him. Amidst great excitement, Speyer, the Banker, went crazy, and it took five men to hold him and everybody lost their head. The W.U. operator came to me and said: "Shake Edison, we are O.K., we hav'nt got a cent. I felt very happy because we were poor. These occasions are very enjoyable to a poor man, but they occur rarely.

[2] A friend of mine was an operator who worked in the back office of Wm. Beldon & Company, 60 Broadway, which were headquarters for Fiske. Mr. Gould was up town in the Erie offices on the Grant Opera House. The other firm on Broad Street, Smith Gould and Martin, was the other branch. All were connected with wires. Gould seemed to be in charge Fiske being the executive. Fiske wore a velvet corduroy coat and very peculiar vest. He was very chipper and seemed to be light-hearted and happy. Sitting around the room were about a dozen fine looking men, all had the complexion of Cadavers. There was a basket of champagne. Hundreds of boys were rushing in paying in checks, all checks being payable at Belden & Company. When Jas. Brown of Brown Bros. & Company broke the corner by selling five million gold, all payments were repudiated at Smith, Gould & Martin, but they continued to receive checks at Beldon & Company for some time, until the street got wind of the game. There was some kind of a conspiracy with Government people, which I could not make out, but I heard messages that opened my eyes to the ramifications of Wall Street. Gold fell to 132 and it took us all night to get the indicators back to that quotation. All night long the streets were full of people. Every broker's office was brilliantly lighted all night and all hands were at work. The Clearing House for gold had been swamped and was all mixed up. No one knew if they were bankrupt or not.

[3] While a newsboy on the Grant Trunk there was a grand dance of the train mens' Fraternal Organization, which I attended. The dance kept up all night. I knew that the irregular freight train would leave about two hours after the regular train and that everybody would be sleepy and so I determined to stop over and see if I could'nt get a chance to run the freight engine a few miles. I was very anxious to do this; I had been allowed to fire and could handle the small sticks of wood expertly and keep steam up, and also now and then I would be allowed to switch a freight car around to the freight shed to unload small batches of freight, but I wanted to get a chance for a long run. Sure enough, both the engineer and fireman were tired out and very sleepy from the good cheer they had taken. I got the engineer to consent to allowing the fireman to run the engine and I to do the firing, the engineer going back to sleep in the caboose. The engine was one of a number leased to the G.T. by the Chicago Burlington & Quincy R.R. It had bright brass bands all over the wood-work beautifully painted, and everything highly polished, which was the custom up to the time old Commodore Vanderbilt stopped it on his road. After running about 15 miles, the fireman could'nt keep his eyes open and he agreed to permit me to run the engine. I took charge, reducing the speed to about 12 miles an hour and brought the train of 7 cars to her destination at the Grant Trunk

Junction safely, but there was something occurred which was very much out of the ordinary. I was very much worried about the water and I knew that if it got low the boiler was likely to explode. I gathered this from hearing conversation among the engineers, so I kept constantly testing the gauge and watching it, keeping her full of water. I had'nt gone more than 20 miles before black damp mud blew out of the stack and covered every part of the engine, including myself. I was about to awaken the fireman to find out the cause of this, when it stopped. Then I approached a station where the fireman always went out to the cowcatcher, open the oil cup on the steam chest and poured oil in. I started to carry out the procedure when upon opening the oil cup, the steam rushed out with a tremendous noise, nearly knocking me off the engine. I succeeded in closing the oil cup and got back in the cab and made up my mind that she would pull thru without oil. I afterwards learned that the engineer always shut off steam when the fireman went out to oil. This point I failed to notice. My powers of observation were very much improved after this occurrence. Just before I reach the Junction another outpour of black mud occurred and the whole engine was a sight, so much so that when I pulled into the yard, everybody turned out to see it, laughing immoderately. I found out the reason of the mud was that I carried so much water, it carried over into the stack and this washed out all the accumulated soot.

[4] One afternoon about a week before Christmas, our train jumped the track near Utica. Four old Michigan Central cars with rotten sills collapsed in the ditch and all went to pieces, distributing figs, raisens, dates and candies all over the track and ditch. I started in to save all I could of this by eating it. Our family doctor had the time of his life with me in this connection.

[5] Just before the war broke out, there came to the train one afternoon in Detroit, two fine looking young men accompanied by a colored servant. They bought tickets for Port Huron, the terminal point for the train. After leaving the Junction, just outside of Detroit, I brought in the evening papers. The train was called the accommodation and there was only one passenger car. When I came opposite the two young men one said—"Boy, what you got"— "I said, "Papers"— All right He took them and threw them out of the window and turning to the colored man said—Nicodemus, pay this boy. I told him the amount and he opened a satchel and paid me. The passengers did'nt know what to make of this transaction. I returned with the illustrated papers and magazines. These were seized and thrown out of the window and I was told to get my money of Nicodemus. I then returned with all the old magazines and novels I had not been able to sell, thinking perhaps this would be too much for them. I was small and thin and the layer reached above my head and all I could

possibly carry. I had prepared a list and knew the amount in case they bit again. When I opened the door, all the passengers roared with laughter. I walked right up to the young man. One asked what I had; I said—magazines and novels—. He promptly threw them out of the window and Nicodemus settled. Then I came in with cracked hickory nuts, then pop corn balls and finally—molasses candy. All went out of the window. I felt like Alexander, the Great. I had no more chance, I had sold all I had. Finally, I put a rope to my trunk, which about the size of a carpenter's chest and started to pull this from the baggage car to the passenger car. It was almost too much for my strength, but finally I got it in front of these men. I pulled off my coat shoes and hat, laid them on the chest, then he asked what have you got boy—I said—everything sir, that I can spare that is for sale—. The passengers fairly jumped with laughter Nicodemus paid me $27. for this last sale and threw the whole out of the door in the rear of the car. These men were from the South and I have already retained a soft spot in my heart for a Southern gentleman.

[6] When I had charge of the Gold Indicator I spent my evenings with the night operator at the French Cable office, which was at that time in Nassau Street. Bunnell was somewhat inventive and we talked on all kinds of electrical applications possible. Bunnell afterwards established a store and factory for the manufacture of telegraph apparatus in N.Y., which was successful and is still in existence, although Bennell has passed away. Bunnell was a fine operator and worked in the U.S. Military Telegraph in Tenn. When I was in the department at Memphis, Bunnell was night operator at Chattanooga Junction. When it was reported that Hood was marching in Nashville, about 11 o'clock one night a jew came into the office in great excitement, having heard the Hood rumor and wanted to send a message to save his goods, he being a large Salter. Bunnell told him it was impossible, that orders had been given to send no private messages. Then the jew wanted to bribe Bunnell, who steadfastly refused, for the reason, as he told the jew he might be courtmartialled and shot. Finally, when the jew got up to $800. Bunnell swore him secrecy and sent the message. Now there was no such order, and the jew finding it out, complained to Captain Van Duzer, the chief of the telegraphs, who investigated the matter, and while he would'nt discharge Bunnell, laid him off indefinitely. Van Duzer was so lenient that if an operator was discharged, all he had to do was to wait three days and then go and sit on the stoop of Van Duzer's office all day and he would be taken back, but Van Duzer swore he never would give in in Bunnell's case. He said if Bunnell had taken the $800. and sent the message at the regular rate, which was 25 cents, it would have been all right, as the jew should be punished, for trying to bribe a military operator, but

when Bunnell took the $800. and then sent the message dead head, he could'nt stand it and he would never relent.

[7] One night I was walking up Broadway and went into Kohn's Museum of Anatomy, which place I have always wished I never entered. In this place were all kinds of bodies and parts thereof moulded in wax, many of them illustrating the eating away of flesh by certain malignant diseases. The place had many urns filled with tuberoses, which gave an overpowering smell. For two days I could'nt eat a thing, and to this day the smell of tuberoses will spoil my appetite.

[8] After the breaking out of the war, there was a regiment of volunteer soldiers quartered at Fort Gratiot, the reservation extending to the boundary line of our house. The barracks were at the end furtherest from the house. Sentries were stationed at various distances along the reservation; nearly every night we would hear a call, such as "Corporal of the Guard No. 1". This would be repeated from sentry to sentry, until it reached the barracks, when Corporal of the Guard No. 1, would come and see who was wanted. Myself and the little Dutch boy, after returning from the town after selling our papers, thought we would take a hand at Military affairs. So one night, when it was very dark, I shouted—"Corporal of the Guard No. 1". The second sentry, thinking it was the terminal sentry that shouted, repeated it to the third and so on. This brought the Corporal along the ½ mile only to find that he was fooled. We tried him three nights, but the third night, they were watching and caught the little Dutch boy and took him to the lock-up at the Fort and shut him up. They chased me to the house. I rushed for the cellar In one small apartment there were two barrels of potatoes and a third one nearly empty. I poured these remnants into the other barrels, sat down and pulled the barrel over my head, bottom up. The soldiers had awakened my father and they were searching the cellar for me with candles and lanterns. The Corporal was absolutely certain I came into the celler and could'nt see how I could have gotten out and wanted to know from my father if there was no secret hiding place. On assurance of my father that there was not, he said it was most extraordinary. I was glad when they left, as I was cramped and the potatoes were rotten that had been in the barrel and violently offensive. The next morning I was found in bed and received a good switching on the legs from my father, the first and only one I ever received from him, although my mother kept a switch behind the old Seth Thomas clock that had the bark worn off. My mother's ideas and mine differed at times, especially when I got experimenting and mussed up things. The Dutch boy was realsed the next day.

[9] When I was a small boy at Milan, and about five years old I and the son of the proprietor of the largest store in the town,

whose age was about the same as mine, went down in a gully in the outskirts of the town to swim in a small creek. After playing in the water a while, the boy with me disappeared in the creek. I waited around for him to come up but as it was getting dark I concluded to wait no longer and went home. Some time in the night I was awakened and asked about the boy. It seems the whole town was out with lanterns and had heard that I was last seen with him. I told them how I had waited and waited, etc. They went to the creek and pulled out his body.

[10] While learning to telegraph I and a boy named Clancy, built a telegraph wire between our houses, about a mile apart, separated by woods. The wire was that used for suspending stove pipes, the insulators were small bottles pushed on ten-penny nails driven in the trees. It worked fine. My father had a neighbor named Jos. Symington, a highly educated Scotchman, and they would talk politics nearly every night until I returned from town, which varied from 11 P.M. to 1 A.M. I would save one paper, but many nights when I wanted to practice I would give the paper to Clancy and then my father would have to get the news over the wire or not get it at all. This generally resulted in going to bed at 3 A.M.

[11] Up to the time of taking up telegraphy, I had a chemical laboratory on the train. A freight car had been fitted up as a baggage car and one end partitioned off as a smoking department. It was only 8 feet long, had a table in the middle and two benches. There was no ventilation and everybody went into the baggage end to smoke, hence I had it all to myself. George Pullman had a small shop in Detroit, working on his sleeping car and he made me a lot of wooden apparatus for my chemicals. After I had done my train work, I would make chemical experiments. One day a bottle containing a stock of phosphorus jarred off on the floor and set it on fire. The baggage master put it out, but the phosphorus would'nt go out and he picked it up to throw it out and some got on his fingers and every time he rubbed it, it exposed a fresh surface and that got on fire. Water would'nt put it out. He got a bad burn and boxed my ears so severely that I got somewhat deaf thereafter.

[12] This deafness has been of great advantage to me in various ways. When in a telegraph office I could only hear the instrument directly on the table at which I sat, and unlike the other operators, I was not bothered by the other instruments. Again, in experimenting on the telephone, I had to improve the transmitter, so I could hear it. This made the telephone commercial, as the telephone receiver of Bell was too weak to be used as a transmitter commercially.

[14] After I became a telegraph operator I practiced for a long time to become a rapid reader of print, and got so expert that I could sense the meaning of a whole line at once. This faculty, I

believe, should be taught in schools, as it appears to be easily acquired. Then one can read two or three books in a day, whereas if each word at time is only sensed, reading is laborious.

[15] In 1873 (?) the owners of the automatic Telegraph Company commenced negotiations with Jay Gould for the purchase of the wires between N.Y. and Washington, and the patents for the system, then in successful operation. Gould at that time controlled the Atlantic & Pacific Telegraph Co. and was competing with the W.U. Co. and endeavoring to depress the W.U. stock on the Exchange. About this time I had invented the Quadruplex, an instrument for sending and receiving four messages simultaneously over a single wire in various directions. I wanted to interest the W.U. Telegraph Company with it, with a view of selling it, but was unsuccessful, until I made an arrangement with the Chief Electrician of the Company, so he could be known a joint inventor and receive a portion of the money. At that time I was very short of funds and needed it more than glory. This electrician appeared to want glory more than money, so it was an easy trade. I brought my apparatus over and was given a separate room with marble tiled floor, which by the way was a very hard kind of floor to sleep on, and started in putting the finishing touches on. After two months of very hard work, I got a detail at regular times of eight operators and we got it working nicely from one room to another over a wire, which ran to Albany and back. Under certain conditions of weather, one side of the quadruplex would work very shaky and I had not succeeded in ascertaining the cause of the trouble. On a certain day when there was a board meeting of the Company I was to make an exhibition test. The day arrived, I had picked the best operators in N.Y. and they were familiar with the apparatus. I arranged it if a storm occurred and the bad side got shaky to do the best they could and draw freely on their imagination. They were sending old messages—about 12 o'clock everything was working, but there was a storm somewhere near Albany, and the bad side got shaky. Mr. Orton, the president and Wm. Vanderbilt, and the other directors came in. I had my heart trying to climb up around my oesophagus. I was paying a sherriff five dollars a day to withhold a judgment which had been entered against me in a case which I had paid no attention to, and if the quadruplex had not worked before the president I knew I was to have trouble and might lose my machinery, but the operators were stars, they pulled me through. The N.Y. times (get it) came out next day with a full account. I was given $5,000. which made me easy and expected the whole thing would be closed up, but Mr. Orton went on an extended tour just about the time I had paid for all the experiments on the quadruplex and exhausted the money and I was again in straits. In the meantime I had introduced the apparatus on the lines of the company where it was very successful. At that time the General

Supt. of the W.U. was Gen'l T. T. Eckert. It seems there was great friction between Eckert and Orton, and Eckert was secretly negotiating with Gould to leave the W.U. and take charge of the Atlantic & Pacific (Gould's Company). One day Eckert called me into his office and made inquiries about money affairs. I told him Mr. Orton had gone off and left me without means and I was in straits. He told me I never would get another cent, but that he knew a man who would buy it. I told him of my arrangement with the Electrician and said I could not sell it to anybody, but if I got enough for it, I would sell all my interest in any share I might get. He seemed to think his party would agree to this. I had a set of quadruplex over in my shop 10 and 12 Ward Street and he arranged to bring him over next evening to see the apparatus. So the next evening Eckert came over with Mr. Jay Gould, and introduced him to me. This was the first time I had ever seen him. I exhibited and explained the apparatus and they departed. The next day Eckert sent for me and I was taken up to Gould's house, which was near the Windsor Hotel. In the basement, he had an office it was in the evening and we went in by the servants' entrance as Eckert probably feared that he was watched. Gould started in at once and asked me how much I wanted. I said—make me an offer—then he said—I will give you $30,000—I said I will sell any interest I may have for that money, which was somewhat more than I thought I could get. The next morning I went with Gould to Sherman & Sterlings' office and received a check for $30,000. with a remark by Gould that I had got the Steamboat "Plymouth Rock" as he had sold her for $30,000 and had just received the check.

[16] There was a big fight on between Gould's Co. and the W.U. and this transaction caused more litigation. The Electrician on account of the testimony lost his glory. The Judge never decided the case, but went crazy a few months afterwards.

[20] After this Gould wanted me to help install the automatic system in the Atlantic & Pacific Co. which General Eckert had been elected president, the company having bought the Automatic Telegraph Company. I did a lot of work for this company, making automatic apparatus in my shop at Newark. About this time, I invented a District Messenger Call Box system and organized a company called the Domestic Telegraph Company and started in to install the system in N.Y. I had great difficulty in getting subscribers, having tried several canvassers, who, one after the other failed to get subscribers. When I was about to give it up, a test operator named Browne, who was on the Automatic Telegraph Company wire between N.Y. and Washington, and which passed through my Newark shop, asked permission to let him try and see if he could'nt get subscribers. I had very little faith in his ability to get any, but thought I would give him a chance as he felt certain of his ability to succeed. He started in

and the results were surprising. Within a month he had procured 200 subscribers and the Company was a success. I have never quite understood why six men should fail absolutely, while the seventh man should succeed. Perhaps hypnotism would account for it. This company was sold out to the Atlantic & Pacific Company.

[21] While engaged in putting in the Automatic System, I saw a great deal of Gould and frequently went up town to his office to give information. Gould had no sense of humor. I tried several times to get off a funny story, but he failed to see any humor in them. I was very fond of stories and had a choice lot always kept fresh and which I could throw a man into convulsions. One afternoon Gould started in to explain the great future of the Union Pacific R.R., which he then controlled. He got a map and had an immense amount of statistics. He kept at it for over four hours and got very enthusiastic. Why he should explain to me a mere inventor, with no capital or standing, I could not make out. He had a peculiar eye and I made up my mind that there was a strain of insanity somewhere. This idea was strengthened shortly afterwards, when the W.U. raised the monthly rental of the stock tickers. Gould had one in his house office, which he watched constantly. This he had removed because the price had been advanced a few dollars and to his great inconvenience. He railed over it. This struck me as abnormal. I think Gould's success was due to abnormal development. He certainly had one trait that all men must have who want to succeed. He collected every kind of information and statistics about his schemes and had all the data. His connection with men prominent in official life of which I was aware of, was surprising to me. His conscience appeared to be atrophied, but that may be due to the fact that he was contending with men that were worse. He worked incessantly until 12 or 1 o'clock at night. He took no pride in building up an enterprise, he was after money and money only. Whether the company was a success or failure, mattered not to him. After he had hammered the W.U. through his opposition Company, and had tired out Mr. Vanderbilt, the latter retired from control and Gould went in and consolidated his company and controlled the W.U. He then repudiated the contract with the Automatic Telegraph people and they never received a cent for their wires or patent, and I lost three years of very hard labor, but I never had any grudge against him, because he was so able in his line and as long as my part was successful the money with me was a secondary consideration. However, suits were brought by the stockholders and have been in the courts for 25 years, and recently the Federal Court decided them against the Gould Estate and the Estate may have to pay for the weapon he used to get control of the W.U. When Gould got the W.U. I knew that no further progress in telegraphy was possible and I went into other lines.

[22] Towards the latter part of 1875 in the Newark shop I invented a device for multiplying copies of letters, which I sold to Mr. A. B. Dick of Chicago, and in the years since it has been universally introduced throughout the world. It is called the Mimeograph. I also invented devices and introduced paraffin paper, now used universally for wrapping up candy, etc.

[23] In 1876, I moved to Menlo Park, N.J. on the P.R.R. several miles below Elizabeth. The cause of this move was due to trouble I had about rent. I had rented a small shop on the top floor of a Padlock factory and had rented it by the month. I gave notice that I would give it up at the end of the month, paid the rent, moved out and delivered the keys; shortly afterwards I was served with a paper, probably a judgment, wherein I was to pay for 9 months rent. There was some law, it seems, that made a monthly renter liable for a year. This seemed so unjust that I determined to get out of a place that permitted such injustice and I moved.

[33] While I was in Newark, I discovered a peculiar phenomenon, which was unexplainable until years after. Beard of N.Y., a prominent scientific man, called it a new force and named it "Etheric Force". The principal phenomenon was shown by breaking the circuit of a magnet when a spark was noticed passing between two carbon points in a dark box, both points being short circuited by a loop of wire and this spark could be obtained anywhere in the vicinity of a magnet, or drawn from gas pipes anywhere in the building. Whatever passed through the isolated circuit would not affect the most delicate detector of electricity. Beard published a description of the phenomenon in the   journal, also in the N.Y. Tribune of   It was published in various electrical journals. The apparatus was shown at the Electrical Exhibition at Paris in   . Lord Kelvin knew of the phenomenon, but could not give an explanation.   (get years) years afterwards, Herz brought out a paper, using the dark box and other devices and clearly explained the whole matter, which opened up the possibility of wireless telegraphy. (Meadowcroft has Beard's article and has found our sketches, etc.)

TD (transcript), NjWOE, Meadowcroft. Because this transcription of Edison's manuscript is presented only as a reference text, typographical errors have not been reproduced or noted.

Appendix 2

Charles Batchelor's Recollections of Edison

The following is one of five reminiscences written by Charles Batchelor concerning his work with Edison. He entered them in one of his daily journals (Cat. 1339) between October 1906 and February 1908. Although they were published in Welch 1972 (app. 1), we have decided to present new transcriptions of them in the appropriate volumes of this work.

The recollection presented below discusses Edison's work on an electromotograph relay which began during the period covered by Volume Two. A draft version of this recollection was also found in the same journal on loose paper. We have included the last paragraph of that draft, which contains information not found in the journal version.

#5 THE ELECTROMOTOGRAPH

In the year 1875 whilst working out a system of rapid automatic telegraphy Mr. Edison discovered a new principle on which he built an instrument to replace the relay in ordinary telegraphy and which could have been built up into a new system of telegraphy if there had been no such thing known as a magnet with its armature

In the above mentioned system of rapid telegraphy a strip of chemically prepared paper was passed rapidly beneath a stylus of metal and received a mark from it each time a current of electricity was sent over the line, the result appeared as a blue mark on a yellow paper or a black mark on a white paper according to which stylus and which chemical were used at the time— To find the best mark and the best paper, he made a great many experiments with a small apparatus that I made for him on which he could try any metal on any solution in a very short time

The experiment worked as follows:— a metal plate was fastened on the table and connected to one pole of a battery, the other pole being connected to a brass disc from whose edges radiated a stylus of every known metal— A piece of paper being saturated with a solution was then laid on the plate and one after another all the different metals were rubbed along the paper whilst the current as being constantly made and broken that passed through both— A paper moistened with ferricyanide of Potassium when put in contact with an iron stylus and the current passed through would receive a blue mark (permanent) on a yellow ground— Another moistened with iodide of potassium and passed under a platinum stylus would receive a brownish or purple mark on a white paper (not permament) In these two cases the stylus was connected to the pole that gives off O.— If the paper was moistened with caustic Potash (weak) and a stylus of Tellurium was used the paper would receive a dense black mark on pure white paper and in this case the stylus would be connected with the pole of the battery at which H. is given off.

Working with this apparatus Mr. Edison discovered that whenever the current passed through the friction of the stylus on the paper was diminished— This action was much more marked with some solutions than others and was exceedingly sensitive to very weak currents so much so that he thought that messages could be received over a wire better and more rapidly than with a relay which is limited as to speed by the charge and discharge of the magnet

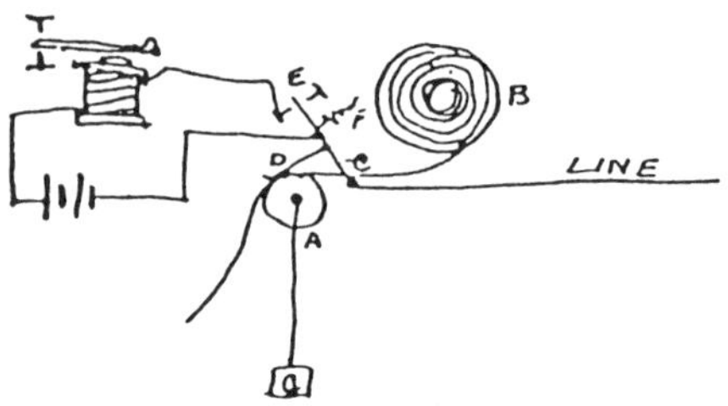

We therefore designed and made an instrument making the moving parts exceedingly light to prevent inertia— A is a metal drum run by clock work over which runs a strip of moistened paper B and under a stylus D that is hinged to an upright lever C. This lever moves between points at E and works a sounder in the usual manner— When current is passed over the line the stylus at D slips forward and makes contact at E and when the current is interrupted the spring F draws it back—

This proved to be a most remarkable instrumment; the relay owing to its inability to charge and discharge its magnet quickly could never get up to more than 60 or 70 words a minute on long lines and then only with careful adjustment; but this worked at that speed so that very little adjustment was necessary and it has been worked up to 400 words a minute— This seems incredible when you think how many movements the little lever must make (about 10000 per min) but we proved it by making it repeat the message on the automatic telegraph from the original punched paper strip

AD, NjWOE, Batchelor, Cat. 1339:256 (*TAEM* 90:590).

ELECTROMOTOGRAPH DRAFT

This was not of great need in telegraphy at the time as the operators could only receive 60 words a minute, and there was only one company to sell it to the West U which had bought out all competitors—[a] A short time [------][b] previous to this[c] however a new company the Mutual Union T Co had[d] appeared in the field and began to cut prices— The W.U. as it had[e] done before raked up some absurd claims about owning the so called Page patent whereby it seems they[f] could prevent any one using an armature ~~before a magnet,~~[g] with a retractile spring before a magnet— This would have put any telegraph Company out of existence if it could be enforced[h] The Mutual Union was owned by Jay Gould—and he had made some advances toward securing the motograph—[i] Mr Edison offered the Motograph relay to the W.U. Tel Co but Mr Orton the Prest[j] informed him that he could not see his way to buy it— He then offered it to Mr Gould for $1000,000 and when there seemed to be a probabllty of this offer being accepted Mr Orton sent for him and made a contract to pay $100,000 for it in 16 yearly payments of $6250 and they locked the instrument up and as far as I know it has never been used since for telegraphic work

AD, NjWOE, Batchelor, Cat. 1339, following p. 258 (*TAEM* 90:593).
[a]"the West U . . . competitors" interlined above. [b]Erased. [c]"previous to this" interlined above. [d]"the Mutual Union had" interlined above. [e]"The W.U. as it had" written over erasure. [f]Repeated as page turn. [g]"~~before a magnet,~~" erased. [h]"This . . . enforced" interlined above. [i]"and he . . . monograph" interlined above. [j]"the Prest" interlined above.

Appendix 3

The Dispute over the Quadruplex

The disposition of Edison's quadruplex patent applications was the focus of litigation in three state and three federal cases, an interference case in the U.S. Patent Office, and two extraordinary administrative proceedings, one in the Patent Office and one in the Secretary of the Interior's Office. It was also a crucial element in another Patent Office proceeding and another federal court case. The core patent itself was not granted until eighteen years after the application.[1] The last appeal in one of the related cases was not ruled upon until 1913. Edison spent a good deal of time attending to these disputes—giving testimony, preparing arguments, drafting amendments, and planning strategy—and the various records fill many thousands of pages. No discussion to date—however brief or lengthy—has managed to be accurate, comprehensive, or even clear about what happened and just what the disputes concerned.[2]

The selections included here present two opposing positions fairly compactly, and they bring out most of the points at issue. They thus introduce the contest in broad terms and indicate where the dispute stood in the first half of 1875. The first document, a legal instrument written early in 1875, presents Western Union's view of the case. In the second, an affidavit generated three months later in connection with the Patent Office–Interior Department proceedings, Edison presents the tale from his standpoint.

The story may be divided conveniently, if somewhat arbitrarily, into seven phases. The first, in which both the technological and contractual roots are found, fell in the years covered by the first volume of Edison's papers.[3] The second phase—that of invention, development, and innovation—ran from the opening of Volume Two through the end of 1874.[4] The third phase extended into March 1876, closing as this volume ends, with Edison moving to Menlo Park.[5] This phase primarily concerned claims about how any patents Edison might get on his quadruplex designs should be initially assigned, and the subsequent action and inaction by successive Secretaries of the Interior regarding the Patent Office decision on those claims.[6] Western Union initiated one legal action during this time.[7] The selections included here were produced during this third phase. The fourth part of the tale focused on the courts. In March 1876 lawyers were preparing to file three cases, two of which had the quadruplex as their chief object.[8] The end of this stage may be placed either in 1877, with a cartel-forming business agreement,[9] or in 1878, with the only verdict directly addressing the central subject.[10]

Subsequent phases involved, fifth, settling Edison's priority dispute with Henry Nicholson (a patent interference case appealed into 1885),[11] and, sixth, further exchanges between Edison and the Patent Office until the basic quadruplex patent finally issued in 1892, some eighteen years after Edison prepared the application.[12] That was not the final phase, however, since the quadruplex was also involved in a legal case that was filed in May 1876 but did not go to trial until 1905.[13] This seventh phase ended only in 1913, when the U.S. Supreme Court set aside a lower court decision on the grounds that the case had been initiated in the wrong jurisdiction, and the remaining parties declined to start over.[14]

Almost every point in the two documents presented here was disputed. It is difficult to identify points of agreement, because these statements were deliberately argumentative, and what may seem to be neutral description often had contentious implications. The fundamental questions addressed a few basic issues. One was whether Edison's business arrangements with George Harrington should be regarded as covering the quadruplex; a second was whether Western Union had established a legal claim to the quadruplex.[15] Each of these raised fundamental legal issues as well as problems in determining the specific facts of the case, and each in turn, however resolved, brought a series of further litigious points in its wake. A third main focus was what, if any, rights to the quadruplex were ever properly held by either George Prescott or Jay Gould.[16] Other contested points included the appropriate policies of the Patent Office and the Interior Department, and possible instances of improper practice by various parties. Eventually even the patentability of the quadruplex technology was questioned.[17]

As with Edison's later autobiographical notes, these selections have been carefully transcribed but have not been given detailed annotation. Some information necessary for the reader's understanding of the pieces is provided at the start of each.

1. U.S. Pat. 480,569.
2. Few authors have treated the topic in any detail, and fewer have

examined any of the related documents. Among biographies of Edison, only Conot 1979 (52–75) and Josephson 1959 (107–27) are based at all on such evidence, as is one history of American electrical communications (Harlow 1936, 405–9). Dyer, Martin, and Meadowcroft 1929 (148–67) relies on Edison's autobiographical notes; other Edison biographies and related works rely mostly on secondary sources when they deal with this (e.g., Clark 1977, 50–55; and Klein 1986, 196–205). All the books significantly misrepresent events and chronology and misinterpret causality and equity. The one reliable paper related to the quadruplex deals only with its comparative international use (Nier 1986).

3. See, for example, Docs. 28, 47–50, 88, 275–78, 285, 294, 300, and 315 for aspects of the technological roots; and Docs. 61, 68, 109, 155, 282, 284, 288, 292, 303, and 316 for much of the contractual background. See also *TAEB* 1:104, 143–44, 493, 507, 513–14, 556, and App. 1.D220.

4. See Docs. 347–48, 356–57, 387, 392, 432, 441–42, 444–46, 449–54, 466–73, 477, 485, 488, 498, 507, 512–13, 515, 517, 519, and 521.

5. See Doc. 734.

6. See, in addition to the material in this appendix, Docs. 526, 531–38, 554, 569, 577, 584, 603, 662–63, 693, 717, and 727. During this time others also advanced claims to the quadruplex; see Chapter 7 introduction, n. 1.

7. The first selection in this appendix initiated that case.

8. Those two were: *Atlantic & Pacific v. Prescott & others* in state court in New York, filed 11 April 1876 (see n. 10); and *Edison & Harrington v. Western Union & others* in federal court in Washington, D.C., filed 11 May 1876 (Quad. TLC.3, 5 [*TAEM* 10:909, 948]). The other court cases centering on the quadruplex question were: *Edison & Harrington v. Western Union & Prescott,* filed 23 October 1876 in federal court in New York (Quad. TLC.4 [*TAEM* 10:926]); *Western Union v. Harrington & others,* filed 21 February 1877 in state court in New York (Quad. TLC.1–2 [*TAEM* 10:800–908]); and *Welch v. Edison,* filed 15 May 1883 in federal court in Massachusetts after being removed from state court, where it had been filed on 1 January 1883.

9. In August 1877 Jay Gould and William H. Vanderbilt reached an agreement to pool patent rights and to divide traffic between Western Union and Atlantic and Pacific, with Western Union buying control of the latter company and Gould acquiring a large share of the former. Harlow 1936, 408–9; Grodinsky 1957, 148–58; Klein 1986, 201–5; Reid 1879, 586–87.

10. The case was *Atlantic & Pacific v. Prescott & others* (see Quad. TLC.2, p. 74; and Quad. 70, 71, 73, passim [*TAEM* 10:857; 9:288–10:797]). The decision, in Western Union's favor, was rendered on 3 June 1878; there was no appeal, the contending parties having ended their competition many months earlier (Quad. 73.16 ([*TAEM* 10:773]).

11. See *Nicholson v. Edison*. A fairly compact review of much of the case is in Gustav Bissing to TAE, 16 May 1892, in Pat. App. 480,567.

12. See Pat. App. 480,567, particularly the historical review in the 16 May 1892 letter mentioned in n. 11.

13. *Harrington v. A&P.*

14. *Harrington v. A&P* was filed on 17 May 1876 in federal court in New York. Since the filing was incorrect, all subsequent findings were

ignored and the appeal was dismissed without comment or prejudice as to the merits of the case. By 1913 Harrington, Gould, and Josiah Reiff—nominally a co-plaintiff, but actually the prime instigator throughout—had all died, and the suit was by Edison against Gould's heirs.

15. Each side claimed its position on more than one basis; any one was sufficient, but no particular one was necessary. Most of the arguments were presented, from various viewpoints, in the briefs and related materials in Quad. 73.1–15 (*TAEM* 10:346–772).

16. Depending on the disposition of various questions, Prescott's share ranged from none to half of the total, and Gould's from none to all.

17. See Pat. App. 480,567.

WESTERN UNION BILL OF COMPLAINT

The Western Union Telegraph Company filed this complaint as a petition in the New Jersey Court of Chancery on 28 January 1875. The company sought an immediate injunction to prevent Edison from selling the quadruplex design to anyone except itself. Western Union claimed that it had a contractual right to the invention, that Edison had reneged, and that it was unlikely that the company could later recover adequate damages by suing him. Since Western Union also insisted on the validity and equity of Edison's arrangement with George Prescott making Prescott joint owner of any quadruplex patents, this complaint and petition named Prescott as a co-defendant. An injunction was granted that day barring Edison (and Prescott) from selling the quadruplex until action was taken by the court following their answering this bill of complaint.

The company's bias is clear throughout the bill. President William Orton's December departure for the Midwest was a simple business trip; Edison's January trip to Port Huron was made solely to escape Western Union. Orton is said to have presented Edison's late December terms of sale to the Executive Committee of the Board of Directors, but no evidence of this has been found in company records.

The complaint presented or cited several documents as evidence. The exhibits included transcriptions of three documents published in this volume: Doc. 517 (Exhibit A), Doc. 519 (Exhibit C), and Doc. 535 (Exhibit D). Exhibits B and E are in Quad. 70.2, p. 12, and Quad. 70.4, p. 19 (*TAEM* 9:315, 334). This complaint also referred to the Edison-Prescott partnership deed of 19 August 1874 (Doc. 466), the patent application cases 94–100 of the same date (Docs. 467–73), and the proferred terms of sale from late December 1874 (Doc. 519). Affidavits of Marshall Lefferts and Gerritt Smith accompanied the bill and were referred to extensively, though without attribution. Smith's is included in this volume as Doc. 538; Lefferts's is described and quoted in annotation to that document and to Doc. 535.

To His Honor, THEODORE RUNYON, *Chancellor of the State of New Jersey:*

Humbly complaining, showeth unto your Honor, your orator, the Western Union Telegraph Company, a corporation and citizen of the State of New York:

That your orator, the Western Union Telegraph Company, is a corporation organized and doing business under the laws of the State of New York, for the purpose of operating Telegraph lines in that and in other States of the United States, and having its principal office in the City of New York.

That on or previous to the month of January, 1874, Thomas A. Edison, who, as your orator is informed and believes to be true, was then and ever since has been and is now a resident of the City of Newark, in the State of New Jersey, entered into an arrangement and agreement with the complainant, through William Orton, its President, that said Edison should endeavor to invent improvements in Stearns' Duplex system of Telegraphy, and to invent other Duplex apparatus, and other processes in connection with Telegraphy, by which more than one message could be sent on one wire at the same time; and should use therefor the lines of your orator, and its men and employees, workshops and materials; and that whatever inventions should so be made by the said Edison, and all patents that might be secured therefor, should be assigned and conveyed to your orator, who should pay such price therefor as was reasonable and just, the amount of compensation to be ascertained either by agreement between the parties or by arbitration.

And your orator further shows, that in pursuance of this arrangement and agreement, said Edison was given the use of the Electrician of the Company, and such other of its employees as he desired; the use of its wires, workshops and materials; and his experiments were continued for the space of several months in endeavoring to invent some improvements upon the Stearns' process, or some new process by which more than one message could be sent over one wire at the same time.

And your orator further shows that, on or about the 19th day of August, 1874, the said Edison associated with himself George B. Prescott, who was then the Electrician of the complainant, and articles of agreement were made and entered into between said Edison and Prescott, of that date, which recited that said Edison had invented certain improvements in Duplex telegraphs, for which he had then executed, or was about to execute applications for Letters Patent of the United States, and that the numbers of such applications were 94, 95, 96, 97, 98, 99 and 100, bearing date August 19, 1874; and further reciting, that said Prescott was entitled to an equal interest in the same; and said agreement further witnessed that, in consideration of the premises and of one dollar, the receipt whereof was thereby acknowledged by said Edison, he, the said Edison, had sold and assigned and did thereby set over and convey unto the said Prescott, one undivided half part of the right, title and interest of every character in, to, under and connected with each and all of the said inventions and letters patent on the same, when the same should be granted; and requested the Commissioner of Patents to issue the said letters patent to said Edison and Prescott, as assignees of said Edison, for the use of said parties and their legal representatives; said agreement further recited that, whereas, the said Edison had also invented other improvements in Duplex telegraphs, the descriptions of which had been lodged with George M. Phelps, for the purpose of models being constructed, it was agreed that such inventions should be included in the said agreement between said Edison and Prescott, and that when the applications for patents therefor should be made, that the said patents should be granted in accordance with the said agreement made with reference to the numbers which were given in said agreement; and that said Edison should sign the required papers therefor; and said agreement further witnessed, that the said transfer to the said Prescott was made on the following terms and conditions, which were thereby made part of the consideration aforesaid, to wit:

First.—That both of the parties should have an equal, undivided interest in all the letters patent of the United States or of any foreign countries, which might be granted for all or any of the said inventions, or of any future improvements thereon made by either party, and of all extensions or re-issues of such letters patent.

Second.—That, whereas, Edison had theretofore expended eleven hundred and twenty-five dollars, for models and patent fees, the said Prescott agreed to pay solely, and without contribution from Edison, all future expense and cost of specifications, drawings, models, Patent Office fees, and Patent Solicitors' and Agents' fees, and all other charges incident to the procuring of letters patent, for any of the said inventions.

Third.—That neither of said parties should sell, assign or otherwise dispose of the whole or any of his interest in said inventions, or letters patent therefor, or any of them, without the written consent thereto first obtained of the other party.

Fourth.—That neither of said parties would himself manufacture, use, or sell, or grant licenses, or the right in any way to any other party to manufacture, use or sell any of the said inventions, or any improvements thereof, or any machine embodying, or article containing, any of said inventions or improvements, or protected by any of the said letters patent, without the written consent first obtained of the other party.

Fifth.—That no sale of any of the said inventions, and no license or right to make or use the same in any way, should be made or given except at a price to which both parties should agree in writing; and all net profits should be equally divided between the said parties.

And your orator further shows, that said agreement between the said Edison and Prescott was received for record in the United States Patent Office, on the 29th day of August, 1874, and recorded in Liber R 18, page 62, of Transfers of Patents.[a]

And your orator further shows, that after the execution of said agreement between said Edison and Prescott, said parties continued to make experiments over the lines of your orator, under the same arrangement, as to them jointly, as had been theretofore made with said Edison individually; and all facilities and assistance required by the said parties to make the said experiments were afforded to them by your orator.

And your orator further shows, that these experiments continued until on or about the 10th day of December, 1874, when your orator was informed by the said Edison and Prescott that the experiments were in such shape as to give a promise of developing important results; and that they, said parties, had discovered valuable improvements in connection with the duplex and quadruplex system of telegraphing; and that said parties expected, soon, to be able to demonstrate the practicability of their inventions. That, at the last mentioned date, the said Edison desired to secure, in advance, a portion of the money which would be due to him, or to said Prescott and Edison, as the price or compensation for such inventions, as aforesaid, and such letters patent of the United States as either or both might obtain for such inventions; and it was thereupon agreed by said Edison with your orator, that the said Prescott and Edison would assign all their right, title and interest in and to all the inventions that they had so made and might make in telegraphy, relating to duplex and quadruplex telegraphy, and all letters patent of the United States which they might obtain therefore, to your orator, upon receiving such compensation as should be thereafter satisfactorily adjusted between your orator and the said Edison and Prescott; and that thereupon, upon the making of the said agreement, and in consideration of the agreement theretofore entered into between your orator, and the said Edison and Prescott, and in execution thereof, your orator paid to the said Edison the sum of five thousand dollars, in part payment of the price so agreed to be paid to said Edison and said Prescott, and thereupon a paper was executed by said Edison, under his hand and seal, and subscribed by him, bearing date the 10th day of December, 1874, a copy of which is hereto annexed as Exhibit A.

And your orator further shows that, at the same time, the said Prescott, being desirous of obtaining a portion of the price or compensation to which he would be entitled, under the arrangement and agreement entered into between your orator and the said Edison and Prescott, requested the payment to him by your orator of the sum of five thousand dollars in part payment of such sum, as it might thereafter be ascertained he was entitled to under the arrangements and agreements aforesaid. That such request was assented to, and the said sum was thereafter paid to him by your orator, and he subscribed and executed, under his hand and seal, a paper bearing date the sixteenth day of January, 1875, a copy of which is hereto annexed, and marked Exhibit B.

And your orator further shows, that on or about the 16th day of December, 1874, negotiations commenced between your orator and the said Edison and Prescott, for the purpose of ascertaining the amount and terms of payment as provided for in Exhibits A and B, by a proposition in writing signed by said Edison and Prescott, a copy of which is hereto annexed, marked Exhibit C; and said negotiations were continued until about the 30th day of December, 1874, when a formal proposition in writing was made by the said Edison and Prescott to your orator, to fix the amount to be received by them for the transfer of all United States patents of said Edison and Prescott, and inventions of said Edison or Prescott, heretofore mentioned, at the sum of twenty-five thousand dollars in cash, and a royalty of $233 per year, for each circuit created. That, at the time this offer was made to the President of your orator, said President submitted the same to the Executive Committee of the Board of Directors, and they appointed the President, in connection with George H. Mumford and Norvin Green, Vice-Presidents of said Company, a committee, with full power to act with reference to said matter, on behalf of your orator. That, at the same time, the President of your orator was about to depart from the State of New York, upon business which would keep him absent some time, which fact was stated to the said Edison and Prescott by him; and, further, that on his return he would take up and consider the offer aforesaid immediately, and reply to the same, and he requested that the offer might be left as made, to await his return, to which said Edison and Prescott then and there assented. That the said President returned on or about the eleventh day of January, 1875, and thereafter, upon a consultation with said Mumford and said Green, and other members of the Executive Committee of said Company, it was agreed that they would accept, on behalf of the Company, the said proposition of the said Edison and Prescott, which had not, at that time, been in any way revoked or modified by them, or either of them.

That, on the 19th day of January, 1875, your orator accepted the offer which the said Edison and Prescott had so made as aforesaid, and embodied said acceptance in a letter by its President, a copy of which letter was delivered to said Edison and said Prescott, on or about the date thereof; a copy of which is hereto annexed, as Exhibit D.

And your orator further shows, upon information and belief, that, on the delivery of said letter to said Edison, he declared to the person who handed him the same that he repudiated all obligation to convey his said inventions, patents, or any patents, or interest therein, to the said Western Union Telegraph Company, and all propositions and agreements theretofore made with that Company relating thereto, and that he should at all times refuse to perform any of said agreements.

The said Prescott, upon receiving said letter, addressed and delivered to your orator a letter of ratification, a copy of which is hereto annexed and marked Exhibit E, and that said Prescott is, as your orator is informed and believes, now ready and willing to execute his portion of said contract and to unite with the said Edison, as required by their agreements above mentioned, to convey to the Western Union Telegraph Company, your orator, the entire interest in all the inventions or letters patent of the United States which may be issued thereon, as above described.

And your orator further shows, that your orator, at all times since its said acceptance of said offer of the said Edison and Prescott, liquidating and fixing, in the manner provided in said agreement, the amount of consideration payable by your orator to the said Edison and Prescott thereunder, has been and now is ready and willing, and hereby offers to pay to the said Edison and Prescott the cash portion thereof remaining due as aforesaid, and to execute all reasonable and proper instruments securing the payment of the said royalty of $233 per annum, for each circuit created, and to fully do and perform all the conditions, terms and obligations of the said agreement and all other acts and things on its part to be performed, done and kept; but that the said Edison, as your orator is informed and believes, claims and charges that the said agreement of sale is void and of no effect, and has prevented your orator from making any tender to him, either of money or documents, and has for that purpose absented himself from his dwelling-house in the said City of Newark, and has kept himself concealed from the agents of your orator, and departed out of the State of New Jersey; and that your orator, after making due and reasonable efforts to find the said Edison, for the purpose of tendering to him the remainder of said consideration in money, and a suitable agreement for the payment of the said royalty, and of again offering to perform the said agreement on its part, and, upon failing so to do by reason of the said concealment and absence of the said Edison, did, on the twenty-eighth day of

January, 1875, at the City of New York, demand of the said Prescott a conveyence to your orator of the matters so agreed to be sold to your orator by the said Edison and Prescott, and did then and there tender to the said Prescott the said money, and offered to execute a proper agreement for securing the payment of the said royalty; and thereupon Prescott declared his willingness to comply with the said demand and execute an assignment, and to accept the said money and agreement for royalty, and to execute a proper assignment to your orator of the interest in the said inventions and letters patent so agreed to be sold, as aforesaid; but that he was unable so to do by reason of the repudiation of said Edison of the said agreement, and the refusal of the said Edison to perform on his part said agreement, and to join with him, said Prescott, in performing the same.

And your orator further shows, that your orator entered into the arrangements and agreements hereinbefore set forth, for the purpose of obtaining the exclusive use of all inventions which might be made by said Edison and Prescott, or either of them, relating to duplex or quadruplex telegraphy, and was induced to give the time of its employees and the use of its works, workshops, and materials and lines, to said Edison and Prescott, and to make the payments to them as aforesaid, solely because of the agreement on the part of the said Edison and Prescott to give to the Company the exclusive benefit and use of such inventions as might be made; in the experiments above mentioned.

That your orator would not have entered into the said arrangement or agreement, or made the expenditures, or advanced the money as aforesaid, had it not relied upon the good faith of the said Edison and Prescott, to faithfully carry out and perform the agreement made by them.

And your orator further shows that, unless your orator secures the exclusive control of the said inventions, and such patents as may be obtained therefor, your orator will be deprived of large gains and profits and advantages—the exact amount of which it is impossible to estimate or ascertain; and if the conveyance of these patent interests and inventions are made to other parties, your orator will be subjected to litigation for infringements of patents, and for actions for damages on behalf of those to whom any such conveyance or transfer may be made by said Edison and Prescott; and that there is nothing now known which will take the place of, or supply the inventions embodied in, the applications made to the Patent Office by the said Edison and Prescott, as aforesaid.

And your orator further shows, that the said Prescott is willing to make the transfer, according to the agreement of said

Edison and Prescott with your orator, in pursuance of his agreement, but the same requires the joining therein of the said Edison, under the terms of the agreement between the said Edison and Prescott, hereinbefore recited, in order to make the same valid and effectual. And your orator is informed and believes, that the refusal of the said Edison to join in such conveyance, and to make the transfer to your orator, as agreed by him, arises solely from the fact that he is now in negotiations with other parties, from whom he expects to obtain an arrangement which will be of greater pecuniary advantage to himself, than that heretofore made with your orator.

And your orator further shows, that said Edison threatens to convey his interest in the inventions and patents to be obtained by him to other parties than your orator, and that, as your orator is informed and believes to be true, the said Edison is insolvent, and of no pecuniary responsibility.

And your orator hoped that said Edison and Prescott would have performed the said agreement on their part respectively, as in justice and equity they ought to have done, respectively.

But now, so it is, may it please your Honor, the said Thomas A. Edison and George B. Prescott, who are the defendants to this Bill, combining and confederating with others at present unknown to your orator, but whose names, when discovered, it prays may be inserted herein, with proper and apt words to charge them herein, absolutely refuse to perform said contract, and sometimes give out and pretend that said agreement was never made; at other times, that your orator had failed to perform its part of the contract, and that the said Edison and Prescott were, and each of them was, always ready, and had offered to perform their part of said agreement; but your orator expressly charges the contrary thereof to be true; all of which actings, doings, refusals and pretences, are contrary to equity and good conscience, and tend to the manifest wrong and injury of your orator in the premises. In consideration whereof, and for as much as your orator can have adequate relief in the premises only in a Court of Equity, where matters of this nature are properly cognizable and relievable; to the end therefore, that said Thomas A. Edison and George B. Prescott, and their, and each of their confederates, when discovered, may, upon their several and respective oaths, true, full, perfect and distinct answers make to all and singular the premises; and that the defendants, their agents, attorneys, solicitors, servants and workmen, and each and every of them, may be perpetually restrained and enjoined from conveying to any other party than your orator, any of the inventions for which the defendant Edison made application for letters patent of the United States, by applications bearing date August 19, 1874, and numbered 94, 95, 96, 97, 98, 99 and 100, or any and all letters patent of the United States, which have been or may be issued for the same inventions, or any of them, or any and all the improvements in duplex telegraphs, the descriptions of which were by the said defendants, or either of them, lodged with

George M. Phelps, for the purpose of the construction of models, on or before the 19th day of August, 1874, or any and all letters patent of the United States, which have been or may be issued for the same, or any of them; or any and all inventions and improvements in duplex and quadruplex telegraphy, made, or to be made, by said defendants, or either of them, and in which the said defendants are, or at any time may be, jointly interested by virtue of their said agreement between them, bearing date August 19, 1874, and received for record in United States Patent Office, on the 29th day of August, 1874, and recorded in Liber K 18, page 62, of Transfers of Patents; or any and all letters patent of the United States, which have been or may be issued for the same, or any of them; or the right to use all or any of said inventions, or letters patent, or the system secured by the same.

And that the defendants may be enjoined and restrained in like manner, until the final decree of this suit.

And that said Edison and Prescott may be decreed to join in the execution of an instrument or instruments effectually conveying and assigning to your orator all the inventions, for which the defendant Edison made application for letters of patent of the United States, by applications, bearing date August 19, 1874, and numbered 94, 95, 96, 97, 98, 99 and 100; and all letters patent of the United States, which have been or may be issued for the same inventions, or any of them; and also all the improvements in duplex telegraphs, the descriptions of which were by the said defendants, or either of them, lodged with George M. Phelps, for the purpose of the construction of models, on or before the 19th day of August, 1874, and all or any letters patent of the United States, which have been or may be issued for the same, or any of them; and also all inventions and improvements in duplex and quadruplex telegraphy, made or to be made by said defendants, or either of them, and in which the said defendants are, or at any time may be jointly interested, by virtue of their said agreement between them, bearing date August 19, 1874, and received for record in the United States Patent Office, on the 29th day of August, 1874, and recorded in Liber K 18, page 62, of Transfers of Patents; and all letters patent of the United States, which have been or may be issued for the same, or any of them; your orator paying to them, upon the delivery of said assignment, the unpaid balance of the said sum of twenty- five thousand dollars, and convenanting and agreeing to pay the annual royalty of $233, for each circuit created; and that your orator may have such other or further relief in the premises, as may be agreeable to equity and good conscience.

May it please your Honor, the premises considered, to grant unto your Orator, not only the State's writ of injunction, issuing out of and under the seal of this Court, directed to the said defendants therein, enjoining and restraining them, and each of them, as aforesaid; but also a writ of subpoena, issuing out of and under the seal of this Court, directed to the defendants, Thomas A. Edison and George B. Prescott, commanding them on a certain day, and under a certain penalty, therein to be inserted, to be and appear before your Honor, in this Honorable Court, then and there to answer all and singular the premises, and to stand to, abide and perform such order and decree therein, as to your Honor shall seem meet, and agreeable to equity and good conscience.

And your Orator will ever pray, &c.

LEON ABBETT,
Solicitor, and of Counsel with Complainant.

PD (transcript), NjWOE, Quad. TLC.7 (*TAEM* 10:984). Oath (by George Mumford), notarization, exhibits, and affidavits by Marshall Lefferts and Gerritt Smith (Doc. 538) omitted. Another, probably later, transcription of this bill (differing on hundreds of inconsequential details) is in Quad. 70.8, p. 17 (*TAEM* 9:772). [a]Other version adds here "as in and by the said agreement remaining of record as aforesaid, more fully and at large appears, and to which or a duly certified copy thereof, your orator, for greater certainty, begs leave to refer."

EDISON'S AFFIDAVIT

Edison and his lawyers (mainly Benjamin Butler and Robert Russell, presumably paid by Jay Gould) prepared this statement, signed and sworn to on 27 April 1875, in connection with arguments made to Secretary of the Interior Columbus Delano. They sought to negate the Patent Commissioner's 20 March decision assigning any patents on Edison's quadruplex to George Prescott and Edison rather than to George Harrington and Edison (see Docs. 569 and 577). This statement overlaps considerably Edison's 20 May 1875 answer to Western Union's petition (above). However, that answer had to give space to legal technicalities, such as formally admitting that Western Union was a corporation, and it devoted comparatively more space to Harrington's claims and less to Edison's story.

This affidavit was phrased throughout to fit legal arguments enhancing Harrington's and Edison's claims (and thus those of Jay Gould and the Atlantic and Pacific Telegraph Co.) as opposed to those of Prescott (and thus of Western Union). For example, various types of telegraph technology were conflated here, and Western Union president Orton was said to have considered duplex telegraphy as a way of increasing "rapidity in telegraphing," thus reinforcing the claim that the quadruplex fell under Harrington's and Edison's agreement regarding "fast" telegraphy.

The affidavit mentioned without adequate identification two people in connection with a lecture at the Polytechnic Association of the American Institute at Cooper Union. Dr. P. H. Vander Weyde, editor of the *Manufacturer and Builder*, had taught at New York University Medical College and at Cooper Union and contributed articles to the *Scientific American* (Borut 1977, 223–24); Hamilton Towle, an engineer from Newark, ran a New York shop with William Unger, formerly of Edison and Unger (see *TAEB* 1:156 n. 1; and Holbrook 1871, 637). Neither the date of the lecture nor press reports about it have been identified.

The affidavit identified by letter five of the nine documents that followed it as exhibits. Those five are included in the present volume: Doc. 451 (Exhibit A), Doc. 466 (Exhibit B), Doc. 519 (Exhibit C), Doc. 517 (Exhibit D), and Doc. 536 (Exhibit E). Doc. 537 was mentioned, but its designation as Exhibit H was not; the same holds for Exhibit I, Edison's 23 January 1875 letter to George Prescott (Quad. 72.16, p. 21 [*TAEM* 9:269]; see also Doc. 537 n. 2). The affidavit also referred to the terms Edison offered in Doc. 521, which Orton quoted in Doc. 535, which in turn was quoted in full in the letter that was Exhibit F (Porter, Lowrey, Soren, and Stone to TAE, 9 Feb. 1875, Quad. 71.2, p. 144 [*TAEM* 10:300]; see also Doc. 537 n. 5). Exhibit G reprinted a duplex design Edison published in 1868 (Doc. 28).

Affidavit of Thomas A. Edison in Regard to his Inventions of Duplex and Quadruplex Telegraphy

Thomas A. Edison, residing in Newark, New Jersey, on oath deposes and says:

I am twenty-eight years old. My business is inventor and electrician. I have been engaged in this business for eighteen years, commencing when a small boy. My attention has principally been turned to the invention of telegraphic apparatus and methods. My earliest perfected invention, for which I made application for a patent, was in 1869 for a printing telegraph; that is to say, a machine for recording telegraphic signals automatically by printing. I have since directed my attention to improving the facilities with which telegraphic signals might be recorded and transmitted, and in 1869, I had a small shop of my own in the city of Boston. Previously thereto I had been employed as a telegraphic operator in Western Union Telegraph offices at Cincinnati, Ohio; Louisville, Kentucky; Memphis, Tennessee, and Indianapolis, Indiana. My employment with this company began in 1864 and continued until 1869, and I was employed with that company while I had the small shop of my own in Boston.

While so employed as a telegraph operator I was engaged in experiments in attempting to invent the requisite machinery and means to carry forward what is now known as duplex telegraphing. I began those experiments as early as 1865, and for that purpose I used the wires of the Western Union Telegraph Company. Those having charge of the telegraph offices over me knew that I was engaged in experiments to develop my idea that two messages might be sent over the same wire at practically the same time, which is what I mean by duplex telegraphing. This was previous to my idea that telegraphic signals might be recorded automatically in print, I supposing that wires might be used in duplex form by hand operation.

In the course of those experiments I satisfied myself that a wire could be used for double transmission; that is to say, by an arrangement of different forces of battery, I could send two messages for all practical purposes at the same time over the same wire. I set myself to work in my little shop for the purpose of developing, improving, and reducing to practical results by machinery these ideas, as well as prosecuting my experiments in printing telegraphs, or recording the signals automatically on paper.

Leaving the further immediate development of duplex machines, I turned my attention more particularly to the invention and perfection of machinery for automatic telegraphy, or the recording of telegraphic signals automatically; and it was upon such machinery that I obtained my first patent, and the result was improvements in the automatic machine, which is used now so largely by bankers and others in self-recording stock quotations. I engaged myself in inventing improvements from time to time and modifications of this machinery, which improvements I covered by a large number of patents. I had no money, except what came from my salary when employed as an operator, and which I received, in not very large returns, for my several inventions applied to automatic machines; but I was engaged from time to time in experiments upon the idea of duplex telegraphy from the time it first occurred to me, while I was perfecting these machines.

In September, 1870, being desirous of continuing my experiments as to duplex telegraphy, and being also desirous of having the means of manufacturing and perfecting my machines for automatic telegraphy, I made an arrangement with George Harrington to go into a joint business, wherein he was to supply tools, machinery, shop, and means, and I was to supply on my part my inventions, already perfected, and my inventions, the idea of which I had conceived, and some of which I was engaged in developing, as my share of the capital. But having sold out interests in the automatic stock-reporting machine, which was used by the stockbrokers and bankers, I was obliged to except and take them out from my proportion of the joint capital.

That arrangement was reduced to writing in a deed which was

executed between myself and Harrington, October 1, 1870, and established a copartnership at Newark, known as the American Telegraph Works. Under that deed machinery and shop and tools were furnished on the one side, and an arrangement by which automatic telegraphy might be further perfected was made by a line between New York and Washington, which passed through our shop.

Under this arrangement I went on perfecting my inventions in printing and automatic telegraphy, and again took up my experiments on the duplex telegraphy. The American Telegraph Works also had in view the making of other machines and devices besides automatic telegraph machines, such as burglar-alarms, printing-telegraph instruments, and electrical instruments for other purposes, and whatever else might grow out of these experiments.

The partnership transaction in regard to the manufacturer of machinery by these telegraphic works was not a business success, and I was not satisfied with its operation. In other words, the manufacturing part of our arrangement was unsuccessful, while the experimental was more successful. We took out a large number of patents for various devices and improvements relative to machinery for utilizing the action of electricity, which were directed to be issued to Harrington and myself jointly, under the arrangement, as they were perfected.

In order to[a] more unity of action another deed, in April, 1871, was made between Harrington and myself, by which I gave him a control of the whole business as the business man of the concern, while I devoted myself to my experiments and inventions.

Meanwhile I had a smaller private shop of my own, disconnected from the manufactory, for the purpose of carrying on experiments, and also for making the machines for the stockbrokers and bankers which I had excepted from the arrangement.

Meanwhile my experiments in duplex telegraphy had led me to conceive the idea of quadruplex telegraphy, but I found that somewhat difficult, and doubted whether I should be able to make it a success. In January, 1873, I had made several duplex machines and one quadruplex machine. It became useful and convenient to have an opportunity of testing those machines upon extended circuits and upon circuits of different lengths, and I applied to Mr. William Orton, president of the Western Union Telegraph Company, through N. C. Miller, at that time treasurer of the Gold and Stock Telegraph Company, for which I was making machines, for leave to put my machines on the wires of the Western Union Telegraph Company when they were not in use between 12 o'clock at night and 7 in the morning, when a portion of its wires were idle, and he (Orton) permitted me do so, saying that he should be glad to have any machine developed which would promote rapidity in telegraphing.

Under that permission I worked some one hundred nights, from twelve at night until the wires were put in use the next morning, and developed the operation of the duplex machines so that I believed them to be a success; and so far tested the quadruplex machines that I believed that by further experiments they could be made a success.

I applied to Mr. Orton to have my duplex machines put on his wires for use, and he expressed a willingness to have it done, but gave no orders to his subordinates by which I was enabled to do it, and whenever I tried to do anything I found that I had no power to have anything done. I, therefore, wrote him asking what I should do under the circumstances to have my machines put upon the wires, and that I would like some orders about it, which letter he received on the eve of his sailing for Europe and did not answer me, so that I was left without any power of testing my machines by practical use on extended circuits as working machines in the business of telegraphy.

Having access, for practical operation, to no extended circuits of wires, Mr. Orton having gone to Europe without giving me the order I sought, I applied, soon after Mr. Orton's return, to Mr. George B. Prescott, who was at that time the electrician of the Western Union Telegraph Company, and stated my difficulty to him, and asked him to aid me in having my duplex instruments put to work on the Western Union lines. To induce him to aid me in so doing, and to allow me further to experiment with my quadruplex machine, I offered a half interest in my inventions, provided he would aid me in what I desired. He agreed so to do if I would agree that he should appear to be a joint inventor with me of these instruments, although up to that time he had done nothing about them, and I had never spoken to him on the subject. I told him that as he had not invented any part of them the only way he could be considered joint inventor was to do some trifling thing about them which would enable him to swear in the Patent Office that he was joint inventor.

In pursuance of that idea a paper, dated July 9th, 1874, was executed, wherein I agreed that we were joint inventors, a copy of which paper is hereto annexed, marked A, and I was to give him one-half of my interest. But, being fearful that he might sell me out, in the first agreement, which was written out by me and submitted to him, I provided that neither should sell without the written consent of the other, guarding the instrument as well as I could, to prevent any possible use of my inventions by Prescott without my consent. That instrument, which I drew, was taken by Mr. Prescott to the attorneys of the Western Union Telegraph Company, Porter, Lowry, Soren & Stone, and was redrafted I think before it was signed, after which it was signed by me and Prescott.

Up to this time Mr. Prescott had done nothing towards the invention, and had paid me no money therefor, and the only consideration for it was that he, as electrician of the Western Union Company, would allow the use of my machines upon their lines, and render me his support and influence; and the only thing that he was to do in the future was to pay the costs of models, solicitors' fees, Patent Office fees, and provide for the necessary apparatus to carry on further experiments, which would be the use of the Western Union Company's lines, and their apparatus for working them.

I consulted with L. W. Serrell, a patent solicitor, as to the legality of the patents if they should be issued in the name of myself and Prescott as joint inventors, stating to him the circumstances, and he advised me that it was very doubtful whether patents issued under those circumstances would be valid, and the question could be raised at any time, in court, in testing the validity of the patents. I came back and stated what Mr. Serrell had told me to Mr. Prescott, and he went up and had a consultation with Mr. Serrell, and received from him substantially the same advice. In consequence of that, it was concluded between us that the patents had better not be taken out in the name of Mr. Prescott and myself as joint inventors.

In the meantime models bearing our joint names, and specifications had been drawn up setting forth that these duplex and quadruplex machines were the joint invention of Mr. Prescott and myself, and these all had to be altered and made out in my own name as sole inventor, and were so filed in the Patent Office, and the agreement between Mr. Prescott and myself was reformed and made, as it appears in the deed of assignment of August 19, 1874. The paper changing the agreement, and which was duly executed, (a copy of which, marked B, is hereto annexed,) was drawn by Mr. Serrell, and for that agreement I received no other or different consideration or payment than that which I had agreed to receive for the other deed for which it was substituted.

In consequence of this agreement I was allowed to put my machines upon the Western Union Company's lines, and to put them at work.

But during the negotiations I had told Mr. Prescott about my quadruplex, and had explained it and its operations to him, so that, if it were a working instrument, it would be so much more rapid and efficacious than the duplex. Mr. Prescott immediately advised me to test that and drop the duplex; and although the duplex machines had been made for the purpose of being put upon the lines, yet, acting under the advice of Mr. Prescott, I dropped the further experiments with the duplex, and had made and perfected a quadruplex instrument, and put it upon the line formed of a return circuit between New York and Philadelphia, and, in the presence of Mr. Prescott, Mr. Orton and others, I worked it all one day. Its working was considered so far successful experimentally that Mr. Orton advised a new set of instruments to be made and put upon the line between Boston and New York, and to have it worked in the actual business of telegraphing, which was done.

This set of instruments was made in the shop of the Western Union Telegraph Company, and I suppose paid for by Mr. Prescott as he was bound to do by his agreement; at any rate, I did not pay for it. My invention proved a success, both in theory and in the working of the machine to carry out the theory, and was worked for some three or four weeks. Then Mr. Orton said to me; "This is a substantial success working on a short circuit, but in order to test its value as a working instrument it will be necessary to have an experiment made with it on a long circuit," (where there must be a translation of signals,) and suggested that the experiment be tried between New York and Chicago.

Thereupon, abandoning the working at Boston, I caused the Chicago wire to be transferred to this instrument instead of the Boston wire, and with another set of machinery for Buffalo, where the translation was to take place, and also a set at Chicago. I remained at the New York end, working the machine, and sent Mr. Gerrit Smith, whom I had instructed in the manner of working the machine, to Buffalo and Chicago to manage that end of it as an expert; and the machines have been working on that line ever since in the service of the Western Union Telegraph Company except while changing into their new building in New York.

During all this time Prescott has never paid me any money for any share in my inventions, nor has he himself personally done anything about them, except, possibly, having paid some solicitor's fee or for some machines.

After my invention was found to be a success Mr. Prescott consulted with me as to what proposition we should make to sell it to the Western Union Telegraph Company. During all these experiments there had been no proposition on the part of the Western Union Company as to what they would pay me, but they had expressed a disposition to buy it. Prescott consulted with me as to the proposition we should make, and I drew out a memorandum, and afterwards a proposition was made to the Western Union Company, signed by both of us, of which the annexed exhibit, marked C, is a copy, and that is the only formal proposition which was made. That proposition was not accepted by the Western Union Company.

I then, being in want of money, submitted other unsigned memoranda of lower rates on my part to Mr. Orton, but he accepted none of them; but, as they were lower in amount than my signed proposition, he said I was doing well, and to keep on going lower and lower. Prescott did not join with me in any of these lower memoranda. I was induced to make these memoranda because I wanted the money badly. I had my last negotiation with Mr. Orton in New York, and then I informally proposed, as a matter of negotiation, to take twenty-five thousand dollars and a royalty, instead of the first proposition, which was as per Exhibit C, and would have amounted to about four hundred and fifty thousand dollars.

After I had made the machine a success between New York and Chicago I needed money to carry on my shop business, and supposing that I should make an arrangement with Mr. Orton for the sale of my invention I applied to him for some money, and he gave me five thousand dollars, and I gave him a receipt, of which exhibit marked D is a copy, and in which Prescott appears as joint inventor, and I received the money, supposing that a satisfactory arrangement would be made.

Prescott knew that I had received the money, but did not sign any receipt or agreement that I should sell. I desired some more money, because Mr. Orton had given me an order for making some twenty sets of my quadruplex machines in my own shop, and I needed the money to pay my help. I called upon him repeatedly, and made known my wants, and he put me off from time to time. At last it was absolutely necessary that I should have the money, and I called upon him to close up the business, and he made me no reply whatever, but treated me with great coldness and indifference, and the next day he went to Chicago, as I learned, leaving me in my difficulties.

That left me in this condition: I had some hundred and twenty men employed, who must have their wages, and I had no property in my own invention which I could sell, being apparently tied up with the agreement with Mr. Prescott; and in the receipt which Mr. Orton wrote me to sign he had put in that the invention was the joint invention of Mr. Prescott and myself, although Mr. Prescott had never done anything, or claimed to have done anything, about the invention after he had consulted with Mr. Serrell, as before stated; and I, therefore, determined to make the best sale of my invention that I could, knowing that the Western Union Telegraph Company had no right, and believing that Mr. Prescott had only taken hold of it for the purpose of getting a portion of it without consideration, and to tie me up so as to put me entirely at the mercy of the Western Union Company whose employee he was; and I made sale of whatever rights I might have to Mr. Jay Gould, in behalf of the Atlantic and Pacific Telegraph Company, so that I might relieve my then pressing necessities, and pay my men. I saw Mr. Harrington very soon after, and communicated the facts to him of my sale, and he approved that sale, under the circumstances, and took such measures to protect his own interests as he was advised, as I am informed.

I have now given a truthful statement of all the facts in relation to these several inventions, transfers, deeds, and negotiations. I felt myself justified in making the sale without reference to the deed to Prescott, because I felt that I had been overreached by him in the bargain; that he had paid no consideration for it; and that he probably was acting simply in the interest of the Western

Union Telegraph Company, because in the course of my negotiations with Mr. Orton when he said that I asked too large a sum of money, I said to him that I was obliged to ask so large a sum because I thought I was to divide with Mr. Prescott, and he replied to me that he had his doubts to whether I ought to divide with Prescott, because, being an employee of the Western Union Telegraph Company, he did not think he had a right to have any; and this statement of Mr. Orton confirmed me in the belief that the Western Union Company had simply employed Prescott to allow the use of their lines for my experiments in order to control my invention, which I still firmly believe.

I was further confirmed in this belief, because the experiments with the quadruplex machine had been kept substantially private; certainly nothing had been given out to the press about it until the agreement of the 9th of July, between myself and Prescott, was signed in the afternoon of that day. I think it was that evening that Mr. Orton had an interview with a reporter from the office of the New York *Times*, and gave to him a glowing account of the experiments that had been made by the Western Union Telegraph Company with the joint inventions of Mr. Edison and Mr. Prescott, the electrician of the company, and that it was to be immediately put in operation on all their lines. The reporter read his notes to me to see if they were correct, and the report read that this was the joint invention of "Edison and Prescott." Mr Prescott was present and heard it read. As the reporter was going out he called him back and spoke to him privately, and when I saw the printed report in the *Times* the next morning, it appeared as the joint invention of "Prescott and Edison." So that I became convinced that not only the profits of my invention were to be got from me, but also any credit which might come to me because of my experiments and labor in making it, which would be due to me.

Again, in this latter opinion I was still further strengthened, because soon afterwards Professor Vander Weyde, who was about to deliver a lecture on duplex and quadruplex telegraphy before the Polytechnic Institute in New York, called on Mr. Prescott to learn of the invention, who gave him a history of it, as I learned afterwards in this way; I went to hear the lecture, being interested in that science, and the Professor gave a brief account of the invention, and spoke of it in an approving manner as being a very valuable scientific invention. He was unable, however, to give a description of the operation of the quadruplex machine, but stated that he had been to the office of the Western Union Telegraph Company and had been shown it in operation between New York and Chicago. The Professor had not been properly instructed in its operation, because he attempted to show, by the movement of a number of balls, how the electrical currents passed each other; whereas in operation, in fact, the currents do not pass each other. At this point in the lecture a gentleman named Hamilton E. Towle, a civil engineer, arose and asked the Professor to whom belonged the credit of the invention of the quadruplex machine, and he answered to Mr. Prescott, the electrician of the Western Union Telegraph Company. Mr. Towle replied that that was not so; that the Professor was mistaken; that the invention belonged to Mr. Edison, and Mr. Towle thereupon took a piece of chalk and wrote the name of "Edison" on the black-board in large letters, and under it wrote the name of "Prescott" in very small writing, saying: "The credit belongs to those two men in proportion to the size of those two names." There was a number of gentlemen there who heard it and it is a little remarkable that that part of the lecture was not reported in the newspapers.

Acting under the belief that I had been wronged by Prescott, and desiring to test his right to any portion of my invention, either as joint inventor or as assignee, I felt it my duty, as well as right, to revoke the request that I had made in my application, that the invention should be issued jointly to Prescott and myself, at the same time being advised and believing that under my former contracts with Harrington, as I have explained, a share of interest in the invention of right belonged to him, because it was the result and perfection of experiments which I had made before my partnership arrangement, and deeds to him were made, and that it passed to him by those deeds. I asked the Commissioner of Patents that the patent might be issued to Mr. Harrington, as per annexed copy of letter dated January, 1875, marked E.

I am now advised that if there is any doubt to whom an interest in this invention belongs, I should be enabled to protect myself and the party to whom it of right belongs by having the patent issue to myself alone, so that whenever either Mr. Prescott or Mr. Harrington ask the courts to assign to them any interest therein, I may call the other into court to interplead with him, and then make the assignment of the proper interest according to the judgment of the court, and thus protect myself from suits for damages brought by either against me.

At this same time I prepared letters addressed to Mr. Geo. B. Prescott and Mr. Wm. Orton, President Western Union Telegraph Company, advising them of my action, and offering to return to them respectively any money I had received from Mr. Orton, president, or that had been expended by Mr. Prescott on account of contracts. (See copies annexed, marked ______.)
(Signed)

THOMAS A. EDISON.

PD (transcript), NjWOE, Quad. 72.16 (*TAEM* 9:259). Notarization and exhibits omitted. Another copy is at MiDbEI. [a]Probably should read "to secure".

Appendix 4

Edison's U.S. Patents, June 1873–March 1876

The following list contains all patents for which Edison executed an application in the period covered by Volume Two. It is arranged in chronological order by execution date, which is the date on which Edison signed the application and the date in the patenting process that comes closest to the time of actual inventive activity. The application date is the date on which the Patent Office received and recorded the application. Edison's American patents from this period are in *TAEM* 1–2; British patents are in Cat. 1321, Batchelor (*TAEM* 92:30–53).

	Exec. Date	*Appl. Date*	*Issue Date*	*Pat. No.*	*Title*
80.	08/25/73	09/02/73	05/26/74	151,209	Automatic Telegraphy and Perforators Therefor
81.	09/29/73	06/01/74	03/02/75	160,402	Solutions for Chemical Telegraph-Paper
82.	09/29/73	06/01/74	03/02/75	160,404	Solutions for Chemical Telegraph-Paper
83.	10/14/73	06/01/74	03/09/75	160,580	Solutions for Chemical Telegraph-Paper
84.	10/29/73	06/01/74	03/02/75	160,403	Solutions for Chemical Telegraph-Paper
85.	04/02/74	05/11/74	09/08/74	154,788	District Telegraph Signal-Boxes
86.	05/22/74	06/01/74	09/21/75	168,004	Printing-Telegraphs
87.	06/01/74	07/25/74	08/17/75	166,859	Chemical Telegraphy
88.	06/01/74	07/25/74	08/17/75	166,860	Chemical Telegraphy
89.	06/01/74	07/25/74	08/17/75	166,861	Chemical Telegraphy
90.	08/07/74	08/13/74	01/19/75	158,787	Telegraph Apparatus
91.	08/07/74	01/15/75	01/18/76	172,305	Automatic Roman-Character Telegraphs

92.	08/07/74	01/15/75	02/22/76	173,718	Automatic Telegraphy
93.	08/14/74	01/15/75	10/05/75	168,465	Solutions for Chemical Telegraphs
94.	08/19/74	09/01/74	05/30/76	178,221	Duplex Telegraphs
95.	08/19/74	09/01/74	05/30/76	178,222	Duplex Telegraphs
96.	08/19/74	09/01/74	05/30/76	178,223	Duplex Telegraphs
97.	08/19/74	09/01/74	08/08/76	180,858	Duplex Telegraphs
98.	08/19/74	09/01/74	09/03/78	207,723	Duplex Telegraphs
99.	08/19/74	09/01/74	08/09/92	480,567	Duplex Telegraph
100.	12/14/74	12/28/74	09/03/78	207,724	Duplex Telegraphs
101.	01/18/75	01/26/75	09/28/75	168,242	Transmitters and Receivers for Automatic Telegraphs
102.	01/18/75	01/26/75	09/28/75	168,243	Automatic Telegraphs
103.	01/18/75	01/26/75	10/05/75	168,385	Duplex Telegraphs
104.	01/18/75	01/26/75	10/05/75	168,466	Solutions for Chemical Telegraphs
105.	01/18/75	01/26/75	10/05/75	168,467	Recording-Points for Chemical Telegraphs
106.	01/18/75	01/27/75	10/02/77	195,751	Automatic Telegraphs
107.	01/19/75	01/27/75	10/02/77	195,752	Automatic Telegraphs
108.	02/11/75	02/16/75	12/21/75	171,273	Telegraph Apparatus
109.	02/24/75	03/23/75	11/16/75	169,972	Electric-Signalling Instruments
110.	02/24/75	03/23/75	10/22/78	209,241	Quadruplex-Telegraph Repeaters
111.	03/07/76	03/13/76	08/08/76	180,857	Autographic Printing

Bibliography

Appleyard, Rollo. 1939. *The History of the Institution of Electrical Engineers (1871–1931).* London: Institution of Electrical Engineers.

Atkinson, E., trans. and ed. 1890. *Elementary Treatise on Physics, Experimental and Applied.* From *Ganot's Eléments de Physique.* 13th ed. New York: William Wood and Co.

Atkinson, E., trans. 1910. *Elementary Treatise on Physics, Experimental and Applied.* Ed. A. W. Reinold. From *Ganot's Eléments de Physique.* 18th ed. New York: William Wood and Co.

Baker, Edward C. 1976. *Sir William Preece, F.R.S.* London: Hutchinson and Co.

Barty-King, Hugh. 1979. *Girdle Round the Earth: The Story of Cable and Wireless and Its Predecessors to Mark the Group's Jubilee, 1929–1979.* London: Heinemann.

Beard, George M., and Alphonso D. Rockwell. 1871. *A Practical Treatise on the Medical and Surgical Uses of Electricity, Including Localized and General Electrization.* New York: William Wood.

Bloxam, Charles L. 1869. *Laboratory Teaching: or, Progressive Exercises in Practical Chemistry.* London: John Churchill & Sons.

Bright, Charles. 1974 [1898]. *Submarine Telegraphs: Their History, Construction, and Working.* New York: Arno Press.

Bruce, Robert V. 1987. *The Launching of Modern American Science, 1846–1876.* Ithaca, N.Y.: Cornell University Press.

Bryant, John H. 1987. "An Engineering and Scientific Look at the Etheric Force Episode." Paper presented at the annual meeting of the Society for the History of Technology, Raleigh, N.C., Oct. 29–Nov. 1.

Bud, Robert, and Gerrylynn K. Roberts. 1984. *Science versus Practice: Chemistry in Victorian Britain.* Manchester: Manchester University Press.

Butrica, Andrew J. 1986. "From *Inspecteur* to *Ingénieur:* Telegraphy and the Genesis of Electrical Engineering in France, 1845–1881." Ph.D. diss., Iowa State University.

Buttrick, John. 1952. "The Inside Contract System." *Journal of Economic History* 12:205–21.

Carlson, W. Bernard. 1984. "Invention, Science, and Business: The Professional Career of Elihu Thomson, 1870–1900." Ph.D. diss., University of Pennsylvania.

Clark, Ronald W. 1977. *Edison: The Man Who Made the Future.* New York: G. P. Putnam's Sons.

Conot, Robert. 1979. *A Streak of Luck.* New York: Seaview Books.

Crookes, William. 1871. *Select Methods in Chemical Analysis (Chiefly Inorganic).* London: Longmans, Green, and Co.

Crowther, J. G. 1962. *British Scientists of the Nineteenth Century.* London: Routledge & Kegan Paul.

Culley, Richard S. 1871. *A Handbook of Practical Telegraphy.* London: Longmans, Green, Reader, and Dyer.

Davies, John D. 1971. *Phrenology: Fad and Science. A Nineteenth-Century American Crusade.* 2d ed. Hamden, Conn.: Archon.

Davis, Charles H., and Frank B. Rae. 1876. *Handbook of Electrical Diagrams and Connections.* New York: Graphic Co.

De Cogan, Donard. 1986. "From Technical Wonder to Profitable Investment: The Economics of Nineteenth-Century North Atlantic Telegraphy." Paper presented at the annual meeting of the Society for the History of Technology, Pittsburgh, Pa., Oct. 23–26.

Dupree, A. Hunter. 1957. *Science in the Federal Government: A History of Policies and Activities.* Cambridge: Harvard University Press, Belknap Press.

Dyer, Frank, Thomas C. Martin, and William H. Meadowcroft. 1929. *Edison: His Life and Inventions.* 2 vols. New York: Harper & Bros.

Englander, Ernest J. 1987. "The Inside Contract System of Production and Organization: A Neglected Aspect of the History of the Firm." *Labor History* 28:429–46.

Fels, Rendigs. 1973 [1959]. *American Business Cycles, 1865–1897.* Westport, Conn.: Greenwood Press.

Finn, William. 1894. "Fast Speed Telegraphy." *Teleg. and Tel.* 14:322–24.

Fischer, Henry C., and William H. Preece. 1877. "Joint Report upon the American Telegraph System." UKLPO.

Ford, William F. 1874. *The Industrial Interests of Newark, N.J.* New York: Van Arsdale & Co.

Fox, Cornelius. 1873. *Ozone and Antozone: Their History and Nature.* London: J. & A. Churchill.

Fox, Edwin M. 1879. "Edison's System of Fast Telegraphy." *Scribner's* 18:840–46.

Franklin Institute. 1874. *Report of the Twenty-seventh Exhibition of American Manufactures, Held in the City of Philadelphia, from October 6th, to November 12, 1874.* Philadelphia: Franklin Institute.

Freedley, Edwin T. 1859. *Philadelphia and Its Manufactures.* Philadelphia: Edward Young.

Friedel, Robert. 1983. *Pioneer Plastic: The Making and Selling of Celluloid.* Madison: University of Wisconsin Press.

Gospill, James, ed. (Printed annually.) *Gospill's Philadelphia City Directory.* Philadelphia: James Gospill.

Granville, Arthur R. 1874. "Duplex and Quadruplex Telegraphy." *Telegr.* 7:229–30.

Gray, Elisha. 1977 [1878]. *Experimental Researches in Electro-Harmonic Telegraphy and Telephony.* In *The Telephone: A Historical Anthology,* ed. George Shiers. New York: Arno Press.

Greenway, John L. 1987. "'Nervous Disease' and Electric Medicine." In *Pseudo-Science and Society in Nineteenth-Century America,* ed. Arthur Wrobel. Lexington: University Press of Kentucky.

Haber, L. F. 1958. *The Chemical Industry during the Nineteenth Century.* London: Oxford University Press.

Haigh, K. R. 1968. *Cableships and Submarine Cables.* London: Adlard Coles Ltd.

Hardaway, Paul A. 1986 [1950]. "J. H. Bunnell & Co.—A Magic Name in Telegraphy." *Dots and Dashes* 14:2.

Harlow, Alvin F. 1936. *Old Wires and New Waves.* New York: D. Appleton-Century Co.

Heaviside, Oliver. 1873. "On Duplex Telegraphy." *Phil. Mag.,* 4th ser., 45:426–32.

Helmholtz, Hermann von. 1875. *On the Sensations of Tone as Physiological Basis for the Theory of Music.* London: Longmans, Green, and Co.

———. 1962 [1881]. *Popular Lectures on Scientific Subjects.* New York: Dover Publications.

History of St. Clair County, Michigan. 1883. Chicago: A. T. Andreas and Co.

Holbrook's Newark City Directory. (Printed annually.) Newark, N.J.: A. Stephen Holbrook.

Hunt, Robert. 1857. "On Electro-Magnetism as a Motive Power." *Proceedings of the Institute of Civil Engineers* 16:386–421.

Israel, Paul B. 1989. "From the Machine Shop to the Industrial Laboratory: Telegraphy and the Changing Context of American Invention, 1830–1920." Ph.D. diss., Rutgers University.

Jehl, Francis. 1937–41. *Menlo Park Reminiscences.* 3 vols. Dearborn, Mich.: Edison Institute.

Jenkins, Reese V., and Keith A. Nier. 1984. "A Record for Invention: Thomas Edison and His Papers." *IEEE Transactions on Education* E-27:191–96.

Jenks, William Lee. 1912. *St. Clair County, Michigan: Its History and Its People.* 2 vols. Chicago: Lewis Publishing Co.

———. n.d. "History of Port Huron Street Railways." William Lee Jenks Collection, MiD-B.

Jones, Francis W. 1875. "The Quadruplex." *Journal of the American Electrical Society* 1:16–29.

Kieve, Jeffrey L. 1973. *The Electric Telegraph: A Social and Economic History.* Newton Abbot, England: David & Charles.

King, W. James. 1962a. "The Development of Electrical Technology in the Nineteenth Century: 1. The Electrochemical Cell and the Electromagnet." *United States Museum Bulletin 228.* Washington, D.C.: Smithsonian Institution.

———. 1962b. "The Development of Electrical Technology in the

Nineteenth Century: 3. The Early Arc Light and Generator." *United States Museum Bulletin 228.* Washington, D.C.: Smithsonian Institution.

Klein, Maury. 1986. *The Life and Legend of Jay Gould.* Baltimore: Johns Hopkins University Press.

Knight, Edward H. 1876–77. *Knight's American Mechanical Dictionary.* 3 vols. New York: Hurd and Houghton.

Lawford, G. L., and L. R. Nicholson. 1950. *The Telcon Story: 1850–1950.* London: The Telegraph Construction & Maintenance Co. Ltd.

[Little, George.] 1873. "The American Automatic Telegraph System." *Telegr.* 9:307–8.

Little, George. 1874a. Letter to the Editor. *Telegr.* 10:182–83.

———. 1874b. Letter to the Editor. *Telegr.* 10:189.

———. 1874c. "Western Union and Automatic Telegraphy.—Mr. Little in Reply to Mr. Orton." *Telegr.* 10:259.

Marchand, John T. 1922. "Corporate History of Western Union." Engineering Library, NjWU.

Maver, William, Jr. 1892. *American Telegraphy: Systems, Apparatus, Operation.* New York: J. H. Bunnell & Co.

Maxwell, James Clerk. 1873a. "On Loschmidt's Experiments on Diffusion in Relation to the Kinetic Theory of Gases." *Nature* 8:298–300.

———. 1873b. "Molecules." *Nature* 8:437–41.

McFeely, William S. 1981. *Grant: A Biography.* New York: W. W. Norton & Co.

McGaw, Judith A. 1987. *Most Wonderful Machine: Mechanization and Social Change in Berkshire Paper Making, 1801–1885.* Princeton, N.J.: Princeton University Press.

Mitts, Dorothy. 1970. "Don't Forget Charles Edison." *Port Huron Times-Herald,* 10 May.

Nevins, Allan. 1957. *Hamilton Fish: The Inner History of the Grant Administration.* Rev. ed., 2 vols. New York: Frederick Ungar.

Nicholson, Henry C. 1874. "Double Duplex and Quadruple Telegraph." *Telegr.* 10:194.

Nier, Keith A. 1987. "The Perplexing Fate of the Quadruplex: Phantom Wires and Technological Myths." *Proceedings of the One Hundred First Annual Meeting of the American Historical Association, 1986.* Ann Arbor: University Microfilms.

Ord-Hume, Arthur W. G. 1977. *Perpetual Motion: The History of an Obsession.* New York: St. Martin's Press.

Pepper, John H. [1869a]. *Cyclopedia of Science Simplified.* London: Frederick Warne and Co.

———. 1869b. *The Playbook of Metals.* London: George Routledge and Sons.

Phillips, Walter Polk. 1897. *Sketches Old and New.* New York: George Munro's Sons.

Pope, Franklin. 1872. *Modern Practice of the Electric Telegraph: A Handbook for Electricians and Operators.* 7th ed. New York: D. Van Nostrand.

———. 1875. "On Quadruplex Telegraphy." *J. Teleg.* 8:337–39.
Post Office London Directory. (Printed annually.) London: Kelly & Co.
Preece, William H., and J. Sivewright. 1876. *Telegraphy.* 2d ed. London: Longmans, Green, and Co.
———. 1891. *Telegraphy.* 9th ed. London: Longmans, Green, and Co.
Prescott, George B. 1863. *History, Theory, and Practice of the Electric Telegraph.* Boston: Ticknor and Fields.
———. 1877. *Electricity and the Electric Telegraph.* New York: D. Appleton and Co.
———. 1879. *The Speaking Telephone, Electric Light, and other Recent Electrical Inventions.* New York: D. Appleton and Co.
———. 1885. *Electricity and the Electric Telegraph.* 6th ed. New York: D. Appleton and Co.
Proudfoot, W. B. 1972. *The Origin of Stencil Duplicating.* London: Hutchinson of London.
Reader, W. J. 1987. *A History of the Institution of Electrical Engineers, 1871–1971.* London: Peter Peregrinus Ltd.
Reid, James D. 1879. *The Telegraph in America.* New York: Derby Bros.
———. 1886. *The Telegraph in America.* Rev. ed. New York: John Polhemus.
Rosenberg, Charles. 1976. *No Other Gods: On Science and American Social Thought.* Baltimore: Johns Hopkins University Press.
Rosenberg, Nathan. 1976. *Perspectives on Technology.* Cambridge: Cambridge University Press.
Sabine, Robert. 1867. *The Electric Telegraph.* London: Virtue Bros. & Co.
———. 1872. *The History and Progress of the Electric Telegraph.* 3d ed. London: Lockwood and Co.
Scheips, Paul J. 1963. "Union Signal Communications." *Civil War History* 9:399–421.
Scott, J. M. 1972. *Extel One Hundred: The Centenary History of the Exchange Telegraph Company.* London: Ernest Benn.
Shaffner, Taliaferro P. 1859. *The Telegraph Manual.* New York: Pudney & Russell.
Simonds, William Adams. 1940 [1934]. *Edison: His Life, His Work, His Genius.* New York: Blue Ribbon Books.
Sinclair, Bruce. 1974. *Philadelphia's Philosopher Mechanics: A History of the Franklin Institute, 1824–65.* Baltimore: Johns Hopkins University Press.
Smith, George David. 1985. *The Anatomy of a Business Strategy: Bell, Western Electric, and the Origins of the American Telephone Industry.* Baltimore: Johns Hopkins University Press.
Smith, Gerritt. 1914. "Looking Backward Sixty Years." *Teleg. and Tel.* 15:140–43, 166–68.
Smith, Willoughby. 1974 [1891]. *The Rise and Extension of Submarine Telegraphy.* New York: Arno Press.
Sobel, Robert. 1968. *Panic on Wall Street: A History of America's Financial Disasters.* New York: Macmillan.

Sprague, John T. 1875. *Electricity: Its Theory, Sources, and Applications.* London: E. & F. N. Spon.

Springfield City Directory and Business Advertiser. 1875. Springfield, Mass.: Clark W. Bryan & Co.

Taltavall, John B. 1893. *Telegraphers of Today.* New York: John B. Taltavall.

Taylor, Jocelyn Pierson. 1978. *Grosvenor Porter Lowrey.* New York: Privately printed.

Thompson, George Raynor. 1958. "Development of the Signal Field Telegraph, 1861–1863." *Signal* 12:28.

Western Electric Manufacturing Co. 1876. *Catalogue of the Western Electric Manufacturing Co.* Chicago: Western Electric Manufacturing Co.

Williams' Cincinnati Directory. (Printed annually.) Cincinnati: Williams Directory Co.

Wilson, H., comp. (Printed annually.) *Trow's Business Directory of New York City.* New York: John F. Trow.

Wilson, James Grant, ed. 1968 [1901]. *Appleton's Cyclopaedia of American Biography.* Detroit: Gale Research Co.

Wilson, Joseph M. 1876–78. *The Masterpieces of the Centennial Exhibition.* Philadelphia: Gebbie & Barrie.

Yates, JoAnne. 1989. *Control Through Communication: The Rise of System in American Management.* Baltimore: Johns Hopkins University Press.

Young, Robert M. 1970. *Mind, Brain, and Adaptation in the Nineteenth Century.* Oxford: Clarendon Press.

Credits

Reproduced with permission of the AT&T Corporate Archive: Doc. 484; illustration p. 120. Reproduced with permission of Dun and Bradstreet and the Baker Library: Docs. 352, 448, 547, 616. Reproduced with permission of the Rare Book and Manuscript Library, Columbia University: Doc. 564. From the collections of the Henry Ford Museum & Greenfield Village: Docs, 349, 458, 504, 721; illustrations on pp. 25, 188, 626, 658. Reproduced with permission of the Historical Society of Pennsylvania: Doc. 672. From the collection of Charles Hummel: Doc. 435. Courtesy of Philip Kunhardt: frontispiece. Reproduced with permission of the Missouri Historical Society: Doc. 710. Reproduced courtesy of the Post Office Archives, London: Docs., 381, 383, 384, 385, 390, 416. Reproduced with permission of the University of Pennsylvania Archives: Doc. 502. Reproduced with permission of the Western Union Corporation: Doc. 509.

Courtesy of Edison National Historic Site (all designations are to *TAEM* reel:frame unless other indicated): pp. 22 (5:796), 23 (5:799), 24 (Rick Mitchell), 41 (6:111), 118 *top* (6:158), 118 *center* (6:159), 118 *bottom* (6:160), 119 *top* (6:160), 122 (90:604), 123 (90:619), 198 (13:104), 208 (27:137), 223 (5:941), 244 *top* (Rick Mitchell), 246 (Rick Mitchell), 286 (5:916), 298 (5:900), 299 (5:909), 404 (8:369), 446 (Rick Mitchell), 465 (13:404), 502 (3:148), 565 (Bradley catalog), 607 (3:199).

Index

Boldface page numbers signify primary references or identifications; italic numbers indicate illustrations. Page numbers refer to headnote or document text unless the reference appears only in an endnote. The profusion, nature, and inconstant nomenclature of chemicals used by Edison makes their indexing fruitless. This is equally true of telegraph components such as relays, transmitting circuits, and batteries, as well as experimental phenomena such as induction and tailing. Only in broad, defining, or exceptional instances do such items appear here.